PHYSICS
FOR SCIENTISTS AND ENGINEERS

AN INTERACTIVE APPROACH

VOLUME 1

Chapters 1 – 18

HAWKES   IQBAL   MANSOUR   MILNER-BOLOTIN   WILLIAMS

1914–2014
Nelson Education celebrates 100 years of Canadian publishing

NELSON
EDUCATION
CELEBRATE LIFELONG LEARNING

NELSON / EDUCATION

ISBN-13: 978-0-17-648384-5
ISBN-10: 0-17-648384-5

Consists of Selections from:

Physics for Scientists and Engineers: An Interactive Approach, 1st Edition
Robert Hawkes,
Javed Iqbal,
Firas Mansour,
Marina Milner-Bolotin,
Peter Williams
ISBN 10: 0-17-659637-2, © 2014

Cover Credit:

© 2012 – Jim Kramer

Contents

ENHANCED WebAssign

Increased Engagement. Improved Outcomes. Superior Service.

Exclusively from Nelson Education and Cengage Learning, **Enhanced WebAssign®** combines the exceptional content that you know and love with the most powerful and flexible online homework solution, **WebAssign**. **Enhanced WebAssign** engages students with immediate feedback, rich tutorial content, and interactive eBooks, helping students to develop a deeper conceptual understanding of subject matter. Instructors can build online assignments by selecting from thousands of text-specific problems or supplementing with problems from any Nelson Education or Cengage Learning textbook in our collection.

With **Enhanced WebAssign**, you can

- Help students stay on task with the class by requiring regularly scheduled assignments using problems from the textbook.

- Provide students with access to a personal study plan to help them identify areas of weakness and offer remediation.

- Focus on teaching your course and not on grading assignments. Use **Enhanced WebAssign** item analysis to easily identify the problems that students are struggling with.

- Make the textbook a destination in the course by customizing your Cengage YouBook to include sharable notes and highlights, links to media resources, and more.

- Design your course to meet the unique needs of traditional, lab-based, or distance learning environments.

- Easily share or collaborate on assignments with other faculty or create a master course to help ensure a consistent student experience across multiple sections.

- Minimize the risk of cheating by offering algorithmic versions of problems to each student or fix assignment values to encourage group collaboration.

Learn more at www.cengage.com/ewa

Iofoto/Shutterstock, Monkey Business Images/Shutterstock, Monkey Business Images/Shutterstock, Bikeriderlondon/Shutterstock, Goodluz/Shutterstock, StockLite/Shutterstock, Shutterstock, Andresr/Shutterstock, Andresr/Shutterstock, Dboystudio/Shutterstock, Pablo Cavlog/Shutterstock, WaveBreakMedia/Shutterstock

NELSON EDUCATION

Preface

THE NEED FOR THIS RESOURCE AND ITS APPROACH

Physics for Scientists and Engineers: An Interactive Approach provides a more relevant approach to the subject to match the Canadian curriculum and will better reflect this fundamental, multidisciplinary, inquisitive, and inspirational science as it applies to Canadian students and instructors. The text also capitalizes on the growing interest in and appreciation of (PER) Physics Education Research. Too many students are either intimidated by physics or not interested in it, and our goal is to simply change that. Physics should be seen as the creative process that it is, and the reader should be helped to feel the thrill of discovery. *Physics for Scientists and Engineers: An Interactive Approach* reflects Canada's revitalized physics curriculum, providing a fresh approach to teaching and more exciting and relevant pedagogy for students. The story is supported by the following key changes that are occurring in the discipline and the market:

- The text is written with Physics Education Research (PER) findings in mind, and in a way to encourage and support instructional strategies based on PER. The author team brings considerable PER expertise to the project, including direct experience with a variety of PER-supported instructional strategies, such as classroom response systems (clickers), peer instruction, studio physics, online simulations, interactive lecture demonstrations, online tutorial systems, collaborative learning, project-based approaches, student reflection, and PSI (personalized system of instruction) approaches.

- While the text encourages PER-based approaches, **it does not support only a single instructional strategy**. Instructors who use traditional lecture + laboratory approaches, those who use audience-response systems in peer instruction, those who favour interactive lecture demonstrations, or indeed use other approaches, will find the text well suited for their needs.

- One idea that spans a number of the PER-supported instructional strategies is **the value of student collaboration**. It is clear that learning is deeper when students develop ideas in collaboration with peers and work together both in brainstorming approaches and in developing solutions. While no book can ensure such approaches, the text has been written in a manner to encourage collaborative learning. For example, the open-ended problems are ideally suited to a group approach. The conceptual problems in each chapter are well suited for use in audience-response peer-instruction approaches, or in studio-style classrooms. In some places we have moved derivations from chapters to problems to encourage student discovery of key relationships. The simulations and experiment suggestions will encourage interactive, engaged learning. If student collaboration is valuable in a classroom, think how powerful collaboration more broadly can be.

- As the Canadian Association of Physicists (CAP), Division of Physics Education (DPE) and others have pointed out, the work of physicists—and the use of physics by other scientists, engineers, and professionals from related fields—is increasingly interdisciplinary. We feel this text is one of the richest in terms of **promoting the interdisciplinary nature of physics** beyond simply having problem applications from various fields. Chapter content is presented with a rich interdisciplinary feel and stresses the need to use ideas from other sciences and related professions. The diverse backgrounds of the author team help create this rich interdisciplinary environment.

- One key differentiator between expert and novice views of physics is that a professional physicist views all of physics as a unified, small set of concepts, principles, and approaches that can be applied to a very diverse set of problems, while the novice sees physics as a huge number of loosely related ideas. Almost all introductory physics texts are much too long and do not adequately illuminate the key principles. This text is **concise in wording and emphasizes the unifying principles and techniques**.

- Closely related to this is the importance of **mastering a number of key concepts**, as opposed to learning incompletely a much broader body of knowledge. As mentioned earlier, the text uses more concise writing in each chapter; in addition, to the degree possible, we have made chapters self-contained so that each instructor can select which content is covered in a course. A carefully selected set of problems, both conceptual and quantitative, helps to reinforce mastery of key concepts.

- Modern computational tools play key roles in the lives of physicists and have been shown to be effective in promoting the learning of physics concepts. Where applicable, we have encouraged explorations through the PhET and similar animation and simulation products. These **allow students to develop their own unifying ideas by manipulating variables in the simulated environments**.

One role of any book is to inspire students to appreciate the beauty of the subject and go on to contribute and become leaders in the field. In order for this to be achieved, students must see the relevance of the subject.

The strong interdisciplinary focus throughout the book will help achieve this goal. But it is also important that **students can see themselves as future physicists**. This is the first broad-market introductory physics text written by a Canadian author team, and we have used Canadian and international examples (discoveries, applications, and current scientists) throughout, including an emphasis on contributions from young Canadians.

Current students place high value on **learning that will help them contribute to society**. For example, service learning is more popular than ever before, and a high number of students set goals of medical or social development careers. Also, there is strong public interest in such fundamental areas as particle physics, quantum mechanics, relativity, string theory, and cosmology.

This book is richer than competitive texts (in the science and engineering physics market) with respect to these topics. We have also employed many examples from the medical physics world, from alternative energy, and from sustainable environments. Thus, the text is richer than most in coverage of areas such as relativity, particle physics, quantum physics, and cosmology. While all physics texts strive to provide **"real-world problems"**, we feel that we have achieved this to a higher degree. Our open problems are modelled on how the world really is: a key part of applying physics is deciding what is relevant, and to make reasonable approximations as needed. Closed-form problems, which in most textbooks are the only type used, portray an artificial situation in which what is relevant—and only that—is given to the student. Our data-rich problems and encouragement of graphing, statistical, and numerical solution software help reinforce realistic situations.

We feel that traditional physics texts have not employed the power of images to the degree possible. Too-early adoption of highly abstract summaries may interfere with learning of concepts through visual images. Physics texts tend to have idealized drawings and a set of images drawn mainly from royalty-free sources. As appropriate, **students are encouraged to explore deeply the messages in rich and compelling visual images, including video**.

Ultimately students must take ownership of their learning; that is essentially the goal of all education. The strong link of objectives, sections, and checkpoint questions will provide an efficient environment for students to achieve this. We view our role in terms of maximizing interest and engagement and eliminating obstacles.

Robert Hawkes
Javed Iqbal
Firas Mansour
Marina Milner-Bolotin
Peter Williams
March 2013

About the Authors

Bob Hawkes

ROBERT HAWKES Dr. Robert Hawkes is a professor of physics at Mount Allison University. In addition to having extensive experience in teaching introductory physics, he has taught upper-level courses in mechanics, relativity, electricity and magnetism, electronics, and astrophysics, as well as education courses in science methods and technology in education. His research program is in solar system astrophysics using advanced electro-optical devices to study atmospheric meteor ablation, as well as complementary lab-based techniques such as laser ablation. Dr. Hawkes received his B.Sc. (1972) and a B.Ed. (1978) at Mount Allison University, and his M.Sc. (1974) and Ph.D. (1979) in physics from the University of Western Ontario. He has won a number of teaching awards, including a 3M STLHE National Teaching Fellow, the Canadian Association of Physicists Medal for Excellence in Undergraduate Teaching, and the Science Atlantic University Teaching Award. He has been an early adopter of a number of interactive teaching techniques, in particular collaborative learning in both introductory and advanced courses. He was the co-editor of the *Physics in Canada* special issue on physics education.

Javed Iqbal

JAVED IQBAL Dr. Javed Iqbal is the director of the Science Co-op Program and an adjunct professor of physics at the University of British Columbia. At UBC he has taught first-year physics for 20 years and has been instrumental in promoting the use of "clickers" at UBC and other Canadian universities. In 2004, he was awarded the Faculty of Science Excellence in Teaching Award. In 2012, he was awarded the Killam Teaching Prize. His research areas include theoretical nuclear physics, computational modelling of light scattering from nanostructures, and computational physics. Dr. Iqbal received his Doctoral Degree in Theoretical Nuclear Physics from Indiana University.

Photo by Hany M. D. Ibrahim

FIRAS MANSOUR As a lecturer in the Department of Physics and Astronomy at the University of Waterloo since 2000, Firas Mansour has gained respect and praise from his students for his exceptional teaching style. He currently teaches first-year physics classes to engineers, life science, and physical science students and has taught upper-year elective physics courses in the past. He is highly regarded for his quality of teaching, his enthusiasm in teaching, and his understanding of students' needs. His dedication to teaching is exemplary, as is his interest in outreach activities in taking scientific knowledge beyond the university boundary. He is a 2012 Distinguished Teaching Award recipient at the University of Waterloo.

Marina Milner-Bolotin

MARINA MILNER-BOLOTIN Dr. Marina Milner-Bolotin is a science educator within the Department of Curriculum and Pedagogy at the University of British Columbia. She specializes in science (physics) teaching and studies ways of using technology to promote student interest in science. For the last 20 years she has been teaching science and mathematics in Israel, the United States (Texas and New Jersey), and Canada. She has taught physics and mathematics to a wide range of students, from elementary gifted students to university undergraduates in science programs and future teachers. She has also led a number of professional development activities for science in-service and pre-service teachers and university faculty—from LoggerPro training workshops, to clicker and tablet training, to physics content presentations at conferences and professional development days. She has been engaged in science education research since 1994. Dr. Milner-Bolotin earned her M.Sc. in theoretical physics at Kharkiv National University, Ukraine, in 1991 and completed her M.A. and Ph.D. in science education at the University of Texas at Austin in 2001. At UT Austin, she investigated how project-based instruction in science courses for future elementary teachers affected their interest in science and their ability to do and teach science. Before joining UBC she was an assistant professor of physics at Ryerson University in Toronto. Dr. Milner-Bolotin is a member of the executive board of the American Association of Physics Teachers (AAPT) and the president and AAPT section representative for the BC Association of Physics Teachers. In 2010, Dr. Milner-Bolotin was awarded the CAP Medal for Excellence in Teaching Undergraduate Physics, the first woman to receive this award.

Deborah Nicholson

PETER WILLIAMS Dr. Peter Williams is dean of the Faculty of Pure and Applied Science and professor of physics at Acadia University and won the 2006 Canadian Association of Physicists (CAP) Medal for Excellence in Teaching. Dr. Williams has developed and taught a great diversity of courses, has shown innovation in the classroom, and has published a number of articles in teaching journals. Dr. Williams received his B.Sc (1985), M.Sc (1988), and Ph.D (1991), from McMaster University. Dr. Williams devotes significant efforts to improving secondary and postsecondary physics education in Atlantic Canada. He is one of those rare individuals who can effectively combine the best of technology-enhanced educational techniques while maintaining a strong personal approach to teaching. He has played a critical role in the development of studio physics modes of instruction at Acadia University and has developed several innovative courses, including most recently a Physics of Music course at Acadia. Furthermore, Dr. Williams has provided regional, national, and international leadership in new modes of physics teaching through his writing and presentations and set an exemplary model for applying research methodology to evaluation of the effectiveness of different modes of physics instruction.

ENRICHED MEDIA

ENHANCED WebAssign Nelson Education's Enhanced WebAssign (EWA) allows you to easily assign, collect, grade, and record homework assignments via the Internet. This proven and reliable system uses pedagogy and content found in Cengage Learning and Nelson Education's best-selling physics textbooks, then enhances them to help your students visualize the problem-solving process and reinforce mathematical concepts more effectively. EWA encourages active learning and time on task, and respects diverse ways of learning.

Used by more than 2.2 million students, Enhanced WebAssign is the leading homework system for math and science. Created by physicists for physics instructors, Enhanced WebAssign is easy to use and works with all major operating systems and browsers. Enhanced WebAssign adds interactive features to go far beyond simply duplicating text problems online. Students can watch solution videos, see problems solved step-by-step, and receive feedback as they complete their homework.

Nelson YouBook

The innovative YouBook available through Enhanced WebAssign allows full instructor customization and robust class-communication tools, including animations, interactive flashcards, and an online glossary to motivate student review, and additional materials only available online.

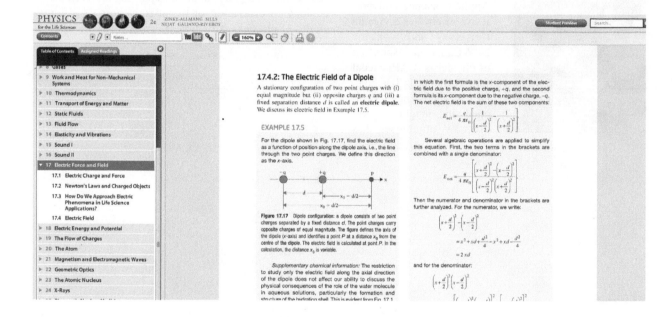

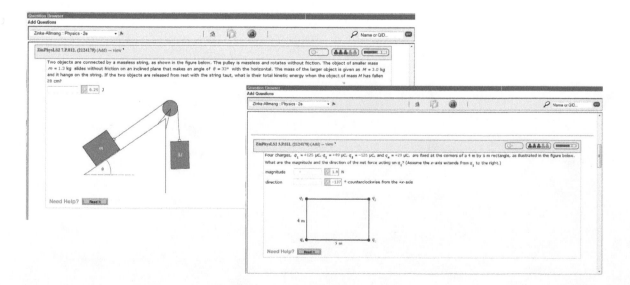

THE BIG PICTURE

Each chapter of *Physics for Scientists and Engineers: An Interactive Approach* is carefully organized according to learning objectives so you can stay focused on the most important concepts. Easy-to-use learning tools point out the topics covered in each chapter, show why they are important, and help you learn the material.

Look Ahead Section These narratives at the beginning of each chapter introduce topics through an interesting and engaging real-life example that pertains to the chapter's subject matter. Each Look Ahead section is illustrated with an engaging photo.

Learning Objectives Learning objectives are brief numbered and directive goals or outcomes that the student should take away from the chapter. Each learning objective corresponds to a major heading within the chapter.

Icons

Major sections are directly tied to **learning objectives**; section elements are visually indicated by individual **icons**.

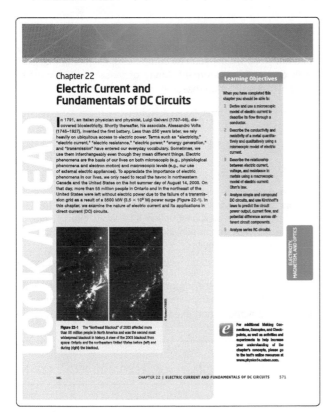

22-4 Analysis of DC Circuits and Kirchhoff's Laws

A **direct current (DC)** circuit is a circuit in which electric current flows in one direction. A simple DC electric circuit consists of circuit elements, such as batteries, resistors, switches, capacitors, and diodes, connected by conducting wires that allow the flow of electric current. We will use the concepts of electric current I, resistance R, power P, and voltage V to describe the operation of these circuits. First, we consider the source of electric potential energy.

Making Connections Making Connections boxes are provided in a narrative format and contain concise examples from international contexts, the history of physics, daily life, and other sciences.

MAKING CONNECTIONS

How Do Touch-Sensitive Lamps Work?

Touch-sensitive lamps commonly have a power-level switch controlled by an RC circuit. Your body has some capacitance. When you touch the metal body of a touch-sensitive lamp, you add capacitance to the control circuit and usually some static charge from your body, as well. These changes in the capacitance trigger the control circuit such that it changes the voltage supplied to the lamp's bulb. Often, touch-sensitive switches cycle through three power levels: off-low-medium-high, then back to off.

Peer to Peer Written by students for students, Peer to Peer boxes provide useful tips for navigating difficult concepts.

PEER TO PEER

I always thought that in a parallel circuit, the current splits equally between the circuit's parallel branches. Then I realized that this is only true when the branches have equal resistance. The current will split in a way such that the more resistance the branch has, the less current will flow through it. For example, if 6 A of current enters the junction and splits between two branches of 2 Ω and 4 Ω, the branch that has twice the resistance will receive half the current. Thus, the current will split into 4 A and 2 A, respectively. Keeping this in mind always helps me determine whether my circuit solution is reasonable.

Online Activity Online Activity boxes provide interactive activities, such as computer simulations, that help with concept development. Many of these are matched to the physics education research-validated PhETs.

ONLINE ACTIVITY

Simple and Not So Simple DC Circuits

The e-resource that accompanies every new copy of this textbook contains an Online Activity using the PhET simulation "Circuit Construction Kit." Work through the simulation and accompanying questions to gain an understanding of DC circuits.

Key Equations It is important for students to differentiate fundamental relationships from equations that are used in steps of derivations and examples. Key equations are clearly indicated.

KEY EQUATION

$$I = I_e \cdot e = (n_e v_d A)e = n_e v_d A (1.6 \times 10^{-19}\,\text{C}) \quad (22\text{-}4)$$

Examples Each example is numbered and corresponds to each major concept introduced in the section. Examples include a statement of the problem, the solution, and "Making sense of the result." Within the example, the authors have modelled desired traits, such as care with units and consideration of appropriate significant figures. "Making sense of the result" is one of the most important features, in which authors model the idea of always considering what has been calculated to determine whether it is reasonable.

EXAMPLE 22-2

Applying Ohm's Law to a Current-Carrying Wire

A potential difference of 100 mV is applied across a 5.0 m-long metal wire that has a diameter of 0.50 mm. The resulting electric current is 66.7 mA.

(a) Find the resistance of the wire.
(b) Find the resistivity of the wire, and suggest what metal it could be made of.

SOLUTION

(a) We assume that the metal wire obeys Ohm's law. Therefore, we can find its resistance, R, using Equation (22-15):

$$R = \frac{V}{I} = \frac{100 \text{ mV}}{66.7 \text{ mA}} = \frac{100 \times 10^{-3} \text{ V}}{66.7 \times 10^{-3} \text{ A}} = 1.50 \ \Omega$$

(b) Rearranging Equation (22-14), we can find the resistivity of the wire:

$$\rho = \frac{RA}{l} = \frac{(1.5 \ \Omega) \cdot \pi (0.25 \times 10^{-3} \text{m})^2}{5.0 \text{ m}} = 5.9 \times 10^{-8} \ \Omega \cdot \text{m}$$

Comparing this value to the resistivities listed in Table 22-2, we find that the wire could be made of zinc.

Making sense of the result:

Resistance and resistivity are different physical quantities. The resistance provides a description of how a particular piece of wire resists the flow of electric current. The resistivity is the resistance of a hypothetical wire that is 1 m long and has a cross-sectional area of 1 m².

Checkpoints Each learning objective has a Checkpoint box to test students' understanding of the material they have just read. Checkpoint boxes include questions in different formats, followed immediately by the answer placed upside-down at the end of the box. While different formats are used, in all cases these checkpoints have a single clear answer so that they can be self-administered and students know whether they have mastered the concept before moving on to dependent material. The close linking of sections, learning objectives, and checkpoints is a major feature of the text.

Putting It All Together The Putting It All Together section at the end of each chapter provides students with an opportunity to review the key concepts discussed in the chapter. Care has been taken to make these concise and yet at the same time cover all core ideas.

 CHECKPOINT

C-22-6 The Relative Brightness of Two Light Bulbs in Series and Parallel Electric Circuits

A 25 W light bulb and a 40 W light bulb are first connected in series. Then they are connected in parallel. Which statement describes their relative brightness?

(a) The 25 W light bulb is brighter in a series circuit and dimmer in a parallel circuit.
(b) The 25 W light bulb is brighter in a parallel circuit and dimmer in a series circuit.
(c) The 25 W light bulb is always brighter.
(d) The 40 W light bulb is always brighter.

C-22-6 (b) The light bulbs are designed to be used in a parallel circuit; therefore, the higher the nominal power (power rating), the less the resistance of the light bulb (higher-rated light bulbs have thicker filaments).

Questions, Problems by Section, and Comprehensive Problems are provided at the end of each chapter to test the students' understanding of the material. For the problems, star ratings are used (*, **, or ***), with more stars indicating more challenging problems.

QUESTIONS

1. Which of the following is *not* a characteristic of physics?
   (a) Physics seeks to explain phenomena in the simplest way possible and in a way that has the widest application.
   (b) While some phenomena may be described qualitatively, in general, physics seeks quantitative descriptions.
   (c) Physics is primarily involved with developing a large number of equations so that there is one equation relevant for any given situation.
   (d) Physics requires creativity.

2. Assume that measurements of a time for some process are normally distributed, with no systematic errors. About two-thirds of the time the value is between 3.4 and 4.8 s. What are the mean and standard deviation of the time measurement?

3. Assume that measurements of a length are normally distributed, with no systematic errors. About 19 times out of 20, the value is between 2.2 and 5.8 m. What are the mean and standard deviation of the length measurement?

4. How many significant figures are in the number 230 100?
   (a) 3
   (b) 4
   (c) 5
   (d) 6

5. Complete the calculation $\frac{12.0 \times 5.0}{4.000}$. According to the rules for significant digits, how should the answer be written?
   (a) 15
   (b) 15.0
   (c) 15.00
   (d) 15.000

6. Which of the following is *not* a base SI unit?
   (a) kg
   (b) s
   (c) N
   (d) cd

7. Which of the following is *not* a base SI unit?
   (a) kg
   (b) m
   (c) J
   (d) K

8. Work (or energy) is equal to force times x. Use Tables 1-2 and 1-3 to identify the units of x.

9. Consider the graph of Figure 1-4. Taking into account the various points made in this chapter, critically evaluate the graphical presentation, indicating all aspects that could be improved.

10. (a) Use a reference source to look up how the length of the metre is currently defined.
    (b) From this definition, what must be the speed of light (to the number of significant figures inherent in the definition)?
    (c) Provide a brief overview of some of the previous ways that have been used to define the length of the metre.

11. Use a reliable reference to determine how the second is currently defined.

12. (a) Use a reliable reference to determine the method currently used to define the kilogram.
    (b) Provide a brief overview of some other ways that have been used to define the kilogram.

13. Describe the current method used to define the ampere. Include a reference to the source of your information.

14. Use a reliable reference source to identify the method currently used to define the candela, as well as the method that was historically used.

15. How many cubic centimetres are in one cubic metre?
    (a) 100
    (b) 1 000
    (c) 10 000
    (d) 1 000 000

16. Two data sets, A and B, have the same mean value, but data set B has a larger standard deviation by a factor of 2. There are 16 data points in data set A. How many data points are needed in set B for the SDOM to be identical for the two sets?

17. If we multiply a force (N) by a velocity (m/s), which of the following will be the units of the resulting quantity?
    (a) kg/s
    (b) J
    (c) Pa
    (d) W

PROBLEMS BY SECTION

For problems, star ratings will be used, (*, **, or ***), with more stars meaning more challenging problems.

Section 1-3   Mean, Standard Deviation, and SDOM

18. * A set of 16 measurements has a mean of 4.525 and a standard deviation of 1.2. How should the mean be written with an uncertainty given by the SDOM?

19. * The SDOM is 1.2, the mean is 4.8, and the standard deviation is 2.2. Approximately how many data points must be in the set?

20. * According to a journal article, experiments suggest that 19 times out of 20, the measurement obtained is within ±0.5 of 10.5. What must be the standard deviation and mean values, assuming that the data is normally distributed?

Section 1-4   Significant Figures

21. * How many significant figures are in each of the following?
    (a) 324.2
    (b) 0.032
    (c) 9.004
    (d) 12 000
    (e) 85.0
    (f) 6000.

22. ** Suppose that the uncertainty in the length 17.386 22 m is ±0.04 m. Round the length to the correct number of significant figures.

Section 1-5   Scientific Notation

23. * Write each of the following in scientific notation.
    (a) 2452
    (b) 0.592
    (c) 12 000
    (d) 0.000 045

24. * Perform the following calculation, and express the final result in scientific notation with the correct number of significant figures.

$$\frac{(3.20 \times 10^5)(1.5 \times 10^4)}{6.400 \times 10^3}$$

MECHANICS

ANCILLARIES

An extensive array of supplemental materials is available to accompany the first edition of *Physics for Scientists and Engineers: An Interactive Approach*. The supplements are designed to make teaching and learning more effective. For more information on any of these resources, please contact your local Nelson Education sales representative or call Nelson Education Ltd. Customer Support at 1-800-268-2222.

Instructor Ancillaries

The **Nelson Education Teaching Advantage (NETA)** program delivers research-based instructor resources that promote student engagement and higher-order thinking to enable the success of Canadian students and educators.

Instructors today face many challenges. Resources are limited, time is scarce, and a new kind of student has emerged: one who is juggling school with work, has gaps in his or her basic knowledge, and is immersed in technology in a way that has led to a completely new style of learning. In response, Nelson Education has gathered a group of dedicated instructors to advise us on the creation of richer and more flexible ancillaries that respond to the needs of today's teaching environments.

The members of our editorial advisory board have experience across a variety of disciplines and are recognized for their commitment to teaching. They include:

Norman Althouse, Haskayne School of Business, University of Calgary

Brenda Chant-Smith, Department of Psychology, Trent University

Scott Follows, Manning School of Business Administration, Acadia University

Jon Houseman, Department of Biology, University of Ottawa

Glen Loppnow, Department of Chemistry, University of Alberta

Tanya Noel, Department of Biology, York University

Gary Poole, Senior Scholar, Centre for Health Education Scholarship, and Associate Director, School of Population and Public Health, University of British Columbia

Dan Pratt, Department of Educational Studies, University of British Columbia

Mercedes Rowinsky-Geurts, Department of Languages and Literatures, Wilfrid Laurier University

David DiBattista, Department of Psychology, Brock University

Roger Fisher, PhD

In consultation with the editorial advisory board, Nelson Education has completely rethought the structure, approaches, and formats of our key textbook ancillaries.

We've also increased our investment in editorial support for our ancillary authors. The result is the Nelson Education Teaching Advantage and its key components: *NETA Engagement*, *NETA Assessment*, and *NETA Presentation*. Each component includes one or more ancillaries prepared according to our best practices, and a document explaining the theory behind the practices.

NETA Engagement presents materials that help instructors deliver engaging content and activities to their classes. Instead of instructor's manuals that regurgitate chapter outlines and key terms from the text, NETA Enriched Instructor's Manuals (EIMs) provide genuine assistance to teachers. The EIMs answer questions like *What should students learn?*, *Why should students care?*, and *What are some common student misconceptions and stumbling blocks?* EIMs not only identify the topics that cause students the most difficulty, but also describe techniques and resources to help students master these concepts. Dr. Roger Fisher's *Instructor's Guide to Classroom Engagement (IGCE)* accompanies every Enriched Instructor's Manual. (Information about the NETA Enriched Instructor's Manual prepared for *Physics for Scientists and Engineers: An Interactive Approach* is included in the description of the IRDVD below.)

NETA Assessment relates to testing materials: not just Nelson's Test Banks and Computerized Test Banks, but also in-text self-tests, Study Guides and web quizzes, and homework programs like CNOW. Under *NETA Assessment*, Nelson's authors create multiple-choice questions that reflect research-based best practices for constructing effective questions and testing not just recall but also higher-order thinking. Our guidelines were developed by David DiBattista, a 3M National Teaching Fellow whose recent research as a professor of psychology at Brock University has focused on multiple-choice testing. All Test Bank authors receive training at workshops conducted by Prof. DiBattista, as do the copyeditors assigned to each Test Bank. A copy of *Multiple Choice Tests: Getting Beyond Remembering*, Prof. DiBattista's guide to writing effective tests, is included with every Nelson Test Bank/Computerized Test Bank package. (Information about the NETA Test Bank prepared for *Physics for Scientists and Engineers: An Interactive Approach* is included in the description of the IRDVD below.)

NETA Presentation has been developed to help instructors make the best use of PowerPoint® in their classrooms. With a clean and uncluttered design developed by Maureen Stone of StoneSoup Consulting, NETA Presentation features slides with improved readability, more multi-media and graphic materials, activities to use in class, and tips for instructors on the Notes page. A copy of *NETA Guidelines for Classroom Presentations* by Maureen Stone is included with each set of PowerPoint slides. (Information about the NETA PowerPoint® prepared for *Physics for Scientists and Engineers: An Interactive Approach* is included in the description of the IRDVD below.)

IRDVD Key instructor ancillaries are provided on the *Instructor's Resource DVD* (ISBN 978-0-17-666168-7), giving instructors the ultimate tool for customizing lectures and presentations. (Downloadable web versions will also be available at **www.physics1e.nelson.com**.) The IRDVD includes

- **NETA Engagement:** The Enriched Instructor's Manual is organized according to the textbook chapters and addresses eight key educational concerns, such as typical stumbling blocks students face and how to address them. Other features include tips on teaching using cases, as well as suggestions on how to present material and use technology and other resources effectively, integrating the other supplements available to both students and instructors. This manual doesn't simply reinvent what's currently in the text; it helps the instructor make the material relevant and engaging to students. The Enriched Instructor's Manual was written by Dan MacQueen of Okanagan College.

- **NETA Assessment**: The Test Bank includes multiple-choice questions written according to NETA guidelines for effective construction and development of higher-order questions. Also included are true/false and completion-type questions. Test Bank files are provided in Word format for easy editing and in PDF format for convenient printing whatever your system.

  The Computerized Test Bank by ExamView® includes all the questions from the Test Bank. The easy-to-use ExamView software is compatible with Microsoft Windows and Mac. Create tests by selecting questions from the question bank, modifying these questions as desired, and adding new questions you write yourself. You can administer quizzes online and export tests to WebCT, Blackboard, and other formats. The NETA Test Bank was written by Naeem Syed Ahmed of Laurentian University.

- **NETA Presentation:** Microsoft® PowerPoint® lecture slides are available for every chapter and include key figures, tables, and photographs from *Physics for Scientists and Engineers: An Interactive Approach*. NETA principles of clear design and engaging content have been incorporated throughout. The NETA Presentation was created by Jason Harlow of the University of Toronto and Dan MacQueen of Okanagan College.

- **Image Library:** This resource consists of digital copies of figures, short tables, and photographs used in the book. Instructors may use these jpegs to create their own PowerPoint presentations.

- **Day One:** Day One—Prof InClass is a PowerPoint presentation that you can customize to orient your students to the class and their text at the beginning of the course.

- **TurningPoint®:** Another valuable resource for instructors is **TurningPoint® classroom response software** customized for *Physics for Scientists and Engineers: An Interactive Approach*. Now you can author, deliver, show, access, and grade, all in PowerPoint ... with no toggling back and forth between screens! JoinIn on Turning Point is the only classroom response software tool that gives you true PowerPoint integration. With JoinIn, you are no longer tied to your computer. You can walk about your classroom as you lecture, showing slides and collecting and displaying responses with ease. There is simply no easier or more effective way to turn your lecture hall into a personal, fully interactive experience for your students. If you can use PowerPoint, you can use JoinIn on TurningPoint! (Contact your Nelson Education publishing representative for details.) These contain poll slides and pre- and post-test slides for each chapter in the text.

- **Instructor's Solutions Manual:** This manual has been independently checked for accuracy. It contains complete solutions to all end-of-chapter problems in the book. The Instructor's Solutions Manual was written by Naeem Syed Ahmed of Laurentian University.

Nelson YouBook The innovative YouBook available through Enhanced WebAssign allows full instructor customization and robust class-communication tools, including videos, animations, interactive flashcards, online glossary, stimulating games to motivate student review, and additional materials only available online.

Student Ancillaries

A groundbreaking homework management system, Enhanced WebAssign (EWA) combines our exceptional physics content with the most flexible online homework solution. Powerful, effective, yet simple to use, EWA provides easy problem selection along with automatic grading and interactive tutorial assistance for students that reinforce the book's pedagogical approach.

EWA engages students with immediate feedback, rich tutorial content, and interactive e-books, helping them develop a deeper conceptual understanding of the subject matter. EWA offers interactive features that go far beyond simply duplicating text problems online. Students can watch solution videos, see problems solved step-by-step, and receive feedback as they complete their homework. Online assignments can be built by selecting from thousands of text-specific problems or supplemented with problems from specific Cengage Learning and Nelson Education textbooks. EWA promotes active learning and time on task, and respects diverse ways of learning.

The more you study, the better the results. Make the most of your study time by accessing everything you need to succeed in one place. Read your textbook, take notes, watch videos, and take practice quizzes—online with Premium Website.

Student Solutions Manual (ISBN 978-0-17-667213-3) The Student Solutions Manual includes solutions to all odd-numbered questions.

Acknowledgements

The production of a "ground up" new physics textbook is a major endeavour and would not have been possible without the expert work and assistance of many people.

The Nelson Education team has been amazing—this book would never have been completed without their expertise, attention to detail, flexibility, and above all emphasis on producing a high quality and innovative text. Particular credit goes to Paul Fam, Publisher—Higher Education who has so enthusiastically guided and supported the project from the earliest days. During the key stages of the project the Senior Developmental Editor was Mark Grzeskowiak. His extensive experience and professionalism were critical in moving us from rough drafts to finished manuscript. Because of the length of the project, in the early stages we worked with two different editors; first Tracy Yan and then Alwynn Pinard. Their leadership helped with critical decisions on the format and composition of the book. In particular we note that it was Alwynn who first suggested that each section of the text should be tied to exactly one learning objective—a key feature of the text.

A text written by a team of physicists who had not previously worked together posed a challenge in making the final book have a common voice and a consistent approach. The success we have achieved in that regard is due in large part to our substantive editor, David Peebles, and our copy editor, Betty Robinson. Words cannot adequately express the debt we owe to these two incredible professionals. David's background in physics, along with his editing experience, helped him to play a major role in suggesting alternative and clearer ways to explain concepts. He helped us to write more concisely and more clearly. Betty demonstrated a care and expertise that improved every single page of the book. As authors we were time and time again amazed at the care to detail she demonstrated. The fact that she is also such an efficient and personable individual made her contribution even more special.

The production stage was more demanding and complex than the authors realized at the outset. We thank the many people who helped us through this process—often under tight deadlines—especially Christine Gilbert, Content Production Manager, who had primary responsibility for overall production issues. Kristiina Paul, our photo researcher, worked hard to get permissions for our first choices for images, and, when they were not available, to find suitable alternatives. As a first edition the task of creating the various figures was challenging. We thank in particular Indhu Gunasekaran, and her team at Integra, and Deborah Crowle of the CrowleArt group. The publisher and the author team would also like to convey their thanks to Santo D'Agostino for his technical edit of the text. Santo's edit ensured consistency in key areas, such as the use of significant digits, accuracy in the figures, tables, and photos, and making sure all of the steps were accounted for in the examples presented.

The text is of course only part of the learning package. We acknowledge the leadership of Roberta Osborne, Senior Development Editor, in that component of the work, and the valuable work of the various ancillary authors. The authors would also like to thank Naeem Syed Ahmed, Mihai Gherase, and Santo D'Agostino for their work on the solutions manual to accompany this text, and the help they provided in reviewing the many end-of-chapter problems contained herein.

Thanks go to those who reviewed the text. Collectively these professionals offered ideas, and occasional corrections, that helped make the book more accurate and improved clarity. The diversity of their views of physics and how it should be taught—while occasionally resulting in that not all suggestions could be incorporated in this printing—provided us with a broader view than that of the five authors alone. We give thanks to the following individuals:

Ben Newling, University of New Brunswick
Syed Ahmed Naeem, Laurentian University
Tetyana Antimirova, Ryerson University
Mark Laidlaw, University of Victoria
Calvin Kalman, Concordia University
Erik Rosolowsky, University of British Columbia—Okanagan
Salam Tawfiq, University of Toronto
Stan Greenspoon, Capilano University
Mahi Singh, University of Western Ontario
Roman Koniuk, York University
Neil Alberding, Simon Fraser University
Ruxandra M. Serbanescu, University of Toronto

The Student Advisory Board (SAB) members helped us remain grounded in what sort of text students wanted and would use, and most of them also contributed to the Peer to Peer boxes ensuring that the material in this text is presented from a truly "student" perspective. The members of the Mount Allison-related SAB were Kevin Alexander, Jeremy Bourque, Emma Driedger, Darin Eddy, James Ehrman, Ashley Faloon, Monica Firminger, Theo Heffler, Kyle Hill, Catherine Hughes, Luyao Li, Ellen Milley, R. Bruce Muir, Andrew Nelles, Ryan Robski, Paras Satja, Ari Silburt, Asmita Sodhi, Sarah Thomas, Evelyn Wainewright, David Watson, and Rory Woods. The members of the Waterloo-based SAB were Andrew Boyko, Xin Jin, Mathew Michael, Ali Ali-Mahmoud, Mukund Vermuri, Yun Zhe Li, James Gerinea, Scott Duncan, David Lu, Calvin Lu, Kurt McCarrell, Aaamir Jawaid, and Lawrence Lu.

We thank Sean Chamberland, Executive Marketing Manager, and Leanne Newell, Marketing Manager at

Nelson Education for their skilled promotion of the book, along with the team of publisher's sales representatives across the country.

To be honest, this book has taken more of our time and energy than any of the authors ever anticipated. All of the authors combined the writing of this text with other career demands, and as a result many weekends and evenings were devoted to this text. We thank our family members for their understanding and encouragement. We thank our colleagues for their support in various ways during the course of this project. The authors would also like to acknowledge Rohan Jayasundera and Simarjeet Saini for enlightening discussions. The authors would like to thank Olga Myhaylovska for her valuable input and advice. The authors would also like to thank Reema Deol and Renee Chu for conducting background research on some of the material used in the text.

The authors would like to thank professor David F. Measday (UBC) for providing valuable feedback for Chapter 33, Introduction to Nuclear Physics.

Daily in our classrooms we learn from our students, and are rejuvenated by their enthusiasm, creativity, and energy. Our view of the teaching and learning of physics owes a great deal to them, probably more than they realize. Similarly, our interactions with colleagues, both at our own institution and beyond, have shown us new and more effective ways to approach difficult concepts, and helped in our own education as physicists. We would like to acknowledge the valuable work done by organizations such as the Canadian Association of Physicists Division of Physics Education (CAP DPE), and the American Association of Physics Teachers (AAPT).

Chapter 1
Introduction to Physics

Why should you study physics? To start with, physics helps us answer amazing questions. For example, physics has provided a remarkably detailed picture of what the universe is like and how it has developed over time. The Hubble Space Telescope produced the image in Figure 1-1. The image shows an area of the sky equivalent to what you would cover if you held a 1 mm square at arm's length. Yet this image shows about 10 thousand galaxies, and each galaxy typically contains 100 billion stars. In the hundred years since the first evidence of the existence of galaxies, observations and theoretical calculations by numerous physicists have provided strong evidence that the evolution of the universe started in a "big bang" about 13.7 billion years ago, when the universe was infinitesimally small, almost infinitely dense, and incredibly hot.

Like cosmology, aspects of research in particle physics, quantum mechanics, and relativity can be fascinating because they challenge our commonsense ideas. However, discoveries in these fields have also led to numerous extremely useful applications. For example, atomic physics underlies medical imaging technologies from simple X-rays to the latest MRIs and CAT scans. In fact, physics concepts are the basis for almost all technologies, including energy production and telecommunications. As you explore this book, each chapter will bring a new answer to the question, "Why should physics matter to me?"

R. Williams (STScI), the Hubble Deep Field Team and NASA

Figure 1-1 A tiny part of the night sky captured with incredible detail by the Hubble Space Telescope

Learning Objectives

When you have completed this chapter you should be able to:

1 Explain the distinguishing characteristics of physics.

2 Differentiate between precision and accuracy, and list the types of possible errors.

3 Calculate the mean, standard deviation, and standard deviation of the mean (SDOM) for data sets, and correctly use $\pm$ notation.

4 Relate uncertainties to significant figures, and state the number of significant figures in expressions.

5 Use scientific notation, and convert quantities to and from scientific notation.

6 Correctly apply significant figure rules to calculated quantities.

7 Use and interpret error bars to express uncertainties in values.

8 State the SI units for physical quantities, and write the units and their abbreviations correctly.

9 Apply dimensional analysis to help determine whether a relationship is correct.

10 Perform unit conversions.

11 Make reasonable order-of-magnitude estimates in solving open problems.

12 Comment on what a professional physicist does and how one becomes a physicist.

 For additional Making Connections, Examples, and Checkpoints, as well as activities and experiments to help increase your understanding of the chapter's concepts, please go to the text's online resources at www.physics1e.nelson.com.

LO 1

1-1 What Is Physics?

We will start this chapter by considering some defining characteristics of physics.

The domain of physics is the physical universe Physics does not seek to answer questions of religion, literature, or social organizations.

Physics is a quantitative discipline Although there are a few topics in physics where our understanding is mainly qualitative, overall, measurement and calculation play critical roles in developing and testing physics ideas.

Physics is strongly model-driven Physicists develop hypotheses and models, use them to make predictions, and then test those predictions with experiments and observations. Through the study of physics you can learn analytical skills that can be applied beyond the sciences and engineering. For example, a number of physicists find employment developing economic and investment models for financial institutions.

Physicists need to be creative Physicists design experiments, find applications for physical principles, and develop new models and theories. Some philosophers of science have asked whether the electron was invented or discovered. Such questioning stresses that, while there is a part of physics that is independent of the observer, the specific models we develop to help understand nature critically depend on the creativity and imagination of physicists. It is not surprising that many physicists also work in other creative areas, such as music and art.

Physics interfaces strongly with other sciences The study of physics is increasingly interdisciplinary, with many physicists working in areas that span physics and other disciplines, for example, medical physics, biophysics, chemical physics, materials science, geophysics, or physical environmental science.

Physics is a highly collaborative discipline Most physicists routinely work with colleagues from other countries, often using international research facilities. Pick up a physics research journal, and you will see that the majority of papers are written by collaborations of scientists from different institutions and countries.

Physics is both deeply theoretical and highly applied Some physicists work exclusively in fundamental areas that have no immediate application, whereas others concentrate on solving applied problems. Sometimes work initially deemed to have little practical use turns out to

MAKING CONNECTIONS

Creative Physicist and Musician

Canadian Diane Nalini de Kerckhove is both a brilliant physicist and a highly acclaimed jazz musician. Growing up, she studied ballet, modern dance, and jazz and considered a career in one of those fields. After an undergraduate degree in physics at McGill University, Montréal, Quebec, she studied at Oxford University, United Kingdom, as a Rhodes Scholar, where she did her graduate work using a powerful analytical instrument called PIXE (proton-induced X-ray emission). At the University of Guelph, her research projects included strain measurements in semiconductors, ion optics, and trace element analysis of human hair samples.

She has performed as a jazz singer on national radio and has released several CDs, including *Kiss Me Like That*, which is based on music created for the International Year of Astronomy (2009), and *Sweet Fire*, which contains Shakespeare sonnets put to music.

Guelph Mercury–Nicki Corrigall/The Canadian Press

Figure 1-2 Physicist and musician Diane Nalini de Kerckhove, a Rhodes Scholar, a researcher in high-resolution nuclear microprobe technology, and an acclaimed jazz musician

have important applications. For example, in 1915, Albert Einstein published a new theory of gravitation called general relativity. When general relativity was developed, it had no practical application. Today, however, the Global Positioning System (GPS) would be hopelessly inaccurate without corrections for the gravitational effects predicted by general relativity theory.

Physics seeks explanations with the greatest simplicity and widest realm of application Physics is not a collection of a huge number of laws. Rather, physics seeks to explain the physical universe and all that it contains, using a limited number of relationships. For example, we

MAKING CONNECTIONS

The CCD: Applied Physics and a Nobel Prize

The 2009 Nobel Prize in Physics was awarded to three scientists: the late Canadian Willard S. Boyle and American George E. Smith for the invention of the charge coupled device (CCD), and Charles Kuen Kao from China for work leading to fibre-optic communication. Born in Amherst, Nova Scotia, Boyle studied at McGill University before working at Bell Laboratories in New Jersey, where he and Smith made the first CCD.

The CCD is a semiconductor device with rows of tiny cells that accumulate an electric charge proportional to the light intensity at each cell. The cells are arranged such that a sequence of voltage pulses applied to the cells causes the charge in each cell to transfer to the next cell in the row. The charge leaving the last cell produces a signal that corresponds to the light that was focused on the cells. This signal is amplified to make an electronic record of the image that was stored as charge on the CCD.

CCD imaging has many advantages over film, including substantially greater sensitivity, which is of great importance to astronomers. CCDs are the heart of the Hubble Space Telescope—indeed, space telescopes would not be feasible without them. CCDs in camcorders

and digital cameras have revolutionized photography, making photographic film almost obsolete. CCDs have also enabled significant advances in medical and scientific instrumentation.

Reprinted with permission of Alcatel-Lucent USA Inc

Figure 1-3   Willard Boyle (left) and George Smith in 1970, shortly after their invention of the charge coupled device (CCD)

only need to invoke four types of interactions to explain all "forces" in physics: gravitation, electromagnetism, weak nuclear, and strong nuclear. In your study of physics it is critical to keep in mind this goal of applying core theoretical ideas in a wide variety of situations. When you have finished this book, perhaps you may be able to list the key ideas of physics on a single page.

Physics uses equations extensively to express ideas You should view physics equations as a shorthand notation for the theories and relationships they represent. Applying physics is not simply selecting from a large pool of

established equations. You will need to develop proficiency in manipulating equations and deriving relationships from basic principles.

LO 2

1-2 Measurement and Uncertainties

Why do we measure? Often we make measurements to test hypotheses. For example, if a technician proposes that a new material expands less in response to temperature changes than some other material, we can use physics measurements to test this claim.

Sometimes, we make measurements to learn about relationships between quantities and to help develop models and theories. For example, in the 1920s, astronomers Edwin Hubble and Milton Humason made measurements of the light from 46 galaxies and found that the light from the more distant galaxies has a greater redshift. (A *redshift* is a shift to longer wavelengths caused by motion away from the observer.) These measurements showed that the relationship between velocity of recession and distance was approximately linear, as shown in Figure 1-4. This relationship, in turn, led to the startling conclusion that the universe is expanding.

✓ CHECKPOINT

C-1-1 What Is Physics?

Which of the following activities is *not* consistent with our description of physics.
(a) testing models for subatomic particles with experiments
(b) collecting observations of interacting galaxies
(c) developing a plan for an experiment to test a new theory of gravity
(d) creating a new mathematics function

C-1-1 (d) Physics should in some way be linked to the physical universe, not simply the world of mathematical definitions (although physicists might later use that mathematical function).

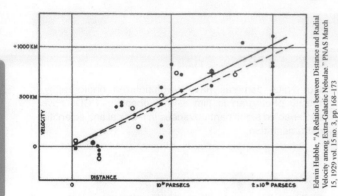

Figure 1-4 The plot of expansion velocity (in km/s) versus distance (in parsecs) published by Hubble and Humason in 1925. A parsec is the distance that light travels in 3.26 years, so the galaxies in the middle of the graph are about 3 million light years away.

Edwin Hubble, "A Relation between Distance and Radial Velocity among Extra-Galactic Nebulae." PNAS March 15, 1929 vol. 15 no. 3, pp. 168–173

Sometimes, we make measurements not to develop a theory, but rather to test its predictions. At other times, we make measurements to obtain precise values for quantities such as the speed of light and the charge on an electron.

Now, suppose that measurements of the expansion of two materials showed that for a given increase in temperature, the length of one material increased by 0.0025% and the length of the other material increased by 0.0030%. Do these measurements mean that the first material has a lower expansion coefficient? The answer depends not only on the measurements, but on the uncertainties in those measurements.

There are three different classes of potential errors in measurements of physical quantities: true errors, systematic errors, and inherent uncertainties.

True errors are mistakes made in the measurement. These errors can in general be identified by conducting more measurements, ideally by experimenters using different techniques. Examples of true error are an experimenter reading a scale on a measurement incorrectly, or transcribing a reading into a notebook incorrectly, or connecting electrical equipment incorrectly so that the measurement made is not the one intended.

Systematic errors result from some bias in the way that the quantity is being measured. For example, if you are measuring an electrical current with a meter that is not accurately calibrated, all the measured values might be somewhat higher (or lower) than the true values.

Inherent uncertainties (also called **random errors**) result from limitations in the precision of the measurements. No matter which technique is used, a measurement cannot be made with absolute precision, so there is always some uncertainty. In Chapter 31, you will learn how the uncertainty principle determines the minimum uncertainty inherent in all measurements.

We need to distinguish between the terms "precision" and "accuracy." **Accuracy** is how close the measurement of some quantity is to the true value. **Precision** is a measure of how close to one-another repeated measurements are. For example, if you measured the length of a box and got 100 values all between 1.259 and 1.260 m, then the precision would be quite good. However, if the actual length of the box is 1.204 m, then the accuracy is rather poor (and you need to look for the cause of the systematic error).

MAKING CONNECTIONS

Becoming a Physicist: Conventional and Unconventional Paths

Edwin Hubble and Milton Humason (Figure 1-5) came to careers in physics by radically different paths. Hubble took an undergraduate degree in astronomy and mathematics and then studied law as one of the first Rhodes Scholars. He practised law in Kentucky but soon abandoned law for a doctorate in astronomy at the University of Chicago. He graduated in 1917. After serving in World War I, Hubble got a job at the Mount Wilson Observatory in California.

Humason dropped out of school at the age of 14. He was a mule driver and hauled materials and equipment up Mount Wilson when the observatory was being built. Humason joined the observatory as a janitor and then volunteered to work as a night assistant. The quality of his work led to an appointment of staff astronomer despite his lack of academic credentials. Among Humason's discoveries is the giant comet C/1961 R1, which is named after him.

Today, the vast majority of physicists follow a path of undergraduate degree followed by graduate work, but there are still a few rare exceptions of individuals who come to physics by unconventional routes, as Humason did.

Figure 1-5 A group of distinguished physicists at Mount Wilson Observatory in 1931. Milton Humason and Edwin Hubble are at the left. Just to the left of Einstein is Albert Michelson, whose work is discussed in Chapters 28 and 29.

The best way to state the uncertainty in a measured quantity is to provide an estimate of the uncertainty. That is, we could write a length measurement as 1.15 ± 0.08 m, which means that the true value is likely between 1.07 and 1.23 m. Often the $\pm$ value is selected so that the true value is within the stated range about two-thirds of the time. In some fields, the criterion is 95% certainty, meaning, in this case, that 95 times out of 100 the true value will be between 1.07 and 1.23 m. We will consider these criteria in more detail in the next section, along with the question of how to decide whether two different mean values are significantly different.

✓ CHECKPOINT

C-1-2 Comparing Means

The mean of a set of length measurements for object A is 1.25 ± 0.05 m, and the mean of similar measurements for the length of object B is 1.32 ± 0.20 m. Can we conclude that B is longer than A? Explain your reasoning.

C-1-2 No. When the range of uncertainties is taken into account, A could be between 1.20 and 1.30 m long, whereas B is between 1.12 and 1.52 m long. Therefore, we cannot conclude that B is longer than A.

1-3 Mean, Standard Deviation, and SDOM

The first step in determining the uncertainty inherent in a set of measurements is to calculate the **mean** (or arithmetic average). For a set of N measurements with values x_i, the mean, μ, is defined as follows:

KEY EQUATION
$$\mu = \bar{x} = \frac{1}{N} \sum_{i=1..N} x_i \qquad (1\text{-}1)$$

Consider the two sets of measurements shown in Table 1-1 and plotted in Figure 1-6. They both have the same mean value (5.0), but they have very different distributions: set A has values with a much smaller spread about the mean value.

Table 1-1 Two Sets of Data with the Same Mean

Set A	Set B
5.5	3.0
5.0	8.5
4.7	7.0
4.8	1.5
4.4	5.5
5.6	4.5
Mean = 5.0	Mean = 5.0

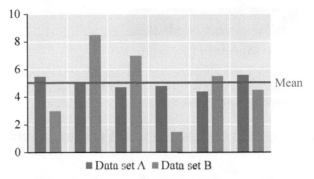

Figure 1-6 The two distributions have the same mean value (5.0), but the data shown with green bars has much more scatter around this mean value.

We can use the standard deviation, σ, to specify how near the values cluster around the mean value. The **standard deviation** is defined as follows:

KEY EQUATION
$$\sigma = \sqrt{\sum_{i=1..N} \frac{\left(x_i - \bar{x}\right)^2}{N-1}} \qquad (1\text{-}2)$$

The smaller the standard deviation, the more closely the measured values cluster around their mean. For example, data set A in Table 1-1 has a standard deviation of 0.47, and data set B has a standard deviation of 2.57.

For many types of measurements, intrinsic uncertainties (random errors) follow a **normal distribution** (also called a Gaussian distribution). This distribution is shown in Figure 1-7. Approximately 68% of the measurements from a normally distributed set lie within ± 1 standard deviation of the mean value, about 95% of the values lie within ± 2 standard deviations of the mean, and 99.7% lie within ± 3 standard deviations of the mean.

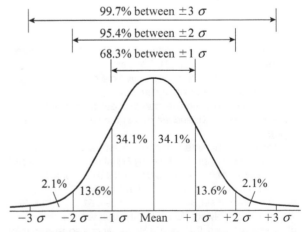

Figure 1-7 An example of a normal distribution (also called a Gaussian distribution)

5

C-1-3 Standard Deviation

The results of a public opinion poll are reported as being accurate to within 4%, 19 times out of 20. Expressed as a percentage, what is the implied standard deviation?

C-1-3 Nineteen times out of 20 is about 95%, so it must correspond to *two* standard deviations. Therefore, one standard deviation is 2% uncertainty.

The more measurements that we take, the closer the mean of the sampled values in our set of measurements will come to the true mean of the distribution. Now, we consider how to determine whether differences between two mean values are significant. You might think we could use the standard deviations of the two sets of data, but these standard deviations are actually quite misleading for comparing different sets of data. If the mean is based on a larger number of observations, its value is more certain, even if the standard deviation of the set is the same as the standard deviation of a smaller set.

Therefore, we use a quantity called the **standard deviation of the mean** (**SDOM**), which is defined in terms of the number of observations (N) used for the mean

and the standard deviation (σ). The SDOM is also called the *standard error of the mean*.

KEY EQUATION
$$\text{SDOM} = \frac{\sigma}{\sqrt{N}} \qquad (1\text{-}3)$$

For example, in the data in Table 1-1, set B has a SDOM of $\dfrac{2.57}{\sqrt{6}}$, or about 1.0.

It is easy to make mistakes when interpreting the standard deviation and the SDOM. Remember that the standard deviation is a measure of the uncertainty of a single value, and the SDOM is a measure of the uncertainty of a mean. For example, about 68% of the time, a single measurement will be within ± 1 standard deviation of the mean value, and about 68% of the time, the mean of a set of measurements will be within ± 1 SDOM of the true value. Therefore, the SDOM is used when you have taken the average value from multiple samples of two different populations, and you want to determine whether the sets are statistically different. Most literature in physics uses the notation with the form mean $\pm$ SDOM when expressing a quantity, but you will sometimes see a notation with the SDOM in parentheses after the mean value.

 EXAMPLE 1-1

Significant Difference?

In an experiment, the decay time of a subatomic particle (the time the particle takes to break down into other particles) was measured repeatedly, producing the following values: 0.45, 0.37, 0.67, 0.84, 0.38, 0.55, and 0.48 μs. A different experiment measured decay times of 0.68, 0.48, 0.88, 0.87, and 0.46 μs. Can we conclude that the decay times in the two experiments are statistically different? Note: In this chapter we will use a criterion of ± 1 standard deviation, unless otherwise stated.

SOLUTION

First, we calculate the mean of each set of measurements using Equation (1-1). For set A (plotted in blue in Figure 1-8), we have a mean decay time of 0.53 μs, and for set B we have a mean decay time of 0.67 μs.

Next, we calculate the standard deviation of each set, using Equation (1-2), and we obtain the values 0.17 μs and 0.20 μs.

Now, we calculate the two SDOM values. Set A has $N = 7$, and set B has $N = 5$. Using Equation (1-3), we obtain SDOM values of 0.06 μs for set A and 0.09 μs for set B.

The mean of set A can be written as 0.53 ± 0.06 μs, and the mean of set B can be written as 0.67 ± 0.09 μs. So, we are reasonably confident that the true value of the mean of A lies between 0.47 and 0.59 μs, and also reasonably confident that the true value of the mean of B lies between

0.58 and 0.76 μs. Since these two ranges overlap slightly, we cannot conclude that the difference between the sets of data is statistically significant.

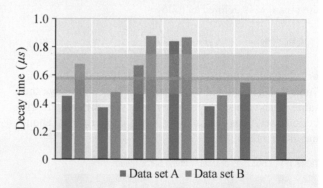

Figure 1-8 Comparison of the two sets of data for particle decay times. The blue bars represent the values from set A, and the green bars represent the values from set B. The shaded rectangles show the areas defined by the mean $\pm$ SDOM for each distribution.

Making sense of the result:

Note that, although we cannot conclude that the differences between the means are significant, we have not proven that the means are the same. If we took more measurements, we might find that the mean decay times are significantly different.

LO 4

1-4 Significant Figures

When uncertainty in values is not stated explicitly using a ±range or standard deviation, we assume that the uncertainty is indicated by the **significant figures** (sometimes called *significant digits*) in the data. For example, two measurements of a time interval are stated as 1.6 and 1.5984 s; the latter measurement implies a much greater precision. The first measurement indicates that the true value is somewhere between 1.55 and 1.65 s; the second measurement implies a much narrower range, between 1.598 35 and 1.599 45 s. We say that there are more significant digits in the second measurement than in the first. It is important not to state a measurement with more significant digits than justified by the true uncertainty in the value.

Unfortunately, conventions for determining the number of significant figures vary somewhat. In this book, we use the following rules:

1. All nonzero digits are always significant. For example, 345.6 has 4 significant digits.

2. Zeros between nonzero digits are always significant. For example, 1204 has 4 significant digits.

3. Zeros at the end of a number are significant only if they are to the right of the decimal point *or* followed by a decimal point. For example, both 450 and 45 000 have 2 significant digits, 450. has 3 significant digits, and both 450.0 and 1.560 have 4 significant digits.

4. Zeros at the beginning of a number are not significant. This rule makes the number of significant digits consistent when unit prefixes are changed. For example, the measurements 0.067 m and 6.7 cm express the same length, and both have 2 significant digits. Example 1-2 demonstrates the application of these rules.

EXAMPLE 1-2

Significant Figures

How many significant figures are in each of the following?
(a) 5.46 m
(b) 0.024 575 m
(c) 320 m
(d) 24.260 m
(e) 204 m

SOLUTION

(a) The measurement 5.46 m has 3 significant figures according to rule 1.

(b) The measurement 0.024 575 m has 5 significant figures (neither 0 is significant according to rule 4).

(c) The measurement 320 m has 2 significant figures according to rule 3.

(d) The measurement 24.260 has 5 significant figures according to rules 1 and 3.

(e) The measurement 204 m has 3 significant figures according to rule 2.

✓ CHECKPOINT

C-1-4 Significant Digits

How many significant digits are in each of the following numbers?
(a) 340.03
(b) 245 000
(c) 12.00

C-1-4 (a) 5 significant digits according to rule 1. (b) 3 significant digits according to rule 3. (c) 4 significant digits according to rule 3.

LO 5

1-5 Scientific Notation

Scientific notation eliminates any confusion over whether zeros at the end of a number are significant digits. Scientific notation also makes it easier to work with the very large and the very small numbers that we frequently encounter in physics.

In scientific notation we write the number 46 580 as 4.658×10^4. To convert a number to scientific notation, we determine the power of 10 factor that will let us express the number in a form with just one nonzero digit to the left of the decimal point. Example 1-3 demonstrates conversions to and from scientific notation.

EXAMPLE 1-3

Scientific Notation

(a) The approximate distance from Earth to the Sun is 150 000 000 000 m. Express this distance in scientific notation.
(b) How would 0.023 be written in scientific notation?
(c) Write the number 2.45×10^{-3} in regular notation.

SOLUTION

(a) We move the decimal point 11 places to the left to get 1.5×10^{11} m.
(b) We move the decimal point 2 places to the right to get 2.3×10^{-2}.
(c) We move the decimal point 3 places to the left to eliminate the multiplier, giving 0.002 45.

C-1-5 Scientific Notation and Significant Digits

How many significant digits are in the number 3.50×10^{-2}?

C-1-5 There are 3 significant digits in the number 3.50×10^{-2}.

LO 6

1-6 Significant Figures and Mathematical Operations

There are two basic rules for dealing with significant figures in calculations.

Multiplication and Division When multiplying and dividing quantities, the number of significant figures in the final answer should be the same as the least precise quantity used in the calculation, as demonstrated in Example 1-4. Note that integers and mathematical constants are assumed to have infinite significant figures. For example, in the expression for the surface area of a sphere, $4\pi r^2$, the factors 4 and π do not affect the number of significant figures in the calculated area.

EXAMPLE 1-4

Significant Figures in Multiplication and Division

Express the answer to the following calculation with the correct number of significant figures:

$$\frac{32.55 \times 0.021}{465}$$

SOLUTION

When we enter these numbers into a calculator, the answer will likely be displayed as 0.001 47 or 0.001 470. However, the number 32.55 has 4 significant figures, the number 0.021 has 2 significant figures, and the number 465 has 3 significant figures. Therefore, the final answer should have only 2 significant figures, and the calculated number should be rounded to 0.0015.

Addition and Subtraction When adding and subtracting quantities, we keep the same number of significant figures after (or in front of) the decimal point as in the least precise of the quantities used in the calculation. For example, the number 2.45 is assumed to have an uncertainty of the order of 0.01. If we subtract the number 2.30 from 2.45, we obtain 0.15—the uncertainty is still of the order of 0.01. This uncertainty stemming from the first measurement would not be reduced if the second number had a higher precision, such as 2.3043. Example 1-5 demonstrates the application of this rule.

EXAMPLE 1-5

Significant Figures in Addition and Subtraction

You measure the total length of two objects together as 2.45 m. You know that one of the objects has a length of 2.248 m. What is the length of the other object?

SOLUTION

A calculator will give an answer of 0.202 m. However, the measurement 2.45 m has 2 significant figures to the right of the decimal point. Therefore, the final answer should be written as 0.20 m.

Making sense of the result:

The measured lengths have 3 and 4 significant digits, but the calculated length has only 2. When we subtract two quantities that are approximately equal, the result will have fewer significant digits than the original quantities.

C-1-6 Computations and Significant Digits

How should the answer be written for the following calculation?

$$\frac{37.45 - 37.020}{124.67}$$

(a) 0.003 449
(b) 0.003 45
(c) 0.0034
(d) 0.003

C-1-6 (c) According to the subtraction rule, the operation in the numerator gives us 2 significant figures. Therefore, the final answer should have the lesser of 2 and 5 significant figures.

Note that the significant figure techniques described above give only an approximation of the uncertainty in measured and calculated quantities. More advanced courses in physics will show you more precise methods for dealing with the uncertainty in calculated quantities.

LO 7

1-7 Error Bars

When graphing data, we can indicate the uncertainty in the data with **error bars**. In Figure 1-9, the error bars are symmetric about the data points, and the uncertainty is the same for all the data values. However, uncertainties are often not symmetric, and they can vary with the data values. In some instances, there are significant uncertainties in the *x*-values as well, in which case the graph would have both vertical and horizontal error bars.

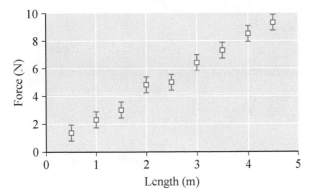

Figure 1-9 The uncertainty in *y*-values of measured quantities is indicated with error bars. When a data point is obtained by finding the mean of values, it is important to realize that we use the SDOM in determining the length of the error bar. Sometimes, however, the error bar is determined from an estimate of the uncertainty rather than a calculation.

✓ CHECKPOINT

C-1-7 Error Bars

A data point on a plot has the *x*-value determined by taking the mean of the numbers 2.4, 2.7, 3.0, 3.3, and 3.6. The *y*-value is obtained by taking the mean of the values 3.0, 4.0, 5.0, 6.0, and 7.0. Describe how the error bars should be drawn for this case.

C-1-7 First, consider the *x*-values, which have a mean of 3.0 and a standard deviation of 0.47. The error bar is drawn using the SDOM, however, which will be 0.47 divided by the square root of *N* or 2.2 in this case. Therefore, the error bar will extend a horizontal distance of 0.2 (in both directions). Similar arguments lead to a mean of 5.0 in the *y*-direction and a SDOM of 0.7, so the error bars in the *y*-direction extend 0.7 (in both directions) from the point.

Error bars are important enough to be included in most experimental results published in physics. These bars give a clear indication of uncertainty in the individual values and how much confidence we can have in any apparent trend in the data.

ONLINE ACTIVITY

Curve Fitting and Error Bars

The e-resource that accompanies every new copy of this textbook contains an Online Activity using the PhET simulation "Curve Fitting." Work through the simulation and accompanying questions to gain an understanding of error bars, curve fitting, and uncertainty.

1-8 SI Units

In this text, we use almost exclusively the International System of Units, or **SI**. The abbreviation comes from the French title, *le Système internationale d'unités*. This system is used worldwide now; in fact, only three countries have not adopted SI for commerce as well as for science: Burma (also known as Myanmar), Liberia, and the United States.

SI is an integrated metric system, designed so that it can be used readily by all the sciences. SI has seven base units, chosen such that all quantities can be expressed in some combination of the base units. SI also has a set of derived units, defined in terms of the base units, to provide convenient measures of particular quantities. Table 1-2 lists SI base units and the quantities they measure.

Table 1-2 The Base SI Units

Unit Name	Unit Symbol	Measures
metre	m	length
second	s	time
kilogram	kg	mass
kelvin	K	temperature
mole	mol	amount of substance
ampere	A	electric current
candela	cd	luminous intensity

The *metre*, *second*, and *kilogram* are units that you likely know well from everyday use. For historical reasons, the kilogram—not the gram—is the SI base unit of mass. The kilogram is the only base unit that has a prefix.

A *kelvin* is equal to a Celsius degree, but the zero point of the Kelvin temperature scale, denoted 0 K, is absolute zero, the lowest temperature possible (approximately −273°C).

A *mole* is the amount of a substance that contains the same number of elementary entities (e.g., atoms or molecules) as exactly 12 g of the isotope carbon-12, about 6.022×10^{23}. In other words, 1 mol contains Avogadro's number of atoms or molecules.

The *ampere* is a measure of electric current, the flow of electric charge past a given point. When you study electromagnetism later in this book, you will see why the designated base unit for electrical quantities is based on current rather than charge.

The *candela* is the unit of luminous intensity, a measure of the light output of a source. As you might guess from the name, the candela is based on the light output of a standard candle with a specified composition and rate of combustion. In fact, a typical candle has brightness in the order of 1 cd. The candela is a precisely defined unit that replaces older units.

As well as the base units, there are a number of fundamental derived SI units. We list the most important ones used in this book in Table 1-3.

There are a number of frequently used units not shown in Table 1-3. Some of these are simply combinations of the units listed. For example, speed and velocity have units of m/s, and acceleration (the rate of change of velocity) has units of m/s^2.

There are a number of commonly used non-SI units, including the minute (min), the solar day (d), the year (a or y), and the light year (ly), which is the distance light travels in one year.

Table 1-3 Derived SI Units Commonly Used in First-Year Physics Courses

Unit Name	Unit Symbol	Quantity	Base Unit Equivalent
newton	N	force	$m \cdot kg \cdot s^{-2}$
joule	J	energy, work	$m^2 \cdot kg \cdot s^{-2}$
watt	W	power	$m^2 \cdot kg \cdot s^{-3}$
hertz	Hz	frequency	s^{-1}
radian	rad	angle	$m \cdot m^{-1}$ (dimensionless)
steradian	sr	solid angle	$m^2 \cdot m^{-2}$ (dimensionless)
pascal	Pa	pressure	$m^{-1} \cdot kg \cdot s^{-2}$
coulomb	C	electric charge	$s \cdot A$
volt	V	electric potential difference	$m^2 \cdot kg \cdot s^{-3} \cdot A^{-1}$
ohm	Ω	electric resistance	$m^2 \cdot kg \cdot s^{-3} \cdot A^{-2}$
farad	F	electric capacitance	$m^{-2} \cdot kg^{-1} \cdot s^4 \cdot A^2$
henry	H	electromagnetic inductance	$m^2 \cdot kg \cdot s^{-2} \cdot A^{-2}$
tesla	T	magnetic field strength	$kg \cdot s^{-2} \cdot A^{-1}$
weber	Wb	magnetic flux	$m^2 \cdot kg \cdot s^{-2} \cdot A^{-1}$
becquerel	Bq	radioactivity	s^{-1}
lumen	lm	luminous flux	cd
lux	lx	luminous flux per unit area	$m^{-2} \cdot cd$

SI Prefixes

SI has a set of prefixes that indicate different powers of 10; for example, a kilometre is 1000 m, and a millimetre is 0.001 m. Table 1-4 lists the names and abbreviations for the most commonly used SI prefixes. Note that the prefix symbols greater than "kilo" are capitalized.

Table 1-4 The Most Common SI Prefixes

Name	Symbol	Multiplication
zetta	Z	10^{21}
exa	E	10^{18}
peta	P	10^{15}
tera	T	10^{12}
giga	G	10^9
mega	M	10^6
kilo	k	10^3
hecto	h	10^2
deca	da	10^1
deci	d	10^{-1}
centi	c	10^{-2}
milli	m	10^{-3}
micro	μ	10^{-6}
nano	n	10^{-9}
pico	p	10^{-12}
femto	f	10^{-15}
atto	a	10^{-18}
zepto	z	10^{-21}

SI specifies rules for writing measurements in SI units. These rules combine aspects of traditional styles used in various countries into a single, unambiguous convention that is recognized worldwide.

■ There is always a space between the number and the unit symbol (e.g., 10 m, not 10m).

■ Abbreviations for units are capitalized when they are named after a person but not otherwise (e.g., N for newton but kg for kilogram).

■ Units are not capitalized, even when named after a person, when written out in full (e.g., newton not Newton). This rule prevents confusion over whether one is referring to the person or the unit.

■ SI unit symbols are not considered abbreviations, so they do not end with a period (e.g., W and cd, not W. and cd.).

■ Unit symbols stand for both single units and multiples of the unit, so we never add an "s" to make a symbol plural (e.g., 10 m means ten metres, but 10 ms means ten milliseconds).

■ The unit symbols are written in normal (roman or upright) type to distinguish them from variables, which are written in italic (e.g., "m" always represents the unit metre, while m usually represents the variable mass).

- When writing numbers with many digits, sets of 3 digits on either side of the decimal point are separated by spaces instead of commas. In a number with 4 digits, the space is optional (e.g., 52 469 450, 7.630 941 23, and 4123 or 4 123).

 CHECKPOINT

C-1-8 Writing SI

Identify all the errors in this statement: "The force applied was 23Ns. and the displacement ...."

C-1-8 One should not use an "s" with the unit symbol for the newton, N. There should be a space between the number and the unit. There should not be a period after the unit symbol. So the statement should be, "The force applied was 23 N and the displacement ...."

When we use SI units consistently and carry the units through in calculations, the results will have appropriate SI units. The website of the National Institute of Standards and Technology (a U.S. body) is a helpful reference for learning more about the definitions, derivations, and uses of SI units (http://physics.nist.gov/cuu/Units/).

 LO 9

1-9 Dimensional Analysis

We can consider the units in a problem to help determine whether a proposed relationship is possible; this is called **dimensional analysis**. Note that dimensional analysis determines only whether the units in a relationship are equivalent; further data is needed to prove that the relationship is actually valid. We demonstrate this technique in Example 1-6.

 EXAMPLE 1-6

Dimensional Analysis for the Period of a Pendulum

Simple observations show that the period, T, of a simple pendulum depends on the length of the pendulum and on the acceleration due to gravity, g (about 9.8 m/s² on Earth's surface). Use dimensional analysis to determine a possible equation for period.

SOLUTION

The period must have units of time, s. The SI unit for length is the metre, and acceleration due to gravity has units of m/s². We can see that the only way to get units of

time by combining the units of length and acceleration is as follows:

$$T = \sqrt{\frac{\ell}{g}} \leftrightarrow \sqrt{\frac{m}{m/s^2}} = \sqrt{s^2} = s$$

In this case, the real expression has a numerical factor (see below).

Making sense of the result:

Observations confirm that a longer pendulum will oscillate more slowly and hence have a longer period, and a faster acceleration due to gravity would make the period shorter. In fact, the actual relationship for the period does have the form we determined above, but the relationship also contains a dimensionless numerical factor:

$$T = 2\pi \sqrt{\frac{\ell}{g}}$$

 CHECKPOINT

C-1-9 Dimensional Analysis

Later in the text, you will use kinematic relationships between displacement (in m), velocity, acceleration, and time. Use dimensional analysis to determine whether the following relationship is correct (x is in m, v_0 is in m/s, and a is in m/s²). If the relationship is not correct, suggest how it could be altered.

$$x (=?) x_0 + v_0 t^2 + at$$

C-1-9 This relationship is not possible. The term $v_0 t^2$ has units of m · s, and the term at has units of m/s. We can rewrite the relationship in the following form, which does meet the dimensional analysis requirements:

$$x (=?) x_0 + v_0 t + at^2$$

As you will see in later chapters, the real expression (for constant acceleration situations) is $x = x_0 + v_0 t + \frac{1}{2} at^2$. Remember that dimensional analysis cannot give you numerical factors, such as the $\frac{1}{2}$.

 LO 10

1-10 Unit Conversion

You can easily do some **unit conversions** in your head; for example, 25 mm is the same as 2.5 cm, and 12 km is the same as 12 000 m. However, for conversions that involve more than a simple power of 10, we recommend that you use the formal unit conversion method outlined in Example 1-7. This method is based on multiplying by factors in which the quantities in the numerator and denominator are equivalent. Since these factors equal 1, the multiplication does not change value of the quantity. We can cancel units to confirm that the conversion is correct.

EXAMPLE 1-7

Unit Conversion

Express the speed 72 km/h in SI base units.

SOLUTION

Speed is a displacement (unit of length) divided by a time, so the SI base units are metres per second.

We do the conversion by writing factors so that our units cancel into the desired final units of m/s, with each factor having quantities in the numerator and the denominator that are equivalent.

$$\frac{72 \text{ km}}{1 \text{ h}} \times \frac{1000 \text{ m}}{1 \text{ km}} \times \frac{1 \text{ h}}{60 \text{ min}} \times \frac{1 \text{ min}}{60 \text{ s}} = 20 \text{ m/s}$$

Making sense of the result:

If we travelled at 1 m/s, we would go 3600 m, or 3.6 km, in an hour. Thus, it makes sense that the speed in m/s will be numerically slightly less than one-third of 72.

 CHECKPOINT

C-1-10 Converting Units

The rate at which a process deposits mass on a surface is given as 25 g/cm². What would this rate be in units of kg/m²?

(a) 2.5
(b) 250
(c) 2500
(d) 25 000

C-1-10 (b) We multiply by two factors: 1 kg/1000 g and (100 cm/1 m)².

 MAKING CONNECTIONS

Costly Mistakes in Unit Conversion

In 1999, the Mars Climate Orbiter (Figure 1-10) provided a spectacular and very expensive demonstration of the critical importance of unit conversions. This spacecraft was intended to orbit Mars and collect data on the planet's climate. However, on September 23, 1999, the orbiter angled far too low into the Martian atmosphere and disintegrated. The reason for the trajectory error was a misunderstanding between NASA staff and contractors on the units used in thrust calculations.

On July 3, 1983, an Air Canada jetliner ran out of fuel while on a flight from Montréal to Edmonton. The fuel gauge system on the aircraft was malfunctioning, so the ground crew and pilots had to calculate the fuel for the flight manually. These calculations involved both the volume and mass of the fuel. The aircraft was a new model calibrated with metric units, but the paperwork for refuelling was based on gallons and pounds, the units used by most aircraft at the time. Although the pilots double-checked the calculations, one of the conversion factors was inverted. Fortunately, the captain was an experienced glider pilot and managed to land with only minor damage on a drag strip at a former military base at Gimli, Manitoba. That aircraft has gone down in aviation history as the Gimli Glider.

Figure 1-10 The ill-fated Mars Climate Orbiter (artist's rendition)

1-11   Fermi Problems

Often in physics (and in engineering and other sciences), we encounter situations in which we do not have enough information to calculate precise values, but we still may be able to make useful approximations. Such estimates often involve reasonable guesses for the values of unknown quantities, probabilities, and dimensional analysis if a relevant physical law is not known. The famous physicist Enrico Fermi (1901–54) skillfully used such techniques, and problems involving such estimates are now called **Fermi problems** in his honour.

When doing Fermi problems, scientific notation is very useful because it separates the magnitude of the number (in exponent) from the precise value. Often, we are just seeking an order-of-magnitude estimate and can ignore everything other than the powers of 10 in coming up with an estimate.

Physicists regularly work in situations in which they cannot necessarily establish even a best estimate for a number, but they can determine a maximum possible value (or minimum possible value).

MAKING CONNECTIONS

Enrico Fermi

Enrico Fermi (Figure 1-11) was an Italian-born physicist who worked in the United States for the last 16 years of his life. He made important contributions to quantum mechanics, nuclear physics, statistical mechanics, and particle physics. He was an experimentalist and a theorist, and he was awarded the Nobel Prize in Physics in 1938. Fermi was a master of "back of the envelope" calculations. In one often-quoted example, he estimated the energy released by the Trinity bomb test in 1945 from a measurement of how much the blast deflected bits of paper at a position well away from the test site. His estimate was accurate to within one order of magnitude.

Department of Energy. Office of Public Affairs

Figure 1-11   Enrico Fermi

EXAMPLE 1-8

The Wheels on the Bus Go Round and Round

How many revolutions do the wheels on a bus make during a trip from Toronto to Montréal?

SOLUTION

We first estimate (or look up) the driving distance from Toronto to Montréal. This distance is about 540 km.

Next, we need to measure, estimate, or look up the radius, r, of a typical bus wheel. Tire sizes for buses vary somewhat, but a reasonable value is 50.0 cm.

With each revolution, the bus tire will move a distance of $2\pi r$ along the road. Using N as the number of revolutions and D as the total distance travelled, then

$$N(2\pi r) = D$$

We rearrange this equation to solve for N. Before we substitute the values for our estimates, we convert the distance and radius estimates to SI units (m) using the techniques of the earlier section. Then we solve for N:

$$N = \frac{D}{2\pi r} = \frac{540\,000 \text{ m}}{2\pi \times 0.50 \text{ m}} = 1.7 \times 10^5$$

Therefore, the wheels on the bus turn about 170 000 times during the trip. Note that our input data is probably only valid to 2 significant figures, so we round the final answer to that precision.

Making sense of the result:

For each kilometre travelled, the wheel makes about 300 revolutions, which seems reasonable. If the wheels are rated 60 000 km between wheel-bearing servicings, this would imply that the wheels turn more than 18 million times between servicings.

✓ CHECKPOINT

C-1-11 Standing Side by Side

Suppose the entire population of the world stood next to each other, forming a square. Calculate the length of the side of the square, in kilometres.

(a) 50 km
(b) 250 km
(c) 500 km
(d) 5000 km

C-1-11 (a) Earth's population is about 7 billion people. If we assume that the area occupied by each person averages about 1 m × 1 m, or 0.3 m², then the total area required is about 2.3 billion square metres. This is about 2300 square kilometres. One side of the square is the square root of this, which is about 48 km. Therefore, (a) is the best choice.

We usually represent a **probability** using a numerical value between 0 and 1, with 0 representing no chance of something happening and 1 indicating that it is certain to happen. For example, if you flip a coin, the probability of heads is 0.5, and so is the probability of tails. Similarly, if you roll a six-sided unbiased die, the probability of any particular number, for example, a 4, will be 1/6, or about 0.17.

If we want to determine the overall probability of some event that depends on multiple independent events, then we simply multiply the probabilities of each of the events. For example, the probability of rolling two 6s with a pair of unbiased dice is (1/6)(1/6), or about 0.028. Note that this technique is valid only if the individual probabilities are independent.

 EXAMPLE 1-9

Habitable Planets

Estimate the number of habitable planets in the universe.

SOLUTION

Start with an estimate of the number of stars in our galaxy, the Milky Way. Informed estimates range from about 100 to 400 billion. We will use 150 billion for our calculation.

Now, consider the image of galaxies shown at the beginning of this chapter. Extrapolating from this sample region, we can estimate that the total number of galaxies in all directions is about 100 billion. It is possible that the image does not show some of the fainter, more distant galaxies, so this estimate may be low.

Therefore, the total number of stars in the visible galaxies is roughly 150 billion times 100 billion, or about 1.5×10^{22}.

Next, we need to estimate what fraction of stars will have planetary systems. Astrophysicists believe that the same process that produces stars from clouds of dust and gas also creates planets. However, some of these stars will be in binary (two-star) or multiple-star systems that may not have planets. With considerable uncertainty, we estimate that one-tenth of the stars have planetary systems, so there could be about 1.5×10^{21} planetary systems.

Now, we must consider the probable number of habitable planets per star. We know that our solar system has

eight major planets. However, only Mars and Earth have conditions that could support life forms that depend on water. We might consider other forms of life, but many experts regard them as unlikely. In any case, we are looking only for an order-of-magnitude estimate. So, we have two habitable planets out of eight. Many stars will have energy outputs much different from that of the Sun, which could substantially reduce the number of planets with habitable temperatures. So, we estimate that there is an average of perhaps 0.2 habitable planets per planetary system.

Then the number of habitable planets in the universe is in the order of

$$0.2 \times 1.5 \times 10^{21} = 3.0 \times 10^{20}$$

Making sense of the result:

This is an almost unimaginably large number. For example, if a human could count one planet per second, and if the entire world population of about 7 billion people counts simultaneously (for 10 h/day, 6 days/week), about

$$365 \times (6/7) \times 10 \times 60 \times 60 \times 7 \times 10^9 = 7.9 \times 10^{16} \text{ planets}$$

could be counted per year. In other words, it would take the entire population about 3800 years to count this many planets at one per second, 6 days a week, 10 h/day.

LO 12

1-12 Becoming a Physicist

We hope that as you study physics in this text you will consider becoming a physicist, or using physics in some other science or engineering career. The Canadian Association of Physicists (CAP) has a useful website (www.cap.ca)

that includes a section on careers in physics. The corresponding organization in the United States is the American Physical Society (APS) (www.aps.org). An international organization with particular strengths in Europe is the Institute of Physics (IOP) (www.iop.org).

It is important to realize that many physicists work as part of interdisciplinary teams and apply physics in areas such as medical physics, astrophysics, biophysics, geophysics, materials science, and many areas of engineering.

MAKING CONNECTIONS

Professional Physicist Certification

PPhys, and its associated logo are registered trademarks of the Canadian Association of Physicists (CAP). The logo is reproduced here with the express permission of the CAP

The Canadian Association of Physicists has a professional certification program called Professional Physicist (PPhys). To qualify for a PPhys certification, you must complete minimum education and experience requirements, agree to a code of ethical conduct, and pass a professional examination that tests your ability to interpret physics in applied situations, as well as your sensitivity to ethical aspects of the profession. See the CAP website (www.cap.ca) for additional information.

Figure 1-12

The typical route to becoming a research physicist is to complete an undergraduate degree (usually an honours degree) in physics or a closely related area, such as applied mathematics, geophysics, or astrophysics, followed by graduate degrees (masters and doctorate). Many physicists then complete several years of postdoctoral work prior to taking a permanent position in research, development, teaching, or administration. Yes, that is a long road. However, graduate students in physics are almost always fully supported by graduate scholarships and research and teaching assistantships, so following your undergraduate degree you will probably be financially independent.

While most research positions require a doctorate, a B.Sc. is sufficient for many careers, including technical positions in research or development labs, application of physics in professional programs such as patent law, teaching at the high school level, marketing of physics-related materials, and science journalism. Others use skills learned in physics courses such as quantitative analysis, problem solving, model formation, and experimental design in a broad array of fields, including administration.

Before finishing this chapter, we want to leave you with some suggestions for doing well in your study of physics. You only really learn physics by mentally interacting with the ideas. Don't skip suggested interactive activities; they are key to your learning. While the chapter summaries are valuable, you should read the entire chapter at least once. Remember that your goal in physics is not to remember a large number of relationships or algorithms for solving problems of a certain type. Rather, it is to get a sound grasp of a surprisingly small number of key concepts. Those, along with expertise in the appropriate mathematical techniques, will allow you to tackle novel problems. Physics can, for all of us, feel frustrating at times, but the more problems you try on your own, the more your expertise will grow. It is important that you try the whole range of questions and problems, including conceptual questions, problems by section, the broader comprehensive problems, and problems that involve data analysis and the open problems that will require you to make reasonable assumptions and decide what is relevant to the situation.

Physics deals with physical phenomena on all scales from subatomic particles to the universe as a whole.

Accuracy is how close a measurement is to its true value, and precision is a measure of how close to one-another repeated measurements are.

For data that is normally distributed, about two-thirds of the data lie within ± 1 standard deviation of the mean, and about 95% of the data lie within ± 2 standard deviations of the mean.

SI is a unified system of units recognized worldwide and applicable for all fields of science and engineering.

Unit conversions are performed by multiplying factors with equivalent expressions in the numerator and the denominator.

Nature of Physics

Physics has a number of distinguishing characteristics, including its quantitative nature, the strong role of models, both theoretical and applied aspects, the need for creativity and a collaborative work environment, and, perhaps most important, the goal of explaining physical phenomena in the simplest way possible.

Precision and Accuracy

Precision specifies how narrowly the values range around a central value. Accuracy refers to how close that central value is to the "correct" value. There are three types of errors: true errors (mistakes), random errors (spread about the central value), and systematic errors (e.g., if the technique overestimated all values).

Mean and Standard Deviation

The mean, μ, and standard deviation, σ, of N measurements are defined by the following:

$$\mu = \bar{x} = \frac{1}{N} \sum_{i=1..N} x_i \qquad \sigma = \sqrt{\sum_{i=1..N} \frac{(x_i - \bar{x})^2}{N - 1}}$$

Standard Deviation of the Mean

When determining whether the mean values from two different data sets are statistically different, we use the standard deviation of the mean (SDOM), also called the standard error of the mean:

$$\text{SDOM} = \frac{\sigma}{\sqrt{N}}$$

Significant Figures

Significant figures indicate the approximate precision of a value. When multiplying and dividing, you should have the same number of significant digits in the result as in the least significant entry number. When adding and subtracting, keep the same number of significant figures after the decimal point as in the least precise of the quantities used in the calculation.

SI Units

SI has seven base units: metre, kilogram, second, ampere, candela, kelvin, and mole. Many other SI units are derived from these base units. If you use only SI units when solving problems, the answer will always be another SI unit.

Unit Conversion

To convert from one unit to another, you multiply by a series of factors, each with a numerator and a denominator that are equivalent. The factors are arranged so that the original units cancel out, leaving the desired final units.

Dimensional Analysis

Dimensional analysis can be used to determine possible relationships among physical quantities. Dimensional analysis can rule out an incorrect equation, but it is not sufficient to prove whether a proposed equation is correct.

Fermi Problems

We can combine realistic estimates, dimensional analysis, and other techniques to obtain order-of-magnitude estimates for values that cannot be calculated precisely.

Combining Probabilities

Independent probabilities can be multiplied to obtain the probability of a combined event.

APPLICATIONS

Applications: determination of the significance of the differences between mean values for any two sets of measurements (e.g., is the average temperature different from an airport location and one near the centre of the city), unit conversion using the multiplication by equal terms ratio method (e.g., converting speeds from km/h to m/s, or areas from square centimetres to square metres), using dimensional analysis to check or propose functional relationships, quantitative approximation techniques to provide estimates for situations in real life (e.g., if I turn down the thermostat in a building by $2°C$, how much energy will I conserve over a year, and what does that mean for greenhouse gas emissions).

Key Terms: accuracy, dimensional analysis, error bars, Fermi problems, mean, normal distribution, precision, probability, random error, SI, significant figures, standard deviation of the mean (SDOM), standard deviation, unit conversions

QUESTIONS

1. Which of the following is *not* a characteristic of physics?
   (a) Physics seeks to explain phenomena in the simplest way possible and in a way that has the widest application.
   (b) While some phenomena may be described qualitatively, in general, physics seeks quantitative descriptions.
   (c) Physics is primarily involved with developing a large number of equations so that there is one equation relevant for any given situation.
   (d) Physics requires creativity.

2. Assume that measurements of a time for some process are normally distributed, with no systematic errors. About two-thirds of the time the value is between 3.4 and 4.8 s. What are the mean and standard deviation of the time measurement?

3. Assume that measurements of a length are normally distributed, with no systematic errors. About 19 times out of 20, the value is between 2.2 and 5.8 m. What are the mean and standard deviation of the length measurement?

4. How many significant figures are in the number 230 100?
   (a) 3
   (b) 4
   (c) 5
   (d) 6

5. Complete the calculation $\dfrac{12.0 \times 5.0}{4.000}$. According to the rules for significant digits, how should the answer be written?
   (a) 15
   (b) 15.0
   (c) 15.00
   (d) 15.000

6. Which of the following is *not* a base SI unit?
   (a) kg
   (b) s
   (c) N
   (d) cd

7. Which of the following is *not* a base SI unit?
   (a) kg
   (b) m
   (c) J
   (d) K

8. Work (or energy) is equal to force times x. Use Tables 1-2 and 1-3 to identify the units of x.

9. Consider the graph of Figure 1-4. Taking into account the various points made in this chapter, critically evaluate the graphical presentation, indicating all aspects that could be improved.

10. (a) Use a reference source to look up how the length of the metre is currently defined.
    (b) From this definition, what must be the speed of light (to the number of significant figures inherent in the definition)?
    (c) Provide a brief overview of some of the previous ways that have been used to define the length of the metre.

11. Use a reliable reference to determine how the second is currently defined.

12. (a) Use a reliable reference to determine the method currently used to define the kilogram.
    (b) Provide a brief overview of some other ways that have been used to define the kilogram.

13. Describe the current method used to define the ampere. Include a reference to the source of your information.

14. Use a reliable reference source to identify the method currently used to define the candela, as well as the method that was historically used.

15. How many cubic centimetres are in one cubic metre?
    (a) 100
    (b) 1 000
    (c) 10 000
    (d) 1 000 000

16. Two data sets, A and B, have the same mean value, but data set B has a larger standard deviation by a factor of 2. There are 16 data points in data set A. How many data points are needed in set B for the SDOM to be identical for the two sets?

17. If we multiply a force (N) by a velocity (m/s), which of the following will be the units of the resulting quantity?
    (a) kg/s
    (b) J
    (c) Pa
    (d) W

PROBLEMS BY SECTION

For problems, star ratings will be used, (✱, ✱✱, or ✱✱✱), with more stars meaning more challenging problems.

Section 1-3   Mean, Standard Deviation, and SDOM

18. ✱ A set of 16 measurements has a mean of 4.525 and a standard deviation of 1.2. How should the mean be written with an uncertainty given by the SDOM?

19. ✱ The SDOM is 1.2, the mean is 4.8, and the standard deviation is 2.4. Approximately how many data points must be in the set?

20. ✱ According to a journal article, experiments suggest that 19 times out of 20, the measurement obtained is within ± 0.5 of 10.5. What must be the standard deviation and mean values, assuming that the data is normally distributed?

Section 1-4   Significant Figures

21. ✱ How many significant figures are in each of the following?
    (a) 324.2
    (b) 0.032
    (c) 9.004
    (d) 12 000
    (e) 85.0
    (f) 6000.

22. ✱✱ Suppose that the uncertainty in the length 17.386 22 m is ± 0.04 m. Round the length to the correct number of significant figures.

Section 1-5   Scientific Notation

23. ✱ Write each of the following in scientific notation.
    (a) 2452
    (b) 0.592
    (c) 12 000
    (d) 0.000 045

24. ✱ Perform the following calculation, and express the final result in scientific notation with the correct number of significant figures.

$$\frac{(3.20 \times 10^5)(1.5 \times 10^4)}{6.400 \times 10^3}$$

Section 1-6 Significant Figures and Mathematical Operations

25. ✱ Perform the following calculation. Express the answer with the correct number of significant figures.

$$\frac{12\,400 \times 0.025\,63}{12.34}$$

26. ✱ Perform the following calculations. Express the answer with the correct number of significant figures.
    (a) 12.456 m − 11.9 m
    (b) 6.542 m + 0.0092 m

Section 1-7 Error Bars

27. (a) ✱✱ Plot the following data points, with error bars, on a graph. Assume that there is no uncertainty in the x-value and that each y-value has an uncertainty of ±1 m: (0,0), (1,2), (2,5), (3,7).
    (b) Is there a line of best fit that passes through the data points, within the uncertainty of each data point? If so, plot it approximately on your graph.
    (c) Add a new data point: (4,6). What must be the error bar for that point in order for the line of best fit to include that point as well?

Section 1-8 SI Units

28. ✱ What are the units of RC, where R is electrical resistance and C is electrical capacitance? (Use Tables 1-2 and 1-3.)

29. ✱✱ Acceleration due to gravity, g, has dimensions of acceleration (m/s²). Assuming that the gravitational potential energy depends only on mass m, g, and height h above Earth h in some way, what must be the form of the equation? Use units to verify this.

30. ✱✱ In thermodynamics, we have the equation $PV = nRT$, where P is pressure, V is volume, n is the number of moles, T is the temperature (in kelvins), and R is a constant. Use the data in Tables 1-2 and 1-3 to determine the units for the constant R.

Section 1-9 Dimensional Analysis

31. ✱✱ Use dimensional analysis to determine a relationship between electric potential difference (voltage), current, and electrical resistance. Consider the units given in Tables 1-2 and 1-3.

32. ✱✱ The aerodynamic drag force, F, for fast relative motion (e.g., a car at highway speeds) has the form $F = -cv^2$, where v is speed (m/s). What must be the units (in terms of SI base units) of the variable c?

33. ✱✱ Aerodynamic drag for slow relative motion can be expressed in the following form: $F = -bv$, where F is a force in newtons and v is a speed (m/s). Use dimensional analysis to determine the units (in terms of SI base units) of the variable b.

Section 1-10 Unit Conversion

34. ✱ Scientists believe that continental drift in the Atlantic happens at the rate of about 1 to 2 cm/year. Convert this rate into nm/s.

35. ✱✱ A cell of baker's yeast is typically 4.0 μm in diameter. (Assume that the cells are spherical, although in reality they have a variety of shapes.) If you had 25 cm³ of yeast (i.e., one-tenth of a measuring cup), about how many cells would it contain?

Section 1-11 Fermi Problems

36. ✱✱ Volunteers are searching a wooded area for a piece of forensic evidence. It is estimated that searchers should cross on a grid with a separation no larger than 4.0 m between searchers and that the searchers average a walking speed of 0.25 m/s. How long would it take a search party of 10 people to search an area of 1 km²?

37. ✱✱ One of the famous Fermi problems is a calculation of how many piano tuners there were in Chicago. (If you are interested in the problem, search the Internet. There are many solutions available.) Using similar methods, estimate the number of goalies there are in Canada at any one time (from all types of hockey teams).

38. ✱✱✱ Using reasonable estimates, determine how many molecules there are in one litre of milk.

COMPREHENSIVE PROBLEMS

39. ✱✱ The HIV virus is approximately spherical, and it has a diameter of about 130 nm. If (reasonably) the mass density is about the same as water, 1000. kg/m³, what is the mass of one HIV virus? Express your answer in pg.

40. ✱✱ An Intel Core 2 Duo processor has the equivalent of 291 million transistors. The dimensions of the integrated circuit chip are approximately 4.0 mm × 4.0 mm. What are the dimensions of one transistor on the integrated circuit?

41. ✱✱ Based on the data in Table 1-5, can we conclude that the teams of the years 1990–99 were significantly better than the teams of the years 2000–09? Clearly explain your reasoning.

Table 1-5 Win–Loss Statistics for the Toronto Blue Jays Baseball Team

Year	Win–Loss Percentage	Year	Win–Loss Percentage
1990	0.531	2000	0.512
1991	0.562	2001	0.494
1992	0.593	2002	0.481
1993	0.586	2003	0.531
1994	0.478	2004	0.416
1995	0.389	2005	0.494
1996	0.457	2006	0.537
1997	0.469	2007	0.512
1998	0.543	2008	0.531
1999	0.519	2009	0.463

42. ✱✱✱ The expansion of the universe mentioned at the beginning of this chapter is often represented by the equation $v = H_0 d$, where v is the speed of recession of the galaxy, d is the distance to the galaxy, and H_0 is the Hubble constant. In this equation, astronomers usually express v in km/s and d in parsecs.

(a) What are the units of H_0 in terms of these units?

(b) The parsec is equal to 3.26 ly (light years). A light year is the distance travelled by light in one year. Write the number of metres in one parsec in scientific notation.

(c) Use unit conversion to write the units of H_0 in SI base units.

43. ✷✷ Scientists estimate that the flux of meteoritic material on the entire Earth is about 20 000 tonnes per year. One tonne is 1000 kg.

(a) Convert this flux to mass per square metre of Earth's surface.

(b) Micrometeorites are meteorites that are small enough to float to Earth without being ablated significantly. They come in a variety of sizes, but for this problem we assume that a typical micrometeorite has a diameter of 25 μm. If 25% of the total meteoritic annual mass was in the form of micrometeorites of this size, how many micrometeorites (on average) would be deposited on each square metre of Earth annually? Assume density of 3400 kg $\cdot$ m^3.

44. ✷✷✷ Look up the values and units for the following three fundamental physical constants: G (the universal gravitational constant), c (the speed of light), and $\hbar$ (Planck's constant divided by 2π).

(a) Use dimensional analysis to determine an expression using only G, c, and h that has the units of length.

Hint: The expression will have a square root, and one of the constants will be to a power, such as squared or cubed. The other two will be just the constant.

(b) The expression that you obtained is called the Planck length. Calculate the numerical value of the Planck length.

(c) Use Internet or other resources to find objects that might have dimensions of the order of the Planck length.

45. (a) ✷✷ Find a combination of the same three physical constants listed in question 44 (G, h, and c) that has units of time. This is called the Planck time. Some physicists consider the Planck time to be the time after the Big Bang, which we can first study using physics.

(b) Calculate a numerical value for the Planck time.

46. ✷✷ A typical oil molecule has an effective diameter of 0.25 nm. If 25 L (one litre is the same as 1000 cm^3) of oil is spilled on a lake, and if the oil is spread so that it is only a single molecule thick, what will be the area of the oil slick? Assume that the oil molecules are just touching. Hint: Determine the cross-sectional area of an oil molecule.

 See the text online resources at www.physics1e.nelson.com for Open Problems and Data-Rich Problems related to this chapter.

Learning Objectives

When you have completed this chapter you should be able to:

1 Determine whether a physical quantity is a vector or a scalar.

2 Describe vector quantities in terms of components and in Cartesian and polar notation.

3 Add and subtract vector quantities, and find the product of a vector and a scalar.

4 Define a dot (scalar) product between two vectors, and use dot products to determine angles between vectors and projections of a vector on arbitrary axes.

5 Define and calculate a cross (vector) product between two vectors using either Cartesian notation or the algebraic definition and the right-hand rule.

Chapter 2
Scalars and Vectors

Philosophy is written in this grand book, the universe which stands continually open to our gaze. But the book cannot be understood unless one first learns to comprehend the language and read the letters in which it is composed. It is written in the language of mathematics, and its characters are triangles, circles and other geometric figures without which it is humanly impossible to understand a single word of it; without these, one wanders about in a dark labyrinth.

Galileo Galilei in *Assayer*

In this chapter, we introduce the concepts of scalars and vectors and develop the mathematical language that will be used throughout the book to describe these concepts.

Have you ever wondered why two different words—speed and velocity—are used in English to describe how fast you are moving? The word "speed" has Old English roots, and the word "velocity" has Latin roots. Speed refers to how fast you are moving. If you are asked about your driving speed, your answer might be 55 km/h. However, if you are asked about your velocity, the 55 km/h value does not suffice: velocity requires additional information about the direction of motion, for example, 55 km/h north on Main St. For many physical quantities, not knowing the direction makes the information incomplete. For example, if you have been flying from Ottawa for 6 h with an average speed of 500 km/h, your final destination could be anywhere within the circle shown in Figure 2-1. To be able to pinpoint the exact destination, you need to know the direction of your flight. If your plane headed directly northwest your destination could be Yellowknife, the capital of Northwest Territories.

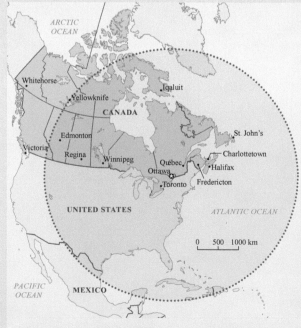

 For additional Making Connections, Examples, and Checkpoints, as well as activities and experiments to help increase your understanding of the chapter's concepts, please go to the text's online resources at www.physics1e.nelson.com.

Figure 2-1 If you were to fly from Ottawa for 6 h with an average speed of 500 km/h, what would your final destination be?

LO 1

2-1 Definitions of Scalars and Vectors

The physical quantities we use throughout this book can be divided into two categories: scalars and vectors.

Scalar quantities, or **scalars,** require only a number (either positive or negative) and a unit for their description. For example, temperature, mass, length, area, volume, time, distance, speed, work, and energy are all scalars. Scalar symbols are denoted using *italics*. For example, the quantities listed above are usually represented by the symbols $T, m, L, A, V, t, d, v, W$, and E.

Vector quantities, or **vectors,** require a positive number, called a **vector magnitude**, a unit, and a direction for their description. For example, displacement, velocity, acceleration, force, momentum, angular momentum, impulse, and magnetic field are all vectors. To describe vectors we need a **frame of reference** (a convention about what is "at rest"), and a coordinate system. For example, a two-dimensional (2-D) coordinate system can consist of an x-axis and a y-axis, as shown in Figure 2-2. A three-dimensional (3-D) coordinate system has an additional axis, the z-axis.

In this textbook, we use primarily rectangular (orthogonal) coordinate systems, where the coordinate axes are perpendicular to each other. In Figure 2-2, vector $\vec{v_1}$ has a magnitude of 30 m/s and a direction of 25° above the horizontal ($\theta_1 = 25°$); the second velocity, $\vec{v_2}$, has a magnitude of 15 m/s and a direction of $\theta_2 = -10° = 350°$. By convention, a positive angle is measured in the counterclockwise sense from the positive x-axis, and a negative angle is measured in the clockwise sense from the positive x-axis.

By convention, the symbols for vector quantities are either italic letters with arrows over them or bold letters. Throughout this textbook, vector quantities will be denoted using the arrow notation, for example, $\vec{v}$ for velocity, $\vec{p}$ for momentum, $\vec{F}$ for force, and $\vec{a}$ for acceleration. The magnitude of a vector is always a positive scalar quantity and can be denoted using the absolute value sign or by using the same letter as the vector without the arrow or bolding. For example, the magnitude of the velocity vector $\vec{v}$ (speed) can be denoted as either $|\vec{v}|$ or v.

Vectors are represented graphically using arrows. The length of the arrow represents the vector's magnitude; the vector's direction coincides with the arrow's direction. Note that a vector is not attached to any particular location in space: moving a vector parallel to itself does not affect it. The end-point of a vector is called the tail of the **vector**, and the arrow-end of a vector is called the **head of the vector** (Figure 2-2). In order to describe a vector, we have to know its magnitude (including its units) and direction. A 2-D vector can be described in terms of a magnitude and an angle measured counterclockwise from the positive x-axis. Such a description is referred to as a **polar notation** or **polar coordinates**. For example, the polar notation for vector $\vec{v_1}$ in Figure 2-2 can be written as 30 m/s with a direction of 25° or as 30 m/s [25°].

✓ CHECKPOINT

C-2-1 Describing Vectors

Which of the following is the most accurate description of a vector quantity?
(a) vector magnitude, direction, and the location of its tail relative to the origin of the coordinate system
(b) vector magnitude, including its units, and direction
(c) vector magnitude, including its units, and location of its head relative to the origin
(d) vector magnitude, including its units, direction, and location of its head relative to the origin
(e) none of the above

C-2-1 (b)

Vectors can also be described in terms of their **components**, which are scalar quantities. A 2-D vector has two components, v_x and v_y, as shown in Figure 2-3. These components are **projections** of the vector onto the x- and y-axes:

$$\begin{cases} v_x = v\cos\theta \\ v_y = v\sin\theta \end{cases} \Rightarrow \begin{cases} v = \sqrt{v_x^2 + v_y^2} \\ \tan\theta = \dfrac{v_y}{v_x} \end{cases} \quad (2\text{-}1)$$

where $|\vec{v}| \equiv v$

Using the Pythagorean theorem, we can see that the magnitude of a vector is equal to the square root of the

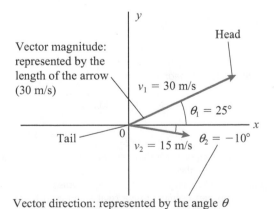

Figure 2-2 Graphical representation of two 2-D velocity vectors

NEL

CHAPTER 2 | **SCALARS AND VECTORS** 21

21

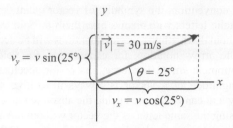

$$v_y = v \sin(25°)$$

$|\vec{v}| = 30$ m/s

$\theta = 25°$

$v_x = v \cos(25°)$

Figure 2-3 Vector components of velocity vector $\vec{v}_1$

sum of the squares of its components. By convention, all angles are measured in a counterclockwise direction from the positive x-axis. However, it is sometimes convenient to use the complementary angle θ^*, which measures the angle clockwise from the vertical y-axis, such as $\theta^* = 90° - \theta$. Hence,

$$\begin{cases} v_x = v \sin\theta^* \\ v_y = v \cos\theta^* \end{cases} \Rightarrow \begin{cases} v = \sqrt{v_x^2 + v_y^2} \\ \tan\theta^* = \dfrac{v_x}{v_y} \end{cases} \quad (2\text{-}2)$$

where $|\vec{v}| \equiv v$

Representing a vector by using its components is often called a **scalar notation**. The vector component in the x-direction is positive when the vector is fully or partially oriented in the positive x-direction. Similarly,

the x-component is negative when the vector is fully or partially oriented in the negative x-direction. The same principle applies to any other axis. Therefore, the notation $v_x = 20$ m/s means that a projection of the object's velocity vector onto the x-axis is +20 m/s. However, an endless number of vectors have this x-component, for example, 20 m/s [0°] and 40 m/s [60°]. We need to provide both components to fully define a 2-D vector. It takes three components to represent a 3-D vector. In general, the description of a vector using components is called **vector resolution into scalar components**. Example 2-1 demonstrates how to resolve a vector into scalar components.

PEER TO PEER

I used to forget to place an arrow above vectors. Leaving the arrow off is confusing because then I didn't remember if I was dealing with a vector or a scalar. For example, it was impossible for me to distinguish if I was talking about speed or velocity. Now I know that keeping track of the arrows eliminates lots of misunderstandings.

EXAMPLE 2-1

Vector Resolution

Force $\vec{F}_1$ has a magnitude of 20.0 N and is directed at a 45.0° angle to the positive x-axis. What will happen to the components of the force when

(a) the magnitude of the force remains the same, but the angle decreases to 30.0°;
(b) the magnitude of the force doubles, but the direction of the force remains the same?

SOLUTION

Let us express the components of force $\vec{F}_1$ in scalar notation (2-1). The magnitudes of forces will be denoted as F_1, F_2, and F_3 respectively:

$$\begin{cases} F_{1x} = F_1 \cos\theta \\ F_{1y} = F_1 \sin\theta \end{cases} \Rightarrow \begin{cases} F_{1x} = (20.0\text{ N}) \cos(45.0°) = 14.14\text{ N} \\ F_{1y} = (20.0\text{ N}) \sin(45.0°) = 14.14\text{ N} \end{cases}$$

(a) Let $\vec{F}_2$ denote the force when the angle decreases to 30.0°:

$$\begin{cases} F_{2x} = F_2 \cos\theta \\ F_{2y} = F_2 \sin\theta \end{cases} \Rightarrow \begin{cases} F_{2x} = (20.0\text{ N}) \cos(30.0°) = 17.32\text{ N} \\ F_{2y} = (20.0\text{ N}) \sin(30.0°) = 10.0\text{ N} \end{cases}$$

We can check that force $\vec{F}_2$ has the same magnitude as the original force:

$$F_2 = \sqrt{F_{2x}^2 + F_{2y}^2} = \sqrt{(17.32\text{ N})^2 + (10.0\text{ N})^2}$$

$$F_2 = 20.0\text{ N}$$

(b) Let $\vec{F}_3$ denote the force when the magnitude of the force doubles while the direction remains the same. Then,

$$\begin{cases} F_{3x} = F_3 \cos\theta_2 \\ F_{3y} = F_3 \sin\theta_2 \end{cases} \Rightarrow \begin{cases} F_{3x} = (40.0\text{ N}) \cos(45.0°) = 28.28\text{ N} \\ F_{3y} = (40.0\text{ N}) \sin(45.0°) = 28.28\text{ N} \end{cases}$$

We can verify that these components describe a vector with twice the magnitude of $\vec{F}_1$.

$$F_3 = \sqrt{F_{3x}^2 + F_{3y}^2} = 40.0\text{ N}$$

Making sense of the result:

Changing the angle such that the vector is closer to the x-axis increases the x-component and decreases the y-component. Doubling the magnitude doubles both components. You can confirm these relationships in a vector diagram.

LO 2

2-2 Vector Addition: Geometric and Algebraic Approaches

From everyday experience, we intuitively know that the sum of two forces is also a force. The sum of two equal forces pulling in opposite directions is zero, and the sum of two equal forces pulling in the same direction is a force with double the magnitude. Therefore, it is not surprising that the sum of two or more vectors depends on both their magnitudes and directions.

The Geometric Addition of Vectors

There are two commonly used rules for the **geometric method for vector addition**: the **parallelogram rule** and the **triangle construction rule**. Figure 2-4 shows how these rules can be used to find the resultant vector when

two students pull on the arms of a friend. Both rules use the fact that a vector can be translated as long as its magnitude and direction are not changed.

Parallelogram Rule Move the vectors so that they are joined at their tails. Then, build a parallelogram with the two vectors forming adjacent sides, as shown in Figure 2-4(b). The vector sum of $\vec{F_1}$ and $\vec{F_2}$ is represented by the diagonal of this parallelogram, $\vec{F_R}$ (the subscript R stands for resultant). The tail of $\vec{F_R}$ coincides with the tails of $\vec{F_1}$ and $\vec{F_2}$, and the head of $\vec{F_R}$ is at the opposite vertex of the parallelogram.

Triangle Construction Rule Move the vectors so that the head of the first vector is joined with the tail of the second vector. Then, connect the tail of the first vector with the head of the second vector, as shown in Figure 2-4(c). The vector sum ($\vec{F_R}$) of forces $\vec{F_1}$ and $\vec{F_2}$ is represented by the third side of the triangle, where the tail of $\vec{F_R}$ coincides with the tail of $\vec{F_1}$, and its head coincides with the head of $\vec{F_2}$. The same method can be applied to add a chain of three or more vectors.

As an exercise, prove that the parallelogram and triangle construction rules produce the same result.

(a)

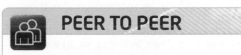

A student is pulled by two forces

(b)

Parallelogram rule

(c)

Triangle construction rule

Figure 2-4 Vector addition. Two forces are exerted on a person. The resultant of these two forces is found using (a) the parallelogram rule and (b) the triangle construction rule.

 ONLINE ACTIVITY

Vector Addition

The e-resource that accompanies every new copy of this textbook contains an Online Activity using the PhET simulation "Vector Addition." Work through the simulation and accompanying questions to gain an understanding of vector addition.

✓ CHECKPOINT

C-2-2 Geometrical Addition of Vectors

Which diagram in Figure 2-5 (on page 24) correctly represents the vector sum of three vectors $\vec{F_1}$, $\vec{F_2}$, and $\vec{F_3}$? Explain.

(a) diagram (a) only
(b) diagram (b) only
(c) diagrams (c) and (d)
(d) diagram (c) only
(e) diagram (d) only
(f) diagrams (b) and (c)
(g) diagrams (a) and (b)

C-2-2 (c) Only diagrams (c) and (d) properly apply the triangle construction rule and the parallelogram rule to vector addition.

👥 PEER TO PEER

I find that adding vectors geometrically helps me get a sense of how the result of vector addition or subtraction should look. Even though drawing by hand might not be exact, it is often helpful for understanding.

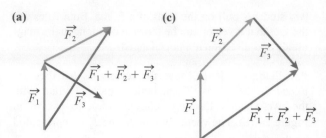

(a)

(c)

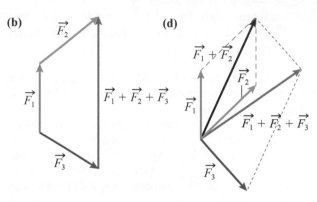

(b)

(d)

Figure 2-5 C-2-2 Geometrical Addition of Vectors

The Algebraic Addition of Vectors

The geometric approach to vector addition is useful for visualizing the relationships between vectors, but it is difficult to draw and measure the diagrams with high precision. Example 2-2 illustrates an algebraic approach to vector addition.

The method for finding the resultant vector for the two perpendicular 2-D vectors in Example 2-2 can be generalized to apply to any number of vectors with arbitrary orientations:

KEY EQUATION

$$\begin{cases} F_{R,x} = F_{1x} + F_{2x} + \cdots + F_{N,x} \\ F_{R,y} = F_{1y} + F_{2y} + \cdots + F_{N,y} \\ F_R = \sqrt{F_{R,x}^2 + F_{R,y}^2} \\ \tan \theta = \dfrac{F_{R,y}}{F_{R,x}} \end{cases} \qquad (2\text{-}3)$$

EXAMPLE 2-2

Finding the Resultant of Perpendicular Forces

Two 40.0 N forces, $\vec{F_1}$ and $\vec{F_2}$, directed at 90.0° to each other act on a hook, as shown in Figure 2-6.

(a) Use the geometric approach to vector addition to find the resultant force acting on the hook.
(b) Find the components of each force along the x- and y-axes.
(c) Find the components of the net force (resultant vector).
(d) Compare your results from parts (b) and (c). What conclusions can you draw?

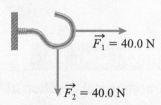

Figure 2-6 Two 40.0 N forces are acting on a hook

SOLUTION

(a) To add the two vectors geometrically, we can use the parallelogram rule, as shown in Figure 2-7. Since the original forces have equal magnitudes, the resultant force, $\vec{F_R}$, is represented by a diagonal of a square. Therefore,

$$\vec{F_R} = \sqrt{2}(40.0 \text{ N})$$
$$\vec{F_R} = 56.6 \text{ N} \qquad \text{and} \qquad \theta = -45.0°$$

(Make sure you can justify these results.)

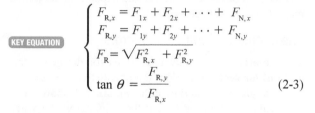

Figure 2-7 Vector addition for Example 2-2

(b) To find the components of forces $\vec{F_1}$ and $\vec{F_2}$, we first choose a coordinate system, then we use Equation (2-1):

$$\begin{cases} F_{1x} = 40.0 \text{ N} \\ F_{1y} = 0 \end{cases} \quad \text{and} \quad \begin{cases} F_{2x} = 0 \\ F_{2y} = -40.0 \text{ N} \end{cases}$$

(c) To find the components of the resultant vector $\vec{F_R}$, we use Equation (2-3):

$$\begin{cases} F_{R,x} = 40.0 \text{ N} \\ F_{R,y} = -40.0 \text{ N} \end{cases} \quad \text{and} \quad F_R = \sqrt{(40.0 \text{ N})^2 + (40.0 \text{ N})^2}$$

$$= \sqrt{2}(40.0 \text{ N}) = 56.6 \text{ N}$$

(d) Comparing the results of (b) and (c), we see that the x-components of the two individual vectors sum to 40.0 N and their y-components sum to −40.0 N. These sums match the components of the resultant vector.

Making sense of the result:

The components of the resultant vector are equal to the sums of the corresponding components of the individual vectors.

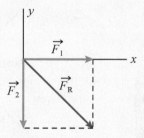

These relationships give us an **algebraic method for vector addition** (Figure 2-8):

1. Resolve the vectors into components.

2. Add components pointing along the same axis to obtain the corresponding components of the resultant vector.

3. Use the components of the resultant vector to determine its magnitude and direction.

By resolving vectors into components, we can use scalar calculations to perform operations with vectors.

The algebraic method can also be extended to 3-D vectors:

KEY EQUATION

$$\begin{cases} F_{R,x} = F_{1x} + F_{2x} + \cdots + F_{N,x} \\ F_{R,y} = F_{1y} + F_{2y} + \cdots + F_{N,y} \\ F_{R,z} = F_{1z} + F_{2z} + \cdots + F_{N,z} \end{cases}$$

$$F_R = \sqrt{F_{R,x}^2 + F_{R,y}^2 + F_{R,z}^2} \qquad (2\text{-}4)$$

Example 2-3 demonstrates algebraic addition of two arbitrary vectors. It also shows how a **free body diagram (FBD)** can help you visualize the forces acting on an object.

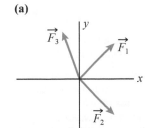
(a)

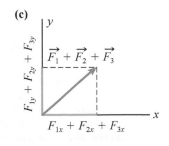
(c)

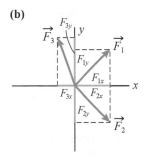
(b)

Figure 2-8 (a) Visual representation of the addition of three vectors (b) Vector resolution of the three vectors (c) Vector addition using vector components

EXAMPLE 2-3

Finding the Resultant Force Acting on a Student Standing on a Ramp

A 51.0 kg student steps onto a slippery 10.0° ramp. Two forces are acting on her: the force of gravity exerted by Earth (her weight), which is $F_g = 500$ N directed down, and the normal force exerted by the ramp, which is $N = 493$ N, directed perpendicular to the ramp, as shown in Figure 2-9.

(a) Find the components of the student's weight along x- and y-axes directed along the incline and perpendicular to the incline, respectively.

(b) Find the components of the normal force along these axes.

(c) Find the resultant force acting on the student.

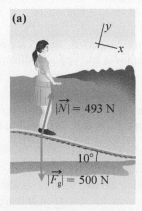
(a)

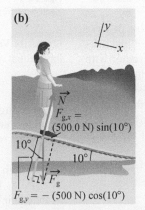
(b)

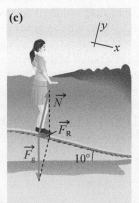
(c)

Figure 2-9 (a) A lifelike problem representation (b) The FBD shows the forces acting on the student. (c) Geometrical representation of the net force acting on the student

(continued)

SOLUTION

When dealing with two or more forces, it is often helpful to draw a detailed FBD, such as the one in Figure 2-9(b).

(a) To find the components of each vector, we have to find the angles that the vector forms with the coordinate axes. The angle between the force of gravity $\vec{F}_g$ and the y-axis equals the angle of the ramp. (Make sure you can see why these two angles are equal.) The force of gravity vector, marked in green, forms the hypotenuse of the right-angled triangle in Figure 2-9(b). In this triangle, the side adjacent to the 10.0° angle represents the y-component of $\vec{F}_g$, and the opposite side represents the x-component. Therefore, the components of the force of gravity are as follows, by Equation (2-1):

$$\begin{cases} F_{g,x} = F_g \sin(10.0°) = (500 \text{ N})(0.1736) = 86.8 \text{ N} \\ F_{g,y} = -F_g \cos(10.0°) = -(500 \text{ N})(0.9848) = -493 \text{ N} \end{cases}$$

The negative sign indicates that the y-component of the force of gravity is directed in the negative y-direction. In this case, it is convenient to measure the angle from the negative y-axis; we compare the direction of the vector with the direction of the positive x- and y-axes to determine the signs on the x- and y-components.

(b) The normal force is directed in the positive y-direction. Consequently, the angle between the normal force and the x-axis is 90.0°, and

$$\begin{cases} N_x = N \cos(90.0°) = (493 \text{ N})(0) = 0 \\ N_y = N \sin(90.0°) = 493 \text{ N} \end{cases}$$

(c) From Figure 2-9(c) we can see that if the vectors are not drawn exactly to scale, or if their directions are slightly inaccurate, we will not be able to get a precise value of the resultant force from measuring it on the diagram. Therefore, we use the algebraic approach. Adding the x- and y-components of the gravitational and normal forces, we get

$$\begin{cases} F_{R,x} = 86.8 \text{ N} + 0 = 86.8 \text{ N} \\ F_{R,y} = -493 \text{ N} + 493 \text{ N} = 0 \end{cases}$$

The resultant force has a nonzero x-component and a zero y-component. Therefore, it points along the x-axis in the positive x-direction—down the ramp. Since the y-component is zero, the magnitude of the resultant force is equal to the x-component, 86.8 N.

Making sense of the result:

The resultant force in this problem is not zero. As we will discuss later in the chapters on Newton's laws, an unbalanced net force (the resultant of all forces acting on the object) means that the object's velocity is changing. Therefore, we should expect the student to slide down the ramp with increasing speed. In this example, we ignored the force of friction between the student and the ramp (since it was a slippery ramp). If present, a friction force would have been directed up the ramp. Choosing an x-axis along the ramp simplified the calculations because the resultant force is directed down the ramp. It is often convenient to have the resultant force directed along one of the axes.

Now let us consider a few consequences of the vector addition rules:

Multiplying a Vector by a Scalar Adding a vector to itself preserves its direction while doubling the magnitude (Figure 2-10(a)). In general, adding N identical vectors means preserving the original direction and multiplying the original magnitude of the vector by N, as shown in Figure 2-10(b). We can generalize this process to a multiplication of a vector by any scalar, not just integers. The product of a scalar and a vector is a vector with a magnitude equal to the product of the magnitude of the original vector and the absolute value of the scalar. The direction of the resulting vector coincides with the original direction of the vector if the scalar is positive, and is opposite to the original vector direction if the scalar is negative:

$$|a\vec{F}| = |a||\vec{F}|$$

$$\begin{cases} \text{if } a > 0, a\vec{F} \text{ has the same direction as } \vec{F} \\ \text{if } a < 0, a\vec{F} \text{ has the opposite direction to } \vec{F} \\ \text{if } a = 0, a\vec{F} \text{ has a magnitude of zero and no direction} \end{cases}$$

$$(2\text{-}5)$$

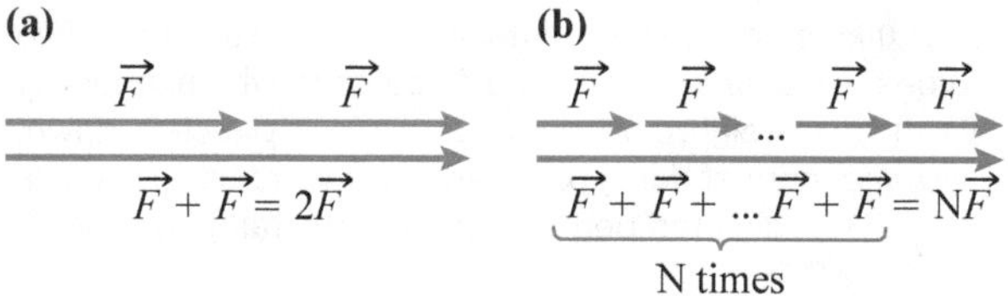

(a)

$$\vec{F} + \vec{F} = 2\vec{F}$$

(b)

$$\underbrace{\vec{F} + \vec{F} + \dots \vec{F} + \vec{F} = N\vec{F}}_{N \text{ times}}$$

Figure 2-10 Multiplication of a vector by a scalar (a) Adding two identical vectors (b) Adding n identical vectors

Opposite Vectors Vectors $\vec{F}_1$ and $\vec{F}_2$ are **opposite** if they have equal magnitudes and opposite directions:

$$\vec{F}_1 = -\vec{F}_2 \qquad (2\text{-}6)$$

Subtracting Vectors We can perform vector subtraction by adding an opposite vector, as shown in Figure 2-11:

$$\vec{F}_2 - \vec{F}_1 = \vec{F}_2 + (-\vec{F}_1) \qquad (2\text{-}7)$$

We can also find the difference between two vectors, $\vec{F}_2 - \vec{F}_1$, by moving the two vectors parallel to themselves so that their tails coincide, and then drawing a vector from the head of $\vec{F}_2$ to the head of $\vec{F}_1$. If you know the magnitudes of the two vectors and the angle between them, you can use the law of cosines to find the magnitude of the difference vector.

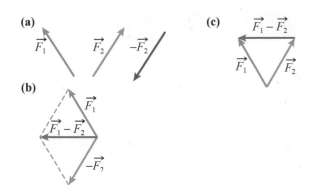

(a)

$\vec{F_1}$ $\vec{F_2}$ $-\vec{F_2}$

(b)

$\vec{F_1}$

$\vec{F_1} - \vec{F_2}$

$-\vec{F_2}$

(c)

$\vec{F_1} - \vec{F_2}$

$\vec{F_1}$ $\vec{F_2}$

Figure 2-11 Vector subtraction. To find $\vec{F_1} - \vec{F_2}$, first construct a vector opposite to $\vec{F_2}$ and then add $\vec{F_1}$ and $(-\vec{F_2})$.

 **CHECKPOINT**

C-2-3 Operations with Vectors

Which, if any, of the following statements is true? Explain.
(a) Opposite vectors always have opposite magnitudes and are directed along the same line.
(b) The sum of two vectors is also a vector directed in the same direction as one of the vectors being added.
(c) The difference between two vectors is a vector that is always perpendicular to the sum of the two vectors.
(d) The magnitude of the sum of two vectors always equals the sum of the magnitudes of the vectors being added.
(e) The magnitude of the product of a vector and a scalar is always larger than the magnitude of the original vector.

C-2-3 All the statements are false.

PEER TO PEER

Leaving the "hat" off a unit vector can be confusing. Sometimes I forget that a unit vector is a vector and not a scalar. Although the magnitude of a unit vector is always one, the direction might vary, and omitting the hat makes it easy to forget that it's a vector.

In Cartesian notation, any 3-D vector can be represented as follows:

$$\vec{F} = F_x \hat{i} + F_y \hat{j} + F_z \hat{k} \qquad (2\text{-}8)$$

where F_x, F_y, and F_z are the usual scalar vector components: projections of the vector $\vec{F}$ onto the x-, y-, and z-axes, respectively.

 **CHECKPOINT**

C-2-4 Cartesian Vector Notation

Which of the following statements are true? Explain.
(a) $3\hat{i} + 4\hat{j} = 7\hat{k}$
(b) $(3\hat{i} + 4\hat{j} + 2\hat{k}) + (2\hat{i} - 2\hat{j} - 2\hat{k}) - 5\hat{i} + 2\hat{j} + 4\hat{k}$
(c) $(3\hat{i} + 4\hat{j}) + (2\hat{i} - 2\hat{k}) = 3\hat{i} + 4\hat{j}$
(d) $(3\hat{i} + 4\hat{j} + 2\hat{k}) + (2\hat{i} - 2\hat{j} - 2\hat{k}) = 5\hat{i} + 2\hat{j}$
(e) $3\hat{i} - 3\hat{j} + 3\hat{k} = 3\hat{k}$

C-2-4 (d) $(3\hat{i} + 4\hat{j} + 2\hat{k}) + (2\hat{i} - 2\hat{j} - 2\hat{k}) = (3\hat{i} + 2\hat{i}) + (4\hat{j} - 2\hat{j}) + (2\hat{k} - 2\hat{k})$
$= (3+2)\hat{i} + (4-2)\hat{j} + (2-2)\hat{k} = 5\hat{i} + 2\hat{j}$

LO 3

2-3 Cartesian Vector Notation

Cartesian vector notation is another method for handling vector components. This notation is based on the concept of **unit vectors**, specifically, unit vectors in the x-, y-, and z-directions. A unit vector has a dimensionless magnitude of 1. Unit vectors in the positive x-, y-, and z-directions are denoted as $\hat{i}$, $\hat{j}$, and $\hat{k}$, respectively (Figure 2-12). Some books use $\hat{x}$, $\hat{y}$, and $\hat{z}$ as an alternative notation for unit vectors. A unit vector in the direction of an arbitrary vector $\vec{F}$ is denoted as $\hat{u}_F$. A "hat" above an italic letter indicates a unit vector.

Cartesian vector notation allows us to simplify vector addition significantly:

KEY EQUATION

$$\vec{F}_R = \vec{F_1} + \vec{F_2} + \vec{F_3} + \cdots + \vec{F_N}$$

$$\vec{F}_R = F_{1x}\hat{i} + F_{1y}\hat{j} + F_{1z}\hat{k} + F_{2x}\hat{i} + F_{2y}\hat{j} + F_{2z}\hat{k}$$
$$+ F_{3x}\hat{i} + F_{3y}\hat{j} + F_{3z}\hat{k} + \cdots + F_{N,x}\hat{i} + F_{N,y}\hat{j} + F_{N,z}\hat{k}$$

$$\vec{F}_R = [F_{1x} + F_{2x} + F_{3x} + \cdots + F_{N,x}]\hat{i}$$
$$+ [F_{1y} + F_{2y} + F_{3y} + \cdots + F_{N,y}]\hat{j}$$
$$+ [F_{1z} + F_{2z} + F_{3z} + \cdots + F_{N,z}]\hat{k}$$

$$\vec{F}_R = \left(\sum_{l=1}^{N} F_{l,x}\right)\hat{i} + \left(\sum_{l=1}^{N} F_{l,y}\right)\hat{j} + \left(\sum_{l=1}^{N} F_{l,z}\right)\hat{k} \quad (2\text{-}9)$$

Equation (2-9) states that the sum of all the x-components of the original vectors equals the x-component of the resultant vector, the sum of all the y-components of the original vectors equals the y-component of the resultant vector, and the sum of all the z-components of the original vectors equals the z-component of the resultant vector. Example 2-4 illustrates the use of Cartesian notation.

(a)

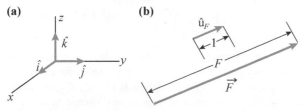

z

$\hat{k}$

$\hat{i}$

$\hat{j}$

y

x

(b)

$\hat{u}_F$

F

$\vec{F}$

Figure 2-12 (a) Cartesian unit vectors: $\hat{i}$, $\hat{j}$, and $\hat{k}$ (b) A unit vector in the direction of the force vector $\vec{F}$

EXAMPLE 2-4

Finding a Unit Vector in the Direction of a Given Force

A force vector is described as $\vec{F} = (3\hat{i} + 4\hat{j} - 5\hat{k})$ N. Express the unit vector in the $\vec{F}$ direction using Cartesian notation $(\hat{i}, \hat{j}, \hat{k})$.

SOLUTION

A unit vector must have a magnitude of 1. Therefore, if we divide the force vector $\vec{F}$ by its magnitude (which is a scalar), the resultant vector will have a magnitude of 1 and the same direction as the force $\vec{F}$. To find the magnitude of the force vector, we use the Pythagorean theorem:

$$F = \sqrt{F_x^2 + F_y^2 + F_z^2} = \sqrt{(3 \text{ N})^2 + (4 \text{ N})^2 + (-5 \text{ N})^2}$$
$$= \sqrt{50} \text{ N} = 5\sqrt{2} \text{ N}$$

Thus, the unit vector in the direction of $\vec{F}$ is

$$\hat{u}_F = \frac{\vec{F}}{F} = \frac{(3\hat{i} + 4\hat{j} - 5\hat{k}) \text{ N}}{(5\sqrt{2}) \text{ N}} = \frac{3}{5\sqrt{2}}\hat{i} + \frac{4}{5\sqrt{2}}\hat{j} - \frac{1}{\sqrt{2}}\hat{k}$$

Making sense of the result:

Since the units divide out, $\hat{u}_F$ is dimensionless, as required for a unit vector and its magnitudes equals 1.

The method used in Example 2-4 can be applied to any vector:

KEY EQUATION

$$\hat{u}_F = \frac{\vec{F}}{F} = \frac{F_x}{F}\hat{i} + \frac{F_y}{F}\hat{j} + \frac{F_z}{F}\hat{k}, \quad \text{which is equivalent to}$$

$$\vec{F} = F\hat{u}_F \tag{2-10}$$

where $\hat{u}_F$ is a dimensionless unit vector, and F is the magnitude of vector $\vec{F}$.

✓ CHECKPOINT

C-2-5 Identifying Unit Vectors

Which of the following vectors is a unit vector? Explain.

(a) $\frac{3}{2\sqrt{2}}\hat{i} + \frac{3}{2\sqrt{2}}\hat{j} - \frac{2}{2\sqrt{2}}\hat{k}$

(b) $3\hat{i} + 3\hat{j} - 5\hat{k}$

(c) $\frac{3}{\sqrt{14}}\hat{i} + \frac{3}{\sqrt{14}}\hat{j} - \frac{2}{\sqrt{14}}\hat{k}$

(d) $\frac{3}{\sqrt{22}}\hat{i} + \frac{3}{\sqrt{22}}\hat{j} - \frac{2}{\sqrt{22}}\hat{k}$

(e) $\frac{4}{\sqrt{12}}\hat{i} - \frac{4}{\sqrt{12}}\hat{j} - \frac{4}{\sqrt{12}}\hat{k}$

C-2-5 (d) The unit vector must have a magnitude of 1.

✓ CHECKPOINT

C-2-6 Adding Vectors Using Cartesian Notation

Which of the following vectors represents the sum of the two vectors $\vec{F}_1 = 3\hat{i} + 3\hat{j} - 5\hat{k}$ and $\vec{F}_2 = -3\hat{i} + 6\hat{j} - 5\hat{k}$? Explain.

(a) $\vec{F}_R = 9\hat{j} - 10\hat{k}$

(b) $\vec{F}_R = 6\hat{i} + 9\hat{j} - 10\hat{k}$

(c) $\vec{F}_R = 6\hat{i} - 3\hat{j}$

(d) $\vec{F}_R = -10\hat{k}$

(e) $\vec{F}_R = -3\hat{i} - 10\hat{k}$

C-2-6 (a) This follows from the vector addition rule, Equation (2-9).

We can specify the direction of a 3-D vector by giving the angles the vector makes with each of the coordinate axes. These angles are called **coordinate direction angles** and are denoted α, β, and γ, as shown in Figure 2-13. In Example 2-5, we will see that these three angles are not independent of each other.

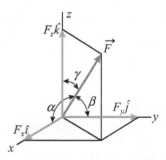

Figure 2-13 The angles describing the orientation of vector $\vec{F}$ relative to coordinate axes x, y, and z

Position Vector A special vector widely used in physics is the **position vector**, $\vec{r}$ (Figure 2-15). A position vector describes the location of a point in space relative to some fixed point (often, the origin). The magnitude of the position vector corresponds to the distance between the origin and point A. The position vector of point A (A_x, A_y, A_z) can be expressed as

$$\vec{r}_A = (A_x - 0)\hat{i} + (A_y - 0)\hat{j} + (A_z - 0)\hat{k}$$
$$= A_x\hat{i} + A_y\hat{j} + A_z\hat{k} \tag{2-15}$$

When the position of an object is described relative to an arbitrary point (not necessarily the origin), for example, from point A (A_x, A_y, A_z) to point B (B_x, B_y, B_z), the position vector is called a **displacement vector** because it refers to how much an object was displaced while moving from point A to point B.

EXAMPLE 2-5

Relationships Between the Coordinate Direction Angles

Prove that the coordinate direction angles for any 3-D vector $\vec{F}$ satisfy the equation

$$\cos^2 \alpha + \cos^2 \beta + \cos^2 \gamma = 1 \qquad (2\text{-}11)$$

SOLUTION

From Figure 2-14 we see that

$$\cos \alpha = \frac{F_x}{F}; \quad \cos \beta = \frac{F_y}{F}; \quad \cos \gamma = \frac{F_z}{F} \qquad (2\text{-}12)$$

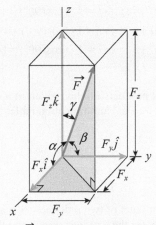

Figure 2-14   Vector $\vec{F}$ expressed in Cartesian vector notation

Substituting expressions (2-12) into the equation for a unit vector, Equation (2-10), gives

$$\hat{u}_F = \frac{\vec{F}}{F} = \frac{F_x}{F}\hat{i} + \frac{F_y}{F}\hat{j} + \frac{F_z}{F}\hat{k}$$

$$= (\cos \alpha)\hat{i} + (\cos \beta)\hat{j} + (\cos \gamma)\hat{k} \qquad (2\text{-}13)$$

The magnitude of the unit vector is 1. Therefore,

$$(\cos \alpha)^2 + (\cos \beta)^2 + (\cos \gamma)^2 = 1$$

Making sense of the result:

This result means that the three directional angles are not independent. In fact, given two of the angles, we can find the absolute value of the cosine of the third angle, which limits the value of the unknown angle to four possible values. This result also applies to 2-D vectors. In the case of 2-D vectors, $\gamma = 90°$, $\cos \gamma = 0$; therefore, $\cos^2 \alpha + \cos^2 \beta = 1$. However, for 2-D vectors, the angles α and β are complementary angles; therefore, $\alpha = 90° - \beta$. Consequently,

$$\cos^2 \alpha + \cos^2 \beta = \cos^2 \alpha + \cos^2(90° - \alpha) = \cos^2 \alpha + \sin^2 \alpha = 1 \qquad (2\text{-}14)$$

This is a well-known result. Thus, expression (2-11) applies to both three- and two-dimensional vectors.

$$\vec{r}_{A \to B} = (B_x - A_x)\hat{i} + (B_y - A_y)\hat{j} + (B_z - A_z)\hat{k} \qquad (2\text{-}16)$$

The magnitude of the displacement vector $\vec{r}_{A \to B}$ corresponds to the distance between point A and point B.

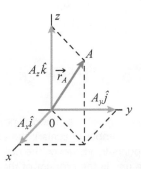

Figure 2-15   Position vector $\vec{r}_A$ represents the location of point A relative to the origin. Its magnitude corresponds to the distance between the origin and point A.

☑ CHECKPOINT

C-2-7 Cartesian Vector Notation and Coordinate Direction Angles

Which, if any, of the following statements do you agree with? Explain.

(a) The sum of the coordinate direction angles is always equal to 1.
(b) A unit vector must always be directed along one of the three coordinate axes.
(c) A position vector and a force vector must always be directed along the same line.
(d) When adding two force vectors, the resultant vector cannot be perpendicular to either one of them.
(e) The magnitude of the unit vector is always 1, and its dimension is the newton (N).
(f) all of the above
(g) none of the above

C-2-7 (g)

MECHANICS

LO 4

2-4 The Dot Product of Two Vectors

Many physics and engineering applications involve the projection of a vector onto a given axis. Although this projection can be done geometrically, it is often more practical to use an algebraic approach. The **dot**, or **scalar**, **product** is a scalar quantity that represents the product of the magnitudes of a vector and the component of another vector projected onto it.

The scalar product of vectors $\vec{A}$ and $\vec{B}$ is defined as the product of their magnitudes and the cosine of the angle between the vectors:

KEY EQUATION $\qquad \vec{A} \bullet \vec{B} = AB \cos\theta$, where $0° \leq \theta \leq 180°$

$$(2\text{-}17)$$

and θ is the angle between the vectors, as shown in Figure 2-16.

The product $\vec{A} \bullet \vec{B}$ is commonly read as "A dot B."

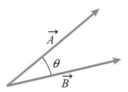

Figure 2-16 The dot product of vectors $\vec{A}$ and $\vec{B}$ is the product of their magnitudes and the cosine of the angle between them.

From the definition above, we can see that a dot product has the following properties:

- A dot product has the same units as the product of the vectors being multiplied.

- A dot product of two perpendicular vectors equals zero because $\cos(90°) = 0$.

- A dot product of two vectors of given magnitudes is greatest when the vectors are parallel to each other because $\cos(0°) = 1$. The dot product is smallest when the vectors are antiparallel to each other because $\cos(180°) = -1$.

- The dot product of an arbitrary vector $\vec{A}$ and $\hat{u}_B$, a unit vector in the direction of vector $\vec{B}$, represents a projection of vector $\vec{A}$ onto the direction of vector $\vec{B}$.

- A dot product is commutative, that is,

$$\vec{A} \bullet \vec{B} = \vec{B} \bullet \vec{A} = AB \cos\theta \qquad (2\text{-}18)$$

- Multiplication of a dot product by a scalar can be expressed as follows:

$$a(\vec{A} \bullet \vec{B}) = (a\vec{B}) \bullet \vec{A} = \vec{A} \bullet (a\vec{B}) = (\vec{A} \bullet \vec{B})a = aAB\cos\theta$$

$$(2\text{-}19)$$

- The dot product is distributive:

$$\vec{A} \bullet (\vec{B} + \vec{C}) = \vec{A} \bullet \vec{B} + \vec{A} \bullet \vec{C}$$
$$= AB \cos(\theta_{A-B}) + AC \cos(\theta_{A-C}), \qquad (2\text{-}20)$$

where θ_{A-B} and θ_{A-C} are the angles between vectors $\vec{A}$ and $\vec{B}$, and $\vec{A}$ and $\vec{C}$, respectively.

A dot product has two important uses: it can be used to find a vector component or a projection of a vector onto a given direction, and it can be used to find the angle between any two given vectors.

The Dot Product and Unit Vectors

The dot products of the various combinations of unit vectors are as follows:

KEY EQUATION

$\hat{i} \bullet \hat{i} = 1 \cdot 1 \cos(0°) = 1$; $\quad \hat{i} \bullet \hat{j} = 1 \cdot 1 \cos(90°) = 0$; $\quad \hat{i} \bullet \hat{k} = 1 \cdot 1 \cos(90°) = 0$

$\hat{j} \bullet \hat{j} = 1 \cdot 1 \cos(0°) = 1$; $\quad \hat{j} \bullet \hat{i} = 1 \cdot 1 \cos(90°) = 0$; $\quad \hat{j} \bullet \hat{k} = 1 \cdot 1 \cos(90°) = 0$

$\hat{k} \bullet \hat{k} = 1 \cdot 1 \cos(0°) = 1$; $\quad \hat{k} \bullet \hat{i} = 1 \cdot 1 \cos(90°) = 0$; $\quad \hat{k} \bullet \hat{j} = 1 \cdot 1 \cos(90°) = 0$

$$(2\text{-}21)$$

These dot products can be written in a more compact form, as a dot (scalar) multiplication table:

KEY EQUATION

$\bullet$	$\hat{i}$	$\hat{j}$	$\hat{k}$
$\hat{i}$	1	0	0
$\hat{j}$	0	1	0
$\hat{k}$	0	0	1

$$(2\text{-}22)$$

The dot product of two unit vectors is at the intersection of the row for the first unit vector and the column for the second unit vector. For example, the dot product of $\hat{i}$ and $\hat{i}$ is 1. The dot products $\hat{i}$ of $\hat{i}$ and $\hat{j}$, and $\hat{i}$ and $\hat{k}$ are 0. Since the dot product is commutative, the table is symmetric about its main diagonal.

Now we are ready to define the dot product for two arbitrary vectors using Cartesian notation:

$$\vec{A} \bullet \vec{B} = (A_x\hat{i} + A_y\hat{j} + A_z\hat{k}) \bullet (B_x\hat{i} + B_y\hat{j} + B_z\hat{k})$$
$$= A_xB_x(\hat{i} \bullet \hat{i}) + A_xB_y(\hat{i} \bullet \hat{j}) + A_xB_z(\hat{i} \bullet \hat{k})$$
$$+ A_yB_x(\hat{j} \bullet \hat{i}) + A_yB_y(\hat{j} \bullet \hat{j}) + A_yB_z(\hat{j} \bullet \hat{k})$$
$$+ A_zB_x(\hat{k} \bullet \hat{i}) + A_zB_y(\hat{k} \bullet \hat{j}) + A_zB_z(\hat{k} \bullet \hat{k})$$
$$= A_xB_x + A_yB_y + A_zB_z \qquad (2\text{-}23)$$

PEER TO PEER

I have to remind myself that the dot product of two vectors is just a number (a scalar); therefore, it does *not* have components. However, the dot product has a unit, and the unit is the product of the units of the two vectors. I find it helpful to call the dot product "the scalar product."

Therefore, the dot product of two vectors can be defined in terms of their components:

KEY EQUATION $\quad \vec{A} \cdot \vec{B} = A_x B_x + A_y B_y + A_z B_z \qquad (2\text{-}24)$

Since the magnitude of a unit vector is 1, the dot product of vector $\vec{A}$ and a unit vector gives the component of $\vec{A}$ along an axis in the direction of the unit vector:

KEY EQUATION $\quad \vec{A} \cdot \hat{u}_l = A_l \cdot 1 = A_l \qquad (2\text{-}25)$

Where $\hat{u}_l$ is a unit vector in an arbitrary direction l, and A_l is the projection of vector $\vec{A}$ onto the direction l.

CHECKPOINT

C-2-8 The Dot Product of Two Vectors

Which of the following expressions correctly represents the dot product of vectors $\vec{A}$ and $\vec{B}$, where $\vec{A} = 3\hat{i} + 4\hat{j} - 5\hat{k}$ and $\vec{B} = -2\hat{i} + \hat{j} - 2\hat{k}$? Explain.

(a) -1
(b) 8
(c) 20
(d) $-6\hat{i} + 4\hat{j} + 10\hat{k}$
(e) other

EXAMPLE 2-6

The Dot Product and Simple Coordinate Transformations

Two forces, $\vec{F}_1$ and $\vec{F}_2$, are described as $\vec{F}_1 = (3.00\hat{i} - 2.00\hat{j})\,\text{N}$ and $\vec{F}_2 = (-1.00\hat{i} + 3.00\hat{j})\,\text{N}$.

(a) Find the angle between the forces.
(b) Verify that the x- and y-components of the forces, F_{1x}, F_{1y}, F_{2x}, and F_{2y}, are their projections onto the x- and y-axes.
(c) Find the components of forces $\vec{F}_1$ and $\vec{F}_2$ in the orthogonal $x'y'$-coordinate system that is rotated by $30°$ counterclockwise about the origin, as shown in Figure 2-17(a).
(d) Calculate the magnitudes of the forces using their components in both coordinate systems, and compare the results.
(e) Suppose that we move the xy-coordinate system in a plane such that the new coordinate axes remain parallel to the original axes, but the origin O moves to a new location (Figure 2-17(b)). Find the components and the magnitudes of forces $\vec{F}_1$ and $\vec{F}_2$ in the $x''y''$-coordinate system.

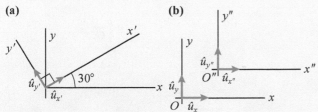

(a)

(b)

Figure 2-17 (a) The *xy*-coordinate system undergoes rotation in a plane (by 30°) around its origin transforming into an *x′y′*-coordinate system. (b) The *xy*-coordinate system undergoes translation in a plane such that the new coordinate axes (*x″*, *y″*) remain parallel to the original coordinate axes (*x*, *y*).

SOLUTION

(a) To find the angle between the forces, we use Equation (2-17):

$$\cos\theta = \frac{\vec{F}_1 \cdot \vec{F}_2}{F_1 F_2} = \frac{F_{1x}F_{2x} + F_{1y}F_{2y}}{\left(\sqrt{F_{1x}^2 + F_{1y}^2}\right)\left(\sqrt{F_{2x}^2 + F_{2y}^2}\right)}$$

$$= \frac{(-3.00 - 6.00)\,\text{N}^2}{\sqrt{13\,\text{N}^2}\sqrt{10\,\text{N}^2}} = \frac{-9.00\,\text{N}^2}{\sqrt{130\,\text{N}^2}} = -0.789$$

$$\theta = \cos^{-1}\left(\frac{-9.00}{\sqrt{130}}\right) = \cos^{-1}(-0.789) = 142°$$

(b) We can use dot products with the $\hat{i}$ and $\hat{j}$ unit vectors to find the projections of the forces onto the x- and y-axes:

$$\begin{cases} F_{1x} = \vec{F}_1 \cdot \hat{i} = [(3.00\hat{i} - 2.00\hat{j}) \cdot \hat{i}]\,\text{N} = 3.00\,\text{N} \\ F_{1y} = \vec{F}_1 \cdot \hat{j} = [(3.00\hat{i} - 2.00\hat{j}) \cdot \hat{j}]\,\text{N} = -2.00\,\text{N} \end{cases}$$

Dot products for $\vec{F}_2$ indicate that its x- and y-projections are -1.00 N and 3.00 N, respectively.

$$\begin{cases} F_{2x} = \vec{F}_2 \cdot \hat{i} = [(-\hat{i} + 3.00\hat{j}) \cdot \hat{i}]\,\text{N} = -1.00\,\text{N} \\ F_{2y} = \vec{F}_2 \cdot \hat{j} = [(-\hat{i} + 3.00\hat{j}) \cdot \hat{j}]\,\text{N} = 3.00\,\text{N} \end{cases}$$

(c) To find the projections of forces $\vec{F}_1$ and $\vec{F}_2$ onto the orthogonal x'- and y'-axes, we define unit vectors $\hat{u}_{x'}$ and $\hat{u}_{y'}$ directed along the x'- and y'-axes, as shown in Figure 2-18(a):

$$\hat{u}_{x'} = \cos(30°)\hat{i} + \sin(30°)\hat{j} = \frac{\sqrt{3}}{2}\hat{i} + \frac{1}{2}\hat{j}$$

$$\hat{u}_{y'} = -\sin(30°)\hat{i} + \cos(30°)\hat{j} = -\frac{1}{2}\hat{i} + \frac{\sqrt{3}}{2}\hat{j}$$

Then, we use Equation (2-25) to find the projections of the forces:

$$F_{1x'} = \vec{F}_1 \cdot \hat{u}_{x'} = \left(3.00\hat{i} - 2.00\hat{j}\right)\left(\frac{\sqrt{3}}{2}\hat{i} + \frac{1}{2}\hat{j}\right)\,\text{N}$$

$$= \left(\frac{3\sqrt{3}}{2} - 1\right)\,\text{N} = 1.60\,\text{N}$$

(continued)

$$F_{1y'} = \vec{F}_1 \cdot \hat{u}_{y'} = (3.00\hat{i} - 2.00\hat{j})\left(-\frac{1}{2}\hat{i} + \frac{\sqrt{3}}{2}\hat{j}\right) N = \left(-\frac{3}{2} - \sqrt{3}\right) N$$

$$= -3.23\, N$$

$$F_{2x'} = \vec{F}_2 \cdot \hat{u}_{x'} = (-\hat{i} + 3.00\hat{j})\left(\frac{\sqrt{3}}{2}\hat{i} + \frac{1}{2}\hat{j}\right) N = \left(-\frac{\sqrt{3}}{2} + \frac{3}{2}\right) N$$

$$= 0.634\, N$$

$$F_{2y'} = \vec{F} \cdot \hat{u}_{y'} = (-\hat{i} + 3.00\hat{j})\left(-\frac{1}{2}\hat{i} + \frac{\sqrt{3}}{2}\hat{j}\right) N = \left(\frac{1}{2} + \frac{3\sqrt{3}}{2}\right) N$$

$$= 3.10\, N$$

(d)

$$\left.\begin{array}{l} F_1 = \left(\sqrt{F_{1x}^2 + F_{1y}^2}\right) = (\sqrt{13})\ N = 3.61\ N \\[2mm] F_2 = \left(\sqrt{F_{2x}^2 + F_{2y}^2}\right) = (\sqrt{10})\ N = 3.16\ N \end{array}\right\}$$ the xy-coordinate system

$$\left.\begin{array}{l} F_1 = \left(\sqrt{F_{1x'}^2 + F_{1y'}^2}\right) = \left(\sqrt{(1.60)^2 + (3.23)^2}\right) N = 3.61\ N \\[2mm] F_2 = \left(\sqrt{F_{2x'}^2 + F_{2y'}^2}\right) = \left(\sqrt{(0.634)^2 + (3.16)^2}\right) N = 3.16\ N \end{array}\right\}$$ the $x'y'$-coordinate system

(e) When a coordinate system is moved parallel to itself—that is, the coordinate system undergoes pure translation—such that its origin is moved to a different location but

the directions of its axes remain unchanged, the components of a vector in the new coordinate system ($x''y''$) will be the same as its components in the old coordinate system (xy). The reason for this is that the magnitude of a vector, as well as the angle between the vectors and the coordinate axes, will not change under this coordinate system transformation (Figure 2-18(b)). Therefore, applying Equation (2-1) we find

$$\begin{cases} F_{1x} = F_{1x''} = 3.00\ N \\ F_{1y} = F_{1y''} = -2.00\ N \end{cases} \text{and} \begin{cases} F_{2x} = F_{2x''} = -1.00\ N \\ F_{2y} = F_{2y''} = 3.00\ N \end{cases}$$

Making sense of the result:

Coordinate systems can undergo transformations, such as coordinate axes translation or rotation. When a coordinate system undergoes pure translation, the coordinates of the vectors remain unchanged. When coordinate axes are rotated about the origin (the coordinate system undergoes a rotational transformation), the components of the vectors change, but the magnitudes of the vectors do not.

✓ CHECKPOINT

C-2-9 Identifying Orthogonal Vectors

Which of the following pairs of vectors are orthogonal to one another? Explain.

(a) $\vec{A} = 3\hat{i} + 4\hat{j} - \hat{k}$ and $\vec{B} = -2\hat{i} + \hat{j} - 2\hat{k}$
(b) $\vec{A} = 2\hat{i} - 3\hat{j} - 5\hat{k}$ and $\vec{B} = -2\hat{i} + \hat{j} - 2\hat{k}$
(c) $\vec{A} = -2\hat{i} - 3\hat{j} - 5\hat{k}$ and $\vec{B} = 4\hat{i} + 6\hat{j} - 2\hat{k}$
(d) $\vec{A} = 2\hat{i} - 2\hat{j} - 2\hat{k}$ and $\vec{B} = -2\hat{i} + 2\hat{j} + 2\hat{k}$
(e) $\vec{A} = 2\hat{i} - 3\hat{j} - 5\hat{k}$ and $\vec{B} = -5\hat{i} + 2\hat{j} - 3\hat{k}$

C-2-9 (a) The dot product of two orthogonal vectors equals zero.

LO 5

2-5 The Cross Product of Vectors

A number of physics and engineering applications involve vectors that are perpendicular to two given vectors. As you will see in later chapters, the angular momentum vector is perpendicular to both the position vector and the velocity vector of an object, and the vector for the magnetic force acting on a charged particle is perpendicular to both the magnetic field vector and the velocity vector of the particle. The **cross**, or **vector**, **product** of two vectors defines a vector that is perpendicular to the two given vectors.

The magnitude of the cross product, $\vec{C}$, of vectors $\vec{A}$ and $\vec{B}$ is defined as the product of the magnitudes of $\vec{A}$ and $\vec{B}$ and the sine of the angle θ between them

(Figure 2-18(a)). The direction of the cross product is determined using the **right-hand rule**: when you curl the fingers of your right hand from vector $\vec{A}$ to vector $\vec{B}$, your thumb will point in the same direction as the cross product, $\vec{C}$ (Figure 2-18(b)).

KEY EQUATION
$$\vec{A} \times \vec{B} = \vec{C}$$

$$C = AB\sin\theta, \text{ where } 0° \leq \theta \leq 180° \qquad (2\text{-}26)$$

(a)

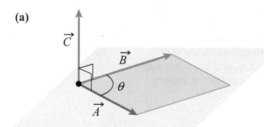

(b)

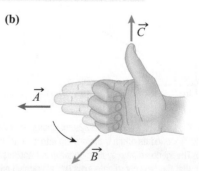

Figure 2-18 (a) Visual representation of the cross product of two vectors, $\vec{A}$ and $\vec{B}$, the area of the parallelogram represents the magnitude of the cross product (b) The right-hand rule for determining the direction of the cross product

PEER TO PEER

I have to remind myself that the cross product of two vectors is also a vector (it has magnitude, direction, and a unit). Its unit is the product of the units of two original vectors. I find it helpful to call a cross product "a vector product" so I remember it.

A cross product has the following properties:

- The cross product is a vector.

- The unit of a cross product is the product of the units of the vectors being multiplied.

- The magnitude of the cross product corresponds to the area of a parallelogram that has vectors $\vec{A}$ and $\vec{B}$ as its sides.

- The cross product of two parallel or two antiparallel vectors equals zero because $\sin(0°) = \sin(180°) = 0$. The cross product of two vectors of a given magnitude is greatest when the vectors are perpendicular to each other because $\sin(90°) = 1$.

- The cross product is anticommutative:

$$\vec{A} \times \vec{B} = -\vec{B} \times \vec{A} \qquad (2\text{-}27)$$

You can verify this property by applying the right-hand rule to the vector products in the equation above.

- The cross product can be multiplied by a scalar. From the definition of the vector product, we can see that

$$a(\vec{A} \times \vec{B}) = (a\vec{A}) \times \vec{B} = \vec{A} \times (a\vec{B}) = (\vec{A} \times \vec{B})a$$
$$= aAB \sin(\theta) \qquad (2\text{-}28)$$

- The cross product operation is distributive:

$$\vec{A} \times (\vec{B} + \vec{C}) = (\vec{A} \times \vec{B}) + (\vec{A} \times \vec{C}) = \vec{A} \times \vec{B} + \vec{A} \times \vec{C} \qquad (2\text{-}29)$$

The Cross Product and Unit Vectors

Consider the vector products of the unit vectors $\hat{i}$, $\hat{j}$, and $\hat{k}$. Since the unit vectors are orthogonal and have a magnitude of 1,

$$\hat{i} \times \hat{i} = 0; \quad \hat{i} \times \hat{j} = \hat{k}; \quad \hat{i} \times \hat{k} = -\hat{j}$$
$$\hat{j} \times \hat{j} = 0; \quad \hat{j} \times \hat{i} = -\hat{k}; \quad \hat{j} \times \hat{k} = \hat{i}$$
$$\hat{k} \times \hat{k} = 0; \quad \hat{k} \times \hat{i} = \hat{j}; \quad \hat{k} \times \hat{j} = -\hat{i} \qquad (2\text{-}30)$$

As with the dot products in Section 2-4, we can write cross products as a cross product multiplication table:

KEY EQUATION

$\times$	$\hat{i}$	$\hat{j}$	$\hat{k}$
$\hat{i}$	0	$\hat{k}$	$-\hat{j}$
$\hat{j}$	$-\hat{k}$	0	$\hat{i}$
$\hat{k}$	$\hat{j}$	$-\hat{i}$	0

$$(2\text{-}31)$$

The cross product of two unit vectors is at the intersection of the row for the first vector and the column for the second vector. Since the cross product is anticommutative, the order in which vectors are multiplied changes the result. Consequently, unlike a dot product multiplication table (see Equation 2-2), this table is *not* symmetrical about its main diagonal.

You can use a **cross-multiplication circle** to remember the signs of the cross products of the unit vectors. Arrange the unit vectors in a counterclockwise order around a circle, as shown in Figure 2-19. To find a cross product of two unit vectors, locate the first vector and then move along the circle to the second vector of the product. If you move counterclockwise (in the direction of the arrows), the result is the third vector along the circle. If you have to move clockwise, the result is the vector opposite the third vector. For example, to find the cross product $\hat{i} \times \hat{j}$, you move counterclockwise along the circle, so the result is $\hat{k}$. However, to find $\hat{j} \times \hat{i}$, you move clockwise, so the result is $-\hat{k}$.

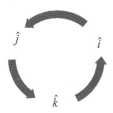

Figure 2-19 The unit cross-multiplication circle

✓ CHECKPOINT

C-2-10 Practising Cross Multiplication

Which of the following represents the result of $(\hat{i} \times \hat{k}) \times \hat{i}$? Explain.

(a) $\hat{i}$
(b) $-\hat{i}$
(c) $\hat{j}$
(d) $-\hat{j}$
(e) $\hat{k}$
(f) $-\hat{k}$
(g) 1
(h) 0

C-2-10 (e) $(\hat{i} \times \hat{k}) \times \hat{i} = -(\hat{j}) \times \hat{i} = -(-\hat{j}) \times \hat{i} = \hat{j} \times \hat{i} = -(-\hat{k}) = \hat{k}$

We can define a cross product for two arbitrary vectors in terms of their components and unit vectors:

$$\vec{A} \times \vec{B} = (A_x\hat{i} + A_y\hat{j} + A_z\hat{k}) \times (B_x\hat{i} + B_y\hat{j} + B_z\hat{k})$$
$$= A_xB_x(\hat{i} \times \hat{i}) + A_xB_y(\hat{i} \times \hat{j}) + A_xB_z(\hat{i} \times \hat{k})$$
$$+ A_yB_x(\hat{j} \times \hat{i}) + A_yB_y(\hat{j} \times \hat{j}) + A_yB_z(\hat{j} \times \hat{k})$$
$$+ A_zB_x(\hat{k} \times \hat{i}) + A_zB_y(\hat{k} \times \hat{j}) + A_zB_z(\hat{k} \times \hat{k})$$
$$= A_xB_y\hat{k} + A_xB_z(-\hat{j}) + A_yB_x(-\hat{k}) + A_yB_z(\hat{i})$$
$$+ A_zB_x(\hat{j}) + A_zB_y(-\hat{i})$$
$$= (A_yB_z - A_zB_y)\hat{i} + (A_zB_x - A_xB_z)\hat{j}$$
$$+ (A_xB_y - A_yB_x)\hat{k} \qquad (2\text{-}32)$$

Therefore, using Cartesian notation the cross product of two vectors can be expressed as follows:

KEY EQUATION
$$\vec{A} \times \vec{B} = (A_y B_z - A_z B_y)\hat{i} + (A_z B_x - A_x B_z)\hat{j}$$
$$+ (A_x B_y - A_y B_x)\hat{k} \qquad (2\text{-}33)$$

The **determinant form of a vector product** is a more elegant way of representing Equation (2-33):

KEY EQUATION
$$\vec{A} \times \vec{B} = \begin{vmatrix} \hat{i} & \hat{j} & \hat{k} \\ A_x & A_y & A_z \\ B_x & B_y & B_z \end{vmatrix} \qquad (2\text{-}34)$$

The first row of the matrix lists the unit vectors, the second row lists the components of the first vector to be multiplied, and the bottom row lists the components of the second vector. Expand the determinant in Equation (2-34) about its first row and you will reproduce Equation (2-33).

✅ **CHECKPOINT**

C-2-11 Vector Cross Products

Which of the following expressions does *not* make sense? Explain.

(a) $(\vec{A} \times \vec{B}) \times \vec{C}$

(b) $(\vec{A} \times \vec{B}) \cdot \vec{C}$

(c) $\vec{A} \times (\vec{B} \cdot \vec{C})$

(d) $\vec{A} \cdot (\vec{B} \times \vec{C}) \cdot \vec{D}$

(e) $\vec{A} \cdot (\vec{B} + \vec{C})$

C-2-11 (c) The vector in (c) is cross multiplied by a scalar, which is not possible. The expression in (d) also makes no sense, because it is the dot product of three vectors, which is not defined.

EXAMPLE 2-7

Finding a Unit Vector Perpendicular to Two Given Vectors

Two forces are described as $\vec{F}_1 = (\hat{i} + 2\hat{j} - 3\hat{k})$ N and $\vec{F}_2 = (-2\hat{i} + 3\hat{j})$ N. Find an expression for a unit vector directed perpendicular to both of them.

SOLUTION

By definition, the cross product of any two vectors is a vector that is perpendicular to both vectors. Therefore, the unit vector we are looking for is directed along the line of the cross product of the two given vectors. Let us first find the cross product:

$$\vec{C} = \vec{F}_1 \times \vec{F}_2 = [(\hat{i} + 2\hat{j} - 3\hat{k})\,\text{N}] \times [(-2\hat{i} + 3\hat{j})\,\text{N}]$$

$$= [3\hat{k} + 4\hat{k} + 6\hat{j} + 9\hat{i}]\,\text{N}^2 = [9\hat{i} + 6\hat{j} + 7\hat{k}]\,\text{N}^2$$

Now, we find a unit vector along the direction of vector $\vec{C}$:

$$\hat{u}_c = \frac{\vec{C}}{C} = \frac{[9\hat{i} + 6\hat{j} + 7\hat{k}]\,\text{N}^2}{[\sqrt{81 + 36 + 49}]\,\text{N}^2} = \frac{1}{\sqrt{166}}(9\hat{i} + 6\hat{j} + 7\hat{k})$$

Making sense of the result:

We can check geometrically that our answer is correct. We can also use scalar products to confirm that the unit vector is perpendicular to the force vectors:

$$\hat{u}_c \cdot \vec{F}_1 = \frac{9\hat{i} + 6\hat{j} + 7\hat{k}}{\sqrt{166}} \cdot (\hat{i} + 2\hat{j} - 3\hat{k})\,\text{N} = \frac{9 + 12 - 21}{\sqrt{166}}\,\text{N} = 0$$

$$\hat{u}_c \cdot \vec{F}_2 = \frac{9\hat{i} + 6\hat{j} + 7\hat{k}}{\sqrt{166}} \cdot (-2\hat{i} + 3\hat{j})\,\text{N} = \frac{-18 + 18}{\sqrt{166}}\,\text{N} = 0$$

Since the dot products are zero, the unit vector is perpendicular to both force vectors. Note that the vector opposite to $\hat{u}$ is also perpendicular to both force vectors.

MECHANICS

FUNDAMENTAL CONCEPTS AND RELATIONSHIPS

Physical quantities used in this book can be categorized as scalars and vectors. Scalars require only a number (either positive or negative) and a unit for their description. Vectors require a number, a unit, and a direction. Vector operations include addition and subtraction, multiplication of a vector by a scalar, and dot (scalar) and cross (vector) products of two or more vectors.

Vector Components

Vector components are scalar quantities that represent vector projections onto the coordinate axes. In the case of 2-D vectors:

$$\begin{cases} v_x = v\cos\theta \\ v_y = v\sin\theta \end{cases} \Rightarrow \begin{cases} v = \sqrt{v_x^2 + v_y^2} \\ \tan\theta = \dfrac{v_y}{v_x} \end{cases}$$

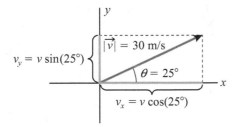

Cartesian Notation

$\vec{F} = F_x\hat{i} + F_y\hat{j} + F_z\hat{k}$, where $\hat{i}$, $\hat{j}$, and $\hat{k}$ are unit vectors

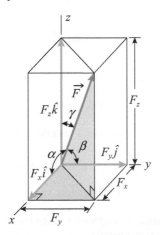

Vector Magnitude and Simple Coordinate System Transformations

When a coordinate system undergoes either pure translation or rotation, the magnitude of a vector in the transformed coordinate system remains unchanged. The vector components in the translated coordinate system do not change either. However, the vector components in a rotated coordinate system will be different from the vector components in the original coordinate system.

Vector Addition and Subtraction

Vector Addition: Vectors can be added using either a geometric or an algebraic approach.

Geometric Approach: triangle construction or parallelogram rule

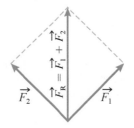

Parallelogram rule   Triangle construction rule

Vector Subtraction: $\vec{F_1} - \vec{F_2} = \vec{F_1} + (-\vec{F_2})$

Algebraic Approach:

$$\begin{cases} F_{R,x} = F_{1x} + F_{2x} + \cdots + F_{N,x} \\ F_{R,y} = F_{1y} + F_{2y} + \cdots + F_{N,y} \\ F_R = \sqrt{F_{R,x}^2 + F_{R,y}^2} \\ \tan\theta = \dfrac{F_{R,y}}{F_{R,x}} \end{cases}$$

Vector Multiplication by a Scalar

$$|a\vec{F}| = |a||\vec{F}|$$
$$\begin{cases} \text{if } a > 0, a\vec{F} \text{ has the same direction as } \vec{F} \\ \text{if } a < 0, a\vec{F} \text{ has the opposite direction to } \vec{F} \\ \text{if } a = 0, a\vec{F} \text{ has a magnitude of zero and no direction} \end{cases}$$

Products of Vectors

Dot Product: $\vec{A} \cdot \vec{B} = AB\cos\theta$,
where $0° \leq \theta \leq 180°$ and

$$\vec{A} \cdot \vec{B} = A_xB_x + A_yB_y + A_zB_z$$

Cross Product: $\vec{A} \times \vec{B} = \vec{C}$

$C = AB\sin\theta$, and $\vec{C}$ is perpendicular to both $\vec{A}$ and $\vec{B}$, where $0° \leq \theta \leq 180°$

$$\vec{A} \times \vec{B} = (A_yB_z - A_zB_y)\hat{i} + (A_zB_x - A_xB_z)\hat{j} + (A_xB_y - A_yB_x)\hat{k}$$

(continued)

Applications: Vectors are used to describe physical quantities that require magnitude, unit, and direction for their description, such as velocity, force, acceleration, and momentum. The dot product is used to determine the projections of vectors onto various directions. The cross product of vectors can be used to define or calculate physical quantities, such as angular momentum, torque, and magnetic force.

Key Terms: algebraic method for vector addition, Cartesian vector notation, coordinate direction angles, cross-multiplication circle, cross (vector) product, determinant form of a vector product, displacement vector, dot (scalar) product, free body diagram, geometric method for vector addition, opposite vectors, parallelogram rule, polar coordinates, polar notation, position vector, reference frame, right-hand rule, scalar notation, scalars, triangle construction rule, unit vectors, vector addition, vector components, vector head, vector magnitude and direction, vector projection, vector resolution into scalar components, vector subtraction, vector tail, vectors

QUESTIONS

1. A given force $\vec{F}$ is located in the horizontal plane. It has a magnitude of 30 N and is directed at a 45.0° angle clockwise from the positive x-axis. Describe $\vec{F}$ using
   (a) polar notation;
   (b) its x- and y-components;
   (c) Cartesian notation.

2. A given force $\vec{F}$ is located in the yz-plane. It has a magnitude of 15 N and is directed at a 30.0° angle clockwise from the positive y-axis. Describe $\vec{F}$ using
   (a) polar notation;
   (b) its y- and z-components;
   (c) Cartesian notation.

3. A force acting on a 5.00 kg object is located in the xy-plane. It has a magnitude of 8.00 N and is directed parallel to the line $y = -3x + 4$. Express this force in Cartesian notation. How many solutions do you have? Explain why.

4. An object is located at the point (3.00 m, –4.00 m, 0 m). A position vector connects this object to the origin. Find the magnitude of the position vector and the angle it makes with the positive x-axis.

5. A displacement vector connects two points, A (–2.00 m, 2.00 m, 0 m) and B (3.00 m, 0 m, –2.00 m). Find the magnitude of the displacement vector.

6. An object travelling in space passes five different points: A (2.00 m, 3.00 m, –1.00 m), B (2.00 m, 6.00 m, –1.00 m), C (2.00 m, 6.00 m, –8.00 m), D (–4.00 m, 6.00 m, –8.00 m), and E (–11.00 m, 6.00 m, –1.00 m). The object returns to its initial location, point A.
   (a) Rank the displacement vectors ($\vec{r}_{AB}$, $\vec{r}_{BC}$, $\vec{r}_{CD}$, $\vec{r}_{DE}$, $\vec{r}_{EA}$) from the vector that has the largest magnitude to the vector that has the smallest magnitude.
      Largest 1____ 2____ 3____ 4____ 5____ Smallest
   (b) Are the displacements the same in all cases?
   (c) Is the displacement zero in all cases?
   (d) If the displacement is the same for two or more cases, clearly indicate it on the ranking scheme. Explain your reasoning.

7. Three vectors are described as $\vec{A}$(1, –3, 4), $\vec{B}$(5, –2, 1), and $\vec{C}$(–2, 3, 6). Find the following vectors:
   (a) $\vec{D} = \vec{A} + \vec{B} + \vec{C}$
   (b) $\vec{E} = -\vec{A} - \vec{B} - \vec{C}$
   (c) $\vec{F} = 2\vec{A} - 4\vec{B} + 3\vec{C}$
   (d) $\vec{G} = 2(\vec{A} - 2\vec{B}) + \vec{C}$

8. Two horizontal forces of magnitude 10 N each are acting on a horizontal force board ring (Figure 2-20). A third force acting on the ring perfectly balances these two forces. What *must be* true about this force?
   (a) The third force must be 10 N, and it must be at an angle of 120 degrees to each one of the forces.
   (b) The magnitude of the third force should be more than 10 N.
   (c) The magnitude of the third force must be less than 10 N.
   (d) The magnitude of the third force must be between –20 N and 0 N.
   (e) The third force must be directed at a 45° angle to the two given forces.

Figure 2-20   Question 8. A bird's eye view of a force board

9. Which of the following statement(s) do you agree with? Explain why.
   (a) Vector components must always be smaller than the vector magnitudes.
   (b) In order to find vector magnitude, one must add the vector components.
   (c) Vector magnitude does not depend on the choice of the coordinate system, even though vector components do.
   (d) Vector components do not depend on the choice of the coordinate system, but vector magnitude does.
   (e) If two vectors have equal magnitudes, they must either be opposite vectors or have the same direction.

10. Which of the following vector pairs are orthogonal (perpendicular to each other)? Explain.
    (a) (2, 3, –6) and (–2, 3, –6)
    (b) (2, 3, –1) and (–2, –1, –7)
    (c) (–2, –3, –1) and (–2, –1, 7)

(d) $(1, 0, -6)$ and $(-2, 3, 0)$

(e) $(2, 3, 0)$ and $(0, -3, 6)$

11. Two students have an argument. Student 1 claims that any two vectors located in orthogonal planes must also be orthogonal, and student 2 disagrees. Who do you agree with and why? (Hint: It is sufficient to have one counterexample to disprove a statement.)

12. Two objects are moving at constant velocities: $\vec{v}_A = 2\hat{i} - 5\hat{j}$ and $\vec{v}_B = -2\hat{i} + 4\hat{j}$.
    (a) Which object is moving faster?
    (b) Find the angle between objects' trajectories.

13. Two vectors are described as $\vec{A}\,(2, 2, -2)$ and $\vec{B}\,(-1, 3, -2)$.
    (a) Find the projections of these vectors on the x-, y-, and z-coordinate axes.
    (b) Find the scalar product of these two vectors.
    (c) Find the vector product of these two vectors.

14. Find the coordinate direction angles α, β, and γ for the position vector $\vec{r} = (2\hat{i} - 5\hat{j} - 3\hat{k})$ m. Verify that expression (2-11) holds true.

15. You overhear your classmate uttering the following statement: "Since the magnitude of a unit vector equals 1, the components of a unit vector must also equal 1." Do you agree or disagree with this statement? Explain why.

16. Three forces are acting on an object: $\vec{F}_1 = (3.00\hat{i} + 4.00\hat{j} - 5.00\hat{k})\,\text{N}; \vec{F}_2 = (\hat{i} - 4.00\hat{j} + 2.00\hat{k})\,\text{N};$ $\vec{F}_3 = (-2.00\hat{i} + 3.00\hat{j} - 3.00\hat{k})\,\text{N}$. Find the expression for the net force acting on it, and then calculate the magnitude of the net force.

17. Complete the following vector addition (the forces are all measured in N), and find the magnitude of each force, as well as the magnitude of the net force:

+	$\hat{i}$	$\hat{j}$	$\hat{k}$
$\vec{F}_1$	1.00	−3.00	1.00
$\vec{F}_2$	−2.00	?	3.00
$\vec{F}_3$	?	2.00	0.00
$\vec{F}_R$	0.00	−5.00	?

18. Prove that the area of a parallelogram that has vectors $\vec{A}$ and $\vec{B}$ for two of its sides can be expressed as $|\vec{A} \times \vec{B}|$.

19. Your classmate claims that, in order to add two vectors using Cartesian notation, you need to add corresponding vector components; therefore, the same can be done using polar notation for vectors. Do you agree with your classmate? Explain why or why not.

20. List the advantages and disadvantages of using polar versus Cartesian vector notation. Illustrate your argument with relevant examples.

21. The head-to-tail rule for adding vectors is as follows: "Start with the first vector in the sum, and then arrange the rest of the vectors so that every vector's tail touches the previous vector's head. The sum can be represented as a vector connecting the tail of the first vector in the sum to the head of the last one." Prove that when you add more than two vectors you can still use the head-to-tail rule.

22. Find the relationship between the coordinate direction angles of any two opposite vectors.

23. Explain why the scalar product is commutative and the vector product is not.

24. If the scalar product of two nonzero vectors equals zero, then the vectors are orthogonal. What does it mean when we say that the vector product of two vectors equals zero? Explain how your result corresponds to the right-hand rule for finding the vector product of two vectors.

25. Force $\vec{F}$ has a magnitude of 5.00 N; its coordinate direction angles are $\alpha = 30.0°$, $\beta = -40.0°$, and $\gamma = -60.0°$. Describe $\vec{F}$ using Cartesian notation.

PROBLEMS BY SECTION

For problems, star ratings will be used, ($\ast$, $\ast\ast$, or $\ast\ast\ast$), with more stars meaning more challenging problems.

Section 2-2 Vector Addition: Geometric and Algebraic Approaches

26. $\ast\ast$ All of the force vectors shown in Figure 2-21 lie in the xy-, yz-, or xz-planes.
    (a) Estimate the x-, y-, and z-components of each force as accurately as you can from the figure.
    (b) Find the angles each force forms with the x- and y-axes.
    (c) Express each force in terms of Cartesian unit vectors.
    (d) Determine the magnitude of each force.
    (e) Write the expressions for the forces $\vec{F}_A, \vec{F}_B, \vec{F}_C,$ and $\vec{F}_D$ in Cartesian notation using an $x'y'z'$ coordinate system that is rotated 40° counterclockwise about the x-axis relative to the original xyz-coordinate system.
    (f) Calculate the magnitude of each force in the $x'y'z'$-coordinate system.
    (g) Compare your results to the previous parts. How do you know if your results make sense? What do they tell you?

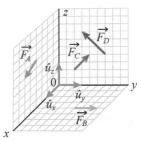

All the vectors are located in the x-y, x-z and y-z planes respectively.

Figure 2-21 Problem 26

27. $\ast$ Find the sum of the forces in Figure 2-21. Express the net force in Cartesian notation. Check your result using the PhET computer simulation "Vector Addition" (http://phet.colorado.edu/simulations/sims.php?sim=Vector_Addition). Explain why it is not practical to add these forces by hand using the geometric method.

28. $\ast$ Four 2-D displacement vectors describing consecutive displacements of an object moving on a flat surface are shown in Figure 2-22.
    (a) Use the geometric approach to vector addition to find the magnitude and direction of the total displacement of the object.

37

(b) Estimate the components of the four displacement vectors as accurately as you can. Then use the algebraic approach to find the total displacement. Compare your results to your results in part (a).

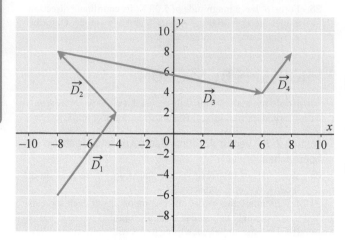

Figure 2-22   Problem 28

29. ✳ A cat travels 10.0 m in the direction of 20.0° [W of N]. Then the cat travels 7.0 m [W] and runs 20.0 m in the direction of 30.0° [W of S]. Find the magnitude of the total displacement of the cat using a geometric method, then check your result using the algebraic approach.

30. ✳✳ A child is sliding down a steep 60.0° water slide. The magnitude of the gravitational force exerted on the child by Earth equals 300 N, and the normal force exerted on the child by the slide is 150 N.
(a) Draw an FBD representing the problem.
(b) Find the components of the gravitational force along the x- and y-axes directed along the slide (x) and perpendicular to it (y).
(c) Find the components of the normal force along the x- and y-axes in (b).
(d) Find the net force acting on the child. What does your result imply?

31. ✳✳ The magnitudes of the vectors shown in Figure 2-23 are $A = 20.0$ units, $B = 15.0$ units, and $C = 25.0$ units, and the corresponding angles are $\theta_1 = -15.0°$, $\theta_2 = 35.0°$, and $\theta_3 = 125°$.
(a) Determine the components of each vector.
(b) Find the sum of the three vectors, and express the sum in both Cartesian and polar notation (magnitude and direction).
(c) Find the magnitude and direction of $\vec{D} = 2\vec{A} - 3\vec{B} + \vec{C}$.

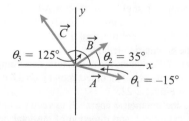

Figure 2-23   Problem 31

Section 2-3   Cartesian Vector Notation

32. ✳✳ Find the magnitude and the coordinate direction angles for the following vectors:
(a) $\vec{F} = (3.00\hat{i} - 5.00\hat{j} + 6.00\hat{k})$ N
(b) $\vec{F} = (-3.00\hat{i} + 5.00\hat{j} - 6.00\hat{k})$ N
(c) Compare your results from (a) and (b). What does your comparison mean?
(d) Find the sum of the squares of the cosines of the corresponding coordinate direction angles found above. How do you know if your result makes sense?

33. ✳✳ Find the Cartesian expression of a unit vector directed
(a) parallel to the force $\vec{F} = (3.00\hat{i} - 5.00\hat{j} + 6.00\hat{k})$ N;
(b) perpendicular to the force $\vec{F} = (3.00\hat{i} - 5.00\hat{j} + 6.00\hat{k})$ N;
(c) parallel to the plane defined by $3x + 2y - 4z = 0$;
(d) perpendicular to the plane defined by $3x + 2y - 4z = 0$.

34. ✳✳ The average velocity of an object as it is moving from position A to position B is defined as $\vec{v}_{avg} = \dfrac{\vec{r}_B - \vec{r}_A}{\Delta t}$, where Δt is the time it takes for the object to move from A to B. The object's positions are described as $\vec{r}_A = 3.00\hat{i} + 2.00\hat{j}$ and $\vec{r}_B = -3.00\hat{i} + 2.00\hat{j}$, and it took 5.00 s to move from A to B.
(a) Find the object's average velocity.
(b) What other information do you need to find the object's average speed?

35. ✳✳ Solve the following vector equations:
(a) $F_x\hat{i} + 3\hat{j} + \sqrt{2}\hat{i} - F_y\hat{j} + F_z\hat{k} - 5\hat{k} = 0$
(b) $3\hat{i} - 5\hat{j} + F_x\hat{i} - 2F_y\hat{j} + F_z\hat{k} - 3\hat{k} = 0$

Section 2-4   The Dot Product of Two Vectors

36. ✳ Calculate the dot product of the following two vectors: $\vec{A}$ (3, 4, –5) and $\vec{B}$ (2, –2, 4).

37. ✳✳ Find a unit vector located in the xy-plane and perpendicular to the given vector $\vec{A} = 3\hat{i} - 2\hat{j} + 5\hat{k}$. How many answers do you have? Explain why.

38. ✳ Solve the following vector equation:
$$(3\hat{i} - 5\hat{j} + F_z\hat{k}) \cdot (4\hat{i} + 4\hat{j} + F_z\hat{k}) = 0$$

39. ✳ Two vectors, $\vec{A}$ and $\vec{B}$, have magnitudes 10 units and 5 units, respectively, and are located in the xy-plane. The angle between them is 20°. Find the dot product of these two vectors.

40. ✳✳ Calculate the following dot products: $\vec{A} \cdot \vec{B}$; $\vec{B} \cdot \vec{C}$; $\vec{A} \cdot \vec{C}$; $\vec{B} \cdot \vec{A}$; $\vec{C} \cdot \vec{B}$ and $\vec{C} \cdot \vec{A}$ for vectors $\vec{A}$, $\vec{B}$, and $\vec{C}$ in Figure 2-24.

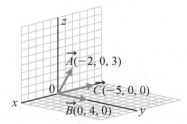

Figure 2-24   Problem 40

41. ✳✳ Find the projection of the 2-D force vector $\vec{F}$ (3.00, 4.00) N onto the line $y = -2x + 5$.

42. ✱✱ Find the angle between the two vectors $\vec{A}$ and $\vec{B}$, where $\vec{A} = -2\hat{i} + 3\hat{j} - 5\hat{k}$ and $\vec{B} = -2\hat{i} + 2\hat{j} - 5\hat{k}$.

Section 2-5 The Cross Product of Vectors

43. ✱ Calculate the vector cross product of the following two vectors: $\vec{A}(3, 4, -5)$ and $\vec{B}(2, -2, 4)$.

44. ✱✱ Three vectors, $\vec{A}(4, 3, 1)$, $\vec{B}(-2, 4, -2)$, and $\vec{C}(-1, -3, 5)$, form the sides of a parallelogram prism.
(a) Find the volume of the prism.
(b) Find the surface area of the prism.

45. ✱ Two vectors, $\vec{A}$ and $\vec{B}$, have magnitudes of 10 units and 5 units, respectively, and are located in the xz-plane. The angle between them is 20°. Find the magnitude and direction of the vector product of these two vectors.

46. ✱ For the vectors $\vec{A}$, $\vec{B}$, and $\vec{C}$ in Figure 2-24, calculate the following cross products:

$$\vec{A} \times \vec{B}; \vec{B} \times \vec{C}; \vec{A} \times \vec{C}; \vec{B} \times \vec{A}; \vec{C} \times \vec{B} \text{ and } \vec{C} \times \vec{A}$$

47. ✱✱ Solve the following vector equations:
(a) $2\hat{k} \times \left(\frac{1}{4} F_x \hat{i} + 7\hat{j} \right) + F_y \hat{i} - \sqrt{2}\hat{j} = 0$
(b) $4\hat{j} \times (F_x \hat{i} + 5\hat{k}) + F_z \hat{k} + \sqrt{3} F_x \hat{i} = 0$

48. ✱✱ Determine the angle between the diagonal of a cube and each one of the edges that shares a vertex with the diagonal.

COMPREHENSIVE PROBLEMS

49. ✱✱ An orthogonal coordinate system, xy, undergoes the following transformation: it moves parallel to itself such that the new location of its origin is $O'(4, 5)$. It is then rotated 60° counterclockwise. The new coordinate system is $x'y'$.
(a) Derive an expression for the coordinates of an arbitrary point under this transformation. Express how the new coordinates depend on the old coordinates.
(b) Derive an expression that describes how vector components change under this transformation. Express how the new vector components depend on the old vector components.

50. ✱✱ For the three vectors defined as $\vec{A} = (a_x, a_y, a_z)$, $\vec{B} = (b_x, b_y, b_z)$, $\vec{C} = (c_x, c_y, c_z)$, prove that

$$\vec{A} \cdot (\vec{B} \times \vec{C}) = \begin{vmatrix} a_x & a_y & a_z \\ b_x & b_y & b_z \\ c_x & c_y & c_z \end{vmatrix}$$

51. ✱✱ Prove that for any two non-zero vectors $\vec{A}$ and $\vec{B}$, both $\vec{A}$ and $\vec{B}$ are perpendicular to their cross product $\vec{A} \times \vec{B}$.

52. ✱✱ Prove the following relationships:
$$\vec{A} \times \vec{B} \times \vec{C} = \vec{C} \times \vec{A} \times \vec{B} = \vec{B} \times \vec{C} \times \vec{A};$$
$$\vec{A} \times \vec{B} \times \vec{C} = -(\vec{A} \times \vec{C} \times \vec{B}) = -(\vec{C} \times \vec{B} \times \vec{A}) = -(\vec{B} \times \vec{A} \times \vec{C})$$

53. ✱✱✱ Prove that the algebraic expression for the cross product of two vectors using the determinant form (2-34) is equivalent to the cross product calculation using the cross product definition (2-26). (Hint: Find the magnitudes of each vector, and then find the angle between them using the dot product.)

54. ✱✱ Prove that the algebraic expression for the cross product of two vectors using the determinant form (2-34) is equivalent to the cross product calculation using Cartesian notation (2.32) and the expressions for the cross product for the unit vectors (2-30).

55. ✱✱ Find the cross product of two vectors $\vec{A} = 3\hat{i} + 2\hat{j} - 5\hat{k}$ and $\vec{B} = 3\hat{i} + 2\hat{j} - 5\hat{k}$ using the determinant form, Equation (2-34).

56. ✱✱ Two straight lines, a and b, are oriented in the xy-plane and described as $a: y = m_1 x + b_1$ and $b: y = m_2 x + b_2$, where neither m_1 nor m_2 is 0. Prove that, in order to be perpendicular to each other, the following must be true: $m_1 m_2 = -1$. (Hint: Prove that any two arbitrary vectors directed along these lines are perpendicular to each other.)

57. ✱✱✱ A triangle is built on the diameter of a circle such that all three of its vertices lie on the circumference of the circle. Prove that the triangle is a right triangle. (Hint: Express the sides of the triangle as vectors, and use a scalar product to prove that these vectors are orthogonal.)

58. ✱✱✱ Show that $|\vec{A} \times \vec{B}| = |\vec{A}|^2 |\vec{B}|^2 - (\vec{A} \cdot \vec{B})^2$.

59. ✱✱✱ The law of cosines states that for any triangle, the following holds true: $C^2 = A^2 + B^2 - 2AB \cos\theta$, where θ is the angle between sides A and B of the triangle. Prove the law of cosines using the triangle construction method of vector addition and the fact that a square of the magnitude of a vector is equal to the dot product of the vector with itself.

60. ✱✱✱ Prove the formula for the differentiation of the scalar product: $d(\vec{A} \cdot \vec{B}) = (d\vec{A}) \cdot \vec{B} + \vec{A} \cdot (d\vec{B})$. Use your proof to find an expression for $d(\vec{v} \cdot \vec{v})$, where $\vec{v}$ is the velocity of a particle.

61. ✱✱✱ The speed of a particle moving in the xy-plane can be described as follows: $\vec{v} = v_{0x}\hat{i} + (v_{0y} - gt)\hat{j}$, where g is a positive constant called the acceleration due to gravity, t is time, and v_{0x} and v_{0y} are the components of the velocity on the x- and y-axes when $t = 0$. For Earth, $g \approx 9.81$ m/s².
(a) Find the expression for the speed of the particle as a function of time.
(b) Draw a diagram that illustrates how the speed of the particle changes with time.
(c) Determine the time when the speed of the particle is the smallest.
(d) This mathematical model is appropriate to describe a common, everyday phenomenon. Suggest what that phenomenon might be.

62. ✱✱ The work of a constant force W_F on an object can be described as a scalar product of the force and the object's displacement, $W_F = \vec{F} \cdot \Delta\vec{r}$, and is measured in joules when the force is measured in newtons and the displacement in metres. A constant force $\vec{F} = (3.00\hat{i} + 4.00\hat{j} - 5.00\hat{k})$ N acts on an object moving along the line $\vec{r} = \vec{a} + t\vec{b}$, with $\vec{a} = 2.00\hat{i} - 3.00\hat{j} - 4.00\hat{k}$ and $\vec{b} = 3.00\hat{i} - 2.00\hat{j} + 4.00\hat{k}$ (t is a parameter represented by a real number: $t \in R$). Find the work done on the object by the force $\vec{F}$ while the object moved from point A (1.00 m, 1.00 m, 0.00 m) to point B (3.00 m, −3.00 m, 6.00 m).

63. ✱✱✱ Newton's second law states that the acceleration of a moving object is proportional to the net force acting on it and inversely proportional to the object's mass:

$\vec{a} = \dfrac{\sum \vec{F}}{m} = \dfrac{\vec{F}_{net}}{m}$. There are three forces acting on a particle:

$\vec{F}_1 = (3.00\hat{i} + 2.00\hat{j} - 5.00\hat{k})$ N; $\vec{F}_2 = (-\hat{i} + 4.00\hat{j} - 2.00\hat{k})$ N;

$\vec{F}_3 = (-2.00\hat{i} + 5.00\hat{j} - 7.00\hat{k})$ N.

The particle's mass is 0.100 kg.
(a) Find the magnitude of the particle's acceleration.
(b) Find the coordinate direction angles of its acceleration vector.
(c) Write the expression for the particle's acceleration in Cartesian notation.
(d) Determine the projections of the particle's acceleration onto each of the coordinate planes xy, xz, and yz.

64. ✸✸✸ Three vectors are described as $\vec{A}(2, 3, -5)$, $\vec{B}(-1, 4, -2)$, and $\vec{C}(1, 2, 1)$. Find the following products:
(a) $\vec{D} = \vec{A} \times \vec{B}$
(b) $\vec{E} = \vec{A} \times \vec{B} + \vec{C}$
(c) $\vec{F} = 2\vec{A} \times \vec{B} + \vec{C}$
(d) $G = (\vec{A} \times \vec{B}) \cdot \vec{C}$

65. ✸✸✸ A trigonometric identity involving the cosine function can be proven using the dot product operation. Imagine a unit circle, which is a circle whose radius is 1 unit. Unit vectors $\hat{u}$ and $\hat{v}$ form angles α and β, respectively, with the horizontal axis. Use the unit circle to prove that $\cos(\beta - \alpha) = \cos\alpha\cos\beta + \sin\alpha\sin\beta$.

66. ✸✸✸ Three vectors are described as $\vec{A}(2, -3, -6)$, $\vec{B}(1, -4, 2)$, and $\vec{C}(1, -2, 3)$. Find all the vectors that are perpendicular to vectors $\vec{A}$ and $\vec{B}$ and have the same magnitude as vector $\vec{C}$.

67. ✸✸ The carbon tetrachloride molecule, CCl_4 (Figure 2-25(a)), has the shape of a tetrahedron. Chlorine atoms are located at the vertices of the tetrahedron, and the carbon atom is located in the centre. The bond angle (θ) is an angle of three connected atoms: Cl–C–Cl. A tetrahedron is formed by connecting alternating vertices of a cube, as shown in Figure 2-25(b). Find the bond angle (θ) of the carbon tetrachloride molecule.

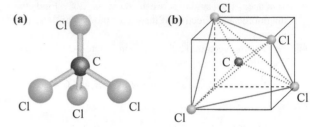

Figure 2-25 Problem 67 (a) A carbon tetrachloride molecule (b) Atom locations in the carbon tetrachloride molecule

See the text online resources at www.physics1e.nelson.com for Open Problems and Data-Rich Problems related to this chapter.

Chapter 3
Motion in One Dimension

Learning Objectives

When you have completed this chapter you should be able to:

1 Define distance and displacement.

2 Define velocity and average velocity, and distinguish between velocity and speed.

3 Define acceleration and average acceleration, and calculate acceleration from velocity and position information.

4 Develop and use kinematics equations for constant acceleration.

5 Define and calculate relative motion quantities in one dimension.

High accelerations can cause serious injuries. During emergency landings, a plane and its passengers can experience high accelerations as the plane comes to a stop over a short distance. In the United States, passenger planes designed between 1952 and 1988 had to have 9 g seats, which means that the seats had to be able to withstand accelerations 9 times the acceleration due to gravity. However, analysis of aircraft crashes equipped with these seats indicated that often the majority of the fatalities and serious injuries in a crash resulted from the seats separating from the tracks on which they were mounted. In 1988, the U.S. Federal Aviation Administration (FAA) introduced regulation requiring commercial aircraft designed after 1988 to be equipped with seats that could withstand accelerations of 16 g. This measure appears to have saved numerous lives during emergency landings and runway overshoots. For example, in 2009, American Airlines flight 331 overshot a runway in Jamaica without any fatalities, even though the Boeing 737 cracked its fuselage (Figure 3-1).

Similarly, there were no fatalities and only 12 serious injuries when an Air France Airbus 340 with 309 people onboard ran off a runway into a ravine at Pearson Airport, Toronto, Ontario, while landing in bad weather in 2005. All passenger aircraft in the United States built after October 27, 2009, are required to have the 16 g seats, and older aircraft are to be retrofitted with 16 g seats by 2016.

ANDREW P. SMITH/Reuters/Landov

Figure 3-1 American Airlines flight 331 crashed on landing at Kingston's Norman Manley International Airport. The Boeing 737 cracked its fuselage, stopping just before it hit the water. There were no fatalities among the 154 passengers.

For additional Making Connections, Examples, and Checkpoints, as well as activities and experiments to help increase your understanding of the chapter's concepts, please go to the text's online resources at www.physics1e.nelson.com.

Kinematics is the study of motion. In kinematics, we analyze how an object's position, velocity, and acceleration relate to one another, and how they change with time. We restrict the discussion in this chapter to motion in one dimension, and extend it to two and three dimensions in the next chapter. We will derive the general relationships for kinematics and the equations that describe motion under constant acceleration.

LO 1

3-1 Distance and Displacement

The **displacement** of an object is the change in its **position**. As shown in Figure 3-2, the displacement is the difference between the final position vector and the initial position vector:

$$\Delta \vec{x} = \vec{x_2} - \vec{x_1} \qquad (3\text{-}1)$$

where $\vec{x_1}$ and $\vec{x_2}$ are, respectively, the initial and final positions of the object as measured from a reference point.

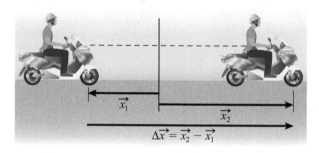

Figure 3-2 Displacement is the change in the position vector.

The SI unit for both distance and displacement is the metre (m). The displacement of an object does not depend on the details of the path taken; it depends only on the initial and final positions. The distance travelled by an object, however, *does* depend on the details of the journey. Distance is a scalar and does not depend on direction.

The distance travelled by an object can be defined as the integral of the lengths of infinitesimal segments of the path:

$$d = \int_{x_1}^{x_2} |d\vec{x}| \qquad (3\text{-}2)$$

Figure 3-3 illustrates the difference between displacement and distance. A person walks from point A to

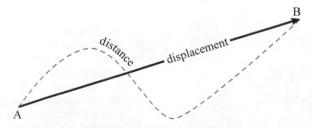

Figure 3-3 Distance (dashed line) and displacement (arrow from A to B)

point B along the path represented by the dashed line—the displacement is the shortest distance between the initial and final positions; however, the distance travelled is the total length of the path. Examples 3-1 and 3-2 highlight the difference between distance and displacement.

 EXAMPLE 3-1

Distance and Displacement

A water taxi starts from a dock on the west bank of a channel, delivers passengers to a dock on the east bank, and then returns to its starting point on the west bank. The two docks are 230 m apart.

(a) Determine the displacement of the water taxi.
(b) Calculate the distance travelled by the taxi.

SOLUTION

(a) At the end of the round trip, the taxi ends up where it started, so its displacement is zero.
(b) The taxi covers 230 m each way, for a total distance travelled of 460 m.

Making sense of the result:

Displacement is not the same as distance travelled.

 EXAMPLE 3-2

Calculating Distance and Displacement

A tarantula is on a wall 20 cm below a nail. It moves down to a point 60 cm below the nail to catch a fly, and then takes the fly straight up to a position 30 cm above the nail, as shown in Figure 3-4.

(a) Find the displacement of the tarantula.
(b) Find the distance covered by the tarantula.

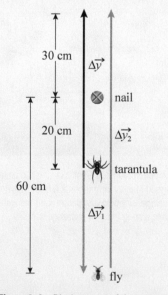

Figure 3-4 Displacement of the tarantula

SOLUTION

(a) We will use the nail as our origin. We define up to be the positive y direction. Therefore, y_1, the initial position of the tarantula as measured from the nail, is negative, and y_2, the final position of the tarantula, is positive. Therefore, the displacement is given by

$$\Delta \vec{y} = \vec{y}_2 - \vec{y}_1 = 30\hat{j} \text{ cm} - (-20)\hat{j} \text{ cm} = +50\hat{j} \text{ cm}$$

(b) On the way down, the tarantula moves from an initial position of 20 cm below the nail to 60 cm below the nail. The distance covered on the way down equals the magnitude of the displacement:

$$\left| \Delta \vec{y}_{\text{down}} \right| = 40 \text{ cm}$$

On the way up, the tarantula moves from an initial position of 60 cm below the nail to a position of 30 cm above the nail, for a total upward distance of

$$\left| \Delta \vec{y}_{\text{up}} \right| = 90 \text{ cm}$$

The total distance covered by the tarantula is the sum of the two distances, 130 cm.

Making sense of the result:

For displacement all we care about are initial and final positions, the distance travelled takes into account the details of the path taken.

Position and displacement are defined as vector quantities: they have a magnitude and a direction. For one-dimensional motion, only two directions are possible. Since these two directions are opposite, we can define one of the directions as positive, and then indicate the direction of vector quantities by simply putting the appropriate positive or negative sign in front of the magnitude.

 CHECKPOINT

C-3-1 Distance and Displacement

When a Formula One race car drives a full lap around a circular track 210 m in radius,

(a) the displacement of the car is the circumference of the track;

(b) the distance travelled by the car is the circumference of the track;

(c) the distance travelled by the car is zero;

(d) the displacement and distance travelled are equal.

C-3-1 (b) displacement is the difference between final and initial positions, and is therefore zero.

 MAKING CONNECTIONS

Range of Distances in Nature and Everyday Life

Table 3-1 Range of Distances in Nature and Everyday Life

Distance	Description
1.6162×10^{-35} m	Planck length, the shortest distance that can theoretically be measured
0.8768 fm = 8.768×10^{-16} m	Radius of a nucleon
12.4 pm = 12.4×10^{-12} m	Upper limit for the wavelength of gamma rays
52.9 pm = 52.9×10^{-12} m	Radius of a hydrogen atom
~0.1 nm	Covalent bond length
~0.2 nm	Hydrogen bond length
0.1–10 nm or 0.1×10^{-9}–10×10^{-9} m	Wavelength of X-rays
580–590 nm	Wavelength of yellow light
30.48 cm	1 ft.
1.609 344 km	1 mile
1.852 km	1 nautical mile
79.8 m	Wingspan of Airbus 380-800
1737 km	Mean radius of Moon
6371 km	Mean radius of Earth
~35.4 solar radii	Radius of R136a1, the most massive known star
1800–2100 solar radii	Radius of VY Canis Majoris, possibly the largest known star,

(continued)

Distance	Description
384 400 km	Mean distance between Earth and the Moon
6.955×10^8 m	Radius of the Sun
149 597 871 km	1 astronomical unit, the mean Earth–Sun distance
4.545 billion km	Mean Neptune–Sun distance
$9.460\ 528\ 4 \times 10^{15}$ m	1 light year, the distance that light travels in one year
100 000 light years	Approximate diameter of the Milky Way galaxy
93 billion light years	Approximate diameter of the observable universe

LO 2

3-2 Velocity and Speed

The terms "velocity" and "speed" are often inter-changeably in everyday life. However, in physics, **speed** is a scalar quantity and **velocity** is a vector. Speed tells us only how fast an object is moving; velocity also tells us the direction of the motion.

Average Velocity

For an object that moves from position $\vec{x_1}$ to position $\vec{x_2}$ during an amount of time Δt, the **average velocity** of the object is given by its displacement divided by the time it takes to complete the displacement:

$$\vec{v}_{\text{avg}} = \frac{\Delta \vec{x}}{\Delta t} = \frac{\vec{x_2} - \vec{x_1}}{\Delta t} \qquad (3\text{-}3)$$

The average velocity does not depend on the details of the journey between two points. A car driver might decide to drive along a highway at a given velocity for some time, stop for a coffee, and then continue to a destination, quite possibly at a different velocity, depending on traffic and road conditions. The total displacement divided by the total elapsed time, including the time for the coffee break, gives us the average velocity for the trip. For someone driving along a straight, flat highway, the average velocity between two points is represented by the slope of the line connecting the corresponding two points on the position-time plot; that is, the rise, Δx, divided by the run, Δt, as shown in Figure 3-5.

The average speed gives us more detail regarding how fast the car was going throughout the trip. **Average speed** is defined as the distance travelled divided by the elapsed time.

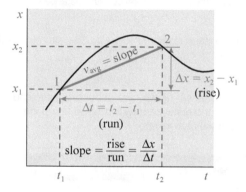

Figure 3-5 On a position versus time plot, the average velocity between two points is given by the slope (rise over run) of the line connecting the two points.

EXAMPLE 3-3

Average Velocity

A plane takes off from Prague in the Czech Republic and lands in Brussels, Belgium, 710 km northwest 1.20 h later. What is the average velocity of the plane?

SOLUTION

The average velocity is given by Equation (3-3):

$$\vec{v}_{\text{avg}} = \frac{\Delta \vec{x}}{\Delta t} = \frac{710\hat{i}\ \text{km}}{1.20\ \text{h}} = 592\hat{i}\ \text{km/h}$$

$$= 592\hat{i}\ \frac{\text{km}}{\text{h}} \left(\frac{1000\ \text{m/km}}{3600\ \text{s/h}} \right) = 164\hat{i}\ \text{m/s}$$

where we have chosen the positive x-axis to be along the direction of motion of the plane.

Making sense of the result:

The magnitude of the average velocity is approximately 600 km/h, which is a reasonable speed for a passenger jet.

EXAMPLE 3-4

Speed and Velocity

A plane flies straight from Stockholm, Sweden, to arrive over Moscow, Russia, 1240 km southeast, 1.70 h later. The pilot is then told to return to Stockholm. The return trip to Stockholm takes 2.10 h, but the plane is then diverted to Oslo, Norway, 425 km to the northwest of Stockholm. The plane finally lands 0.900 h later. The three cities lie along a straight line, as shown in Figure 3-6.

(a) Find the average velocity of the plane.
(b) Find the average speed of the plane for the entire trip.

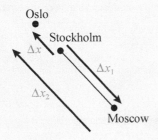

Figure 3-6 Flight path of the plane in Example 3-4

SOLUTION

(a) The average velocity depends on the net displacement and the total elapsed time:

$$\vec{v}_{avg} = \frac{\Delta\vec{x}}{\Delta t} = \frac{425\hat{i}\ \text{km}}{1.70\ \text{h} + 2.10\ \text{h} + 0.900\ \text{h}} = 90.4\hat{i}\ \text{km/h}$$

(b) The average speed is the total distance travelled divided by the total time:

$$v_{avg} = \frac{d}{\Delta t} = \frac{1240\ \text{km} + 1240\ \text{km} + 425\ \text{km}}{1.70\ \text{h} + 2.10\ \text{h} + 0.900\ \text{h}}$$

$$= \frac{2905\ \text{km}}{4.70\ \text{h}} = 618\ \text{km/h}$$

where we have chosen the positive x-axis to point from Moscow to Oslo.

Making sense of the result:

Unlike the average speed, which tells us how fast the plane was moving in the air, the average velocity depends only on the starting and ending positions.

✓ **CHECKPOINT**

C-3-2 Average Speed

You make a bungee jump off a 300 m-high bridge. After 20 s, you have fallen 270 m and bounced 230 m back up. What is your average speed?

(a) 2 m/s
(b) 3.5 m/s
(c) 25 m/s
(d) none of the above

Instantaneous Velocity

If we want to know details of how an object's position is changing at a given location along its trajectory, we consider the **instantaneous velocity**, the velocity of the object at a given moment. To calculate the instantaneous velocity, we find the limit of the average velocity as the change in time goes to zero. In other words, the instantaneous velocity is the derivative of the position with respect to time:

$$\vec{v} = \lim_{\Delta t \to 0} \vec{v}_{avg} = \lim_{\Delta t \to 0} \frac{\Delta\vec{x}}{\Delta t} = \frac{d\vec{x}}{dt} \qquad (3\text{-}4)$$

Similarly, **instantaneous speed** is the magnitude of the instantaneous velocity. On a plot of position versus time, the instantaneous speed corresponds to the absolute value of the slope of the tangent to the plot at that point (Figure 3-7).

Note, however, that the average speed is not necessarily equal to the magnitude of the average velocity as demonstrated by Example 3-4 above.

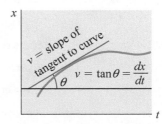

Figure 3-7 Instantaneous velocity is given by the slope of the tangent on the position versus time plot.

MAKING CONNECTIONS

Range of Speeds in Nature and Everyday Life

Table 3-2 Range of Speeds in Nature and Everyday Life

Speed	Description
~1 m in 10^{10} y	Theoretical rate of flow of glass at room temperature
~1 m in 800 y	Theoretical rate of flow of glass at 414°C
~1 cm/y	Speed of continental drift
0.2 cm/s	Speed of a small earth worm
1 cm/s	Speed of a molecule in a Bose–Einstein condensate (occurs at very low temperatures)
1.852 km/h	1 knot, the customary unit for the speed of ships
~10 m/s	Sprint speed in a 100 m race
41 km/h	Highest speed ever achieved by a sailing ship
52 km/h	Top running speed of a bear
82.78 km/h	Top speed of K222, an Anchar (Papa) class Soviet submarine
112 km/h	Top speed of cheetahs and sailfish
325 km/h	Top swooping speed of a peregrine falcon
360 km/h	Top speed of Formula One cars at Monza, Italy
431.07 km/h	Top speed of a Bugatti Veyron race car
574.8 km/h	Speed record for the V150 version of Alstom's TGV high-speed passenger train
581 km/h	Speed record for the MLX01 magnetic levitation train (Japan, 2003)
945 km/h	Maximum operating speed of an Airbus 380 at cruising altitude
330 m/s	Maximum rate of climb of a Mig-29 fighter plane
340 m/s	Speed of sound at sea level at 15°C
1040 km/h	Top speed of the Yokosuka MXY7 Ohka attack plane
1352 m/s	Root mean square speed of a helium atom at room temperature
465 m/s or 1674 km/h	Equatorial rotation speed of Earth
2172 km/h	Top speed of the Concorde supersonic jet at cruising altitude
Mach 2.8	Speed of the Brahmos antiship missile
3630 km/h	Maximum speed of a Mig 25 fighter plane
1 km/s or 3600 km/h	Speed of the Moon around Earth
2 km/s	Equatorial rotation speed of the Sun
Mach 7	Speed of the Brahmos II missile
7275 km/h	Top speed of the X-15, the fastest piloted rocket plane
12 144 km/h	Maximum speed of the X-43, an unpiloted experimental hypersonic jet
28 000 km/h	Speed of the space shuttle on reentry
16.26 km/s or 58 536 km/h	Speed of NASA's *New Horizons* spacecraft

Speed	Description
29.77 km/s	Speed of Earth around the Sun
220 km/s	Speed of the Sun around the galaxy centre
~600 km/s	Speed of the Milky Way galaxy (relative to cosmic microwave background)
2100 km/s	Equatorial rotation speed of the Crab Pulsar
3×10^8 m/s	Speed of light

 EXAMPLE 3-5

Velocity and Speed Plots

Figure 3-8(a) shows a position versus time graph for a car. Use the graph to plot the velocity vs. the time t, and the speed versus time for the time intervals shown.

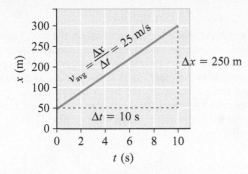

Figure 3-8(a) Position versus time for a car

SOLUTION

The slope of the position graph is constant, indicating that the velocity and the speed are constant. Hence, the instantaneous velocity is equal to the average velocity for the 10 s interval shown on the graph. Since the displacement is 250 m in the positive direction during this time,

$$\left|\vec{v}_{avg}\right| = \frac{\Delta x}{\Delta t} = \frac{250 \text{ m}}{10 \text{ s}} = 25 \text{ m/s}$$

A plot of the instantaneous velocity and speed is simply a horizontal line with a y-intercept of 25, as shown in Figure 3-8(b).

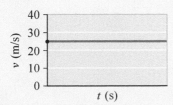

Figure 3-8(b) Velocity versus time graph for the car

Making sense of the result:

The x vs t plot is linear, which means the speed is constant.

 ## MAKING CONNECTIONS

Fastest Planes in World War II

Toward the end of World War II, the major powers on both sides began to produce jet- and rocket-powered warplanes. The German Messerschmitt Me262 was the first operational jet fighter (Figure 3-9). It had a maximum speed of 900 km/h. The Allies' first operational jet was the British Gloster Meteor, which had a top speed of 965 km/h. The fastest of these new aircraft was the Yokosuka MXY7 Ohka (Figure 3-10), used by Thunder God Corps of the Japanese air force. The Ohka was essentially a rocket-powered warhead designed for suicide attacks on Allied warships. When diving with its three rocket engines at full power, the Ohka could reach speeds as high as 1040 km/h. However, the speed of the MXY7 in level flight was 649 km/h, and it had a range of only 36 km.

Figure 3-9 The Me 262 in the collection of the National Air and Space Museum, Steven F. Udvar-Hazy Center, Washington, D.C.

Figure 3-10 The Yokosuka MXY7 Ohka on display in the Air & Space Gallery, Museum of Science & Industry, Manchester, U.K.

EXAMPLE 3-6

Average and Instantaneous Velocity

An object moves in a straight line such that its position (in metres) is given by $x = 0.250 t^3$, where t is measured in seconds.

(a) Plot the position versus time for the interval $t = 0$ s to $t = 2.00$ s.
(b) Find the average velocity during the interval $t = 1.00$ s to $t = 2.00$ s, and include the corresponding line on your graph.
(c) Find the average velocity during the interval 1.25 s to 1.75 s, and include the corresponding line on your graph.
(d) Find the instantaneous velocity at $t = 1.50$ s, and include the corresponding line on your graph.

SOLUTION

(a) To create the plot, we can pick convenient values along the t-axis, calculate the corresponding x-coordinates, and sketch a smooth curve from the resulting points, as shown in Figure 3-11.
(b) We can apply Equation (3-3) to find the average velocity between $t = 1$ s and $t = 2$ s:

$$v_{avg} = \frac{\Delta x}{\Delta t} = \frac{x_2 - x_1}{\Delta t} = \frac{0.25(2^3) \text{ m} - 0.25(1^3) \text{ m}}{2 \text{ s} - 1 \text{ s}} = +1.75 \text{ m/s}$$

This velocity is illustrated by an orange line on the graph. The average velocity between $t = 1$ s and $t = 2$ s

is actually the slope of the line connecting the points at $t = 1$ s and $t = 2$ s on the position curve.

(c) The average velocity between $t = 1.25$ s and 1.75 s is

$$v_{avg} = \frac{\Delta x}{\Delta t} = \frac{x_2 - x_1}{\Delta t} = \frac{0.25(1.75^3) \text{ m} - 0.25(1.25^3) \text{ m}}{1.75 \text{ s} - 1.25 \text{ s}}$$

$$= +1.70 \text{ m/s}$$

This velocity corresponds to the slope of the line connecting the points at $t = 1.25$ s and $t = 1.75$ s also shown in orange.

(d) To get the instantaneous velocity at $t = 1.50$ s, we use Equation (3-4):

$$v = \frac{dx}{dt} = \frac{d(0.25t^3)}{dt} = 3(0.25)t^2 = 0.75(1.5)^2 = +1.69 \text{ m/s}$$

This velocity corresponds to the slope of the tangent to the x versus t graph at $t = 1.5$ s shown in green.

Making sense of the result:

We see that as the time, Δt, separating the two points gets smaller, the line representing the average velocity approaches the instantaneous velocity given by the slope of the tangent to the graph, until in the limit as Δt goes to zero, the two points separated by Δt merge, and we get the tangent to the line at the point of interest, who's slope is given by the derivative of x as a function of t.

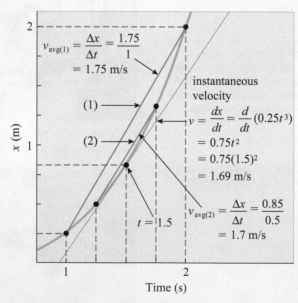

Figure 3-11 The closer that two points are on a position versus time plot, the closer the average velocity gets to the instantaneous velocity.

C-3-3 Average Velocity

The average velocity of an object

(a) must be greater than its instantaneous velocity;

(b) must be less than its instantaneous velocity;

(c) cannot be equal to its instantaneous velocity;

(d) none of the above.

C-3-3 (d) the average velocity can be greater or less than or equal to the instantaneous velocity.

 MAKING CONNECTIONS

Cosmic Cannonball

The Chandra X-ray Observatory detected a neutron star, RX J0822-4300, which is moving away from the centre of Pupis A, a supernova remnant about 7000 ly away (Figure 3-12). Believed to be propelled by the strength of the lopsided supernova explosion that created it, this neutron star is moving at a speed of about 4.8 million km/h, putting it among the fastest-moving stars ever observed. At this speed, its trajectory will take it out of the Milky Way galaxy in a few million years. Astronomers were able to estimate its speed by measuring its position over a period of 5 years.

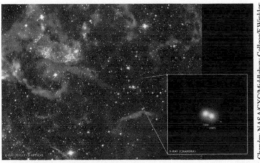

Chandra: NASA/CXC/Middlebury College/F. Winkler; ROSAT: NASA/GSFC/S. Snowden et al.; Optical: NOAO/CTIO/Middlebury College/F. Winkler et al.

Figure 3-12 Supernova remnant RX J0822-4300

 LO 3

3-3 Acceleration

The rate of change of the velocity with time is called **acceleration**. The **average acceleration** of an object is given by the change in its velocity divided by the time over which it changes:

$$\vec{a}_{avg} = \frac{\Delta \vec{v}}{\Delta t} = \frac{\vec{v}_2 - \vec{v}_1}{\Delta t} \tag{3-5}$$

The average acceleration of an object between two points depends on only the final and initial velocities and the elapsed time. The details of the object's motion between the two points do not affect the average acceleration.

Taking the limit of the average acceleration as Δt goes to zero effectively gives us the **instantaneous acceleration** at a specific point. Thus, the instantaneous acceleration is the derivative of the velocity as a function of time:

$$\vec{a} = \lim_{\Delta t \to 0} \frac{\Delta \vec{v}}{\Delta t} = \frac{d\vec{v}}{dt} \tag{3-6}$$

Since the instantaneous velocity is equal to the derivative of the position with respect to time, the instantaneous acceleration is equal to the *second* derivative of position with respect to time:

$$\vec{a} = \frac{d^2\vec{x}}{dt^2} \tag{3-7}$$

 CHECKPOINT

C-3-4 Acceleration

If the acceleration vector of a moving object points in the same direction as the motion but is decreasing in magnitude, the speed of the object

(a) is increasing;

(b) is decreasing;

(c) is constant;

(d) could be either increasing or decreasing, depending on the magnitude of the velocity.

C-3-4 (a) there is non zero forward acceleration, the object's speed will increase.

 ONLINE ACTIVITY

The e-resource that accompanies every new copy of this textbook contains an Online Activity using the PhET simulation "Moving Man." Work through the simulation and accompanying questions to gain an understanding of the relationships between the quantities that describe motion.

MAKING CONNECTIONS

Acceleration and Safety Standards

Until the end of World War II, the prevalent belief among military aircraft engineers was that subjecting the human body to accelerations in excess of 18 g would be fatal. Therefore, airplane cockpits and seats did not have to be built to withstand accelerations greater than 18 g. However, research led by Colonel John Stapp (Figure 3-13(a)) of the U.S. Army Air Force showed that pilots could survive much greater accelerations if the seats and restraining harnesses could withstand the forces involved. The most notable tests used a rocket sled called the "Gee Whiz" (Figure 3-13(b)) to subject test dummies and human volunteers to accelerations comparable to those during a crash landing.

Stapp often rode the Gee Whiz himself and sustained broken bones and burst blood vessels in some of the tests. In 1954, he survived 46.2 g, likely still the world record for such tests. As a result of Stapp's research, new standards were adopted to require fighter jet seats to be able to withstand accelerations of 32 g. Stapp also developed improved harnesses for pilots and paratroopers, and advocated strongly for the installation of seat belts in cars. His work led to the 1966 law that made the provision of seat belts mandatory for cars sold in the United States.

Figure 3-13 (a) Col. John P. Stapp secures Capt. Edward Magid into "the vertical accelerator." (b) The Gee Whiz, at the International Space Hall of Fame Museum, New Mexico

MAKING CONNECTIONS

Noteworthy Accelerations

Table 3-3 Noteworthy Accelerations

0.033 73 m/s^2	Radial acceleration at Earth's equator due to Earth's rotation
9.81 m/s^2	Typical acceleration due to gravity at Earth's surface at sea level (g)
1.7 g	Typical transverse acceleration of a Formula One car on a curve
~4 g	Acceleration experienced by astronauts during takeoff and reentry for both the space shuttle and the Soyuz
5.2 g	Typical acceleration of a luge sled at Whistler, B.C., during the Vancouver Olympics
5.3 g	Typical acceleration of a Top Fuel drag racer
6 g	Lateral acceleration, Formula One, Suzuka circuit; Maximum deceleration, Formula One
8–9 g	Tolerable sustained acceleration during inside loop for pilots
12 g	Threshold acceleration for sustained inside loop (vertical), with G-suit
16 g	Commercial aircraft seat standard as of 1988
18 g	Injury threshold for upward acceleration for a duration of 100 ms
20–25 g	Spinal injury threshold for upward acceleration for a duration of 100 ms
20 g	Injury threshold for lateral acceleration
32 g	Tolerance standard for U.S. military aircraft seats
~214 g	Highest acceleration survived during a crash
18 000 g	Approximate maximum acceleration of a golf ball in competition (Jason Zuback, 2009)
2 000 000 g	Radial acceleration in an ultracentrifuge
3.3 × 10^{10} g	Radial acceleration at the equator of the fastest known spinning neutron star[1]
~7 × 10^{11} g	Gravitational acceleration on the surface of a neutron star

[1]At 716 Hz (716 rotations per second), PSR J1748-2446ad is the fastest *confirmed* pulsar.

EXAMPLE 3-7

Using Derivatives to Calculate Acceleration

The position in metres of a particle on the x-axis is given by $x = -0.400t^4 + 4.00t^2 - 16.0t$, where time is measured in seconds. Find

(a) the velocity of the particle at $t = 2.00$ s;
(b) the acceleration of the particle at $t = 1.00$ s;
(c) the average acceleration of the particle between $t = 0.900$ s and $t = 1.10$ s;
(d) the maximum velocity the particle will reach.

SOLUTION

(a) To find the velocity from the position, we use Equation (3-4):

$$v = \frac{dx}{dt} = -4(0.4)t^3 + 2(4)t - 16 = -1.6t^3 + 8t - 16$$

At $t = 2$ s,

$$v = \frac{dx}{dt} = -1.6(2^3) + 8(2) - 16 = -12.8 \text{ m/s}$$

(b) To find the acceleration, we take the derivative of the velocity with respect to time:

$$a = \frac{dv}{dt} = \frac{d^2x}{dt^2} = -12(0.4)t^2 + 2(4) = -4.8t^2 + 8$$

At $t = 1$ s,

$$a = -4.8(1)^2 + 8 = +3.20 \text{ m/s}^2$$

(c) For the average acceleration, we use Equation (3-6):

$$a_{\text{avg}} = \frac{\Delta v}{\Delta t} = \frac{v_2 - v_1}{\Delta t}$$

Here, v_2 is the velocity at 1.1 s, and v_1 is the velocity at 0.9 s, so

$$v_2 = -1.6(1.1)^3 + 8(1.1) - 16 = -9.3296 \text{ m/s}$$

$$v_1 = -1.6(0.9)^3 + 8(0.9) - 16 = -9.9664 \text{ m/s}$$

$$a_{\text{avg}} = \frac{v_2 - v_1}{\Delta t} = \frac{-9.3296 - (-9.9664)}{0.2} = +3.18 \text{ m/s}^2$$

(d) The maximum (or minimum) speed occurs when the derivative of the velocity is zero, which occurs when the acceleration is zero:

$$a = -4.8t^2 + 8 = 0$$

$$\Rightarrow t = \sqrt{\frac{8}{4.8}} = 1.291 \text{ s}$$

At this time, the velocity is

$$v = -4(0.4)(1.291)^3 + 2(4)(1.291) - 16 = -9.115 \text{ m/s}$$

The maximum velocity is therefore -9.12 m/s.

Making sense of the result:

In the interim calculations we keep more significant digits than is required for the question. For parts b and c, since the time interval is so small, the average acceleration between $t = 0.9$ s and $t = 1.1$ s is very close to the instantaneous acceleration at $t = 1$ s. Note that the maximum velocity does not correspond to the maximum speed.

 CHECKPOINT

C-3-5 Average Acceleration

An object is thrown straight down with a speed of 10 m/s. The object then bounces back up from the ground, reaching its initial location 4 s later with the same speed. The average acceleration of the object is

(a) 9.8 m/s²;
(b) zero;
(c) 5 m/s²;
(d) None of the above.

C-3-5 (c) the change in velocity is 20 m/s, the time elapsed is 4 s.

LO 4

3-4 Kinematics Equations

General Framework for Kinematics Equations

In the previous section, we obtained velocity and acceleration from position information for a given object. Conversely, if we know how the acceleration of an object

varies with time, we can determine how the velocity and the position of the object change with time. If the acceleration is given by a continuous smooth function $a(t)$, then we can rearrange Equation (3-6) and integrate to find the velocity of the object:

$$d\vec{v} = \vec{a}\,dt \tag{3-8}$$

$$\Delta\vec{v} = \vec{v}(t) - \vec{v}_0 = \int_0^t \vec{a}\,dt \tag{3-9}$$

or

$$\vec{v}(t) = \vec{v}_0 + \int_0^t \vec{a}\,dt \tag{3-10}$$

where $\vec{v}_0$ is the velocity of the object at time $t = 0$, that is, the initial velocity of the object.

According to Equation (3-10), $\vec{v}(t)$, the velocity of the object as a function of time, is given by the vector sum of the initial velocity of the object and the change in velocity due to the acceleration term.

Similarly, rearranging Equation (3-4) and integrating gives $\vec{x}(t)$, the position of the object as a function of time:

$$d\vec{x} = \vec{v}\,dt \tag{3-11}$$

$$\Delta x = \vec{x}(t) - \vec{x_0} = \int_0^t \vec{v}\, dt \qquad (3\text{-}12)$$

or

$$\vec{x}(t) = \vec{x_0} + \int_0^t \vec{v}\, dt \qquad (3\text{-}13)$$

where $\vec{x_0}$ is the initial position of the object.

EXAMPLE 3-8

Applying Kinematics Equations

A Formula One car is moving at a speed of 98.0 m/s along a straight track when the driver brings the car to a stop, gradually applying the brakes such that the acceleration increases linearly with time. The car takes 3.40 s to stop.

(a) Express the car's acceleration as a function of time.
(b) Calculate the speed of the car 3 s after the driver starts braking.
(c) Find the total distance covered by the car while slowing.
(d) Find the highest acceleration of the car.

SOLUTION

(a) Since the acceleration changes linearly with time and has an initial value of zero, we can write the acceleration as

$$a(t) = ct \qquad (1)$$

where c is the *rate of change* of the acceleration (often called *jerk* or *jolt*).
Substituting this acceleration function into Equation (3-10), we have

$$v(t) = v_0 + \int_0^t a(t)dt = v_0 + \frac{1}{2}ct^2 - \frac{1}{2}c(0)^2 = v_0 + \frac{1}{2}ct^2 \quad (2)$$

Choosing the direction of motion as positive, and seeing that the speed is zero at the end of 3.4 s,

$$0 = v_0 + \frac{1}{2}ct^2 \Rightarrow c = \frac{-2v_0}{t^2} = \frac{-2(98 \text{ m/s})}{(3.4 \text{ s})^2} = -16.96 \text{ m/s}^3 \quad (3)$$

Combining Equations (1) and (3):

$$a(t) = (-17.0 \text{ m/s}^3)t$$

to 3 significant figures.

(b) For the speed of the car 3 s into the acceleration, we use Equation (3):

$$v(t) = v_0 + \frac{1}{2}ct^2 = 98 \text{ m/s} + \frac{1}{2}(-16.96 \text{ m/s}^3)(3 \text{ s})^2 = +21.7 \text{ m/s}$$

(c) For the distance covered by the car, we use Equation (3-12):

$$\Delta x = x(t) - x_0 = \int_0^t v\,dt = \int_0^t \left(v_0 + \frac{1}{2}ct^2\right)dt$$

$$= (v_0 t - v_0(0)) + \left(\frac{1}{6}ct^3 - \frac{1}{6}c(0)\right) = v_0 t + \frac{1}{6}ct^3$$

taking the direction of motion to be positive,

$$\Delta x = v_0 t + \frac{1}{6}ct^3 = 98(3.4) + \frac{1}{6}(-16.96)(3.4)^3 = 222.1 \text{ m}$$

(d) From Equation (1), the highest acceleration is

$$a = -16.96 \text{ m/s}^3 \times 3.4 \text{ s}$$

$$= -57.7 \text{ m/s}^2$$

$$a = -5.88 \text{ g}$$

Making sense of the result:

The constant c has the units of m/s^3, and the acceleration as a function of time is the product of that constant with time, leaving us with m/s^2 as the units for the acceleration, as expected. The initial speed of the car is about 353 km/h, so the fairly long stopping distance seems reasonable. The maximum acceleration is equal to 5.88 g, not unheard of in Formula One racing.

The body's tolerance to acceleration depends on the direction of the acceleration as well as the rate of change of the acceleration, the convention illustrated in Figure 3-14 is applied. This is discussed in more detail in the text online resources.

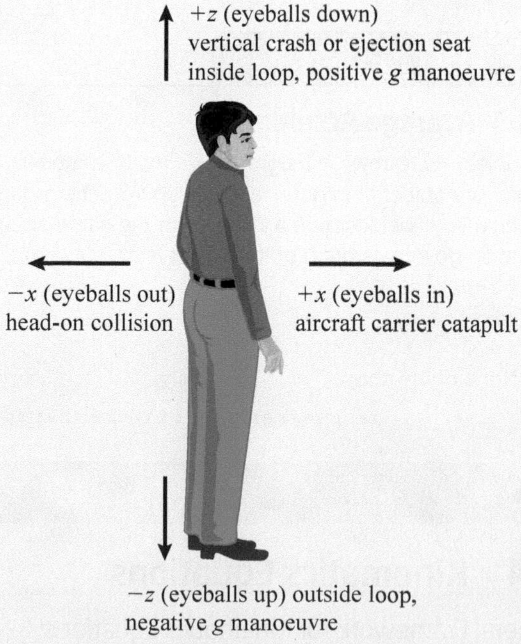

$+z$ (eyeballs down)
vertical crash or ejection seat
inside loop, positive g manoeuvre

$-x$ (eyeballs out)
head-on collision

$+x$ (eyeballs in)
aircraft carrier catapult

$-z$ (eyeballs up) outside loop,
negative g manoeuvre

Figure 3-14 Conventions used for discussing the tolerance of the human body to acceleration

✓ CHECKPOINT

C-3-6 Acceleration-Time Graph

Figure 3-15 shows an acceleration versus time graph for an object. Positive is taken to be right. At $t = 3$ s, the object is

(a) moving to the right;
(b) moving to the left;
(c) not moving at all.
(d) Any of the above could be correct.

C-3-6 (d) Answer could be any of the above.

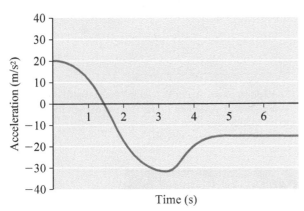

Figure 3-15 Acceleration-Time Graph

⊞ MAKING CONNECTIONS

EMAS Runway Extensions

Modern jet aircraft require longer runways than propeller aircraft. However, at some older airports, runway expansion is not feasible. Some of the older runways are long enough for large jets under normal conditions, but do not meet safety standards, which must allow for conditions such as high winds, storms, and brake failure. One solution used at some airports with serious space constraints is a runway extension made with an engineered material arresting system (EMAS), such as "foamcrete." These materials are designed to crumble under the weight of an aircraft to slow the plane down enough to prevent it from running off the end of the extension. In a number of incidents, EMAS materials have stopped aircraft from running off the end of a runway while causing few injuries to the passengers and limited damage to the aircraft. Figure 3-16 shows a Bombardier CRJ 200 that ran off a runway into an EMAS installation when the takeoff was aborted. None of the 34 occupants were injured.

(continued)

(continued)

Figure 3-16 A PSA Bombardier CRJ200 jet stopped in the EMAS installation at Charleston, South Carolina, on January 19, 2010.

✓ CHECKPOINT

C-3-7 Velocity-Time Graph

Figure 3-17 shows a velocity versus time graph for a speedboat. For the time when the velocity curve is below the x-axis,

(a) the average speed is zero;
(b) the average acceleration is zero;
(c) the instantaneous acceleration is always negative.
(d) None of the above.

C-3-7 (b) change in velocity for interval is zero

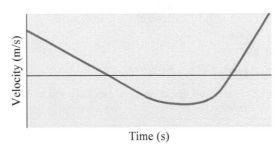

Figure 3-17 Velocity versus time graph for a speedboat

Kinematics Equations for Constant Acceleration

Although most situations in life do not involve constant acceleration, the acceleration we experience due to gravity is very nearly constant as long as we stay close to Earth's surface. When you let a ball drop, it falls with almost constant acceleration. Air drag can slow the acceleration, but this effect is negligible at slow speeds. The acceleration due to gravity at Earth's surface is about 9.81 m/s². The force of gravity is discussed in detail in Chapter 11.

When acceleration is constant, Equation (3-10) becomes

$$\vec{v}(t) = \vec{v}_0 + \int_{t_0}^{t} \vec{a}\, dt = \vec{a}(t - t_0) + \vec{v}_0 \qquad (3\text{-}14)$$

If we start at $t = 0$, Equation (3-14) can be written as follows.

KEY EQUATION First time-dependent kinematics equation:

$$\vec{v}(t) = \vec{a}t + \vec{v}_0 \qquad (3\text{-}15)$$

Combining Equations (3-13) and (3-15) gives us the position as a function of time:

KEY EQUATION
$$\vec{x}(t) = \vec{x}_0 + \int_{t_0}^{t} \vec{v}\, dt = \vec{x}_0 + \int_{t_0}^{t} (\vec{a}t + \vec{v}_0)\, dt \qquad (3\text{-}16)$$

or

KEY EQUATION Second time-dependent kinematics equation:

$$\vec{x}(t) = \frac{1}{2}\vec{a}t^2 + \vec{v}_0 t + \vec{x}_0 \qquad (3\text{-}17)$$

which can also be written as

$$\vec{x}(t) - \vec{x}_0 = \frac{1}{2}\vec{a}t^2 + \vec{v}_0 t \qquad (3\text{-}18)$$

or

$$\Delta \vec{x} = \frac{1}{2}\vec{a}t^2 + \vec{v}_0 t \qquad (3\text{-}19)$$

EXAMPLE 3-9

Gravitational Acceleration

A child throws a marble straight up in the air with a speed of 21.0 m/s. Find the speed of the marble after

(a) 1.00 s;
(b) 3.00 s.

SOLUTION

(a) Choosing up as our positive direction, and applying Equation (3-15) with $t = 1$ s, we have

$$\vec{v}(t) = \vec{a}t + \vec{v}_0 = 21\hat{j}\text{ m/s} + (-9.81\hat{j}\text{ m/s}^2)(1\text{ s}) = +11.2\hat{j}\text{ m/s}$$

(b) Similarly, after 3 s,

$$\vec{v}(t) = 21\hat{j}\text{ m/s} + (-9.81\hat{j}\text{ m/s}^2)(3\text{ s}) = -8.43\hat{j}\text{ m/s}$$

Making sense of the result:

Every second the velocity changes by 9.81 m/s. the difference in speed between $t = 1$s and $t = 3$s should be 19.62 m/s, which it is, within the accuracy of the data.

EXAMPLE 3-10

Speed of a Falling Object

You drop a ball from a point 16 m above the ground. Find the speed of the ball just before it hits the ground.

SOLUTION

We know the acceleration, the initial speed (zero), and the displacement but not the elapsed time. The time it takes for an object to fall to the ground from rest depends on the height from which it falls. Choosing up as the positive direction and applying Equation (3-19), we have

$$\Delta \vec{y} = \frac{1}{2}\vec{a}t^2 + \vec{v}_0 t$$

$$-16.0\hat{j}\text{ m} = \frac{1}{2}(-9.81\hat{j}\text{ m/s}^2)t^2$$

$$t = \sqrt{\frac{2 \times 16.0\text{ m}}{9.81\text{ m/s}^2}} = 1.81\text{ s}$$

Now that we know the elapsed time, we can apply Equation (3-15):

$$\vec{v}(t) = \vec{v}_0 + \vec{a}t = 0 - (9.81\hat{i}\text{ m/s}^2)(1.81\text{ s}) = -17.8\hat{i}\text{ m/s}$$

Making sense of the result:

The speed of an object in free fall increases by 9.81 m/s every second. After just under 2 s, we expect the speed to be slightly less than 19.6 m/s.

✓ CHECKPOINT

C-3-8 Relative Speed

You throw two balls from the top of a building at the same time: a red one with a speed of 7 m/s straight up, and a grey one with a speed of 14 m/s straight down. After 2 s,
(a) the relative speed of the two balls has increased;
(b) the relative speed of the two balls has decreased;
(c) the relative speed of the two balls has not changed;
(d) the grey ball is accelerating faster than the red ball.

C-3-8 (c) the two objects have the same acceleration and hence zero relative acceleration.

In Example 3-10, we found the time it took the ball to fall first, then we applied Equation (3-15). Since the question did not explicitly ask about time, one wonders if there is a way to find how fast the object is falling without having to calculate the elapsed time. We do know that the greater the distance an object falls, the higher its speed at impact will be.

In problem 64 you will be asked to substitute Equation (3-15) into Equation (3-19) in order to obtain the third kinematics equation. The third kinematics equation is time-independent and relates displacement to velocity and acceleration:

KEY EQUATION Third kinematics equation:

$$v^2 - v_0^2 = 2\vec{a} \cdot \Delta \vec{x} \qquad (3\text{-}20)$$

where v is the final speed of the object.

Note that the term on the right side of Equation (3-20) is a scalar product. Unlike the two time-dependent kinematics, Equations (3-15) and (3-19), the third kinematics equation is not a vector equation. In problem 119, you will be asked to prove that Equation (3-20) holds for nonconstant acceleration as well.

The last three examples each applied one of the kinematics equations. Next, we look at more involved examples.

EXAMPLE 3-11

Launching Lunch

A construction worker standing on a beam asks his friend on the ground 9.00 m below to throw his lunch bag up to him. The friend wants to throw the bag straight up such that it has a speed of 0.500 m/s when it reaches the worker on the beam. At what speed should the friend throw the lunch bag?

SOLUTION

Here, we know the distance covered, the acceleration, and the final speed. We need to find the initial speed. We can apply Equation (3-20) here. Note that the acceleration and the initial velocity have opposite directions.

$$v^2 - v_0^2 = 2\vec{a} \cdot \Delta \vec{y} = |\vec{a}| |\Delta \vec{y}| \sin(180°)$$

$$(0.50 \text{ m/s})^2 - v_0^2 = 2(9.81 \text{ m/s}^2)(9 \text{ m})(-1)$$

$$v_0 = \sqrt{(0.5 \text{ m/s})^2 + 2(9.81 \text{ m/s}^2)(9 \text{ m})} = 13.3 \text{ m/s}$$

where we have used Equation 2-18 for the scalar product of $\vec{a}$ and $\Delta \vec{y}$.

Making sense of the result:

The minus sign in the brackets indicates that the acceleration is opposite to the displacement.

EXAMPLE 3-12

Falling Firecrackers

A boy reaches over the railing around the top of a building and throws a firecracker straight up in the air, with a speed of 17.0 m/s.

(a) Find the time it takes for the firecracker to reach the ground 24.0 m below.
(b) Find the speed of the firecracker just before it touches the ground.
(c) How would the answer to (b) change if the boy threw the firecracker straight down at the same speed of 17.0 m/s?

SOLUTION

(a) We know the displacement, the acceleration, and the initial velocity of the firecracker, and we need to find the elapsed time. Therefore, Equation (3-19) is ideal for this part.

$$\Delta \vec{y} = \frac{1}{2} \vec{a} t^2 + \vec{v_0} t$$

Choosing up as the positive direction, we can write

$$-24 \hat{j} \text{ m} = \frac{1}{2}(-9.81 \text{ m/s}^2)t^2 + (17 \text{ m/s})t$$

We now have a quadratic equation for t. We rearrange this equation in the form $ax^2 + bx + c = 0$, and then use the formula for the roots of a quadratic equation:

$$4.905t^2 - 17t - 24 = 0$$

$$t = \frac{-b \pm \sqrt{b^2 - 4ac}}{2a} \quad \text{where } a = 4.905, b = 17, \text{ and } c = 24$$

the roots are:

$$t = 4.54 \text{ s} \quad \text{or} \quad t = -1.08 \text{ s}$$

We discard the negative root, so the firecracker takes 4.54 s to reach the ground.

(b) We have the initial speed, the displacement, and the acceleration. So we can use either Equation (3-15) or Equation (3-20) to find the speed of the firecracker before it hits the ground. Applying Equation (3-20) and noting that the acceleration and the displacement are both directed downward,

$$v^2 - v_0^2 = 2\vec{a} \cdot \Delta \vec{y}$$

$$v^2 = v_0^2 + 2\vec{a} \cdot \Delta \vec{y} = (17 \text{ m/s})^2 + 2(9.81 \text{ m/s}^2)(24 \text{ m})(1)$$

$$v = \sqrt{17.0^2 + 2(9.81)(24.0)(1)} = 27.6 \text{ m/s}$$

where we have used Equation 2-18 for the scalar product of $\vec{a}$ and $\Delta \vec{y}$.

(c) The expression for v^2 above depends on the square of the initial speed, and not on the initial velocity. Provided the initial speed is the same, the final speed will not be affected by whether the firecracker is thrown straight up or straight down.

Making sense of the result:

If you throw the firecracker upward, it will reach a maximum height and then come back down toward you. When it reaches your position again, it will have the same speed downward as its initial speed upward (by conservation of energy). It then has the same motion as if it had been thrown downward at the same speed.

CHAPTER 3 | MOTION IN ONE DIMENSION 55

EXAMPLE 3-13

Rocket Sled

Figure 3-18 shows an acceleration versus time graph for a rocket sled equipped with thrusters on both ends. Take right to be the positive x-direction. The magnitude of the acceleration is shown in the figure and the sled is slowing down at $t = 0$.

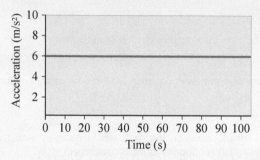

Figure 3-18 Acceleration versus time graph for a rocket sled

(a) Plot the velocity versus time graph until the sled comes to a stop for the first time, knowing that at $t = 0$ the velocity was 250 m/s to the right.
(b) For what value of t will $v = 0$?
(c) Plot the position versus time graph for the rocket sled as measured from the observaion tower, knowing that at $t = 0$ the sled was 700 m to the left of the tower. Your plot should show the points where $x = 0$ and $v = 0$.

SOLUTION

(a) The graph shows that the acceleration is constant at 6 m/s². Since the acceleration is constant, we can use Equation (3-15) to find an expression for the velocity as a function of time:

$$\vec{v}(t) = \vec{v}_0 + \vec{a}t = +250\hat{i}\ \text{m/s} + (-6\hat{i}\ \text{m/s}^2)t \qquad (1)$$

The velocity versus time plot is linear, as shown in Figure 3-19.

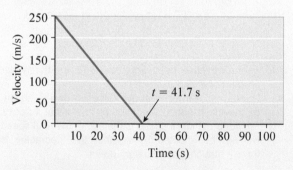

Figure 3-19 Velocity versus time plot

(b) The change in velocity is the area under the acceleration curve. For v to be zero, that area must be 250 m/s. Using Equation (3-16), we then have

$$at = 250 \Rightarrow t = \frac{250\ \text{m/s}}{6\ \text{m/s}^2}$$

$$t = 41.7\ \text{s} \qquad (2)$$

we report this as 41.7 s but keep more digits for the next part.

(c) The dependence of the position on time is given by Equation (3-17):

$$\vec{x}(t) = \frac{1}{2}\vec{a}\,t^2 + \vec{v}_0 t + \vec{x}_0 \qquad (3)$$

or

$$\vec{x}(t) = \frac{1}{2}(-6\hat{i}\ \text{m/s}^2)t^2 + (250\hat{i}\ \text{m/s})t - 700\hat{i}\ \text{m} \qquad (4)$$

The graph of this equation is a parabola. At $t = 0$, $x = -700$ m. Let us pick a few more points to plot the position. The position x will be zero when the expression in Equation (4) is 0. The roots of this quadratic are given by

$$t = \frac{-b \pm \sqrt{b^2 - 4ac}}{2a} = \frac{-250 \pm \sqrt{(+250)^2 - 4(3)(700)}}{-6} \qquad (5)$$

So, $t = 2.90$ s or 80.4 s when $x = 0$.

Since the velocity is the derivative of the position, a maximum or minimum for x will occur at the point of zero velocity. From part (b), the velocity is zero when $t = 41.7$ s. At this time,

$$\vec{x}(t) = ((-3\ \text{m/s}^2)(41.67\ \text{s})^2 + (250\ \text{m/s})(41.67\ \text{s}) - 700\ \text{m})\hat{i}$$

$$= 4510\hat{i}\ \text{m} \qquad (6)$$

A plot of $x(t)$ is shown in Figure 3-20.

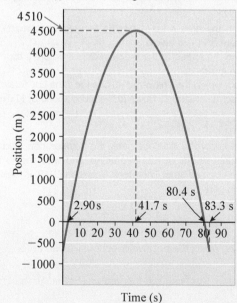

Figure 3-20 Position versus time for the rocket sled

Making sense of the result:

The rocket sled will go past the control tower at $t = 2.90$ s. The speed of the sled will continue to drop until it stops momentarily at $t = 41.7$ s. When the sled stops, it will be 4510 m to the right of the tower. The sled will then reverse direction, reaching the tower again at $t = 80.4$ s.

LO 5

3-5 Relative Motion in One Dimension

When you are driving, trees and other objects beside the road appear to be moving past you. Although these objects are not moving relative to the ground, they do have motion relative to you. Similarly, oncoming traffic seems to be moving toward you at a much faster speed. In this section, we introduce methods for describing such **relative motion**. First, a note about notation in relation to relative motion: we use double subscripts to indicate relative velocities. For example, $\vec{v}_{1g}$ is the velocity of object 1 with respect to the ground, and $\vec{v}_{12}$ is the velocity of object 1 with respect to object 2.

Relative Velocity

In general, the velocity of object 1 with respect to object 2 is given by

$$\vec{v}_{12} = \vec{v}_{1g} - \vec{v}_{2g} \tag{3-21}$$

Here, we have used the ground as our frame of reference, but we could choose any reference frame instead.

EXAMPLE 3-14

Relative Velocities

You are driving east on the Trans-Siberia highway at 100 km/h. Take east to be the positive x-direction.

(a) Express the velocity of a tree on the side of the highway with respect to you, using double-subscript notation.
(b) Find the speed of a tree on the side of the highway with respect to you.
(c) Find your velocity with respect to the tree.
(d) Find the velocity, with respect to you, of a bus driving west at 115 km/h. Find the velocity of your car with respect to the bus.

SOLUTION

We will use the subscripts c, t, and b for your car, the tree, and the bus, respectively. We choose east as the positive x-direction.

(a) For the velocity of a tree relative to your car, we use the relative velocity Equation (3-21):

$$\vec{v}_{tc} = \vec{v}_{tg} - \vec{v}_{cg} = 0 - 100\hat{\imath} \text{ km/h} = -100\hat{\imath} \text{ km/h}$$

(b) The speed of the tree relative to you equals the magnitude of the relative velocity found in part (a): 100 km/h.

(c) The velocity of your car with respect to the tree is

$$\vec{v}_{ct} = \vec{v}_{cg} - \vec{v}_{tg} = 100\hat{\imath} \text{ km/h} - 0 = 100\hat{\imath} \text{ km/h} \tag{3}$$

(d) The velocity, with respect to you, of the bus moving in the opposite direction is

$$\vec{v}_{bc} = \vec{v}_{bg} - \vec{v}_{cg} = -115\hat{\imath} \text{ km/h} - (+100\hat{\imath} \text{ km/h})$$
$$= -215\hat{\imath} \text{ km/h}$$

Similarly, your velocity with respect to the bus is

$$\vec{v}_{cb} = \vec{v}_{cg} - \vec{v}_{bg} = 100\hat{\imath} \text{ km/h} - (-115\hat{\imath} \text{ km/h}) = 215\hat{\imath} \text{ km/h}$$

Making sense of the result:

Vehicles coming in the direction opposite to yours often seem to be moving twice as fast.

✓ CHECKPOINT

C-3-9 Relative Velocity

You are in the free-fall stage of a skydive when a skydiver below you opens her parachute. A few seconds later, she passes you at a relative velocity of 50 km/h upward. She has a velocity with respect to the air of 80 km/h downward. What is your velocity with respect to the air?

(a) 30 km/h down;
(b) 30 km/h up;
(c) 130 km/h down;
(d) 130 km/h up.

C-3-9 (c) your velocity is 50 km/h down with respect to her, her velocity is 80 km/h relative to the air.

Relative Acceleration and Position

The equation for the relative velocity of objects is easily adapted to apply to other vector quantities that describe motion. We simply differentiate Equation (3-21) to find relative acceleration (see problem 96):

$$\vec{a}_{12} = \vec{a}_{1g} - \vec{a}_{2g} \tag{3-22}$$

Similarly, integrating Equation (3-21) gives relative position (see problem 96):

$$\vec{x}_{12} = \vec{x}_{1g} - \vec{x}_{2g} \tag{3-23}$$

For displacement, we have

$$\Delta\vec{x}_{12} = \Delta\vec{x}_{1g} - \Delta\vec{x}_{2g} \tag{3-24}$$

A comprehensive treatment of relative acceleration involves a discussion on frames of reference that is beyond the scope of this text. Equation 3-22 on relative acceleration certainly applies in the case of two objects moving in the same inertial frame of reference, such as two cars on the road.

MECHANICS

C-3-10 Acceleration and Relative Velocity

You throw a ball straight down with a speed of 20 m/s, wait 1.0 s, and throw another ball down with a speed of 20 m/s. Use down as the positive direction. The velocity of the second ball relative to the first ball will

(a) always be −9.8 m/s;
(b) always be +9.8 m/s;
(c) change with time.
(d) None of the above.

C-3-10 (a) after 1s the speed of first ball has increased by 9.8 m/s. Then both speeds increase at same rate.

 ## EXAMPLE 3-15

Relative Position

At the annual auto show in Frankfurt, Germany, an Aston Martin is 7.0 m to left of a post and a Maserati is 12 m to the left of the same post. Find the position of the Aston Martin relative to the Maserati.

SOLUTION

We use Equation (3-23) and choose right as the positive x-direction. We will use the post as our origin; therefore, left of the post is the negative direction. Both x_A and x_M are negative because they are both left of the origin (Figure 3-21).

$$\vec{x}_{AM} = \vec{x}_A - \vec{x}_M$$

Here we drop the subscript g for simplicity as both positions are taken with respect to ground by default.

$$\vec{x}_{AM} = -7.0\hat{i}\text{ m} - (-12\hat{i}\text{ m}) = 5.0\hat{i}\text{ m}$$

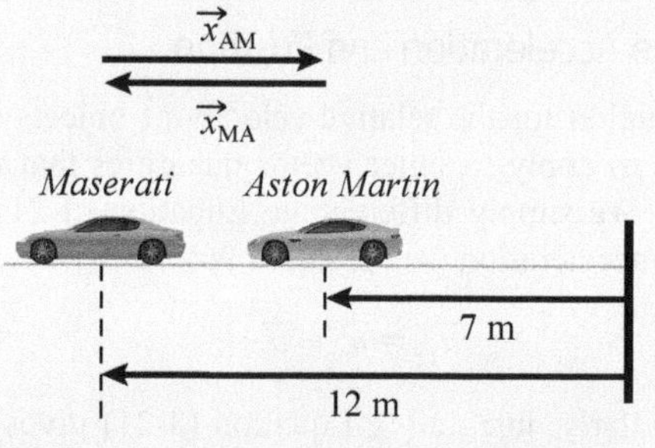

Figure 3-21 The position of one car relative to the other

Making sense of the result:

The Aston Martin is to the right of the Maserati, hence its position relative to the Maserati should be positive.

Derivation of the General Kinematics Equations for Relative Motion

We now derive the general kinematics equations for one object with respect to another when both objects have constant accelerations. Here again we drop the subscript g, as all quantities are taken relative to ground.

Relative Position The position of object 1 with respect to object 2 is given by Equation (3-23):

$$\vec{x}_{12} = \vec{x}_1 - \vec{x}_2$$

The position of each object as a function of time is given by Equation (3-17):

$$\vec{x}(t) = \frac{1}{2}\vec{a}\,t^2 + \vec{v}_0 t + \vec{x}_0$$

Combining these two equations, we get the following:

$$\vec{x}_{12}(t) = \frac{1}{2}\vec{a}_1 t^2 + \vec{v}_{01} t + \vec{x}_{01} - \left(\frac{1}{2}\vec{a}_2 t^2 + \vec{v}_{02} t + \vec{x}_{02}\right)$$

$$\vec{x}_{12}(t) = \left(\frac{1}{2}\vec{a}_1 - \frac{1}{2}\vec{a}_2\right)t^2 + (\vec{v}_{01} - \vec{v}_{02})t + (\vec{x}_{01} - \vec{x}_{02})$$

$$\vec{x}_{12}(t) = \frac{1}{2}\vec{a}_{12} t^2 + \vec{v}_{012} t + \vec{x}_{012} \qquad (3\text{-}25)$$

Relative Velocity We begin with Equation (3-21):

$$\vec{v}_{12} = \vec{v}_1 - \vec{v}_2$$

 ## EXAMPLE 3-16

Relative Acceleration

You are driving a pickup truck when the driver of a motorcycle in front of you suddenly brakes to slow down for a bump on the road. Being a smaller vehicle with good brakes, the motorcycle can slow down at an acceleration of 19 m/s². Your heavy pickup truck slows at a rate of 12 m/s².

(a) What is the acceleration of the motorcycle with respect to the pickup truck?
(b) What is the acceleration of the pickup truck with respect to the motorcycle?

SOLUTION

For this example, we use Equation (3-22), and choose the direction of motion as the positive x-direction. Therefore, both accelerations are negative. We use the subscript p for the pickup truck and m for the motorcycle.

(a) The acceleration of the motorcycle with respect to the pickup truck is

$$\vec{a}_{mp} = \vec{a}_{mg} - \vec{a}_{pg} = -19\hat{i}\text{ m/s}^2 - (-12\hat{i}\text{ m/s}^2) = -7.0\hat{i}\text{ m/s}^2$$

(b) The acceleration of the pickup truck with respect to the motorcycle is

$$\vec{a}_{pm} = \vec{a}_{pg} - \vec{a}_{mg} = -12\hat{i}\text{ m/s}^2 - (-19\hat{i}\text{ m/s}^2) = +7.0\hat{i}\text{ m/s}^2$$

Making sense of the result:

Since the motorcycle is slowing down faster than the pickup truck, from the point of view of the pickup truck the motorcycle is accelerating toward it. The acceleration of the motorcycle relative to the truck will point towards the truck which is the negative direction as per our choice of coordinate system.

Then using Equation (3-15) for the velocity of an object moving with constant acceleration, we have

$$\vec{v}_{12}(t) = \vec{a}_1 t + \vec{v}_{01} - (\vec{a}_2 t + \vec{v}_{02})$$

$$\vec{v}_{12}(t) = (\vec{a}_1 - \vec{a}_2)t + \vec{v}_{01} - \vec{v}_{02}$$

$$\vec{v}_{12}(t) = \vec{a}_{12} t + \vec{v}_{012} \qquad (3\text{-}26)$$

EXAMPLE 3-17

Relative Motion with Constant Acceleration

At the same time, you throw a red ball straight up with a speed of 21.0 m/s and a blue ball straight down with a speed of 17.0 m/s. Find the acceleration, speed, and position of the blue ball with respect to the red ball 3.00 s later. Assume that air resistance is negligible.

SOLUTION

Once thrown, both balls have the same constant acceleration of 9.81 m/s².

We calculate the relative speed using Equation (3-26). We use the subscript r for the red ball and the subscript b for the blue ball:

$$\vec{v}_{br}(t) = \vec{a}_{br} t + \vec{v}_{0br} \qquad (1)$$

Using Equations (3-21) and (3-22), we can rewrite (1) as

$$\vec{v}_{br} = (\vec{a}_b t - \vec{a}_r t) + (\vec{v}_{0b} - \vec{v}_{0r}) = \vec{v}_{0b} - \vec{v}_{0r} \qquad (2)$$

The terms in the first parentheses cancel because the two balls have the same constant acceleration. Consequently, the relative velocity of the two balls is constant because it depends on only their initial velocities. Taking down to be the positive y-direction, we can write

We leave it as an exercise in problem 72 to apply relative motion notation to the time-independent kinematics Equation (3-20) to show that

$$v_{12}^2 - v_{012}^2 = 2\vec{a}_{12} \cdot \Delta \vec{x}_{12} \qquad (3\text{-}27)$$

$$v_{br} = v_{0b} - (-v_{0r}) = v_{0b} + v_{0r} = 17 \text{ m/s} + 21 \text{ m/s} = 38.0 \text{ m/s} \quad (3)$$

where the variables in Equation (3) denote magnitdues of the quantities in Equation (2)

For relative position we use Equation (3-24):

$$\vec{y}_{12}(t) = \frac{1}{2}\vec{a}_{12} t^2 + \vec{v}_{012} t + \vec{y}_{012} \qquad (4)$$

We use Equations (3-21) and (3-23) to rewrite (4) as

$$\vec{y}_{br} = \left(\frac{1}{2}\vec{a}_b - \frac{1}{2}\vec{a}_r\right)t^2 + (\vec{v}_{0b} - \vec{v}_{0r})t + (\vec{y}_{0b} - \vec{y}_{0r})$$

$$= (\vec{v}_{0b} - \vec{v}_{0r})t \qquad (5)$$

Again, the acceleration terms cancel out. The last two terms also cancel because the initial position of the two balls is the same. Since we have chosen down to be positive,

$$y_{br} = (v_{0b} - (-v_{0r}))t = (17 \text{ m/s} + 21 \text{ m/s})3 \text{ s} = 114 \text{ m} \qquad (6)$$

where the variables in Equation (6) denote magnitudes of the vectors in Equation (5)

Making sense of the result:

Since gravity contributes to the change in velocity and position of both objects identically, the acceleration due to gravity makes no contribution to the relative quantities.

EXAMPLE 3-18

Applying the Kinematics Equations to Relative Motion

A Mercedes Formula One car is moving at a speed of 330 km/h, 50.0 m directly behind a Renault moving at a speed of 220 km/h. Preparing for a sharp turn ahead, the Renault driver slows down with an acceleration 2.62 times the acceleration due to gravity.

(a) Find the minimum acceleration that the Mercedes must have to avoid running into the Renault. Assume that the Mercedes brakes at the same time as the Renault, and assume constant acceleration for both cars.

(b) How long does it take the Mercedes to reach the Renault if the Mercedes does brake with the minimum acceleration needed to avoid a collision? How far did each car move during this time?

SOLUTION

Since both cars are accelerating, it could be quite involved to calculate the displacement and velocity of each car, and then compare them. However, the solution is much easier if we think in terms of the relative motion of the two cars.

(a) To avoid just colliding with the Renault, the Mercedes must reach the same speed as the Renault a moment before catching up to it. In other words, the relative speed of the two cars must be zero by the time the Mercedes closes the distance between them.

We choose the direction of motion as the positive x-direction, which makes the accelerations for both cars negative. Since the Mercedes is initially moving faster than the Renault, the Mercedes will have to slow more rapidly (i.e., have greater acceleration in the negative direction) than the Renault in order to match

(continued)

its speed. Therefore, the relative acceleration of the Mercedes, with respect to the Renault, $\vec{a}_{MR} = \vec{a}_M - \vec{a}_R$, also points in the negative direction.

The initial speed of the Mercedes with respect to the Renault is

$$\vec{v}_{0MR} = \vec{v}_{0M} - \vec{v}_{0R} = 330\hat{i} \text{ km/h} - 220\hat{i} \text{ km/h}$$

$$= 110\hat{i} \text{ km/h} = 110\hat{i} \text{ km/h} \times \frac{1000 \text{ m}}{3600 \text{ s}} = 30.56\hat{i} \text{ m/s}$$

From the point of view of the Renault, the Mercedes is getting closer. Hence the direction of the displacement of the Mercedes relative to the Renault, $\Delta\vec{x}_{MR}$, as per our choice of coorindate system, is positive.

Applying the third kinematics equation for relative motion, Equation (3-27), we have

$$v_{MR}^2 - v_{0MR}^2 = 2\vec{a}_{MR} \cdot \Delta\vec{x}_{MR}$$

where the subscript MR refers to the Mercedes with respect to the Renault.

The two vectors in this equation have opposite directions, so using Equation 2-18

$$v_{MR}^2 - v_{0MR}^2 = 2a_{MR}\Delta x_{MR}(\cos(180°)) = 2a_{MR}\Delta x_{MR}(-1)$$

$$\Rightarrow 0 - (30.56 \text{ m/s})^2 = 2a_{MR}(50 \text{ m})(-1)$$

$$a_{MR} = 9.336 \text{ m/s}^2 = 0.952g$$

$$a_M - a_R = 0.952g \Rightarrow a_M = a_R + 0.952g$$

$$= 2.62g + 0.952g = 3.57g$$

(b) To find the time it takes for the relative velocity to be zero, we use Equation (3-26):

$$\vec{v}_{MR}(t) = \vec{a}_{MR}t + \vec{v}_{0MR}$$

$$0 = (\vec{a}_M - \vec{a}_R)t + \vec{v}_{0M} - \vec{v}_{0R}$$

With our choice of positive direction, both accelerations are negative and both initial velocities are positive. In the approach we take below, a_M and a_R are introduced as variables that denote the magnitudes of the accelerations. This approach is more adequate for the level of complexity of the problem. Therefore,

$$(-a_M - (-a_R))t\hat{i} + (v_{0M} - v_{0R})\hat{i} = 0$$

Rearranging and using the values calculated in part (a), we have

$$t = \frac{v_{0R} - v_{0M}}{a_R - a_M} = \frac{-30.56 \text{ m/s}}{-9.336 \text{ m/s}^2} = 3.273 \text{ s}$$

During that time, the Mercedes would have moved by

$$\Delta\vec{x} = \frac{1}{2}\vec{a}t^2 + \vec{v}_0 t$$

$$= \frac{1}{2}(-35.039\hat{i} \text{ m/s}^2)(3.273 \text{ s})^2 + (91.667\hat{i} \text{ m/s})(3.273 \text{ s})$$

$$= 112.348\hat{i} \text{ m} = 112\hat{i} \text{ m}$$

to 3 significant figures. For the Renault,

$$\Delta\vec{x} = \frac{1}{2}\vec{a}t^2 + \vec{v}_0 t$$

$$= \frac{1}{2}(-25.702\hat{i} \text{ m/s}^2)(3.273 \text{ s})^2 + (61.111\hat{i} \text{ m/s})(3.273 \text{ s})$$

$$= 62.349\hat{i} \text{ m} = 62.4\hat{i} \text{ m/s}$$

to 3 significant figures.

Making sense of the result:

The distance covered by the Mercedes should be 50 m longer than the distance covered by the Renault. The result does agree to within the accuracy of the given data.

Although the relative motion Equations (3-21) to (3-24) have been developed using one-dimensional examples, these vector equations are readily applied to two- and three-dimensional situations, as we will see in Section 4-4.

 CHECKPOINT

C-3-11 Relative Acceleration

You are slowing your car, which at a given instant is moving at 34 m/s with an acceleration of 23 m/s^2. Behind you is a truck moving in the same direction as you at a speed of 43 m/s and slowing down with an acceleration of 16 m/s^2. At that instant, the truck's acceleration relative to you is

(a) parallel to the truck's relative displacement with respect to you;

(b) antiparallel to the truck's relative position with respect to you at that moment;

(c) parallel to the truck's relative velocity with respect to you;

(d) antiparallel to the truck's relative velocity with respect to you.

C-3-11 (c) truck has larger speed than you, and a smaller acceleration than you in opposite direction to motion. Look at equations, draw diagrams if in doubt.

FUNDAMENTAL CONCEPTS AND RELATIONSHIPS

Kinematics is the study of motion. In kinematics, we study the relationships between an object's position, displacement, velocity, and acceleration and their dependence on time. We also examine relative motion.

Position, Velocity, Acceleration, and Time

The displacement of an object is the change in its position vector:

$$\Delta \vec{x} = \vec{x}_2 - \vec{x}_1 \tag{3-1}$$

Distance is the path length covered by an object:

$$d = \int_{x_1}^{x_2} |d\vec{x}| \tag{3-2}$$

The average velocity is the displacement divided by the time elapsed:

$$\vec{v}_{\text{avg}} = \frac{\Delta \vec{x}}{\Delta t} = \frac{\vec{x}_2 - \vec{x}_1}{\Delta t} \tag{3-3}$$

The average speed is the distance covered divided by the time elapsed The instantaneous velocity is the derivative of the position with respect to time:

$$\vec{v} = \lim_{\Delta t \to 0} \frac{\Delta \vec{x}}{\Delta t} = \frac{d\vec{x}}{dt} \tag{3-4}$$

The instantaneous speed is the magnitude of the instantaneous velocity.

The average acceleration is the change in the velocity divided by the time elapsed:

$$\vec{a}_{\text{avg}} = \frac{\Delta \vec{v}}{\Delta t} = \frac{\vec{v}_2 - \vec{v}_1}{\Delta t} \tag{3-5}$$

The instantaneous acceleration is the derivative of velocity with respect to time:

$$\vec{a} = \frac{d\vec{v}}{dt} = \frac{d^2\vec{x}}{dt^2} \tag{3-6), (3-7}$$

Kinematics Equations

The velocity of an object can be obtained by integrating the acceleration with respect to time:

$$\vec{v}(t) = \vec{v}_0 + \int_{t_0}^{t} \vec{a}\, dt \tag{3-9}$$

The position (and displacement) of an object can be obtained by integrating velocity with respect to time:

$$\vec{x}(t) = \vec{x}_0 + \int_{t_0}^{t} \vec{v}\, dt \tag{3-13}$$

For constant acceleration, the velocity of an object as a function of time is given by

$$\vec{v}(t) = \vec{a}t + \vec{v}_0 \tag{3-15}$$

The position of an object moving with constant acceleration is given by

$$\vec{x}(t) = \frac{1}{2}\vec{a}t^2 + \vec{v}_0 t + \vec{x}_0 \tag{3-17}$$

The time-independent kinematics equation is given by

$$v^2 - v_0^2 = 2\vec{a} \cdot \Delta \vec{x} \tag{3-20}$$

Relative Motion

The relative velocity of object 1 with respect to object 2 is

$$\vec{v}_{12} = \vec{v}_{1g} - \vec{v}_{2g} \tag{3-21}$$

The relative acceleration is given by

$$\vec{a}_{12} = \vec{a}_{1g} - \vec{a}_{2g} \tag{3-22}$$

The relative displacement is given by

$$\Delta \vec{x}_{12} = \Delta \vec{x}_{1g} - \Delta \vec{x}_{2g} \tag{3-24}$$

The relative position is expressed as

$$\vec{x}_{12} = \vec{x}_1 - \vec{x}_2$$

APPLICATIONS

Applications: airline safety, motor vehicle safety, travel times, runway construction, relative motion

Key Terms: acceleration, average acceleration, average speed, average velocity, displacement, instantaneous acceleration, instantaneous speed, instantaneous velocity, kinematics, position, relative motion, speed, velocity

QUESTIONS

1. Figure 3-22 shows a plot of the position versus time for two objects.
   (a) Which object has the higher average velocity for the interval shown?
   (b) Which object has the higher speed at $t = 30$ s? Which has the higher speed at $t = 90$ s?
   (c) Do the objects ever have the same speed? If so, at what time do you estimate this will happen?

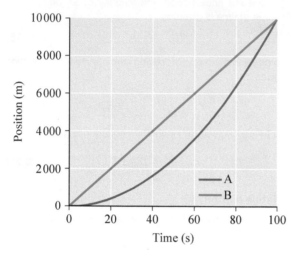

Figure 3-22   Question 1

2. Figure 3-23 shows a plot of the position versus time for two objects, A and B.
   (a) Which object has the higher average speed for the interval shown?
   (b) Estimate the average speeds of object A and object B for the interval shown.
   (c) At the points indicated by the arrows, do the objects have the same speed? The same acceleration?
   (d) Which object has the higher speed at the position indicated by arrow 1? Is the speed of object A ever higher than the speed of object B?
   (e) Which object has the higher speed at the position indicated by arrow 2?

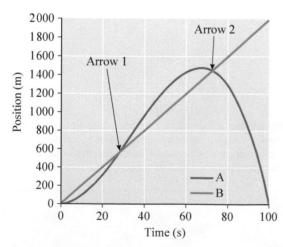

Figure 3-23   Question 2

3. A plot of velocity versus time for two objects is shown in Figure 3-24.
   (a) Starting at the same point, are the two objects ever at the same position again?
   (b) Do the two objects have the same speed? If so, when does that occur?
   (c) Which object has the higher acceleration for the interval shown? Estimate the magnitude of the maximum acceleration for each object.

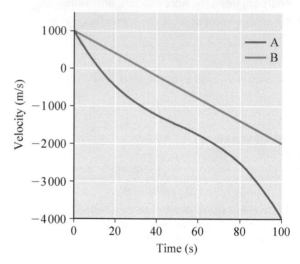

Figure 3-24   Question 3

4. Figure 3-25 shows a plot of the speed versus time for two boats.
   (a) For the interval shown, which boat covers more distance?
   (b) Do the boats ever have the same acceleration?

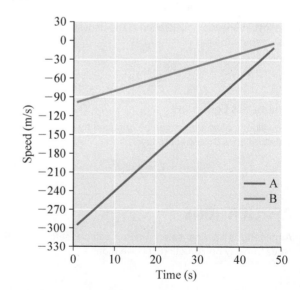

Figure 3-25   Question 4

5. Figure 3-26 shows a velocity versus time plot for two objects; they are at the same point at $t = 0$.
   (a) For the interval shown, which object has the higher average acceleration?

(b) Which object covers more distance?

(c) Which object has the higher magnitude of acceleration at the point of intersection for the two curves?

(d) Which object is ahead of the other at the point of intersection of the two curves?

(e) Do the objects ever have the same displacement and/or acceleration? If so, where does that happen?

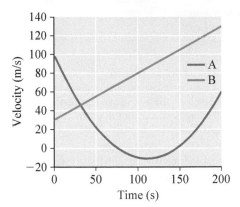

Figure 3-26   Question 5

6. Figure 3-27 show a plot of an object's position versus time.
   (a) Where does this object have its highest speed?
   (b) What is the object's lowest speed?
   (c) Infer where the object's acceleration is zero.
   (d) Is the object ever at rest?
   (e) Does the object cover more distance between $t = 0$ and $t = 140$ s than it does between $t = 140$ s and $t = 200$ s? Explain your reasoning.

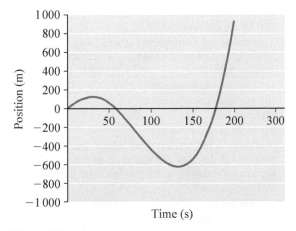

Figure 3-27   Question 6

7. The velocity versus time graph for an object is shown in Figure 3-28.
   (a) Is the acceleration of this object ever zero?
   (b) Where is this object's magnitude of acceleration the highest? The lowest?

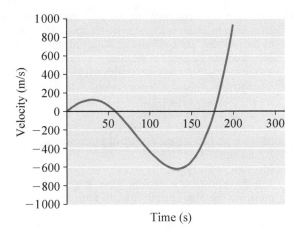

Figure 3-28   Question 7

8. Two balls are thrown simultaneously, one up and one down, both with an initial speed of 27 m/s. They pass each other 3 s after they are thrown. What is the speed of one ball with respect to the other ball as they pass each other?
   (a) 27 m/s
   (b) 54 m/s
   (c) It depends on the distance the balls travel.
   (d) None of the above.

9. You drop a tennis ball from a hot-air balloon that is going upward at a speed of 4 m/s. Relative to ground, the ball
   (a) falls straight down;
   (b) goes up a short distance, stops, and then goes down;
   (c) hangs in the air for a short time, then goes down.
   (d) None of the above.

10. You are standing in an open construction elevator at a site. The elevator is going up at a constant speed v. You throw a stone up at speed $2v$ with respect to the elevator, while your friend on the ground throws a stone up at speed $2v$ relative to ground. Who will get their stone back first?
    (a) you
    (b) your friend
    (c) You both get them back at the same time.
    (d) The answer cannot be determined without a value for v.

11. You throw two stones at a speed of 24 m/s: a small one straight up and a larger one at an angle of 42° above the horizontal. The stones are launched from and land at the same height. Which one lands with the greater speed? Ignore air resistance.
    (a) the small one
    (b) the large one
    (c) The speeds will be the same.
    (d) The answer cannot be determined from the given data.

12. You are piloting a plane at a speed of 980 km/h relative to the ground. Another plane has a speed of zero with respect to you. Relative to the ground, the other plane is
    (a) stationary;
    (b) moving at 980 km/h in the opposite direction;
    (c) moving at 980 km/h in some arbitrary direction.
    (d) None of the above.

13. When a Formula One race car completes two laps around a track,
    (a) the distance covered by the car is zero;
    (b) the car's displacement is zero;
    (c) the car's displacement is twice the circumference of the track.
    (d) None of the above.

63

14. Evaluate this statement: When you throw two balls from a 10 m-high building, one with twice the speed of the other, by the time they get to the ground, the speed of the one will still be double the speed of the other.

15. You throw two balls down, one with a speed of 10 m/s and the other with a speed of 6 m/s. As long as both balls are in the air, the speed of one relative to the other
    (a) depends on the height from which they fall;
    (b) stays the same as it was initially;
    (c) changes because the faster ball gains more speed than the slower ball.
    (d) None of the above.

16. You throw two balls from the same height, one straight down and the other straight up at the same speed v. Which of the following are correct?
    (a) As long as both balls are in the air, the ball you throw straight down will have twice the speed of the ball you throw straight up.
    (b) As long as both balls are in the air, the relative speed of the two balls is $2v$.
    (c) The balls will hit the floor with the same speed.
    (d) Both (b) and (c) are correct.

17. If a rocket sled has an acceleration to the right three times the acceleration due to gravity,
    (a) the rocket sled must be moving;
    (b) the rocket sled must be moving to the left;
    (c) the rocket sled is stationary.
    (d) All the above are possible.

18. An object moves with an acceleration that increases in magnitude. If the object is not stationary, its speed
    (a) will increase;
    (b) will decrease;
    (c) can be zero.
    (d) All the above are possible.

19. Can an object's average acceleration be equal to its instantaneous acceleration? Explain.

20. Can an object's average speed be equal to the magnitude of its instantaneous velocity? Explain.

21. Can an object's instantaneous velocity be equal to its average velocity? Explain.

22. When an object moves to the right and its acceleration vector points to the right, then
    (a) the object's speed will decrease;
    (b) the object's speed will increase.
    (c) Both (a) and (b) are possible.
    (d) None of the above.

23. An object moves to the right with increasing speed. Define left to be the positive direction. The acceleration of the object
    (a) is negative;
    (b) is positive;
    (c) could be either negative or positive.
    (d) None of the above.

24. If an object's velocity increases, then
    (a) its speed must increase;
    (b) its speed must decrease.
    (c) Both (a) and (b) are possible.
    (d) None of the above.

25. An object moves to the right, but its acceleration vector is directed to the left. Define left to be the positive direction. Which of the following is true?
    (a) The object's velocity increases.
    (b) The object's velocity decreases.
    (c) The object's speed increases.
    (d) More one than of the above is possible.

26. Can your speed be zero when your velocity is nonzero? Explain.

27. Can your speed be zero when your acceleration is nonzero? Explain.

28. Can your acceleration be zero when your speed is nonzero? Explain.

29. When an object's instantaneous velocity is zero, its acceleration must be
    (a) positive;
    (b) negative;
    (c) zero.
    (d) All the above are possible.

30. A person's displacement over a given interval of time is zero. Therefore, the average speed must be zero. Which of the following is true about this claim?
    (a) true
    (b) false
    (c) It depends on the start and finish speeds.
    (d) It depends on the distance that the person moved during the interval.

31. You start from rest at point A, then go to point B, where you stop. Which of the following is true?
    (a) Your average speed is zero.
    (b) Your average velocity is zero.
    (c) Your average displacement is zero.
    (d) None of the above are true.

32. If your net displacement is zero, then
    (a) the distance you travel must be zero;
    (b) your average speed must be zero;
    (c) your average velocity is zero.
    (d) All the above are possible.

33. If your average speed over a given interval is zero, then
    (a) you did not move;
    (b) you returned to the same place;
    (c) your displacement can be nonzero.
    (d) All the above are possible.

34. You drop a ball from the top of a building. Then 3 s later you drop another ball. While the two balls are in the air, the distance between them
    (a) increases as the square of t;
    (b) increases linearly with t;
    (c) stays constant.
    (d) None of the above.

35. You throw a ball straight down with a speed of 20 m/s. It bounces off the floor and comes up to the same height at the same speed with which you launched it. Which of the following is true?
    (a) The average acceleration is 9.8 m/s.
    (b) The average speed is 20 m/s.
    (c) The average velocity is zero.
    (d) None of the above are true.

36. Two friends start from points that are 30 m apart and change positions. It takes them 6 s to do so. Which of the following is true?
    (a) Their relative displacement is zero.
    (b) Their average relative velocity is zero.
    (c) Their average relative velocity is 10 m/s.
    (d) Their relative displacement is 30 m.

PROBLEMS BY SECTION

For problems, star ratings will be used, (✶, ✶✶, or ✶✶✶), with more stars meaning more challenging problems.

Section 3-1 Distance and Displacement

37. ✶ You walk from your house 1100 m down the street, then you reverse direction and walk 1300 m up the street.
(a) What is the distance you covered?
(b) What is your total displacement? Choose down the street as the positive direction.

38. ✶ Starting from a trail 3 m above water level, you climb straight up a cliff to a point 11 m above the water. Then you dive into the water and climb back to the trail. Find your displacement as well as the distance you covered.

39. ✶ A grasshopper jumps straight up from a leaf that is 80 cm above the ground, reaches a maximum height of 75 cm above the leaf, and lands on the ground. Find the distance covered by the grasshopper as well as its displacement.

Section 3-2 Velocity and Speed

40. ✶ A barge takes a load of processed fish from a fishing boat to the harbour 3.2 km away, and then doubles back to the fishing boat along the same direction. In the meantime, the fishing boat has moved 2.0 km from the dock. The whole journey takes 20 min. Find
(a) the displacement of the barge;
(b) the average velocity of the barge;
(c) the average velocity of the fishing boat;
(d) the average speed of the barge;
(e) the average speed of the fishing boat.

41. ✶ On a distant planet, a ball is dropped from rest from a height of 21 m. It hits the ground and bounces back up until it is 1 m below its original position. The trip down and back up takes 10 s. Find the ball's average speed, displacement, and average velocity.

42. ✶✶ The sailfish has a maximum speed of 112 km/h. A sailfish moves in a straight line at a maximum speed for 3.00 s, slows down, and then comes to a stop 2.00 s later, having covered a total distance of 125 m.
(a) Find the acceleration during the last 2 s, assuming constant acceleration.
(b) Find the average acceleration during the entire 5 s.
(c) Find the average velocity during the entire 5 s.
(d) The sailfish now takes 8.00 s to return to its starting point, accelerating to its maximum speed as it does so. Find its average speed and average velocity for the round trip.

43. ✶✶ A stone is released at the top of a 1.2 m-tall barrel of liquid. The stone reaches a speed of 1.7 m/s just before it hits the bottom of the barrel 0.90 s later. Assuming constant acceleration, find the stone's average speed and average velocity.

44. ✶✶ Find the magnitude of the average velocity of Earth with respect to the Sun over one month. Also find the average speed and average acceleration. Assume that Earth's orbit is circular.

45. ✶ The maximum speed of an Airbus 380 at cruising altitude is 945 km/h. How long would this plane take to go from Paris to New York at that speed?

46. ✶ The fastest known rotation of a neutron star is an equatorial speed that is 0.24 of the speed of light. How long is a "day" on this star? Assume a radius of 16 km.

47. ✶ The distance between the Sun and Earth is one astronomical unit, 149 597 871 km. Calculate the speed of Earth around the Sun, assuming that Earth has a nearly circular orbit and that the Sun is at the centre of Earth's orbit.

48. ✶ The speed of the Moon around Earth is approximately 1 km/s. Calculate the distance between the Moon and Earth.

49. ✶ Lightning flashes in your neighbourhood, and 3.00 s later you hear thunder. How far away was the lightning flash? Assume the speed of sound to be 345 m/s.

50. ✶ The speed of the space shuttle in low Earth orbit is about 28 000 km/h. When the altitude of the shuttle was 300 km, how long did it take to orbit Earth once?

51. ✶ A rocket car covers a distance of 10 km in 1.0 min, turns around, which takes 5.0 min, and goes back halfway to its starting position in 1.5 min.
(a) Find the displacement of the rocket car.
(b) Find the distance covered by the rocket car.
(c) Find (i) the average velocity and (ii) the average speed of the rocket car.

52. ✶ The maximum speed of the Concorde at cruising altitude was 2172 km/h. How long would it take the Concorde to travel from London, England, to Montréal, Québec, at that speed?

53. ✶✶ Figure 3-29 shows a velocity versus time plot for a motorcycle.
(a) Find the displacement of the motorcycle between $t = 0$ s and $t = 3$ s, between $t = 3$ s and $t = 7$ s, and between $t = 7$ s and $t = 9$ s.
(b) What is the acceleration of the motorcycle between $t = 2$ s and $t = 3$ s?

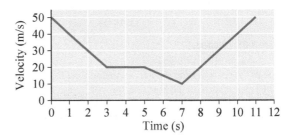

Figure 3-29 Question 53

54. ✶✶ Figure 3-30 shows a position versus time plot for a speedboat.

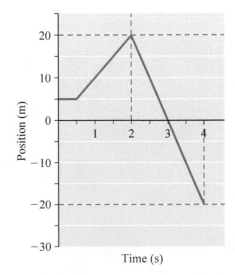

Figure 3-30 Problem 54

(a) Find the average speed of the speedboat over the entire interval shown.

(b) Find the average velocity of the speedboat over the entire interval.

(c) Find the velocity of the speedboat between $t = 1$ s and $t = 2$ s.

55. ✱ (a) How long does light take to reach us from the Sun?

(b) Roughly how long does it take light to reach us from the star Sirius?

(c) How long would it take light to cross the diameter of the Milky Way galaxy?

Section 3-3 Acceleration

56. ✱ The space shuttle had a takeoff acceleration approximately 3.5 times the acceleration due to gravity. What was the speed of the space shuttle 100 s after takeoff when it maintained that acceleration?

57. ✱✱ Figure 3-31 shows a position versus time plot for a particle. Find the speed and velocity of the particle.

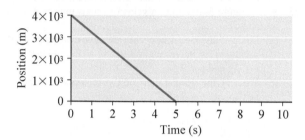

Figure 3-31 Problem 57

58. ✱✱ A sunspot is on the Sun's equator. Find the velocity of this spot relative to the Sun's centre.

59. ✱✱ Starting with a speed of 4 m/s, a stone falls toward a container of molasses, reaching a speed of 21 m/s by the time it hits the surface of the molasses. Over 2 s, the molasses slows the stone down to a speed of 4 m/s, its terminal velocity. Find the average speed of the stone during this fall.

60. ✱✱ Figure 3-32 shows a velocity versus time plot for a car.

(a) Find the average acceleration of the car over the first 4 s.

(b) Find the acceleration at $t = 11$ s.

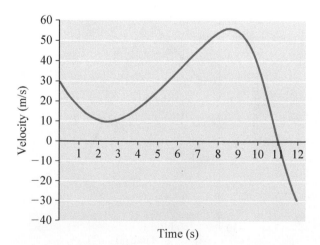

Figure 3-32 Problem 60

(c) Find the average acceleration of the car over the entire interval shown.

(d) Estimate the distance covered by the car between $t = 0$ s and $t = 12$ s.

61. ✱✱✱ Figure 3-33 shows an acceleration versus time plot for a particle undergoing a form of cyclic motion called simple harmonic motion. The velocity is zero when $t = 0$.

(a) Find the average speed of the particle between $t = 2$ s and $t = 3$ s.

(b) Find the speed of the particle at $t = 3$ s.

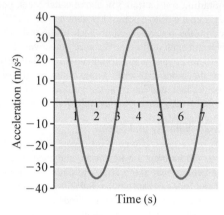

Figure 3-33 Problem 61

62. ✱ Using SI base units, a particle's position as a function of time is given by $x(t) = 7t^3 - 4t^2 + 6$. At what time will the particle's acceleration be zero? Find the speed of the particle at that time.

Section 3-4 Kinematics Equations

For this section some data is given as obtained from NASA. Convert all units to SI units and report your answers in SI units. mph = miles per hour ft/s = feet/s.

63. ✱✱ During the second-stage burn of an *Apollo* launch, the rocket's speed increased from 7882.9 ft./s to 21 377.0 ft./s. This burn lasted 384.22 s. Assuming constant acceleration and a straight path, find

(a) the average acceleration

(b) the distance travelled during this second-stage burn.

64. ✱✱ Use Equations (3-15) and (3-19) to derive Equation (3-20).

65. ✱✱ Show that Equation (3-27) is valid.

66. ✱✱✱ You throw two balls down from a height of 90 m, 3.0 s apart. You throw the first ball at a speed of 8.0 m/s. At what speed must you throw the second ball if it is to reach the ground

(a) with the same speed as the first ball;

(b) at the same time as the first ball?

67. ✱✱ You throw a stone straight down at a speed of 8.0 m/s from a height of 70 m. What is the average acceleration of the stone while it is in the air?

68. ✱ Typically, a golf club is in contact with a golf ball for about 0.5 ms. Jason Zuback is reported to have hit a golf ball with a speed of 357 km/h. Assuming constant acceleration, estimate the acceleration of Zuback's golf ball and the distance his club travelled while in contact with the ball.

69. ✳ A Formula One car has an acceleration 1.45 times the acceleration due to gravity. Find the time the car takes to go from 0 to 100 km/h, assuming constant acceleration.

70. ✳✳ You throw a marble at a speed of 10 m/s straight down from the top of a 30 m-tall building, and follow it with a second marble 0.50 s later. At what speed must you throw the second marble so that the two marbles arrive at the same time?

71. ✳✳ Your throw a ball straight up in the air with an initial speed of 32 m/s, and catch it at the moment it has the same speed with which you launched it. Find the ball's average acceleration, average speed, and average velocity.

72. ✳✳ A peregrine falcon flies in a straight line at 78 km/h for 3.0 s and spots a mouse on the ground 310 m below. The falcon dives at an angle of 30° to the vertical, reaching a maximum speed of 325 km/h in 4.0 s. The falcon maintains that speed as it makes a circular arc of radius 73 m to fly horizontally for another 2.0 s at 1.0 m above the ground. Find the average speed and the magnitude of the average velocity of the falcon during this flight.

73. ✳✳ The Bugatti Veyron can go from 0 to 100 km/h in 2.46 s (Figure 3-34). It takes another 7.34 s to reach the speed 240 km/h. Assume constant acceleration.
(a) Calculate the acceleration of the car over the first 2.46 s.
(b) How much distance does the Veyron have to cover to reach a speed of 240 km/h?
(c) The speed of the car is limited to 415 km/h to protect the tires. When the car maintains the same acceleration that it had at 240 km/h, how long would it take the car to reach 415 km/h?

© culture-images GmbH/Alamy

Figure 3-34 Problem 73

74. ✳✳ The greatest acceleration experienced by Col. Stapp on the Gee Whiz rocket sled was 46.2 g as the sled braked from a speed of 632 mph. Over what distance would the sled come to a stop if the acceleration was constant? How long would the sled take to come to a stop?

75. ✳✳ You throw a ball up straight in the air, and your friend, who is 14 m above you, catches it when its speed is 3 m/s on the way back down.
(a) How long does it take for the ball to pass your friend on the way up?
(b) What is the average speed of the ball between the time you let it go and the time your friend catches it?
(c) What is the average speed of the ball from launch to maximum height, and what is that maximum height?

76. ✳✳ Engineered material arresting systems (EMASs) are sometimes used to stop airplanes when they overshoot a runway. Consider a worst-case scenario in which an Airbus 380 moving at its full landing speed of 250 km/h hits the EMAS. What is the minimum stopping distance that will keep the acceleration below the 16 g rating for the seats on the plane?

77. ✳✳ Crumple zones in cars increase the time and the distance over which a car stops during a collision. The length of one crumple zone is 1.6 m. Find the maximum speed that a car could have when hitting a wall head-on without subjecting the passenger to an acceleration of more than 46.2 g. How long would that collision last?

78. ✳✳ Kenny Brack survived a 214 g car crash during an Indy car race in Fort Worth, Texas, in 2003. His speed is reported to have been 354 km/h when he crashed into a wall. If his final speed was zero, how long did the crash last and over what distance did he stop? Assume constant acceleration during the crash.

79. ✳ Some ejection seats have an acceleration of 14 g upward. How fast would the pilot be going relative to the plane at the end of a 1.3 s ejection burst?

80. ✳✳ The Peugeot EX1 electric sports car set a few speed records in 2010. Starting from rest, it covered the first 201.2 m (1/8 mile) in 8.89 s.
(a) Assuming constant acceleration, find the time the EX1 takes to go from 0 to 100 km/h.
(b) The car took a total of 16.81 s to travel 500 m starting from rest. Find the average acceleration for the period after it covered the first 1/8 mile.

81. ✳✳ A rocket is launched such that for the first 40.0 s its acceleration is upward at 5.00 g. For the following 25.0 s, its acceleration is upward at 3.50 g. The final stage lasts 90.0 s with an acceleration of 3.20 g.
(a) What is the speed of the rocket at the end of the second stage ($t = 65.0$ s)?
(b) What is the total distance covered by the rocket in the third stage?
(c) What is the total distance covered by the rocket during the three stages?

82. ✳✳ A lunar lander 200 m above the surface of the Moon has a downward speed of 50 m/s. Find the constant acceleration required for the lander to arrive at the surface with a speed of 1.1 m/s.
(a) What must the acceleration of the lander be so that it arrives at the surface with a speed of 1.1 m/s?

83. ✳ A rocket is launched vertically such that its acceleration upward is 4.50 g. If it can maintain that acceleration, how much time would it take to reach the orbit of the International Space Station 300 km above Earth's surface?

84. ✳✳✳ A stone is tossed straight down from the top of a building. During the third second of its flight, the stone covered a distance of 45.0 m. How much distance did the stone cover during the first 2.00 s? What was the stone's initial speed?

85. ✳✳ You throw a ball straight up in the air and catch it 7.00 s later. Find the average speed and average acceleration of the ball, the total distance covered by the ball, and its displacement.

86. ✳✳ You are standing on top of a building and are challenged by a friend to throw a ball such that it covers the 1.7 m distance between the top and the bottom of his window in 0.3 s. The top of the window is 13 m below the point where you release the all. Is this throw possible? If so, what is the maximum amount of time during which the ball can cover the distance?

Section 3-5 Relative Motion in One Dimension

87. ✱✱✱ A child standing on a balcony drops a marble from a height of 16 m above the ground at the same time that another child throws a tennis ball up from a height of 1 m above the ground.
    (a) What must be the initial speed of the tennis ball for the two to meet at the highest point in the ball's trajectory?
    (b) What was the speed of the ball relative to the marble just before the marble hit the ground?

88. ✱✱ Derive the relative velocity, acceleration, displacement, and position for two objects moving with constant accelerations.

89. ✱ A Mini Cooper moving at 135 km/h passes a Mustang moving in the same direction at 90 km/h. Calculate the velocity of the Mustang relative to the Mini, and the speed of the Mustang relative to the Mini.

90. ✱✱ Two young men on a balcony are trying to attract the attention of a young lady on the ground 11 m below. One young man drops a flower. The other young man wants his flower to land first. After 0.7 s, how fast should he throw it down?

91. ✱✱ A ball is dropped from the top of a 56.0 m-tall building. After 2.00 s, another ball is thrown upward from the ground. When the two balls pass the same point, they have the same speed. How fast was the upward-bound ball thrown?

COMPREHENSIVE PROBLEMS

92. ✱✱ A stone is dropped from a high cliff such that its speed just before impact on the ground below is 190 m/s. How much distance did the stone cover during the interval $5.0 \text{ s} \leq t \leq 15.0 \text{ s}$? How far did the stone travel in its last second of flight?

93. ✱✱✱ You are riding in an elevator with an open roof. You are moving down at a speed of 4.00 m/s when you throw an apple up into the air at a speed of 6.00 m/s relative to the elevator.
    (a) How long does it take for you to catch the apple?
    (b) What is the maximum height the apple reaches above the elevator floor? How does this height change if the elevator is moving up?
    (c) A friend standing on the ground throws an apple up at a speed of 6.00 m/s. How long does it take for that apple to return to his hand?

94. ✱✱ (a) A Formula One race car can go from 0 to 100 km/h in 1.70 s, from 0 to 200 km/h in 3.80 s, and from 0 to 300 km/h in 8.60 s. Assuming constant acceleration during the intervals 0 s to 1.70 s, 1.70 s to 3.80 s, and 3.80 to 8.60 s, find the distance that the car covers in the first 1.70 s and between 3.80 s and 8.60 s.
    ✱✱ (b) A Formula One race car starts from rest, reaches a speed of 100 km/h in 1.70 s, reaches a speed of 325 km/h over the next 7.30 s, then slows with an acceleration of 4.00 g for the next 100 m. Find the average speed, average acceleration, and average velocity of the car for the entire interval.

95. ✱✱ How much distance does a Formula One race car initially travelling at 360 km/h cover while it brakes at 6.00 g for 1.20 s?

96. ✱✱ You are sitting well back from the windows in the library at the Bamfield Marine station on Vancouver Island. A bald eagle hovers overhead with a fish in its grip. The fish wriggles loose and falls straight down. You happen to be looking out the window and notice that the fish is in sight for 0.14 s. The windows are 1.2 m high. How high was the eagle when it dropped the fish? What is the speed of the fish as it passes the window sill?

97. ✱✱ A particle's position is given by $x(t) = x_0 \cos(\omega t + \phi)$, where x_0 and ϕ are constants. At $t = 0$, the particle is at $x = 0$.
    (a) Find the possible values of ϕ.
    (b) Given that $x_0 = 20$ cm, plot the position of the particle over the interval $t = 0$ s to 10 s with (i) $\omega = \pi$ and $\phi = \pi$, (ii) $\omega = \pi/2$ and $\phi = 0$, and (iii) $\omega = \pi/2$ and $\phi = \pi/2$.
    (c) Write an expression for the particle's velocity as a function of time.
    (d) Write an expression for the particle's acceleration as a function of time.
    (e) Deduce a relationship between the acceleration and the position of a particle executing this motion (called simple harmonic motion).

98. ✱✱ Find the maximum speed and the magnitude of the maximum acceleration of the particle in problem 97. Use mathematical equations to show where these maxima occur.

99. ✱✱ An ocean liner approaches a pier at 20 km/h. When the ocean liner is 10 km away, the captain sends a speedboat to the pier to pick up the harbour pilot, who will help dock the ocean liner. The speedboat takes 12 min to get to the pier.
    (a) How long would the speedboat take to return to the ocean liner if it maintains the same speed?
    (b) Find the displacement of the speedboat during the round trip and the distance covered by the speedboat during the round trip.

100. ✱✱ A speedboat accelerates from rest at the start line of a race to reach a speed of 30 km/h over a distance of 70 m, then continues with the same constant acceleration for another 11 s before slowing down to a speed of 25 km/h over a distance of 100 m. Another speedboat begins to accelerate from rest at the start line and catches up to the first boat within 30 s.
    (a) Find the acceleration of the second speedboat.
    (b) Find the time it takes the first speedboat to cover the first 70 m.
    (c) Find the speed of the second speedboat when it catches up to the first speedboat.

101. ✱✱ You are in a hot-air balloon that is accelerating upward at 4.00 m/s². The balloon is at a height of 47.0 m when its speed upward is 9.00 m/s. You drop an apple from the side. Find the maximum height the apple will reach above the ground, and the speed of the apple when it hits the ground.

102. ✱✱✱ A child has a tennis ball and a baseball. He lets go of the tennis ball from the top of a high building, waits, and then throws the baseball straight down with a speed such that the two balls have a relative speed of 10 m/s when they meet 40 m below. How long did the child wait before throwing the baseball?

103. ✱✱✱ You throw a ball up in the air with a speed such that it just reaches your friend on a balcony 30.0 m above. At the same time, your friend throws another ball down to you such that it reaches you with twice the speed with which you threw your ball. What is the relative velocity of the two balls when they pass each other?

104. ✶✶ Two children are playing a game in which one of them throws a baseball down from the top of his 6.0 m-high tree house and the other is supposed to hit the baseball with a tennis ball. If the child in the tree house throws the baseball with a downward speed of 1.2 m/s, what is the speed the child on the ground has to throw his baseball with from a height of 1.1 m so the baseballs meet halfway?

105. ✶✶✶ Two children are playing a game in which one of them throws a volleyball up in the air and the other, standing very close to her friend, throws a tennis ball up to hit the volleyball as it reaches its highest point. The first child throws the volleyball at a speed of 12 m/s, and the second child throws the tennis ball at a speed of 25 m. How long after the volleyball is thrown should the second child wait to throw the tennis ball?

106. ✶✶✶ Show that Equation (3-20) is valid for nonconstant acceleration.

107. ✶✶✶ Two friends live beside a 25.0 m-high cliff, one at the top of the cliff and the other directly below, at the bottom. While they are making dinner, the friend at the top of the cliff throws a loaf of bread down with a speed of 7.0 m/s at the same time that the other friend throws a lemon up with a speed of 30.0 m/s. Find the relative speed of the bread and the lemon when they pass one another.

108. ✶✶ Madrid, Spain, and Beijing, China, both lie on the 40th parallel. Beijing is at 116°23′ E, and Madrid is at 3°40′ W. Find the average speed of a plane that takes 7.0 h to travel from Beijing to Madrid.

109. ✶✶✶ Figure 3-35 shows data for the velocity of a meteor fragment entering the atmosphere as a function of altitude.
(a) Estimate the rate of change of the velocity at altitudes of 120 km and 92 km.
(b) Assuming that gravity and friction with the atmosphere are the principal forces acting on the meteor fragment, what does the graph reveal about the relative strength of the two forces?
(c) What does the shape of the velocity graph suggest about the upper atmosphere?

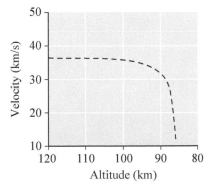

Figure 3-35 Problem 109

110. ✶✶✶ Figure 3-36(a) shows a graph of altitude versus time for the *Soyuz* launch vehicle.

(a) Estimate the radial speed (i.e., the magnitude of the vertical component of velocity) of the vehicle at (i) 100 s, (ii) 200 s, (iii) 300 s, and (iv) 400 s.
(b) Find the mean radial velocity between (i) 0 s and 100 s and (ii) between 300 s and 400 s.

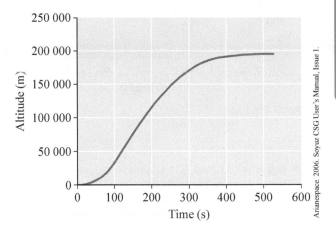

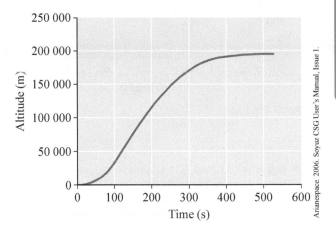Arianespace. 2006. *Soyuz CSG User's Manual*, Issue 1.

Figure 3-36 (a) Problem 110

© ITAR-TASS Photo Agency/Alamy

Figure 3-36 (b) Problem 110

111. ✶✶✶ Figure 3-37 shows a velocity versus time plot for the *Soyuz* launch vehicle.
(a) Estimate the peak acceleration at $t = 120$ s, $t = 280$ s, and $t = 540$ s.
(b) What is the average acceleration of the vehicle for the entire flight time shown?
(c) Find the average acceleration for each of the three stages ending at 120 s, 280 s, and 540 s.

MECHANICS

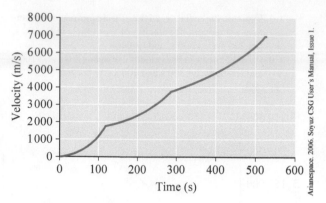

Arianespace. 2006. Soyuz CSG User's Manual, Issue 1.

Figure 3-37   Problem 111

112. ✖✖✖   Use Figure 3-37 to estimate the distance covered by the *Soyuz* vehicle in the first 120 s. How much distance did the vehicle cover by the end of 280 s? Compare these results to the graph of Figure 3-36(a), and explain the difference.

113. ✖✖✖   Use the velocity versus time graph for the *Soyuz* launch vehicle (Figure 3-37) to construct an acceleration versus time graph. Use two scales to show your results in m/s² and in multiples of g. Compare your graph to Figure 3-38 below.

114. ✖✖✖   Figure 3-38 shows an acceleration versus time plot for the *Soyuz* launch vehicle. Using the graph, calculate the speed of the vehicle at each of the three peak accelerations shown, and construct a velocity versus time graph for the vehicle. Compare your results with Figure 3-37.

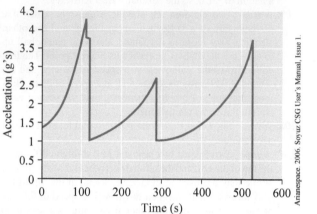

Arianespace. 2006. Soyuz CSG User's Manual, Issue 1.

Figure 3-38   Problem 114

 See the text online resources at www.physics1e.nelson.com for Open Problems and Data-Rich Problems related to this chapter.

Chapter 4
Motion in Two and Three Dimensions

The equatorial speed of a star, also called rotational speed, can vary significantly from one star to another. The Sun's equatorial speed is only a few kilometres per second. The equatorial speed for some stars can exceed 200 km/s. Some neutron stars, which are vastly smaller than stars like our Sun, have rotational speeds as high as 38 000 km/s. With an equatorial speed of approximately 600 km/s, the fastest spinning full-sized star observed to date is VFTS 102. VFTS 102 was discovered by astronomers using the Very Large Telescope at the European Southern Observatory in Paranal, Chile (Figure 4-1).

VFTS 102 is 25 times more massive and 100 times brighter than the Sun. The star is located in the Tarantula nebula within the Large Magellanic Cloud, about 160 000 ly from Earth. The high radial acceleration resulting from the high equatorial speed has VFTS 102 close to tearing itself apart. The rotational and translational speeds of this star suggest that it was ejected from a binary star system when its companion exploded in a supernova. If VFTS 102 does tear itself apart, it is expected to produce one of the brightest stellar explosions in the universe.

Learning Objectives

When you have completed this chapter you should be able to:

1. Express displacement, velocity, and acceleration in two and three dimensions.
2. Describe and analyze the motion of a projectile.
3. Derive the radial acceleration vector for an object moving in a circle.
4. Apply relative motion principles to two and higher dimensions.

ESO/M.-R. Cioni/VISTA Magellanic Cloud survey. Acknowledgment: Cambridge Astronomical Survey Unit

Figure 4-1 VFTS 102, the fastest spinning star found to date, is at the centre of the Tarantula nebula in the Large Magellanic Cloud.

For additional Making Connections, Examples, and Checkpoints, as well as activities and experiments to help increase your understanding of the chapter's concepts, please go to the text's online resources at www.physics1e.nelson.com.

LOOK AHEAD

4-1 Displacement, Velocity, and Acceleration in Two and Three Dimensions

We can use the vector algebra tools from Chapter 2 to extend our analysis of motion to two and three dimensions. As described in Section 2-2, two- and three-dimensional vectors can be written in terms of components. For example, in two dimensions the position vector $\vec{r}$ can be expressed as

$$\vec{r} = \vec{r}_x + \vec{r}_y = x\hat{i} + y\hat{j} \qquad (4\text{-}1a)$$

In three dimensions, the position vector is given by

$$\vec{r} = x\hat{i} + y\hat{j} + z\hat{k} \qquad (4\text{-}1b)$$

Equation 4-1a and 4-1b give the position vector as the sum of its scalar components, multiplied by the corresponding unit vectors. This will, in general, be the approach we take to representing vectors in this text. However, occasionally it proves to be more elegant to represent a given vector as the sum of its vector components. In three dimensions the position vector given in equation 4-1b can be written as

$$\vec{r} = \vec{x} + \vec{y} + \vec{z} \qquad (4\text{-}1c)$$

Similarly, the displacement vector between two points can be written as

$$\Delta\vec{r} = x\hat{i} + y\hat{j} \qquad (4\text{-}2a)$$

and in three dimensions

$$\Delta\vec{r} = x\hat{i} + y\hat{j} + z\hat{k} \qquad (4\text{-}2b)$$

The average and instantaneous velocities of an object are then given by

$$\vec{v}_{\text{avg}} = \frac{\Delta\vec{r}}{\Delta t} = v_{\text{avg},x}\hat{i} + v_{\text{avg},y}\hat{j} \qquad (4\text{-}3a)$$

in two dimensions and

$$\vec{v}_{\text{avg}} = \frac{\Delta\vec{r}}{\Delta t} = v_{\text{avg},x}\hat{i} + v_{\text{avg},y}\hat{j} + v_{\text{avg},z}\hat{k} \qquad (4\text{-}3b)$$

in three dimensions; and

$$\vec{v} = \lim_{\Delta t \to 0} \frac{\Delta\vec{r}}{\Delta t} = \frac{d\vec{r}}{dt} = v_x\hat{i} + v_y\hat{j} \qquad (4\text{-}4a)$$

in two dimensions and

$$\vec{v} = \lim_{\Delta t \to 0} \frac{\Delta\vec{r}}{\Delta t} = \frac{d\vec{r}}{dt} = v_x\hat{i} + v_y\hat{j} + v_z\hat{k} \qquad (4\text{-}4b)$$

in three dimensions.
Alternatively, one can write

$$\vec{v} = \vec{v}_x + \vec{v}_y + \vec{v}_z \qquad (4\text{-}4c)$$

The average and instantaneous accelerations are written as

$$\vec{a}_{\text{avg}} = \frac{\Delta\vec{v}}{\Delta t} = a_{\text{avg},x}\hat{i} + a_{\text{avg},y}\hat{j}$$

$$\vec{a} = \lim_{\Delta t \to 0} \frac{\Delta\vec{v}}{\Delta t} = \frac{d\vec{v}}{dt} = \vec{a}_x + \vec{a}_y$$

in two dimensions, and

$$\vec{a}_{\text{avg}} = \frac{\Delta\vec{v}}{\Delta t} = a_{\text{avg},x}\hat{i} + a_{\text{avg},y}\hat{j} + a_{\text{avg},z}\hat{k} \qquad (4\text{-}5)$$

and

$$\vec{a} = \lim_{\Delta t \to 0} \frac{\Delta\vec{v}}{\Delta t} = \frac{d\vec{v}}{dt} = a_x\hat{i} + a_y\hat{j} + a_z\hat{k} \qquad (4\text{-}6)$$

in three dimensions.
Alternatively, one can write

$$\vec{a} = \vec{a}_x + \vec{a}_y + \vec{a}_z \qquad (4\text{-}4c)$$

Although we will use the scalar form for the components of a vector in general, there will be occasions where it will prove more elegant, convenient, and advantageous to treat the components of a vector as vectors. These occurrences will be clearly pointed out.

EXAMPLE 4-1

Average Speed and Average Velocity

An airplane flying at a constant speed takes 1.20 h to go from Amsterdam, the Netherlands, to Berne, Switzerland, 630 km to the southwest. The airplane is then diverted to Paris, France, 432 km northwest. It takes 0.900 h to get there, travelling at the same constant speed.

(a) Find the average speed of the airplane during the trip.
(b) Find the average velocity of the airplane, given that Amsterdam and Paris are 440 km apart.
(c) In a diagram, show the difference between the displacement of the airplane and the distance covered by the airplane throughout the journey.

SOLUTION

(a) The average speed is the total *distance* covered by the airplane divided by the total time elapsed. Since distance is a scalar quantity, the total distance is the sum of the distances in each part of the journey:

$$d = d_1 + d_2 = 630 \text{ km} + 432 \text{ km} = 1062 \text{ km}$$

The total time taken is 1.2 h + 0.9 h = 2.1 h. Therefore, the average speed is

$$v_{\text{avg}} = \frac{1062 \text{ km}}{2.1 \text{ h}} = 506 \text{ km/h}$$

(b) Since the airplane started over Amsterdam and ended over Paris 440 km away, the displacement is 440 km pointing directly from Amsterdam to Paris. The average velocity is given by the displacement divided by the elapsed time:

$$\vec{v}_{avg} = \frac{\Delta x \hat{i}}{\Delta t} = \frac{440 \hat{i}\ km}{2.1\ h} = 210 \hat{i}\ km/h$$

where the direction from Amsterdam to Paris is chosen as the positive x-axis

(c) The distance covered by the plane is shown in Figure 4-2 with the dashed line that follows the airplane's path. The displacement is the vector connecting the start and finish points.

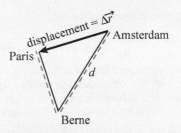

Figure 4-2 The difference between distance covered (dashed line) and the displacement of the plane

Making sense of the result:

Unlike the average speed, which tells us how fast the airplane is moving, the average velocity depends only on the direct distance between the start and finish points and the time elapsed on the way there.

 CHECKPOINT

C-4-1 Distance and Displacement

When you walk halfway around a circle with a radius of 1.2 m,

(a) your displacement is 1.2πm;

(b) you travel a distance of 2.4 m;

(c) your displacement is 2.4 m directed straight across the circle;

(d) your displacement and the distance travelled are equal.

C-4-1 (c) Displacement depends on initial and final positions only.

 EXAMPLE 4-2

Components of Velocity

At a practice, football players run 100 m at top speed in one direction, make a 90° turn, and run another 120 m, again at top speed (Figure 4-3). One player runs the first leg in 12.0 s and the second leg in 18.0 s.

(a) Ignoring the time taken to change directions, find the player's average velocity. Express your answer in polar and Cartesian components.

(b) Calculate the average velocities during the first and the second legs of the run.

(c) Calculate the player's average speed for the entire run.

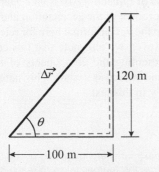

Figure 4-3 The path distance (dashed line) and the displacement (vector arrow) for the football player in Example 4-1

SOLUTION

(a) Let us pick the positive x-axis to lie along the first 100 m and the positive y-axis to lie along the second direction.

The total displacement is

$$\Delta \vec{r} = \Delta x \hat{i} + \Delta y \hat{j} \qquad (1)$$

Since the angle between the two legs of the trip is 90°,

$$\Delta r = \sqrt{\Delta x^2 + \Delta y^2} = \sqrt{100^2 + 120^2} = 156.2\ m \qquad (2)$$

to 3 significant figures, this is 156 m, but for the calculations below we keep the value from Equation 2.

The total time elapsed is $12\ s + 18\ s = 30\ s$, so the magnitude of the average velocity is

$$v = \frac{\Delta r}{\Delta t} = \frac{156.2}{30} = 5.21\ m/s$$

The Cartesian components of the average velocity are

$$v_{avg,x} = v_{avg} \cos\theta = 5.21 \cos(50.19°) = 3.34\ m/s$$

$$v_{avg,y} = v_{avg} \sin\theta = 5.21 \sin(50.19°) = 4.00\ m/s$$

(b) The velocity during the first leg of the run is the displacement along the x-axis divided by the time elapsed:

$$\vec{v}_1 = \frac{\Delta x \hat{i}}{\Delta t} = \frac{100 \hat{i}\ m}{12.0\ s} = 8.33\ m/s\ in\ the\ +x\text{-direction}$$

(continued)

During the second leg of the run,

$$\vec{v}_2 = \frac{\Delta y \hat{j}}{\Delta t} = \frac{120 \hat{j} \text{ m}}{18.0 \text{ s}} = 6.67 \text{ m/s in the } +y\text{-direction}$$

(c) The average speed for the entire run is the total distance (not the displacement) divided by the time elapsed. The distance covered is 100 m + 120 m = 220 m, and the total time is 30 s. Therefore,

$$v_{\text{avg}} = \frac{d}{\Delta t} = \frac{220 \text{ m}}{30 \text{ s}} = 7.33 \text{ m/s}$$

Since displacement is not the same as distance, we get different answers for parts (a) and (b).

✓ CHECKPOINT

C-4-2 Components of Velocity

Why are the components of the average velocity for the entire run in Example 4-2 not equal to the average velocities along each direction for each leg of the run?

C-4-2 The average velocities along each direction are not the components of the overall average velocity vector.

PEER TO PEER

The average speed is the total distance travelled divided by the time elapsed; however, the average velocity is the displacement vector divided by the time elapsed.

LO 2

4-2 Projectile Motion

Figure 4-4 shows the trajectory of a projectile under the influence of gravity only. Other factors, such as the effect of air resistance, are ignored. The launch point is used as the origin of the coordinate system. The projectile was launched with an initial speed v_0, making an angle θ above the horizontal. The initial velocity can be resolved into a horizontal component equal to $v_0 \cos \theta$ and a vertical component equal to $v_0 \sin \theta$.

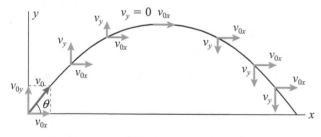

Figure 4-4 Typical projectile trajectory

Since there is no acceleration in the horizontal direction, the horizontal component of the velocity is constant throughout the motion, as indicated by the length of the horizontal arrows representing this component in Figure 4-4. In the vertical direction, the force of gravity is

acting on the projectile, so the vertical component of the velocity decreases until it reaches zero momentarily at the top of the trajectory, and then increases in the downward direction. The overall speed of the projectile therefore reaches a minimum at the top of the trajectory.

The horizontal distance covered by the projectile depends on the speed in the horizontal direction and the time of flight:

$$\Delta x = v_x t \qquad (4\text{-}7)$$

In the y-direction, we can apply the results we obtained for motion in one dimension under constant acceleration. Using Equation (3-19), the vertical displacement can be written as

$$\Delta \vec{y} = \frac{1}{2} \vec{a} t^2 + \vec{v}_{0y} t \qquad (4\text{-}8)$$

Similarly, Equation (3-15) gives a vertical velocity of

$$\vec{v}_y(t) = \vec{a} t + \vec{v}_{0y} \qquad (4\text{-}9)$$

and the time-independent kinematics Equation (3-20) gives

$$v_y^2 - v_{0y}^2 = 2\vec{a} \cdot \Delta \vec{y} \qquad (4\text{-}10)$$

Note that we are copying the approach we have taken in Chapter 3 in dealing with velocity, displacement, and acceleration as vectors. The dot product form used on the right-hand side of Equation 4-10 is an elegant way of representing the product of the acceleration and the displacement. We keep the vector format here for relevance to later chapters. However, we also note that this can equally be done by representing the components of the quantities along the x- and y-axes as scalars. This latter approach is used frequently in this text.

EXAMPLE 4-3

Football Trajectory

A football is kicked from a level field at a velocity of 19.0 m/s at an angle of 41.0° above the horizontal.

(a) Find the horizontal distance covered by the football.

(b) Find the maximum height reached by the football.
(c) Find the time the football takes to reach its maximum height.
(d) Find the minimum speed of the football during its flight.

SOLUTION

The trajectory of the soccer ball is shown in Figure 4-5. We choose up and right as the positive directions.

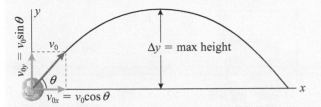

Figure 4-5 Example 4-3

(a) The velocity in the horizontal direction is constant. Therefore, the horizontal distance covered by the football is given by $\Delta x = v_x t$. The horizontal component of the velocity, v_x, is

$$v_x = v_0 \cos\theta = (19 \text{ m/s})(\cos(41°))$$

Now we need to find the time, t, that the football is in the air. At the end of this time, the ball hits the ground, so its height, y, is zero. From Equation (4-8) we have

$$\Delta \vec{y} = \frac{1}{2}\vec{a}t^2 + \vec{v}_0 t$$

$$0 = \frac{1}{2}(-9.81\hat{j}) \text{ m/s}^2 (t^2) + (19 \sin(41°)\hat{j}) \text{ m/s} (t)$$

This quadratic equation gives us two solutions: $t = 0$ and $t = 2.5413$ s. Clearly, the later time is the landing time for the football. Using this time in the expression for v_x gives

$$x = v_0 \cos(41°)t = (19 \cos(41°) \text{ m/s})(2.5413 \text{ s}) = 36.4 \text{ m}$$

(b) To find the maximum height reached by the ball, we apply Equation (4-10) to the moment the ball reaches the top of its trajectory:

$$v_y^2 - v_{0y}^2 = 2\vec{a} \cdot \Delta\vec{y}$$

At the maximum height the vertical speed is zero. We know that the initial speed in the y-direction is $v_{0,y} = 19 \sin(41°)$. Since the acceleration vector points downward as the ball moves upward, the vector dot product $\vec{a} \cdot \Delta\vec{y}$ is negative. Therefore, using Equation 2-18 for the scalar product of $\vec{a}$ and $\Delta\vec{y}$

$$0 - (19 \sin(41°) \text{ m/s})^2 = 2(9.81 \text{ m/s}^2)(\Delta y)(-1)$$

$$\Delta y = 7.92 \text{ m}$$

(c) To find the time the football takes to reach the maximum height, we use Equation (4-9):

$$\vec{v}_y(t) = \vec{v}_{0y} + \vec{a}t$$

$$0 = (19 \sin(41°)\hat{j}) \text{ m/s} + (-9.81\hat{j}) \text{ m/s}^2 (t)$$

$$\Rightarrow t = \frac{19 \sin(41°) \text{ m/s}}{9.81 \text{ m/s}^2} = 1.27 \text{ s}$$

(d) The minimum speed occurs at the top of the trajectory, where the vertical component of the velocity is zero. We only have the horizontal velocity $v_x = 19 \cos(41°) = 14.339$ m/s, which is 14.3 m/s to three significant figures.

Making sense of the result:

The height reached by the football is reasonable, and the time it takes for the football to reach the maximum height is half its flight time, which makes sense due to symmetry.

 CHECKPOINT

C-4-3 Minimum Speed of a Projectile

What is the minimum speed of a projectile launched with a velocity of 200 m/s at a 60° angle above the horizontal?

(a) zero
(b) 100 m/s
(c) $100\sqrt{3}$ m/s
(d) none of the above

C-4-3 (b) min speed is the speed in the x direction $= 200 \cos(60°) = 100$ m/s.

 ONLINE ACTIVITY

Projectile Motion

The e-resource that accompanies every new copy of this textbook contains an Online Activity using the PhET simulation "Projectile Motion." Work through the simulation and accompanying questions to gain an understanding of projectile motion.

 MAKING CONNECTIONS

Giant Artillery

One of the world's largest artillery pieces was "Dora," a railway-mounted siege gun built for the German army during World War II (Figure 4-6). This cannon had a calibre (bore diameter) of 800 mm, and could fire a 4800 kg shell up to 48 km. Because of air resistance, the firing angle had to be slightly greater than 45° to get the maximum range.

Figure 4-6 The Dora rail cannon

EXAMPLE 4-4

Projectile Motion

You are standing on a roof 13.0 m above the ground, and you throw a tennis ball to your friend, who is on the ground 20 m from the building. Your hand is 2.00 m above the roof when you release the ball. Your friend reaches up such that her hand is 2.10 m above the ground. The ball leaves your hand with a velocity of 16.0 m/s at an angle of 15.0° below the horizontal.

(a) By how much vertical distance does the ball miss your friend's hand?

(b) What is the speed of the ball as it passes your friend's hand?

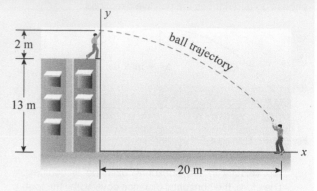

Figure 4-7   Trajectory of the ball in Example 4-4

SOLUTION

(a) Since we are given the horizontal distance and the horizontal speed, we can calculate the time the ball takes to cover the distance. We can then find the height of the ball at this time. We choose up and right as the positive directions.

From Equation (4-7) and Figure 4-7,

$$\Delta x = v_x t = (v \cos\theta) t$$

$$t = \frac{\Delta x}{v \cos\theta} = \frac{20 \text{ m}}{16 \cos(15°) \text{ m/s}} = 1.2941 \text{ s}$$

The vertical displacement is given by Equation (4-8):

$$\Delta \vec{y} = \frac{1}{2}\vec{a}t^2 + \vec{v}_{0,y} t$$

$$\Delta y = \frac{1}{2}(-9.81\hat{j} \text{ m/s}^2)(1.2941^2) - (16\sin(15°))\hat{j} \text{ m/s} (1.2941 \text{ s})$$

$$= -13.57\hat{j} \text{ m}$$

The ball has therefore dropped a total of 13.57 m from its launch height, which was 15 m above the ground.

Thus, the ball is at a height of 1.43 m when it gets to your friend. Since her hand is 2.1 m above the ground, the ball misses by 67.0 cm.

(b) As the ball passes your friend, its velocity has a constant horizontal component of $v_0 \cos(15°)$ and a vertical component given by

$$\vec{v}(t) = \vec{a}t + \vec{v}_0$$

$$v_y(t) = -9.81 \text{ m/s}^2(1.2941 \text{ s}) - 16\sin(15°) \text{ m/s} = -16.836 \text{ m/s}$$

$$v = \sqrt{v_x^2 + v_y^2} = \sqrt{(16\cos(15°) \text{ m/s})^2 + (16.836 \text{ m/s})^2}$$

$$= 22.9 \text{ m/s}$$

Making sense of the result:

For the distances given, the amount by which your friend misses the ball is reasonable.

LO 3

4-3   Circular Motion

Uniform Circular Motion

An object moving with **uniform circular motion** moves along a circular path of fixed radius at a constant speed. Since the direction of motion of the object is constantly changing, its velocity is also constantly changing. Therefore, the object must be continuously accelerating.

Figure 4-8 shows an object moving around a circle at a constant speed. The velocity at point A is $\vec{v}_A$, and the velocity at point B is $\vec{v}_B$ The velocities have exactly the same magnitude, v. The position vector at point A is $\vec{r}_A$, and the position vector at point B is $\vec{r}_B$. Both position vectors have magnitude r, as measured from the centre of the circle. At every point on the circle, the velocity of the object is tangential to the circle and perpendicular to the position vector. The average acceleration is

$$\vec{a}_{avg} = \frac{\Delta\vec{v}}{\Delta t}$$

Therefore, the direction of average acceleration is parallel to $\Delta\vec{v}$.

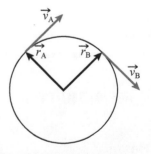

Figure 4-8   An object in uniform circular motion

Now, we redraw the position and velocity vectors using the principle that a vector can be translated as long as its magnitude and direction do not change. First, we construct a triangle formed by the two position vectors and the

Uranium Enrichment and Centrifugation

Highly enriched uranium fuels are used in nuclear reactors. Gas centrifuges are often used to enrich the uranium by separating uranium's various isotopes. In a gas centrifuge, vaporized material is spun at very high speeds to separate components with different densities. As early as the end of the 19th century, chemists used simple hand-cranked centrifuges to help precipitate fine particles from liquids. Ultracentrifuge rotors spin at very high speeds, producing accelerations up to 2 million times the acceleration due to gravity.

Theodor Svedberg invented the first ultracentrifuge in 1925, and he received the Nobel Prize for his research on colloids and proteins with his invention. The Russians, using the expertise of Austrian scientist Gernot Zippe, were the first to mass-produce enriched uranium using centrifuge technology (Figure 4-9). The same centrifuge principles used to produce fuel for nuclear reactors can be used to produce the more highly enriched weapons-grade uranium.

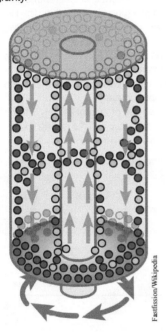
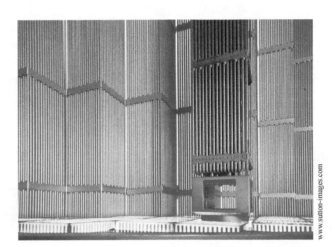

Figure 4-9 (a) A schematic of the Zippe centrifuge (b) A cascade of gas centrifuges used to successively increase the concentration of the fissionable isotope of uranium

change in position Δr, as shown in Figure 4-10. We then construct a triangle formed by the two velocity vectors and the change in velocity. Since the velocity vector is perpendicular to the position vector at either location, we can conclude that the change in velocity and, consequently, the average acceleration vector, are perpendicular to Δr:

Next, we take the limit of the average acceleration as Δt goes to zero to get the instantaneous acceleration:

$$\vec{a} = \lim_{\Delta t \to 0} \frac{\Delta \vec{v}}{\Delta t} = \frac{d\vec{v}}{dt}$$

As the time interval Δt shrinks, so does the displacement between A and B. In the limit, the two points merge. As the position triangle collapses, so does the velocity triangle. However, the acceleration which is parallel $\Delta \vec{v}$ remains perpendicular to $\Delta \vec{r}$, which is a chord of the circle. The acceleration vector, being perpendicular to that chord, points along the radial to the centre of the circle. In the limit as Δt goes to zero, the instantaneous acceleration

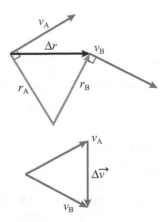

Figure 4-10 The change in velocity and, therefore, the acceleration vector, are perpendicular to the change in the position vector.

points along the radius. Therefore, an object moving in a circle must continuously accelerate toward the centre of the circle. Acceleration toward the centre of the circle is called **radial acceleration**.

To find the magnitude of the acceleration, consider the following ratios of the two similar triangles formed by the position and velocity vectors:

$$\frac{\Delta v}{v} = \frac{\Delta r}{r}$$

$$\Delta v = \frac{v\Delta r}{r}$$

Dividing both sides by Δt, the change in time, gives

$$\frac{\Delta v}{\Delta t} = \frac{v\Delta r}{r\Delta t}$$

As Δt goes to zero,

$$\frac{dv}{dt} = \frac{vdr}{rdt}$$

KEY EQUATION

$$a = \frac{v^2}{r} \qquad (4\text{-}11)$$

We have thus established both the direction and the magnitude for the radial acceleration of an object moving around a circle at a constant speed.

 CHECKPOINT

C-4-4 Uniform Circular Motion

For an object moving in uniform circular motion,
(a) the velocity of the object is constant;
(b) the speed of the object is constant.
(c) Both (a) and (b) are true.
(d) None of the above are true.

C-4-4 (b) the speed must be constant for uniform circular motion, but the velocity is not constant.

ONLINE ACTIVITY

The e-resource that accompanies every new copy of this textbook contains an Online Activity using the PhET simulation "Circular Motion." Work through the simulation and accompanying questions to gain an understanding of circular motion.

Nonuniform Circular Motion

An object moving in a circle does not necessarily maintain a constant speed. For example, if you let a marble roll down the inside of a hemispherical bowl, the marble will speed up while travelling down the circular surface. When the speed of an object changes while travelling along a circle, then the velocity of the object changes

along the direction of motion, which is tangential to the circle. Such an object is said to have **tangential acceleration**. Therefore, the **total acceleration** of this object is the *vector sum* of its tangential and radial accelerations (Figure 4-12).

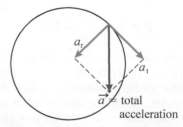

Figure 4-12 The total acceleration for an object moving in a circle at a changing speed is the vector sum of the radial and tangential acceleration vectors.

MAKING CONNECTIONS

Sustained versus Sudden Acceleration

The magnitude of the acceleration that the human body is subjected to is not the only factor in determining the effects of acceleration on the body: the direction and duration of the acceleration are also important. The body reacts differently and has different tolerance thresholds to abrupt as opposed to sustained acceleration. The resulting symptoms can be quite different. The body can cope with rather high accelerations applied over a short time. For example, the body can tolerate peak vertical accelerations up to 25 g in the +z-direction over a period of 100 ms without incurring spinal injury.

However, much smaller accelerations can produce adverse effects if applied over a longer time. For example, pilots can sustain the short period of high acceleration in an ejection (Figure 4-13) seat without injury (ejection seat accelerations can reach 13 g). Accelerations as small as 5 g, however, can result in blackouts if sustained for a few seconds in manoeuvres such as an inside loop. The effects of sustained acceleration depend, in part, on the rate of change of the acceleration, called *onset*. For example, if the acceleration during an inside loop manoeuvre were to increase from zero to about 5 g over a period of 5 s, it would produce loss of consciousness. It would take longer for symptoms to occur if the onset were lower.

Figure 4-13 A pilot experiences +z-acceleration when deploying an ejection seat.

EXAMPLE 4-5

Nonuniform Circular Motion

Starting from rest, an experimental mini-drone (remote-controlled aircraft) flies around a circle of radius 2.00 m three times in 3.00 s. The mini-drone has constant tangential acceleration. Find the magnitude of the mini-drone's acceleration at the end of 0.50 s.

SOLUTION

Since the mini-drone is accelerating along the circumference of the circle, its acceleration has both tangential and radial components. We know the initial speed, the total distance travelled in the three circuits, and the time taken. We pick the direction of motion as positive.

To find the tangential acceleration, we apply Equation (3-19):

$$d = \frac{1}{2}a_t t^2 + v_0 t$$

$$3(2\pi)2 \text{ m} = \frac{1}{2}a_t(3 \text{ s})^2 + 0$$

$$a_t = \frac{8\pi}{3} = 8.378 \text{ m/s}^2$$

From Equation (4-11), we know that the radial acceleration depends on speed. Since the speed is increasing continuously, the magnitude of the radial acceleration of the mini-drone is also increasing. The speed at the end of 0.50 s is given by Equation (3-15):

$$v = a_t t + v_0 = 8.378 \text{ m/s}^2 (0.50 \text{ s}) + 0 = 4.189 \text{ m/s}$$

Substituting this speed into Equation (4-11) gives

$$a_r = \frac{v^2}{r} = \frac{(4.189 \text{ m/s})^2}{2 \text{ m}} = 8.773 \text{ m/s}^2$$

The total acceleration in this particular case is the vector sum of the radial and the tangential accelerations, which are perpendicular to each other:

$$\vec{a} = \vec{a}_r + \vec{a}_t$$

$$a = \sqrt{a_r^2 + a_t^2} = \sqrt{(8.773 \text{ m/s}^2)^2 + (8.378 \text{ m/s}^2)^2} = 12.1 \text{ m/s}^2$$

CHECKPOINT

C-4-5 Total Acceleration

An object under the influence of gravity and frictional forces is moving at a constant speed along the hemispherical bowl shown in Figure 4-14. The direction of the total acceleration is

(a) toward the centre of the circle;

(b) straight down;

(c) in the direction of the vector sum of the gravitational acceleration and the radial acceleration;

(d) impossible to determine without knowing the directions of the frictional forces.

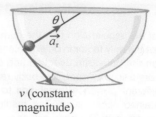

Figure 4-14 The object is moving in a bowl at a constant speed.

C-4-5 (a) since the object has constant speed the only acceleration it has is radial.

 MAKING CONNECTIONS

G-LOC, Grey-outs, and Red-outs

G-LOC, induced loss of consciousness caused by acceleration, was first reported toward the end of World War I. Pilots who underwent accelerations of approximately $4.5\ g$ often experienced G-LOC for about 20 s. During a greyout, the pilot is conscious but has partial loss of vision. When the body is accelerated upward, blood is forced away from the brain, resulting in reduced consciousness. Upward acceleration results from positive G-manoeuvres, such as the inside loop shown in Figure 4-15(a). Before the onset of G-LOC there are other symptoms, such as blurry vision, tunnel vision, and grey-outs. An outside loop (Figure 4-15(b)) produces $-z$-acceleration and can cause a red-out, which is a reddening in the visual field as excess blood is forced into the brain. Human tolerance to $-z$-acceleration is less than human tolerance to $+z$-acceleration. For high-speed aerobatic manoeuvres, pilots wear a G-suit, which is a suit that is designed to restrict blood flow from the head to the body. With a G-suit, the threshold for vertical acceleration during an inside loop is about $12\ g$, and $8\ g$ accelerations can be sustained over relatively long periods.

 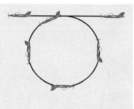

Figure 4-15 Acceleration during (a) an inside loop and (b) an outside loop

---

LO 4

4-4 Relative Motion in Two and Three Dimensions

A ferry crosses the English Channel from Calais, France, to Hastings, England, 110 km west-southwest. However, there is a strong flood tide flowing northeast. If the captain simply sets a course straight for Hastings, the ferry will be too far north when it reaches the English coast. To make the crossing in the least amount of time, the captain must adjust the course to allow for the tidal current. In Section 3-5, we introduced the concept of relative motion needed to deal with such problems. Since we developed the equations in vector form, we can readily adapt them to relative motion in two and three dimensions.

Equation (3-23) was written for the relative position along the x-axis, but we can also apply it to the components of the relative position of the two objects in three dimensions:

$$\vec{x}_{12} = \vec{x}_{1\mathrm{g}} - \vec{x}_{2\mathrm{g}} \qquad (4\text{-}12\mathrm{a})$$

$$\vec{y}_{12} = \vec{y}_{1\mathrm{g}} - \vec{y}_{2\mathrm{g}} \qquad (4\text{-}12\mathrm{b})$$

$$\vec{z}_{12} = \vec{z}_{1\mathrm{g}} - \vec{z}_{2\mathrm{g}} \qquad (4\text{-}12\mathrm{c})$$

Adding the three equations above gives

$$\vec{r}_{12} = \vec{r}_{1\mathrm{g}} - \vec{r}_{2\mathrm{g}} \qquad (4\text{-}13\mathrm{a})$$

$$\vec{r}_{12} = x_{12}\hat{i} + y_{12}\hat{j} + z_{12}\hat{k} \qquad (4\text{-}13\mathrm{b})$$

Similarly, the formulas for relative velocity and relative acceleration for two dimensions are

$$\vec{v}_{12} = \vec{v}_{1\mathrm{g}} - \vec{v}_{2\mathrm{g}} \qquad (4\text{-}14)$$

$$\vec{a}_{12} = \vec{a}_{1\mathrm{g}} - \vec{a}_{2\mathrm{g}} \qquad (4\text{-}15)$$

and are easily extended to motion in three dimensions.

EXAMPLE 4-6

Flight Plan

A plane is scheduled to fly from Casablanca, Morocco, to Lisbon, Portugal, 581 km [13° W of N]. A Poniente wind is blowing from Gibraltar at 34 km/h [15° N of W]. When the plane's airspeed is 560 km/h, and the pilot flies on a direct bearing to Lisbon, how long does the flight take?

SOLUTION

If the pilot were to point the plane along the straight line from Casablanca to Lisbon, the wind would push the plane off course to the west. To compensate, the pilot needs to point the plane somewhat to the east of the direct heading to Lisbon.

Applying Equation (4-14) to find the velocity of the plane (p) with respect to the wind (w) gives

$$\vec{v}_{pw} = \vec{v}_{pg} - \vec{v}_{wg}$$

$$\vec{v}_{pg} = \vec{v}_{pw} + \vec{v}_{wg}$$

Thus, the velocity of the plane with respect to the ground equals the sum of the velocity of the plane in the air and the velocity of the air with respect to the ground (i.e., the wind velocity). Let us construct a diagram of these velocities (Figure 4-16).

Although we do not know yet in which direction the plane should be oriented, we do know that the sum of the velocities of the plane and the wind point along the line connecting the two cities. Since the wind speed is small compared to the airspeed of the plane, the airspeed and the speed of the plane with respect to ground will not be very different, as we can see from our scale diagram.

We label the angles between the velocities α, β, and γ as shown. In Figure 3-16, v_{pg} makes an angle of 13° with the vertical, and v_{wg} makes an angle of 15° with the horizontal. Then,

$$\alpha = 90° - (13° + 15°) = 62°$$

Applying the sine law to the velocity triangle, we have

$$\frac{\sin\alpha}{v_{pw}} = \frac{\sin\beta}{v_{wg}}$$

$$\sin\beta = \frac{v_{wg}(\sin\alpha)}{v_{pw}} = \frac{34 \text{ km/hr } (\sin(62°))}{560 \text{ km/hr}} = 0.0536$$

$$\beta = 3.07°$$

Therefore,

$$\gamma = 180 - (\alpha + \beta) = 180° - (3.07° + 62°) = 114.93°$$

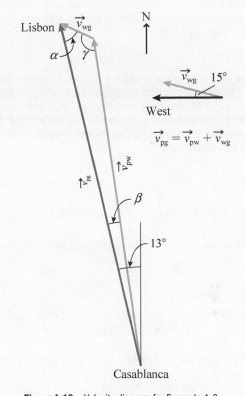

Figure 4-16 Velocity diagram for Example 4-6

To three significant digits the value is 115°.

Using the sine law once again, we have

$$\frac{\sin\alpha}{v_{pw}} = \frac{\sin\gamma}{v_{pg}} \Rightarrow v_{pg} = \frac{v_{pw}\sin\gamma}{\sin\alpha} = \frac{(560 \text{ km/hr}) \sin(114.93°)}{\sin(62°)}$$

$$= 575.16 \text{ km/hr}$$

The time it takes the plane to arrive at its destination is

$$t = \frac{581 \text{ km}}{575.14 \text{ km/h}} = 1.01 \text{ h}$$

Making sense of the result:

The plane's speed is very large compared to speed of the wind, so the small heading correction of 3.07° is reasonable.

EXAMPLE 4-7

Military Mud Ball

A French patrol on a cliff 50.0 m above the St. Lawrence River during the 1759 Battle for Québec spots a British scout boat leaving the seemingly undefended beach below.

Not wanting to give away their position by firing guns, the French soldiers prepare to launch a large mud ball from an improvised catapult that operates almost silently. The catapult launches mud balls at a velocity of 35.0 m/s at an

(continued)

angle of 41.0° above the horizontal. The British scout boat starts 60.0 m from the point directly below the French patrol, and has a speed of 4.00 m/s. The French take 5.00 s to aim the catapult. How much longer should they wait before releasing the catapult?

SOLUTION

Figure 4-17 shows the trajectory needed to hit the British boat. We know the launch velocity and the change in height, but we need to determine how long the mud ball will be in the air. This transit time is determined by the y-component of the motion. Choosing up as the positive direction, we apply Equation (4-8):

$$\Delta \vec{y} = \frac{1}{2}\vec{a}t^2 + \vec{v}_{0y}t$$

$$-50\hat{j}\ \text{m} = \frac{1}{2}(-9.81\hat{j}\ \text{m/s}^2)t^2 + (35\sin(41°))\hat{j}\ \text{m/s}\ (t)$$

$$0 = \frac{1}{2}(9.81\ \text{m/s}^2)t^2 - (35\sin(41°))\ \text{m/s}\ (t) - 50$$

$$t = \frac{-b \pm \sqrt{b^2 - 4ac}}{2a}$$

$$= \frac{35\sin(41°)\ \text{m/s} \pm \sqrt{(35\sin(41°)\ \text{m/s})^2 + 4(4.905\ \text{m/s}^2)(50\ \text{m})}}{2(4.905\ \text{m/s}^2)}$$

The two roots are $t = 6.2995$ s and $t = -1.6182$ s. We take the positive root.

The distance that the mud ball travels in the x-direction is given by Equation (4-7):

$$x = v_x t = v\cos\theta t = 35\cos(41°)\ \text{m/s}\ (6.2995\ \text{s}) = 166.40\ \text{m}$$

Initially, the British soldiers are 60 m from the cliff. By the time the catapult is ready to fire, the additional distance covered by the British soldiers is

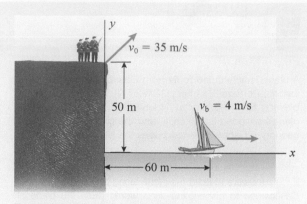

Figure 4-17 Schematic for Example 4-7

$$x = v_x t = 4\ \text{m/s}\ (5\ \text{s}) = 20.0\ \text{m}$$

Thus, they are a distance of $60\ \text{m} + 20\ \text{m} = 80.0\ \text{m}$ from the cliff by the time the French are ready to fire. In the time the mud ball takes to reach the river, the British will travel an additional distance of

$$x = 6.2995\ \text{s}\ (4\ \text{m/s}) = 25.198\ \text{m}$$

Now the British boat will be a total horizontal distance of $80\ \text{m} + 25.198\ \text{m} = 105.198\ \text{m}$ from the cliff. If the French fire when ready, the mud ball will overshoot the boat by $166.40\ \text{m} - 105.198\ \text{m} = 61.203\ \text{m}$. Before releasing the catapult, the French should wait for

$$t = \frac{61.203\ \text{m}}{4\ \text{m/s}} = 15.3\ \text{s}$$

Making sense of the result:

If the catapult fires only at the given angle then the French Patrol has to wait. The final results are reported to three significant figures. However, in the intermediate values needed to obtain the final answer, more significant figures are kept.

EXAMPLE 4-8

Relative Speed

Two children launch marbles from spring-loaded toy cannons placed side by side. The slower of the two cannons fires a marble at a velocity of 17.0 m/s at an angle of 50.0° above the horizontal; the other cannon fires a marble at a velocity of 21.0 m/s at an angle of 45.0° above the horizontal. Find the relative speed of the two marbles 1.70 s into the flight.

SOLUTION

We can find the relative speed from the components of the relative velocity. The relative velocity is given by Equation (4-14):

$$\vec{v}_{12} = \vec{v}_{1g} - \vec{v}_{2g}$$

In the x-direction,

$$\vec{v}_{12,x} = \vec{v}_{1g,x} - \vec{v}_{2g,x}$$

Taking the x-direction of motion of the faster marble to be positive,

$$\vec{v}_{12,x} = 21\cos(45°)\hat{i}\ \text{m/s} - 17\cos(50°)\hat{i}\ \text{m/s} = 3.922\hat{i}\ \text{m/s}$$

In the y-direction, the vertical relative velocity between the two marbles remains constant because both marbles undergo the same acceleration due to gravity. Thus, we can find the vertical relative velocity from the vertical components of the initial velocities:

$$\vec{v}_{12,y} = \vec{v}_{1g,y,0} - \vec{v}_{2g,y,0}$$

Choosing up to be positive,

$$v_{12y} = 21\sin(45°)\ \text{m/s} - 17\sin(50°)\ \text{m/s} = 1.826\ \text{m/s}$$

Therefore, the relative speed is given by

$$v_{12} = \sqrt{(v_{12,x})^2 + (v_{12,y})^2} = \sqrt{(3.922\ \text{m/s})^2 + (1.826\ \text{m/s})^2}$$

$$= 4.33\ \text{m/s}^2$$

Making sense of the result:

The relative speed between the two marbles will be constant, since the acceleration contribution is the same for both.

EXAMPLE 4-9

Torpedo and Boat

In a naval tactics competition, a contestant is launching a torpedo at another contestant's boat. The boat and torpedo launch from positions 70.0 m apart on a straight shore that runs east to west, with the torpedo launched east of the boat. The boat starts from rest with a constant acceleration of 0.60 m/s² in a direction 21.0° west of north. At the same time, the torpedo is launched at an angle of 31.0° to the shore with an initial speed of 3.00 m/s. What constant acceleration must the torpedo have to hit the boat?

SOLUTION

We will break up the motion into x- and y-components and choose an x-axis parallel to the shore, as shown in Figure 4-18. The distance between the torpedo (designated t) and the boat (designated b) is given by Equation (4-13):

$$\vec{r}_{tb} = \vec{r}_t - \vec{r}_b \tag{1}$$

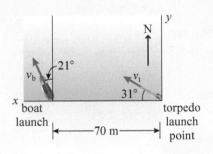

Figure 4-18 Paths of the torpedo and boat in Example 4-9

The x-component is given by Equation (4-12a):

$$\vec{x}_{tb} = \vec{x}_t - \vec{x}_b \tag{2}$$

From Equation (3-25) we can express this component as

$$\vec{x}_{tb}(t) = \frac{1}{2}\vec{a}_{tb,x}t^2 + \vec{v}_{tb,0,x}t + \vec{x}_{tb,0} \tag{3}$$

We choose west as the positive direction. The boat is 70 m west of the torpedo at the time of launch. Therefore, $x_{0tb} = -70$ m. The x-component of the relative velocity is

$$\vec{v}_{tb,0,x} = \vec{v}_{t,0,x} - \vec{v}_{b,0,x} \tag{4}$$

The initial speed of the boat is zero. Since the torpedo is headed in a westerly direction, the x-component of its velocity is positive and is equal to $v_{t,0,x} = v_{t,0}\cos(31°)$, as we can see from Figure 4-18.

$$\vec{v}_{tb,0,x} = (v_{t,0,x} - v_{b,0,x})\hat{i} = v_{t,0}\cos(31°)\hat{i} - v_{b,0}\sin(21°)\hat{i}$$

$$= 3\cos(31°)\hat{i}\text{ m/s} - 0 = 2.572\hat{i}\text{ m/s} \tag{5}$$

The acceleration in the x-direction is given by

$$\vec{a}_{tb,x} = \vec{a}_{t,x} - \vec{a}_{b,x} \tag{6}$$

Since both accelerations point west, their x-components will be positive and

$$\vec{a}_{tb,x} = a_t\cos(31°)\hat{i} - 0.6\sin(21°)\hat{i} \tag{7}$$

Substituting (5) and (7) into (3), we have

$$x_{tb} = \frac{1}{2}(a_t\cos(31°) - 0.215\text{ m/s}^2)t^2 + 2.572t - 70 = 0 \tag{8}$$

In the y-direction, we can apply Equation (4-12b) and Equation (3-25):

$$\vec{y}_{tb} = \vec{y}_t - \vec{y}_b \tag{9}$$

$$\vec{y}_{tb}(t) = \frac{1}{2}\vec{a}_{tb,y}t^2 + \vec{v}_{tb,0,y}t + \vec{y}_{tb,0} \tag{10}$$

Similarly, using Figure 4-18, we can write

$$y_{tb} = \frac{1}{2}(a_t\sin(31°) - a_b\cos(21°))t^2$$

$$+ (v_{t,0}\sin(31°) - v_{b,0}\cos(21°))t + 0 = 0 \tag{11}$$

Equations (8) and (11) have two unknowns: the acceleration of the torpedo and the time it takes to reach the boat. We can find the acceleration in terms of the time from Equation (11) and substitute into Equation (8). Equation (11) is a quadratic equation, but the constant term (c) in the quadratic formula) is zero, so there is only one root:

$$a_t = -2\frac{(v_{t,0}\sin(31°) - v_{b,0}\cos(21°))}{t\sin(31°)} + a_b\frac{\cos(21°)}{\sin(31°)}$$

$$= -\frac{2(3\text{ m/s}\sin(31°))}{t\sin(31°)} + \frac{0.6\text{ m/s}\cos(21°)}{\sin(31°)}$$

$$= -\frac{6\text{ m/s}}{t} + 1.088\text{ m/s}^2 \tag{12}$$

Substituting into Equation (8), we have

$$x_{tb} = \frac{1}{2}\left[\left(-\frac{6\text{ m/s}}{t} + 1.088\text{ m/s}^2\right)\cos(31°) - 0.215\text{ m/s}^2\right]t^2$$

$$+ 2.572\text{ m/s }t - 70\text{ m} = 0$$

$$x_{tb} = -2.572\text{ m/s }t + 0.466\text{ m/s}^2t^2 - 0.1075\text{ m/s}^2t^2$$

$$+ 2.572\text{ m/s }t - 70\text{ m} = 0$$

$$t = \sqrt{\frac{70\text{ m}}{0.3585\text{ m/s}^2}} = 13.97\text{ s}$$

Substituting this value for t into Equation (12), we get

$$a_t = -\frac{6\text{ m/s}}{13.97\text{ s}} + 1.088\text{ m/s}^2 = 0.658\text{ m/s}^2$$

Making sense of the result:

Both x and y components of the motion need to be accounted for as the torpedo approaches the boat. While the torpedo has the advantage of an initial speed on the boat, it has an extra 70 m to travel, which explains why it has to have a slightly higher acceleration than the boat.

FUNDAMENTAL CONCEPTS AND RELATIONSHIPS

Position, Velocity, Acceleration, and Time in Three Dimensions

Position:

$$\vec{r} = x\hat{i} + y\hat{j} + z\hat{k} \tag{4-1}$$

Displacement:

$$\Delta\vec{r} = \Delta x\hat{i} + \Delta y\hat{j} + \Delta z\hat{k} \tag{4-2}$$

Distance covered = path length
Average velocity:

$$\vec{v}_{avg} = \frac{\Delta\vec{r}}{\Delta t} = v_{avg,x}\hat{i} + v_{avg,y}\hat{j} + v_{avg,z}\hat{k} \tag{4-3}$$

Instantaneous velocity:

$$\vec{v} = \lim_{\Delta t \to 0} \frac{\Delta\vec{r}}{\Delta t} = \frac{d\vec{r}}{dt} = v_x\hat{i} + v_y\hat{j} + v_z\hat{k} \tag{4-4}$$

Average acceleration:

$$\vec{a}_{avg} = \frac{\Delta\vec{v}}{\Delta t} = a_{avg,x}\hat{i} + a_{avg,y}\hat{j} + a_{avg,z}\hat{k} \tag{4-5}$$

Instantaneous acceleration:

$$\vec{a} = \lim_{\Delta t \to 0} \frac{\Delta\vec{v}}{\Delta t} = \frac{d\vec{v}}{dt} = a_x\hat{i} + a_y\hat{j} + a_z\hat{k} \tag{4-6}$$

Projectile Motion

Horizontal distance for a projectile:

$$\Delta x = v_x t \tag{4-7}$$

Vertical displacement:

$$\Delta\vec{y} = \frac{1}{2}\vec{a}t^2 + \vec{v}_{0y}t \tag{4-8}$$

Vertical velocity:

$$\vec{v}_y(t) = \vec{a}t + \vec{v}_{0y} \tag{4-9}$$

Time-independent kinematics equation for motion in the vertical direction:

$$v_y^2 - v_{0y}^2 = 2\vec{a}\cdot\Delta\vec{y} \tag{4-10}$$

Circular Motion

Radial acceleration of an object moving in a circle of radius r at speed v:

$$a = \frac{v^2}{r} \tag{4-11}$$

Total acceleration of an object moving in a circle:

$$\vec{a} = \vec{a}_r + \vec{a}_t$$

Relative Motion

The relative position of object 1 as seen by object 2:

$$\vec{r}_{12} = \vec{r}_{1g} - \vec{r}_{2g} \tag{4-13}$$

Relative velocity:

$$\vec{v}_{12} = \vec{v}_{1g} - \vec{v}_{2g} \tag{4-14}$$

Relative acceleration:

$$\vec{a}_{12} = \vec{a}_{1g} - \vec{a}_{2g} \tag{4-15}$$

Applications: projectile motion, centrifuges, race car acceleration, effects of acceleration on the human body, military tactics

Key Terms: nonuniform circular motion, radial acceleration, tangential acceleration, total acceleration, uniform circular motion

QUESTIONS

1. A projectile is launched horizontally from a cliff at a very high speed with a horizontal acceleration that slows the projectile down but decreases with time. Which trajectory in Figure 4-19 best describes the motion of the projectile?

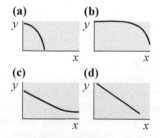

Figure 4-19 Question 1

2. A missile is launched horizontally with a low speed. The missile's engine accelerates it in the horizontal direction only. The acceleration provided by the engine increases with time. Which path in Figure 4-20 best describes its trajectory?

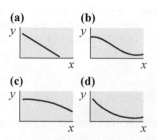

Figure 4-20 Question 2

3. You release a marble at the top edge of a bowl. The marble rolls down the inside of the bowl and continues up the opposite side. What is the direction of the marble's acceleration when the marble comes to a momentary stop at its highest point on the side of the bowl?
   (a) down
   (b) normal to the bowl
   (c) tangent to the bowl
   (d) The acceleration is zero, so it has no direction.

4. A rocket is launched horizontally off a cliff. The rocket engine provides a constant horizontal acceleration with a magnitude greater than the magnitude of the acceleration due to gravity. Which path in Figure 4-21 best corresponds to the trajectory of the rocket?

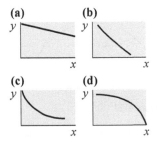

Figure 4-21 Question 4

5. A missile is launched from a plane travelling horizontally. The missile leaves the plane in a horizontal direction, with an acceleration that increases linearly with time. Which path in Figure 4-22 best describes its trajectory?

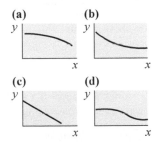

Figure 4-22 Question 5

6. You intend to cross a river that flows due south. Your speed with respect to the water is twice the speed of the river current. In which direction relative to the current should you swim to cross in the shortest time?
   (a) at 30° to the river
   (b) at 60° to the river
   (c) at 45° to the river
   (d) perpendicular to the river

7. You have a battery-powered model boat that has a fixed rudder so that it always moves straight ahead. You launch the boat in a pond, aiming straight at a point 60 m away on the opposite side of the pond. The boat reaches the shore 12 m north of the target point because of a current in the pond. On the next run, you aim the boat 12 m south of your target point. The boat will
   (a) arrive at the target;
   (b) arrive north of the target;
   (c) arrive south of the target.
   (d) Cannot answer without more information.

8. Two objects, A and B, travelling along the same line have speeds v_A and v_B, respectively. The speed of object A with respect to object B is
   (a) $v_B - v_A$;
   (b) $v_A - v_B$;
   (c) $v_A + v_B$.
   (d) Could be any of the above.

9. Two objects move along the same circular trajectory with the same speed. They start at the same time, one on the x-axis and the other on the y-axis. When the objects move in the same direction around the circle,
   (a) their relative position vector will always point in the same direction;
   (b) their relative acceleration vector will not change;
   (c) the magnitude of their relative acceleration will always equal $\frac{v^2}{r}$.
   (d) None of the above.

10. An object falls straight down while another object moves in a vertical circular trajectory at a speed that is not necessarily constant. The acceleration vector of the free-falling object can be equal to
    (a) the radial acceleration vector of the object in the circle;
    (b) the linear acceleration vector of the object in the circle;
    (c) the total acceleration vector of the object in the circle.
    (d) Any of the above.

11. You slide a marble along your kitchen counter toward a friend who stands 1 m from the end of the counter. Your friend is holding a marble in her hand at the height of the counter. She lets go of her marble just as your marble leaves the countertop. The speed of your marble is such that it will travel 1 m horizontally before your friend's marble hits the floor, as shown in Figure 4-23. Which of the following statements is true?
    (a) The two marbles will never collide.
    (b) Your marble will pass over your friend's marble.
    (c) Your marble will pass below your friend's marble.
    (d) The marbles will collide before they hit the floor.

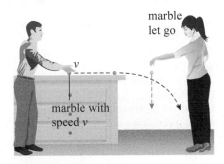

Figure 4-23 Question 11

12. A projectile is fired at an angle of 30° with an initial speed of 340 m/s. It lands at the same elevation from which it was fired. While in the air,
    (a) the projectile's average acceleration is zero;
    (b) the projectile's average acceleration is 9.81 m/s²;
    (c) the projectile's average velocity in the y-direction is zero.
    (d) Both (b) and (c) are true.

13. An object completes 2.25 revolutions around a circle of radius $4/\pi$ m. Which of the following statements is true?
    (a) The distance covered by the object is 4 m.
    (b) The displacement of the object is $A = \dfrac{2\sqrt{2}}{\pi}$.
    (c) Both (a) and (b) are correct.
    (d) None of the above are true.

14. An object travelling at a constant speed takes 3 s to complete three full cycles around a circle of radius 1 m. Which of the following statements is true?
    (a) The object's average velocity is zero.
    (b) The object's average speed is 2π m/s.
    (c) The object's average acceleration is $4\pi^2$ m/s².
    (d) Both (a) and (b) are true.

15. A sharpshooter has a rifle that fires bullets at a speed of 1100 m/s. A trainer 100 m away throws an apple 30 m straight up in the air. The sharpshooter tracks the apple and fires directly at it when it is at its maximum height. Which of the following statements is true?
    (a) The bullet will hit the apple at the apple's maximum height.
    (b) The bullet will pass under the apple on its way down.
    (c) The bullet will pass over the apple on its way down.
    (d) The bullet will hit the apple on its way down.

16. Two archers aim for the same coconut on a tree 14 m above the ground. One archer stands 50 m away and shoots an arrow at 60 m/s. The other archer stands 70 m away and shoots an arrow at 87 m/s. They both aim directly at the coconut and release their arrows at the moment the coconut begins to fall from the tree. Which of the following statements is true?
    (a) The two arrows will hit the coconut at the same time.
    (b) Neither arrow will hit the coconut.
    (c) The faster arrow will hit the coconut but not the slower one.
    (d) Both arrows will hit the coconut but at different times.

17. Two cannons, side by side on a cliff, fire projectiles at the same speed, one horizontal and one at an angle of 30° above the horizontal. Which of the following statements is true?
    (a) The vertical distance between the two projectiles is $v\sin(30°)t$.
    (b) The vertical distance between the two projectiles is $v\sin(30°)t - 4.9t^2$.
    (c) The horizontal distance between the two projectiles is constant.
    (d) Both (a) and (c) are correct.

18. NASA's 20 g centrifuge is made to spin at 17 g. Which of the following statements is true?
    (a) The acceleration is constant.
    (b) The magnitude of the acceleration is constant.
    (c) The velocity is constant.
    (d) None of the above statements are true.

19. If you throw a tennis ball straight up (relative to yourself) while running at a constant speed, the ball will land
    (a) in your hand;
    (b) slightly ahead of you;
    (c) slightly behind you.
    (d) I cannot tell.

20. While a projectile is in the air, can its velocity and acceleration vectors ever be parallel or perpendicular? Explain.

21. One of Robin Hood's men shoots an arrow horizontally at a target, with a speed of 75 m/s. At the same time, the shooter drops an apple he is holding while shooting the arrow. Ignoring air resistance, which object will hit the ground first? Explain.

22. A projectile is launched at speed v, making an angle θ with the horizontal. What is the projectile's acceleration
    (a) right after the launch;
    (b) at the top of its trajectory?

23. A bead sits on top of an overturned hemispherical bowl and begins to slide down, as shown in Figure 4-24. What are the directions of the tangential acceleration and radial acceleration of the bead when the bead's height changes by one-quarter of the radius of the bowl?

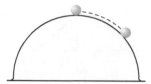

Figure 4-24   Question 23

24. An astronaut gets bored while on an extravehicular activity outside a space station. She happens to have a piece of putty, which she throws at the station, with a speed of 20 m/s. The putty covers a distance of 50 m and sticks to the wall of the station. The recoil of the station is negligible. Which of the following is true?
    (a) The average speed of the putty is 10 m/s.
    (b) The average acceleration of the putty is 2 m/s².
    (c) The average acceleration of the putty is zero.
    (d) None of the above are true.

PROBLEMS BY SECTION

For problems, star ratings will be used, (✷, ✷✷, or ✷✷✷), with more stars meaning more challenging problems.

Section 4-1   Displacement, Velocity, and Acceleration in Two and Three Dimensions

25. ✷   A fireman hears the station alarm, runs 3.0 m straight to the fire station pole in 1.7 s, and descends 4.0 m down the pole in 2.0 s.
    (a) Find the magnitude of the fireman's overall average velocity.
    (b) Find the fireman's overall average speed.
    (c) Find the fireman's speed during the run.
    (d) Find the fireman's speed during the descent down the pole.
    (e) Can the quantities from (c) and (d) be considered components of the fireman's overall average velocity? Explain why or why not.

26. ✷✷   A cheetah chases a gazelle at a speed of 112 km/h for 3.00 s, then turns 42° from its original direction, runs at 78.0 m/s for 2.00 s, and finally tackles the gazelle, coming to a stop in 11.0 m.
    (a) Find the cheetah's average speed.
    (b) Find the cheetah's average velocity.
    (c) Find the cheetah's average acceleration.
    (d) How long does it take the cheetah to stop in the last 11.0 m? Assume constant acceleration.

27. ⁎⁎ A particle has the coordinates $x = r\cos(8t)$ and $y = r\sin(8t)$.
(a) Find the x- and y-components of the velocity of this particle at $t = 0.20$ s.
(b) Find the velocity at $t = 0.20$ s.
(c) Describe the trajectory of the particle. Express the position, velocity, and acceleration of the particle as a function of time using Cartesian notation.

28. ⁎⁎ A particle has coordinates given by $x = 11\cos(4t)$, $y = 11\sin(4t)$, and $z = 0.7t$.
(a) Describe the trajectory of the particle.
(b) Find the speed of the particle at $t = 4.0$ s.
(c) Express the particle's position, velocity, and acceleration as a function of time, using Cartesian notation.
(d) When will the acceleration vector make an angle of 30° with the x-axis? Is that time affected by the z-motion? Explain.

29. ⁎⁎ A particle has coordinates given by $x = 8t$ and $y = \frac{1}{16t}$. Describe the trajectory of the particle, and find the particle's speed and acceleration at $t = 3$ s.

30. ⁎⁎ A particle has coordinates given by $x = 20\cos(5t)$, $y = 5\sin(0.5t)$, and $z = 7t$.
(a) Express the particle's position, velocity, and acceleration as a function of time using Cartesian notation.
(b) What is the speed of the particle at $t = 0$?
(c) Find the velocity of the particle at (i) $t = \pi$ s, (ii) $t = 2\pi$ s, and (iii) $t = 3\pi$ s.
(d) When will the velocity make an angle of 45° with the positive x-axis? Does that happen only once? Explain, and describe the trajectory of the particle.

31. ⁎⁎ Starting from rest, a race car moves straight up an incline with a 37.0° angle at a constant acceleration for 5.00 s such that its speed reaches 23.0 m/s. The road becomes horizontal, and the car slows down with constant acceleration as it follows a circular curve that changes its direction by 90°. The radius of the curve is 121 m, and the car is moving at a speed of 7.00 m/s at the end of the curve.
(a) Find the total displacement of the race car.
(b) Find the average velocity of the race car from the start until the end of the circular curve.
(c) Find the average acceleration of the race car during this trip.

Section 4-2 Projectile Motion

32. ⁎ A projectile is fired from and lands at ground level 43.0 km away 50.0 s later. Find the projectile's launch speed, minimum speed, angle of launch, average speed, average acceleration while in the air, and average velocity.

33. ⁎ A child flings a marble such that it leaves the horizontal surface of a table that is 70 cm high. The marble lands with a speed of 7.0 m/s. How fast did the child fling the marble?

34. ⁎ A particle is launched at an initial speed v making an angle θ above the horizontal. Prove that the trajectory of the particle is a parabola.

35. ⁎⁎ A stunt car driver drives off a 7.0 m-high cliff into a lake with a horizontal speed of 12 m/s. The car needs to clear an 8.0 m-long ledge that is 5.0 m below the edge of the cliff, as shown in Figure 4-25. The car just misses the ledge on the way down.
(a) With what speed does the car hit the water?

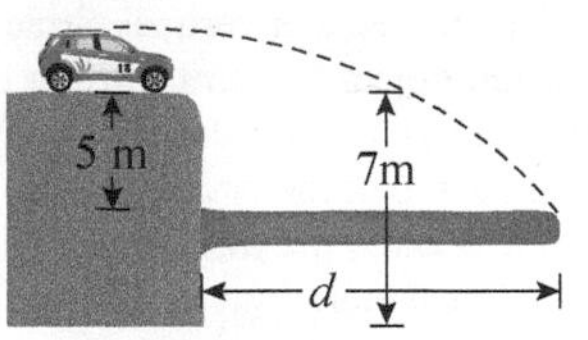

Figure 4-25 Problem 35

Section 4-3 Circular Motion

36. ⁎⁎ If Earth had no atmosphere, what would be the speed of a satellite in a circular orbit that just clears the 8848 m summit of Mount Everest?

37. ⁎⁎ The distance between the cities of Libreville, Gabon, and Kampala, Uganda, is 2570 km. Both cities are almost at the equator. The pilot requires passengers stop their clocks during the flight.
(a) What is the time zone difference between the two cities?
(b) You board a plane that takes off from Kampala toward Libreville. How fast does the plane have to travel so you do not have to adjust your clock when you arrive?

38. ⁎⁎ The distance between São Tomé off the coast of Africa and the city of Quito, Ecuador, is approximately 9500 km. Both cities are almost on the equator.
(a) How long does it take Earth to rotate to cover the distance between the two cities?
(b) You board a plane in São Tomé and fly west to Quito at a speed of 970 km/h, landing 12 h later. By how many hours do you need to adjust your watch? What is your radial acceleration when the plane's altitude is 10 000 m?

39. ⁎ Pilots can sustain accelerations as high as 13 g during an inside loop manoeuvre. When a plane is moving at a speed of 830 km/h, what is the minimum radius for an inside loop turn the plane can fly without its acceleration exceeding 13 g?

40. ⁎ Calculate the radial acceleration of a point on the Sun's equator.

41. ⁎⁎ A bead on top of an overturned hemispherical bowl of radius 32 cm starts sliding down the side of the bowl, as shown in Figure 4-24. By the time the bead has moved by 30°, its speed is 0.92 m/s. Find the bead's tangential acceleration, radial acceleration, and total acceleration.

42. ⁎ At the Istanbul Grand Prix, drivers sustain lateral accelerations (sideways) about five times the acceleration due to gravity while driving around curves at speeds of 280 km/h. Assuming a circular curve, find the radius of that section of the track.

43. ⁎ An ultracentrifuge produces accelerations that are 2 million times the acceleration due to gravity. Find the speed of rotation for the ultracentrifuge, which has a radius of 4.5 cm.

44. ⁎ The P2 centrifuge, used to enrich uranium, has a diameter of approximately 15 cm, a length of 1.0 m, and a peripheral speed of approximately 500 m/s.
(a) What is the acceleration of a point on the circumference of a P2 centrifuge?
(b) What is the acceleration of a point midway between the axis and the circumference of a P2 centrifuge?
(c) How many turns does the P2 centrifuge complete in 1 s?

MECHANICS

45. ✱✱ An object moves in a vertical circle of radius 4.0 m. The speed of the object is 12 m/s when the object is 30° above the lowest point along the circle, and 8.0 m/s when the object is at the top of the circle. The tangential acceleration is uniform. What is the total acceleration of the object when its speed is 8.0 m/s?

46. ✱✱ A particle moves along a circular path of radius r at a constant speed v. The particle is at the angular position ϕ above the positive x-axis, as shown in Figure 4-26. Without using calculus, write expressions for
(a) the projection of the particle's position onto the x-axis as a function of time;
(b) the velocity of the projection as a function of time;
(c) the acceleration of the projection as a function of time.

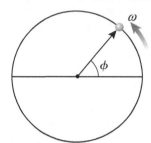

Figure 4-26 Problem 46

47. ✱✱ A steel ball that is accelerated on a circular track of radius 72 cm reaches a speed of 4.0 m/s over three circuits, starting from rest. How long after the ball starts to move does its radial acceleration equal its tangential acceleration? How many turns does the ball take for that to happen? Assume constant tangential acceleration.

48. ✱✱✱ An object moves on a circular track of radius 29 m such that the object's radial acceleration doubles every 2.0 s. The object completes its first circuit in 4.0 s. The object starts from rest, at which point its radial acceleration is zero.
(a) Write the equation for the object's speed as a function of time.
(b) What is the total acceleration of the object 6.0 s into the motion?
(c) What is the distance covered as a function of time?

Section 4-4 Relative Motion in Two and Three Dimensions

49. ✱✱ A parade car moving at 10 km/h shoots an acrobat out of a cannon that is mounted at 90° to Earth's surface. The cannon fires the acrobat with a speed of 14 m/s. Where will the acrobat land relative to the parade car? Find the magnitude of the displacement of the acrobat relative to the ground. Ignore air resistance.

50. ✱✱ A child has a cannon mounted on a battery-powered car. While the car is moving at a speed of 2.00 m/s, the cannon fires a marble with a speed of 7.00 m/s with respect to the car, at an angle of 27.0° above the horizontal. The car's speed does not change noticeably.
(a) What angle does the marble make with the horizontal when it lands on the ground?
(b) How far will the marble land from the position from which it was launched?
(c) How far is the marble from the car when it lands?

51. ✱✱ A swimmer swims in a river against the current for 5 min, and then swims with the current for another 5 min, and ends up 500 m away from where he started. Find the speed of the current in the river if the swimmer's speed relative to the water is 2.6 m/s.

52. ✱✱✱ Two cannons at the same location launch balls at a speed of 450 m/s at the same time. One cannon is aimed east and fires its ball at an angle of 31° above the horizontal, and the other is aimed north and fires its ball at an angle of 21° above the horizontal.
(a) Find the relative speed of the two cannonballs 2.0 s and 5.0 s after firing.
(b) Find the distance between the two cannonballs 3.0 s and 23.0 s after firing.

53. ✱✱ A director wants to produce a scene in which an archer shoots a watermelon that is dropped from a 30 m-high bridge. To compensate for gravity, the director asks the archer to aim 5.0 m above the watermelon, which is released at the same time that the archer releases his arrow. The initial speed of the arrow is 160 m/s, and the distance to the watermelon is 90 m. How close will the arrow get to the watermelon?

54. ✱✱ A child has mounted a spring-loaded cannon on a toy car such that the cannon fires backward from the car at an angle of 42° above the horizontal. While the car is moving forward at 1.1 m/s, the cannon fires a marble with speed v with respect to the car. The marble lands at an angle of 89° with the level ground. What is the speed of the marble with respect to the toy car?

55. ✱✱ A parade car has a cannon that shoots teddy bears into the crowd. As the car moves forward at 2.7 m/s, the cannon fires horizontally at an angle of 37° from the direction of motion of the car. The speed of the teddy bears with respect to the cannon is 6.0 m/s, the cannon is 3.0 m above the ground, and the recoil of the cannon is negligible. Find the landing speed of the teddy bear with respect to the ground.

56. ✱✱✱ During a clay pigeon shooting match, a clay pigeon is fired with an initial speed of 82 km/h at an angle of 39° above the horizontal. Then 1.7 s later, the shooter fires a shot that has a speed of 404 m/s. When the shooter is standing straight behind the clay pigeon launcher, at what angle should she aim?

57. ✱✱✱ You notice an eagle soaring at an altitude of 25 m, eyeing a rabbit. You want to scare the eagle away, so 1.0 s after the eagle flies directly overhead at 34 km/h, you throw a stone at an angle of 38° above the horizontal. You want the stone to pass 50 cm in front of the eagle. With what speed should you throw the stone? Is this doable at all? Assume your hand to be 2.0 m above the ground when you throw the stone.

58. ✱✱✱ (Graphical solution) Your opponent is operating a model helicopter in a simulated battle. You have a gun that fires paintballs at a speed of 40 m/s. The helicopter is hovering 12 m above the ground and is 17 m from you. Just before you fire, the helicopter dives at twice the acceleration due to gravity. You take 0.2 s to correct your aim. What angle must your gun make with the horizontal to hit the helicopter on its way down?

59. ✱✱✱ A bird flies 25 m above a hunter at a speed of 39 km/h. The hunter's gun fires a shot with a speed of 378 m/s. The hunter waits 4.0 s to judge the speed of the bird before firing. At what angle above the horizontal should the hunter aim?

COMPREHENSIVE PROBLEMS

60. ✷   A spinning neutron star with a radius of 16 km completes 712 rev/s.
    (a) Find the average speed of a point on the star's equator over one-third of a revolution.
    (b) Find the average acceleration of a point on the star's circumference over three-quarters of a revolution.
    (c) Find the distance covered by a point on the star's equator in 1 s.
    (d) Find the displacement of the point in part (c).

61. ✷✷   A monkey is 14 m above the ground in a tree and drops a papaya. At the same time, two other monkeys, one 11 m from the base of the tree and the other 16 m from the base of the tree, throw small stones at the papaya such that all three objects collide 0.9 s before the papaya hits the ground. Find the speed of each stone as it hits the papaya. Assume both stones were thrown from ground level.

62. ✷✷✷   Starting from rest, a particle moves under constant acceleration in a horizontal line such that its speed 3.00 s later is 1900 m/s. The particle is deflected by 37.0° in the horizontal plane without changing its speed. In this new direction, the particle accelerates such that it reduces its speed by half in 1.00 s. The particle is then deflected straight up, and immediately accelerates uniformly such that its velocity after 3.00 s is 9000 m/s straight up. For the whole 7.00 s interval, find the particle's total displacement, average velocity, average speed, and average acceleration. Express these quantities using magnitude and angles measured with respect to the positive coordinate axes.

63. ✷✷   A missile is launched with an initial speed of v, making an angle θ above the horizontal. Radar tracking shows that the altitude of the missile is given by $60t^2 - 7t^3 + 4t + 120$, and the horizontal distance from the point of observation is given by $1400 + 3t$.
    (a) Determine the maximum height reached by the missile.
    (b) Find the speed and horizontal displacement of the missile just before it hits the ground 120 m below the point of launch.
    (c) Where does the maximum acceleration of the missile occur?

64. ✷✷   A sounding rocket has an engine that allows it to maintain a constant horizontal speed of 190 m/s while providing it with a constant vertical acceleration of 32 m/s² until it reaches a height of 60 km. The engine then shuts off, and the rocket continues under the influence of just the gravitational force. Assume that air resistance and the change in gravitational acceleration with altitude are negligible.
    (a) What is the maximum height reached by the rocket, and how long does it take to reach that height?
    (b) What angle does its velocity make with the horizontal just before the landing?

65. ✷✷   A projectile is fired at speed 140 m/s such that the projection of its velocity onto the xy-plane makes an angle of 40° with the positive x-axis. The initial velocity vector makes an angle of 29° with the positive z-axis. Express the x-, y-, and z-coordinates of the projectile as a function of time, taking the launch point as the origin (0, 0, 0).

66. ✷✷   A fireworks rocket is fired horizontally from a height of 17 m above the ground. The rocket has a horizontal acceleration that is twice as large as the acceleration due to gravity.

    (a) What trajectory will the rocket follow?
    (b) What will its speed be just before it hits the ground?
    (c) Find the horizontal distance between the launch and landing points.

67. ✷   A child fires a tomato from an improvised catapult from a tree house 4.00 m above the ground. The tomato leaves the catapult at a speed of 35.0 m/s, making an angle of 37° with the horizontal.
    (a) Find the maximum height reached by the tomato.
    (b) How far from the tree house will the tomato land?
    (c) How fast is the tomato moving just before it lands?

68. ✷✷   A plane is approaching a remote community in northern Manitoba to drop a crate of supplies when clear turbulence buffets the plane, pushing it suddenly upward. The crate tears loose at an altitude of 160 m, leaving the plane with a velocity of 290 km/h directed at an angle of 23° above the horizontal. What is the speed of the crate just before it lands?

69. ✷✷   The location of Toronto's Lester B. Pearson International Airport is 43°40′ N latitude and 79°36′ W longitude, and the Beijing Capital International Airport is situated at 40°44′ N and 116°36′ E. A plane takes you from Toronto to Beijing in 11.0 h.
    (a) How fast is the plane?
    (b) By how much do you have to adjust your watch?

70. ✷   How fast would a plane have to fly at an altitude of 13 000 m above the equator so that passengers would not have to adjust their watches? Would the plane be heading east or west? The pilot requires batteries be removed from all watches during the flight.

71. ✷✷   You are onboard a plane flying west at an altitude of 600 m above the 32nd parallel. You fly over a given point at 12:00 p.m. local time, and arrive at an airport seven time zones away such that you only have to adjust your watch by 1 h. Estimate the airspeed of the plane. Your watch was stopped during the entire flight.

72. ✷   Express y as a function of x for a projectile that is launched with speed v_0 at an angle θ and lands at the same level from which it was launched. Derive expressions for the angle for which the maximum distance is covered, the distance at which the maximum height occurs for that angle, and the maximum height.

73. ✷✷   A particle's position is given as $(2 + (0.5)\sin(0.4t))\hat{i} + (3 + (0.5)\sin(0.4t))\hat{j} + 2t^2\hat{k}$.
    (a) What is the position of the particle at $t = 0$?
    (b) Find the speed of the particle at $t = 0.20$ s.
    (c) Find the average velocity of the particle between $t = 0.20$ s and $t = 0.4$ s.
    (d) Find the average acceleration of the particle between $t = 0.2$ s and $t = 0.4$ s.
    (e) Find the displacement of the particle between $t = 0$ s and $t = 1.1$ s.
    (f) What is the trajectory of the particle?

74. ✷✷   Two disks of radius 32 cm are placed side by side, and the two closest points on the disks are painted red. The disks are then made to spin about fixed axes in opposite directions at 10 rpm.
    (a) Write an expression for the relative acceleration of one red point with respect to the other.
    (b) Write expressions for the relative velocity and relative speed of the two points.
    (c) Given that the two points are in contact at the start, write an expression that describes their relative position.

75. ✷✷ A ring lies in a horizontal plane and spins about its symmetry axis at a rate of 3 rev/s. Find the position, acceleration, and velocity of a point A on the ring with respect to a point B that is in the same plane and is a distance $3r$ from A, given that the distance between A and B is $2r$ at $t = 0$.

76. ✷✷ A Boeing 747 is flying 790 km/h at an altitude of 20 km in a direction 11° [N of W]. A smaller jet at the same altitude is 12 km away 17° [N of E] from the Boeing. The velocity of the smaller jet is 430 km/h 23° [W of S]. Find the distance of closest approach between the planes and the time they take to reach that distance.

77. ✷✷ NASA reports that the position of a space shuttle during the first few minutes of its flight can be represented by the following equations:
altitude: $h(t) = 658.8 - 0.154t^3 + 18.3t^2 - 345t$
down-range distance: $R(T) = 1426.5e^{0.029t}$
Find the position, speed, and acceleration of the shuttle 1 min after takeoff. Coordinates are in m, and t is in s.

78. ✷✷ A test pilot is placed in a high-g training pod that follows a horizontal path of radius 20 m. The pod accelerates from rest to achieve an acceleration six times the acceleration due to gravity in the radial direction over five turns. The pod has constant tangential acceleration during these five turns. The pod then continues at a constant speed for 2.0 s, and then undergoes constant deceleration, coming to a stop in the next 6.0 s.
(a) How long does the pod take to finish the first five turns?
(b) What is the total acceleration of the pod 3.0 s after it begins to decelerate?
(c) What is the total number of turns the pod makes?
(d) Find the magnitude of the average acceleration between 3.0 s and 6.0 s.
(e) Find the magnitude of the average velocity between 4.0 s and 6.0 s.
(f) Find the magnitude of the displacement of the pod between $t = 2.0$ s and $t = 5.0$ s.

79. ✷✷ You are in the passenger seat of a car driving on a country road at 43 km/h. You throw an acorn out the window at an angle of 26° above the horizontal with a speed of 4.0 m/s relative to the car. The projection of the acorn's velocity onto the horizontal makes an angle of 34° with the velocity of the car.
(a) How far ahead of the car is the acorn 1 s later?
(b) What is the speed of the acorn in flight with respect to the ground?
(c) What angle with respect to the horizontal does the acorn's velocity make just before the acorn lands? What angle does the projection of the acorn's velocity make with the road?

80. ✷✷ Two monkeys standing 20 m apart are both trying to use stones to knock down a coconut hanging 7.0 m up in a tree that is directly between them. One monkey is 7.0 m from the tree, and the other is 13 m from the tree. They aim straight for the coconut, not accounting for the acceleration due to gravity. The close monkey throws a stone at speed v, and the far monkey throws a stone at $2v$. The stone thrown by the close monkey strikes the ground 3.0 s after it leaves her hand. By what vertical distance do the two stones miss each other?

81. ✷✷✷ In a robot demolition derby, your friend's robot, Sir Burnalot, is stuck and is being charged by an opponent's more powerful robot, Sir Crunchalot. You have a cannon that fires a heavy ball at a speed of 7.0 m/s. Crunchalot is moving at 1.1 m/s on course for a head-on collision with Burnalot 11 m in front of it. Your cannon is 4.0 m behind Burnalot, which is 50 cm tall. At what angle must you fire the ball to knock over the opposing robot before it gets to Burnalot?

82. ✷✷ A particle's position along the x-axis is given by $x = 20\cos(0.4t)$, and its position along the y-axis is given by $y = 20\sin(0.4t)$.
(a) Describe the trajectory of the particle.
(b) Find the time it takes the particle to finish a complete cycle. (This time is the period of the motion.)
(c) Find the tangential speed of the particle.
(d) Find the magnitude of the acceleration of the particle at a given time t.

83. ✷✷ Given that the particle in problem 82 is at $x = 20$ at $t = 0$, find
(a) the velocity of the particle at $t = 12$ s;
(b) the average velocity and the average acceleration of the particle between $t = 12$ s and $t = 20$ s.

84. ✷✷✷ You are playing an innovative fortress defence game. You build a 4.0 m-high fortress out of wood and cover the front wall with a thin layer of playdough. Your opponents build their fortress 21 m from yours and mount a home-built cannon at its edge, 7.0 m above the ground. Their cannon fires baseballs with a speed of 15.0 m/s. They fire a baseball at your fortress at an angle of 19° above the horizontal. The opponents score points with every mark their baseball leaves on your playdough. You have a defensive cannon that shoots golf balls at a speed of 52.0 m/s. If your cannon fires from ground level 0.80 s after their cannon fires, at what angle must you fire your golf ball to hit the incoming baseball?

85. ✷✷ A pilot points her plane in the direction of an airfield that is 300 km east and 210 km north of her current location. Since the winds are light and varying, the pilot does not attempt to correct her bearing for wind speed. Near the end of her flight, she finds that her course will take her 34 km north of the airfield. The plane's airspeed is 270 km/h, and the flight would take 98 min by the most direct route. Find the bearing that would have taken her straight to the airfield.

86. ✷✷ In a marine-rescue exercise, a small boat moves at velocity 13 km/h [37° N of E]. The rescue boat has a speed of 30 km/h and is positioned 60 m [42° S of E] from the small boat.
    In what direction must the rescue boat head to intercept the small boat?

87. ✷✷✷ A pheasant flying at 28 m/s at a height of 20 m passes directly over a hunter. The hunter's shotgun fires pellets at a speed of 332 m/s. At what angle should the hunter aim to hit the pheasant if he fires just as it passes overhead? Assume that the effects of wind and air resistance are negligible.

88. ✷✷✷ Figure 4-27 is a speed versus time graph for a meteoroid fragment entering the atmosphere. Focus on the region where the speed drops from 45 to 35 km/s. Use the graph to construct approximate distance versus time and acceleration versus time graphs.

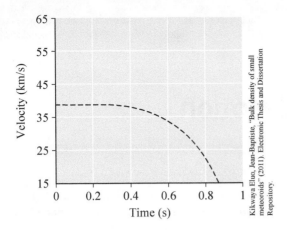

Kikwaya Eluo, Jean-Baptiste. "Bulk density of small meteoroids" (2011). Electronic Thesis and Dissertation Repository.

Figure 4-27 Problem 88

89. ✷✷ Figure 4-28 shows an altitude versus time plot for the *Arianne 5*'s ascent to space.
(a) Estimate the average radial velocity (i.e., vertical component of velocity) of the rocket between launch time and the separation of the second rocket stage (point H2 on the graph).
(b) Estimate the radial speed at $t = 200$ s, $t = 400$ s, $t = 1500$ s, and $t = 2000$ s.
(c) Describe the radial acceleration of the Arianne 5 after point H3 on the graph (separation of stage 3 of the rocket).
(d) Why does the slope of the graph in Figure 4-28 not correspond to the total speed of the rocket?

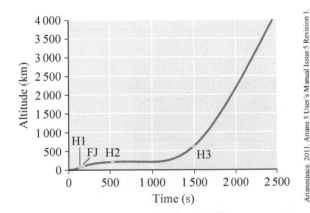

Arianespace. 2011. Ariane 5 User's Manual Issue 5 Revision 1.

Figure 4-28 Problem 89

90. ✷✷ Figure 4-29(a) shows the acceleration curve for a Ferrari Scuderia (Figure 4-29(b)).
(a) What is the maximum acceleration on the curve?
(b) Calculate the acceleration at $t = 7$ s.

(c) What is the average acceleration between $t = 0$ and $t = 14$ s?
(d) How far does the car go before reaching maximum speed?
(e) How far does the car go between $t = 14$ and $t = 16$ s?
(f) What is the total distance covered by the car during the whole interval shown?

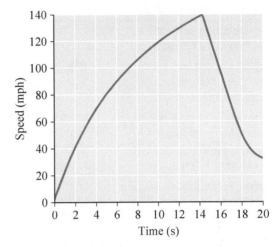

Figure 4-29(a) Problem 90

© John Lamm / TRANSTOCK/Transtock/Corbis

Figure 4-29(b) Problem 90

91. ✷ You want a golf ball on its way to the hole to clear a 12 m-tall tree 40 m away. You estimate that the golf ball will leave the ground at an angle of 47°. What initial speed must the golf ball have to just clear the tree? Is there only one answer to this problem? Explain.

 See the text online resources at www.physics1e.nelson.com for Open Problems and Data-Rich Problems related to this chapter.

Learning Objectives

When you have completed this chapter you should be able to:

1 Represent forces as vectors, and resolve and add forces.

2 Explain Newton's laws and the relationship between force, mass, and motion.

3 Draw free body diagrams to resolve forces into components, and write and solve equations of motion.

4 Explain weight, *g* forces, and weightlessness.

5 Incorporate normal forces into free body diagrams.

6 Differentiate between tension and compression, and explain the action of tension in a string.

7 Distinguish between static and kinetic friction, and define the maximum force of static friction.

8 Explain Hooke's law and the forces exerted by an ideal spring.

9 Identify the forces contributing to the circular motion of an object.

10 Explain fictitious forces and the difference between inertial and noninertial reference frames.

e For additional Making Connections, Examples, and Checkpoints, as well as activities and experiments to help increase your understanding of the chapter's concepts, please go to the text's online resources at www.physics1e.nelson.com.

Chapter 5
Forces and Motion

During the reentry of a space capsule into Earth's atmosphere (Figure 5-1), the two main forces acting on the capsule are gravity and drag resulting from collisions with air molecules. Reentry capsules are shaped to provide some lift to slow their descent if they enter the atmosphere at the correct angle. If the trajectory is too steep, the drag force, or the force of air resistance, can overheat the hull of the capsule and decelerate the capsule so rapidly that the crew may be harmed. On April 19, 2008, a malfunction caused the *Soyuz TMA-11* capsule to enter the atmosphere at an abnormally steep angle. Although the three astronauts onboard survived, they experienced forces as high as 11 times the force of gravity on Earth's surface.

© Sergei Remezov/Reuters/Corbis

Figure 5-1   Image of Soyuz TMA 11 after re-entry

LO 1

5-1 Force and Net Force

We first discuss the basic properties of forces and then their relation to mass and motion in the context of Newton's laws. Later in the chapter we outline and apply strategies for solving problems involving the more common types of mechanical forces.

Forces Are Vectors

Forces are vectors; therefore, they have both magnitude and direction. When you push against a wall, you are exerting a force. When you pull a door handle, you are also exerting a force. How hard you push or pull determines the magnitude of the force you exert. The direction in which you push or pull determines the direction of the force vector. The application of force on an object results in the motion of the object, the deformation of the object, or both. In this chapter, we will deal primarily with objects that do not deform.

Fundamental and Nonfundamental Forces

In this chapter, for the most part we will focus on forces that require physical contact between objects, except for the force of gravity. Such contact forces include friction, tension, elastic forces, normal forces, and applied forces. By "applied force" we specifically refer to the force applied by you, me, or a donkey, for example, on a crate, a box, or some other object. These forces are typically called **nonfundamental forces**. In contrast, the fundamental forces in nature – gravitational attraction, the electromagnetic force, and the two nuclear forces – do not require contact between objects and can act at a distance. Although the exact nature of the four fundamental forces is not fully known, they are real, identifiable, and measurable. What we sometimes refer to as nonfundamental forces are, in fact, macroscopic manifestations of the fundamental forces. The fundamental forces underlie the nonfundamental forces.

Forces must have real physical identifiable causes. They must be one of the fundamental or nonfundamental forces, a few of which are mentioned above.

Net Force

The net force acting on an object is the sum of all the forces acting on that object. Forces are vectors, so by "sum" we mean the "vector sum."

✓ CHECKPOINT

C-5-1 Net Force of "Squeeze"

You squeeze a baseball between your hands with a force of 210 N from each hand. Assuming no deformation, the net force acting on the ball is

(a) 420 N;
(b) zero;
(c) 210 N.
(d) None of the above.

C-5-1 (b) the two forces add up to zero.

 EXAMPLE 5-1

Adding Forces

Calculate the net force on the object in Figure 5-2(a). Give the magnitude of the resultant force and the angle it makes with the direction of $\vec{F}_1$.

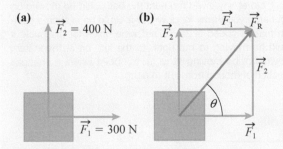

Figure 5-2   Example 5-1

SOLUTION

$$\vec{F}_R = \vec{F}_1 + \vec{F}_2$$

Since the forces act at 90° to each other, we can apply Pythagoras' theorem:

$$F_R^2 = F_1^2 + F_2^2 = (300 \text{ N})^2 + (400 \text{ N})^2 = 250000 \text{ N}^2$$

$$F_R = 500 \text{ N}$$

giving us a magnitude of 500 N for F_R.

From Figure 5-2(b), we can see that the angle that the resultant makes with $\vec{F}_1$ is given by

$$\tan\theta = \frac{400 \text{ N}}{300 \text{ N}} \Rightarrow \theta = 53.1°$$

Making sense of the result:

The forces make a 3;4;5 triangle, therefore the value of 500 N is expected.

EXAMPLE 5-2

Calculating Net Force

Calculate the net force acting on the object in Figure 5-3.

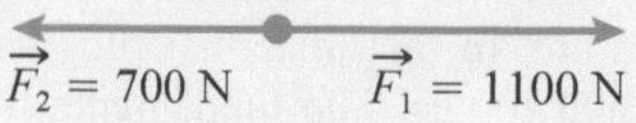

$$\vec{F_2} = 700 \text{ N} \qquad \vec{F_1} = 1100 \text{ N}$$

Figure 5-3 Example 5-2

SOLUTION

$$\vec{F_R} = \vec{F_1} + \vec{F_2} \qquad (1)$$

we will pick the direction of $\vec{F_1}$ to be positive. Therefore,

$$F_R = F_1 - F_2 = 1100 \text{ N} - 700 \text{ N} = 400 \text{ N} \qquad (2)$$

where the quantities in Equation 2 are variables that denote the magnitudes of the forces in Equation (1).

Making sense of the result:

The vector sum results in the magnitudes of the forces being subtracted.

LO 2

5-2 Newton's Laws

Newton's First Law

When a hockey player's stick strikes a puck, the puck moves in a straight line on the ice, and its speed does not change significantly as it travels across the rink if the ice is smooth. In fact, if there were no friction between the puck and the ice, the velocity of the puck would not change at all. If you want to change the speed or the direction of the moving puck, you have to exert a force on it. In addition, if you were to set the puck gently on the horizontal ice surface, the puck would remain where you placed it. The puck will remain at rest unless a force is exerted on it.

These examples demonstrate **Newton's first law:** If no net force is exerted on an object, the object's velocity will not change.

✓ CHECKPOINT

C-5-2 Dropped Ball

A ball is dropped from rest and bounces off the floor. Which of the following statements applies to the ball on its way up?
(a) The ball's acceleration is upward because it is moving upward.
(b) There must be a force causing the ball to continue to move upward.
(c) The ball is only under the influence of the force of gravity as it moves upward.
(d) None of the above apply.

C-5-2 (c) the only force acting on the ball, in the absence of air resistance, is the force of gravity.

Newton's Second Law

If a hockey puck is at rest on the ice and you push it with a horizontal force, the puck will move in the direction of the force, and the speed of the puck will continue to increase as long as you apply the force. The response of the puck to the force you apply to it is given by Newton's second law, which is often presented in equation form:

KEY EQUATION
$$\vec{F}_{net} = m\vec{a} \qquad (5-1)$$

EXAMPLE 5-3

Which Way Does the Puck Move?

The puck in Figure 5-4 is on frictionless ice and is acted upon by the forces as shown. What can we say about the motion of the puck: is the puck moving to the left, to the right, or is it stationary?

SOLUTION

The direction of the net force on an object does not necessarily dictate the direction in which the object is moving. We do not know which way the puck is moving. For example, when you throw a ball up in the air, it is under the influence of only the force of gravity as it moves up. The net force on the ball is downward as the ball moves up. The force of gravity will eventually cause the ball to slow down and move downward, but at any given moment the ball could be moving up (right after you throw it), down, or it could be in such a position that its speed is zero. The same holds for the puck: it could be moving to the right, to the left, or, at the instant shown, not be moving at all (at the point where it changes direction of motion from left to right).

$$\vec{F_1} = 60 \text{ N} \qquad \vec{F_2} = 110 \text{ N}$$

Figure 5-4 Example 5-3

Equation (5-1) states that the acceleration of an object is directly proportional to the net force exerted on it. If you double the force on the puck in the above example, the acceleration will double. If you double the mass while keeping the force constant, the acceleration will drop by a factor of 2. It is much easier to push a hockey puck across the ice than push a more massive object such as a hockey player. Ultimately, how massive an object is depends on how much material it contains.

EXAMPLE 5-4

Force and Motion

An object moving to the right is under the effect of a force that points to the right but whose magnitude decreases linearly with time. What can we say about the speed of the object: is it decreasing or increasing as a result of this force?

SOLUTION

Despite the fact that the magnitude of the force is decreasing with time, at any given point the force is non-zero and pointing to the right. According to Newton's second law in Equation (5-1), the acceleration is to the right; therefore, the speed is increasing. Of course, this applies as long as the force does not reach a value of zero, at which point the acceleration will be zero and the puck's speed will be constant.

Making sense of the result:

The fact that the magnitude of the force decreases with time means that the acceleration decreases with time, but as long as the acceleration is not zero, the puck will continue to accelerate and its speed will continue to increase.

Inertia refers to an object's resistance to being accelerated. The quantitative measure of inertia is mass. A net force of 1 N will cause a 1 kg mass to accelerate at 1 m/s^2:

$$1 \text{ N} = 1 \text{ kg m/s}^2$$

Point Mass A point mass is an object that has mass but whose dimensions are infinitesimally small. The concept of the point mass does not take into account the dimensions and shape of the body. For all intents and purposes, all the objects in this chapter will be treated as point masses. This will serve us well as long as we do not consider any rotation of these objects that could be caused by the forces acting on them. In the diagrams we draw, however, the objects will be shown as having some dimensions and not just as point masses for clarity and sometimes ease of perception.

CHECKPOINT

C-5-3 Net Force

A cat of mass 2.4 kg enters an elevator, which moves upward with an acceleration of 4 m/s^2. The net force on the cat is

(a) 33.1 N;

(b) 23.5 N;

(c) 13.9 N;

(d) 9.6 N.

C-5-3 (d) By Newton's second law, to get the net force on an object one multiplies its mass with its acceleration.

EXAMPLE 5-5

Force and Acceleration

A Formula One race car is at rest at the start line. When the race starts, the car attains a speed of 135 km/h in 4.50 s. Find the net force on the driver, who has a mass of 67.0 kg. Assume constant acceleration.

SOLUTION

Newton's second law relates the net force on the driver and his acceleration:

$$F_{net} = ma$$

The driver's acceleration is the same as the car's acceleration, and from Equation 3-15

$$v(t) = at + v_0$$

The initial speed is zero, and the final speed is

$$v(t) = 135 \text{ km/h} = 135\left(\frac{1000 \text{ m}}{3600 \text{ s}}\right) = 37.5 \text{ m/s}$$

Hence,

$$a = \frac{v(t) - v_0}{t} = \frac{37.5 \text{ m/s} - 0}{4.5 \text{ s}} = 8.33 \text{ m/s}^2$$

Therefore,

$$F_{net} = 67.0 \text{ kg} (8.33 \text{ m/s}^2)$$
$$F_{net} = 558 \text{ N}$$

Making sense of the result:

All we need to know is an object's mass and its acceleration for the net force acting on it.

EXAMPLE 5-6

Braking Distance

A 120 tonne metro train is brought to a stop from a speed of 23.0 m/s using a constant braking force of 400 kN. Over what distance does the train come to a stop? Assume that the only force stopping the train is the force from the brakes.

SOLUTION

Since the only force acting on the train is from the brakes, we choose the positive x-axis to be the direction of motion of the train.

$$F_{net} = ma \Rightarrow a = \frac{F_{net}}{m} = \frac{400\,000 \text{ N}}{120\,000 \text{ kg}} = 3.33 \text{ m/s}^2 \quad (1)$$

We can use the time-independent kinematic equation for constant acceleration to find the distance that the train travels while braking:

$$v_f^2 - v_i^2 = 2\vec{a} \cdot \Delta\vec{x} \quad (2)$$

The final speed is zero. Since the train is slowing down, the acceleration and the displacement have opposite directions. Therefore, using Equation 2-18, $\vec{a} \cdot \Delta\vec{x} = |\vec{a}||\Delta\vec{x}| \cos(180°)$ and Equation (2) becomes

$$0 - (23 \text{ m/s})^2 = -2(3.33 \text{ m/s}^2)\Delta x \Rightarrow \Delta x = 79.43 \text{ m} = 79.4 \text{ m}$$

where Δx denotes the magnitude of the displacement

Making sense of the result:

400 KN is a large force, but the train is also very massive.

EXAMPLE 5-7

Force in Circular Motion

You are sitting in the rear seat of a car, which is going around a track with a radius of 120 m at a speed of 26.0 m/s. The track is banked at an angle of 21°. The coefficient of static friction between the car and the road is 0.34. The car is on the verge of sliding up the bank. A friend sitting next to you pushes you against the door with a force of 290 N. Another friend, sitting in the seat in front of you, pulls you forward, squeezing you against the back of her seat with a force of 360 N. Your mass is 68 kg. Calculate the net force acting on you.

SOLUTION

From the statement of the problem, you are evidently not moving relative to either of your friends or the car; therefore, your acceleration is the same as the car's acceleration.

According to Newton's second law, the net force on you is simply your mass times your acceleration, and since the car is moving in a circle, that acceleration, as we saw in Chapter 4, is equal to v^2/r.

$$F_{net} = ma = \frac{mv^2}{r} = \frac{(68 \text{ kg})(26.0 \text{ m/s})^2}{120 \text{ m}} = 383 \text{ N}$$

Making sense of the result:

Here, Newton's second law allows us to calculate the net force without considering the individual forces. There is extra information given in the question, but that information is not needed. The point here is that all you need is the mass and the acceleration to determine the net force.

MAKING CONNECTIONS

Newton's Laws Before Newton?

Before Isaac Newton (1642–1727), Western civilization's perception of motion was dominated by Aristotle's view dating back to the 4th century BCE. According to Aristotle, in the absence of an external force, or "motive power," an object would come to rest. Consequently, an object moving at a constant velocity would have a net force acting on it. Once that force is removed, the object will begin to slow down until it comes to rest.

According to Newton's second law, when a net force acts on an object, the object's velocity will not be a constant; the object will accelerate. Newton's first law states that an object will move at a constant velocity only if no net force acts on it.

Statements summarizing the principle contained within Newton's first law date as far back as the 5th century BCE in China and the 11th century CE in parts of the Islamic world. Newton first published his laws of motion in 1687 as part of his three-volume work *Philosophiæ Naturalis Principia Mathematica*, or *Principia*. *Principia* also includes his theory of universal gravitation and a mathematical derivation of Kepler's laws of planetary motion. Newton's laws of motion laid the foundation for classical mechanics and dominated physicists' view of the universe for the next three centuries. Go to the text's online resources for an interesting story on the connections between coffee, Newton's Laws and a dutch spyglass.

Newton's Third Law

Now consider what you will feel if you stand firmly on a nonslippery surface beside an ice rink and push with your hand on a friend who is standing on the ice. You will feel a force from your friend on your hand, and your friend will feel the force of your push, which causes your friend to slide away from you. This interaction demonstrates

Newton's third law: Whenever you exert a force on an object, that object will exert a force on you that is equal in magnitude and opposite in direction to the force you exert on it, whether or not motion is involved. In vector form, one can say $\vec{F}_{12} = -\vec{F}_{21}$.

EXAMPLE 5-8

Who Is Pushing Harder?

A polar bear cub and his mother are standing still on a frictionless icy patch in the Gulf of Boothia, Nunavut. The 50.0 kg cub pushes the 400 kg mother bear with a constant force. What happens to the cub?

SOLUTION

According to Newton's third law, the force exerted by the cub on the mother must be equal in magnitude and opposite in direction to the force exerted by the mother on the cub. The acceleration of each is given by Newton's second law

$$a = \frac{F_{net}}{m}$$

Since they both experience forces of equal magnitude but have different masses, their accelerations will not be the same. In fact, when the cub pushes his mother on the ice, he will end up moving backward at a much higher speed than she will end up moving in the direction he pushed her.

Making sense of the result:

In everyday life, when a lighter person pushes a heavier person, the lighter person does not expect to move backward. Enough friction is often present to prevent the lighter person from sliding backwards.

EXAMPLE 5-9

The Force of a Serve

A volleyball player serves a 249 g volleyball at a speed of 90.0 km/h. Assuming that the player's hand contacts the ball for 42.0 ms, find the average force exerted by the ball on the player's hand during that time.

SOLUTION

We have no knowledge of the mass of the player's hand or its change in speed during the collision. However, Newton's third law tells us that the magnitude of the force on the player's hand equals the magnitude of the force on the volleyball. The average acceleration of the volleyball is given by

$$a_{avg} = \frac{\Delta v}{\Delta t} = \frac{25 \text{ m/s}}{0.042 \text{ s}} = 595.2381 \text{ m/s}^2$$

$$F_{avg} = ma_{avg} = (0.249 \text{ kg})(595.2381 \text{ m/s}^2) = 148 \text{ N}$$

Making sense of the result:

The force on the player's hand is roughly sixty times the weight of the volleyball, which is reasonable because serving the volleyball takes a lot more effort than holding it stationary. Also note how the final answer is kept to 3 significant figures, whereas the interim value of a, needed to calculate F, is kept to more than just 3. Ideally, for an interim calculation you keep as many digits as possible.

C-5-4 When Are the Forces Equal?

When you are standing in an elevator that is accelerating upward,

(a) the force that the elevator exerts on you is less than the force that you exert on the elevator;

(b) the force that the elevator exerts on you is more than the force that you exert on the elevator;

(c) the force that the elevator exerts on you is equal to the force that you exert on the elevator;

(d) the forces will be equal only when the acceleration is constant.

C-5-4 (c) as given by Newton's third law.

LO 3

5-3 Free Body Diagrams, Acceleration, and Equations of Motion

Free Body Diagrams

Once the forces acting on an object have been identified, the next step is to draw a **free body diagram (FBD)**, which is a diagram that shows only the object and the forces acting on it.

EXAMPLE 5-10

Drawing FBDs

In an FBD, show all the forces acting on a crate that is sliding down a frictionless incline.

SOLUTION

The forces are shown in the FBD in Figure 5-6. We have the force of gravity, acting vertically downward, and the normal force from the surface of the incline acting normal (perpendicular) to the surface. We will discuss these forces in detail below.

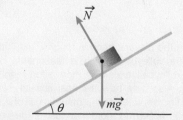

Figure 5-6 The FBD of a crate on a frictionless incline

The Direction of Acceleration and the Choice of Coordinate System

Often, we deal with forces that act in different directions, like those in Figure 5-6. To solve problems, we need to resolve these forces into components. The Cartesian system of coordinates is generally the most convenient for this purpose. Any set of perpendicular axes can be used to resolve vectors into Cartesian components. The choice of axes as such does not affect the physics of the problem. However, we can simplify the calculations considerably by having *one of the Cartesian axes point in the same direction as the acceleration of the object.*

For example, when an object slides down an incline, the acceleration is directed down the incline; therefore, we choose an x-axis along the incline and a y-axis normal to the incline, as shown in Figure 5-7(a). (Example 5-10 demonstrates the advantage of this choice of axes.) Figure 5-7(b) shows a car going around a frictionless banked curve at just the right speed to avoid sliding up or down the incline. Here, the acceleration is a horizontal radial acceleration, so it is better to have a horizontal axis instead of an axis aligned with the incline.

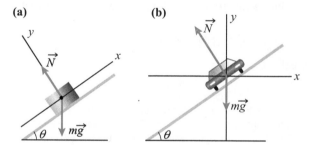

Figure 5-7 Comparing choice of axes

Equations of Motion

When the forces are resolved and broken into components, we can write equations of motion by simply applying Newton's second law along the directions of each axis in the coordinate system. The equations of motion can then be solved individually or simultaneously to extract the desired information, as will be demonstrated in examples throughout this chapter.

A Problem-Solving Strategy for Applying Newton's Laws

We can combine the graphical and mathematical techniques described above into a powerful strategy for analyzing and solving problems involving Newton's laws:

Step 1. Identify all the forces acting on the object.

Step 2. Draw a free body diagram.

Step 3. Establish the direction of acceleration.

Step 4. Set up a coordinate system with one axis oriented along the direction of the acceleration.[1]

Step 5. Resolve the forces into components along the axes of the coordinate system.

Step 6. Write the equations of motion along each axis.

Step 7. Solve the equations of motion.

C-5-5 Choosing Coordinate Axes

Figure 5-8 shows a FBD for a polar bear cub sliding down an igloo. The polar bear is not on the verge of leaving the igloo's surface. Which of the following statements is true?
(a) The axes should be vertical and horizontal.
(b) The normal should point in the opposite direction to the one shown.
(c) The acceleration is along the direction of the force of gravity.
(d) none of the above.

C-5-5 the axis should be as shown, the normal points as it should, and the acceleration points to centre of circle, so answer is d).

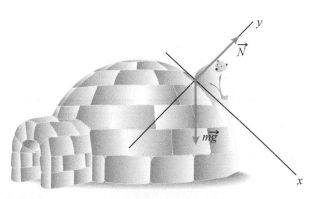

Figure 5-8 A FBD of a polar bear sliding down an igloo. The axes shown would be used typically to resolve the forces into components.

LO 4

5-4 The Force of Gravity

Among Newton's greatest contributions was the formulation of the law of universal gravitational attraction, which is detailed in Chapter 11. For this chapter, we will consider only the gravitational attraction between Earth and objects at or near its surface. The attractive force that Earth exerts on an object is called the weight of the object and is given by

$$\vec{W} = \vec{F} = m\vec{g}$$

where m is the mass of the object and g is the gravitational acceleration (a measure of the strength of Earth's gravitational field). The value of g varies somewhat with altitude and with latitude because the force of gravity depends on the distance from Earth's centre. For calculations in this chapter, we will assume that g has a constant value of 9.81 m/s[2] near Earth's surface. This approximation is reasonably accurate, especially for objects near sea level.

By Newton's third law, the gravitational force that Earth exerts on an object is equal in magnitude and opposite in direction to the gravitational force that the object exerts on Earth.

C-5-6 The Force of Gravity

A 45 kg mountain goat is drinking water from Shunda Lake in Nordegg, Alberta. The acceleration due to gravity at that elevation (1378 m) is approximately 9.8024 m/s[2]. What is the force with which the goat pulls Earth?
(a) negligible
(b) M_E(9.8024 m/s[2]), where M_E is the mass of Earth
(c) (45 kg) (9.8024 N)
(d) There is no way that a goat can pull Earth.

C-5-6 (c)

MAKING CONNECTIONS

The *g* Force

The **g force** can be defined as the sum of nongravitational forces acting on an object divided by the object's mass. Often, the nongravitational force of concern is the normal force. For example, an astronaut in a space shuttle before launch experiences a *g* force of 1 (from the normal force exerted by the astronaut's seat). If the shuttle takes off with an acceleration that is equal to three times the acceleration due to gravity, or 3 *g*, the *g* force experienced by the astronaut in the shuttle is 4 *g*.

LO 5

5-5 The Normal Force

When you are standing indoors, two forces act on you: the force of gravity, directed downward (toward the centre of Earth), and a **normal force**, directed upward, exerted by the floor under your feet. Since you are not accelerating, the vector

[1]Sometimes, it is convenient to decompose the acceleration of an object into two components, for example, in the case of an object moving freely in a vertical circle. In this case, one axis can point along the direction of one of the components of the acceleration, and the other axis will then point along the direction of the second component.

sum of the two forces is zero – the floor must be pushing upward on your body with a force equal to your weight.

When two objects are in physical contact, often they exert a normal force on each other. Newton's third law states that this pair of forces is equal in magnitude and opposite in direction. Therefore, the normal force that the objects exert on each other are equal in magnitude and opposite in direction. Further, the normal force acts in a direction that is perpendicular to the surface of contact between the two objects. The normal force is sometimes called the contact force, or the force of contact. We use the term "normal force" to avoid confusion with friction, which also results from contact between objects. However, frictional forces are directed *parallel* to the surface of contact.

 CHECKPOINT

C-5-7 Newton's Third Law

You are standing on the floor in your study space. Your mass is m. Which of the following statements is true?

(a) Earth pulls you, and you pull Earth with a force of the same magnitude, mg, by Newton's third law.

(b) You push the floor, and the floor pushes you with a force of the same magnitude by Newton's third law.

(c) Since the sum of the forces on you is zero, $N = mg$.

(d) All the above statements are true.

C-5-7 (d) all the statements are true.

EXAMPLE 5-11

Apparent Weight in an Elevator

A person with a mass of 61.0 kg stands on a scale in an elevator that is accelerating upward at 3.00 m/s². Find the reading of the scale, in newtons.

SOLUTION

The scale reads the force that the person exerts on it. This force is the normal force exerted by the person on the scale and has the same magnitude, N, as the force exerted by the scale on the person. The FBD in Figure 5-9 shows the forces acting on the person.

Figure 5-9 The FBD of a person in an elevator

According to Newton's second law

$$\sum \vec{F} = m\vec{a} \qquad (1)$$

Summing forces in the y direction

$$\sum F_y = N_y + mg_y = ma_y \qquad (2)$$

Taking up as the positive direction,

$$N - mg = ma \qquad (3)$$

Therefore, the reading of the scale is

$$N = mg + ma = m(g + a) = 61.0(3.00 \text{ m/s}^2 + 9.81 \text{ m/s}^2)$$

$$= 781.41 \text{ N} = 781 \text{ N}$$

Please note in the approach above, we have written the sum of the forces in the y direction in scalar form. This was done in Equation 2. Another important aspect of the approach above that we will use frequently is in step 3. In step 3, the symbols are variables that denote magnitudes, or absolute values, of the quantities in step (2). Alternatively, the solution could be done using full vector representation, using unit vectors, in which case, Newton's second law can be expressed as

$$\vec{N} + m\vec{g} = m\vec{a}$$

which can then be written as

$$N\hat{j} - mg\hat{j} = ma\hat{j}$$

Making sense of the result:

The scale supports the person's weight and transfers the force from the elevator to accelerate the person upward. Thus, it makes sense that the force exerted by the scale is greater than mg. You have most likely experienced the increase in the force on your feet as an elevator accelerates upward, making you feel heavier.

How would the scale reading in Example 5-11 change if the elevator were accelerating downward? We can show that the magnitude of the normal force is given by

$$N = mg - ma$$

If the acceleration of the elevator equals g (i.e., the elevator is free-falling), the normal force on the passenger inside is zero. This normal force is often called apparent weight, as it is the value shown by the scale. When there is no normal force acting on us, we experience a feeling of weightlessness. It is important to remember that our weight in the elevator is not affected by the elevator's acceleration, but the force we are pushing down on a scale (apparent weight) is. (On certain amusement park rides, this feeling of apparent weightlessness is often accompanied by loud screams.)

MAKING CONNECTIONS

Osteoporosis and Microgravity

Osteoporosis, or "porous bones," is a condition that afflicts a significant fraction of the aging population, especially women. Osteoporosis weakens the bones, increasing the risk of fractures and a host of related complications. Among the various causes of osteoporosis are physical inactivity and reduced mobility. Exposure to the microgravity of space also results in a significant loss of bone mass. NASA reports an average bone mass loss of about 1% per month for astronauts during space missions. There are various remedies for earthbound osteoporosis, depending on the cause. Subjecting the body to higher forces generally increases the loading on the bones so as to promote bone mass increase. In order to mitigate the effects of space bound bone mass loss, astronauts are required to exercise for two hours per day using various devices; however the current measures are not sufficient. One promising proposed remedy is to have the astronauts endure vibrations for 20 to 30 minutes per day, thus subjecting their bodies to higher accelerations, and subsequently more force for the purpose of stimulating bone growth in the body (Figure 5-10).

Cary Wolinsky and Trillium Studios

Figure 5-10 Bioengineering Professor Dr. Clinton Rubin, of the State University of New York, sitting opposite a turkey on a vibration plate used for experiments on bone loss.

EXAMPLE 5-12

Horizontal Normal Forces

Block 1 in Figure 5-11(a) has a mass of 2.00 kg, and block 2 has a mass of 8.00 kg. The two blocks are sitting on a horizontal frictionless surface. You push block 1 to the right with a force of 100 N.

(a) Find the normal force between the two blocks.
(b) Verify that the normal forces exerted by the two blocks have equal magnitudes, as predicted by Newton's third law.
(c) You now push block 2 to the left with a 100 N force. Find the normal force between the blocks.

(a)

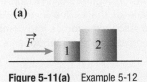

Figure 5-11(a)   Example 5-12

SOLUTION

(a) Let us begin by drawing an FBD of the two blocks combined (Figure 5-11(b)).
     In the horizontal direction, the only force acting on the system of two blocks is the 100 N force directed right. This will be the direction of the positive x-axis. Hence the acceleration of the system is

(b)

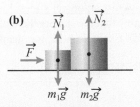

Figure 5-11(b)   The FBD of block 1 and block 2 together

$$\sum F_x = F_x = m_T a_x$$

$$a = \frac{F}{m_T}$$

According to the choice of the coordinate axis we made above,

$$a = \frac{F}{m_T} = \frac{100\ N}{10\ kg} = 10.0\ m/s^2$$

We now draw the FBD of the larger block, block 2, shown in Figure 5-11(c). The only force acting on block 2 in the horizontal direction is the normal force from the 2.0 kg block. Denoting the magnitude of that normal force as F_{12}, we get

$$\sum F_x = F_{12,x} = m_2 a_x = (8\ kg)(10\ m/s^2) = 80.0\ N$$

(continued)

(c)

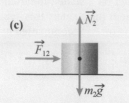

Figure 5-11(c) The FBD of block 2

(b) Now we draw the FBD of block 1 (Figure 5-11(d)).

(d)

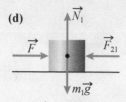

Figure 5-11(d) The FBD of block 1

$$\sum F_x = F_x + F_{21,x} = m_1 a_x$$

Choosing right as the positive direction,

$$F - F_{21} = m_1 a$$

$$F_{21} = F - m_1 a = 100 - 20 = 80.0 \text{ N}$$

We note here that F, F_{21}, and a are variables that denote the magnitudes of the forces and the acceleration.

Thus in magnitude,

$$F_{21} = F_{12}$$

Recall however that these forces are an action-reaction pair, and though they are equal in magnitude they are opposite in direction.

(c) Now we are pushing m_2, which in turn pushes m_1. Since the total mass is unchanged and the magnitude of the applied force is the same, the acceleration of the two-mass system is still 10.0 m/s^2, to the left. The normal force between the two masses is the only force acting on the smaller mass (Figure 5-12). We can show that by summing the forces on m_1 in the x-direction,

$$\sum F_x = F_{21,x} = m_1 a_x = (2 \text{ kg})(10 \text{ m/s}^2) = 20.0 \text{ N}$$

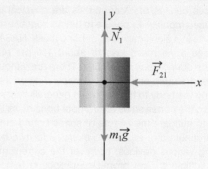

Figure 5-12 The FBD of block 1, now being pushed to the left

Making sense of the result:

The ratio of the net forces on the blocks equals the ratios of their masses: 1 : 4.

Inclined Surfaces

What dictates how fast an object slides down an incline? Intuitively, we know that the steeper the incline, the higher the acceleration. In Example 5-13, we quantify the motion of an object on an inclined surface. This section deals with frictionless surfaces, but later in the chapter we will consider motion with friction.

 EXAMPLE 5-13

Motion on a Frictionless Incline

A mass is placed on a frictionless surface inclined at an angle θ above the horizontal (Figure 5-13). Determine the acceleration of the mass when it is allowed to slide freely, and calculate the magnitude of the normal force acting on the mass.

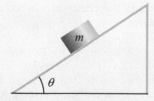

Figure 5-13 Example 5-13

SOLUTION

Here, we apply the analytical strategy outlined in Section 5-3.

Step 1. Identify all the forces acting on the object.

The forces acting on the mass are the force of gravity, which pulls the mass downward, and the normal force, which acts perpendicular to the incline.

Step 2: Draw an FBD.

The FBD is shown in Figure 5-14. Notice that we have shown the surface of the incline in the FBD. This will make it easier to resolve the forces into components and determine the angles and the components; in general,

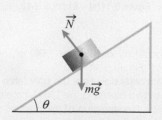

Figure 5-14 The FBD of the mass on an incline in Example 5-13

showing the surface also aids with the aesthetics of the diagram.

Steps 3 and 4: Establish the direction of the acceleration, and set up a coordinate system with one axis oriented along the direction of this acceleration.

The mass will accelerate down the incline. Therefore, one of the axes should be oriented in that direction. We refer to that axis as the x-axis (to the motion). We will refer to the second axis as the y-axis. Alternatively, these axes can be referred to as the parallel and perpendicular axes, respectively. We choose the centre of the object as the origin for the coordinate system.

Step 5: Resolve the forces into components along the chosen axes.

The normal force, $\vec{N}$, from the incline on the block points along the perpendicular axis (or y-axis), so the y-component is equal to the magnitude of the normal force, and the x-component is zero.

To resolve the force of gravity on the block into components, we draw two lines from the tip of the mg vector, one parallel to the incline (along the x-axis) and one perpendicular to the incline (along the y-axis). These lines intersect the axes at the tips of the components of mg. We designate these components mg_x and mg_y, respectively (Figure 5-15). Vector mg is perpendicular to the horizontal, and the perpendicular component of mg is perpendicular to the incline. The angle between mg and mg_y, then, is the same as the angle between the horizontal and the incline, θ.

Step 6: Write the equations of motion along each axis.

The sum for the forces along the direction normal to the incline (i.e., along the y-axis) is

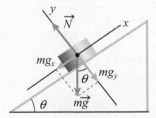

Figure 5-15 Components of the forces along the chosen axes

$$\sum F_y = ma_y \qquad (1)$$

or

$$N_y + mg_y = ma_y \qquad (2)$$

To analyze the motion along the incline, we sum the forces in the direction of the incline (parallel to the x-axis):

$$\sum F_x = ma_x$$

Therefore,

$$mg_x = ma_x \qquad (3)$$

Step 7: Solve the equations of motion.

The components of $m\vec{g}$ parallel to the incline and, therefore, to the motion of the mass on the incline depend on θ. In the right-angled triangle formed by the force of gravity and its components, mg_x is opposite the angle θ, mg_y is adjacent to the angle θ, and mg is the hypotenuse. Therefore, in terms of magnitudes,

$$|mg_x| = mg \sin\theta \qquad (4)$$

and

$$|mg_y| = mg \cos\theta \qquad (5)$$

Therefore, from Equation (2), keeping in mind that the N points along the positive y-direction, and mg_y points along the negative y-direction we have

$$N - mg \cos\theta = 0 \qquad (6)$$

or

$$N = mg \cos\theta$$

And from Equations (2) and (3) we get

$$a = g \sin\theta \qquad (7)$$

where we use the variable a to denote the magnitude of the acceleration.

Making sense of the result:

When the mass is placed on a horizontal surface, the acceleration is zero. An incline angle of 90° means that the mass is effectively placed against a vertical wall, in which case it will simply free-fall with an acceleration equal to g, as given by Step 6. According to step 5, the normal will be zero for a vertical incline.

✓ CHECKPOINT

C-5-8 The Normal Force

For the box in Figure 5-16, the magnitude of the normal force

(a) equals mg;

(b) is greater than mg;

(c) is less than mg;

(d) may be less than, equal to, or greater than mg, depending on the value of θ.

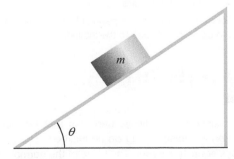

Figure 5-16 C-5-8

C-5-8 (c) box sits freely, $N = mg \cos\theta$.

EXAMPLE 5-14

MECHANICS

Horizontal Force Acting on an Object on an Inclined Plane

Figure 5-17 shows a worker, wearing special boots for added traction, pushing a crate of mass m up an incline with a slope of angle θ using a horizontal force F. Find the normal force from the incline on the crate, and determine the acceleration of the crate. Assume that the friction between the crate and the incline is negligible.

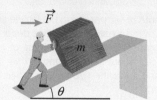

Figure 5-17   Example 5-14

SOLUTION

Here, again, we invoke our strategy for solving problems involving Newton's laws.

Step 1.   Identify the forces acting on the object.

There are three forces acting on the crate: the force of gravity, the normal force exerted by the incline on the crate, and the force applied by the worker.

Step 2.   Draw an FBD for the crate.

The FBD is shown in Figure 5-18.

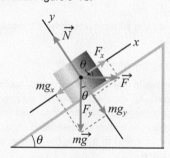

Figure 5-18   The FBD of the crate

Step 3.   Determine the direction of the acceleration.

The acceleration of the crate is along the incline.

Step 4.   Choose your system of axes.

We choose the x-axis to be parallel to the incline and the y-axis to be perpendicular to the incline.

Step 5.   Resolve the forces into components along the chosen axes.

Here, again, the normal force points along the y-axis and has an x-component of zero. As in Example 5-13, the parallel component of the gravitational force is $mg \sin\theta$, and the perpendicular component is $mg \cos\theta$. In Figure 5-18, in the triangle formed by F we can see the hypotenuse F and the components F_x and F_y, where F_x is the adjacent side and F_y is the side opposite the angle θ. Therefore, in terms of magnitudes,

$$|F_x| = F\cos\theta \quad \text{and} \quad |F_y| = F\sin\theta$$

Step 6.   Write the equations of motion along each axis. Along the x-axis we have

$$\sum F_x = ma_x \Rightarrow F_x + mg_x = ma_x \tag{1}$$

If we choose up the incline as the positive direction, then

$$F\cos\theta - mg\sin\theta = ma \tag{2}$$

where F, g and a in Equation (2) are used as variables to denote magnitudes of the quantities in Equation (1).

Along the y-axis we have

$$\sum F_y = ma_y \Rightarrow F_y + mg_y + N_y = 0 \tag{3}$$

Taking the direction of the normal force to be positive,

$$-F\sin\theta - mg\cos\theta + N = 0 \tag{4}$$

Step 7.   Solve the equations of motion.

We can solve Equation (4) for the normal force:

$$N = F\sin\theta + mg\cos\theta \tag{5}$$

Note that here the normal force is not equal to $mg \cos\theta$. To determine the normal force, we had to sum the y-components of the forces.

We can find the acceleration of the crate from Equation (2):

$$a = \frac{F\cos\theta - mg\sin\theta}{m} \tag{6}$$

Making sense of the result:

Since the worker is pushing the crate to a certain degree against the face of the incline, the y-component of the applied force contributes to the normal force. Consequently, the magnitude of the normal force is greater than $mg \cos\theta$. The expression for the acceleration is the net force divided by the mass, as we would expect.

 PEER TO PEER

A misconception I always had was that the normal force on an object sitting on an inclined surface was always equal to $mg \sin\theta$. The value of the normal force can vary, depending on the details of the problem, and should be obtained by solving the equations of motion.

LO 6

5-6   Tension

When you hold a string stationary with a weight attached to it, the string is under **tension** because the weight is pulling the lower end of the string downward while you are pulling the top end upward to prevent it from moving down.

The string is pulling the weight up and pulling your hand down at the same time. The force of tension in a string (or rope) pulls the weight toward the string. Generally, as you pull objects with a string, the string exerts a force of equal magnitude on you and on the object, in accordance with Newton's third law.

Similarly, if you grip both ends of a pen and try to pull the ends apart, the pen is under tension. Your hands are being pulled toward one another by the tension in the pen. Alternatively, you can say that the force of tension prevents your hands from moving apart. If you press on both ends of the pen, the force of **compression** in the pen prevents your hands from coming toward each other, and under compression. However, you cannot have compression in a rope: if you try to push the two ends of a rope together, the rope will buckle. In other words, for almost all practical applications you cannot push on a flexible rope.

The force of tension at either end of an accelerating string will not be the same; neither will the force of compression on the above pen be the same, if the string or the pen is accelerating. The net force, whose magnitude is the difference between the magnitudes of the forces at either end, is, by Newton's second law, equal to the mass of the string or pen multiplied by its acceleration. Therefore, for the tension or compression forces at either end to be equal, we require that either the string (or pen) have zero acceleration or zero mass. In this chapter, we will deal almost exclusively with massless strings.

MAKING CONNECTIONS

Tensile Strength

The ultimate tensile strength (UTS) indicates the maximum tension force per unit area that a material can withstand without structural failure. Structural steel has a UTS of about $400 \ MN/m^2$. Although concrete is widely used in elements under compression, its UTS is relatively weak, with a typical value of approximately $3 \ MN/m^2$. Silk has 2.5 times the tensile strength of structural steel, and some multiwalled carbon nanotubes have been made with a UTS exceeding $60 \ GN/m^2$. For a more detailed discussion of this topic, please see chapter 10.

Certain armed-forces safety regulations require that, when a vehicle is being winched, all nonessential personnel stay at least a cable length away. This rule reduces the risk of injuries should the winching cable snap, because the amount of tension that develops in a towing line can be quite significant.

CHECKPOINT

C-5-9 Pulling on a Rope

A 40 kg seal corners you ($m = 80 \ kg$) on a frozen shelf in the Davis Strait. He grabs a rope with his mouth and brings it to you, insisting with repeated head nods that the way out for you is to play a game where you pull the rope he is holding with his teeth. You pull the rope and notice that the seal accelerates with respect to the frictionless ice at $3 \ m/s^2$. Which of the following statements is true? Assume you experience no friction.

(a) The net force on you is twice the force on the seal.
(b) Your acceleration is the same as the seal's acceleration.
(c) The net force on you is 240 N.
(d) The net force on you 480 N.
(e) None of the above are true.

C-5-9 (e) force on you equals that on seal, none of the answers adds up to that.

EXAMPLE 5-15

Towing Tension

Three masses ($m_1 = 20.0 \ kg$, $m_2 = 50.0 \ kg$, and $m_3 = 30.0 \ kg$) resting on a horizontal frictionless surface are connected by ropes as shown in Figure 5-19. A force of 200.0 N pulls m_3 to the right. Find the tension in the rope connecting m_1 and m_2. The mass of the ropes is negligible.

Figure 5-19 Example 5-15

SOLUTION

From the FBD for m_1 in Figure 5-20, we see that the only force acting on m_1 in the x-direction is the tension in the rope connecting it to m_2. Using T_1 to denote the magnitude of this force, we have

$$\sum F_x = T_{1x} = m_1 a_x \Rightarrow T_1 = m_1 a$$

Figure 5-20 The FBD of m_1

Since m_1 is known, we can solve for T_1 if we have a value for a. The acceleration of all three masses must be the same, as we can see from the FBD of the whole system in Figure 5-21.

(continued)

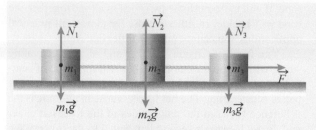

Figure 5-21 The FBD of the three masses in Example 5-15

We take the positive x-axis to point along the line of action of the force F. Since the only force acting on the three-mass system in the x-direction is the 200 N applied force,

$$\sum F_x = F_x = m_T a_x \Rightarrow F = m_T a$$

and

$$a = \frac{F}{m_T} = \frac{200 \text{ N}}{100 \text{ kg}} = 2.00 \text{ m/s}^2$$

Therefore,

$$T_1 = m_1 a = 20 \text{ kg} \ (2 \text{ m/s}^2) = 40.0 \text{ N}$$

we note here that the variables F and a denote magnitudes of the force and the acceleration.

Making sense of the result:

We are able to treat the three masses like one mass because they are connected by taut ropes (which we assume do not stretch). Although the applied force acts directly on m_3 only, in effect, the force is pulling all three masses.

 CHECKPOINT

C-5-10 How Much Tension?

What is the tension in the string in Figure 5-22?

(a) 196 N
(b) 98 N
(c) zero
(d) We need to know whether or not the system is moving.
(e) None of the above.

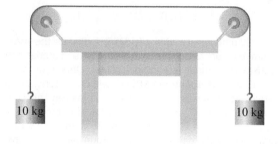

Figure 5-22 C-5-10

C-5-10 (b) for either block, $mg = T = 98$ N, tension is same in rope everywhere.

 **EXAMPLE 5-16**

Simple Rope and Pulley System

An 8.00 kg mass (m_1) is resting on a horizontal frictionless surface. A rope that runs over a massless pulley connects m_1 to a 6.00 kg mass (m_2), as shown in Figure 5-23. The two-mass system is initially at rest. Find the acceleration of m_1 and the tension in the rope.

Figure 5-23 Example 5-16

SOLUTION

Here we have two unknowns, the tension force and the acceleration, so we will need two equations to solve this problem. We begin by drawing a FBD for each object (Figure 5-24).

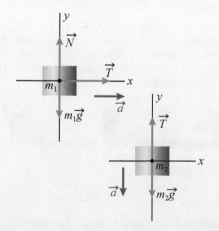

Figure 5-24 The FBD of the two masses in Example 5-16

Since the only force acting on m_1 in the x-direction is the tension in the rope,

$$\sum F_x = T_x = m_1 a_x \Rightarrow T = m_1 a \qquad (1)$$

From the FBD of m_2, we get

$$\sum F_y = T_y + m_2 g_y = m_2 a_y$$

Choosing up as our positive direction gives

$$T - m_2 g = -m_2 a \qquad (2)$$

Note that we have introduced a variable, a, to denote the magnitude of the acceleration, the same is true for T and g.

Substituting Equation (1) into Equation (2), we get

$$m_1 a - m_2 g = -m_2 a \Rightarrow a = \frac{m_2 g}{m_1 + m_2}$$

$$= \frac{(6\ \text{kg})\,(9.81\ \text{m/s}^2)}{(6 + 8)\ \text{kg}} = 4.20\ \text{m/s}^2$$

To find the tension in the rope, we substitute this value for a into either Equation (1) or Equation (2). From Equation (1) we get

$$T = m_1 a = 8\ \text{kg}\ (4.2\ \text{m/s}^2) = 33.6\ \text{N}$$

Making sense of the result:

The acceleration of the system is less than g because the weight of m_2 is pulling both m_1 and m_2. Note that the tension is not equal to $m_2 g$, as it would have been were m_2 not accelerating. The tension in the horizontal and the vertical segments of the rope is the same (but would not be if the pulley has mass, as we will see in Chapter 8). We also note than when $m_1 = 0$, the acceleration of m_2 is equal to g.

EXAMPLE 5-17

Train Acceleration

A bored but observant child on a moving train ties a ball to one end of a string and ties the other end to a light fixture on the ceiling of the train. When the train slows down, the child notices that the string is no longer hanging straight down but instead makes an angle of $19°$ with the vertical. The mass of the ball is 175 g.

(a) Will the string tilt forward or backward? Find the acceleration of the train.

(b) What is the tension in the string?

SOLUTION

(a) The string tilts forward. As the train slows down, the only forces acting on the ball are the tension in the string and the force of gravity. The acceleration is opposite to direction of motion of the ball (and the train). So, we choose a horizontal x-axis and a vertical y-axis. The FBD in Figure 5-25 shows the forces resolved into components along these axes.

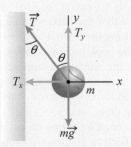

Figure 5-25 The FBD of the suspended ball in Example 5-17

Summing the forces on the ball in the y-direction, we get

$$\sum F_y = T_y + mg_y = 0$$

Choosing up as our positive direction gives

$$T \cos\theta - mg = 0, \quad \text{or} \quad T \cos\theta = mg \qquad (1)$$

Here, we have used the fact that the vertical component of the force of tension is the side adjacent to θ in the triangle formed by T, T_x, and T_y in Figure 5-25. Since the acceleration of the ball is in the horizontal direction,

$$\sum F_x = T_x = ma_x \Rightarrow T \sin\theta = ma \qquad (2)$$

We have two unknowns—the tension and the acceleration—and we now have two equations involving these unknowns. Dividing Equation (2) by Equation (1), we get

$$\tan\theta = \frac{a}{g} \Rightarrow a = g\tan\theta = (9.81\ \text{m/s}^2)\tan(19°) = 3.38\ \text{m/s}^2$$

(b) For the tension in the rope we can use either Equation (1) or Equation (2). Substituting for the known values in Equation (1) gives

$$T = \frac{mg}{\cos\theta} = \frac{0.175\ \text{kg} \times 9.81\ \text{m/s}^2}{\cos(19°)} = 1.82\ \text{N}$$

Making sense of the result:

The tension is the only force acting to slow the ball down. The tension is not much greater than mg because the acceleration is relatively small, as evidenced by the small angle the string makes with the vertical.

107

C-5-11 Which Mass Is Faster?

Which of the following statements is true about the accelerations of the two masses in Figure 5-26?
(a) They both have the same acceleration.
(b) The acceleration of m_1 is twice the acceleration of m_2.
(c) The acceleration of m_2 is greater than the acceleration of m_1.
(d) None of the above.

C-5-11 (b) If the vertical mass moves by 2d, the horizontal mass will move by d.

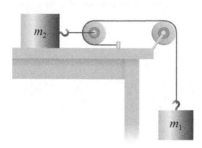

Figure 5-26 C-5-11

LO 7

5-7 Friction

If you push gently on the side of a heavy textbook lying on a table, the book does not move even though you are exerting a horizontal force on it. According to Newton's second law, the net force on the book must be zero because the acceleration is zero. Therefore, another force must be acting on the book to exactly balance the horizontal force you are applying. If you decrease the force you are applying, increase it slightly, or change which side of the book you push on, the book still does not move. The force that is preventing the book from moving is changing to remain exactly opposite to the force you apply. The force that prevents the book from moving is **friction**. Friction that acts between objects that are not moving relative to each other is called **static friction**.

If you now gradually increase the horizontal force you apply to the book, you will find that the static friction reaches a maximum just before the book starts to move relative to the table. The maximum force of static friction between two objects in contact with each other always occurs when the objects are on the verge of moving relative to each other.

If you were to place a second, identical book on top of the first book and gradually increase the horizontal

force you apply to the bottom book, you will notice that the force you have to exert on the bottom book to bring it to the verge of moving has increased. In fact, if you pull the book with a spring scale, you will find that the force required to bring the book to the verge of moving is doubled when you place the second book on top of it. The maximum force of static friction between two surfaces is directly proportional to the normal force between the two surfaces:

KEY EQUATION $$f_s^{max} = \mu_s N \qquad (5\text{-}2)$$

where μ_s is the coefficient of static friction.

The force of static friction therefore takes on values between zero and the maximum force of static friction:

KEY EQUATION $$0 \le f_s \le f_s^{max} \qquad (5\text{-}3)$$

 CHECKPOINT

C-5-12 The Force of Static Friction

What is the minimum value for the force of static friction?

C-5-12 In general, zero.

 CHECKPOINT

C-5-13 Force of Static Friction

You place a book on the horizontal desk in front of you. The book's weight is 20 N, and the coefficient of static friction between the desk and the book is 0.25. You push the book with a horizontal force of 2 N. What is the force of static friction on the book?
(a) zero
(b) 2 N
(c) 5 N
(d) none of the above

C-5-13 (b) book is not moving so the force of friction is equal to the force you apply.

When you apply a force to the book discussed above that is slightly more than the maximum force of static friction, the book will begin to move and continue to accelerate. As soon as the book begins to move, you will notice a sudden drop in the force that opposes you. To keep the book moving at a constant speed, you will have to decrease the applied force. The force of friction decreases somewhat when an object starts moving.

Friction that acts on a moving object is called **kinetic friction**. Kinetic friction is directly proportional to the normal force between an object and the surface on which it moves:

KEY EQUATION $$f_k = \mu_k N \qquad (5\text{-}4)$$

where μ_k is the coefficient of kinetic friction.

For a given object and surface, the coefficient of kinetic friction is usually less than the coefficient of static friction.

ONLINE ACTIVITY

Friction

The e-resource that accompanies every new copy of this textbook contains an Online Activity using the PhET simulation "Motion." Work through the simulation and accompanying questions to gain an understanding of the force of friction.

EXAMPLE 5-18

Coefficient of Kinetic Friction

Your friend is sitting on a sled on a level snow-covered surface. The sled has steel runners. You give the sled one push, which causes it to move away from you at a speed of 2.10 m/s. The sled stops 7.00 s later. Find the coefficient of kinetic friction between the sled and the snow-covered surface.

SOLUTION

Since we know the initial speed and elapsed time, we can calculate the acceleration, which will give us information about the forces acting on the sled.

The FBD of the sled in Figure 5-27 shows that the only horizontal force acting on the sled after you stop pushing it is the force of kinetic friction, f_k. Therefore, using Equation 5-4,

$$\sum F_x = m_1 a_x \Rightarrow f_k = ma \Rightarrow \mu_k N = ma \qquad (1)$$

For the normal force, we sum the forces in the vertical direction. Choosing up as our positive direction, we get

$$\sum F_y = ma_y \Rightarrow N - mg = 0$$
$$\Rightarrow N = mg \qquad (2)$$

Substituting Equation (2) into Equation (1), we get

$$\mu_k mg = ma \Rightarrow a = g\mu_k \qquad (3)$$

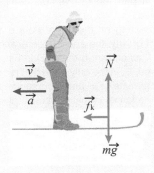

Figure 5-27 The FBD for the sled in Example 5-18

Choosing the direction of the sled displacement as positive, we have

$$\vec{v}(t) = \vec{a}t + \vec{v}_0$$

$$0 = -a(7s) + 2.1\,\frac{m}{s} \Rightarrow a = \frac{2.1\ \text{m/s}}{7\ \text{s}} = 0.300\ \text{m/s}^2 \qquad (4)$$

note that a in the line above is a variable that denotes the magnitude of the acceleration.

Substituting Equation (4) into Equation (3) gives

$$\mu_k = \frac{a}{g} = \frac{0.3\ \text{m/s}^2}{9.81\ \text{m/s}^2} = 0.0306$$

Making sense of the result:

A sled with steel runners slides easily on snow, so a low value for μ_k seems reasonable. We also note that μ_k has no units as expected.

EXAMPLE 5-19

Static or Kinetic?

You are pushing a 26.0 kg crate on a horizontal floor, where the coefficient of static friction is 0.170 and the coefficient of kinetic friction is 0.130. You apply a 140 N force at an angle of 27.0° below horizontal.

(a) Will the crate move?
(b) If the crate does move, find its acceleration.

SOLUTION

First, we draw the FBD of the crate; see Figure 5-28.

(a) The only force that opposes the motion of the crate is the force of static friction. If the horizontal component of the applied force is greater than the maximum force of static friction, then the crate will move. The maximum force of static friction depends on the normal force between the surface and the crate, which

(continued)

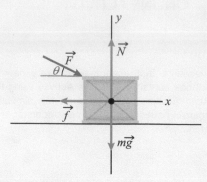

Figure 5-28   The FBD of the crate in Example 5-19

depends, in part, on the vertical component of the applied force.

The forces acting on the crate are the normal force, the force of gravity, the applied force, and the force of friction ($\vec{f}$), as shown in the FBD in Figure 5-28. The applied force, $\vec{F}$, is the only force that requires calculations to determine its x- and y-components.

Summing the forces in the x-direction gives

$$\sum F_x = m_1 a_x \Rightarrow F_x + f_x = ma_x \qquad (1)$$

Choosing right as our positive direction,

$$F\cos\theta - f = ma \qquad (2)$$

Next, we sum forces in the y-direction:

$$\sum F_y = F_y + mg_y + N_y = 0 \qquad (3)$$

Choosing up as our positive direction, we get

$$-F\sin\theta - mg + N = 0 \qquad (4)$$

$$N = mg + F\sin\theta \qquad (5)$$

$$N = F\sin\theta + mg \qquad (6)$$

where the variables in Equations (2) and (4) denote magnitudes of the quantities in Equations (1) and (3) respectively.

The maximum force of static friction is, by Equation 5-2,

$$f_s^{\max} = \mu_s N = \mu_s(F\sin\theta + mg)$$

$$= 0.17\,\text{N}\,[140\sin(27°) + (26\,\text{kg})(9.81\,\text{m/s}^2)]$$

$$= 54.2\,\text{N} \qquad (7)$$

The x-component of the applied force is

$$F_x = F\cos(27°) = (140\,\text{N})\cos(27°)$$

$$F_x = 125\,\text{N}$$

Since the x-component of the applied force is greater than the maximum force of static friction, the crate will move.

(b)   We use the force of kinetic friction in the calculation of the acceleration. Summing the forces in the x-direction and using Equation (2) we can write

$$F\cos\theta - \mu_k N = ma \qquad (8)$$

$$F\cos\theta - \mu_k(F\sin\theta + mg) = ma$$

Therefore,

$$a = \frac{F\cos\theta - \mu_k(F\sin\theta + mg)}{m}$$

$$= \frac{140\cos(27°) - 0.13[140\,\text{N}\sin(27°) + (26\,\text{kg})(9.81\,\text{m/s}^2)]}{26\,\text{kg}}$$

$$= 3.20\,\text{m/s}^2$$

Making sense of the result:

Since the vertical component of the applied force contributes to the normal force, the magnitude of the normal force is greater than mg. This, in turn, affects the magnitude of the force of friction. The units for acceleration make sense. In limit of no friction, the acceleration reduces to the force applied in the direction of motion divided by the mass.

 EXAMPLE 5-20

Which Way Will the System Move?

A 4.20 kg mass, m_1, hangs from a rope that runs over a massless pulley and is connected to a 6.70 kg mass, m_2, resting on the surface at an incline of 31.0°, as shown in Figure 5-29(a). The coefficient of static friction between the incline and m_2 is 0.09, and the coefficient of kinetic friction

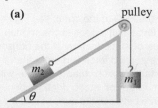

(a)   pulley

Figure 5-29(a)   Example 5-20

is 0.07. Initially, the system is at rest. Find the acceleration of the masses and the tension in the string.

SOLUTION

First, we determine which way the masses will move. On one side we have the weight of m_1 at 41.2 N [(4.2 kg)(9.81 m/s²)]; on the other side we have the component of the weight of m_2 along the incline at 33.9 N [(6.7 kg)(9.81 m/s²)(sin(31°))]. The weight of m_1 is opposed by the weight of m_2 and the force of friction. The force of friction, as a quick check, is given by $N\mu_k$. Mass m_2 is on the incline, so we can say that the normal force between m_2 and the incline is $N = m_2 g\cos\theta$. Therefore, $f_k = m_2 g\cos\theta\mu_k = 5.07\,\text{N}$. The sum of the component of the weight of m_2 along the incline and the force for friction is less than the weight of m_1, which means that m_1 will move down and m_2 will move up the incline.

In problems involving friction, it is best to do the above check because the direction of motion will affect the form of the equations of motion. This is not necessary in the absence of friction.

For the FBD for m_1 we have the force of gravity and the force of tension acting in the y-direction (Figure 5-29(b)).

(b)

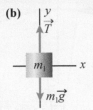

Figure 5-29(b)   The FBD for m_1

$$\sum F_y = T_y + m_1 g_y = ma_y \qquad (1)$$

Choosing up as our positive direction, we get

$$T - m_1 g = -m_1 a \qquad (2)$$

The acceleration of m_1 shows up as negative because up is our choice for a positive direction and the mass is moving down. Note that we have introduced 'a' as a variable to denote the magnitude of the acceleration.

The forces acting on m_2 are the force of friction, the normal force from the incline, the tension in the string, and the force of gravity. Figure 5-29(c) shows these forces resolved into components along the coordinate axes, one of which, the parallel, or the x-axis, points along the direction of the incline, the direction of the acceleration. Summing the forces in the perpendicular, or y-direction, we get

$$\sum F_y = N_y + m_2 g_y = 0 \qquad (3)$$

Choosing the direction of the normal force to be positive, we get

$$N - m_2 g \cos\theta = 0, \text{ hence } N = m_2 g \cos\theta \qquad (4)$$

Summing the forces in the parallel, or x-, direction yields

$$\sum F_x = T_x + f_{kx} + m_2 g_x = ma_x \qquad (5)$$

Choosing up the incline as our positive direction,

(c)

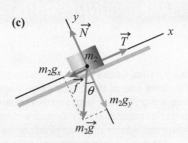

Figure 5-29(c)   The FBD for m_2

$$T - f_k - m_2 g \sin\theta = m_2 a \qquad (6)$$

where the variables in Equations (4) and (6) denote magnitude of the quantities in Equations (3) and (5) respectively.

We multiply Equation (2) by -1 and add it to Equation (6) to get

$$m_1 g - f_k - m_2 g \sin\theta = m_1 a + m_2 a \qquad (7)$$

Therefore,

$$a = \frac{m_1 g - m_2 g \mu_k \cos\theta - m_2 g \sin\theta}{m_1 + m_2}$$

$$= \frac{(4.2)(9.81) - (6.7)(9.81)(0.07)\cos(31°) - (6.7)(9.81)\sin(31°)}{4.2 + 6.7}$$

$$(8)$$

Substituting the values, we get $a = 0.313$ m/s², and

$$T = m_1 g - m_1 a = m_1(g - a) = 4.2 \text{ kg } (9.81 - 0.313) \text{ m/s}^2$$
$$= 39.9 \text{ N} \qquad (9)$$

Making sense of the result:

Since the acceleration is rather small, we would expect the value of the tension in the rope to be close to $m_1 g$. In the limiting case, where the system is not moving (or moving at a constant speed), the tension would be exactly equal to $m_1 g$.

It is worthwhile to determine whether the system will move when let go initially. For this, we have to look at the maximum force of static friction and see whether it is large enough to prevent the system from moving. In this case it is not. You can verify that yourself by using the value of μ_s provided.

ONLINE ACTIVITY

Ramp: Forces and Motion

The e-resource that accompanies every new copy of this textbook contains an Online Activity using the PhET simulation "Motion." Work through the simulation and accompanying questions to gain an understanding of the forces acting on an object and the object's resulting motion.

MAKING CONNECTIONS

The Power of Friction

While mooring a ship, sailors can exert a relatively small tension force on their end of the mooring line, while the other end sustains a significantly larger tension due to the pull by the ship being moored. This difference in tension results from the tremendous force of friction that can be generated by wrapping the rope a few times around a bollard. The force of friction between the bollard and the rope increases exponentially with the angle subtended by the rope as it hugs the bollard. (See Example 5-21 below.) For a coefficient of friction of 0.5 between the bollard and the rope, if the sailor wraps the rope only one time around the bollard, the force of tension on the other end will be over 20 times that with which the sailor is pulling. If, however, the sailor manages to wrap the rope 4 times around the bollard, the force on the other end can be over 12 000 times larger than the force with which the sailor pulls. Bollards are located on the docks or on the ship's deck.

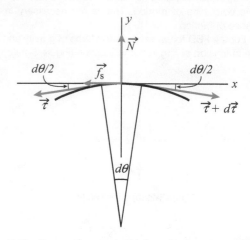

Figure 5-30   Forces acting on rope. This is used in Example 5-21 below.

EXAMPLE 5-21

Mooring a Ship: Rope and Bollard

The problem of calculating the effect of the force of friction between the rope and the bollard is complicated by the fact that the force of friction depends on the tension in the rope in the first place. For a given number of turns, the force of friction increases as the tension increases. A sailor moors a ship by wrapping one of its mooring lines around a bollard. Develop a relationship between the tension forces on each side of the bollard.

SOLUTION

Let us consider a small segment of the bollard subtended by the rope. On one side, the tension is $\tau_1 = \tau$, and it increases by an amount of $d\tau$ over an angle $d\theta$, giving us a value of $\tau_2 = \tau + d\tau$ on the other side (See Figure 5-30). The other force is the normal force exerted by the bollard on the rope. It can be shown that the angle between either end of the rope segment and the tangent to the bollard at the midpoint of the segment is $\dfrac{d\theta}{2}$.

$$N - \tau \sin \frac{d\theta}{2} - (\tau + d\tau) \sin \frac{d\theta}{2} = 0 \qquad (1)$$

For a very small angle $d\theta$, $\sin d\theta \approx d\theta$. Therefore, Equation (1) becomes

$$N - \tau \frac{d\theta}{2} - (\tau + d\tau) \frac{d\theta}{2} = 0 \qquad (2)$$

The product of the two infinitesimal quantities $d\theta$ and $d\tau$ becomes negligible compared to the other quantities in the problem. In the limit, we can ignore it. Therefore,

$$N \sim 2\tau \frac{d\theta}{2} = \tau d\theta \qquad (3)$$

The maximum force of static friction occurs when the rope is on the verge of slipping on the bollard. For this limiting case we have

$$f_s^{\max} = \mu_s N = \mu_s \tau d\theta \qquad (4)$$

Next, we sum the forces in the x-direction. Choosing right as positive we have

$$(\tau + d\tau) \cos \frac{d\theta}{2} - f_s^{\max} - \tau \cos \frac{d\theta}{2} = 0 \qquad (5)$$

where, for an infinitesimally small length of the rope, we can say that the force of friction acts in the x-direction.

We will use the small-angle approximation here again, for which $\cos d\theta \approx 1$. Using Equation (4), we get

$$(\tau + d\tau) - \mu_s \tau d\theta - \tau = 0 \qquad (6)$$

or

$$d\tau = \mu_s \tau d\theta \Rightarrow \frac{d\tau}{\tau} = \mu_s d\theta \qquad (7)$$

Integrating we get

$$\int_{\tau_1}^{\tau_2} \frac{d\tau}{\tau} = \mu_s \int_{\theta_1}^{\theta_2} d\theta$$

or

$$\ln \frac{\tau_2}{\tau_1} = \mu_s \Delta\theta \qquad (5\text{-}5a)$$

or

KEY EQUATION
$$\frac{\tau_2}{\tau_1} = e^{\mu_s \Delta\theta} \qquad (5\text{-}5b)$$

Making sense of the result:

The force of friction increases tremendously with each turn.

LO 8

5-8 The Force of a Spring

The force required to extend or compress an "ideal spring" is directly proportional to the displacement of the free end of the spring. This relationship is called **Hooke's law** and is often expressed as

$$F_s = -kx \qquad (5\text{-}6)$$

where k is the spring constant.

The greater the value of k, the more force required to stretch the spring by a given amount and, therefore, the stiffer the spring will be. Hooke's law is a useful approximation for many elastic materials.

✓ CHECKPOINT

C-5-14 Mass on a Vertical Spring

A spring with a spring constant k hangs vertically from a ceiling. You attach a 2 kg mass to the free end of the spring, pull the mass down, and then release it. When will the acceleration of the mass be zero?

(a) Never, because the mass is always under the effect of gravity.

(b) When the mass stops momentarily at the maximum extension of the spring.

(c) When the mass returns to the top and stops momentarily.

(d) When the extension of the spring is mg/k.

C-5-14 (d) This is the point where the net force is zero.

MECHANICS

EXAMPLE 5-22

Hooke's Law

A spring hangs freely from the ceiling of an elevator. While the elevator is accelerating upward at 2.90 m/s², you attach a 300 g mass to the spring and find that the spring stretches by 7.00 cm. Calculate the spring constant.

SOLUTION

The forces acting on the mass are the force of the spring and the force of gravity, as shown in the FBD in Figure 5-31. Summing the forces in the vertical direction we get

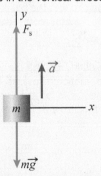

Figure 5-31 The FBD of the mass in Example 5-22

$$\sum F_y = F_{sy} + mg_y = ma \qquad (1)$$

Choosing up as the positive direction, we have

$$F_s - mg = ma, \quad \text{or} \quad F_s = mg + ma \qquad (2)$$

where the variables in Equation 2 denote magnitudes of the force of the spring, the acceleration due to gravity and the acceleration of the mass.

Applying Hooke's law, we get

$$kx = mg + ma$$

$$k = \frac{mg + ma}{x} = \frac{0.3 \text{ kg} (9.81 \text{ m/s}^2 + 2.9 \text{ m/s}^2)}{0.070 \text{ N/m}} = 54.5 \text{ N/m}$$

Making sense of the result:

The force of the spring counters the force of gravity acting on the mass and transmits the acceleration of the elevator to the mass. Recall that Hooke's law, $F = -kx$, means that the spring force points in the opposite direction to that of the stretch or compression. The actual sign of the force (kx) in the solution above depends on our choice of positive and negative directions.

MAKING CONNECTIONS

Inertial Navigation

An inertial navigation system uses mechanical forces to calculate the speed and position of a moving object, such as a ship or an airplane. Force sensors disclose data on the acceleration of an object. The acceleration information is then integrated to calculate the object's speed and location relative to its starting point. Since inertial navigation systems do not require signals from navigation beacons or satellites, such systems are immune to signal reception problems and electronic jamming. A rather simple motion (or acceleration) sensor can, in principle, be built using springs and masses.

LO 9

5-9 Circular Motion and Centripetal Forces

In Chapter 3, we saw that an object moving with speed v along a circular trajectory of radius r has a radial acceleration with a magnitude of

$$a_r = \frac{v^2}{r} \tag{5-7}$$

According to Newton's second law, the object must then be experiencing a net force that points toward the centre of the circle, as well. This net force is the vector sum of all the forces acting on the object in the radial direction. This net force is often called a centre-seeking force, or a **centripetal force**.

✓ CHECKPOINT

C-5-15 Centripetal Force

You swing a mass at the end of a string in a vertical circle such that the string remains taut at all times. Which of the following forces is the centripetal force acting on the mass when it is at the bottom of the circular trajectory?
(a) the force of gravity
(b) the tension in the string
(c) the vector sum of the force of gravity and the tension
(d) a force other than the force of gravity and the tension

C-5-15 (c)

👥 PEER TO PEER

I always thought that there was a separate physical entity called the "centripetal force" that existed in addition to the real and actual forces acting on an object moving in a circular trajectory. However, the centripetal force is really the net force, or the sum of all the actual physical forces, acting along the radial direction. Without these real physical forces, such as the force of tension, the normal force, and other real forces such as the ones discussed in this chapter, there would be no centripetal force.

➕➖ EXAMPLE 5-23

Traffic Circle Speed Limit

A car is going around a traffic circle (roundabout) that has a radius of 42.0 m. The traffic circle is not banked, and the coefficient of static friction between the tires and the road is 0.270.
(a) Find the maximum speed that the car can travel on the traffic circle without sliding.
(b) Comment on the centripetal force in this problem.

SOLUTION

The FBD in Figure 5-32 shows the forces as viewed from in front of the car. The force of friction, as shown, points to the centre of the circle.

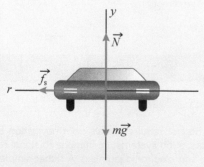

Figure 5-32 The FBD of the car in Example 5-23

(a) Since the direction of the acceleration is toward the middle of the traffic circle, we choose a horizontal x-axis or r-axis. We choose the direction pointing to the centre of the circle as the positive r-axis. Summing the forces in the horizontal (or radial) direction, we get

$$\sum F_r = ma_r \tag{1}$$

The only force acting on the car is the force of friction, so static friction provides all the centripetal force. Since the car is on the verge of slipping, the force of static friction is at its maximum value, using Equation 5-2,

$$f_s^{max} = N\mu_s = \frac{mv^2}{r} \tag{2}$$

To solve for v, we need to calculate the value for N. Summing the forces in the y-direction

$$\sum F_y = N_y + mg_y = ma_y = 0 \tag{3}$$

Choosing up as our positive direction, we get

$$N - mg = 0 \Rightarrow N = mg \tag{4}$$

Combining Equations (2) and (3), we get

$$mg\mu_s = \frac{mv^2}{r} \Rightarrow \tag{5}$$

$$v^2 = rg\mu_s = (42 \text{ m})(9.81 \text{ m/s}^2) \times 0.27 = 111.24 \text{ m}^2/\text{s}^2$$

$$v = 10.5 \text{ m/s}$$

(b) The centripetal force in this case is the force of friction because it is the only physical force pointing toward the centre of the circle.

Making sense of the result:

When the coefficient of friction is zero, the speed of the car, for it not to slip, must be zero! The car would not be able to turn, no matter how low its speed is.

EXAMPLE 5-24

Staying in the Loop

You swing a 310 g metal ball at the end of a 1.20 m-long string in a vertical circle. The mass has a speed of 17.0 m/s at the lowest point in the circle.
(a) Find the tension in the string.
(b) Comment on the centripetal force in this problem.

SOLUTION

(a) The FBD of the ball in Figure 5-33 shows that only the tension in the string and the force of gravity act on the mass.

Summing the forces in the vertical, or radial, direction gives

$$\sum F_y = T_y + mg_y = ma_y \quad (1)$$

Taking the direction toward the centre of the circle to be positive, we get

$$T - mg = ma \Rightarrow T = mg + \frac{mv^2}{r}$$
$$= 0.310 \text{ kg}\left(9.81 \text{ m/s}^2 + \frac{17^2}{1.2 \text{ m}} \text{ m}^2/\text{s}^2\right) = 77.7 \text{ N} \quad (2)$$

where the variables T, g, and a, in Equation (2) denote magnitudes of the tension, the acceleration due to gravity and the acceleration of the ball repsectively.

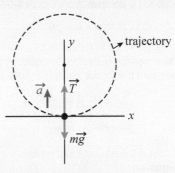

Figure 5-33 The FBD for the mass in Example 5-24

(b) Note that labelling the vertical axis as the r-axis would have also been appropriate. In vector form, the centripetal force is the vector sum of the force of gravity and the tension force. The magnitude of the centripetal force is given in Equation (2).

Making sense of the result:

The tension in the string is much greater than the weight of the ball as it supports the weight of the ball and provides its acceleration. When the speed of the ball is zero, the tension will be equal to mg, as expected.

EXAMPLE 5-25

Banked Racetrack

A race car is on a slippery circular track of radius 190 m. The racetrack is banked at an angle of 27° with respect to the horizontal.
(a) Find the speed at which the race car will not slide up or down the track, assuming that friction with the track surface is negligible.
(b) Identify the centripetal force in this problem.

SOLUTION

(a) Step 1: Identify all the forces acting on the object.

The forces acting on the race car are the force of gravity and the normal force from the incline.

Step 2: Draw an FBD.

Figure 5-34 shows the forces as viewed from in front of the car.

Steps 3 and 4: Establish the direction of the acceleration, and set up a coordinate system with one axis pointing along the direction of the acceleration.

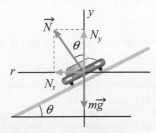

Figure 5-34 The FBD for the car on a banked curve in Example 5-25

The car will move in a horizontal circle. The acceleration is therefore directed horizontally toward the centre of the circle. So we choose a horizontal axis and a vertical y-axis. Either x-axis or r-axis is an appropriate label for the horizontal axis. However, since centripetal or radial acceleration is involved, we will refer to this as the r-axis, for radial.

Step 5. Resolve the forces into components along the chosen axes.

(continued)

115

The force of gravity points along the vertical axis. We use simple trigonometry to resolve the normal force from the incline as a radial component (N_r) and a vertical component (N_y), as shown in Figure 5-34.

Step 6. Write equations for the motion along each axis. The sum of the forces along the vertical direction is

$$\sum F_y = N_y + mg_y = 0 \qquad (1)$$

Taking up to be positive, from Figure 5-34 we have

$$N\cos\theta - mg = 0 \qquad (2)$$

where the variables in Equation (2) are used to denote magnitudes of the quantities in Equation (1).
Next, we sum the forces along the radial direction:

$$\sum F_r = N_r = ma_r \Rightarrow$$

$$N\sin\theta = ma_r = m\frac{v^2}{r} \qquad (3)$$

Step 7. Solve the equations of motion.
From Equation (2),

$$N\cos\theta = mg \qquad (4)$$

We have three unknowns: the mass of the car, m, the speed of the car, v, and the normal force, N. If we divide Equation (3) by Equation (4), we get

$$\tan\theta = \frac{v^2}{rg} \Rightarrow v = \sqrt{rg\tan\theta} = \sqrt{(190 \text{ m})(9.81 \text{ m/s}^2)\tan(27°)}$$

$$v = 30.8 \text{ m/s}$$

(b) The centripetal force is the vector sum of all the forces acting along the radial direction pointing toward the centre. In this case, it is the horizontal component of the normal force, or $N\sin\theta$.

Making sense of the result:

Notice the difference between this example and Example 5-13 where the direction of the acceleration was along the surface of the incline for a mass sliding down the incline. Therefore in that example, one of the axes pointed along the incline. We also note here that the normal force from the track on the car is given by Equation (4) as $N = \dfrac{mg}{\cos\theta}$. For a banking angle of zero, the car, without friction, will not be able to move along the track without sliding.

EXAMPLE 5-26

Friction on a Banked Curve

A car is travelling on a curve of radius r banked at an angle θ. The coefficient of static friction between the car and the road is μ_s.
(a) Derive an expression for the maximum speed that the car can travel without slipping.
(b) Derive an expression for the normal force, and comment on the expression you get for the normal force in this example compared to the expressions you obtained for the normal force in Examples 5-13, 5-14, and 5-25.
(c) Identify the centripetal force acting on the car.

SOLUTION

(a) Steps 1 and 2: Identify all the forces acting on the object, and draw an FBD.
The forces acting on the car are the force of gravity, the normal force from the incline, and the force of friction, as shown in Figure 5-35.
Steps 3 and 4: Establish the direction of the acceleration, and set up a coordinate system with one axis parallel to the acceleration.

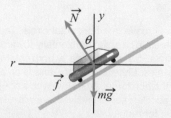

Figure 5-35 The FBD for the car as it goes around a banked curve for Example 5-26. Notice the direction of the force of friction

☑ CHECKPOINT

C-5-16 Friction Direction

Why is the force of friction acting downward on the incline?

As in Example 5-25, the car moves in a horizontal circle, and the acceleration points toward the centre of the circular path. Therefore, we again choose a horizontal (radial) r-axis and a vertical y-axis.

Step 5: Resolve the forces into components along the axes of the coordinate system.

In Figure 5-36, we can see that the components of the normal force and the force of friction can be found using the sine and cosine of the incline angle θ, see step 6 for details.

Step 6. Write the equations of motion along each axis.

The sum for the forces along the vertical direction gives us

$$\sum F_y = N_y + mg_y + f_y = ma_y$$

Taking up as positive

$$N\cos\theta - mg - f\sin\theta = 0 \qquad (1)$$

If the car is travelling at the maximum speed possible without slipping, then it is on the verge of slipping, and

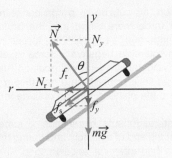

Figure 5-36 The FBD for the car in Example 5-26 with the components of the forces decomposed along chosen axes

the force of static friction is at its maximum value of $N\mu_s$. Therefore, using Equation 5-2, we have

$$N\cos\theta - mg - N\mu_s\sin\theta = 0 \quad (2)$$

Next, we sum the forces along the radial direction:

$$\sum F_r = N_r + f_r = ma_r$$

Taking the radial direction toward the centre of the circle to be positive, we have

$$N\sin\theta + N\mu_s\cos\theta = \frac{mv^2}{r} \quad (3)$$

Alternatively, one can also say that we have taken the positive r-axis to point to the centre of the circle.

Step 7: Solve the equations of motion.
Rearranging Equation (2) gives

$$N\cos\theta - N\mu_s\sin\theta = mg \quad (4)$$

We divide Equation (3) by Equation (4) to eliminate N:

$$\frac{\sin\theta + \mu_s\cos\theta}{\cos\theta - \mu_s\sin\theta} = \frac{v^2}{rg}$$

$$v^2 = \frac{rg(\sin\theta + \mu_s\cos\theta)}{\cos\theta - \mu_s\sin\theta} \quad \text{or} \quad v = \sqrt{\frac{rg(\sin\theta + \mu_s\cos\theta)}{\cos\theta - \mu_s\sin\theta}}$$

note that the variables N, g and f in Equations (1), (2) and (3) denote magnitudes.

(b) As for the normal force as given by Equation (4), we can write

$$N = \frac{mg}{\cos\theta - \mu_s\sin\theta} \quad (5)$$

Each of the examples involving inclined planes (Examples 5-13, 5-14, 5-20, 5-25, and 5-26) gives us a different value for the normal force, depending on the situation involved. It is a common error to assume that the normal force is always equal to mg or $mg\cos\theta$. Now you can see that most often it is not. See the examples for details.

(c) The centripetal force is the sum of all the forces acting along the radial direction. In vector form, the sum is given by $\sum \vec{F_r} = \vec{N_r} + \vec{f_r}$. The magnitude of the sum is given by Equation (3) as $N\sin\theta + N\mu_s\cos\theta$.

Making sense of the result:

When the coefficient of friction is zero, the value of the normal force reduces to $mg/\cos\theta$ and v^2 reduces to $rg\tan\theta$ as we have seen the previous example, (example 5-25).

 PEER TO PEER

I was under the impression that the normal force for a car going around a banked curve always equals $mg/\cos\theta$. This expression does not give the normal force in all cases, although it does apply for a car going around a banked curve where there is no friction. The correct way to deal with the normal force is to treat it as an unknown and obtain its value if needed by solving the relevant equations of motion.

LO 10

5-10 Frames of Reference and Fictitious Forces

A Thought Experiment

A passenger is standing still in a stationary bus. The bus then accelerates before the passenger can sit down. At that moment, the passenger is not holding on to any of the handles or the support poles on the bus. You happen

 MAKING CONNECTIONS

Maximum *g* Forces

A *g* force is a force expressed as a multiple of the force of gravity at Earth's surface. Aerobatics pilots can experience considerable centripetal forces, especially when flying at jet speeds. For an inside loop, the plane flies in a vertical circular path loop with the canopy (and the pilot's head) facing toward the centre of the loop (Figure 5-37). The *g* forces experienced during an inside loop are called positive *g* forces. Positive *g*

forces drive blood away from the brain. During a high positive *g* manoeuvre, pilots can experience grey-outs or blackouts as a result of reduced oxygen in the brain. With special training and a G-suit, which helps maintain blood flow to the brain, a pilot can withstand a positive *g* force of $9g$.

During an outside loop, the top of the plane faces away from the centre of the vertical loop. Depending on

(continued)

Photography Perspectives – Jeff Smith/Shutterstock.com

Figure 5-37 A plane in a positive *g* loop

the acceleration, an outside loop can cause negative *g* forces, which force more blood into the brain and are often more dangerous than positive *g* forces. Generally, the maximum tolerable negative *g* force is about −3*g*. For a high *g* manoeuvre, a pilot will position the plane to experience a positive rather than a negative *g* force, for example, rolling the plane to turn an outside loop into an inside loop.

In contrast, astronauts experience *g* forces mostly during liftoff, at the end of a launch, and during reentry, due to linear rather than rotational accelerations. The blood is "driven" either to the front or to the back of the body. As the body accelerates, the blood's inertia will cause it to lag behind the body parts, owing to blood's fluid nature, which would cause the blood to be pushed toward the front or the back relative to the accelerating body. The *g* forces experienced in this way are called horizontal *g* forces and can be as high as 12*g* or 17*g* for short periods of time, depending on the direction of the force causing the acceleration. In 1954, Colonel John Stapp is reported to have sustained a *g* force of 46.2*g* in a rocket sled as it was slowing down. Higher values have been reported during accidents as well. Refer to Chapters 3 and 4 for a more detailed and in-depth discussion.

to be standing outside the bus while a friend of yours is already sitting comfortably inside the bus. Let us say for the purpose of demonstration that the floor of the bus is so slippery that there is effectively no friction between the bus floor and the shoes of the standing passenger.

The bus accelerates, and from your perspective, as it moves, the passenger simply stays still with respect to you. Since the passenger is not holding on to anything, and since there is no friction between his shoes and the floor of the bus, the net force on the passenger in the direction of motion of the bus is zero. Therefore, that person will not move.

Your frame of reference is called an **inertial frame of reference**. For the passenger to accelerate, he would have to be acted upon either by friction from the floor or by the force from the pole.

In an inertial frame of reference, an object will not accelerate unless there is a real, physical, identifiable force acting on it.

How do things look from the point of view of your friend who is sitting comfortably in her seat on the bus? Suppose that the acceleration is not too great and that she does not feel much of a force from the seat as the bus accelerates. What she will notice, however, is the standing passenger accelerating backward with respect to her. It would then only be natural for her to assume that he is

under the influence of a force pulling him backward. The magnitude of the force will be, naturally, to her, the person's mass times his acceleration. The force she perceives to be there is called a **fictitious force** and only exists with respect to her, in her frame of reference. It is a fictitious force because there is no identifiable agency, or cause, for that force. In her frame of reference the passenger is accelerating without any real, physical identifiable forces acting on him, forces such as the fundamental force (e.g., gravity) or the nonfundamental forces (e.g., tension and the normal force). A frame of reference in which objects accelerate in the absence of real, identifiable physical forces, and in which one must account for fictitious forces to explain physical phenomena is called a **noninertial frame of reference**.

When the bus accelerated, the passenger was standing still. This makes sense from your perspective because there are no forces to move him forward. However, as soon as your friend stretched her arm back to grab the passenger, the passenger started to move forward. From your perspective, this also makes sense. Suddenly, there is a real net force acting on the passenger in the direction of motion of the bus; this is the force your friend is exerting on the passenger and is the only reason the person moves forward from your perspective is that there is a very real force pulling him, and that is the force exerted by your friend's arm on him.

Forces are vectors. Forces are caused by physical inter-actions between objects, such as gravitational attraction between you and Earth. There are fundamental and nonfundamental forces in nature. With the exception of the force of gravity, all the forces reviewed in this chapter are nonfundamental forces that require physical contact. A net force acting on an object causes the object to accelerate, deform, or both. Cartesian axes are often the most convenient type for analyzing forces and motion. Problems become simpler to solve when one of the axes is chosen to lie along the direction of the acceleration.

Forces

1. Forces are vectors that have magnitude and direction.

2. Forces have real, physical causes.

3. The four fundamental forces underlie all the nonfundamental forces.

4. Identifying and quantifying the forces acting on an object is necessary to predict the object's trajectory.

Newton's Second Law

Mass is a measure of an object's inertia. Inertia determines the ability of an object to resist being accelerated. The mass of an object depends on the amount of material contained within that object.

The net force acting on an object results in the acceleration of the object:

$$\vec{F}_{net} = m\vec{a} \tag{5-1}$$

Acceleration and Free Body Diagrams

A free body diagram shows all the forces acting on a single object.

Force vectors can be resolved into components.

When solving problems, the object's acceleration dictates the choice of coordinate system.

Newton's second law can be used to generate equations of motion which can be solved and used to predict trajectories.

APPLICATIONS

Applications: *g*-forces and medical conditions, car racing, driving safety, effects of friction, tension safety, inertial navigation

Key Terms: centripetal force, compression, fictitious force, force, free body diagram, friction, fundamental force, *g* force, Hooke's law, inertia, inertial frame of reference, kinetic friction, mass, Newton's first law, Newton's second law, Newton's third law, nonfundamental force, noninertial frame of reference, normal force, static friction, tension

QUESTIONS

Note that more than one answer can be true

1. A polar bear cub pushes his mother on an icy, frictionless surface in Lancaster Sound, Nunavut. The cub and his mother are originally at rest, and the cub's weight is 1/6 of the mother's weight. She moves away from the cub at 1.1 m/s, and the cub then grabs her arm just before she gets out of his range, and hangs on. What is the final speed of the cub and his mother?
   (a) zero
   (b) 1.1/7 m/s
   (c) The mother stops, and the cub moves at 6.6 m/s.
   (d) The mother slows down slightly while the cub's speed increases.

2. Figure 5-38 shows the FBD of an object. What is the direction of motion of the object?
   (a) The object is moving to the left.
   (b) The object is moving to the right.
   (c) The object is not moving.
   (d) There is not enough information to answer the question.

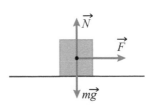

Figure 5-38   Question 2

3. An object moves to the right while subject to a force whose magnitude increases with time but points to the left. Which of the following statements is true?
   (a) The particle will reverse direction as soon as the force is applied.
   (b) The particle will slow down and eventually reverse direction.
   (c) The particle will not slow down because the magnitude of the force decreases with time.
   (d) The particle will not reverse direction.

4. A puck is moving freely on frictionless ice along the positive x-axis when a force is applied to the puck. The direction of the force is maintained along the positive x-axis, but its magnitude drops exponentially with time. Which of the following statements is true?
   (a) The puck's speed increases with time.
   (b) The puck's speed decreases with time.
   (c) The puck's acceleration increases with time.
   (d) None of the above are true.

5. A particle moving to the right is acted upon by a constant net force pointing to the left. The particle moves along the x-axis, and the force is applied at $t = 0$, when the particle's position is to the right of the origin (positive). We can say that the magnitude of the particle's position with respect to time
   (a) changes like a parabola that opens upward;
   (b) changes like a parabola that opens downward;
   (c) decreases linearly;
   (d) increases linearly.

6. You ($m = 76$ kg) are riding in a car moving at a constant speed of 12 m/s on a curved road of radius 10 m. What is the net force acting on you?
   (a) The net force cannot be determined without knowing the force of friction between the road and the car.
   (b) The net force cannot be determined without knowing the force of friction between you and the car.
   (c) The net force cannot be determined without knowing the normal force from the seat cushion on you.
   (d) None of the above.

7. You pull your friend with a horizontal rope such that the force you apply to the rope is 160 N. Your friend stands his ground without pulling back. Neither of you move. Which of the following is true?
   (a) The net force on your friend is 160 N.
   (b) The force of friction on your friend is 160 N.
   (c) The tension in the rope is 320 N.
   (d) None of the above are true.

8. In Figure 5-39, the combined mass of the monkey, bananas, and platform is m. In order to keep the platform still, the monkey has to pull the rope with which force?
   (a) mg
   (b) $\dfrac{mg}{2}$
   (c) $\dfrac{mg}{4}$
   (d) $2mg$

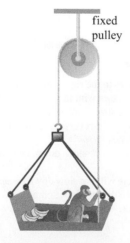

Figure 5-39   Question 8

9. What can be said about the initial accelerations of the two masses in Figure 5-40? How are the accelerations related?

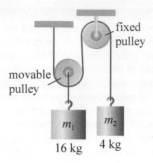

Figure 5-40   Question 9

10. What can be said about the initial accelerations of the three masses in Figure 5-41? How are the accelerations related?

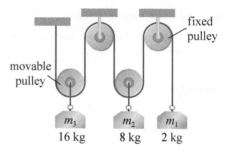

Figure 5-41   Question 10

11. You are on a frictionless skating rink. Your mass is 90 kg. You give a rope to a friend whose mass is 60 kg. You pull the rope, and your friend hangs on as she accelerates toward you at 3 m/s² without pulling on the rope. Which of the following statements is true?
    (a) You do not move.
    (b) The force acting on you is 270 N.
    (c) Your acceleration is 3 m/s².
    (d) The force acting on you is 180 N.

12. Two friends ($m_1 = 52$ kg and $m_2 = 87$ kg) pull one another using a rope. They each apply a force of 230 N to the horizontal rope, but neither move. Which of the following statements is true?
    (a) The tension in the rope is 230 N.
    (b) The tension in the rope is 460 N.
    (c) The force of friction on each friend is 460 N.
    (d) The heavier friend experiences a larger force of friction as they both pull.

13. Two friends are standing on an ice rink, one of them pulling the other with a rope with a force of 145 N. The friend being pulled ($m_1 = 50$ kg) moves toward the friend doing the pulling ($m_2 = 75$ kg) at a constant speed of 6 m/s. Which of the following statements is true?
    (a) The speed of the friend doing the pulling must be 4 m/s.
    (b) The force of kinetic friction on the friend being pulled is 145 N.
    (c) The net force on the friend being pulled is 145 N.
    (d) The force of static or kinetic friction on the friend doing the pulling is 145 N.

14. Two children are riding in the back of a truck. The truck comes to an emergency stop and both children slide with respect to the truck. Assuming the same coefficient of friction between each child and the back of the truck, which child will slide first—the lighter or the heavier child ?

15. Figure 5-42 shows the FBD of an object. The object must
(a) be moving up;
(b) be accelerating up
(c) momentarily stationary
(d) be accelerating down

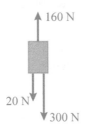

Figure 5-42 Question 15

16. Two boxes, one twice the mass of the other, slide down the same incline, starting from rest at the same height. The coefficient of friction between each of the boxes and the incline is the same. Compare the times the boxes take to go down the incline.

17. You are standing in an elevator that is accelerating upward. The force that you exert on the elevator
(a) is larger than the force that the elevator exerts on you;
(b) is smaller than the force that the elevator exerts on you;
(c) is equal to the force that the elevator exerts on you.
(d) none of the above.

18. You are sitting in a 1000 kg truck, which is slowing down at 10 m/s². Your mass is 50 kg. Which of the following is true?
(a) The force that you exert on the truck is 500 N.
(b) The net force on the truck is 9500 N.
(c) The force that the truck exerts on you is 500 N.
(d) The vertical force that the truck exerts on you is 490 N.

19. A crate is sitting on a horizontal surface where the coefficient of static friction is 0.6. A worker pushes horizontally on the crate, which has a mass of 45 kg. The crate does not move, and there are no other forces acting on it. Which of the following statements is true?
(a) The force of friction on the crate is $mg\mu_s$.
(b) The force of friction on the crate is greater than the force that the worker exerts on the crate.
(c) The force exerted on the worker by the crate is equal in magnitude to the force of friction exerted by the crate on the surface.

20. Two friends are pushing a book in opposite directions, each pushing with a force of 320 N. One friend says the net force on the book is 640 N. Is this correct? Explain.

21. Two friends push a sled on snow in opposite directions. One pushes with a force of 190 N, and the other pushes with a force of 130 N. The sled does not move. What is the net force on the sled?
(a) 320 N
(b) 60 N
(c) zero
(d) The net force cannot be determined.

22. You are sitting in a car that is rounding a 40 m-radius curve banked at 21°, with a speed of 12 m/s. Which of the following statements is true?
(a) The horizontal force exerted on you by the car is larger than the horizontal force exerted on the car by you.
(b) The net force on the car is $\dfrac{Mv^2}{r} - \dfrac{mv}{r}$.
(c) The net force on you is $\dfrac{mv^2}{r}$.
(d) The net force on the car is $\dfrac{mv^2}{r}$.

23. A lion trainer is in a train car looking after his 120 kg lion named Pouncer. Pouncer is not in the mood to play and is sitting still. The trainer prods Pouncer gently with a force of 200 N in a direction toward the front of the train, but it is not enough to get Pouncer to move. The train is slowing down at an acceleration of 5 m/s². The net force on Pouncer is
(a) 720 N;
(b) 600 N;
(c) 480 N;
(d) 120 N;

24. A rope connects M_1 hanging vertically to M_2 resting on top of a horizontal frictionless surface. The system is originally at rest.
(a) Is the tension in the rope larger than, smaller than, or equal to $M_1 g$?
(b) If the surface that M_2 rests on provides enough friction to prevent M_2 from moving, what is the tension in the rope?

25. You are riding in a car moving at a speed of 20.0 m/s around a curve of radius 40.0 m. The mass of the car is 1300 kg, and your mass is 60.0 kg. Find the magnitude of
(a) the force that you exert on the car
(b) the net force on the car
(c) the net force on you
(d) the horizontal force that the car exerts on you
(e) the force that the car exerts on the road

26. Two buckets, each of mass m, are suspended from a rope that goes around the fixed massless frictionless pulley in Figure 5-43. The system is at rest. What is the tension in the rope?
(a) mg
(b) $2mg$
(c) zero
(d) mg^2

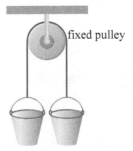

Figure 5-43 Question 26

27. Two buckets of equal mass are suspended vertically from a rope that goes around the fixed massless frictionless pulley as shown in Figure 5-43. The pulley rotates clockwise so that the masses are moving at a constant speed. Which of the following statements is true?

(a) The tension in the rope is equal to mg.

(b) The tension in the rope is equal to $2mg$.

(c) The tension in the rope is zero.

(d) The tension cannot be determined without knowing the masses.

28. The bucket (mass m) hanging from the right side of the massless frictionless pulley in Figure 5-43 is heavier than the bucket hanging from the rope on the left side of the pulley. The tension in the left side of the rope is

(a) mg;

(b) greater than mg;

(c) less than mg;

(d) zero.

29. A centipede of mass m is climbing a thread that is hanging vertically from the edge of a table. When the centipede begins its climb, its acceleration is 2 m/s² upward. The tension in the rope is

(a) mg;

(b) ma;

(c) $mg + ma$;

(d) $mg - ma$.

30. A monkey uses a fixed rope to climb down from a coconut tree. The monkey races down with an acceleration of $1.5g$ downward to avoid a coconut dropped by another monkey. Taking up as the positive direction, the tension in the fixed rope is

(a) $0.5mg$;

(b) $1.5mg$;

(c) $2.5mg$;

(d) $-0.5mg$.

31. A child swings a string with three beads attached to it in a vertical circle. The bead closest to his hand (distance l away) has a mass of m, the bead after that (distance $2l$ away) has a mass of $2m$, and the farthest bead from his hand (distance $3l$) has a mass of $3m$. The tension in the rope is the highest in the section

(a) closest to the child's hand;

(b) between m and $2m$;

(c) between $2m$ and $3m$.

32. Mass m_1 hangs from the massless movable pulley shown in Figure 5-44. One end of the rope supporting the pulley is attached to the ceiling; the other goes over the fixed pulley and is attached to the mass on the frictionless surface. When the system is let go, the acceleration of m_1 is

(a) half the acceleration of m_2;

(b) double the acceleration of m_2;

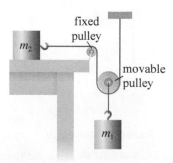

Figure 5-44   Question 32

(c) the same as the acceleration of m_2.

(d) The relationship between the two accelerations depends on the masses.

33. Mass m_1 hangs vertically from one end of a rope as shown in Figure 5-45. The rope goes over the fixed pulley and around the movable pulley that is attached to m_2. The other end of the rope is fixed to the protrusion that is attached to the horizontal surface. As the system moves, the acceleration of m_1 is

(a) twice the acceleration of m_2;

(b) half the acceleration of m_2;

(c) the same as the acceleration of m_2.

(d) More information is needed to answer the question.

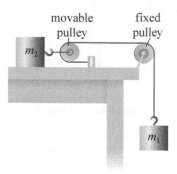

Figure 5-45   Question 33

34. Four blocks rest on top of each other such that $m_1 = 2m_2 = 4m_3 = 8m_4$. The coefficients of friction between the blocks are $\mu_{s3} = 2\mu_{s2} = 4\mu_{s1}$. The bottom block is the heaviest of the four blocks. It is pushed with a horizontal force strong enough that at least one pair of the blocks slips. Which block will be the first to slide with respect to the block beneath it?

(a) m_4

(b) m_2

(c) m_3

(d) They will all slide simultaneously.

35. A bug is standing stationary on a spinning turntable. Some oil is poured on the turntable to render the surface frictionless. As soon as the surface becomes frictionless, the bug will

(a) continue to move in a circle, but the radius of the circle will get larger and larger;

(b) move in a straight line tangent to the circular trajectory it was on;

(c) move radially outward with respect to the turntable;

(d) move radially outward toward the edge of the turntable;

36. A child is swinging a string with a ball attached to its end in a circle that is nearly horizontal. Ignoring gravity, if the string is cut, in which direction will the ball go?

PROBLEMS BY SECTION

For problems, star ratings will be used, (✶, ✶✶, or ✶✶✶), with more stars meaning more challenging problems.

Section 5-1   Force and Net Force

37. ✶ (a) Resolve each force in Figure 5-46 into Cartesian components.

(b) Calculate the net force in Figure 5-46.

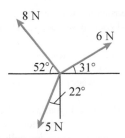

Figure 5-46 Problem 37

38. ✳ Find the magnitude and direction of the forces listed below. Report the direction as an angle measured counterclockwise with respect to the *x*-axis.
   (a) $6\hat{i} + 2\hat{j}$ N
   (b) $-21\hat{i} + 16\hat{j}$ N
   (c) $7\hat{i} - 4\hat{j}$ N

39. ✳ Add the following forces. Express the resultant forces in Cartesian form with unit vectors.
   (a) $-4\hat{i} - 2\hat{j} + 0.5\hat{k}$ N, $3\hat{j} - 4\hat{k}$ N, $3\hat{i} + 2\hat{j}$ N
   (b) 14.0 N [57°], $(3.0\hat{i} - 4.0\hat{j})$ N, 8.0 N [236°]
   (c) $(4\hat{i}, 2\hat{j}, -5\hat{k})$, $(-9\hat{i}, 14\hat{j}, 11\hat{k})$
   (d) $r\theta\phi$ notation, (21 N, 32°, 27°), (17 N, 160°, 295°)

40. ✳ Add the following forces, which are all in newtons. Express your answer in both polar and Cartesian notation.
   (a) $(4\hat{i}$ N, $2\hat{j}$ N), $(6\hat{i}$ N, $-4\hat{j}$ N), $(10\hat{i}$ N, $-4\hat{j}$ N)
   (b) (16 N, 7° N of E), (5 N, 289°), (24 N, −25°)
   (c) $(5\hat{i}$ N, $-2\hat{j}$ N), (2 N, 230°)

Section 5-2 Newton's Laws

41. ✳ A constant force of 130 N is applied to a block of mass 9.00 kg, which is sitting on a frictionless horizontal surface. Find the speed of the mass after it has moved 14.0 m, starting from rest.

42. ✳ The last part of a roller coaster's track is horizontal. A roller coaster is brought to rest using a constant braking force of 4300 N. It takes 9.00 s to bring the roller coaster to a stop from a speed of 23.0 m/s. Find the mass of the roller coaster.

43. ✳ Assume that a particular crash test expert can withstand 36 times the acceleration due to gravity. Find the shortest distance over which her car can be brought to a stop from a speed of 70 km/h without the acceleration exceeding 36*g*.

44. ✳ A car of mass 2100 kg is brought to an emergency stop from a speed of 43 m/s over a distance of 21 m. What is the average net force experienced by the car during this stop?

45. ✳ You are sitting in the passenger seat of a car driving on the highway at a speed of 110 km/h when the driver (*m* = 79 kg) exits on a circular ramp of radius 63.0 m. The mass of the car is 1300 kg, and your mass is 71 kg. Find
   (a) the horizontal force that the road exerts on the car;
   (b) the horizontal force that you exert on the car;
   (c) the net horizontal force on the car.

46. ✳ You are sitting in the back seat of a car moving on the highway, when the driver hits the brakes. The net force on you is 320 N. Find the net horizontal force that you exert on the car.

Section 5-3 Free Body Diagrams, Acceleration, and Equations of Motion

47. ✳✳ For each of the following scenarios, determine the direction of the acceleration. Assign your system of axes, resolve the forces into components, and write the equations of motions for each scenario.
   (a) A child slides down a frictionless incline with a slope of angle θ.
   (b) A child slides down an incline with a slope of angle θ.
   (c) A conical pendulum.
   (d) A man pushes his daughter of mass *m* up the frictionless surface of a slide at an angle θ using a horizontal force *F*. Draw the FBD of the daughter. Assume she is moving up the incline as a result of being pushed by her father.
   (e) A worker pulls a crate of mass *m* on a horizontal surface, where the coefficient of friction is μ_k. She uses a force of magnitude *F* directed at an angle θ above the horizontal. Assume the crate is moving as a result of being pushed.
   (f) A mass moving at a constant speed at the end of a string rotated in a vertical circle when the mass is
      (i) at the top;
      (ii) at the bottom;
   (g) A car banking a frictionless curve with radius *r* and banking angle θ.
   (h) A car banking a curve with radius *r*, banking angle θ, and coefficient of static friction μ_s.

Section 5-4 The Force of Gravity

48. (a) ✳ If your weight is 670 N, what is your mass?
   (b) How much would you weigh on the Moon, where the acceleration due to the Moon's gravity is approximately 1/6 the acceleration due to gravity at Earth's surface?
   (c) If you can jump 60 cm straight up on Earth, how high can you jump on the Moon?

49. ✳✳ You (m_1 = 81 kg) are standing in an elevator (m_2 = 2300 kg) that is moving downward with a speed of 9.0 m/s. The elevator comes to a stop with a constant acceleration over a distance of 4.0 m. Find
   (a) the net force on you;
   (b) the net force on the elevator;
   (c) the force that you exert on the elevator;
   (d) the gravitational force on the elevator.

Section 5-5 The Normal Force

50. ✳✳ Four blocks of masses 20 kg, 30 kg, 40 kg, and 50 kg are stacked on top of each other in an elevator in order of decreasing mass, with the lightest mass on top. The elevator moves down with an acceleration of 3.2 m/s². Find the contact force between the 30 kg and the 40 kg blocks.

51. ✳✳ A woman is standing in an elevator carrying a 2.7 kg cat. The elevator begins to accelerate upward at 3.2 ms². Find the force that the cat exerts on the woman.

52. ✳✳ You are pushing four blocks of masses 4.00 kg, 7.00 kg, 11.00 kg, and 17.00 kg positioned next to each other on a frictionless surface as shown in Figure 5-47 (Page 124). Find the force of contact between the 7.00 kg and 11.00 kg blocks when
   (a) you push on the lightest mass with a 190 N force to the left;
   (b) you push on the heaviest block with a 190 N force to the right.

123

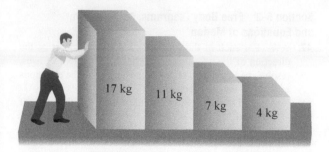

Figure 5-47 Problem 52

53. ✷✷ Starting from rest, a car skids 15 m down an ice-covered road, where the friction is negligible. The speed of the car at the end of 15 m is 3.2 m/s. Find the slope angle of the road.

54. ✷✷ Children are enjoying a constant slope water slide at an amusement park. They run toward the top of the 6 m-long slide and lie flat on the slide surface. A child who arrives at the top of the slide with a speed of 1.4 m/s reaches a speed of 3.6 m/s at the bottom of the slide. Find the slope angle of the slide, assuming that its friction is negligible.

55. ✷ A polar bear cub slides 31 m toward his mother down a snow-covered 37° slope starting from rest. At the end of the 31 m, the cub's speed is 4.7 m/s. Find the coefficient of friction between the cub and the snow-covered surface.

56. ✷✷ You have just had the apples in your orchard harvested into a large number of boxes with masses of 130 kg each. You wish to lift these boxes 12.0 m up to the fourth floor of your packaging facility. Since you have only one person to help, you construct a slippery surface that is inclined 30° above the horizontal, and then fix a pulley at the top corner as shown in Figure 5-48. You hang from the rope at the top of the incline and let the force of gravity pull the crates up the ramp. The friction between the crates and the inclined surface is negligible.
(a) What must your minimum mass be for this system to work?
(b) If you do not wish to hit the ground with a speed greater than 2.0 m/s, what must your maximum mass be?
(c) How long will it take to lift one crate to the fourth floor if you have the mass calculated in part (b)?

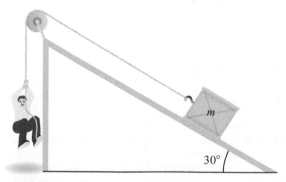

Figure 5-48 Problem 56

Section 5-6 Tension

57. ✷✷ Mass M_1, sitting on a horizontal frictionless surface, is connected by a rope that runs over a massless pulley to mass M_2, which is hanging vertically. The system is originally at rest and is then let go.

(a) Find the tension in the rope when $M_2 > M_1$.
(b) How would your answer to part (a) change if the masses were interchanged?
(c) Obtain numerical values for parts (a) and (b) using $M_1 = 2.0$ kg and $M_2 = 12.0$ kg.

58. ✷✷ Four food packages ($m_1 = 7.00$ kg, $m_2 = 5.00$ kg, $m_3 = 8.00$ kg, and $m_4 = 3.00$ kg) are connected to each other using ropes as shown in Figure 5-49. The four masses rest on a frozen lake surface in Labrador. A curious husky pulls on a rope that is attached to m_4 and pulls the four masses to the right.
(a) The tension in the rope between m_1 and m_2 is 40.0 N. Calculate the tension in the rope connecting m_2 and m_3.
(b) What is the magnitude of the force with which the husky is pulling on m_4?

Figure 5-49 Problem 58

59. ✷✷ You are holding a special pendulum with two masses, m_1 and m_2, instead of one, connected by a rope as shown in Figure 5-50. You lower the pendulum such that the tension in the rope connecting the two masses is half the weight of the bottom mass.
(a) Find the acceleration with which you lower the pendulum.
(b) For what tension in the rope between you and m_1 would the tension in the rope between the two masses be zero?

Figure 5-50 Problem 59

Section 5-7 Friction

60. ✷ A hockey puck is shot across an ice rink at a speed of 3 m/s and comes to a stop over a distance of 11 m. Find the coefficient of friction between the puck and the ice.

Section 5-8   Force of a Spring

61. ✷   In order to stretch a spring by 10 cm, you have to apply a force of 120 N. How much would the spring stretch if a 10 kg mass were to hang from it vertically?

62. ✷   The left side of a mass resting freely on a frictionless surface is attached to an unstretched spring. The mass is pulled 20 cm to the right and then released. Write an expression for the acceleration of the mass when it has moved 15 cm back toward its original position.

63. ✷   A spring is suspended from an elevator, and a 12.0 kg mass is then attached to the spring. When the elevator moves down at an acceleration of 3.00 m/s², the net elongation of the spring is 30.0 cm. Find the spring constant.

Section 5-9   Circular Motion and Centripetal Forces

64. ✷   You swing a marble attached to the end of a string in a horizontal circle as shown in Figure 5-51. The angle that the string makes with the vertical is 37°.
    (a) Find the speed of the marble when the string is 44.0 cm long.
    (b) Write an expression for the tension in the string.

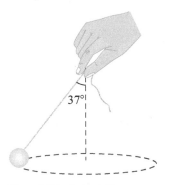

Figure 5-51   Problem 64

65. ✷   A child of mass 20 kg stands on a merry-go-round 3.0 m from the centre. You push the merry-go-round until the child is about to slip on the merry-go-round surface. The coefficient of static friction between the child and the surface is 0.45.
    (a) Find the child's speed at the moment she is about to slip.
    (b) What is the net force acting on the girl at that instant?

66. ✷   You are testing a remote-controlled toy truck in an empty parking lot. The coefficient of static friction between the surface and the truck is 0.34. Find the radius of the smallest circle the truck can move in without slipping at its maximum speed of 33 km/h.

67. ✷✷   You are driving on a circular ramp of radius 72.0 m on an icy January day. The ice-covered ramp is banked at a 23.0° angle, and you anticipate that the ramp is not going to provide any significant friction to help keep your car on the road.
    (a) What must your speed be so that your car does not slide off the road?
    (b) What are the direction and magnitude of the net force acting on you at that point? Your mass is 72.0 kg.

68. ✷✷   Your niece finds her father's watch. The light watch chain has a length of 32 cm, and the mass of the watch is 210 g. Your niece swings the watch in a vertical circle. Find the tension in the chain when it makes an angle of 43° with respect to the vertical if the speed of the watch is 2.8 m/s.

Section 5-10   Frames of Reference and Fictitious Forces

69. ✷✷   You are sitting comfortably in a train. A passenger standing beside you is loading her luggage into the overhead compartment when the train starts moving forward with an acceleration of 5.0 m/s². Find the coefficient of static friction that is necessary to prevent the person from sliding with respect to the train. Solve the problem by considering the scenario from the train's frame of reference, and then again using the train station's frame of reference.

70. ✷✷   Derive the equations of motion for a conical pendulum using the pendulum mass's frame of reference and then again using a stationary frame of reference.

71. ✷✷   A 30 kg bag of mortar mix is sitting in the back of a truck rounding a traffic circle of radius 23 m at a speed of 67 km/h. Using the truck's frame of reference, find the force of contact between the bag and the side of the truck. Assume that the floor at the back of the truck is frictionless.

72. ✷✷   An astronaut sits in a spacecraft orbiting Earth. Explain the feeling of weightlessness that the astronaut experiences. Is it true that an astronaut living on the International Space Station (ISS) is too far from Earth to experience the force of gravity? What is the altitude of the ISS's orbit? Research the value of the force of gravity at that altitude.

COMPREHENSIVE PROBLEMS

73. ✷✷   A curious child finds a rope hanging vertically from the ceiling of a large storage hangar. The child grabs the rope and starts running in a circle. The length of the rope is 17.0 m. When the child runs in a circle of radius 6.0 m, the child is about to lose contact with the floor. How fast is the child running at that time?

74. ✷✷✷   A massless spring with a spring constant $k = 490$ N/m is suspended from a hook on the ceiling of a cargo truck moving in a straight line on the highway. A child reaches over from her seat and suspends a 1.2 kg toy on the free end of the spring. The child notices that the spring is now 4.0 cm longer than when it was hanging freely. Find the acceleration for the truck.

75. ✷✷   A cruise ship moving at a constant speed of 30 km/h is equipped with a gymnasium. A gymnast is climbing a rope suspended from the ceiling of the gym. The captain of the ship puts the ship into a circular turn of radius 150 m to get closer to a pod of whales. Find the angle that the rope makes with the vertical.

76. ✷✷   A child sits in the back of a car with a yo-yo freely suspended from his fingers. The car is leaving the highway at a speed of 78 km/h on a ramp of radius 210 m banked at a 23° angle.
    (a) Find the angle that the yo-yo makes with the vertical.
    (b) Find the angle that the yo-yo makes with the normal to the roof of the car.

77. ✷✷✷   The truck in Figure 5-52 (page 126) is going around a circular track of radius 72 m, banked at a 60° angle. A spider rests on the inside wall of the truck as shown. The coefficient of static friction between the truck wall and the spider is 0.91. Find the maximum speed that the truck can have before the spider begins to slip down the wall.

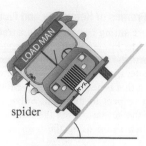

spider

Figure 5-52 Problem 77

78. ✷✷ A toy car is made to run the bowl-shaped track shown in Figure 5-53. As the car's speed increases, the car moves farther and farther up the track walls. In the end, the car is able to round the vertical walls of the track, such that its height from the bottom of the track does not change. The radius of the bowl is 1.9 m. Find minimum speed that the car needs to not slip when rounding the vertical walls. The coefficient of static friction between the car and the wall is 0.39.

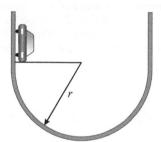

Figure 5-53 Problem 78

79. ✷✷ Two masses (m_1 = 12.0 kg and m_2 = 14.0 kg) are hanging from a rope that passes over a massless pulley, as shown in Figure 5-54. The two masses are originally held in place and then let go.
(a) Find the speed of each mass after one has moved 4.00 m.
(b) What is the tension in the rope?

Figure 5-54 Problem 79

80. ✷✷ A roller coaster car is upside down at the top of a vertical circular loop with a radius of 21 m. The speed of the car is 24 m/s, and the coefficient of kinetic friction between the car and the track is 0.16. Find the radial acceleration of the car.

81. ✷✷ A marble slides down a hemispherical bowl starting from rest as shown in Figure 5-55 starting from rest. When the marble is halfway to the bottom, its speed is 2.3 m/s. Write an expression for the magnitude of the total acceleration of the marble.

Figure 5-55 Problem 81

82. ✷✷✷ A block of mass m is held in place on a ramp of mass M with a slope of angle θ, which in turn is held in place as shown in Figure 5-56. The ramp is free to move on the surface beneath it. There is no friction anywhere. The block and the ramp are then released. Derive an expression for the acceleration of the ramp.

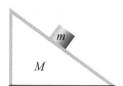

Figure 5-56 Problem 82

83. ✷✷✷ A block of mass m is held in place on a ramp of mass M with a slope of angle θ, which in turn is held in place as shown in Figure 5-57. The ramp is free to move on the surface beneath it. There is no friction anywhere. The ramp is then pushed with a horizontal force so that the block does not slide on the ramp. Derive an expression for the magnitude of this force.

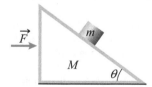

Figure 5-57 Problem 83

84. ✷✷✷ The coefficient of static friction between the inclined face of a wedge (mass M) and the mass m sitting on its inclined edge is μ_s. The coefficient of kinetic friction is μ_k. Derive an expression for the maximum horizontal force that you can apply to the vertical face of the wedge so that the mass m does not move with respect to the wedge.

85. ✷✷✷ The coefficient of kinetic friction between the inclined face of a wedge of mass M and a block of mass m sitting on its inclined face is μ_k (Figure 5-58). The vertical face of the wedge is pushed with a horizontal force such that m slides up the inclined surface with an acceleration a with respect to the inclined surface. Derive an expression for the magnitude of the force.

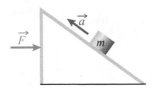

Figure 5-58 Problem 85

86. ✷✷✷ Block A of mass $m_A = 25$ kg rests on a frictionless surface as shown in Figure 5-59. Block B is suspended from a rope that runs over the massless pulley and is then attached to block A. Block C of mass $m_C = 6$ kg rests on top of block A. The coefficient of static friction between blocks A and C is 0.23. The system is initially held in place. Find the maximum value for the mass of block B so that when the system is let go, block C does not slide on block A.

Figure 5-59 Problem 86

87. ✷✷✷ Mass m rests on mass M as shown in Figure 5-60. Mass m is connected with a rope that runs over a massless pulley to mass $3m$. There is no friction anywhere, and the system is originally held in place. When the system is let go, mass M is pushed with a horizontal force such that $3m$ does not move vertically. Derive an expression for the magnitude of the force.

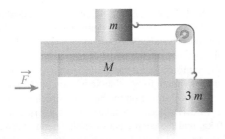

Figure 5-60 Problem 87

88. ✷✷✷ Block B ($m_B = 1.00$ kg) is held in contact with block A ($m_A = 3.00$ kg) as shown in Figure 5-61. When block B is let go, block A is pushed to the right with force of magnitude 160 N. The coefficient of static friction between the two blocks is 0.50.
(a) Find the force of friction on block B.
(b) Find the normal force exerted by block A on block B.
(c) Find the vertical force exerted by block B on block A.

Figure 5-61 Problem 88

89. ✷✷✷ Block B ($m_B = 4.00$ kg) sits on a horizontal frictionless surface. Block A ($m_A = 2.00$ kg) sits on top of block B and is attached to a rope that runs over the massless pulley as shown in Figure 5-62. Block C ($m_C = 1.00$ kg) hangs from the rope. There is no friction anywhere.
(a) With what horizontal force $\vec{F}$ must you push block B so that block C rises at an acceleration of 3.00 m/s²?
(b) For part (a), find the net force on blocks B, A, and C. Why is the net force on block B different from the force used to push block B?

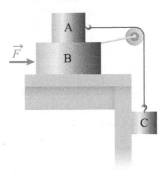

Figure 5-62 Problem 89

90. ✷✷ You swing a 1.2 m-long string with a metal ball attached to its end in a vertical circle, such that the speed of the ball does not change but the rope is always taut. The tension in the string when the ball is at the bottom of the circle is 60 N more than the tension when the ball is at the top.
(a) Find the mass of the ball.
(b) How does the difference in tension change when the speed is doubled?

91. ✷✷✷ The driver of a race car of mass M wishes to test a wide, banked turn at angle θ on a circular race track of radius R. First, the surface of the banked curve is covered with ice. The driver places his car at the top of the bank and observes it as it slides down the bank. The engine at this stage is off, and the car is originally at rest.
(a) Derive an expression for the magnitude of the normal force on the car as it slides downhill.
(b) With the bank still covered in ice, the driver next drives around the track at the exact speed needed to keep the car from sliding up or down the bank. Derive an expression for the magnitude of the normal force acting on the car.
(c) For the last test, the driver asks for the ice to be melted, and he drives his car around the track at the maximum speed that will not cause the car to slide. The coefficient of static friction is μ_s. Derive an expression for the magnitude of the normal force on the car from the track.

92. ✱✱✱ Four blocks stacked vertically have masses as follows: $M_A = 11.0$ kg, $M_B = 7.0$ kg, $M_C = 5.0$ kg, and $M_D = 3.0$ kg. The coefficient of static friction between blocks A and B is 0.12, the coefficient of static friction between blocks B and C is 0.16, and the coefficient of static friction between blocks C and D is 0.18. There is no friction between block A and the surface on which it sits. A horizontal force, F, is applied to block A. Find the maximum value of F for which none of the blocks slide with respect to one another.

93. ✱✱ An astronaut-training device is made of a ramp connected to heavy, 10 m long rope of linear density of 3 kg/m. The other end of the rope is fixed. The ramp is made to spin in a horizontal circle 10 m in radius. The astronaut sits on the ramp facing the centre of the circle. The coefficient of static friction between the astronaut and the ramp is 0.25. Find the maximum tension in the rope as the ramp spins so that the astronaut does not slip on the ramp. There is no friction between the ramp and the surface beneath it.

94. ✱✱ You want to push your 53 kg friend up the 33° ramp of a slide in a nearby park. You do so by pushing your friend with a 510 N horizontal force. The coefficient of kinetic friction between your friend and the slide is 0.17. Find the acceleration of your friend up the slide. What is the normal force from the ramp on your friend?

95. ✱✱✱ The face of block M in Figure 5-63 is shaped like a semicircular bowl of radius R. A mass m is placed at the top-left corner of the bowl and then let go. Find the acceleration of block M when m is a distance of $0.2R$ from the bottom of the bowl. There is no friction between M and m, or between M and the surface on which it sits.

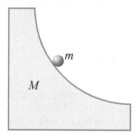

Figure 5-63   Problem 95

96. ✱✱✱ The face of block M is shaped like a semicircular bowl of radius R. There is no friction between M and m, or between M and the surface on which it sits. Derive an expression for the force you need to apply to the vertical face so that a mass that placed on the semicircular face a distance of $0.5R$ above the bottom of the bowl does not slide with respect to M.

97. ✱✱✱ You carve out a piece of ice in the shape of a circular dome of mass M and radius R. You then place the carved piece on an ice surface, and place an ice cube on your dome, at a height h above the ice surface. Derive an expression for the force you need to apply to this dome so that the cube does not slide on it.

98. ✱ Mass m_1 hangs from a massless movable pulley as shown in Figure 5-64. The rope rounding the movable pulley passes over the fixed pulley and is attached to mass m_1. There is no friction between m_2 and the surface supporting it. The system is originally held at rest and then let go. Derive expressions for the acceleration of m_1 and the tension in rope.

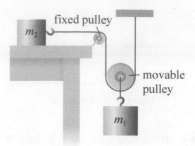

Figure 5-64   Problem 98

99. ✱✱ Figure 5-65 shows two friends testing a stunt ride they invented. One friend ($m_1 = 56.0$ kg) sits on a seat hanging from a cliff above water, and the other friend ($m_2 = 78.0$ kg) stands on a frictionless surface at the top of the cliff while holding a moveable pulley. The upper end of the rope that passes over the fixed pulley and the movable pulley is fastened to a shock-absorbing block and the lower end to the suspended seat as shown. They start with the seat at rest 19.0 m above the water. Find the acceleration of each friend, and the speed at which the seat hits the water.

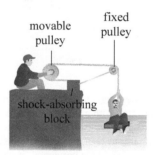

Figure 5-65   Problem 99

100. ✱ A 1300 kg car takes a 44 m-radius roundabout at a speed of 7.0 m/s. The road is not banked. The coefficient of static friction between the car and the road is 0.41. Find
(a) the force of friction on the car;
(b) the net force on the car;
(c) the maximum force of static friction on the car.

101. ✱✱ You are sitting on the back of a truck in the Prairies when the driver decides to take you on a joy ride. Your mass is 69 kg, and the coefficient of static friction between you and the truck is 0.34. The truck follows a circular path of radius 54 m at a speed of 11 m/s. Find
(a) the force of friction on you;
(b) the force that you exert on the truck;
(c) the net force on you;

102. ✱✱ Blocks A (20.0 kg), B (10.0 kg), C (40.0 kg), and D (30.0 kg) are in contact beside each other on a horizontal frictionless surface as shown in Figure 5-66. Block A is being pushed to the left with force F such that the force of contact between the C and D blocks is 60.0 N. Find the force of contact between blocks B and C. Also find the force F and the acceleration of each block.

Figure 5-66 Problem 102

103. ✳✳ You place a book on the seat next to you, and you then drive your car at a speed of 110 km/h on the highway.

You suddenly have to slow down. The book is just beginning to move on the seat. You stop over a distance of 120 m. What is the coefficient of friction between the seat and the book?

 See the text online resources at www.physics1e.nelson.com for Open Problems and Data-Rich Problems related to this chapter.

Learning Objectives

When you have completed this chapter you should be able to:

1 Explain the work–kinetic energy theorem, and define the work done by a force.

2 Calculate the work done by a constant force in one or more dimensions.

3 Calculate the work done by a variable force.

4 Calculate the work done by a spring.

5 Differentiate between conservative and nonconservative forces, and relate the change in potential energy to the work done by conservative forces.

6 Explain the principle of conservation of mechanical energy, and compare the work–energy approach to solving problems to the approach used when solving problems using Newton's laws.

7 Relate potential energy to reference points.

8 Define equipotential surfaces, and explain their relationship to conservative forces.

9 Define power and calculate power output.

Chapter 6
Energy

Most of the energy we use on Earth originates from the Sun. The rivers that run hydroelectric generators are fed by precipitation, which results from water being evaporated by the Sun's energy. When we burn fossil fuels, we use chemical energy from compounds originally created by photosynthesis millions of years ago, another process powered by the Sun.

An unintended consequence of the massive use of fossil fuels and some large-scale industrial processes is a significant increase in the quantities of greenhouse gases, such as carbon dioxide and methane, in the atmosphere. As a result, a greater proportion of the energy radiated by the Sun is absorbed by Earth.

The consensus among scientists is that human activity is having a measurable and accelerating effect on the climate. For example, satellite images from 2007 show that the extent of Arctic ice in the summer had shrunk by over one-third from the average size measured from 1979 to 2000.

Figure 6-1 is a satellite image taken by NASA in 2003. The red line indicates a possible route for the long-sought-after Northwest Passage through the Canadian Arctic. Due to pack ice, this route was considered navigable only by icebreakers and other specially equipped vessels. However, in August 2007, enough pack ice melted to make a route feasible for ordinary cargo vessels. It is conceivable that the increased absorption of energy from the Sun could open the Northwest Passage year-round before the end of the century. While a boon for shipping, this climate change could have disastrous consequences elsewhere.

In this chapter, we will discuss the transformation of several types of energy and solve problems related to energy.

For additional Making Connections, Examples, and Checkpoints, as well as activities and experiments to help increase your understanding of the chapter's concepts, please go to the text's online resources at www.physics1e.nelson.com.

Figure 6-1 A possible route for the Northwest Passage through the Canadian Arctic

NASA

MAKING CONNECTIONS

Going Carbon-Neutral

On November 7, 2007, the residents of the town of Eden Mills, Ontario, launched an initiative to make their town the first carbon-neutral community in North America. The project includes measuring greenhouse emissions and taking active steps to reduce these emissions at both the town and the household levels. The residents of Eden Mills hope that this project will pave the way for other communities to see that going carbon-neutral can be a reality. It is also a testimony to future generations that they have done their share, or a satisfactory portion of it, in this regard.

Figure 6-2 shows the pond at Eden Mills. Waterpowered mills such as the one located at Eden Mills pond rely on a

Mark Hamilton, 2007. www.fourthline.ca

Figure 6-2   The pond at Eden Mills, Ontario

direct conversion of the gravitational potential energy of the water–Earth system to kinetic energy to power processes such as sawing and finishing lumber and grinding grain into flour.

LO 1

6-1   Work and Kinetic Energy for Constant Forces

Work converts energy from one form to another. For example, as water stored behind a hydroelectric dam falls to activate generator turbines, the work done by the force of gravity transforms the gravitational potential energy into kinetic energy.

To the best of our knowledge energy is conserved. We have yet to see this principle violated. The conservation of energy is one of the fundamental laws of modern science. Energy takes on different forms and is manifested in various ways, but it is always conserved.

Consider an object being pushed by a constant net force. The object's acceleration is constant by Newton's second law. The time-independent kinematics equation for motion in one dimension with constant acceleration relates the speed, acceleration, and displacement of an object:

$$v^2 - v_0^2 = 2\vec{a} \cdot \Delta \vec{x} \qquad (3\text{-}18)$$

Dividing each side of the equation by two, and multiplying by the mass of the object gives

$$\frac{1}{2} mv^2 - \frac{1}{2} mv_0^2 = m\vec{a} \cdot \Delta \vec{x} \qquad (6\text{-}1)$$

By Newton's second law, $m\vec{a}$ is the net force acting on the object, $\vec{F}_{\text{net}}$. Therefore,

$$\frac{1}{2} mv^2 - \frac{1}{2} mv_0^2 = \vec{F}_{\text{net}} \cdot \Delta \vec{x} \qquad (6\text{-}2)$$

Since $\vec{F}_{\text{net}}$ is the sum of all the forces acting on an object, the term on the right-hand side of Equation (6-2)

is called the **total work**, W_{T}, done on the accelerating the object by the net force acting on it. It can also be thought of as the sum-total of the work done by all the forces acting on the object.

KEY EQUATION
$$W_{\text{T}} = \vec{F}_{\text{net}} \cdot \Delta \vec{x} \qquad (6\text{-}3)$$

We define the **kinetic energy** of an object of mass m moving at speed v as

$$K = \frac{1}{2} mv^2 \qquad (6\text{-}4)$$

Therefore, Equation (6-2) can be rewritten as follows:

KEY EQUATION   The total work–kinetic energy theorem:

$$W_{\text{T}} = \Delta K = K_{\text{f}} - K_{\text{i}} \qquad (6\text{-}5)$$

Equation (6-5) is a statement of the **work–kinetic energy theorem**, which simply states that the *total* work done on an object (i.e., the work done by *all* the forces acting on the object) is equal to the change in its kinetic energy. Since work is the scalar product of the force vector and the displacement vector, work is a scalar and can be positive or negative.

For the work done by a single constant force, we can write

KEY EQUATION
$$W_F = \vec{F} \cdot \Delta \vec{x} \qquad (6\text{-}6)$$

We note here that for Equation (6-6), we have chosen the x-axis as our default direction. However, the same expression applies to the y-direction, the z-direction, and any other direction.

Expanding the scalar product gives

$$W_F = |\vec{F}| |\Delta \vec{x}| \cos\theta \qquad (6\text{-}7)$$

The first two terms on the right-hand side of Equation (6-7) are magnitudes and are therefore always positive. The term that determines the sign for the work done is therefore $\cos\theta$, where θ is the angle between the force vector and the displacement vector in the tail-to-tail representation. For example, when the force points in the same direction as the displacement the angle between the force and the displacement is $\theta = 0$, since $\cos\theta = 1$, the work is positive. When the force points in the direction opposite to the displacement, the angle between the force and the displacement is $180°$, hence $\cos\theta = -1$ and the work done by the force is negative.

We can also see that work can be positive or negative in light of Equation (6-5), the total work–kinetic energy theorem, as illustrated in Example 6-1.

CHECKPOINT

C-6-1 Work and Total Work

What is the difference between the quantity W_T in Equation (6-5) and the quantity W_F in Equation (6-6)?

SI Energy Units

The SI unit for work and energy is the joule (J), which is equal to a newton metre:

$$1\,\text{J} = 1\,\text{N}\cdot\text{m} = 1\,\text{kg m}^2/\text{s}^2$$

EXAMPLE 6-1

Work and Kinetic Energy

You push a crate of mass 22.0 kg on a horizontal frictionless surface over a distance of 8.00 m using a constant horizontal force of 100 N starting from rest.

(a) Find the work done you as you push the crate.
(b) Find the speed of the crate at the end of the 8 m push.

SOLUTION

(a) Since the force you use is constant, we can use Equation (6-6):

$$W_F = \vec{F}\cdot\Delta\vec{x} = F\Delta x \cos\theta$$

The crate will naturally move in the direction of your push; therefore, the angle between the force and the displacement vectors is zero, and the work for the force you apply can be expressed as

$$W_F = F\Delta x\,(1) = 100\,\text{N}\,(8\,\text{m}) = 800\,\text{J}$$

(b) To calculate the speed of the crate, we use Equation (6-5). The total work done is the work done by you because there are no other forces doing work on the crate. Here, we also note that the normal force and the force of gravity do no work. This is covered in more detail in the next few sections.

$$W_T = W_F = \Delta K = K_f - K_f$$

$$\Rightarrow W_F = F\Delta x = \frac{1}{2}mv^2 - 0 \Rightarrow v^2 = \frac{2W_F}{m}$$

Since the crate starts from rest, its initial kinetic energy is zero and

$$v^2 = \frac{2W_F}{m} \Rightarrow v = \sqrt{\frac{2W_F}{m}} = \sqrt{\frac{2(800\,\text{J})}{22\,\text{kg}}} = 8.53\,\text{m/s}$$

Making sense of the result:

The work done by you is positive, since the crate moves in the same direction in which you push.

EXAMPLE 6-2

Work Against Gravity

You lift your 2.20 kg laptop a total vertical distance of 70.0 cm above the table on which it rests. How much work do you do?

SOLUTION

We do not know if the force you applied was constant, so we cannot assume that the force is equal to mg. Therefore, we cannot use Equation (6-6). However, we can still use Equation (6-5):

$$W_T = \Delta K = 0 \qquad (1)$$

We are assuming that the laptop starts at rest and ends at rest, so the change in its kinetic energy is zero.

The total work in Equation (1) is the work done by all the forces involved; this is the work done by you or the force you apply and the work done by the force of gravity:

$$W_T = W_F + W_g = 0 \qquad (2)$$

Figure 6-3 shows the laptop being lifted upward with the forces acting on it.

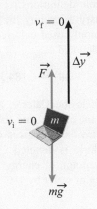

Figure 6-3 Example 6-2

For the work done by the force of gravity, we use Equation (6-6):

$$W_g = m\vec{g} \cdot \Delta\vec{y} = (mg)(\Delta y)\cos\theta \qquad (3)$$

The laptop is lifted upward and the force of gravity points downward, so the angle between the two vectors is $180°$ (Figure 6-3). Therefore,

$$W_g = m\vec{g} \cdot \Delta\vec{y} = (mg)(\Delta y)(-1) = -(mg)(\Delta y) \qquad (4)$$

From Equation (2):

$$W_F = -W_g = mg\Delta y = (2.2 \text{ kg})(9.81 \text{ m/s}^2)(0.7 \text{ m}) = 15.1 \text{ J}$$

Making sense of the result:

The work done by gravity is negative. This is the case whenever the force points in the direction opposite to the displacement. We will discuss this in more detail in the next example.

EXAMPLE 6-3

Negative Work

In a "strong man contest," a car with a mass of 1200 kg rolls toward a contestant at a speed of 0.80 m/s. The contestant pushes against the car with a constant force (Figure 6-4) and brings it to a stop over a distance of 0.5 m.

(a) Find the work that the contestant does on the car.
(b) Find the force exerted by the contestant.

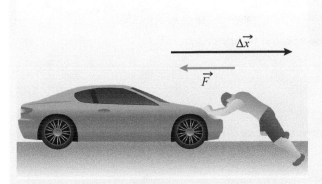

Figure 6-4 In a "strong man contest," a contestant pushes against a car.

SOLUTION

(a) The contestant is the only agent doing work. The normal force and the force of gravity act normal to the direction of motion, and the work done by each of these forces is zero. The total work done on the car is the work done by the contestant. From Equation (6-5),

$$W_c = W_T = \Delta K = K_f - K_i = 0 - \frac{1}{2}mv^2$$

$$= -\frac{1}{2}(1200 \text{ kg})(0.8^2) \text{ m}^2/\text{s}^2 = -384 \text{ J} \qquad (1)$$

(b) From Equation (6-7), the work done by the contestant is given by

$$W_c = |\vec{F}||\Delta\vec{x}|\cos\theta \qquad (2)$$

Since the contestant pushes in the direction opposite to that of the displacement of the car, the angle between the force and the displacement is $180°$. Thus,

$$W_c = |\vec{F}||\Delta\vec{x}|\cos\theta = F\Delta x(-1) \qquad (3)$$

Combining Equations (1) and (3), we have

$$F\Delta x(-1) = -384 \text{ J} \Rightarrow F(5 \text{ m})(-1) = -384 \text{ J}$$

$$F = \frac{-384 \text{ J}}{-0.5 \text{ m}} = 768 \text{ N}$$

Making sense of the result:

The final kinetic energy of the car ($K_f = 0$) is less than the kinetic energy that the car had to begin with. Therefore, the difference, $\Delta K = K_f - K_i$, is negative, and the work done on the car by the contestant is also negative by Equation 6-5.

We can also look at the sign of the work done in light of the scalar product, given by Equation (6-7). To slow the car down, the contestant will naturally have to push in a direction that is opposite to the direction of motion of the car; hence $\theta = 180°$. Since $\cos(180°) = -1$, the work done by the contestant is negative. A force of 768 N is what one would use to lift a weight of ~78 Kg.

Other Energy Units

The joule is the internationally recognized SI unit for energy, but there are several pre-SI units for energy that are still in common use.

One such traditional unit is the calorie (cal), defined as the energy required to raise the temperature of 1 g of water by 1°C. This unit is also called the gram calorie and the small calorie. However, the energy required for a 1°C increase in the temperature of a given quantity of water varies slightly with the starting temperature. Various definitions of the calorie give it a value ranging from 4.182 J to 4.204 J. (SI was developed to eliminate this sort of confusion over unit definitions.) For precise measurements it is necessary to specify the type of calorie. One of the more commonly used calorie units is the thermochemical calorie, denoted cal_{th}:

$$1 \ cal_{th} = 4.184 \ J$$

A related unit is the Calorie, also called the large calorie, the kilogram calorie, and the nutritionist calorie. (However, the nutritional sciences calorie is defined as 4.182 J.) The Calorie is denoted Cal or kcal; this unit is defined as the amount of energy required to raise the temperature of 1 kg of water by 1°C.

Table 6.1 Appreciating the Joule: Energy Orders of Magnitude

Energy (J)	Object
6.068×10^{-21}	Average kinetic energy of a molecule of an ideal gas at room temperature
1.6×10^{-19}	1 eV (electron volt)[1]; the kinetic energy of an electron that has been accelerated through a potential difference of 1 V
2.614×10^{-19} to 5.28×10^{-19}	Energy range of a photon of visible light
8.187×10^{-14}	0.511 MeV (mega electron volt); the rest mass energy of an electron[2]
3.205×10^{-6}	20 TeV (terra electron volt); the proposed beam energy in the Superconducting Supercollider[3]
1	Approximate kinetic energy gained by a 1 kg weight falling a distance of 10 cm
4.184	Thermochemical calorie; the energy required to raise the temperature of 1 g of water by 1°C
7×10^{2}	Approximate kinetic energy of an average person after 1 m of free fall
1.0551×10^{3}	1 BTU (British thermal unit); the amount of energy needed to raise the temperature of 1 lb of water by 1°F[4]
3.6×10^{3}	1 W·h (watt hour)[5]; see Section 6-9
4.184×10^{3}	1 $kcal_{th}$; the energy required to raise the temperature of 1 kg of water by 1°C
3.858×10^{5}	Kinetic energy of a 1000 kg vehicle moving at 100 km/h
5–50 kJ/mol	Range of hydrogen bond energies
~23 kJ/mol	Hydrogen bond energy in water
413 kJ/mol	Hydrogen–carbon covalent bond energy
4.1868×10^{10}	1 toe; the energy released by burning 1 T of crude oil (1 tonne of oil equivalent, or 1 toe)
3.6×10^{6}	1 kW·h (kilowatt hour)
4.184×10^{12}	1 kilotonne of TNT, the energy released in the detonation of 1000 tonnes of TNT
5.1×10^{20}	12 267.38 Mtoe; the world's primary energy supply for 2008
4×10^{23}	Energy of the meteor impact that formed the Chicxulub crater

[1] The electron volt is commonly used as an energy unit in quantum physics. See Chapter 31 for details.

[2] The rest mass energy of a particle is the energy equivalent of the mass of the particle, in effect, the energy required to create the particle, using Einstein's famous equation, $E = mc^2$. See Chapter 29.

[3] The Superconducting Supercollider was to be the largest particle collider built during the 20th century. The project was scheduled to start operating in 2019 but was cancelled by the U.S. Congress in 1993.

[4] The value presented here is the international table BTU.

[5] Unit commonly used in power production and consumption.

MAKING CONNECTIONS

Walking It Off

An average fast food burger without cheese has 590 Cal (kcal) and provides 52% of the daily limit of fat for an average recommended daily caloric intake of 2000 Cal. A medium-sized cola drink (473 mL) has 210 Cal, all from sugar, and a medium order of fries has 453 Cal and provides 33% of the daily fat limit.

On average, a 70 kg person burns 390 Cal/h during a very brisk 7 km/h walk. At this rate, it would take close to 3 h (21 km) to burn off the calories in the fast food described above.

EXAMPLE 6-4

Work Against Friction

A fisherman pushes a crate full of fish ($m = 123$ kg) from the edge of a fishing boat to the middle of the boat with a constant force of 560 N over a total distance of 7.00 m. Over this distance, the speed of the crate changes from zero to 4 m/s. Find the coefficient of kinetic friction between the crate and the boat surface.

SOLUTION

Figure 6-5 shows the FBD for the crate. We can use the work–kinetic energy theorem, Equation (6-5):

$$W_T = \Delta K = K_f - K_i \tag{1}$$

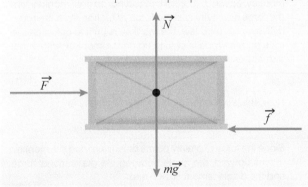

Figure 6-5   Example 6-4

Neither the force of gravity nor the normal force do any work because they both act normal to the direction of motion. The total work done is the work done by the force of the fisherman and the work done by the force of friction. The initial kinetic energy of the crate is zero, then

$$W_F + W_f = K_f - 0 \tag{2}$$

The work done by each force is given by Equation (6-6). The force of the fisherman acts in the same direction as the force of the crate's motion, and the angle between the force and the displacement vectors is zero:

$$W_F = \vec{F} \cdot \Delta \vec{x} = F \Delta x \cos\theta = F \Delta x \tag{3}$$

The force of friction acts opposite to the direction of motion, so the angle between the force and the displacement vectors is 180°:

$$W_f = \vec{f_k} \cdot \Delta \vec{x} = f_k \Delta x \cos(180°) = -f_k \Delta x \tag{4}$$

Putting Equations (3) and (4) into Equation (2) yields

$$F\Delta x - f_k \Delta x = K_f = \frac{1}{2}mv^2 \tag{5}$$

Therefore,

$$f_k \Delta x = F\Delta x - \frac{1}{2}mv^2 \tag{6}$$

The force of friction between the boat's surface and the crate is given by Equation (5-4) as $f_k = \mu_k N$. From Figure 6-5, the normal force on the crate is equal to its weight. Therefore, $f_k = \mu_k mg$, and Equation (6) becomes

$$\mu_k mg\Delta x = F\Delta x - \frac{1}{2}mv^2$$

or

$$\mu_k = \frac{F\Delta x - \frac{1}{2}mv^2}{mg\Delta x} = \frac{(560\ \text{N})(7\ \text{m}) - \frac{1}{2}(123\ \text{kg})(4\ \text{m/s})^2}{(123\ \text{kg})(9.81\ \text{m/s}^2)(7\ \text{m})} = 0.348$$

Making sense of the result:

The coefficient of kinetic friction has no units, as expected. In addition, a value between 0 and 1 is expected.

PEER TO PEER

Something I often ran into trouble with is the concept of kinetic energy. I have to remind myself that kinetic energy is always a positive scalar and it doesn't depend on the direction of motion. If an object of mass m moves along the positive x-axis with speed v,

its kinetic energy is $\frac{1}{2}mv^2$. If that object then moves along the negative x-axis with the same speed, its kinetic energy is the same, $\frac{1}{2}mv^2$. The difference in kinetic energy between the two cases is zero.

✅ CHECKPOINT

C-6-2 Work and Kinetic Energy

You are standing on a flat, horizontal icy stretch in the Arctic, when a baby seal comes sliding slowly toward you. You are wearing shoes with cleats that prevent you from sliding. You place your hand gently against the seal to slow it down and then push it back toward its mother at the same speed it was coming toward you, 0.5 m/s. The seal's mass is 8 kg. Assume that you use a constant force and that you are in contact with the seal for a total of 1.1 s. What is the total work done by you on the seal during the entire process?

(a) 1 J
(b) 2 J
(c) −2 J
(d) −1 J
(e) none of the above

C-6-2 (e) The incoming speed is the same as the outgoing speed, the change in kinetic energy is zero, hence the total work done is zero.

EXAMPLE 6-5

Work and Gravity

You lift a 420 g glass of iced coffee from your desk to a shelf 35.0 cm above the desk. It takes you 0.700 s to lift the glass, and the maximum speed that the glass reaches on its way up is 0.900 m/s. What is the total work done on the glass of coffee during its journey upward?

SOLUTION

Since the glass starts from rest and ends up at rest, the change in its kinetic energy is zero. Therefore, the total work done on the glass is zero according to Equation (6-5).

Making sense of the result:

The work you do is positive, and the work done by the force of gravity is negative. These two quantities comprise the total work and up to zero.

EXAMPLE 6-6

Eager Worker

You are training with two friends at the gym. One friend throws a 9.00 kg medicine ball at you at a speed of 2.00 m/s. You stop the ball over a distance of 30.0 cm, and throw it to your other friend in a direction that makes an angle of 120° with the incident direction. The ball leaves your hands with the same speed it had when you intercepted it. Determine the total work done by you on the ball.

SOLUTION

You slide the ball with the same speed that you receive it, so the overall change in ball's kinetic energy is zero.

Therefore, the total work done on the ball is zero. Since you are the only agency doing work, the work done by you is the total work done on the ball, which is zero.

Making sense of the result:

Although you exert a force on the ball in two different directions, the *total* work done by you is zero. When you slow the ball, you are doing negative work. When you slide the ball, you are doing positive work. The sum total of the work done is zero.

EXAMPLE 6-7

What Is Doing Work?

A 17.0 kg monkey climbs up a rope up to a coconut tree, starting from ground level. The monkey is originally at rest. When the monkey is 6.00 m above the ground, his speed is 2.30 m/s. What force moves the monkey upward? How much work is done by that force?

SOLUTION

The monkey climbs by holding the rope and pulling down on it. The force of friction from the rope pushes the monkey upward. The forces acting on the monkey are the force of gravity and the force of friction from the rope; by Newton's third law, the monkey is pulling down on the rope and the rope is pulling up on the monkey with equal and opposite force. Using Equation 6-5

$$W_T = W_f + W_g = \Delta K \qquad (1)$$

Now, the work done by the force of gravity is

$$W_g = m\vec{g} \cdot \Delta\vec{y} \qquad (2)$$

Since the force of gravity points downward and the monkey moves upward, the angle between the gravitational force and the displacement is 180° and

$$W_g = |m\vec{g}||\Delta\vec{y}|\cos(180°) = |m\vec{g}||\vec{h}|\cos(180°) = -mgh \quad (3)$$

We can rewrite Equation (1) as

$$W_f - mgh = \Delta K = \frac{1}{2}mv^2 - \frac{1}{2}mv_0^2 \qquad (4)$$

Since the monkey starts from rest, $v_0 = 0$ and

$$W_f = mgh + \frac{1}{2}mv^2 = (17\text{ kg})(9.81\text{ m/s}^2)(6\text{ m})$$
$$+ \frac{1}{2}(17\text{ kg})(2.3\text{ m/s}) = 1040\text{ J} \qquad (5)$$

Making sense of the result:

The work done by the force of friction has to account for the work needed to lift the monkey against the force of gravity (mgh) as well as for the increase in the monkey's kinetic energy.

C-6-3 Who Is Doing the Work?

For Example 6-7, which of the following statements is true?
(a) The work is done by the force of static friction.
(b) The work is done by the force of kinetic friction on the monkey because the monkey is moving.
(c) The work is done by the monkey on the monkey to lift the monkey up.
(d) None of the above are true.

C-6-3 (b) Of course, we are assuming that the monkey's hands do not slip on the rope.

MAKING CONNECTIONS

Impact Mass Extinction Events

During a mass extinction event, a significant number of species become extinct. A large meteor impact is perhaps the most dramatic type of such an event. A meteor impact releases enormous amounts of energy, which can cause massive forest fires and gigantic tsunamis. The smoke and dust released into the atmosphere prevent the process of photosynthesis, thus catastrophically disrupting food chains through a domino effect. There is strong evidence that the dinosaurs became extinct as a direct result of a 10 to 20 km-wide meteor striking the Yucatán peninsula about 65 million years ago (Figure 6-6). This impact is estimated to have released an amount of energy equivalent to 100 billion megatonnes of TNT. The impact threw trillions of tonnes of dust into the atmosphere, which may have caused global climate changes in addition to the food chain disruptions mentioned above. To put this in perspective, the atomic bomb that destroyed the city of Hiroshima in 1945 had a yield of only 13 to 18 kilotonnes.

D. VAN RAVENSWAAY/SCIENCE PHOTO LIBRARY

Figure 6-6 An artist's rendition of what the 180 km-wide Chicxulub impact crater on the Yucatán Peninsula would have looked like shortly after it was created.

MECHANICS

LO 2

6-2 Work Done by Constant Forces in Two and Three Dimensions

✓ CHECKPOINT

C-6-4 Work Done by the force of Gravity

A satellite of mass m completes half a turn around Earth's axis in low Earth orbit. Use R as the radius of Earth, h as the altitude of the satellite, and g as the value of the acceleration due to gravity at that altitude. How much work is done by the force of gravity on the satellite as it completes the half turn?

(a) $mg(R + h)$

(b) $2mg(R + h)$

(c) zero

(d) none of the above

C-6-4 (c) The change in the kinetic energy is zero hence the total work, which in this case the is the work done by the force of gravity is zero.

Thus far, we have considered examples where the force and the displacement are directed along the same direction, so our calculations have involved only one dimension. However, forces are frequently not parallel to the direction of motion, as in the example shown in Figure 6-7(a). The work done by a constant force that is not parallel to the displacement can be defined as

$$W_F = \vec{F} \cdot \Delta \vec{s} \qquad (6\text{-}8)$$

where $\Delta \vec{s}$ denotes displacement in more than one dimension. Work, as given in equation 6-8, is the scalar product between the two vectors $\vec{F}$ and $\Delta \vec{s}$.

Note that this equation applies only when the force is constant and when the angle between the force and the displacement remains constant.

(a)

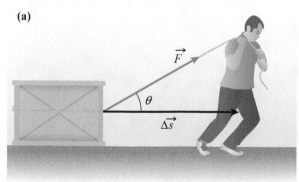

Figure 6-7(a) A crate being pulled on the floor. The crate moves along a horizontal surface, and the force acts in a direction above the horizontal, making an angle θ with the horizontal.

We will highlight three ways to calculate the work done in situations involving two or more dimensions. We emphasize here that the expression for the work done by

a constant force is always given by Equation (6-8) above. The three methods that we develop below are simply three ways of looking at the scalar product of Equation (6-8).

Approach 1. For an object moving in a straight line, expanding the scalar product in Equation (6-8) gives

$$W_F = |\vec{F}| |\Delta \vec{s}| \cos \theta \qquad (6\text{-}9)$$

We can then calculate the work done by using values for the magnitude of the force, the magnitude of the displacement, and the cosine of the angle between the force and displacement vectors, as illustrated in Figure 6-7(b).

(b)

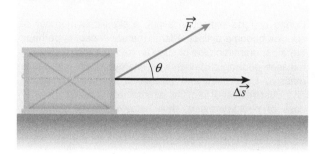

Figure 6-7(b) Calculating the work done by force $\vec{F}$

Approach 2. In the second approach, we rewrite Equation (6-9) as

$$W_F = \{ |\vec{F}| \cos \theta \} |\Delta \vec{s}| = F_\parallel \Delta s \qquad (6\text{-}10a)$$

We can derive this equation more formally by writing the force vector as the sum of two components: one parallel to the displacement and the other perpendicular to the displacement. Since the scalar product of perpendicular vectors is zero, the expression for the work done by force $\vec{F}$ becomes

$$W_F = (\vec{F}_\parallel + \vec{F}_\perp) \cdot \Delta \vec{s} = \vec{F}_\parallel \cdot \Delta \vec{s} = \{ |\vec{F}| \cos \theta \} |\Delta \vec{s}| \qquad (6\text{-}10b)$$

Thus, only the component of $\vec{F}$ parallel to the displacement does work. The work will be positive when $\vec{F}_\parallel$ and $\Delta \vec{s}$ point in the same direction and negative when they point in opposite directions (Figure 6-7(c)).

(c)

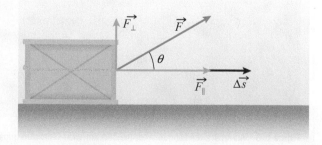

Figure 6-7(c) Only the component of $\vec{F}$ parallel to the displacement does work.

(d)

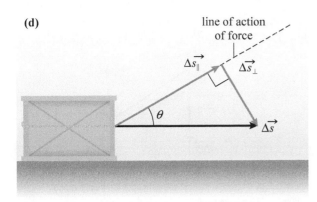

Figure 6-7(d) Only the component of *F* parallel to the displacement does work.

Approach 3. In the third approach (Figure 6-7(d)), we rewrite Equation (6-9) as

$$W_F = |\vec{F}| \{|\Delta\vec{s}| \cos\theta\} = F\Delta s_\parallel \qquad (6\text{-}11\text{a})$$

As with the force vector, we can express the displacement vector as the sum of a component parallel to the force and a component perpendicular to the force:

$$W_F = \vec{F} \cdot (\Delta\vec{s_\parallel} + \Delta\vec{s_\perp}) = \vec{F} \cdot \Delta\vec{s_\parallel} \qquad (6\text{-}11\text{b})$$

EXAMPLE 6-8

Work Done by the Force of Gravity

Starting from rest, a block slides down a frictionless ramp of slope angle 32.0°, as shown in Figure 6-8. Use the three representations introduced above to find the speed of the block at the bottom of the ramp.

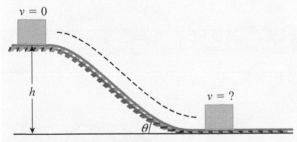

Figure 6-8 A block slides down a frictionless incline.

SOLUTION

The normal force does no work because the angle between the force and the displacement is 90°. Since the force of gravity is the only force doing work,

$$W_g = W_T = \Delta K$$

Once we find the work done on the block, we can use its kinetic energy to find its speed.

APPROACH 1.

Using the scalar product of the force and displacement vectors, from Equation 6-9:

C-6-5 A Sliding Bear

A polar bear cub runs toward an icy hill and slides freely up the icy slope. Which of the following statements is accurate?
(a) The force of gravity does positive work on the cub.
(b) The force of gravity does negative work on the cub.
(c) Work is a scalar, not a vector, and cannot be positive or negative.
(d) There is no work done because there is no friction.

C-6-5 (b) the vertical displacement is up while the force of gravity points down.

Thus, only the component of the displacement that is parallel to the force contributes to the work done. Here, again, when $\vec{F}$ and $\Delta\vec{s_\parallel}$ point in the same direction, the work done by the force is positive, and when they point in opposite directions, the work is negative.

Equations (6-9), (6-10), and (6-11) offer three representations of the same principle. Depending on the problem being solved, using one of these representations might be more convenient than another.

Example 6-8 demonstrates that the work done by the force of gravity does not depend on the path taken; only the vertical component of the displacement matters.

$$W_g = m\vec{g} \cdot \Delta\vec{s} = |m\vec{g}| |\Delta\vec{s}| \cos\beta$$

where β is the angle between the displacement vector and the force vector.

From Figure 6-9(a), $\beta = 90° - \theta$, $\cos\beta = \sin\theta$, and $\sin\theta = h/\Delta s$. Therefore,

$$W_g = |m\vec{g}| |\Delta\vec{s}| \sin\theta = mgh$$

(a)

Figure 6-9(a) Using the scalar product to find the work done

APPROACH 2.

Using the parallel component of the force, from Equation 6-10b:

$$W_g = m\vec{g_\parallel} \cdot \Delta\vec{s}$$

From Figure 6-9(b), the parallel component of *mg* is $mg \sin\theta$. Therefore,

$$W_g = mg \sin\theta(\Delta s) = mgh$$

(continued)

(b)

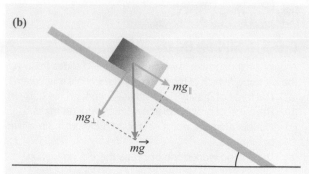

Figure 6-9(b) Only the component of mg along the incline contributes to the work done by gravity.

APPROACH 3.

Using the parallel component of the displacement, from Equation 6-11b:

$$W_g = m\vec{g} \cdot \Delta \vec{s}_\parallel$$

Since $m\vec{g}$ is vertical, the parallel component of Δs is also vertical. From Figure 6-9(c), this component is $\Delta s (\sin \theta)$. Therefore,

$$W_g = mg\Delta s (\sin \theta) = mgh$$

All three approaches give the same expression for W_g, as expected. We now solve for the speed at the bottom of the ramp:

(c)

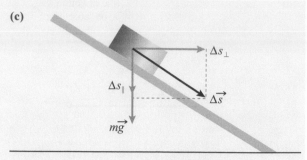

Figure 6-9(c) Only the vertical component of the displacement contributes to the work done by the force of gravity.

$$W_g = \Delta K$$

The work done by the force of gravity by all three approaches is mgh:

$$mgh = \frac{1}{2}mv^2 \Rightarrow v^2 = 2gh$$

Making sense of the result:

This problem can also be solved using Newton's second law, but the work–kinetic energy approach solves it more easily. The result obtained is the same as the result obtained using kinematics equations for an object that falls straight down a distance h.

ONLINE ACTIVITY

Work and Energy

The e-resource that accompanies every new copy of this textbook contains an Online Activity using the PhET simulation "The Ramp." Work through the simulation and accompanying questions to gain an understanding of the concepts of work and energy.

MAKING CONNECTIONS

Turbines on Kites

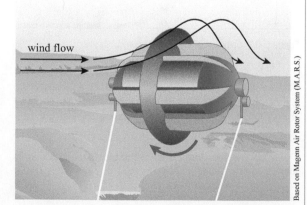

Figure 6-10 (a) An offshore wind farm. Wind turbines convert kinetic energy to electric energy. (b) An artist's rendition of a high-altitude wind turbine.

(continued)

Wind energy reportedly has the potential to provide the world's total need for energy many times over. Although renewable, the amount of energy generated from terrestrial and offshore wind turbines (Figure 6-10(a)) remains relatively modest due to the limited wind speed at low altitudes. Research is being conducted into developing technologies to harvest wind energy at high altitudes (Figure 6-10(b)). Due to the tremendous wind speeds at higher altitudes, the energy available is much higher than what is available closer to Earth's surface. The work done by the wind on the turbine goes into the turbine's kinetic energy, which generates electricity. Figure 6-10(b) shows an artist's rendition of a proposed airborne high-altitude wind turbine. While the European Union seems to be leading the way in terms of harvesting energy from wind, Canada and Russia have the largest potential in this regard. Germany is on schedule to have 80% of its electrical power needs met by renewable energy by 2050, 44% of which is anticipated to come from wind energy.

EXAMPLE 6-9

Pulling a Crate

The worker in Figure 6-11(a) pulls on a rope with a force of 610 N to drag a 90 kg crate along a horizontal surface. The rope makes a 41° angle with the horizontal. The coefficient of kinetic friction between the surface and the crate is 0.24. What is the speed of the crate after having been pulled a total distance of 4.3 m?

(a)

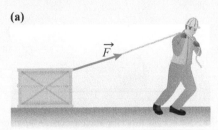

Figure 6-11(a) A worker pulls a crate on a rough surface.

SOLUTION

This is another example where we can use Newton's second law and the kinematics equations to solve the problem. However, since the work–kinetic energy theorem combines both, it is ideal to use it to solve this problem.

The total work done on the crate comes from the work done by the applied force and the work done by friction:

$$W_F + W_f = \Delta K \qquad (1)$$

There is no term in Equation (1) for either the work done by the force of gravity or the work done by the normal force because these forces act normal to the direction of motion. Therefore, by Equations (6-8) and (6-9), the work done by these forces is zero.

Using Equation (6-9), examining Figure 6-11(b), and considering that the two forces act over the distance Δs, we have

$$F\Delta s \cos\theta - f_k d = \frac{1}{2}mv^2 \qquad (2)$$

The work done by F is positive because the horizontal component of F points along the same direction as the displacement. The force of friction points in a direction that is opposite to the displacement's direction. Therefore, the work done by friction is negative.

(b)

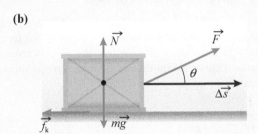

Figure 6-11(b) Forces acting on the crate being pulled

To calculate v from Equation (2), we need to determine the force of friction. We can use $f_k = N\mu_k$ because we can find N, the normal force on the crate, by summing forces in the vertical direction:

$$N + F\sin\theta - mg = 0$$

Therefore,

$$N = mg - F\sin\theta$$

Substituting into Equation (2) gives

$$Fd\cos\theta - (mg - F\sin\theta)\mu_k d = \frac{1}{2}mv^2 \qquad (3)$$

$$v^2 = \frac{2d}{m}\{F\cos\theta - (mg - F\sin\theta)\mu_k\}$$

$$= \frac{2(4.3 \text{ m})}{90 \text{ kg}}\{(610 \text{ N})(\cos(41°))$$

$$- [(90 \text{ kg})(9.81 \text{ kg m/s}^2) - (610 \text{ N})(\sin(41°))](0.24)\}$$

and

$$v = 5.74 \text{ m/s}$$

Making sense of the result:

Here, we had to keep in mind that the magnitude of normal force was not simply equal to mg; rather, it had to be obtained from the summation of the forces in the vertical direction. Without the force of friction, we would have $Fd\cos\theta = \frac{1}{2}mv^2$ and the speed would have been 6.63 m/s. Friction dissipates some of the kinetic energy, resulting in a lower speed.

6-3 Work Done by Variable Forces

So far, we have dealt with work done by constant forces. However, most of the forces we encounter in daily life are not constant. Standing up, sitting down, starting and stopping a car, skiing, and canoeing all involve forces that vary. The fundamental forces in nature—gravity, the electrostatic force, and the two nuclear forces—all vary with distance. How is the calculation of work done by a constant force different from the calculation of work done by a variable force?

Graphical Representation of Work

Consider an object being pushed over a distance Δs by a constant force that points in the direction of the motion. Since this force does not change, the force versus displacement curve is simply a horizontal line, as shown in Figure 6-12.

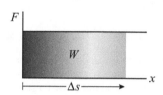

Figure 6-12 Force versus distance graph for a constant force

From Equation (6-12a), the work done by a constant force parallel to the displacement is $F\Delta s$. In Figure 6-12, this quantity is represented by the area of the rectangle under the F versus Δs graph.

Figure 6-13 shows a force versus distance graph for a force with a continuously changing magnitude. As in Figure 6-12, the area under the curve represents the work done by the force. If the force does not have the same direction as the displacement, we can graph just the component of the force parallel to the displacement. Then the area under the graph will correspond to the work done by the force. Alternatively, we can graph each component of the force against its corresponding component of the displacement and sum all the areas to calculate the work done.

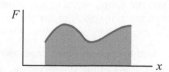

Figure 6-13 The work done by the force over the displacement is the area under the curve.

We can break up the area under the curve into thin rectangles, or trapezoids, and then sum their areas (Figure 6-14):

$$W \approx \sum \vec{F} \cdot \Delta \vec{x} \qquad (6\text{-}12)$$

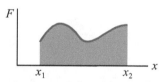

Figure 6-14

The narrower the width of the rectangles, the more accurate the result will be. As Δx approaches zero, that is, for an infinitesimally small value of Δx (denoted dx), the area calculated becomes infinitely accurate. The sum $\sum \vec{F} \cdot \Delta \vec{x}$ then becomes the integral of the force over the displacement. The area obtained using integration is exactly equal to the work done by the variable force over the displacement from x_1 to x_2 in Figure 6-15:

$$W_F = \int_{x_1}^{x_2} \vec{F} \cdot d\vec{x} \qquad (6\text{-}13a)$$

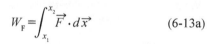

Figure 6-15 The integral gives the exact area under the curve.

For more than one dimension, the expression becomes

$$W_F = \int_a^b \vec{F} \cdot d\vec{s} \qquad (6\text{-}13b)$$

where $d\vec{s}$ is the infinitesimal displacement along the trajectory of the object acted upon by force $\vec{F}$ while moving from point a to point b.

The Work–Kinetic Energy Theorem for Variable Forces

We introduced the total work–kinetic energy theorem with Equations (6-2) and (6-5). Although Equation (6-5) is universal, Equation (6-2) holds only for constant forces. We now present a proof that the work–kinetic energy theorem does hold for variable forces.

In three dimensions, the definition of the total work done on an object by variable forces can be expressed as

$$W_T = \int_{r_1}^{r_2} \vec{F}_{net} \cdot d\vec{s} = \int_{r_1}^{r_2} F_s \, ds \qquad (6\text{-}14)$$

Where F_s is the component of the force parallel to the displacement. Since $F_s = ma_s$,

$$W_T = \int_{r_1}^{r_2} ma_s \, ds$$

MECHANICS

Where r_1 and r_2 are the initial and final positions of the object.

Using the chain rule for integration yields

$$a = \frac{dv}{dt} = \frac{dv}{ds}\frac{ds}{dt} = v\frac{dv}{ds}$$

Therefore,

$$W_T = \int_{r_1}^{r_2} m\left(v\frac{dv}{ds}\right)ds$$

and

$$W_T = \int_{v_1}^{v_2} mv\,dv = \frac{mv^2}{2}\bigg|_{v_1}^{v_2} = \frac{1}{2}mv_2^2 - \frac{1}{2}mv_1^2$$

Thus, the total work–kinetic energy theorem applies for both constant and variable forces.

LO 4

6-4 Work Done by a Spring

A simple example of work done by a variable force is the work done by the force of a spring. As described in Chapter 5, the force exerted by a spring is given by Hooke's law:

$$\vec{F_s} = -k\vec{x} \qquad (6\text{-}15)$$

where $\vec{x}$ is the elongation of the spring from its unstretched length, and k is the spring constant.

Hooke's law is a good approximation for springs within the elastic limit. (Springs stretched beyond this limit become nonlinear and may permanently deform.) The larger the value of k, the stiffer the spring, that is, the more force it takes to stretch the spring by a given amount.

Consider the spring resting on the horizontal surface in Figure 6-17. The left end of the spring is fixed. If we move the right end by a distance x to the right of the unstretched (equilibrium) position, the spring will exert a force on us to the left. If we double the distance by which the spring is stretched, the force that the spring exerts on us doubles. The negative sign indicates that if you move one end of a spring away from the unstretched position, the spring will exert a force on you to pull you back. Thus, the force of a spring is a **restoring force**.

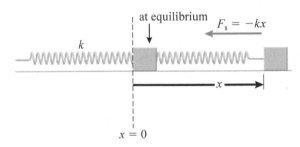

Figure 6-17 The restoring nature of the spring force: when stretched to the right, the spring pulls to the left

✓ CHECKPOINT

C-6-6 Work Done by a Spring

When is the work done by a spring positive?
(a) as the spring is being compressed from equilibrium
(b) as the spring is being stretched from equilibrium
(c) as the spring returns to equilibrium
(d) as the spring is being moved away from equilibrium

C-6-6 (c) this is the only instance when the force of the spring points in the direction of motion, and not opposite to it.

Work Done by a Spring

Let us move the free end of a spring such as the one in Figure 6-17, whose other end is fixed, from position x_1 to position x_2, both relative to the position where the spring is unstretched. Since the force of the spring is not a constant, we use calculus to calculate the work done by a spring:

$$W_s = \int_{x_1}^{x_2} \vec{F_s} \cdot d\vec{x} \qquad (6\text{-}16)$$

The force that the spring exerts is given by

$$\vec{F}_s = -k\vec{x}$$

Substituting for $\vec{F}_s$ in the expression for the work done gives

$$W_s = \int_{x_1}^{x_2} -k\vec{x} \cdot d\vec{x}$$

Since $\vec{x}$ and $d\vec{x}$ have the same direction,

$$W_s = -k \int_{x_1}^{x_2} x\, dx = \left(-\frac{kx_2^2}{2}\right) - \left(-\frac{kx_1^2}{2}\right)$$

and

$$W_s = \frac{1}{2}kx_1^2 - \frac{1}{2}kx_2^2 \qquad (6\text{-}17)$$

While deriving Equation (6-17), we did not need to specify whether the spring was being compressed or stretched. Nor did we consider whether the spring might have already been stretched at position x_1 and then either stretched farther to position x_2, or perhaps moved through

equilibrium to become compressed at position x_2. The initial and final positions, as measured from the unstretched position, determine the work done by the spring, regardless of *how* the spring moved between these positions.

 CHECKPOINT

C-6-7 $(\Delta x)^2$ or $\Delta(x^2)$?

A spring of spring constant $k = 690$ N/m is held at a position such that it is compressed by 10.0 cm from its unstretched position. The spring is then compressed by another 90.0 cm. The work done by the spring is

(a) 221 J;
(b) −221 J;
(c) 276 J;
(d) −276 J;
(e) none of the above.

C-6-7 (d) the first two answers use $(0.8)^2$, and are wrong, the work should be negative.

 ## EXAMPLE 6-10

Collision with a Spring

A 17.0 kg block is sliding on a frictionless horizontal surface at a speed of 3.80 m/s when it strikes a spring that is already compressed by 21.0 cm. Given that the spring constant is 710 N/m, find the distance by which the spring is further compressed during the collision.

SOLUTION

The force of the spring and, therefore, the acceleration of the mass are not constant. The kinematics equations we developed in Chapter 3 are not applicable here because they apply only for constant acceleration. However, we easily apply a work–energy approach because the work done by the spring depends only on the initial and final positions.

The only force doing work on the block is the force of the spring; therefore, $W_T = W_s = \Delta K$. Since the final speed of the block is zero at maximum spring compression,

$$W_s = \frac{1}{2}kx_1^2 - \frac{1}{2}kx_2^2 = 0 - \frac{1}{2}mv^2 \qquad (1)$$

Then,

$$\frac{1}{2}kx_2^2 = \frac{1}{2}mv^2 + \frac{1}{2}kx_1^2 \qquad (2)$$

and

$$x_2^2 = \frac{1}{k}(mv^2 + kx_1^2)$$

$$= \frac{1}{710}(17\text{ kg }(3.80\text{ m/s})^2 + 710\text{ N/m }(0.21\text{ m})^2)$$

$$x_2 = 0.624\text{ m}$$

The distance by which the spring compresses after the block strikes it is the difference between the final and initial positions:

$$\Delta x = x_2 - x_1 = 0.624\text{ m} - 0.210\text{ m} = 0.414\text{ m}$$

Making sense of the result:

If the spring had not been compressed initially, the result would have been

$$\frac{1}{2}kx_2^2 = \frac{1}{2}mv^2 \quad \text{and} \quad \Delta x = 0.588\text{ m}$$

The more compressed a spring is, the harder it becomes to compress it further.

EXAMPLE 6-11

Spring Trampoline

While testing springs in your backyard, you jump from a height of 4.00 m down onto a platform sitting on top of a spring, as shown in Figure 6-18. The spring constant is 501 N/m. Find the maximum compression of the spring. Your mass is 72.0 kg, and assume that the mass of the platform is negligible.

SOLUTION

The work–energy approach is once again ideal to use to solve this problem involving springs. Two forces do work on you: the force of gravity and the force of the spring. You start from rest and are momentarily at rest again at maximum spring compression. Therefore, the change in kinetic energy of the platform is zero, and

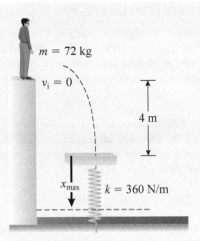

$$m = 72 \text{ kg}$$

$$v_i = 0$$

4 m

$$x_{max}$$

$$k = 360 \text{ N/m}$$

Figure 6-18 Example 6-11

$$W_T = W_g + W_s = \Delta K = 0$$

Since the spring is originally unstretched, $x_1 = 0$. Substituting into Equation (6-17) gives

$$W_s = \frac{1}{2} kx_1^2 - \frac{1}{2} kx_2^2 = -\frac{1}{2} kx_2^2$$

The force of gravity does work on you as long as you are moving downward. Thus,

$$W_g = mgh + mgx_2$$

The work done by the force of gravity is positive as given by Equations (6-8) and (6-9) because the force of gravity points along the same direction as your displacement before and during spring compression. Therefore,

$$mgx_2 + mgh - \frac{1}{2} kx_2^2 = 0$$

$$\Rightarrow 72 \text{ kg } (9.81 \text{ m/s}^2)x_2 + 72 \text{ kg } (9.81 \text{ m/s}^2)4 - 250.5x_2^2 = 0$$

Solving this quadratic equation gives a maximum compression of $x_{max} = x_2 = 5.05$ m.

Making sense of the result:

The key here is to keep in mind that the force of gravity does work as the spring is being compressed. If we neglect the mgx_2 term (the work done by the force of gravity while the spring is being compressed), we get a maximum compression of 3.36 m much less than the actual value.

6-5 Conservative Forces and Potential Energy

Suppose you lift a book a distance d straight up from a table and then let the book drop. The book will fall back to its original position, hitting the table with a certain speed. Now suppose you slide the book the same distance d horizontally along the table and then release it. The book does not return to its original position. What is the difference between the two cases?

In the first case, you had to do work against the force of gravity to lift the book. In the second case, you did work against the force of friction. The work you did against friction was largely converted into thermal energy, which, for practical purposes, you cannot recover. When you let the book drop after lifting it, not only does it return to its original position, but it gains some kinetic energy. We can show that you can recover all the work that you put in when you work against the force of gravity.

Conservative and Nonconservative Forces

Forces in the physical universe can be either **conservative** or **nonconservative**. For example, the force of friction is a dissipative, or a nonconservative, force because the work you do against friction is dissipated, primarily as thermal energy (we use this fact to warm our hands by rubbing them together). The dissipated energy is "lost" and cannot be recovered as kinetic energy. Common examples of nonconservative forces are applied forces, which are forces that we would apply, for instance, to push a crate on an incline, compress a spring, or lift a weight.

The work you do against conservative forces can be recovered in the form of kinetic energy. When you stretch a spring and release it, it "springs" back to its equilibrium position. In the case of charged particles, the closer you bring two opposite charges together, the more kinetic energy they will have when they move away from each other after being released. The more work you do in bringing them together, the more energy is stored in the system that can be recovered in the form of kinetic energy. This stored energy is called **potential energy**.

Potential Energy

When you lift a book a distance h as shown in Figure 6-19, beginning and ending with the book at rest, the change in kinetic energy is zero, and

$$W_T = \Delta K = 0$$

The work, W_F, that you do on the book as you lift it upward is positive because the force that you exert has the same direction as the displacement. However, the *total* work done on the book also includes the work done by the force of gravity, W_g, which is negative:

$$W_T = W_F + W_g = \Delta K = 0$$

Therefore, $W_F + W_g = 0$

and $W_F = -W_g$

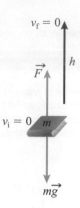

Figure 6-19 The work done by $\vec{F}$ is positive. The work done by gravity is negative

We can use Equation (6-9) to find the work done by the constant force of gravity:

$$W_g = \vec{F} \cdot \vec{h} = m\vec{g} \cdot \vec{h} \qquad (6\text{-}18)$$

Expanding the scalar product, we get

$$W_g = \vec{F} \cdot \vec{h} = m|\vec{g}||\vec{h}|\cos\theta \qquad (6\text{-}19)$$

Since the angle between the vectors $\vec{g}$ and $\vec{h}$ is 180°,

$$W_g = mgh \cos(180°) = -mgh$$

and

$$W_F = -W_g = mgh$$

Gravitational Potential Energy

When you lift an object of mass m a height h, the gravitational potential energy of the object–Earth system increases by mgh. The conventional symbol for potential energy is U. Thus, in this situation,

$$\Delta U = mgh$$

Note that the quantity mgh is also the negative of the work done by the force of gravity. In fact, the change in gravitational potential energy is defined in terms of the work done by the force of gravity:

KEY EQUATION
$$\Delta U_g = -W_g \qquad (6\text{-}20a)$$

A common misconception is that the change in potential energy is given by the work done by you as you lift the book. The only reason the work done by you in this case is equal to the change in potential energy is that you started and ended with the book at rest. If this is not the case, the work done by you would not be equal to the change in potential energy. We look at this in detail shortly.

If we move an object up, the work done by the force of gravity is negative, and the change in potential energy is positive. The gravitational potential energy increases as the height increases. Conversely, if we move the object down, the change in potential energy is negative.

MAKING CONNECTIONS

Green Cars: Hybrids, Biofuels, and Electric Cars

According to the International Energy Agency, the global transportation sector used approximately 61.4% of the oil consumed worldwide and produced around 20% of global carbon dioxide (CO_2) emissions in 2008. Alternatives to fossil fuel–powered cars are being developed worldwide. Hybrid vehicles driven in the city do reduce fuel consumption and reduce CO_2 emissions; however, the bulk of the energy to drive the hybrid vehicle still comes from fossil fuels.

Biofuel, in the form of bioethanol and biodiesel, is another alternative. In Brazil, ethanol use already exceeds gasoline use as a car fuel. Theoretically, biofuels put no more CO_2 back into the atmosphere than what the crops use to produce the biofuels; unlike fossil fuels, biofuels are renewable. The use of biofuel, however, is questionable because of competition with food crops and limitations in the capacity to meet global demand.

Electric cars remain a viable alternative, depending on circumstances. In the United States and China, the bulk of electricity is produced by burning coal. One issue here is that burning coal releases more CO_2 than burning gasoline. Considering the efficiency limits inherent in the energy-conversion stages required to put an electric car in motion (from thermal energy from burning the coal, to electrical to electrochemical energy in the car's battery, to the mechanical energy of the moving car), do you think it would be advisable to operate an electric car in a region in which electricity is produced by burning fossil fuels?

Canada is the largest producer of hydroelectricity in North America and the third largest producer in the world after China and Brazil. In 2008, close to 70% of Canada's electric power came from hydroelectricity. It would be advantageous to operate an electric car in a region of the country that is powered by a hydroelectric generating plant. Figure 6-20 shows the Tesla Roadster 2.5, a premium, fully electric car with impressive performance.

Figure 6-20 This Tesla Roadster 2.5 boasts a 288 horsepower output; it accelerates from 0 to 100 km/h in 3.7 s and has a range per charge of approximately 400 km.

EXAMPLE 6-12

Work Done by a Variable Force

You push a block of mass m up a quarter-circle track as shown in Figure 6-21(a) using a variable force. You start from rest such that the mass arrives at the top of the track with a speed of zero. Calculate the work done by the force of gravity and the change in the block's potential energy.

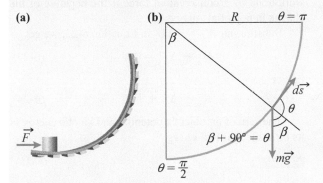

Figure 6-21 (a) You push a block up a semicircular track. (b) The FBD of the block moving up the track

SOLUTION

From Equation (6-20), the change in potential energy of the block is the negative of the work done by the force of gravity. The force of gravity is a constant and points downward. The displacement is directly along the track. Therefore, the angle θ between the force of gravity vector and the displacement changes continuously (Figure 6-21(b)). Since the

angle is not constant, we have to use integration to calculate the work done by the force of gravity:

$$W_g = \int_{s_1}^{s_2} \vec{F} \cdot d\vec{s} = \int_{s_1}^{s_2} Fds\cos\theta \qquad (1)$$

When the block is at the bottom of the track, θ is 90°; at the top of the track, θ is 0°. We will use β for the angular position of the block with respect to the vertical, as measured from the centre of the circular track. From Figure 6-21(b), we can see that

$$\beta = \theta - 90° \qquad (2)$$

The infinitesimal arc length ds is given by

$$ds = Rd\beta = Rd\theta \qquad (3)$$

Substituting for ds in the integral gives

$$W_g = \int_{\pi/2}^{\pi} mgRd\theta \cos\theta = mgR\int_{\pi/2}^{\pi} \cos\theta\, d\theta$$

$$W_g = mgR\left[\sin(\pi)\right] - \{mgR\left[\sin(\pi/2)\right]\} = -mgR$$

The change in potential energy is the negative of the work done by the force of gravity. Therefore using Equation 6-20a,

$$\Delta U = mgR$$

Making sense of the result:

The change in the height of the mass is equal to R, the radius of the track. So, this example illustrates the fact that the change in gravitational potential energy, ΔU, is path-independent and depends only on the endpoints of the path.

The key relationship given in Equation (6-20a) applies for any conservative force:

The change in potential energy is defined as the negative of the work done by the conservative force:

KEY EQUATION $$\Delta U_c = -W_c \qquad (6\text{-}20b)$$

The Change in Potential Energy Is Path-Independent

In Section 6-3, we described the work done by the force of gravity as path-independent because it is dependent only on the change in vertical position. Equation (6-20) illustrates that the change in gravitational potential energy is path-independent and depends only on the height of the initial and final positions. In fact, the change in potential energy associated with all conservative forces, and not just the force of gravity, depends on the initial and final states and not on the path taken between these states. The chapters on thermodynamics discuss the concepts of "states" and "state functions" in more detail.

Potential Energy of a Spring

The work done by the force of a spring was derived in Section 6-4-1:

$$W_s = \frac{1}{2}kx_1^2 - \frac{1}{2}kx_2^2 \qquad (6\text{-}17)$$

where x is the distance from the equilibrium position.

Using the definition of the change in potential energy from Equation (6-20b), we conclude that the change in the potential energy of a spring is given by

$$\Delta U_s = -W_s = \frac{1}{2}kx_2^2 - \frac{1}{2}kx_1^2 \qquad (6\text{-}20c)$$

The change in the potential energy of the spring depends only on the initial and final positions, and is not affected by the path taken to get from one state to another. We can define the potential energy stored in a spring as

$$U_s = \frac{1}{2}kx^2 \qquad (6\text{-}21)$$

Then, Equation (6-20c) becomes

$$\Delta U_s = U_{s_2} - U_{s_1}$$

CHECKPOINT

C-6-8 Zero of Potential Energy

Compare the expression for the elastic potential energy of a spring in Equation (6-21) and the expression for the gravitational potential energy in Equation (6-20a). Why is there a uniquely identifiable zero for the elastic potential energy of a spring but not for the force of gravity?

C-6-8 The potential energy for a spring is zero when it is not stretched. The zero for gravitational potential energy near Earth can be set arbitrarily.

LO 6

6-6 Nonconservative Forces and Mechanical Energy

When you carry a box up a ladder, the change in the potential energy of the box might not equal the work done by you. If the final speed of the box does not equal the initial speed, you do work to change the box's kinetic energy in addition to the work you do against the force of gravity.

Now, let us apply the work–kinetic energy theorem to a situation in which you lift a stationary basketball straight up from the floor and prepare to throw it. The initial speed of the basketball is zero. You accelerate the basketball such that its speed upward when it reaches chest height is nonzero. Then,

$$W_T = \Delta K = W_F + W_g$$

where W_F is the work done by you and W_g is the work done by gravity.

We have already established that the force of gravity is a conservative force. Let us consider why the applied force is nonconservative. If you do work against a spring, by compressing it, you can get the work you do back when the spring goes back to equilibrium. If you do work by lifting a book, you can get the work you did back in the form of kinetic energy when you let the book go. Both the spring force and the force of gravity are conservative forces. If however, you were to bend your friend's arm, or push a book on a rough surface, the work you do is not recoverable in the form of mechanical energy. Hence

 EXAMPLE 6-13

Stretching a Spring

A block of mass 7.20 kg lies on a horizontal frictionless surface. It is attached to a horizontal unstretched spring with a spring constant $k = 190$ N/m. The other end of the spring is fixed. You apply a constant force $\vec{F}$ to the mass to pull it away from the spring. When the mass has moved by 40 cm, its speed is 3.10 m/s. Find the force $\vec{F}$. Assume that the force acts in the direction of motion of the mass.

the force of friction and the applied force (force exerted by your friend, or by you) are nonconservative. The force applied by you is a dissipative, or nonconservative, force. In general, since all forces can be classified as conservative or nonconservative, we can express the statement of the total work–kinetic energy theorem as follows:

$$W_T = W_{nc} + W_c = \Delta K \qquad (6\text{-}22)$$

where the subscripts nc, and c, stand for nonconservative and conservative respectively. From Equation 6-20b, the work done by a conservative force is the negative of the change in potential energy.

Substituting $W_c = -\Delta U_c$ in Equation 6-22, we get

$$W_{nc} - \Delta U_c = \Delta K \qquad (6\text{-}23)$$

or

$$W_{nc} = \Delta K + \Delta U_c \qquad (6\text{-}24)$$

The sum of an object's potential and kinetic energy is its **total mechanical energy**:

KEY EQUATION
$$E_m = U + K \qquad (6\text{-}25)$$

Equation (6-24) can be rewritten as follows:

KEY EQUATION
$$W_{nc} = \Delta E_m \qquad (6\text{-}26)$$

Therefore, the action of a nonconservative force changes the total mechanical energy of an object. When we do positive work on an object, its mechanical energy increases. Examples are pushing a car (increasing its kinetic energy) and lifting a book (increasing its potential energy). When we do negative work on an object, its mechanical energy decreases. Examples are slowing down an object, such as when applying the brakes in a moving car, and taking a book down from a high shelf.

 CHECKPOINT

C-6-9 Lifting Weights

How much work do you have to do to lift a 1.00 kg dumbbell from the floor by 1.00 m so that it has a speed of 3.00 m/s at that height?

C-6-9 $W_F = \frac{1}{2} mv^2 + mgh = \frac{1}{2}(9)(1) + 1(9.81)(1) = 14.3$ J

SOLUTION

Since the force you apply is a nonconservative force, we use Equation (6-24):

$$W_{nc} = \Delta K + \Delta U_c \qquad (1)$$

The change in the kinetic energy of the block is

$$\Delta K = K_f - K_i = \frac{1}{2}mv^2 - 0 \qquad (2)$$

The change in the spring's potential energy is given by Equation (6-20c):

$$\Delta U_s = \frac{1}{2} kx_2^2 - \frac{1}{2} kx_1^2 = \frac{1}{2} kx^2 - 0 \qquad (3)$$

The work done by the force you apply is

$$W_{nc} = W_F = \vec{F} \cdot \vec{x} = Fx \cos(0°) = Fx \qquad (4)$$

where x is the distance over which the force acts.
Putting together the above equations, we have

$$Fx = \frac{1}{2} mv^2 + \frac{1}{2} kx^2 \qquad (5)$$

$$F = \frac{\frac{1}{2} mv^2 + \frac{1}{2} kx^2}{x}$$

$$= \frac{\frac{1}{2}(7.2 \text{ kg})(3.1 \text{ m/s})^2 + \frac{1}{2}(190 \text{ Nm}^{-1})(0.4 \text{ m})^2}{0.4 \text{ m}} = 124 \text{ N}$$

Making sense of the result:

The plus sign in Equation (5) indicates that the work done by the applied force is responsible for both stretching the spring and increasing the kinetic energy of the block, as expected.

Conservation of Mechanical Energy

Equation (6-26) has a profound consequence: if there are no nonconservative forces (or the work of the nonconservative forces sums to zero), then the change in the object's mechanical energy will be zero:

$$\Delta E_m = 0 \quad \text{if} \quad W_{nc} = 0 \qquad (6\text{-}27)$$

Since the mechanical energy does not change in the absence of the work done by nonconservative forces, we can say the mechanical energy of an object is conserved:

$$E_m = constant \qquad (6\text{-}28)$$

Consequently, the mechanical energy of an object at any point on its trajectory is the same:

$$E_{m1} = E_{m2} = E_{m3} = \cdots \qquad (6\text{-}29)$$

Since the mechanical energy of an object is the sum of its potential and kinetic energies, the **law of conservation of mechanical energy** can also be stated as follows:

In the absence of nonconservative forces,

$$0 = \Delta K + \Delta U_c \qquad (6\text{-}30)$$

Therefore,

$$\Delta K = -\Delta U_c \qquad (6\text{-}31)$$

Equation (6-31) indicates that for an object under the influence of conservative forces only, whenever potential energy is lost, it is converted into kinetic energy, and vice versa.

☑ CHECKPOINT

C-6-10 What Is Conserved?

Which of the following statements is true?
(a) Potential energy is not conserved.
(b) Kinetic energy is conserved.
(c) Mechanical energy is always conserved.
(d) None of the above are true.

C-6-10 (d) Mechanical energy is only conserved when the work of non conservative forces is zero.

EXAMPLE 6-14

Block on an Incline

A 1.20 kg block slides down a frictionless incline with a slope angle of 42°, starting from a height of 2.30 m above the bottom of the incline, as shown in Figure 6-22. The incline meets a frictionless horizontal surface, at the end of which is an unstretched spring ($k = 460$ N/m) used to stop the block. Find the maximum compression of the spring.

SOLUTION

Since there are no nonconservative forces, we can apply the principle of conservation of mechanical energy. We begin with Equation (6-31):

$$\Delta K = -\Delta U_c$$

The block starts at rest, and when the spring is at maximum compression, the block is also momentarily at rest.

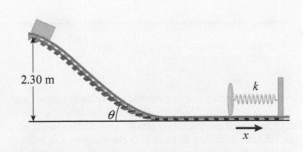

Figure 6-22 Example 6-14

Therefore, the change in kinetic energy, ΔK, is zero. Therefore, the change in potential energy, ΔU, is also zero.

The change in potential energy has two components: the change in the elastic potential energy of the spring and

(continued)

149

the change in the gravitational potential energy of what is referred to as the block–Earth system.

$$\Delta U_s + \Delta U_g = 0$$

When the block reaches the bottom of the incline, the change in gravitational potential energy is $\Delta U_g = -mgh$. The change in the spring's potential energy is given by

$$\Delta U_s = \frac{1}{2} kx_2^2 - \frac{1}{2} kx_1^2 = \frac{1}{2} kx_2^2$$

Since the spring is originally unstretched,

$$\frac{1}{2} kx^2 - mgh = 0$$

and

$$x^2 = \frac{2mgh}{k}$$

Therefore,

$$x = \sqrt{\frac{2(1.2 \text{ kg})(9.81 \text{ m/s}^2)(2.3 \text{ m})}{460 \text{ N/m}}} = 0.343 \text{ m}$$

Making sense of the result:

The gravitational potential energy is converted into kinetic energy, which is converted into spring elastic potential energy. However, there was no net change in kinetic energy, so we were able to bypass the kinetic energy step altogether.

ONLINE ACTIVITY

The e-resource that accompanies every new copy of this textbook contains an Online Activity using the PhET simulation "Energy Skate Park." Work through the simulation and accompanying questions to gain an understanding of the concepts of the conservation of energy.

EXAMPLE 6-15

Marble on a Track

The marble in Figure 6-23 starts from rest and slides down a frictionless incline from a height h above the ground, and then goes up the circular track of radius r as shown. Derive an expression for the minimum value of the height, h, so that the marble stays in contact with the circular track at the highest point on the track.

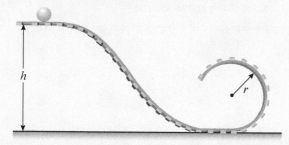

Figure 6-23   Example 6-15

SOLUTION

We are demanding that the marble barely be in contact with the track at the top. The force of contact between the marble and the top of the track is equal to zero. So, we begin with an FBD of the marble at the top of the track (Figure 6-24). Applying Newton's second law to the marble, we have, in the vertical direction,

$$\vec{N} + m\vec{g} = m\vec{a} \qquad (1)$$

Choosing down as our positive direction, we get

$$N + mg = \frac{mv^2}{r} \qquad (2)$$

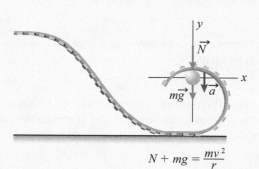

$$N + mg = \frac{mv^2}{r}$$

Figure 6-24   The FBD of the marble

Since N is zero,

$$v^2 = rg \qquad (3)$$

For the marble to have this speed at the top of the track, some of the gravitational potential energy was converted to kinetic energy. Since there are no dissipative forces, we can apply the conservation of mechanical energy:

$$\Delta K = -\Delta U_g \qquad (6\text{-}32)$$

The change in height is $h - 2r$. Since the marble ends up at a lower height, the gravitational potential energy decreases, and the change in potential energy is negative:

$$\Delta U_g = -mg(h - 2r) \qquad (4)$$

Since the marble starts from rest, the change in its kinetic energy is

$$\Delta K = \frac{1}{2} mv^2 - 0$$

where v is the speed of the marble at the top of the track. Therefore,

$$\frac{1}{2}mv^2 = mg(h - 2r) \tag{5}$$

Using the result from Equation (3), we get

$$\frac{1}{2}mrg = mg(h - 2r) \tag{6}$$

and, therefore,

$$h = 2.5\,r \tag{7}$$

Making sense of the result:

The speed at the top of the circular track cannot be zero because the marble needs enough speed to make it across the top without falling. Therefore, the initial height of the marble should be higher than the height of the loop ($2r$).

EXAMPLE 6-16

Sliding on a Spherical Surface

A polar bear cub sits on top of an igloo. The surface of the igloo has a spherical curvature and is smooth enough to have negligible friction. The cub begins to slide down the igloo. Determine where the cub will leave the surface of the igloo.

SOLUTION

As the bear cub slides down, the Earth-cub's potential energy is converted to kinetic energy and the cub gains speed. Let us draw the FBD of the cub at the point where the cub leaves the surface of the igloo. We will measure the angular position of the bear from the vertical. The forces acting on the bear are the force of gravity and the normal force. Since the cub's motion is circular, the principal acceleration is radial. Thus, we choose our axes so that one axis points along the radial direction and the other axis is tangent to the igloo. We then resolve the forces into components along these axes, and write the equation for motion along each axis.

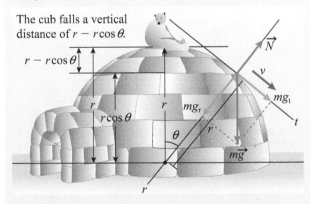

The cub falls a vertical distance of $r - r\cos\theta$.

Figure 6-25 Example 6-16

In the radial direction we have

$$N_r + mg_r = ma_r \tag{1}$$

Choosing the inward direction as positive gives

$$mg\cos\theta - N = \frac{mv^2}{r} \tag{2}$$

where N used in Equation (2) denotes the magnitude of the normal force.

When the cub is on the verge of leaving the surface, the normal force is zero and

$$mg\cos\theta = \frac{mv^2}{r} \tag{3}$$

We have two unknowns—the speed, v, and the angle, θ, at which the cub leaves the surface. Since we need one more relevant equation, we now consider using the mechanical energy of the cub.

As the cub slides down the surface, some potential energy is converted to kinetic energy. From Equation (6-32),

$$\Delta K = -\Delta U_g \tag{4}$$

As we can see from Figure 6-25, the change in height is $r - r\cos\theta$. Since the bear moves to a lower height, the change in potential energy is negative:

$$\Delta U_g = -mg(r - r\cos\theta) \tag{5}$$

The cub starts from rest, so the change in kinetic energy is

$$\Delta K = \frac{1}{2}mv^2 - 0 \tag{6}$$

Substituting Equations (5) and (6) into Equation (4), we have

$$\frac{1}{2}mv^2 = mg(r - r\cos\theta) \tag{7}$$

Using the result from Equation (3), we get

$$\frac{1}{2}mgr\cos\theta = mg(r - r\cos\theta) \tag{8}$$

$$\cos\theta = \frac{2}{3} \tag{9}$$

$$\theta = 48.19°$$

Thus, the cub will leave the surface at an angle of about 48.2° from the vertical.

Making sense of the result:

This example again demonstrates the effectiveness of combining mechanical energy considerations with Newton's laws. The normal force is zero when the cub is about to leave.

EXAMPLE 6-17

Pendulum Tension

The small marble in Figure 6-26 is attached to a massless string, which is free to rotate about the pivot at its other end. The marble is initially held such that the string is horizontal and then let go. Find the angle that the string makes with the vertical when the tension in the string is equal to the weight of the marble.

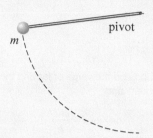

Figure 6-26   Example 6-17

SOLUTION

When the string is horizontal and the mass is at rest, the tension is zero because the gravitational force is directed straight downward, and thus has no horizontal component. As the mass moves down, its speed increases. At the bottom of the mass's path, the string both supports the mass and provides the centripetal acceleration, making the tension greater than mg. Therefore, at some angle θ between $0°$ and $90°$ from the vertical, the tension will be equal to mg.

As shown in the FBD in Figure 6-27, the forces acting on the mass are tension and the force of gravity. Since the pendulum swings in a circular arc, the principal acceleration

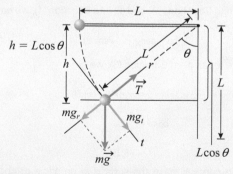

Figure 6-27   The FBD of the marble

is along the radial direction. Therefore, we choose one axis along the radial direction and the other axis perpendicular to it and tangent to the path of the mass.

If we sum the forces along the radial direction r-axis, we get

$$T_r + mg_r = ma_r \tag{1}$$

Taking the radial direction (direction of the tension) as the positive r-axis, we have

$$T - mg\cos\theta = \frac{mv^2}{r} \tag{2}$$

When the tension equals mg,

$$mg - mg\cos\theta = \frac{mv^2}{L} \tag{3}$$

We see that the mass cancels out, thus, we have two unknowns: the speed of the mass and the angle of the string.

The mass gains speed as it falls because gravitational potential energy is being converted into kinetic energy. Since there are no dissipative forces, mechanical energy is conserved and, using Equation 6-31

$$\Delta K = -\Delta U_g \tag{6-32}$$

As the pendulum swings down, ΔK is positive and ΔU_g is negative. From Figure 6-27 we can see that $h = L\cos\theta$, where h is the vertical distance from the initial position of the mass and L is the length of the string. Therefore,

$$\Delta K = \frac{1}{2}mv^2 = -(-mgL\cos\theta) \tag{5}$$

and

$$mv^2 = 2mgL\cos\theta$$

Substituting for mv^2 in Equation (2) we have

$$mg - mg\cos\theta = 2mg\cos\theta$$

$$\cos\theta = \frac{1}{3}$$

$$\theta = 70.53°$$

Making sense of the result:

As a quick check setting θ to zero in Equation (5) means that the marble has fallen a total vertical distance of L. The speed of the marble would be $v = \sqrt{2gL}$, as one would expect for an object falling straight down a total distance of L.

✓ CHECKPOINT

C-6-11 $T = mg$?

Why is the tension not equal to mg when the string is vertical?

C-6-11 The tension needs to support the weight and provide radial acceleration.

EXAMPLE 6-18

Using the Work-Energy Approach to Find Acceleration

A block of mass M rests on top of a table with a horizontal frictionless surface. The block is attached to a hanging mass m with a rope that passes over the massless, frictionless pulley as shown in Figure 6-28.

(a) Find the speed of mass M after m has fallen a total distance h, when the system starts from rest.
(b) Use your result from part (a) to find the acceleration of the system.
(c) Repeat parts (a) and (b) when the coefficient of friction between M and the surface of the table is μ_k.
(d) Compare this approach to the approach used in Chapter 5, Example 5-19.

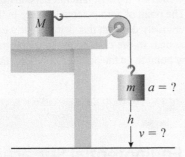

Figure 6-28   Example 6-18

SOLUTION

(a) Since there is no friction, we can use the work–kinetic energy theorem or one of the expressions based on the principle of conservation of mechanical energy. We will use the latter here. Since the total mechanical energy is conserved, the gravitational potential energy is converted as m drops into kinetic energy of the system.

$$\Delta K = -\Delta U_c \qquad (6\text{-}32)$$

ΔK is the change in the kinetic energy of *both* masses:

$$\Delta K = \Delta K_1 + \Delta K_2 = \frac{1}{2}Mv_f^2 - \frac{1}{2}Mv_i^2 + \frac{1}{2}mv_f^2 - \frac{1}{2}mv_i^2 \qquad (1)$$

Since the system starts at rest, $v_i = 0$:

$$\Delta K = \frac{1}{2}Mv_f^2 + \frac{1}{2}mv_f^2 \qquad (2)$$

The only change in potential energy comes from m descending a distance h (Figure 6-28). So,

$$\Delta U = -mgh \qquad (3)$$

and

$$\frac{1}{2}Mv^2 + \frac{1}{2}mv^2 = mgh \qquad (4)$$

We note here that both masses move at the same speed.

Solving Equation (4) for v, we obtain

$$v = \sqrt{\frac{2mgh}{M + m}}$$

(b) To determine the acceleration, we can take the derivative of Equation (4) with respect to time. Using the chain rule gives

$$\frac{d}{dt}v^2 = \left(\frac{d}{dv}v^2\right)\left(\frac{dv}{dt}\right) = 2va$$

Since the derivative of the height with respect to time is $dh/dt = v$,

$$Mva + mva = mgv \qquad (5)$$

$$Ma + ma = mg \qquad (6)$$

$$a = \frac{mg}{M + m} \qquad (7)$$

(c) Since friction is a nonconservative force, we use Equation (6-26):

$$W_{nc} = \Delta E_m = \Delta U + \Delta K \qquad (8)$$

$$W_{nc} = -mgh + \frac{1}{2}Mv^2 + \frac{1}{2}mv^2 \qquad (9)$$

As M moves to the right, the force of friction from the surface on M is directed to the left and therefore, by Equation 6-7, does negative work:

$$W_{nc} = \vec{f_k} \cdot \Delta \vec{x} = -f\Delta x = -N\mu_k h = -Mg\mu_k h \qquad (10)$$

Therefore,

$$-Mg\mu_k h = -mgh + \frac{1}{2}Mv^2 + \frac{1}{2}mv^2 \qquad (11)$$

Solving for v, we obtain

$$v = \sqrt{\frac{2mgh - 2Mg\mu_k h}{M + m}} \qquad (12)$$

To get the acceleration, we take the derivative of Equation 11 with respect to time, as in part (b). The resulting expression for the acceleration is

$$a = \frac{mg - Mg\mu_k}{M + m}$$

(d) If we used the approach that we used in Chapter 5, Example 5-19, we would have to deal with a system of two equations for the unknowns, a and T.

Making sense of the result:

The system is propelled to move by the weight of the hanging mass, and is slowed down by the force of friction on the horizontal mass. It makes sense for the numerator in the expression for the acceleration to be the net force, or the difference between the two forces and for the denominator to be the total mass.

6-7 Potential Energy and Reference Points

As we saw in Section 6-5, the elastic potential energy of the spring can be expressed as

$$U_s = \frac{1}{2}kx^2$$

where x is the distance by which the spring is stretched or compressed as measured from its unstretched length.

The elastic potential energy stored in a spring is position-dependent. The spring's potential energy is zero when $x = 0$. *This point is unique and cannot be arbitrarily assigned.* It is the point where the spring is unstretched. The work done by a spring and, therefore, the elastic potential energy stored in a spring depend on both the starting and ending positions.

Clearly, in the previous example stretching the spring by the same amount did not result in the same amount of change in potential energy. The change in potential energy depends, of course, on the amount by which the spring is stretched. However, it also depends on the starting point.

Now, suppose you carry out an exercise similar to the one in Example 6-19, but this time, we lift an object against the force of gravity. Say you lift a 1 kg book a distance of 50 cm. First, you start from the surface of a table in front of you. Then you start from the roof of a 100 m-tall building. Which exercise involves a greater change in the gravitational potential energy of the book–Earth system?

The answer, of course, is that the change in the potential energy is the same in both cases.

For the book being lifted against the force of gravity, all that matters is the change in height; the starting point is immaterial. However, the change in potential energy of a spring that is stretched by a given amount depends on the starting position. The point of zero potential energy in the case of the spring is uniquely defined: this is the point where the spring is not stretched. In the case of the force of gravity, we can choose any point to be the point of zero potential energy.

EXAMPLE 6-19

Spring Stiffness

You are given a spring of spring constant $k = 1000$ N/m and asked to stretch it 10 cm from the unstretched length. You then repeat the exercise with the same spring, but this time the spring is already stretched by 30 cm before you stretch an extra 10 cm. Calculate the change in the spring's potential energy in each case. Which case involves a larger change in the spring's potential energy?

SOLUTION

The change in the potential energy of the spring is given by Equation (6-20c):

$$\Delta U_s = -W_s = \frac{1}{2}(kx_2^2 - kx_1^2) \qquad \text{(6-20c)}$$

We begin with the initially unstretched spring, $x_1 = 0$, and $x_2 = 10$ cm:

$$\Delta U_s = \frac{1}{2}(1000 \text{ N/m}(0.1 \text{ m})^2 - 0) = 5 \text{ J}$$

When the spring is already stretched by 30 cm, this will be our x_1, and when we stretch it another 10 cm, x_2 is (30 cm + 10 cm) = 40 cm. Then $\Delta U_s = \frac{1}{2}(1000 \text{ N/m}(0.4 \text{ m})^2 - 1000 \text{ N/m }(0.3 \text{ m})^2) = 35$ J.

Making sense of the result:

The spring that is already stretched will require more work to stretch it farther even though the amount by which the distance is stretched is the same in both cases.

6-8 Equipotential Surfaces and Field Lines

We have defined the change in potential energy as the negative of the work of the conservative force:

$$W_c = -\Delta U_c \qquad \text{(6-20b)}$$

For a constant conservative force, the difference in potential energy between two points $\Delta \vec{r}$ apart is

$$\vec{F}_c \cdot \Delta \vec{r} = -\Delta U_c$$

If the conservative force is not constant, we can use the more general form of this relationship:

$$\int_{r_1}^{r_2} \vec{F}_c \cdot \Delta \vec{r} = -\Delta U_c$$

We can then express the conservative force as a vector derivative, in the elegant language of calculus:

KEY EQUATION $$\vec{F}_c = -\nabla U = -grad\,(U) \qquad \text{(6-33a)}$$

The vector differential operator ∇ is called "del." In this context, del denotes the *gradient* of the scalar potential energy. In Cartesian form, the expression in Equation (6-33) is written as follows:

KEY EQUATION $$\vec{F}_c = -\frac{\partial U}{\partial x}\hat{\imath} - \frac{\partial U}{\partial y}\hat{\jmath} - \frac{\partial U}{\partial z}\hat{k} \qquad \text{(6-33b)}$$

The *partial derivative* operator ∂ indicates that we are taking the derivative of a multivariate function (a function that depends on more than one variable) with respect to one of the variables while keeping the other variables constant.

In one dimension, say, along the x-axis, Equation (6-33b) reduces to

$$\vec{F_c} = -\frac{dU}{dx}\hat{\imath} \qquad (6\text{-}34)$$

EXAMPLE 6-20

Force of a Spring

Use the potential energy function for a spring to derive an expression for the force exerted by a spring.

SOLUTION

In Section 6-5, we found that the potential energy stored in a spring is $U = \frac{1}{2}kx^2$. Since the expression depends on only one variable, we can use Equation (6-34):

$$\vec{F_c} = -\frac{dU}{dx}\hat{\imath} = -\frac{d}{dx}\left(\frac{1}{2}kx^2\right)\hat{\imath} = -kx\,\hat{\imath}$$

Making sense of the result:

The derived expression for the force is the same as the equation for Hooke's law.

EXAMPLE 6-21

Using a Gradient to Determine Force

A particle is in a region in which its potential energy is given by $U - (3x^2 - 2xy^2 + zx)$ J. Find an expression for the force acting on the particle, and determine the magnitude of the force at point A (0.5 m, −1 m, 0.5 m).

SOLUTION

Here, we use Equation (6-33b):

$$\vec{F_c} = -\frac{\partial U}{\partial x}\hat{\imath} - \frac{\partial U}{\partial y}\hat{\jmath} - \frac{\partial U}{\partial z}\hat{k}$$

which gives

$$\vec{F_c} = ((-6x + 2y^2 - z)\hat{\imath} + 4xy\hat{\jmath} - x\hat{k})\text{J/m}$$

Substituting the coordinates of point A gives

$$\vec{F_c} = ((-1.5)\hat{\imath} - 2\hat{\jmath} - 0.5\hat{k})\ \text{J/m}$$

The magnitude of the force at point A is

$$|\vec{F_c}| = \left(\sqrt{(-1.5)^2 + (-2)^2 + (-0.5)^2}\right)\text{J/m} = \sqrt{6.5}\ \text{N} = 2.5\ \text{N}$$

Equipotential Surfaces and Field Lines

An **equipotential surface** is a surface along which the potential energy does not change. For example, the surface of Earth is an approximate gravitational equipotential surface. If you walk from point A to point B without any change in elevation, the potential energy of the interaction between you and Earth does not change. In the absence of dissipative forces, we do not need to do work to move along an equipotential surface.

One of the consequences of Equation (6-33) is that a conservative force must be perpendicular to the equipotential surface (see Example 6-22).

Around Earth, the gravitational equipotential surfaces are (almost) perfectly spherical shells with Earth at their centre. As we know from everyday experience, the force of gravity points radially toward Earth's centre, as shown in Figure 6-29. Thus, the force of gravity is perpendicular to these equipotential surfaces.

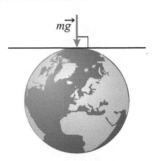

Figure 6-29 The force of gravity points radially toward Earth's centre.

Further, the gravitational acceleration, or field strength, $\vec{g}$, is parallel to the force of gravity because $\vec{F} = m\vec{g}$. We thus conclude that the gravitational field is also perpendicular to the gravitational equipotential surface. This fundamental relationship does not apply to gravity only; it applies to any equipotential surface and its corresponding conservative force: *conservative force vectors are perpendicular to equipotential surfaces.*

EXAMPLE 6-22

Force and Equipotential Surfaces

Prove that the conservative force lines are perpendicular to a straight equipotential line. Assume that the conservative force producing the equipotential line is constant.

SOLUTION

Consider the equipotential surface in the field of a conservative force, as shown in Figure 6-30. We will assume

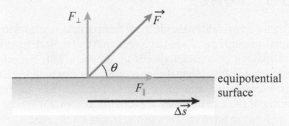

Figure 6-30 Conservative force pointing in an arbitrary direction relative to equipotential surface.

(continued)

that the conservative force (F) points in some arbitrary direction relative to the equipotential surface.

The force vector can be resolved into one component parallel to the equipotential line and another component perpendicular to the equipotential line. The force makes an angle θ with its projection onto the equipotential surface as shown. The work, W_c, done by the conservative force is an object moves along the equipotential line from point A to point B is given by Equations 6-8 and 6-10a

$$W_c = \vec{F} \cdot \Delta \vec{s} = F_\parallel \Delta s$$

where $\Delta \vec{s}$ is the distance between points A and B. The change in potential energy between points A and B is given by Equation 6-20a

$$\Delta U = -W_c = -F_\parallel \Delta s$$

However, since points A and B lie on the same equipotential surface, the potential energy at A must be equal to the potential energy at B, and the potential energy difference between the two points must be zero. Then $F_\parallel$ must be zero. Since the conservative force cannot have a component in a direction parallel to the equipotential surface, the force must be perpendicular to the equipotential surface.

Making sense of the result:

This example also shows that the field of a conservative force is perpendicular to an equipotential surface generated by that force or field for the geometry we chose.

 CHECKPOINT

C-6-12 The Conservative Force

Is the following statement true or false? A conservative force points in the direction of decreasing potential energy.

C-6-12 True

LO 9

6-9 Power

Consider two otherwise identical dump trucks loaded with equal amounts of oil sand in Northern Alberta. Both trucks are moving at constant speeds up the same hill, but one truck moves faster than the other. We know that both trucks are exerting the same magnitude of force on the road, by virtue of the fact that they are moving at constant speeds. The force exerted by each of the trucks on the road must be equal to $mg\sin\theta$ to result in zero acceleration. Further, we know that the two trucks, by the time they deliver their loads at the top of the hill, will have done the same amount of work, namely, mgh, where h is the height of the hill. However, since one of the trucks is moving faster than the other, the *rate* at which they do work is different.

Power is the rate at which work is done. Average power is defined as

$$P_{avg} = \frac{W}{\Delta t} \tag{6-35}$$

The limit of the average power as Δt approaches zero is the instantaneous power:

KEY EQUATION $$P = \frac{dW}{dt} \tag{6-36}$$

where dW is the infinitesimal amount of work done during the infinitesimal amount of time dt.

The power delivered by a variable force acting in one dimension between points x_1 and x_2 can be written as

$$P = \frac{d}{dt}\left(\int_{x_1}^{x_2} \vec{F} \cdot d\vec{x} \right) \tag{6-37}$$

If the force is constant, the expression for dW reduces to $\vec{F} \cdot d\vec{x}$, and the expression for instantaneous power becomes

$$P = \vec{F} \cdot \frac{d\vec{x}}{dt}$$

For forces acting in more than one dimension,

$$P = \vec{F} \cdot \frac{d\vec{s}}{dt}$$

or

$$P = \vec{F} \cdot \vec{v} \tag{6-38}$$

The SI unit for power is the watt (W), which is defined as

$$1 \text{ W} = 1 \text{ J/s}$$

Another commonly used unit to measure power is horsepower (hp). Unfortunately, the horsepower (like the calorie) has a number of different definitions. One commonly used standard for horsepower is

$$1 \text{ hp} = 746 \text{ W}$$

 MAKING CONNECTIONS

The Kilowatt Hour

A kilowatt hour is the amount of energy consumed in one hour at the rate of 1000 W:

$$1 \text{ kW} \cdot \text{h} = 1000 \text{ W} (3600 \text{ s}) = 3.6 \times 10^6 \text{ J} = 3.6 \text{ MJ}$$

In the power industry, the kilowatt hour is often used for measuring consumption and for billing. Large-scale electricity generation and consumption are commonly reported in terawatt hours:

$$1 \text{ TW} \cdot \text{h} = 1 \times 10^9 \text{ kW} \cdot \text{h} = 3.6 \times 10^{15} \text{ J} = 3.6 \text{ PJ}$$

TW stands for terawatt and PJ stands for petajoule.

EXAMPLE 6-23

Water Power

At 14 GW, the Itaipu Dam and generating plant has the second largest installed power generation capacity in the world, after the Three Gorges Dam in China. Itaipu also holds the world record for the most power generated in one year (2008: 94.86 TW). Located on the Upper Paraná River at the border between Paraguay and Brazil, the dam is 196 m high and has an average annual river flow of 12 000 m³/s. Calculate the theoretical maximum possible power generation for the water flowing through the dam.

SOLUTION

As shown in Example 6-8, the work done by the force of gravity as an amount of water of mass m falls through a height h is

$$W_g = mgh$$

The average power from Equation (6-35) is given by

$$P_{avg} = \frac{W_g}{\Delta t}$$

The average amount of water flowing every second through the dam is 12 000 m³. Using 1000 kg/m³ for the density of water, we have 12×10^6 kg of water flowing every second. The work done by the force of gravity is then

$$W_g = mgh = 12 \times 10^6 \text{ kg } (9.81 \text{ m/s}^2)\,(196 \text{ m}) = 2.3073 \times 10^{10}\text{J}$$

This amount of work is done by the force of gravity in one second. Therefore,

$$P_{avg} = \frac{W_g}{\Delta t} = \frac{2.3049 \times 10^{10} \text{ J}}{1 \text{ s}} = 23.0 \text{ GW}$$

Making sense of the result:

This is not much larger than the installed capacity. Hydro plants often run short of the maximum theoretical power generation for various reasons. One of these reasons is the "installed capacity," which generally refers to the infrastructure in place to harness the available energy. Another reason is the issue of the inherent efficiency of a given device or system in converting energy from one form to another. This efficiency is rarely 100%.

The efficiency of a given system or device is measured as the ratio of the power output to the power input:

$$e\ (\%) = \frac{P_{out}}{P_{in}} \times 100\%$$

Potential Energy and Hydroelectricity

As water accelerates in a waterfall, potential energy is converted to kinetic energy through the work done by the force of gravity. This kinetic energy can be harnessed as a clean and renewable source of energy for running electric generators. The water in the Robert Bourassa Reservoir powers the LG-2 (Figure 6-31) and LG-2-A generating stations on the La Grande River in Québec, with a combined output capacity of 7.326 GW. This is the largest installed capacity in North America, and the fifth largest in the world. To date, the Three Gorges Dam in China is the most powerful hydroelectric generating station in the world, with an installed capacity of 18.2 GW, expandable to over 22 GW. However, this power comes at the expense of extensive flooding, loss of cropland, and environmental damage.

Photo Hydro-Québec

Figure 6-31 The main hall at the LG-2 generating station (Robert Bourassa) on La-Grande River in Québec

✓ CHECKPOINT

C-6-13 The Work Done by an Escalator

An escalator takes people up a few floors at the airport. A steward stands still on the escalator, and a pilot with the same mass as the steward walks up the escalator. The pilot reaches the top in less time than the steward. Which of the following statements is true?

(a) The escalator does more work to move the steward upward than it does to move the pilot.

(b) The escalator does the same amount of work to move both the pilot and the steward upward.

(c) The escalator does less work to move the steward than it does to move the pilot.

(d) None of the above are necessarily true.

C-6-13 (c) the pilot does work against the force of gravity to go up.

Forces are classified as conservative and nonconservative. The work you do against conservative forces, such as the force of gravity and the force of a spring, is stored in the form of potential energy. The work you do against nonconservative forces, such as friction, is dissipated and cannot be recovered in the form of kinetic energy. In the absence of nonconservative forces, the mechanical energy of an object is conserved. The work of a nonconservative force on an object results in a change in its total mechanical energy. The work–mechanical energy approach can be a very powerful tool for solving problems and is often more convenient than the approach using Newton's second law.

The Work–Kinetic Energy Theorem

The total work done on an object is equal to the change in its kinetic energy:

$$W_T = \Delta K \qquad (6\text{-}5)$$

Work done by a force:
in one dimension:

$$|\vec{F}||\Delta\vec{x}|\cos\theta \qquad (6\text{-}5),\ (6\text{-}7)$$

$$W_F = \int_{x_1}^{x_2} \vec{F}\cdot d\vec{x} \qquad (6\text{-}13\text{a})$$

Total work done by a constant force:

$$W_T = \vec{F}_{net}\cdot\Delta\vec{x} \qquad (6\text{-}3)$$

Total work done by a varying force:

$$W_T = \int_{x_1}^{x_2} \vec{F}_{net}\cdot d\vec{x}$$

In more than one dimension

$$W_F = \int_{a}^{b} \vec{F}\cdot d\vec{s} \qquad (6\text{-}13\text{b})$$

Work done by a spring:

$$W_s = \frac{1}{2}kx_1^2 - \frac{1}{2}kx_2^2 \qquad (6\text{-}17)$$

Power is the rate at which work is delivered:

$$P = \frac{dW}{dt} \qquad (6\text{-}36)$$

Conservative Forces, Potential Energy, and Mechanical Energy

The work you do against a conservative force is stored as potential energy. The change in potential energy is defined as the negative of the work done by the conservative force:

$$\Delta U_c = -W_c \qquad (6\text{-}20\text{b})$$

The force can be obtained from the potential energy by using

$$\vec{F}_c = -\frac{dU}{dx}\hat{i} \qquad (6\text{-}34)$$

The potential energy is the same everywhere on an equipotential surface.

The force lines are perpendicular to the equipotential surface.

The change in gravitational potential energy is given by

$$\Delta U_g = mg\Delta h$$

The change in the elastic potential energy of a spring is given by

$$\Delta U_s = \frac{1}{2}kx_2^2 - \frac{1}{2}kx_1^2 \qquad (6\text{-}20\text{c})$$

The Conservation of Mechanical Energy

The work done by nonconservative forces on an object results in a change in the total mechanical energy of the object:

$$W_{nc} = \Delta E_m$$

In the absence of nonconservative forces, mechanical energy is conserved.

$$\text{If } W_{nc} = 0, \quad \text{then} \quad \Delta E_m = 0$$

The conservation of mechanical energy can be expressed as

$$E_m = \text{constant} \qquad (6\text{-}28)$$

or

$$\Delta K = -\Delta U_c \qquad (6\text{-}31)$$

Applications: conservation of energy, alternate fuel sources, efficient energy use, renewable energy, the global energy crisis, power generation, energy conversion, energy storage, earthquake engineering, shock absorbers, auto safety, engines, space flight, metabolism, electrostatic interactions, gravitation, friction losses, air resistance, nuclear interactions, thermodynamics, quantum mechanics.

Key Terms: conservation of mechanical energy, conservative force, equipotential surface, kinetic energy, mechanical energy, nonconservative force, potential energy, power, restoring force, total work, total work–kinetic energy theorem, work

QUESTIONS

1. Does the work–kinetic energy theorem hold for forces that are not constant? Explain.

2. When you drop a stone from the top of a building, does the instantaneous power delivered by gravity increase, decrease, or stay the same as the stone falls?

3. True (T) or false (F): A spring is already stretched 10 cm when you decide to stretch it another 10 cm. The spring constant k is 1000 N/m. The work you do is positive.

4. The total work done as you stretch a massless spring is
   (a) the change in the potential energy of the spring;
   (b) the negative of the work done by the spring;
   (c) parts (a) and (b);
   (d) none of the above.

5. You stretch a spring that is already stretched. Is the work done by the spring positive or negative?

6. A spring is compressed from equilibrium. Is the work done by the spring positive or negative?

7. Can the work done by a force ever be negative? Explain.

8. Can the total work done on an object ever be negative? Explain.

9. When you do positive work on a moving car, does the car's speed increase or decrease? Explain.

10. You ride on a Ferris wheel. Your mass is 73 kg, and the radius of the Ferris wheel is 45 m. What is the total work done on you during your journey from the lowest to the highest points? What is the work done by gravity?

11. What is the total work done by Earth on a 22 000 kg satellite as it moves around Earth by 180°, at a constant altitude? What is the work done by gravity during that time?

12. You spin the bob (mass m) at the end of a pendulum of length L in a vertical circle. What is the work done by the tension in the string as the bob is brought from the lowest to the highest position?
   (a) $-2mgL$
   (b) $2mgL$
   (c) zero
   (d) The answer could be either (a) or (b), depending on the choice of positive direction.

13. A hockey player intercepts a 170 g puck moving toward him on ice at 23 m/s and sends it at the same speed toward the opponent's goalie. Find the work done by the hockey player.

14. Your friend is holding on to a rope that is suspended from a tree branch, such that his feet are 60 cm off the ground. You push him with a horizontal force of variable magnitude such that he moves at a constant speed as his feet end up being lifted by an extra 40 cm off the ground. Your friend's mass is 69 kg. Find the work done by your variable force.

15. You drop a 2.0 kg weight from a height of 43 cm above a scale. The weight comes to rest after falling 48 cm. Find the total work done on the weight during its downward journey.

16. You compress a spring from equilibrium by an amount Δx. We can say that the change in potential energy of the spring is equal to
   (a) $\frac{1}{2}k(\Delta x)^2$;
   (b) $-\frac{1}{2}k(\Delta x)^2$;
   (c) part (a) or (b), depending on our choice of a positive direction.
   (d) We need to know either the initial or final length of the spring to answer this question.

17. You stretch a spring from an already stretched position. Which of the following statements is true?
   (a) The work you do is negative.
   (b) The work done by the spring is positive.
   (c) The change in the spring's potential energy is negative.
   (d) None of the above are true.

18. You intercept a hockey puck shot at you at a speed of 15 m/s by kicking it such that it deflects at the same speed at an angle of 65° to the incoming direction. Assume the average force that you exert on the puck is 90 N and that it acts over a total distance of 12 cm in one direction. What is the total work done on the puck by you? Ignore gravity.
   (a) 10.8 J
   (b) −10.8 J
   (c) 5.04 N
   (d) −5.04 N
   (e) none of the above

19. A spring cannon fires a projectile at an angle of 47° above the horizontal. The cannon is fired at $t = 0$, and the projectile reaches its maximum height a time Δt later. Which of the following statements is true?
   (a) During the time Δt, the work done by the force of gravity is larger than the work done by the spring.
   (b) During the time Δt, the work done by the spring is larger than the work done by gravity.
   (c) During the time Δt, the work done by the spring is equal to and opposite in sign to the work done by the force of gravity.
   (d) None of the above are true.

20. You lift your backpack of mass m a vertical distance h. What is the definition of the change in potential energy?
   (a) the work done by you
   (b) the negative of the work done by you
   (c) the negative of the work done by the force of gravity
   (d) the work done by the force of gravity

21. The work done by a spring as it is being compressed from its unstretched position is
   (a) positive;
   (b) negative;
   (c) either positive or negative, depending on where you start;
   (d) depends on how fast you are moving;
   (e) none of the above because work is a scalar that cannot be referred to as positive or negative.

22. A spring is already stretched 20 cm from its equilibrium position (Figure 6-32). You pull it so that it is now stretched 40 cm. While the spring is being stretched,
   (a) you do negative work;
   (b) the spring does positive work;
   (c) the sign of the work done by the spring on you depends on whether we choose right or left to be positive;
   (d) the work done by the spring is always positive because of the term $(dx)^2$;
   (e) none of the above.

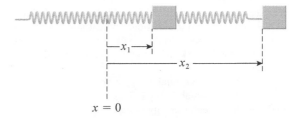

Figure 6-32   Question 22

23. A child moves down a slide at a park at a constant speed. Find a relation between the slope angle of the slide and the coefficient of kinetic friction between the child and the slide's surface. What can be said about the relationship between the work done by gravity and the work done by friction?

24. You slide down a snowy hill on a toboggan from a height of 7 m. The hill has a constant slope such that your speed remains constant at 5 m/s throughout the entire journey. Air resistance is negligible. Which of the following statements is true?
    (a) The magnitude of the work done by gravity is larger than the magnitude of the work done by friction.
    (b) The work done by friction is equal to the change in the gravitational potential energy.
    (c) The work done by friction cannot be determined.
    (d) Friction is not significant because you actually move.
    (e) None of the above is true.

25. A compressed spring is used to launch a marble onto a surface that starts out horizontal and then slopes downward at a constant angle. When the marble moves down the incline at a constant speed,
    (a) the total work done on the marble is zero;
    (b) the work done by the spring is greater than the work done by gravity;
    (c) the work done by friction is equal to the work done by gravity;
    (d) the work done by the spring is greater than the work done by friction.
    (e) This motion is impossible.

26. You move a spring such that the work done by the spring is negative. This means that you
    (a) stretched the spring from equilibrium;
    (b) compressed the spring from equilibrium;
    (c) stretched an already stretched spring;
    (d) any of the above;
    (e) none of the above.

27. You move a spring such that the work done by the spring is negative. This means that
    (a) you allow the spring to move toward equilibrium;
    (b) the potential energy of the spring decreases;
    (c) you do negative work.
    (d) There is not enough information to answer this question.
    (e) None of the above.

28. A spring is compressed by 10 cm from its equilibrium position. When you move the spring so that it is now stretched by 10 cm,
    (a) the work that you do is positive;
    (b) the total work done on the spring is positive;
    (c) the work that you do is negative;
    (d) the work done by the spring is negative;
    (e) none of the above.

29. You are sitting at your desk. You decide to set your desktop as the point of zero gravitational potential energy, then you lift a book from a distance h below the desk to a distance h above the desk. Which of the following statements is true?
    (a) The change in potential energy is zero.
    (b) The final potential energy is $2mgh$.
    (c) The final potential energy is mgh.
    (d) The change in potential energy is $2mgh$.

30. A spring is already stretched 10 cm when you decide to stretch it another 10 cm. The spring constant is 200 N/m. As you stretch the spring,
    (a) the work done by the spring is −1 J;
    (b) the work done by the spring is −2 J;
    (c) the work done by the spring is −3 J;
    (d) the work done by the spring is −4 J.
    (e) None of the above.

31. You drop a 2 kg mass from a height of 7 m above a spring whose spring constant is 100 N/m at its equilibrium position. During the object's downward journey,
    (a) the magnitude of the work done by gravity is less than $14g$;
    (b) the magnitude of the work done by the spring is less than $14g$;
    (c) the total work done on the mass is zero.
    (d) I really don't think this makes sense.
    (e) None of the above.

32. Which of the statements below is true when you catch a baseball that comes at you horizontally? More than one statement might apply.
    (a) The work that you do is positive.
    (b) The work that you do is negative.
    (c) Work is not a vector and cannot be referred to as positive or negative.
    (d) Work is a scalar product expressed in terms of magnitudes, so it is never negative.
    (e) Work is expressed in terms of kinetic energy, which must be positive.

33. A spring is used to launch a marble up a frictionless incline such that the marble reaches the top of the incline with a speed of zero. Which of the following statements is true?
    (a) The total work done on the marble is positive.
    (b) The total work done on the marble is negative.
    (c) The work done on the marble by the spring is equal in magnitude and opposite in sign to the change in the gravitational potential energy.
    (d) The change in the gravitational potential energy is smaller than the change in elastic potential energy.
    (e) None of the above are true.

PROBLEMS BY SECTION

For problems, star ratings will be used, (✶, ✶✶, or ✶✶✶), with more stars meaning more challenging problems.

Section 6-1   Work and Kinetic Energy for Constant Forces

34. ✶   You throw a stone with a mass of 215 g straight up in the air. It reaches a maximum height of 12 m, and then you catch it at the same level from which you threw it.
    (a) Find the work done by the force of gravity on the stone on its way up.
    (b) Find the work done by the force of gravity on the stone on its way down.
    (c) Find the total work done by the force of gravity over the entire trip.

35. ✶   You lift your 400 g cup of coffee from the table a total distance of 45 cm and hold it there.

(a) Find the work done by the force of gravity.

(b) What is the work done by you on the cup?

(c) What is the total work done on the cup?

36. ✳ You push a 10.0 kg wagon with a horizontal force of 145 N over a total distance of 7.00 m.

(a) How much work is done by you?

(b) Now you move so that you are opposing the motion of the wagon. You stop the wagon by applying a force that stops it in 2.00 m. How much work do you do as you slow the wagon?

37. ✳ While ice fishing, your friend slides a 5.0 kg sausage bucket over the ice surface to you at a speed v. You stop the bucket over a distance of 90 cm by applying a force of 20 N on the bucket. How much work did you do?

38. ✳ A tennis player strikes an incoming tennis ball of mass m and speed v such that the ball leaves the racket at speed v as well. How much work was done by the tennis player to reverse the direction of the ball? Explain. Did the tennis player do any negative work?

39. ✳ You throw an 8.0 kg medicine ball straight up in the air and then catch it on the way down. Just before it hits your hands, it is moving at a speed of 3.0 m/s. You stop it over a distance of 50 cm.

(a) How much work did you do to stop it?

(b) How much work was done by the force of gravity while you stopped the ball?

40. ✳ A worker tosses a 1.5 kg lunch box to a friend who is on a platform 12 m above. The lunch box is in contact with the worker's hands for a total distance of 35 cm. How much work did the worker do to accelerate the lunch box?

41. ✳✳ An athlete picks up a 20.0 kg sandbag from the ground and throws it straight up in the air. It leaves her hands 1.50 m above the ground and reaches a height of 7.00 m.

(a) How much work did the athlete do?

(b) How much work was done by the force of gravity while the bag was in contact with the athlete?

42. ✳ A block of mass 26 kg rests on a frictionless inclined surface with a slope of angle 37°. A horizontal force of 510 N is used to push the block a distance of 4.6 m along the inclined surface, starting from rest, as shown in Figure 6-33. Find the work done by the force $\vec{F}$.

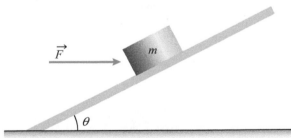

Figure 6-33 Problem 42

43. ✳✳ Two masses, $M = 16.8$ kg and $m = 4.8$ kg, are attached to each other with a rope that winds around a massless hanging pulley as shown in Figure 6-34. Mass m is attached to one end of an unstretched spring whose other end is attached to the ground. The system is initially at rest.

(a) Derive an expression for the speed of mass m after M has fallen 45 cm.

(b) Find the maximum speed of m.

(c) Find the speed of m when $k = 96$ N/m.

Figure 6-34 Problem 43

44. ✳ A child starts from rest and slides down a snow-covered hill with a slope of angle 47° from a height of 1.7 m above the bottom of the hill. The speed of the child at the bottom of the hill is 3.2 m/s. Find the coefficient of friction between the hill and the child.

45. ✳ An object of mass m is pushed along a frictionless horizontal surface by a force F, over a distance d, such that the speed is constant.

(a) Find an expression for the coefficient of kinetic friction between the object and the surface.

(b) Find the work done by F.

(c) Friction is removed, and the force F is now applied along an incline to cause the mass to move up the incline at a constant speed. Does the force F do more or less work over the same distance?

(d) How does your answer to part (c) change if the mass were to accelerate up the incline? (Hint: We make the incline angle smaller.)

(e) How does your answer to part (c) change if friction were brought back such that the object still moved up the incline?

Section 6-2 Work Done by Constant Forces in Two or Three Dimensions

46. ✳ A worker leans on a crate and pushes it on a horizontal surface with a force of 230 N in a direction that makes an angle of 67° below the horizontal over a distance of 12 m. How much work is done by the worker?

47. ✳✳ A 22 kg sled comes loose during a sled race on ice and is stopped by an observer over a horizontal distance of 1.3 m. The observer is pushing down on the sled with a force that makes an angle of 42° with the horizontal. The sled's speed is 4.0 m/s. How much work does the observer do? What is the magnitude of the force used by the observer?

48. ✳ A child pulls a 3.0 kg wooden bus on a horizontal surface over a distance of 3.0 m using a rope that makes a 47° angle above the horizontal starting from rest. The speed of the bus at the end of the pull is 5.0 m/s. Determine the work done by the child and the force exerted by the child. How much work is done by the force of gravity during the pull?

49. ✳✳ A child uses a rope to pull her brother on a toboggan on a horizontal, snow-covered surface. The combined mass of her brother and the toboggan is 34 kg. The rope makes an angle of 41° with the surface. The horizontal projection of the rope makes an angle of 27° with the direction of motion. The child is pulling the toboggan with a force of 60 N, and the toboggan moves at a constant speed. Find the coefficient of friction between the toboggan and the snow.

50. ✳✳ A person applies a force $F = 4\hat{i} - 2\hat{j} + 7\hat{k}$ N to move a crate on a track in the xz-plane that angles 32° above

the positive x-axis. Find the work done by the person over the first 3.0 m of the motion.

51. ✹✹ A particle moves on a track between points A $(3, 2, -1)$ and B $(5, -4, 3)$ as a result of being pushed by a force of magnitude 170 N that rises 32° above the xy-plane such that its projection makes an angle of 195° with the positive x-axis.
   (a) The mass of the particle is 1.1 kg, and the particle starts from rest. Find the work done by the force and find the particle's speed at the end of its trajectory.
   (b) Repeat part (a), but include gravity and use the coefficient of kinetic friction between the particle and the track, $\mu_k = 0.15$.

52. ✹ You pull a crate of mass 7.00 kg a distance of 3.00 m up a frictionless incline with a slope of angle 32.0° by applying a force of magnitude 300 N along the incline.
   (a) Find the amount of work you do.
   (b) You now push the same mass on a frictionless horizontal surface for the same distance as in part (a), using a force of the same magnitude as in (a) but now pointing horizontally in the same direction of motion. Determine the work done by you.

Section 6-3   Work Done by Variable Forces

53. ✹✹ A marble of mass m sits at the bottom of a bowl of radius R. The surface of the bowl is frictionless, and the bowl is not allowed to move. You push the marble in the horizontal direction using a force of variable magnitude until the marble is a height h above the bottom of the bowl, such that $h < R$. You bring the marble to a stop. Find the work done by the variable force.

Section 6-4   Work Done by a Spring

54. ✹✹ Two identical springs, each of spring constant 720 N/m, are set up in parallel horizontally such that they are compressed by 32.0 cm. The two springs are used to eject a 500 g mass onto a frictionless surface. The mass then moves 1.90 m up a frictionless incline with a slope of angle 32.0°, where it encounters a combination of springs identical to the springs used to eject it, lying along the incline. How high above the horizontal surface is the mass when it is finally stopped?

55. ✹✹ A 10.0 kg block slides 3.00 m down a frictionless surface inclined 30° above the horizontal, before being stopped by a spring of spring constant $k = 340$ N/m secured to the inclined surface. Find the maximum compression of the spring.

56. ✹✹✹ A perforated bead of mass 37 g moves on a frictionless horizontal circular track of radius $R = 0.65$ m. The track runs through the bead (Figure 6-35). When the bead is at rest in the horizontal position, it is tied to a spring ($k = 230$ N/m), the other end of which is fixed at a pivot at $R/2$ from the bottom of the track as shown. The unstretched length of the spring is $R/3$. The bead is given a gentle nudge.
   (a) Find the speed of the bead when it has moved by 180°.
   (b) Find the speed of the bead when it has moved by 90°.
   (c) Find the work done by the spring to move the bead by 180°.

(d) Find the work done by the spring to move the bead by 90°.

Figure 6-35   Problem 56

Section 6-5   Conservative Forces and Potential Energy

57. ✹✹ You push a crate of mass 27 kg 15 m up a ramp that is sloped 19° above the horizontal. The coefficient of friction between the ramp and the crate is 0.15, and the speed of the crate, which starts from rest, is 1.7 m/s at the end of the 15 m. To accomplish the task, you use a variable force whose magnitude and direction both change.
   (a) Calculate the change in the crate's gravitational potential energy (crate-Earth system).
   (b) Calculate the work done by the force.

Section 6-6   Nonconservative Forces and Mechanical Energy

58. ✹✹ A mass ($m = 1.20$ kg), originally at rest, sits on a frictionless surface. It is attached to one end of an unstretched spring ($k = 790$ N/m), the other end of which is fixed to a wall (Figure 6-36). The mass is then pushed with a constant force to stretch the spring. As a result, the system comes to a momentary stop after the mass moves 14.0 cm. Find the
   (a) work done by the constant force;
   (b) change in potential energy of the spring;
   (c) change in mechanical energy of the system;
   (d) total work done by nonconservative forces in the problem;
   (e) speed of the mass 4.00 cm into the motion;
   (f) new equilibrium position of the system;
   (g) maximum speed of the mass and where that occurs.

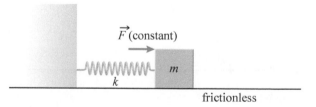

Figure 6-36   Problem 58

59. ✹✹ A mass ($m = 900$ g) rests on a 63.0° incline, where it is attached to one end of a spring, also resting on the incline as shown in Figure 6-37. The other end of the spring is fixed. The mass is held in place so that the spring is unstretched. The mass is then released. The coefficient of kinetic friction between the surface and the mass is 0.0800, and the spring constant is 110 N/m. Find the maximum compression of the spring.

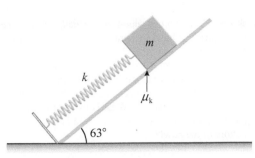

Figure 6-37 Problem 59

Section 6-7 Potential Energy and Reference Points

60. ✱✱✱ Newton's law of gravitation states that the force of attraction between two bodies is proportional to the product of their masses and is inversely proportional to the square of the distance between them. The force is given by $\frac{Gm_1m_2}{r^2}$, where G is a constant. (See Chapter 11.)

(a) Find an expression for the gravitational potential energy between two objects.

(b) Is the expression positive or negative?

(c) What would your escape velocity have to be if you weighed 800 N on Earth's surface? The escape velocity is the velocity with which you have to be launched so you do not come back to Earth. Ignore losses due to air resistance.

(d) Find the speed of a 1000 kg satellite orbiting Earth at an altitude of 1000 km.

(e) Find the total mechanical energy for the satellite in part (d).

(f) What total mechanical energy does the satellite need to just break free from Earth's gravitational field?

(g) The binding energy of a satellite is the additional energy needed for it to escape Earth's gravity. Find the binding energy for the satellite in part (d).

Section 6-8 Equipotential Surfaces and Field Lines

61. ✱✱ The force of gravitational attraction on an object beneath Earth's surface is proportional to r, the distance from the centre. Beyond Earth's surface, the force of attraction becomes inversely proportional to r^2. Plot the gravitational potential energy for an object as it moves from the centre of Earth to a point that is three times the radius of Earth as measured from Earth's centre.

62. ✱✱ A particle enters a region where the potential energy function is given by $\alpha x^{-2} + \beta xy - \gamma z^2 x$. Find an expression for the force in Cartesian unit vector notation.

63. ✱✱✱ The electrostatic potential energy between two charges is given by KQq/r. Charge q is brought to within a distance R of charge Q and then released from rest. Charge Q is fixed in space.

(a) Ignoring gravity, derive an expression for the speed of charge q after it has moved a distance of Δr from Q.

(b) Using $m_q = 51$ g, $q = 16$ μC, $Q = 47$ μC, $K = 8.99 \times 10^9$ Nm/C^2, $\Delta r = 1$ cm and $R = 1.1$ cm, find a value for the speed in part (a).

64. ✱✱ The potential energy of an interaction between two objects is given by $-A(r^3 + r)$, where A has a positive value.

(a) Is the force between these two objects repulsive or attractive? How can you tell?

(b) Plot a potential energy versus r graph for $A = 120$ N/m, over the range $r = 0$ to 10 m.

(c) Derive an expression for the force between the two objects as a function of r.

(d) One of the objects is held at rest, while the other is placed a distance of 20.0 m from the fixed object, and then released from rest. The mass of the moving object is 11.0 kg. Find its speed after it has moved 3.00 m.

Section 6-9 Power

65. ✱ The Tupolev Tu-144 supersonic passenger jet had four Kolesov RD-36-51 afterburning turbojets that each provide a maximum thrust of 200 kN. The maximum flying speed of the Tu-144 was Mach 2.2 at an altitude of 120 000 ft. Find the force of air resistance on the airplane at that speed.

66. ✱✱ A tractor of mass 5400 kg pulls a 1400 kg crate up a hill with a slope of angle 12.0° above the horizontal at a constant speed of 11.0 m/s.

(a) What acceleration would the tractor have on a level road at that speed? Ignore any frictional losses.

(b) Repeat part (a), but now consider that the coefficient of friction between the crate and the ground is 0.3. Assume that all frictional losses come from the friction between the crate being pulled and the ground, and not from the tractor.

67. ✱ A sports car has a maximum speed of 320 km/h and a maximum operating engine power output of 490 hp. Find the force of air resistance on the car at that speed. Ignore all other sources of friction and resistance to the motion.

68. ✱✱ You kick a 200 g rock from the edge of a cliff while hiking along the edge of the Grand Canyon.

(a) Ignoring air resistance, find the average power delivered by gravity during the first 4 s of the fall.

(b) Find the instantaneous power delivered by gravity at the end of the fifth second of the fall.

COMPREHENSIVE PROBLEMS

69. ✱✱✱ A bead slides down a frictionless vertical helical track as shown in Figure 6-38. The helical track has 1 turn/cm of length of the helical axis, and the loops have a radius of 17.0 cm each. There are 7 turns in the helix.

Figure 6-38 Problem 69

(a) Estimate the slope angle experienced by the bead along the helix.

(b) Find the speed of the bead at the bottom of the helix.

70. ✸✸ You swing a massless string of length L with a mass m attached to its end in a vertical circle, such that the mass has more than enough speed to keep the string taut all the time. You then hold the end of the string steady while the other end with the mass attached to it continues to swing freely. Derive an expression for

(a) the difference between the maximum and the minimum tensions in the string as the mass moves in its circular path.

(b) the work done by the tension during a full revolution.

71. ✸✸ A marble slides on the frictionless ramp starting at rest from height h shown in Figure 6-39. The marble then moves to the circular track of radius r. Write an expression for the minimum value of the height h so that the normal force on the marble is three times its weight when it is at the same height as the centre of the circular track.

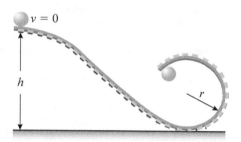

Figure 6-39 Problem 71

72. ✸✸✸ You happen upon what looks like an abandoned hemispherical igloo of radius R. You climb the igloo and sit at the top (Figure 6-40). A few moments later you feel a broom very gently prodding you off the roof. The push you receive from the disgruntled igloo dweller is just enough to cause you to begin to move.

(a) Find the angle that your position vector as measured from the centre of the igloo makes with the vertical at the point where you slide off the surface of the igloo.

(b) at what angle is the normal force equal to 1/3 of you weight.

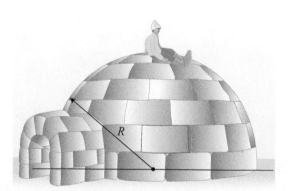

Figure 6-40 Problem 72

73. ✸✸✸ You happen upon a perfectly shaped ceramic semi-spherical bowl at home, with a smooth, frictionless surface. You place a small bead at the top of the inside edge of the bowl and let it slide on the inside of the bowl under the effect of gravity (Figure 6-41). Locate the point along the bowl where the normal force from the bowl on the bead equals the bead's weight. What is force of contact between the bead and the bowl when the bead is passing across the bottom of the bowl?

Figure 6-41 Problem 73

74. ✸✸ The simple pendulum in Figure 6-42 is made to spin in a vertical circle such that it has sufficient kinetic energy to just stay taut at the top of its trajectory. When the string is in the vertical position shown, it is intercepted by a peg at a distance of $2L/3$ from the centre of the vertical circle, and the mass continues along the new trajectory as shown.

(a) Find the speed of the mass(m) at the lowest point of the new trajectory.

(b) Find the speed of the mass at the top of the new trajectory.

(c) Find the tension in the string when it makes an angle of $40°$ with the vertical before hitting the peg.

(d) Find the tension in the string at the top of the new trajectory.

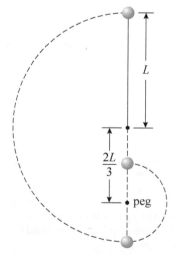

Figure 6-42 Problem 74

75. ✸✸ A 200 g cube slides down a ramp starting from rest as shown in Figure 6-43. The ramp has a $46°$ slope. After falling a distance of 92 cm, the cube strikes a spring of spring constant $k = 20$ N/m. Find the maximum compression of the spring when the coefficient of kinetic friction between the cube and the ramp is 0.17.

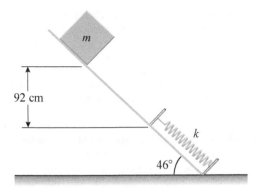

Figure 6-43 Problem 75

76. ✶✶ You are sitting in a roller coaster car that begins its descent, from rest, down the main track from a height of 62.0 m. The roller coaster then climbs up the inside of a vertical circle of radius 20.0 m to a point two-thirds of the circle's radius above the bottom of the circle.
(a) Find the normal force exerted on you by the seat if your mass were 57.0 kg. The normal direction is taken to be perpendicular to the track.
(b) What is the total force exerted on you by the seat?

77. ✶✶ The string in Figure 6-44 is wound around the nail as shown. The end of the string is attached to a mass m that is released from the horizontal position, starting from rest. Derive an expression for
(a) the tension in the string when it makes an angle of $23°$ with the vertical.
(b) the total work done to bring the string to the position in part (a).
(c) the work done by the tension.

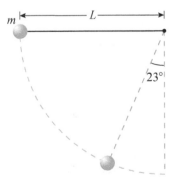

Figure 6-44 Problem 77

78. ✶✶✶ A block of mass M sits on top of a horizontal frictionless bench. One end of the block is connected to an originally unstretched spring of spring constant k. The other end is connected to a vertically hanging mass m by a string that runs over a massless pulley. Find an expression for
(a) the maximum distance that m will fall, starting from rest.
(b) the speed of mass m after it has fallen a distance h

79. ✶✶ An unstretched spring of spring constant k hangs vertically from the ceiling. You now attach a mass m to it and let it extend slowly until the mass stops.
(a) By how much does the spring stretch?
(b) Next, you take the mass back up to the original unstretched position and let it go. How far will the mass fall? What is the maximum speed of the mass?

80. ✶✶ A spring of spring constant $k = 710$ N/m is suspended vertically from a ceiling. You attach a 32 kg mass to the spring and let it drop. Find the speed of the mass when it is three-quarters of the way down to the position where it momentarily stops.

81. ✶✶✶ A block of mass $m = 3.2$ kg rests on top of another block of mass $M = 16.4$ kg, which rests on top of a horizontal frictionless surface. The coefficient of static friction between the two blocks is 0.23, and the coefficient of kinetic friction is 0.17. Block M is pushed with a horizontal force of 37 N. Find the energy lost to friction during the time it takes M to move 4.0 m, starting from rest.

82. ✶✶ You throw a 240 g orange straight up in the air from a height of 90 cm above the ground starting from rest. You let go of the orange when it is 140 cm above the ground and it reaches a maximum height of 14 m.
(a) How much work is done by you?
(b) How much work is done by the force of gravity while your hand is in contact with the orange?
(c) Find the average power delivered by the force of gravity during the last second before the orange reaches it maximum height.

83. ✶✶✶ A block of mass $M = 37$ kg rests on top of a horizontal bench, where the coefficient of friction between the block and the surface is $\mu_k = 0.20$. A string attached to M runs over a massless, frictionless pulley and is attached to the vertically hanging mass ($m = 11$ kg). The system is originally held at rest and then released.
(a) Find the speed of M after m has fallen a distance of 60 cm.
(b) Using energy considerations, derive an expression for the acceleration of the system as a function of time.

84. ✶✶ A block of mass $M = 17$ kg rests on top of a frictionless horizontal bench. A string attached to M runs over a massless frictionless pulley and is attached to a vertically hanging mass ($m = 4.0$ g). The system is originally held at rest and then released. Find the speed of M after m has fallen a distance of 30 cm. Using energy considerations, find the acceleration of the system.

85. ✶✶✶ The force between two particles is measured to be $F = -Ar^{-3}$.
(a) Develop an expression for the change in the potential energy of interaction between the two particles as the distance between them changes from r_1 to r_2.
(b) You are doing work to change the distance between the two particles. Develop an expression for the work done by you when the distance changes at a constant speed.
(c) What is the expression for the work done by the force between the two charges?
(d) Is the force between the particles attractive or repulsive?
(e) One of the particles is fixed in space, and the other is let go from a distance of 110 cm from the fixed particle. The mass of the moving particle is 12 g. Find its speed when it has moved 70 cm from the position indicated above. The experiment is conducted on a horizontal frictionless surface.

86. ✳✳✳ The worker in Figure 6-45 has to pull a crate over a concrete surface over a distance of 9.0 m, as shown.
   (a) Find the angle that will maximize the effectiveness of the worker in pulling the crate. (Hint: Minimize the time for the task.)
   (b) The crate's mass is 340 kg, and the coefficient of kinetic friction between the crate and the floor is 0.20. For a given angle θ, what is the minimum force needed to cause the crate to start moving when the coefficient of static friction between the crate and the surface is 0.50?
   (c) Once the block starts moving, the worker applies just enough force to keep it moving at a constant speed. Find the work done by the worker.
   (d) Find the work done by friction.
   (e) Find the work done by gravity.
   (f) Find the total work done on the crate.

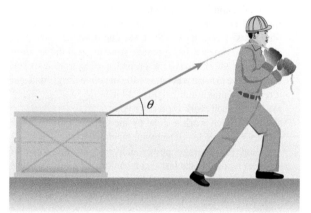

Figure 6-45 Problem 86

87. ✳✳ (Numerical solution) A particle enters a field where the potential energy is given by $U = \dfrac{-7}{r^2 + 6}$ J. Derive an expression for the force experienced by this particle in this region as a function of r. Find the speed of the particle 3 s into the field. Use initial values, $t = 0$, $x = 1$ m, $y = 0$, $vx = 0$, $vy = 0.5$ m/s.

88. ✳✳✳ A 67 kg skydiver free-falls for the first 3 min, reaching 99% of her terminal speed of 210 km/h in 15 s.
   (a) Find the work done by gravity during the last 2 min of the fall.
   (b) Find the work done by the air resistance during the last 2 min of the fall.
   (c) Find the power delivered by gravity during that time.
   (d) Find the power delivered by air resistance during that time.
   (e) Find the average force of air resistance on the skydiver during the first 15 s of the fall.

89. ✳✳✳ A particle with a mass of 310 g, moving in one dimension enters a progressively damped region where the force it experiences is given by $F = -kx^3$. Find how far the particle enters the region when it enters with a speed of 120 m/s. The force constant is 320 N/m³.

90. ✳✳✳ A test particle's position as a function of time is given by $x = -\alpha t^3 + \beta t^2 + \gamma t^{-4}$ as a result of being under the influence of a variable force, F.

(a) Find the position of the particle at $t = 0$ 1 s.
(b) Find the position of the particle at $t = 3.5$ s for $\alpha = 0.70$ m/s³, $\beta = 1.1$ m/s², and $\gamma = 1.5$ ms⁴.
(c) Find the value of the force at $t = 0$.
(d) Find the value of F at $t = 3.5$ s.
(e) Evidently, the acceleration is not constant. Can we still use the work–kinetic energy theorem for this particle? Explain.

91. ✳✳✳ Construct a proof to show that the force lines or the field lines, as given by Equation (6-34), follow the path of steepest descent between equipotential surfaces.

92. ✳✳✳ (Tough) Repeat problem 71 with friction when the coefficient of kinetic friction between the marble and the ramp is 0.15 and the mass of the marble is 120 g.

93. ✳✳✳ Repeat problem 72 when the coefficient of kinetic friction between you and the igloo is 0.2, and your mass is 72 kg.

94. ✳✳✳ The air resistance on a free-falling stone is proportional to the square of the speed of the stone. The stone falls from rest and reaches 99% of its terminal velocity in the first 15 s of the fall. Find the coefficient of proportionality between the resistive force and the square of the speed. The terminal speed is 230 km/h. Find the work done by air resistance during the first 2 s of the fall.

95. ✳ The Avro Arrow jet fighter had a maximum takeoff weight of 62 430 pounds and an operating ceiling of 58 500 feet. Calculate an estimate of the change in gravitational potential energy when a fully loaded Avro Arrow climbs to its operating ceiling. List any assumptions you make for your estimate.
   (a) Find the combined power output of its two Pratt & Whitney turbojet engines, the thrust of which was 23 450 lbs each when cruising at maximum speed.
   (b) Using the fact that the force of air resistance on an object is proportional to the square of its speed, find the maximum acceleration of the plane when at a cruising speed of 701 mph.
   (c) Assuming a constant climb rate, find the work done by the engines to get the plane to its operating ceiling, moving at its maximum speed.

96. ✳✳✳ A typical banana nut muffin or a sesame seed bagel with cream cheese has around 400 Cal. Determine the height of a mountain you would have to climb to do that much work against gravity. Use a mass of 72 kg. How many kilometres of walking does that equal if you are climbing a 20° hill? This initial estimate does not take into account the fact that you actually burn calories walking on a perfectly horizontal surface. A 72 kg person walking briskly at 6.0 km/h will burn approximately 285 Cal on level ground in 1.0 h. How many kilometres do you have to walk if you factor that fact in?

See the text online resources at www.physics1e.nelson.com for Open Problems and Data-Rich Problems related to this chapter.

Chapter 7
Linear Momentum, Collisions, and Systems of Particles

In cars, the combination of crumple zones and strengthened passenger compartments is a direct application of Newton's laws that has proved to be a major contribution to passenger safety (Figure 7-1). The crumple zone is a part of a car that is controllably weakened in such a way that it deforms during a collision, which maximizes the dissipation of the kinetic energy from the collision. The crumpling also increases the time over which the car loses speed and, hence, decreases the acceleration and the force experienced by the passengers during an impact. First introduced by Mercedes Benz in the 1950s, crumple zones are generally installed at both the front and rear of modern cars. Cars with crumple zones generally have a hardened passenger cabin, which is designed to maintain its integrity during a collision and protect passengers against intrusions from the crumpling exterior.

Older cars have rigid bodies that do not deform much during a collision. Such a design makes these cars more likely to rebound from each other during a collision. There is minimum damage to the car bodies but there are high rebound accelerations. High rebound accelerations cause more serious passenger injuries than the reduced collision accelerations experienced in cars with crumple zones.

cla78/Shutterstock.com

Figure 7-1 A car's crumple zone

Learning Objectives

When you have completed this chapter you should be able to:

1 Define linear momentum, and calculate and compare momenta of various real-life objects.

2 Express Newton's laws in terms of rates of change of linear momentum.

3 Define and calculate impulse, and establish the conditions for the conservation of linear momentum.

4 Find the centre of mass of a system of particles.

5 Distinguish between internal and external forces on a given system, and explain how they affect the motion of the system's centre of mass.

6 Distinguish between elastic and inelastic collisions in terms of energy and momentum conservation.

7 Apply the laws of conservation of momentum and energy to elastic collisions.

8 Derive the ideal rocket equation, and explain the basics of jet and rocket propulsion.

MECHANICS

For additional Making Connections, Examples, and Checkpoints, as well as activities and experiments to help increase your understanding of the chapter's concepts, please go to the text's online resources at www.physics1e.nelson.com.

LO 1

7-1 Linear Momentum

In Chapter 6, we stated that in the absence of nonconservative forces, the mechanical energy of an object is conserved.

In this chapter, we introduce another conserved quantity: **linear momentum**, $\vec{p}$. The linear momentum of an object is defined as the product of its mass and its velocity:

$$\vec{p} = m\vec{v} \tag{7-1}$$

Momentum is a vector quantity and has units of kilogram metres per second.

Table 7-1 lists momenta for different objects. The table gives an indication of the wide range of possible values for momentum.

Table 7-1 Appreciating Momentum

Momentum (kg · m/s)	Object
~1.66×10^{-29}	An atom in a Bose–Einstein condensate (a very low-energy state of matter)
3.67×10^{-24}	A free neutron at 290 K
2.123×10^{-23}	An atom of argon at room temperature
2.733×10^{-23}	An electron moving at 1/10 the speed of light
5.01×10^{-20}	A proton moving at 1/10 the speed of light
0.014	A 7.0 g marble moving at 2.0 m/s
3.0	A 7.5 g bullet moving at 400 m/s
2.8×10^4	A 1.0 tonne car moving at 100 km/h
14.5×10^6	A fully loaded Avro Arrow flying at 2100 km/h
1.6×10^9	Super Carrier (100 000 tonnes at full speed, 54 knots)
1.776×10^{29}	Earth as it orbits the Sun

EXAMPLE 7-1

Rocket Momentum

Find the horizontal and the vertical components of the momentum of a space shuttle at the point following its takeoff where the mass of the shuttle (including fuel) is 1.90×10^6 kg and its velocity is 4200 km/h at an angle of $11.0°$ from the vertical.

SOLUTION

The magnitude of the space shuttle's momentum at that point is given by

$$p = mv = 1.90 \times 10^6 \text{ kg} \times 4200 \left(\frac{1000 \text{ m}}{3600 \text{ s}} \right)$$

$$= 2.22 \times 10^9 \text{ kg m/s}$$

The expression for the momentum vector is given by

$$\vec{p} = \vec{p}_x + \vec{p}_y$$

Using Figure 7-2 we can express the momentum as

$$\vec{p} = p\sin(11°)\hat{i} + p\cos(11°)\hat{j}$$

$$= (4.23 \times 10^8 \hat{i} + 2.18 \times 10^9 \hat{j}) \text{ kg m/s}$$

Making sense of the result:

The vertical component is significantly greater than the horizontal component, as we would expect, because the angle from the vertical is relatively small.

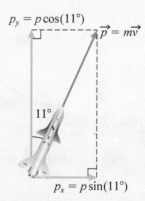

Figure 7-2 The horizontal and vertical components of the linear momentum of a space shuttle

✓ CHECKPOINT

C-7-1 Speed versus Momentum

A car enters a roundabout at a constant speed and exits in the direction opposite to the one in which it entered the roundabout. The change in momentum of the car is

(a) zero;
(b) twice the initial momentum;
(c) twice the final momentum.
(d) None of the above.

C-7-1 (c) $\Delta p = p_f - p_i = (-p_i) - p_i = 2p_f$.

MECHANICS

EXAMPLE 7-2

Change of Momentum

A marble with a mass of 10.0 g is sliding at a constant speed of $v = 2.00$ m/s along a circular, horizontal frictionless track of radius $r = 0.90$ m, as shown in Figure 7-3(a). Find the change in the marble's momentum as it goes from point A to point B. Express the change in vector form, and find the magnitude of the momentum change. Illustrate your solution with a diagram.

SOLUTION

The change in momentum is the difference between the final and the initial momenta. When the marble is in position A, all of its momentum is in the y-direction. Therefore,

$$\vec{p}_A = 0\hat{i} + (mv)\hat{j} = 0\hat{i} + (0.010 \text{ kg})(2 \text{ m/s})\hat{j}$$
$$= (0.020 \text{ kg m/s})\hat{j} \tag{1}$$

At position B, the magnitude of the momentum has not changed, but it is now pointing in the x-direction:

$$\vec{p}_B = (mv)\hat{i} + 0\hat{j} = (0.020 \text{ kg m/s})\hat{i}$$

The change in momentum as a vector is equal to

$$\Delta\vec{p} = \vec{p}_B - \vec{p}_A = 0.020\hat{i} - 0.020\hat{j}$$

The change in momentum can be expressed as

$$\Delta\vec{p} = \Delta\vec{p}_x + \Delta\vec{p}_y = 0.020\hat{i} - 0.020\hat{j}$$

The magnitude of the momentum vector is

$$\Delta p = \sqrt{(\Delta p_x)^2 + (\Delta p_y)^2} = \sqrt{0.020^2 + 0.020^2} = 0.028 \text{ kg m/s}$$

The diagram for the Solution is shown in Figure 7-3(b).

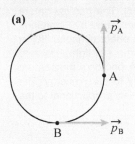 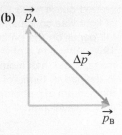

Figure 7-3 (a) The linear momentum has the same magnitude but different directions at A and B. (b) The change in momentum is shown as a difference of vectors.

Making sense of the result:

The change in momentum must point down and to the right to give us the final momentum which points straight to the right.

✓ **CHECKPOINT**

C-7-2 Change in Momentum

A high-g training capsule of mass m moves counterclockwise around a circular track at a constant speed v. What is the change in the capsule's momentum by the time it has completed one-quarter of a turn if it starts in the positive y-direction?
(a) $mv\hat{i} + mv\hat{j}$
(b) $-mv\hat{i} - mv\hat{j}$
(c) $mv\hat{i} - mv\hat{j}$
(d) $-mv\hat{i} + mv\hat{j}$

C-7-2 (b) horizontal momentum goes from zero to $-mv$, vertical momentum goes from $+mv$ to zero.

Momentum and Kinetic Energy

The kinetic energy of a particle is given by

$$K = \frac{1}{2}mv^2 \tag{6-4}$$

From the definition of momentum (Equation (7-1)), we get

$$\vec{v} = \frac{\vec{p}}{m}$$

Then,

$$K = \frac{p^2}{2m} \quad \text{or}$$

$$p = \sqrt{2mK} \tag{7-2}$$

Thus, an object that has momentum also carries kinetic energy.

✓ **CHECKPOINT**

C-7-3 Kinetic Energy and Momentum

When the momentum of a particle doubles, and its mass remains constant, its kinetic energy
(a) doubles;
(b) quadruples;
(c) increases by $\sqrt{2}$.
(d) Both (a) and (b) are possible.

C-7-3 (b).

MAKING CONNECTIONS

Momentum Without Mass

Light has a dual nature: it is both a particle and a wave. Its wave nature can be demonstrated through interference and diffraction measurements, the most famous of which is Young's double-slit experiment, carried out in 1803 (Chapter 28). The particle nature of light was conclusively demonstrated by Einstein in his paper on the photoelectric effect (Chapter 30) in 1905. Although a photon (a particle of light) has zero mass, it still has momentum. A photon's momentum is defined in terms of its ability to carry energy. The momentum of a photon is inversely proportional to the light's wavelength.

Figure 7-4 The remains of the *Titanic* on the floor of the Atlantic Ocean

© Ralph White/CORBIS

LO 2

7-2 Rate of Change of Linear Momentum and Newton's Laws

MAKING CONNECTIONS

Fatal Momentum

The *Titanic* received six warnings of "imminent ice ahead" before its fateful collision with an iceberg. However, the captain maintained full speed, in accordance with the shipping industry's standard operating procedure of maintaining full cruising speed until ice is actually sighted. Up until this time, this procedure had proven efficient and safe. This procedure was reasonable for smaller and more manoeuvrable ships; however, *Titanic* was a 52 000 tonne leviathan. Its large momentum could not be changed quickly, and the corresponding kinetic energy made the collision with the iceberg more damaging than it would have been for a smaller ship travelling at the same speed. By the time the ice was sighted, it was too late to change course enough to save the ship. Ironically, it is believed that had the *Titanic* maintained its course and hit the iceberg head on, the ship might not have sunk, because the bow structure would have distributed the force of the collision lengthwise along the ship's hull. However, the angled impact along the side of the ship broke a line of hull plates loose at their riveted joints, dooming the ship.

Taking the derivative of Equation (7-1) with respect to time gives

$$\frac{d\vec{p}}{dt} = \frac{d}{dt}(m\vec{v}) \qquad (7\text{-}3a)$$

This equation looks similar to Newton's second law, but its form is different from the one we saw in Chapter 5. For a constant mass, however, Equation (7-3) does reduce to the familiar form of Newton's second law. When the mass is constant, it can be taken out of the differentiation and

$$\frac{d\vec{p}}{dt} = m\frac{d}{dt}\vec{v} = m\vec{a} \qquad (7\text{-}3b)$$

Equation (7-3a) is the form of Newton's second law that we must use when mass varies. An example in which a mass can vary is the squid: a squid ejects water as a means of propulsion, so the total mass of the squid decreases as it releases the water. Similarly, jets and rockets lose fuel mass as they propel themselves.

We therefore use a form of Newton's law that can address the acceleration of an object that has changing mass:

$$\vec{F}_{net} = \frac{d\vec{p}}{dt} \qquad (7\text{-}4a)$$

Here, the net force acting on an object is defined as the rate of change of the momentum of the object. The greater the change in momentum, the greater the applied force. Equation (7-4a) also lets us define the average net force acting on an object in terms of the total change in its momentum:

$$\vec{F}_{net}^{avg} = \frac{\Delta\vec{p}}{\Delta t} \qquad (7\text{-}4b)$$

Equation (7-4a) gives us the instantaneous force acting on an object, and Equation (7-4b) gives us the average net force.

MAKING CONNECTIONS

Cushioning the Blow: Airbags and Seat Belts

A critical factor for passengers in a car collision is the time over which their momentum changes, which determines the net force experienced as the car slows down. The dashboard is often padded with a firm rubber foam, which deforms when a passenger hits it, reducing the rate of change of the passenger's momentum. Seat belts are also used to restrain the passengers and prevent them from colliding with the harder components on the car's interior. An essential element of a seat belt is the elasticity, or deformability, which helps increase the time over which the momentum of the passenger changes during a collision. Emergency tension-limiting devices are also used to give the seat belt a controlled amount of "slack" during the collision to help reduce injuries. Sensors in the car seat measure the weight of the passenger and adjust the rate at which the built-in air bag inflates and then deflates to optimize the time over which the passenger's momentum changes, again reducing the net force experienced by the passenger.

EXAMPLE 7-3

Forces in a Car Crash

A 1700 kg car moving at a speed of 90.0 km/h runs into a tree, which stops the car in 100 ms.

(a) Find the average force of impact on a 75.0 kg passenger.
(b) Calculate the acceleration experienced as a result of this force, express in multiples of g.
(c) At what rate should an airbag deflate to keep the force experienced by the passenger below $17g$ (commonly regarded as fatal)?

SOLUTION

We choose the x-axis to be along the direction of motion.
(a) The average force of impact:

$$\vec{F}_{net}^{avg} = \frac{\Delta \vec{p}}{\Delta t} = 0 - \frac{(75 \text{ kg})(90)\left(\frac{1000}{3600}\right)\hat{i} \text{ m/s}}{0.1 \text{ s}} = -18750\hat{i} \text{ N}$$

(b) The average acceleration that the car undergoes is given by Newton's second law, in magnitude:

$$a_{avg} = \frac{F_{avg}}{m} = \frac{18750}{75} = 250 \text{ m/s}^2 = \frac{250}{9.81} g = 25.5 \, g$$

(c) Working backward, we need a maximum g value of 17:

$$a_{avg} = 17 \times 9.81 \text{ m/s}^2 = 166.77 \text{ m/s}^2$$

The average force experienced by the passenger is then

$$F_{avg} = ma_{avg} = 75 \text{ kg } (166.77 \text{ m/s}) = 12508 \text{ N}$$

Therefore, the time the airbag takes to deflate must be at least

$$\Delta t = \frac{\Delta p}{F_{net}^{avg}} = \frac{(75 \text{ kg})(90 \text{ km/h})\left(\frac{1000 \text{ m}}{3600 \text{ s}}\right)}{12508 \text{ N}} = 0.150 \text{ s} = 150 \text{ ms}$$

Making sense of the result:

In an actual collision with a modern car, deformation of the car body would make the duration of the collision longer than 100 ms, thus reducing the force of the impact. Airbags often deploy within 60 ms, and then deflate within approximately 120 ms following full inflation. The average force during a collision is not the maximum force (as we will see below); the time over which the actual maximum force is in effect is very small. Therefore, the average force during the collision is taken to be representative of the force during a collision.

 CHECKPOINT

C-7-4 Bouncing Ball

When a ball hits a wall horizontally and bounces back with the same speed,

(a) the average net force on the ball is zero;
(b) the momentum of the ball does not change;
(c) the force on the ball is parallel to the ball's incident momentum.
(d) None of the above.

 CHECKPOINT

C-7-5 The Force on a Satellite

A satellite of mass m orbits Earth at speed v in an orbital period of T. What is the magnitude of the average force on the satellite during half an orbit?

(a) zero
(b) mv/T
(c) $2mv/T$
(d) $4mv/T$

C-7-4 (d) momentum does change, force is not zero and is parallel to final momentum.

C-7-5 (c) change in momentum is 2 mv.

CHAPTER 7 | LINEAR MOMENTUM, COLLISIONS, AND SYSTEMS OF PARTICLES 171

7-3 Impulse and the Conservation of Linear Momentum

In comic books and movies, Superman, Spiderman, and the Flash sweep people off their feet and pick them up while in mid-fall to save them from peril. What is often missed is the peril that lies in the details of the rescue: It is not uncommon to see a victim falling from the top of a skyscraper, only to have the superhero swoop up from below to reverse the direction of motion almost instantaneously. In real life, changing the speed or the momentum at this rate might well be fatal. Recent movies show the action more realistically, with the superhero gently bringing a falling person or vehicle to a stop before reversing the direction of motion.

In this section, we examine in more detail the behaviour of the force on an object during an impact or collision. We begin by rewriting Equation (7-3) as

$$d\vec{p} = \vec{F}_{net}\, dt \qquad (7\text{-}5)$$

According to Equation (7-5), the infinitesimal change in the momentum of an object equals the product of the net force acting on that object and the infinitesimal amount of time over which the force acts on the object. Integrating Equation (7-5) gives the change in momentum, which is called the **impulse, I**:

$$\vec{I} = \Delta \vec{p} = \int_{t_1}^{t_2} \vec{F}_{net}\, dt \qquad (7\text{-}6a)$$

We can also say that

$$\vec{I} = \Delta \vec{p} = \vec{F}_{net}^{\,avg}\, \Delta t \qquad (7\text{-}6b)$$

where $\vec{F}_{net}^{\,avg}$ is the average net force acting on the object during the time Δt.

When the net force on an object is zero, the impulse is also zero. The fact that an object's linear momentum

---

✔ CHECKPOINT

C-7-6 Direction of Impulse

A 17 g marble moving at a constant speed on a horizontal surface strikes a peg and deflects 37° below its original direction of motion without losing any of its kinetic energy. If the marble was originally moving to the right, the impulse that it receives

(a) has a magnitude of zero;
(b) is nonzero and points downward to the right;
(c) is nonzero and points straight downward;
(d) is nonzero and points downward and the left.

C-7-6 (d) this can be seen by drawing the momentum diagram.

EXAMPLE 7-4

Impulse of a Falling Object

Find the impulse experienced by a 200 g stone when it falls freely for 10.0 s near Earth's surface.

SOLUTION

The force acting on the stone has a constant magnitude, $F = mg$. Taking down as the positive direction and applying Equation (7-6b), we get

$$\Delta \vec{p} = \vec{F}_{net}\, \Delta t \Rightarrow p_f - p_i = mg\, \Delta t$$

Since the stone falls from rest, the initial momentum is zero and

$$\Delta p = mg\, \Delta t = (200 \times 10^{-3}\text{ kg})\,(9.81\text{ m/s}^2)\,(10\text{ s})$$

$$\Delta p = 19.6\text{ kg m/s}$$

Making sense of the result:

The force of gravity changes the momentum of the stone.

---

remains constant in the absence of a net force acting on the object is in keeping with Newton's first law.

The Force of Impact

What takes place when a tennis ball bounces off a wall? For simplicity, we will consider a ball travelling horizontally when it hits a vertical wall. We also assume that the collision is elastic, (neither the wall nor the tennis ball is irrecoverably deformed). Before the impact, the wall does not exert a force on the ball. As the ball strikes the wall, the force of the wall on the ball increases sharply to a peak value. During this first phase of the impact, the ball deforms somewhat. We assume that the wall is hard enough to not deform at all. The second phase of the impact starts after the force reaches its peak. The ball begins to reverse direction as it recovers from the deformation sustained during the collision. If the collision is completely elastic, the recovery of the ball will be complete. This profile for the behaviour of the magnitude of the force with time is typical of elastic collisions.

The area under the force versus time curve corresponds to the integral of the force over the time during the collision. According to Equation (7-6a), the integral is the total change in the momentum of the ball.

Linear Approximation for the Force of Impact

Since the variation of force with time during a collision is complex, integration of real-time data like those shown in Figure 7-5(a) is not easy, so the graph in Figure 7-5(a) is of limited practical use. However, we can often use

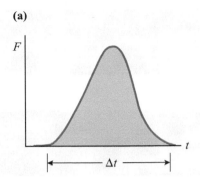

(a)

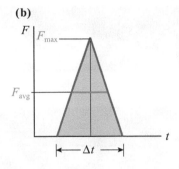

(b)

Figure 7-5   (a) Actual force versus time graph for a collision (b) Linear model for force versus time

the relatively simple linear model to calculate reasonable approximations.

Criteria The two parameters for our linear model are the duration of the impact and the total change in momentum. Our simplified model should produce a reasonable

estimate of the average force of the impact. This model is based on a triangular graph, as shown in Figure 7-5(b). Using the properties of similar triangles, we can show that in this model, the maximum force is twice as great as the average force. Example 7-5 will give us some sense of the accuracy of this approximation.

EXAMPLE 7-5

Linear Model of Impact Force

A golf ball with a mass of 46.0 g strikes a wall horizontally at a speed of 130 km/h, and bounces straight back from the wall with the same speed. The collision takes 0.900 ms.

(a) Find the impulse delivered to the ball by the wall.
(b) Find the average force experienced by the ball during the collision.
(c) Assuming that the force changes linearly with time, plot a force versus time graph.
(d) Estimate the maximum value of the force of impact.

SOLUTION

The situation is depicted in Figure 7-6(a). The initial momentum of the golf ball points to the right; after the collision, the golf ball moves to the left.

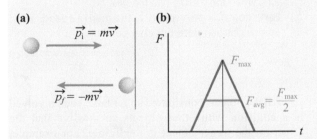

Figure 7-6   (a) The golf ball before and after the collision (b) The linear model for the behaviour of force versus time

(a) For convenience, we take left as our positive direction and convert the speed to metres per second. Then,

$$\vec{p}_f = m\vec{v} = -\vec{p}_i, \text{ then} \tag{1}$$

$$\Delta p = p_f - (-p_i) = 2mv$$

$$= 2(46 \times 10^{-3}) \text{ kg } (130) \left(\frac{1000 \text{ m}}{3600 \text{ s}} \right) = 3.32 \text{ kg m/s} \tag{2}$$

where the variables used for momentum in Equation (2) denote magnitudes

(b) From Equation (7-6b), we have

$$\Delta p = F_{net}^{avg} \Delta t$$

$$\Rightarrow F_{net}^{avg} = \frac{\Delta p}{\Delta t} = \frac{3.32 \text{ kg m/s}}{0.9 \times 10^{-3} \text{ s}} = 3.69 \text{ kN} \tag{3}$$

(c) Figure 7-6(b) shows the force as a function of time, taken to be linear.
(d) The average force is half the maximum value for the linear approximation in Figure 7-6(b); therefore, the maximum value of the force of impact is 7.38 kN.

Making sense of the result:

The force calculated is tremendous—equivalent to the weight of a three-quarter tonne truck. However, this force exists for only a very short time.

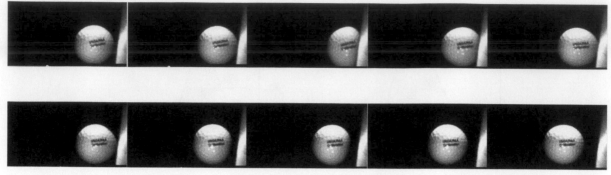

Figure 7-7 High-speed footage of a golf ball colliding with a wall. The frame-by-frame images (taken at small intervals) reveal details that human vision cannot otherwise detect.

The degree to which an object can deform during a perfectly elastic collision can be quite spectacular. The sequence of images shown in Figure 7-7 shows a golf ball during various stages of impact with a metal plate.

☑ CHECKPOINT

C-7-7 Collision Duration

Figure 7-8 shows the normal force versus time for a 90.0 g sponge ball as it hits the ground at a speed of 7.00 m/s. Find the speed with which the ball leaves the ground. The peak force is 1150 N, and the duration of the collision is 2.00 ms.

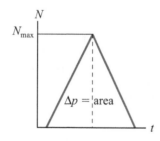

Figure 7-8 Simplified model showing the linear behaviour of force versus time.

LO 4

7-4 Systems of Particles and Centre of Mass

From Newton's third law, we know that when a golf ball strikes a wall, the force exerted by the wall on the golf ball, N, is equal in magnitude and opposite in direction to the force exerted by the golf ball on the wall; we will call this force N for normal. If the wall were free-standing on a frictionless surface, then the wall would accelerate away from the ball during the collision. When the initial velocity of the ball is normal to the wall, the change in momentum and speed for the wall are as follows:

$$\Delta \vec{p} = m\Delta \vec{v}$$

$$\Delta \vec{v} = \frac{\vec{N}\Delta t}{m} \qquad (7-7)$$

Since the mass of the wall is much greater than the mass of the golf ball, the recoil speed of the wall would be much less than the recoil speed of the ball. However, an actual wall is most likely firmly attached to the ground, so the change in momentum will result in an immeasurably small acceleration or deformation of the wall, its foundations, and the ground around it.

☑ CHECKPOINT

C-7-8 Earth-Moving Marble

When you drop a marble from the top of a tall building, and the marble lands on sand without bouncing,
(a) after the marble lands, Earth will gain some speed;
(b) whatever speed Earth gains as the marble falls, it will lose when the marble hits the ground;
(c) Earth does not gain speed as the marble speeds up;
(d) Earth's position does not change at all.

All the objects considered so far have been involved in a collision with other objects so massive that the effect on the larger object is negligible. For example, when a 20 kg dog moves on Earth's surface, the force that causes him to move also acts on Earth's surface, pushing Earth's surface back. We can see the dog move, but the corresponding motion of Earth is far too small for us to detect. What about the motion of objects compared with objects much less massive than Earth? For example, when the same dog walks along the length of a 30 kg canoe, however, the canoe's motion is visible despite the friction from the water around it. Also, when two cars of comparable mass collide, the changes in momentum and velocity for the cars will often be obvious.

Systems of Particles: From Point Masses to Rigid Bodies

We now examine the principles of mechanics as they apply to systems of particles, or objects. So far, we have treated all objects as point masses, or particles. A **point mass** is an object with a nonzero mass but with very small dimensions, or, ideally, zero size. Although this approach was not realistic, it provided a simple platform for the application of physical principles. Now that we have mastered the basics of Newton's mechanics and the work–mechanical energy approach, we are in a good position to move on to a more realistic description of physical objects. Here, we will consider a system of particles, or discrete distributions of mass. Systems of particles are in effect a step in the transition from dealing with point masses to rigid objects. In the next chapter, we will deal with continuous distributions of mass in various rigid objects.

Centre of Mass The first tool in our study of systems of particles is the concept of the **centre of mass**, the point corresponding to the mean position of the mass in a body or a system. For many calculations, the mass of a system or an object can be treated as a point mass, located at the centre of mass. Interactions between the particles or objects in a system do not change the total momentum of the system.

In one dimension, the position of the centre of mass of a system of n particles is given by

$$M_T \vec{x}_{cm} = m_1 \vec{x}_1 + m_2 \vec{x}_2 + \cdots + m_n \vec{x}_n \qquad (7\text{-}8a)$$

where m_i is the mass of each particle, $\vec{x}_i$ is the position vector of each particle, and M_T is the total mass of all the particles in the system: $M_T = m_1 + m_2 + \cdots + m_n$

Therefore, the location of the centre of mass of the system is

$$\vec{x}_{cm} = \frac{m_1 \vec{x}_1 + m_2 \vec{x}_2 + \cdots + m_n \vec{x}_n}{M_T} \qquad (7\text{-}8b)$$

Similarly, for systems in more than one dimension,

$$M_T \vec{r}_{cm} = m_1 \vec{r}_1 + m_2 \vec{r}_2 + \cdots + m_n \vec{r}_n \qquad (7\text{-}9a)$$

and

$$\vec{r}_{cm} = \frac{\sum_{i=1}^{n} m_i \vec{r}_i}{\sum_{i=1}^{n} m_i} \qquad (7\text{-}9b)$$

In three dimensions, the $\vec{r}$ vector has independent components in the x-, y-, and z-directions. Thus, we can have equations with the y- and z-components of $\vec{r}_{cm}$ with the same form as Equation (7-8).

The expression for the centre of mass of a *continuous* distribution of mass is

$$\vec{r}_{cm} = \frac{\int_V \vec{r}\, dm}{\int_V dm} \qquad (7\text{-}9c)$$

where the integration limit V indicates that the integral is taken over the volume of an object or a continuous distribution of particles.

Equations (7-8) and (7-9) are all vector equations, so it is important to define a frame of reference and choose an appropriate coordinate system when dealing with centre of mass, as the next two examples demonstrate.

EXAMPLE 7-6

Centre of Mass of Two Particles

Find the centre of mass of the system of two particles of masses $m_1 = 5\,m$ and $m_2 = m$ located a distance L apart as shown in Figure 7-9(a).

(a)

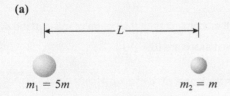

$m_1 = 5m$ $m_2 = m$

Figure 7-9(a) Two masses a distance L apart

SOLUTION

Our first step is to choose a coordinate system. For convenience, we choose the line connecting the two masses as the x-axis. We use the heavier of the two masses, m_1, as our origin (see Figure 7-9(b)) and choose right as the positive direction.

Using Equation 7-8(b) we have

$$\vec{x}_{cm} = \frac{m_1 \vec{x}_1 + m_2 \vec{x}_2}{m_1 + m_2} = \frac{5m(0)\hat{i} + m(L)\hat{i}}{5m + m} = \frac{L\hat{i}}{6}$$

We will now solve this same problem using the lighter of the two masses, m_2, as our origin (Figure 7-9(c)), but still taking right as the positive direction. Then,

(continued)

(b)

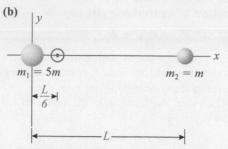

Figure 7-9(b) Mass m_1 is the origin of the coordinate system. The centre of mass is $L/6$ to the right.

$$x_{cm} = \frac{5m(-L)\hat{i} + m(0)\hat{i}}{5m + m} = -\frac{5L\hat{i}}{6}$$

Here, the negative value for x_{cm} tells us that the centre of mass is to the left of the origin.

(c)

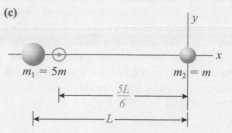

Figure 7-9(c) Mass m_1 is the origin of the coordinate system.

Making sense of the result:

As we would expect, the centre of mass is between the two objects and closer to the more massive object. Whether we place our origin at the heavier or the lighter mass, the centre of mass will be a distance of $L/6$ from the heavier mass and $5L/6$ from the lighter mass.

MAKING CONNECTIONS

What Is Orbiting What?

Earth does not really orbit the Sun, nor does the Moon orbit Earth. In a system consisting of just Earth and the Sun, both objects orbit the centre of mass of the system. The gravitational force that holds the Earth–Sun system together is internal to the system. In the absence of external forces, the centre of mass remains stationary. The centre of mass is much closer to the centre of the Sun than to the centre of Earth.

EXAMPLE 7-7

Three Equally Spaced Masses

Find the centre of mass of three particles of masses m_1, m_2, and m_3 located respectively at the left, right and top vertices of an equilateral triangle as shown in Figure 7-10(a). The length of each side of the triangle is L. All three particles have the same mass, m.

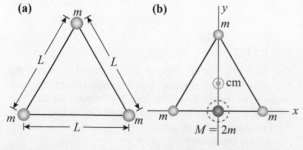

Figure 7-10 (a) Three equal masses at the corners of an equilateral triangle (b) The centre of mass of the system of three equal masses

SOLUTION

First, we find the centre of mass of the two masses at the base of the triangle. From the symmetry of the triangle, we can see this centre of mass is located in the middle of the base. To find the centre of mass of the whole system, we can replace masses m_1 and m_2 with a combined mass $M = 2m$ at the middle of the base of the triangle. We pick this location as the origin of the coordinate system (Figure 7-10(b)).

Now, we have only two objects to deal with: mass $M = 2m$, located at the origin, and mass m_3, located at the top of the triangle. From Figure 7-9(b) we can see that the distance between m_3 and the origin is

$$y = L\cos(30°) = L\frac{\sqrt{3}}{2}$$

Applying Equation (7-1) in the y-direction, we get

$$y_{cm} = \frac{2m(0) + mL\frac{\sqrt{3}}{2}}{3m} = \frac{\sqrt{3}}{6}L$$

Thus, the coordinates of the centre of mass of the system of three masses are $\left(0, \frac{\sqrt{3}}{6}L\right)$.

Making sense of the result:

It makes sense for the centre of mass to be closer to the base than the apex.

In Example 7-7, our first step was to use symmetry to locate the centre of mass of the two masses at the base, m_1 and m_2. Our second step was to choose that location as our origin of coordinates for the problem. Alternatively, our first step could have been to pick the location of our origin, perhaps to coincide with one of the masses, and then calculate the location of the centre of mass of the system.

LO 5

7-5 Newton's Laws Applied to Systems of Particles

We begin this discussion by using momentum to define the velocity, $\vec{v}_{cm}$, of the centre of mass of a system of particles. Taking the derivative of Equation (7-8a) with respect to time gives

$$M_T \vec{v}_{cm} = m_1 \vec{v}_1 + m_2 \vec{v}_2 + \cdots + m_n \vec{v}_n \quad (7\text{-}10a)$$

or

$$\vec{p}_{cm} = \vec{p}_1 + \vec{p}_2 + \cdots + \vec{p}_n \quad (7\text{-}10b)$$

From Equation (7-10a), we see that the velocity of the centre of mass of a system is really the weighted average of the velocities of the particles making up the system. Taking the derivative of Equation (7-10b) with respect to time, we get

$$\frac{d\vec{p}_{cm}}{dt} = \frac{d\vec{p}_1}{dt} + \frac{d\vec{p}_2}{dt} + \cdots + \frac{d\vec{p}_n}{dt} \quad (7\text{-}11a)$$

For a system with constant mass, this is equivalent to taking the derivative of Equation (7-10a):

$$M_T \frac{d^2\vec{r}_{cm}}{dt^2} = m_1 \frac{d^2\vec{r}_1}{dt^2} + m_2 \frac{d^2\vec{r}_2}{dt^2} + \cdots + m_n \frac{d^2\vec{r}_n}{dt^2} \quad (7\text{-}11b)$$

In both Equations (7-11a) and (7-11b), the right-hand side is the sum of all the individual net forces acting on the individual particles of the system. This sum is the net force acting on the system:

$$\frac{d\vec{p}_{cm}}{dt} = \vec{F}_{net} \quad (7\text{-}12a)$$

The significance of Equation (7-12a) is that, in the absence of a net force on a system of particles, the acceleration of the centre of mass of these particles is zero.

Internal Forces and Systems of Particles

For a system of particles, the internal forces acting between particles do not contribute to the net force acting on the system as presented in Equations (7-12a). **Internal forces** between any pair of particles are action–reaction pairs. Action–reaction forces are, by definition, equal in magnitude and opposite in direction, so when

✓ CHECKPOINT

C-7-9 Walking on Water

You are standing on the left end of a dugout canoe as it floats on a still lake in Sleeping Giant Provincial Park, Ontario. The mass of the canoe is twice your mass. When you have moved to the other end of the canoe,

(a) the centre of mass of the canoe does not move;
(b) the centre of mass of the canoe moves to the left;
(c) the centre of mass of both you and the canoe moves to the left;
(d) the centre of mass of both you and the canoe does not move at all.

C-7-9 (d) as dictated by conservation of momentum.

we add all the forces acting on the individual particles in a system, the total is zero. Therefore, internal forces do not affect the acceleration of the centre of mass of a system, and Equation (7-12a) can be written unambiguously as

$$\frac{d\vec{p}_{cm}}{dt} = \vec{F}_{ext} \quad (7\text{-}12b)$$

The implication of Equation (7-12b) is that where the **external force** is zero, the momentum of the system of particles does not change:

$$if \ \vec{F}_{ext} = 0, \ then \ \frac{d\vec{p}_{cm}}{dt} = 0 \Rightarrow \vec{p}_i = \vec{p}_f \quad (7\text{-}12c)$$

Later in the chapter, we will apply this concept to collisions.

System Identification

It is important to clearly define the system we are working with, and to identify which of the forces acting on the system are external and which are internal.

During a collision between two cars, the force of the collision changes the momentum of each car. Therefore, the momentum of each car is not conserved during the collision, although the *total* momentum of the two cars is conserved. The forces exerted by the cars on each other are internal forces, which have no effect on the total momentum of the system that is the two cars, here the two cars are the system.

For the linear momentum of a system to be conserved, there can be no unbalanced external forces acting on it. For problems involving conservation of momentum, we will therefore choose systems such that the forces involved are internal to the systems whenever possible. The total linear momentum of such a system is conserved.

For example, when a dog walks in a canoe, the forces between the dog and the canoe are internal to the dog–canoe system and do not affect the system's total momentum. Similarly, when an astronaut takes a walk in the International Space Station, both the astronaut and the space station move relative to the centre of mass of the astronaut–station system, but the momentum of the system does not change as a result of the force they exert on one another.

However, an external force acting on a system will change its total momentum, as indicated by Equation (7-12a). For example, the total momentum changes in the case when two cars collide on a surface where friction is not zero, and the momentum of the astronaut–space station system changes continuously as the space station (and the astronaut) is held in an orbit around Earth by the force of gravity.

CHECKPOINT

C-7-10 Jumping off a Sled

You are standing on a dogsled, which sits on an icy surface just outside of Oslo, Norway, in early December. You jump off the back of the sled to tend to your dogs. The sled, not tethered to the dogs, moves forward. Ignoring the friction between the sled and the ice, the centre of mass of the system consisting of you and the sled
(a) remains still;
(b) moves forward;
(c) moves backward;
(d) may move either forward or backward depending on whether you or the sled is more massive.

C-7-10 (a) conservation of momentum.

EXAMPLE 7-8

Stranded Astronaut

An astronaut's manoeuvring unit fails while he is inspecting the nose of a space shuttle in orbit. Another astronaut uses the orbiter's (shuttle's) robotic arm to move the stranded astronaut from the nose section to the airlock. Derive an expression for how far the orbiter moves as a result of this action.

SOLUTION

The force that the robotic arm exerts on the astronaut (and, therefore, the force that the astronaut exerts on the arm and the orbiter) is internal to the astronaut–orbiter system. According to Equation (7-9), the acceleration of the centre of mass of the astronaut–orbiter system is zero, because the external, or net, force acting on the system is zero. Since the centre of mass of the two objects is originally stationary (in their orbiting frame of reference), the centre of mass of the orbiter-astronaut system will remain stationary.

Let us draw a diagram of the orbiter and the astronaut before and after the astronaut moves (Figure 7-11).

We can use Equation (7-8b) to locate the centre of mass of the astronaut and the orbiter before and after the robotic arm moves the astronaut:

$$M_T \vec{x}_{cm} = m_a \vec{x}_a + m_o \vec{x}_o \tag{1}$$

$$M_T \vec{x}_{cm}' = m_a \vec{x}_a' + m_o \vec{x}_o' \tag{2}$$

Subtract Equation (1) from Equation (2):

$$M_T \vec{x}_{cm}' - M_T \vec{x}_{cm} = m_a \vec{x}_a' - m_a \vec{x}_a + m_o \vec{x}_o' - m_o \vec{x}_o$$

$$M_T(\vec{x}_{cm}' - \vec{x}_{cm}) = m_a(\vec{x}_a' - \vec{x}_a) + m_o(\vec{x}_o' - \vec{x}_o)$$

$$M_T \Delta \vec{x}_{cm} = m_a \Delta \vec{x}_a + m_o \Delta \vec{x}_o \tag{3}$$

We have already established that the centre of mass of the system does not move; therefore $\Delta \vec{x}_{cm} = 0$. We know that

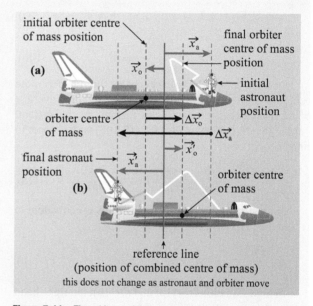

Figure 7-11 The orbiter and the astronaut (a) before and (b) after the astronaut moves.

the orbiter will move by an amount $\Delta \vec{x}_o$ when the astronaut moves by an amount $\Delta \vec{x}_a$. We do not know what these displacements are yet, but they are related by Equation (3).

In principle, we have two unknowns, $\Delta \vec{x}_a$ and $\Delta \vec{x}_o$, and one equation. The relative displacement of the astronaut with respect to the orbiter is given by Equation (3-24):

$$\Delta \vec{x}_{ao} = \Delta \vec{x}_a - \Delta \vec{x}_o \tag{4}$$

The relative displacement of the astronaut with respect to the orbiter is equal to the distance from the tail to the airlock, L. We pick right, as viewed in our diagram, as the positive x-direction. The orbiter moves to the right, so its displacement is positive. The astronaut moves to the left, so his displacement is negative. Since the displacement of the astronaut with respect to the orbiter is to the left, $\Delta\vec{x}_{ao}$ is negative. Therefore,

$$\Delta\vec{x}_{ao} = -L\hat{i}$$

Equations (3) and (4) can be written as

$$M_T \Delta x_{cm} = 0 = -m_a \Delta x_a \hat{i} + m_o \Delta x_o \hat{i} \quad \text{(3a)}$$

$$\Delta x_{ao}\hat{i} = -\Delta x_a \hat{i} - \Delta x_o \hat{i} \quad \text{(4a)}$$

Where the Δx's in Equations (3a) and (4a) indicate magnitudes of the displacements.

From Equation (3a) we have

$$\Delta x_a = \frac{m_o}{m_a} \Delta x_o$$

Substituting into Equation (4a) gives

$$-\Delta x_{ao} = \frac{m_o}{m_a} \Delta x_o + \Delta x_o$$

$$L = \frac{m_o}{m_a} \Delta x_o + \Delta x_o$$

$$\Delta x_o = \frac{L}{m_o/m_a + 1}, \quad \text{and} \quad \Delta x_a = \frac{L}{m_a/m_o + 1}$$

Making sense of the result:

Since the mass of the orbiter is much greater than the mass of the astronaut, $m_0 \gg m_a$, and $\Delta x_o \ll \Delta x_a \approx L$.

CHECKPOINT

C-7-11 Sun Orbits Earth

If we assume that the external forces acting on the Sun–Earth system are negligible, we can say that, as Earth orbits the Sun,

(a) the Sun remains stationary with respect to the centre of mass of the Sun–Earth system;

(b) the centre of mass of the Sun–Earth system moves back and forth;

(c) the Sun orbits the centre of mass of the Earth–Sun system;

(d) the motion of Earth does not affect the Sun's motion.

C-7-11 (c) both Earth and Sun orbit the centre of mass.

PEER TO PEER

In Example 7-8, it is important to distinguish between the displacement of the astronaut, $\Delta\vec{x}_a$, and the displacement of the astronaut with respect to the orbiter, $\Delta\vec{x}_{ao} = \Delta\vec{x}_a - \Delta\vec{x}_o$. The former is what an external observer in the centre of mass frame of reference would see—the actual distance covered by the astronaut relative to that frame of reference. The latter is the distance covered by the astronaut relative to the orbiter. Since the orbiter also moves within the centre of mass frame of reference, the value for $\Delta\vec{x}_{ao}$ depends on both the motion of the orbiter and the motion of the astronaut. For example, if the astronaut moves by 3.0 m and the orbiter moves 0.1 m in the opposite direction, then the astronaut would have moved by 3.1 m with respect to the orbiter.

EXAMPLE 7-9

Applying Conservation of Momentum

A bear cub ($m = 21.0$ kg) is sitting on a wagon ($M = 31.0$ kg), which is resting on a horizontal surface. A larger bear pushes the wagon with the bear cub on it until the wagon is moving at a speed of $v = 2.00$ m/s. The cub then jumps backward off the wagon with a speed of 3.00 m/s with respect to the wagon. Ignoring the effects of friction, find

(a) the speed of the cub with respect to the ground;

(b) the speed of the wagon with respect to the ground.

SOLUTION

The problem involves conservation of momentum, so we will consider the system consisting of the bear cub and

the wagon, just before and just after the cub jumps off, as shown in Figure 7-12.

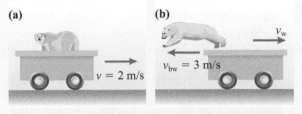

Figure 7-12 The wagon and the bear cub (a) just before the cub jumps and (b) just after the cub jumps.

(continued)

CHAPTER 7 | LINEAR MOMENTUM, COLLISIONS, AND SYSTEMS OF PARTICLES   179

(a) When the bear cub jumps off the wagon, the force that the cub exerts on the wagon and the force that the wagon exerts on the cub are internal to the cub–wagon system. Since we are ignoring the effects of friction and there are no other external forces acting on the system at this time, the linear momentum of the cub–wagon system is conserved as the bear cub jumps off the wagon:

$$\vec{p}_i = \vec{p}_f \qquad (1)$$

Initially, the bear cub and the wagon are both moving with the same velocity $\vec{v}$, so

$$(m + M)\vec{v} = m\vec{v}_b + M\vec{v}_w \qquad (2)$$

where $\vec{v}_b$ and $\vec{v}_w$ are, respectively, the velocities of the bear cub and the wagon after the cub jumps off. Both quantities are unknown, and so far we have one equation only, we need another.

The relative velocity of the bear cub with respect to the wagon after the bear cub jumps is (Equation (3-21)):

$$\vec{v}_{bw} = \vec{v}_b - \vec{v}_w \qquad (3)$$

After the bear cub jumps off the wagon, the velocity of the bear cub with respect to the wagon is the difference between the velocity of the bear cub and the velocity of the wagon (both with respect to the ground). We have used the result for relative velocity from Equation (3-21).

Let us pick right as the positive x-direction. The wagon is originally moving to the right, so v is positive. The bear cub jumps to the left, which is expected to increase the speed of the wagon as it moves to the right. Therefore, v_w should be positive. When the bear cub is in the air, he is moving to the left with respect to the wagon, so v_{bw} will be negative. We will have to guess whether or not the bear cub is moving to the right or to the left with respect to the ground. We emphasize here that *the assumption we make will not make a difference.* If our choice is incorrect, we will end up with a negative value for the speed of the bear cub, but the magnitude will be correct. Let us assume that the bear cub is moving to the left with respect to the ground.

Applying our choice of positive direction and our assumption for the direction of v_b to Equation (2), we get

$$(m + M)v\hat{i} = -mv_b\hat{i} + Mv_w\hat{i} \qquad (4)$$

And Equation (3) becomes

$$-v_{bw}\hat{i} = -v_b\hat{i} - v_w\hat{i} \text{ or}$$
$$v_w = -v_b + v_{bw} \qquad (5)$$

Again, the variables used in Equations (4) and (5) denote magnitudes of the velocity vectors in Equations (2) and (3) respectively. Substituting Equation (5) into Equation (4):

$$(m + M)v = -mv_b + M(-v_b + v_{bw}) = -mv_b - Mv_b + Mv_{bw}$$

The speed of the cub is given by

$$v_b = \frac{(m + M)v - Mv_{bw}}{-(m + M)}$$

$$= \frac{(21 \text{ kg} + 31 \text{ kg})(2 \text{ m/s}) - 31 \text{ kg} (3 \text{ m/s})}{-(21 \text{ kg} + 31 \text{ kg})} = -0.212 \text{ m/s}$$

Since the answer is negative, our choice for the direction of the velocity of the bear cub with respect to ground was incorrect. The bear cub is then actually moving to the right relative to the ground. We will keep than in mind and use v_b as a negative value next.

Had we assumed that the bear cub was moving to the right with respect to the ground, we would have ended up with the same number, but with a negative sign telling us we picked the incorrect direction.

(b) To get the speed of the wagon, we use Equation (5):

$$v_w = -v_b + v_{bw}$$
$$v_w = -(-0.212 \text{ m/s}) + 3 \text{ m/s} = 3.21 \text{ m/s}$$

Note that we can also use unit vector notation to solve this problem. The signs in the vector equations would then indicate the direction relative to the direction chosen for the x-axis unit vector.

Making sense of the result:

As a quick check, the change in momentum for the bear should be equal to and opposite to the change in momentum of the wagon. In this case they both work out to ~37.5 kg·m/s in magnitude. It is not always easy to predict the direction of motion of the bear cub relative to ground. We also note that a correct prediction is not necessary.

✓ **CHECKPOINT**

C-7-12 Leaping Bear

In Example 7-9,

(a) the system had more kinetic energy after the bear cub jumped;

(b) the system had more kinetic energy before the bear cub jumped;

(c) the system's kinetic energy does not change when the frictional forces are negligible;

(d) the bear cub exerts more force on the wagon than the wagon exerts on the bear cub.

C-7-12 (a) Bear does positive work.

LO 6

7-6 Inelastic Collisions

An **inelastic collision** is a collision during which the colliding objects lose kinetic energy. The energy is lost in the form of thermal energy, sound, and irreversible deformation of the objects. In a completely inelastic collision, the two colliding objects stick together. Although kinetic energy is lost during the collision, the momentum of the system is conserved. The less elastic a collision is, the greater the proportion of the kinetic energy lost.

CHECKPOINT

C-7-13 Adding Momentum

Two cars of equal mass (1100 kg) and speed (36 km/h) collide head-on in a completely inelastic collision. What is the total momentum of the system of two cars after the collision?

(a) 11 000 kg·m/s
(b) 22 000 kg·m/s
(c) zero
(d) none of the above

C-7-13 (c) the vector sum of the momenta is zero.

EXAMPLE 7-10

Vehicles in a Collision

In a collision-testing facility, a car with mass $m_1 = 1150$ kg moving at speed of 10.0 m/s collides head-on with a truck of mass $m_2 = 2800$ kg moving at speed of 7.00 m/s. The collision leaves the two vehicles joined in a heap. Find

(a) the speed of the vehicles after the collision;
(b) the kinetic energy lost in the collision.

SOLUTION

(a) The forces that exist between the vehicles during the collision are internal to the two-vehicle system. Therefore, we can say that the total vector momentum of the system is conserved during the collision:

$$\vec{p_i} = \vec{p_f} \quad \text{or} \quad (1)$$

$$m_1\vec{v_1} + m_2\vec{v_2} = (m_1 + m_2)\vec{v_f} \quad (2)$$

The two vehicles were initially moving separately, and after the collision they moved together as one object of mass.

$$M = m_1 + m_2$$

Assume that the car is moving to the right, and choose right to be the positive x-direction. (Note that the choice of direction does not affect the physics of the problem.) Then, Equation (2) can be written as

$$m_1v_1\hat{i} - m_2v_2\hat{i} = (m_1 + m_2)v_f\hat{i} \quad (3)$$

where the velocity symbols without the vector signs are variables that indicate the magnitudes of the velocity vectors. We will assume that the two objects move to the right after the collision; so we write v_f as a positive quantity in Equation (3):

$$v_f = \frac{m_1v_1 - m_2v_2}{m_1 + m_1}$$

$$= \frac{1150 \text{ kg}(10 \text{ m/s}) - 2800 \text{ kg}(7 \text{ m/s})}{(1150 + 2800) \text{ kg}}$$

$$= -2.05 \text{ m/s} \quad (4)$$

(b) For the energy lost in the collision, we will compare the initial and final kinetic energies of the system. The sum of the kinetic energies of the two vehicles before the collision is the initial kinetic energy of the system. The final kinetic energy of the system is the kinetic energy of the heap they form after the collision.

The energy lost is then

$$\frac{1}{2}m_1v_1^2 + \frac{1}{2}m_2v_2^2 - \frac{1}{2}(m_1 + m_2)v_f^2$$

$$= \frac{1}{2}1150 \text{ kg}(10 \text{ m/s})^2 + \frac{1}{2}2800 \text{ kg}(7 \text{ m/s})^2$$

$$- \frac{1}{2}(1150 \text{ kg} + 2800 \text{ kg})(2.05 \text{ m/s})^2$$

$$= 118\,000 \text{ J}$$

Making sense of the results:

The minus sign for the speed of the two vehicles together in Equation (4) indicates that our assumption that the heap is moving to the right is incorrect; the heap is moving to the left. The energy is lost to thermal energy, deformation, and sound.

MAKING CONNECTIONS

Cosmic Collisions

Stars readily form in the arms of a spiral galaxy because these regions have relatively high densities and strong gravitational forces (Chapter 11). Spiral galaxies are sub-classified by their structure. As shown in Figure 7-13, the Whirlpool galaxy (catalogued as NGC 5194 and as M51) has only two prominent spiral arms, which makes it a Grand Design spiral galaxy, in contrast to spiral galaxies that have many arms and often a less-defined spiral structure.

(continued)

The clear spiral structure of the Whirlpool galaxy was likely enhanced by an interaction with the smaller nearby galaxy, NGC 5195. The two galaxies may have undergone two collisions, one a few hundred million years ago and the other about 100 million years ago, as NGC 5195 crossed the disk of the Whirlpool galaxy. The interaction between the two galaxies created regions of even higher gravitational attraction, promoting spectacular star formation near the lagging edge of the arms of the spirals of the Whirlpool galaxy. The Whirlpool galaxy is about 31 million light years away, in the constellation Canis Venatici (the Hunting Dogs). Its orientation with respect to Earth and its relative closeness have given astronomers a great opportunity to study the birth of stars within spiral galaxies.

Robert Gendler Astrophotography

Figure 7-13 One of the sharper images of NGC 5194 and 5195 taken in January 2005 with the Advanced Camera for Surveys (ACS) aboard the Hubble Space Telescope. The pink colour coding represents intense star-forming regions.

EXAMPLE 7-11

Inelastic Collision

Children are playing with a small ball of steel of mass m_b suspended on a string such that the ball hangs just above a slippery plastic mat. The children slide a magnet of mass m_M along the mat to strike the steel ball. When the magnet strikes the steel ball at speed v_M, the magnet sticks to the ball, causing it to swing in an arc from its rest position. Assume that all friction in the system is negligible.

(a) Derive an expression for the height, h, that the ball and the magnet reach after the collision.
(b) Calculate the height using the following values: $m_M = 92.0$ g, $m_b = 190$ g, and $v_M = 7.00$ m/s.

SOLUTION

(a) We would be tempted to say that the kinetic energy of the incoming magnet is completely converted to the potential energy of interaction between the rising ball and magnet and Earth. However, since the magnet sticks to the steel ball, the collision is by definition completely inelastic. Kinetic energy is lost during an inelastic collision. Momentum, however, is conserved.

As Far as the collision is concerned, the system of concern to us is the magnet and the steel ball. The forces that occur as a result of the collision are internal to the system, so the total linear momentum of the system is conserved during the collision.

During the collision, the magnet and the steel ball effectively become one object of mass $M = m_M + m_b$ moving with velocity $\vec{v}$. Therefore,

$$\vec{p}_i = \vec{p}_f \tag{1}$$

$$m_M \vec{v}_M + m_b \vec{v}_b = (m_M + m_b)\vec{v} \tag{2}$$

choosing the direction of motion as the positive x-axis,

$$m_M v_M \hat{i} = (m_M + m_b)v\hat{i} \tag{3}$$

After the collision, the only force doing work on the ball-magnet system is the force of gravity. Since no nonconservative forces are doing work on the system, the total mechanical energy is conserved:

$$\Delta E_m = \Delta U + \Delta K = 0 \tag{4}$$

The kinetic energy of the system is converted into gravitational potential energy as the ball with the magnet attached swings upward:

$$\Delta U = -\Delta K \tag{5}$$

$$(m_M + m_b)gh = -(K_f - K_i) = -\left(0 - \frac{1}{2}(m_M + m_b)\right)v^2$$

$$= \frac{1}{2}(m_M + m_b)v^2 \tag{6}$$

$$h = \frac{v^2}{2g} \tag{7}$$

(b) Substituting the given values into Equation (3) gives

$$v = \frac{m_M v_M}{(m_M + m_b)} = \frac{(0.092 \text{ kg})(7 \text{ m/s})}{(0.092 + 0.190) \text{ kg}} = 2.284 \text{ m/s} \tag{8}$$

and

$$h = \frac{v^2}{2g} = \frac{(2.284 \text{ m/s})^2}{2(9.81 \text{ m/s}^2)} = 0.266 \text{ m} \tag{9}$$

Making sense of the result:

The mechanical energy is not conserved during the collision. Had energy been conserved, then

$$h = \frac{1}{2} m_M (v_M)^2 / (m_M + m_m)g = 0.815 \text{ m}$$

EXAMPLE 7-12

Shock-Absorber Test

In one particular collision test, a 1450 kg car is driven at 48.0 km/h toward a stationary car of mass 1780 kg. A spring with spring constant 88.0 kN/m is used to simulate the effect of shock-absorbing features that could be added to the vehicles, and such a spring is attached to the stationary car. Assuming that friction with the ground can be ignored, find

(a) the maximum compression in the spring;
(b) the maximum force on either car during the collision;
(c) the maximum acceleration experienced by the lighter car during the collision.

SOLUTION

(a) As the incoming car strikes the spring attached to the stationary car, the spring begins to compress, thus exerting a force on each car. As a result, the incoming car slows down, and the stationary car accelerates until both cars reach the same speed, v. At the instant the two cars have the same speed the distance between the two cars does not change, and the spring reaches its maximum compression.

The system is comprised of the two cars and the spring. Since the forces during the collision are internal to this system, the momentum is conserved, and

$$\vec{p}_i = \vec{p}_f$$

where $\vec{p}_i$ is the momentum of the system before the collision, and $\vec{p}_f$ is the momentum at the instant when the spring is at maximum compression. A speed of 48 km/hr reads 13.3 m/s.

Then

$$m_1\vec{v}_1 = (m_1 + m_2)\vec{v}$$

$$v = \frac{m_1 v_1}{(m_1 + m_2)} = \frac{(1450 \text{ kg})(13.3 \text{ m/s})}{1450 \text{ kg} + 1780 \text{ kg}} = 5.99 \text{ m/s}$$

Since the spring cushions the impact and the spring force is conservative, the total mechanical energy of the system is conserved. Some kinetic energy changes into elastic potential energy in the spring, but this energy is not lost from the system. Thus, when the cars are moving at the same speed, we can say

$$\frac{1}{2} m_1 v_1^2 = \frac{1}{2} kx^2 + \frac{1}{2}(m_1 + m_2)v^2$$

$$x = \sqrt{\frac{m_1 v_1^2 - (m_1 + m_2)v^2}{k}}$$

$$= \sqrt{\frac{(1450 \text{ kg})(13.3 \text{ m/s})^2 - (1450 \text{ kg} + 1780 \text{ kg})(5.99 \text{ m/s})^2}{88\,000 \text{ kg/s}^2}}$$

$$= 1.27 \text{ m}$$

(b) The maximum force exerted by one car on the other equals the force of the spring at maximum compression. The magnitude of the force can be obtained using Hooke's law:

$$F = kx = (88\,000 \text{ N/m})(1.27 \text{ m}) = 112 \text{ kN}$$

(c) The maximum acceleration experienced by the lighter car is

$$a = \frac{kx}{m} = \frac{(88\,000 \text{ N/m})(1.27 \text{ m})}{1450 \text{ kg}} = 77.1 \text{ m/s}^2$$

Making sense of the result:

The maximum acceleration experienced by a passenger would be about 8 g, which is high but not lethal in a head-on collision. Note that at the maximum compression both cars are moving, the spring will eventually release its potential energy and the two cars will go at different speeds.

 CHECKPOINT

C-7-14 Conserved Momentum

You throw a bouncy ball directly at a vertical wall (at 90° to the wall), and the ball bounces back at the same speed with which it struck the wall. Which of the following statements is the most correct?

(a) The momentum of the ball is conserved as it bounces back.
(b) As the ball moves toward the wall, the wall's position with respect to Earth's mantle remains constant.
(c) Earth's crust recoils slightly on the mantle as a result of the impact of the ball with the wall, which is rigidly attached to the crust
(d) The surface of Earth is too massive to be affected.

C-7-14 (c) the system's momentum is conserved, and none of the other statements are true.

 LO 7

7-7 Elastic Collisions

During an **elastic collision**, both the total linear momentum and the total kinetic energy of the system are conserved. We demonstrate a strategy for analyzing elastic collisions by considering an elastic collision in one dimension.

Suppose a marble of mass m moving to the right with speed v collides elastically with a marble of mass M moving to the left at speed u (Figure 7-14(a)). Depending on the masses and the initial speeds, after the collision, the two marbles will bounce away from one another (Figure 7-14(b)); both move to the left, or both move to the right. Since linear momentum and velocity are vectors, we need to establish a coordinate system. We take right as positive.

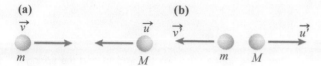

(a) $\vec{v}$ **(b)** $\vec{u}$ $\vec{v'}$ $\vec{u'}$

m M m M

Figure 7-14 (a) The two marbles head toward each other. (b) The two marbles bounce away from each other.

Conservation of Momentum

The total linear momentum of the system of two marbles is the vector sum of their individual momenta:

$$\vec{p_i} = m\vec{v} + M\vec{u}$$

After the collision, the expression of the final momentum of the system is

$$\vec{p_f} = m\vec{v'} + M\vec{u'}$$

where the primed variables indicate the final values.

The forces involved during the collision between the two marbles are internal to the two-marble system. Since there are no external forces involved, the total linear momentum of the system is conserved during the collision, and

$$\vec{p_i} = \vec{p_f} \quad or$$

$$m\vec{v} + M\vec{u} = m\vec{v'} + M\vec{u'} \qquad (7\text{-}13)$$

We could try to predict which of the three scenarios will occur after the collision, but the magnitude of the speeds we calculate after the collision will be the same no matter which assumption we make. Only the sign of the quantity will change depending on whether or not we choose the "correct" scenario. A negative value for a quantity indicates that the direction of the corresponding vector is opposite to what we had assumed.

We choose right to be the direction of the positive *x*-axis. The quantities in Equation (7-13) that point to the right will be positive, and the quantities that point to the left will be negative. Therefore,

$$mv\hat{i} - Mu\hat{i} = -mv'\hat{i} + Mu'\hat{i} \qquad (7\text{-}14)$$

The velocity symbols in Equation 7-14 denote the magnitudes of the velocity vectors.

In Equation (7-14), we see that we have two unknowns: the speeds of the two masses. Notice here that we say speeds and not velocities. The quantities that appear in Equation (7-13) are velocities. In principle, these speeds should be positive. However, if they turn out to be negative, it is because we made the "incorrect" assumption about the post-collision directions of motion. The signs depend on our assumptions, but the magnitudes of these quantities do not change.

To determine v' and u' from the masses and the initial velocities, we need a second equation relating v' and u'.

Conservation of Energy

Since the collision is elastic, kinetic energy is conserved. Therefore, the kinetic energy of the system before the collision is equal to the kinetic energy after the collision:

$$\frac{1}{2}mv^2 + \frac{1}{2}Mu^2 = \frac{1}{2}mv'^2 + \frac{1}{2}Mu'^2 \qquad (7\text{-}15)$$

This equation involves the squares of the final speeds. Although we can solve for the final speeds, it would be simpler to deal with a system of two linear equations. Gathering the terms in Equation (7-15) by mass, we have

$$m(v^2 - v'^2) = M(u'^2 - u^2) \qquad (7\text{-}16)$$

Factoring the differences of squares, we get

$$m(v - v')(v + v') = M(u' - u)(u' + u) \qquad (7\text{-}17)$$

Now we regroup the terms in Equation (7-14) in the same way:

$$m(v + v') = M(u' + u) \qquad (7\text{-}18)$$

If we divide Equation (7-17) by Equation (7-18), we have our second linear equation:

$$v - v' = u' - u \qquad (7\text{-}19)$$

Between Equations (7-18) and (7-19) we now have two equations with our two unknowns. We multiply Equation (7-19) by *m* and add it to Equation (7-18). Expanding all the brackets, we get

$$mv + mv' + mv - mv' = Mu' + Mu + mu' - mu \qquad (7\text{-}20)$$

$$u' = \frac{2mv + (m - M)u}{M + m} \qquad (7\text{-}21)$$

Similarly, solving for v' gives

$$v' = \frac{2Mu + (M - m)v}{M + m} \qquad (7\text{-}22)$$

The above steps are useful for analyzing an elastic collision. Equations (7-16), (7-17), and (7-18) are universal and should look the same for every solution involving elastic collisions. The steps relating to momentum conservation, however, depend on the assumptions we make about the directions of the unknown quantities and our choice of a positive direction. Ultimately, the final results in Equations (7-21) and (7-22) also depend on our choices; these two equations are not universal formulas.

MECHANICS

CHECKPOINT

C-7-15 Something's Got to Give

In a collision between a tennis ball and a tennis racquet, is it possible for the tennis player to hold the racquet hard enough so that the racquet does not recoil at all relative to the player?

(a) No, the racket must have a recoil speed depending on its mass.

(b) Yes, it is possible for the racquet to stay in place. There is no need for recoil if the player holds the racquet with enough force in which case the player must recoil.

(c) If the player does not recoil then the ground she stands on must recoil.

(d) b or c can be correct.

C-7-15 (d) If momentum is to be conserved then something has to give.

ONLINE ACTIVITY

Collisions

The e-resource that accompanies every new copy of this textbook contains an Online Activity using the PhET simulation "Collision Lab". Work through the simulation and accompanying questions to gain an understanding of collisions.

CHECKPOINT

C-7-16 What Went Wrong?

A student carries out the solution for an elastic collision between two billiard balls as outlined above, and makes assumptions about the unknown speeds of the two balls after the collision. The student obtains negative values for the speeds of both balls after the collision. Which of the following statements is correct?

(a) The student made a mistake when picking the negative direction.

(b) The student made an incorrect assumption about the direction of motion of one ball, but not both balls after the collision.

(c) The student made an incorrect assumption about the direction of motion of both balls after the collision.

(d) None of the above statements are correct.

C-7-16 (c) according to the approach we highlight above, a negative value for an unknown speed means the wrong direction was assumed for the velocity.

MAKING CONNECTIONS

Laser Cooling of Atoms

A direct application using the momentum of a photon is the laser cooling of atoms. The higher the temperature of a gas, the more kinetic energy its atoms will have and, hence, the higher their speed. Similarly, gas molecules move slower at lower temperatures. Since photons have momentum, a laser can be used to cool a gas by slowing down its atoms. The laser is fired at atoms in the gas. Incoming photons at the correct wavelength are absorbed by atomic electrons that are moving toward the laser. The momentum of the absorbed photon transfers to the atom, thus slowing the atom. The target atom cools further every time it absorbs a photon. The electron that absorbs the photon moves to a higher state of energy. If the photon is subsequently re-emitted by the excited atom, the momentum of the photon has a random direction. Energy is conserved in this process, and the collision is elastic. Carefully calibrated lasers (such as those in Figure 7-15) can cool atoms to a few tenths of a microkelvin and slow them down to the point where they are effectively "trapped". Researchers began cooling atoms with lasers in 1978 and reached temperatures below 40 K. They achieved temperatures a million times colder just 10 years later. This technology eventually led to better atomic clocks and the observation of an ultracold state of matter. The 1997 Nobel Prize in Physics was awarded to Steven Chu, Claude Cohen-Tannoudji, and William D. Phillips in 1997 for their work in developing technologies to use laser light to cool and trap atoms and the 2001 Nobel prize in physics was awarded to Eric A. Cornell, Wolfgang Ketterie and Carl E. Wieman for the achievement of Bose Einstein condensation.

The lowest temperaturs ever recorded to date is 1/2 of a nano Kelvin reached by an MIT research team lead by Ketterie, Leanhardt and Pritchard.

Figure 7-15 A cloud of cold sodium atoms (bright spot at centre) floats in a trap at the National Institute of Standards and Technology.

7-8 Variable Mass and Rocket Propulsion

Rocket propulsion provides an example of a system with variable mass. Here, the *system* consists of the rocket, its fuel, and the exhaust gas. In the rocket engine, the burning fuel turns into hot exhaust gases, primarily carbon dioxide and water vapour. When this exhaust is directed out the back of the rocket, the rocket accelerates forward. The more fuel that the rocket burns, the more mass it loses. The forces generated by the combustion of the fuel are internal to the system, so the linear momentum of the system is conserved as the fuel burns.

Consider a rocket that fires its engine in deep space. Assume that the engine burns fuel at a constant rate R, and that the exit (or effective) speed, v_{er}, of the exhaust gas relative to the rocket is also constant. Since the rocket is in deep space, we do not have to consider external forces, such as air drag, which would be significant in a launch from Earth. We also assume that gravitational forces are negligible.

At a given instant, the total mass of the rocket including the remaining fuel is M as shown in Figure 7-16(a). Figure 7-16(b) shows the rocket at a time dt later. The speed of the rocket has increased to $v + dv$, the mass of the exhaust released during the interval dt is dm, and the velocity of the exhaust gas is $\vec{v_e}$. The mass of the rocket plus fuel has decreased by dm to $M - dm$. The choice for the direction of $\vec{v_e}$ in Figure 7-16 (b) will be explained shortly. We note here that we assume the mass loss rate is a constant.

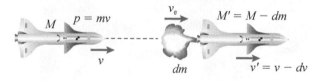

Figure 7-16 (a) A rocket moving at time t (b) The rocket moving at time $t + dt$

Since there are no external forces acting on the rocket–fuel–exhaust system and all the forces generated by burning the fuel are internal to the system, the total momentum of the system is conserved. Therefore, the momentum at time $t + dt$ at time t is

$$\vec{P_{(t)}} = \vec{P_{(t+dt)}} \qquad (7\text{-}23)$$

At time t, the momentum of the rocket and unburned fuel is Mv. After the interval dt, the mass of the rocket and the remaining fuel is $M' = M - dm$, and their velocity is $\vec{v'} = \vec{v} + d\vec{v}$. During the same interval, the mass of exhaust gas increases by dm. Using these expressions to find the momenta in Equation (7-23) gives

$$M\vec{v} = (M - dm)(\vec{v} + d\vec{v}) + dm(\vec{v_e}) \qquad (7\text{-}24)$$

Assuming that we know the initial mass and speed of the rocket system, the unknowns are the new rocket velocity, $\vec{v'} = \vec{v} + d\vec{v}$, and the velocity of the exhaust $\vec{v_e}$. The change in mass can be determined from the given rate of fuel consumption. However, to solve for the two unknown quantities we need another equation.

The equation can be found by considering the relative velocities of the components of the system. We know the exit speed of the exhaust relative to the rocket, $\vec{v_{er}}$. Applying Equation (3-21), we can readily see that

$$\vec{v_{er}} = \vec{v_e} - \vec{v'} = \vec{v_e} - (\vec{v} + d\vec{v}) \qquad (7\text{-}25)$$

Now, we choose the direction of the acceleration of the rocket (right) as the positive direction. The velocity of the exhaust with respect to the rocket, $\vec{v_{er}}$, is undoubtedly negative because it points to the left. However, the direction of the velocity of the exhaust, with respect to a stationary observer, depends on how fast the rocket is moving to the right and how fast the exhaust moves to the left with respect to the rocket.

Rockets in deep space reach speeds of up to 30 times the speed of sound near sea level. When the speed of the rocket is low, a stationary observer would expect to see the rocket move to the right and the exhaust gas move to the left. However, when the speed of the rocket is greater than the exit speed of the exhaust, both the rocket and the exhaust move to the right relative to a stationary observer. We assume that the exhaust is indeed moving to the right with respect to a stationary observer. We can consider the direction of motion of the rocket to align with the positive x-axis. Therefore, we can rewrite Equations (7-24) and (7-25) as

$$Mv\hat{i} = (M - dm)(v + dv)\hat{i} - dm(v_e)\hat{i} \qquad (7\text{-}26)$$

$$-v_{er}\hat{i} = -v_e\hat{i} - v'\hat{i} \quad \text{or} \quad -v_{er} = -v_e - (v + dv) \qquad (7\text{-}27)$$

$$v_e = v_{er} - (v + dv) \qquad (7\text{-}28)$$

Where the velocity symbols in the above three equations denote magnitudes.

Substituting the value of v_e from Equation (7-28) into (7-26) gives

$$Mv = (M - dm)(v + dv) - dm(v_{er} - v - dv) \qquad (7\text{-}29)$$

$$Mv = Mv + Mdv - vdm - dmdv - dmv_{er} + vdm + dmdv \qquad (7\text{-}30)$$

$$Mdv = dmv_{er} \qquad (7\text{-}31)$$

The decrease in the mass of the rocket and the remaining fuel, M, is the same as the increase in the total mass of exhaust, m. Thus

$$dM = -dm$$

Substituting the above into Equation (7-31), we get

$$Mdv = -dMv_{er} \qquad (7\text{-}32)$$

Equation (7-32) gives a relation between the incremental increase of the mass and the rate of change of the speed. Rearranging this equation gives

$$dv = -\frac{dM}{M}\, v_{er} \qquad (7\text{-}33)$$

To find the change in speed of the rocket after a given amount of fuel has been burned, we integrate Equation (7-33) with the initial and final mass as our integral limits:

$$\Delta v = \int_{v_i}^{v_f} dv = -v_{er}\int_{M_i}^{M_f}\frac{dM}{M} \qquad (7\text{-}34a)$$

Since

$$\Delta v = v_f - v_i$$

$$v_f - v_i = -v_{er}\ln\left(\frac{M_f}{M_i}\right) = v_{er}\ln\left(\frac{M_i}{M_f}\right) \qquad (7\text{-}34b)$$

Equation (7-34b) is called the **ideal rocket equation**. The change in speed is proportional to the logarithm of the ratio of the masses. Since the initial mass of the rocket with fuel is greater than the final mass, the final speed will be faster than the initial speed. Among other simplifications, the ideal rocket equation does not include the effects of gravity or air resistance.

The speed of the rocket depends on the mass of the remaining fuel, which is a function of time, since

$$\frac{dM}{dt} = R \qquad (7\text{-}35)$$

Thus, we can also drive an equation for the time dependence of the speed of rocket.

To find the instantaneous acceleration of the rocket, we take the derivative with respect to time of both sides of Equation (7-31):

$$M\frac{dv}{dt} = \frac{dm}{dt}\, v_{er} \qquad (7\text{-}36a)$$

$$Ma = Rv_{er} \qquad (7\text{-}36b)$$

The term Rv_{er} is called the thrust of the rocket engine. For a given mass of the rocket, the instantaneous acceleration is given by Equation (7-36b).

MAKING CONNECTIONS

Rockets and the Space Age

In the spring of 1241, the Mongolian empire defeated the kingdom of Hungary at the battle of Mohi. It is reported that Mongols used gun powder–propelled rockets during that battle, as well as during the rest of their 13th-century invasion of Europe. The Mongols adopted rocket technology from the Chinese, who had been reportedly using rockets for various purposes as early as the 14th century BCE. Chinese rocket experts were captured/hired by the Mongols to supply the rocket power for military conquest.

Before and during World War II, major advancements in rocket technology took place, mostly due to military demands. The advancements in rocket technology continued after World War II during the full-fledged rocket and space conquest race between the East and the West. Some landmark events that resulted include sending a human into space and landing on the Moon. Rocket technology made space exploration a reality and space colonization a possibility. Not free of tragedy, the story of the space age boasts many triumphs, not the least of which is the transformation of what started as an international quest for space supremacy into an endeavour that brings out the best of human achievement through collaboration. Collaborative efforts among the leading nations in the world continue to reach new heights as is evidenced by the International Space Station and other feats.

EXAMPLE 7-13

Applying the Ideal Rocket Equation

A space probe has a total mass of 1300 kg, 520 kg of which is fuel. The probe is near the edge of the solar system. The probe fires its rocket thrusters to make a final course correction, burning the fuel at a constant rate until it is gone.

The burn lasts 100 s and produces a constant thrust of 11.0 kN. Estimate

(a) the initial acceleration of the probe;
(b) the speed of the probe 80.0 s into the burn;
(c) the maximum speed of the probe;
(d) the maximum acceleration of the probe.

(continued)

SOLUTION

If we assume that all external forces acting on the probe are negligible, we can apply the ideal rocket equation. To do so, we need to find the rate of fuel consumption, R, and the effective exhaust speed, v_{er}. Since the rate of fuel consumption is constant and the rocket burns 520 kg of fuel in 100 s,

$$R = \frac{dm}{dt} = \frac{520 \text{ kg}}{100 \text{ s}} = 5.20 \text{ kg/s} \qquad (1)$$

We can determine the effective exhaust speed from the thrust, T:

$$T = Rv_{er} \Rightarrow 11\ 000 \text{ N} = (5.20 \text{ kg/s})(v_{er})$$

$$v_{er} = \frac{11\ 000}{5.20} = 2.115 \text{ km/s}$$

(a) Assume the probe is so far away from any celestial bodies such that the only significant force acting on the probe is the thrust from the rocket. Therefore,

$$F = Rv_{er} = ma$$

Since the initial mass of the rocket and fuel is 1300 kg, the initial acceleration is

$$a = \frac{F}{m} = \frac{Rv_{er}}{m} = \frac{11\ 000 \text{ N}}{1300 \text{ kg}} = 8.46 \text{ m/s}^2$$

(b) The mass of the probe and the remaining fuel after 80.0 s is

$$M_f = M_i - Rt = 1300 \text{ kg} - 5.20 \text{ kgs}^{-1}(80.0) = 884 \text{ kg} \qquad (2)$$

Substituting this mass gives

$$v_f - v_i = v_{er} \ln\left(\frac{M_i}{M_f}\right) = 2115 \ \ln\left(\frac{1300}{884}\right) = 816 \text{ m/s}$$

(c) The maximum speed of the rocket occurs when the rocket has burned all its fuel, at which point the mass of the probe is 1300 kg − 520 kg = 780 kg. Then,

$$v_f - v_i = v_{er} \ln\left(\frac{M_i}{M_f}\right) = 2115 \ \ln\left(\frac{1300}{780}\right) = 1080 \text{ m/s}$$

(d) The probe's maximum acceleration occurs just as the rocket runs out of fuel. At this instant, the mass of the probe is the least it can be:

$$a = \frac{F}{m} = \frac{Rv_{er}}{m} = \frac{11\ 000 \text{ N}}{780 \text{ kg}} = 14.1 \text{ m/s}^2$$

Making sense of the result:

The acceleration is about ~1.5 g, which is reasonable for rockets sent into space. As the mass drops the acceleration will increase, since the thrust is assumed to be constant.

MAKING CONNECTIONS

Flight of the Komet

Unlike its contemporary jet-powered counterparts, of which there were a few during World War II, the Messerschmitt ME 163 (the Komet, seen in Figure 7-17) was a full-fledged rocket plane, powered solely by rocket fuel. While not the first rocket plane to be built, it was the first to become operational during World War II. The Komet's rocket engines gave it a few minutes of powered flight after which it went into a glide. The performance was extraordinary to the point that plans to have a few of these as point defence interceptors against bombers in various locations were put in place, but never implemented. There were, however, some serious drawbacks, mostly due to the high speed of the plane relative to its bomber targets, which allowed the pilot a small window of opportunity to fire on Allied bombers. While almost impossible to hit during its fuel burst, the Komet reportedly made an easy target for fighters during its landing and gliding manoeuvres. The speeds reached by the Komet would only be surpassed in 1947.

Figure 7-17 An ME 163 rocket plane at an airfield in the 1940s

FUNDAMENTAL CONCEPTS AND RELATIONSHIPS

Momentum

Momentum is a vector:

$$\vec{p} = m\vec{v} \qquad (7\text{-}1)$$

The net force on an object results in a change in the object's momentum:

$$\vec{F}_{net} = \frac{d\vec{p}}{dt} \qquad (7\text{-}4a)$$

or

$$\vec{F}_{net}^{\,avg} = \frac{\Delta\vec{p}}{\Delta t} \qquad (7\text{-}4b)$$

Impulse is equal to the change in an object's momentum:

$$\vec{I} = \Delta\vec{p} = \int_{t_1}^{t_2} \vec{F}_{net}\, dt \qquad (7\text{-}5b)$$

or

$$\vec{I} = \Delta\vec{p} = \vec{F}_{net}^{\,avg}\Delta t \qquad (7\text{-}6b)$$

When the net force on an object is zero, the object's momentum does not change.

$$\text{if } \vec{F}_{net} = 0 \quad \text{then} \quad \Delta\vec{p} = 0 \Rightarrow \vec{p}_i = \vec{p}_f \qquad (7\text{-}6c)$$

Systems of Particles

The centre of mass in one dimension is given by

$$M_T\vec{x}_{cm} = m_1\vec{x}_1 + m_2\vec{x}_2 + \cdots + m_n\vec{x}_n \qquad (7\text{-}8a)$$

or

$$\vec{x}_{cm} = \frac{m_1\vec{x}_1 + m_2\vec{x}_2 + \cdots + m_n\vec{x}_n}{M_T} \qquad (7\text{-}8b)$$

The centre of mass in more than one dimension is given by

$$\vec{r}_{cm} = \left(\frac{\sum_{i=1}^{n} m_i\vec{r}_i}{\sum_{i=1}^{n} m_i} \right) \qquad (7\text{-}9b)$$

Continuous Distribution

$$\vec{r}_{cm} = \frac{\int_v \vec{r}\, dm}{\int_v dm} \qquad (7\text{-}9c)$$

External Force and Centre of Mass

$$\frac{d\vec{p}_{cm}}{dt} = \vec{F}_{net} \qquad (7\text{-}12a)$$

or

$$\frac{d\vec{p}_{cm}}{dt} = \vec{F}_{ext} \qquad (7\text{-}12b)$$

Internal forces do not affect the momentum of a system of particles:

$$\text{if } \vec{F}_{ext} = 0, \quad \text{then} \quad \frac{d\vec{p}_{cm}}{dt} = 0 \Rightarrow \vec{p}_i = \vec{p}_f \qquad (7\text{-}12c)$$

Rocket Propulsion

The ideal rocket equation is

$$v_f - v_i = -v_{er}\ln\left(\frac{M_f}{M_i}\right) = v_{er}\ln\left(\frac{M_i}{M_f}\right) \qquad (7\text{-}34b)$$

The mass flow rate is

$$\frac{dM}{dt} = R \qquad (7\text{-}35)$$

Rocket engine thrust is given by

$$Ma = Rv_{er} \qquad (7\text{-}36b)$$

COLLISIONS

Inelastic Collisions

The momentum of a system comprised of colliding particles is conserved:

$$\vec{p}_i = \vec{p}_f$$

Kinetic energy is not conserved in an inelastic collision.

Elastic Collisions

The momentum of a system comprised of colliding particles is conserved:

$$\vec{p}_i = \vec{p}_f$$

Kinetic energy is conserved in an elastic collision:

$$K_i = K_f$$

Applications: rocket propulsion, collisions

Key Terms: centre of mass, conservation of momentum, elastic collision, external forces, ideal rocket equation, inelastic collision, impulse, internal forces, linear momentum, point mass, rocket propulsion

QUESTIONS

For all questions and problems involving springs, assume that the masses of the springs are negligible unless otherwise stated.

1. To double the momentum of a particle, you could
   (a) double the particle's mass;
   (b) double the particle's speed;
   (c) quadruple the particle's kinetic energy.
   (d) All of the above.

2. A bullet is stopped by a wall. Which of the following statements is true?
   (a) The change in the bullet's momentum is opposite in direction to the incoming momentum.
   (b) The change in the bullet's momentum is equal to the incoming momentum.
   (c) Both (a) and (b) are correct.
   (d) None of the above.

3. A car makes a three-quarter turn around a traffic circle at a constant speed. The car is originally moving in the positive x-direction. Which of the following indicates the car's change in momentum?
   (a) $mv\hat{i} + mv\hat{j}$
   (b) $mv\hat{i} - mv\hat{j}$
   (c) $-mv\hat{i} + mv\hat{j}$
   (d) $-mv\hat{i} - mv\hat{j}$

4. A hockey puck of mass m moves at speed v on ice. A hockey player hits the puck such that its speed stays the same, but the puck is now moving at 60° with respect to its incident direction. What is the magnitude of the impulse?
   (a) zero
   (b) $mv \sin(30°)$
   (c) mv
   (d) $mv \cos(30°)$

5. If the puck in question 4 were moving along the positive x-axis before the shot, then the final momentum can be written as
   (a) $\dfrac{mv}{2}\hat{i} - \dfrac{\sqrt{3}mv}{2}\hat{j}$
   (b) $\dfrac{\sqrt{3}mv}{2}\hat{i} + \dfrac{mv}{2}\hat{j}$
   (c) $\dfrac{\sqrt{3}mv}{2}\hat{i} - \dfrac{mv}{2}\hat{j}$
   (d) None of the above.

6. For the puck in question 5, the impulse is
   (a) $\dfrac{mv}{2}\hat{i} + \dfrac{\sqrt{3}mv}{2}\hat{j}$
   (b) $-\dfrac{mv}{2}\hat{i} + \dfrac{mv}{2}\hat{j}$
   (c) $-\dfrac{mv}{2}\hat{i} + \dfrac{\sqrt{3}mv}{2}\hat{j}$
   (d) $\dfrac{mv}{2}\hat{i} - \dfrac{\sqrt{3}mv}{2}$

7. A bouncy ball of mass $m = 100$ g and speed $v = 10$ m/s is incident on a wall at an angle of 30°. It then bounces off the wall at the same speed, still making an angle of 30° with the wall after the collision. The collision takes 1 ms. The average force experienced by the ball during the collision is

   (a) 1000 N into the wall;
   (b) 1000 N out of the wall;
   (c) zero because the speed does not change;
   (d) $\sqrt{3}(1000)$ N parallel to the wall.

8. A bouncy ball of mass m and speed v impacts on a wall at an angle of 60°. The component of its velocity perpendicular to the wall does not change, and the angle it makes with the wall after the collision is 30°. Is this possible? Explain.

9. A slightly deflated ball hits a wall at an angle of 60° and leaves at an angle of 30°. The ratio of the outgoing momentum to the incoming momentum is
   (a) $\dfrac{\sqrt{3}}{2}$
   (b) $\dfrac{1}{\sqrt{3}}$
   (c) $\dfrac{2\sqrt{13}}{13}$
   (d) None of the above.

10. A sponge ball strikes a wall and leaves with a speed that is $\dfrac{\sqrt{13}}{4}$ of its incoming speed. If the incident ball makes an angle of 30° with the wall, then the component of its momentum perpendicular to the wall changes by factor of
    (a) $\dfrac{\sqrt{3}}{2}$
    (b) $\dfrac{\sqrt{13}}{4}$
    (c) $\dfrac{1}{2}$
    (d) None of the above.

11. You balance a firecracker on the tip of a needle and then light it. What happens to the firecracker's centre of mass when it explodes?

12. When you drop a tennis ball onto Earth's surface, the collision is completely elastic and lasts a few milliseconds. Will the ball bounce back to the same height? Explain.

13. Assuming Earth does not move at all during the fall of a bouncy ball and the subsequent collision with Earth's surface, would you expect the bouncy ball to go back to the same height it fell from if the collision is completely elastic? Explain.

14. How can you explain that a bouncy ball, in a completely elastic collision with Earth's surface, bounces back to exactly the same height it fell from despite the fact that the force of gravity acts on the ball during the collision? Would the statement that the ball has a lesser rebound speed than the speed it impacted with be true?

15. You drop a 30 g bouncy ball from a height of 11 m onto a flat, horizontal surface. The ball is in contact with the surface for 20 ms. Find the average force Earth exerts on the ball if the ball rebounds with the same speed. When would the ball rebound with a different speed?

16. Suppose Earth were the only planet in the solar system. As Earth orbits the Sun,
    (a) the Sun would remain stationary;
    (b) the Sun would always move in the same direction as Earth;
    (c) the Earth–Sun centre of mass would move back and forth;
    (d) both Earth and the Sun would orbit the stationary Earth–Sun centre of mass.

17. If one of the stars in a binary system explodes without directly damaging the second star, immediately after the explosion
    (a) the centre of mass of the two stars will move closer to the star that remained intact;
    (b) the centre of mass of the two stars will be at the centre of the intact star;
    (c) the centre of mass will shift, depending on the explosion;
    (d) the orbit of the intact star will not be affected by the explosion if the intact star does not does not absorb any material from the other star.

18. If one of the stars in a binary star system explodes, and the intact star absorbs a good portion of the material from the first star, then
    (a) the centre of mass of the exploding star is not affected by the explosion;
    (b) the centre of mass of the two-star system is not affected by the explosion;
    (c) the centre of mass of the two-star system will move closer to the intact star as the material from the exploding star gets absorbed by intact star.

19. A spring is held compressed between two masses, one having 10 times the mass of the other. The spring is then released. When the spring is at maximum extension,
    (a) the smaller mass is moving toward the larger mass with 9 times the speed of the larger mass;
    (b) the larger mass is moving away from the smaller mass;
    (c) the two masses are stationary;
    (d) only the larger mass is stationary.

20. When you throw a ball straight down so that it bounces perfectly elastically from the surface of Earth, the duration of the impact will dictate how high the ball goes. True or false? Explain.

21. A flamingo lands vertically on the back of a floating hippopotamus, then takes a few steps along the hippopotamus's back toward its head. Which of the following statements is true?
    (a) The hippopotamus will move in a direction opposite to the flamingo's motion.
    (b) The hippopotamus will not move.
    (c) The centre of mass of the hippopotamus–flamingo system will move in the same direction that the hippopotamus moves.
    (d) The centre of mass of the hippopotamus–flamingo system will move in the same direction as the flamingo's motion.

22. A fish is stuck in a closed bottle that is half-filled with water. The bottle floats on its side in a still lake. The fish swims to the left. Which of the following is true?
    (a) The bottle will move to the left.
    (b) The bottle will move to the right.
    (c) The bottle will not move at all.
    (d) The centre of mass of the fish, water, and bottle will move in the same direction as the fish.

23. A bar is held vertically on ice. You let the bar drop such that friction prevents its rough bottom edge from moving. When the bar is horizontal there is little friction between the smooth side of the bar and the ice. After the bar falls down completely,
    (a) it moves on the ice;
    (b) it is stationary;

    (c) its centre of mass has not moved at all.
    (d) None of the above.

24. A child fits a spring on the front of a stationary block of wood and places it on a frictionless surface. He then throws another block at it such that the spring cushions the blow. After a few moments, the two blocks are moving at the same speed. Which of the following statements is true?
    (a) The collision is elastic; therefore, the kinetic energy of the system does not change.
    (b) Since a spring is involved, we cannot say whether the momentum of the two masses is conserved during the collision.
    (c) The kinetic energy of the system changes during the collision.
    (d) The total mechanical energy of the system is not conserved.

25. A child has a spring-loaded cannon that launches small marbles horizontally. The cannon is mounted on little skis and is placed on an icy surface in the backyard during the winter. When the marble is in the air, the cannon moves as a result of firing. Which of the following statements is true?
    (a) The sum of the kinetic energy of the cannon and the flying marble is not equal to the energy that was stored in the spring, so the collision is inelastic.
    (b) The sum of the kinetic energies of the cannon and the marble is equal to the energy stored in the spring.
    (c) The spring will exert a greater force on the marble than it does on the cannon.
    (d) None of the above are true.

26. You are standing on a sled in the Arctic and decide to jump forward off the sled such that your horizontal speed with respect to the sled is 4 m/s. Which of the following statements is correct? (Hint: Distinguish between speed and velocity.)
    (a) Your speed relative to the sled is the sum of both speeds.
    (b) Your velocity vector with respect to the sled is defined as the sum of both velocity vectors.
    (c) Your velocity relative to the ground is the same as your velocity relative to the sled.
    (d) Your velocity relative to the ground is the sum of the velocity of the sled and your velocity with respect to the sled.

27. A toy car is moving forward. The car has a mounted cannon that fires horizontal projectiles backward. Which of the following statements is true?
    (a) The projectiles must be moving backward relative to the ground.
    (b) The projectiles must move forward with respect to the ground.
    (c) Could be (a) or (b), depending on the specifics.
    (d) None of the above are true.

28. You go for a sprint across a soccer field, heading east. Assuming that you are the only mobile object on Earth's surface, Earth's rotation would be
    (a) slower than usual;
    (b) faster than usual;
    (c) unaffected;
    (d) faster or slower, depending on the frame of reference.

29. A cannon fires a probe into the Moon such that the probe leaves Earth's surface at 90° to the surface. Which of the following statements is true?
    (a) Earth will not be affected at all.
    (b) Earth will recoil slightly.

(c) Since the force of gravity is involved, momentum is not conserved.

(d) None of the above are true.

30. A child jumps onboard a stationary wagon. The losses due to friction between the wagon and the floor and in the wagon's wheels are negligible. Neither the child's body nor the wagon sustains an irrecoverable deformation, and the child and the wagon move as one. Which of the following statements is true?

(a) The collision is elastic because there is no irrecoverable deformation.

(b) The two objects stick to one another, so the collision must be inelastic.

(c) The kinetic energy of the child is reduced by an amount that is smaller than the kinetic energy gained by the wagon.

(d) The child's momentum is reduced by an amount that is larger than the amount by which the wagon's momentum increases.

31. For question 30, is the following statement true (T) or false (F): The only mechanism for the kinetic energy to be lost is friction between the child and the wagon. (Ignore losses to sound.)

32. A man standing on a wagon that is originally at rest begins to run. If there are no losses in the wagon motion to friction,

(a) the momentum of the man relative to the ground is the same in magnitude as the momentum of the wagon with respect to the ground;

(b) the momentum of the man with respect to the wagon is the same in magnitude as the momentum of the wagon with respect to the ground;

(c) the speed of the man with respect to the ground is greater than the speed of the man with respect to the wagon;

(d) it depends on the friction between the man and the wagon.

33. A well-behaved cat goes on a canoeing trip with her owners. She is left in the canoe as they set up camp. She sees a fish jump in front of the canoe and leaps forward to try to catch it. The canoe is originally at rest, and we take the direction of motion of the canoe with respect to ground to be negative. Which of the following statements is true?

(a) The speed of the cat with respect to the ground is positive.

(b) The speed of the cat with respect to the ground is negative.

(c) Whether (a) or (b) is correct depends on how fast the canoe moves.

(d) Whether (a) or (b) is correct depends on the ratio of the cat's mass to the canoe's mass.

34. A sled is moving on horizontal frictionless ice when a dog onboard leaps onto the ice, leaving the sled horizontally backward. The direction of motion of the sled is taken to be positive. Whether the velocity of the dog with respect to the ground is positive or negative depends on

(a) the masses of the sled and the dog;

(b) the original speed of the sled and the relative speed of the dog with respect to the sled;

(c) both (a) and (b).

(d) None of the above are true.

35. A rocket is fired vertically upward from Earth's surface. If you were to choose a system in which momentum is conserved during the process, the most correct system to pick would be

(a) rocket + fuel/exhaust;

(b) rocket + fuel/exhaust + atmosphere;

(c) rocket + fuel/exhaust + atmosphere + Earth;

(d) rocket + fuel/exhaust + atmosphere + Milky Way galaxy.

36. A heavy cannon fires a shell horizontally. Which of the following statements is true?

(a) It is possible for the cannon not to recoil.

(b) If the cannon does not recoil at all, then the ground it is sitting on must move a little.

(c) If the ground does not move, then Earth's crust must move on the mantel.

(d) Both (b) and (c) are correct.

37. A canoeist notices her canoe drifting toward a waterfall. She has a battery-powered water pump fitted with a hose that she can use to direct the water from the pump. The pump draws water vertically up from the river. Which of the following statements is true?

(a) If the pump shoots out enough water at a high enough speed, the canoeist can angle the hose so that she can move away from the waterfall.

(b) The pump cannot propel the canoe because the momentum of the system must be conserved.

(c) Conservation of momentum does not apply here because the pump has a power source.

(d) None of the above are true.

PROBLEMS BY SECTION

For problems, star ratings will be used, (∗, ∗∗, or ∗∗∗), with more stars meaning more challenging problems.

Section 7-1   Linear Momentum

38. ∗ Calculate the linear momentum of a space shuttle of mass 2040 tonnes moving at a speed of 27 000 km/h.

39. ∗ Find the momentum of a 60 g stone that starts from rest and falls to the ground from a height of 20 m.

Section 7-2   Rate of Change of Linear Momentum and Newton's Laws

40. ∗∗ The engine of a fire rescue boat gets damaged during a rescue operation. The captain decides to use the fire-fighting water-pumping system to push the boat back to the harbour. The pump is mounted on the deck and draws the water vertically out of the ocean. The crew points the nozzle of the hose toward the rear of the boat. The mass of the boat is 180 000 kg, and the pump sprays water at a rate of 330 kg/min with a muzzle speed of 20 m/s. Ignoring water resistance, find the acceleration of the rescue boat.

41. ∗∗ An athlete of mass 74 kg is running inside a train car of mass 2100 kg, at a speed of 9.0 m/s with respect to the ground. While the athlete is running, the car is stationary. The athlete then comes to a stop over 0.20 s.

(a) Find the speed of the train car when the athlete comes to a full stop on the car.

(b) The athlete comes to a stop over 0.2 s. During this time, the force of kinetic friction is the only force between her feet and the train car. Find the average force exerted on the train car by the athlete.

Section 7-3 Impulse and the Conservation of Linear Momentum

42. ✷ A soccer player kicks a 425 g soccer ball straight on with an average force of 78.0 N. The speed of the soccer ball right after the collision is 41.0 m/s. Find the duration of the collision. Ignore the effects of gravity.

43. ✷ Tennis balls, whose individual masses are 58 g, are fired at a wall by a tennis-ball launcher such that they impact the wall horizontally in perfectly elastic collisions at a speed of 117 km/h.
 (a) The launcher fires the balls at a rate of 55 balls/min. Find the average force exerted on the wall as a result.
 (b) The average collision time is 25 ms. Find the average force per collision.

44. ✷✷ A somewhat deflated volleyball ($m = 220$ g) falls on the floor from a height of 2.0 m, starting from rest. The collision between the ball and the floor takes 0.30 s and is perfectly elastic.
 (a) Find the speed of the ball on the way up after the impact.
 (b) Write an expression for the net force acting on the ball using the linear behaviour model.
 (c) Sketch the net force acting on the ball as a function of time, using the linear behaviour model.
 (d) Find the maximum value for the net force.

45. ✷✷ Starting from rest, a 25 g bouncy ball falls from a height of 10 m onto a flat, horizontal surface and rebounds with 99% of its incident speed. Find the amount of time that the ball was in contact with the ground if the average net force on the ball during the collision is 45 N.

46. ✷ Figure 7-18 shows the net force versus time profile for an object colliding with a horizontal surface. The object's mass is 300 g, and its incident speed is 13 m/s. Find the recoil speed.

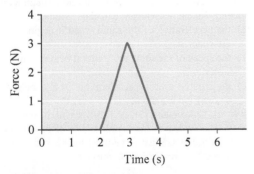

Figure 7-18 Problem 46

Section 7-4 Systems of Particles and Centre of Mass

47. ✷ Sixteen identical point masses are placed at equal intervals such that they outline the circumference of a circle of radius R. By how much does the centre of mass shift when one of those masses is removed?

48. ✷ Find the centre of mass of the particles shown in Figure 7-19.

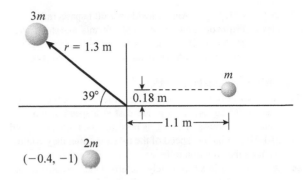

Figure 7-19 Problem 48

49. ✷ Masses m, $2m$, $3m$, and $4m$ are located at the corners of a square of side L as shown in Figure 7-20. Locate the centre of mass of this system.

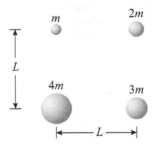

Figure 7-20 Problem 49

50. ✷ A pendulum is fashioned out of a thin bar of length 0.55 m and mass 1.9 kg, welded to a sphere of radius 0.11 m and mass 0.86 kg. Find the centre of mass of the composite object.

Section 7-5 Newton's Laws Applied to Systems of Particles

51. ✷✷ A 4.0 kg dog runs at constant speed from one end of a 21 kg canoe to the other, a distance of 6.7 m, in 3.1 s. Assuming negligible water resistance, how far does the canoe move?

52. ✷✷ An astronaut finds herself stranded on a 43 kg platform designed for short hops around a space station, when the platform loses its propulsion system. The platform is moving away from the space station at a speed of 2 m/s and is oriented such that the astronaut's head points toward the station as she stands straight. She then decides to kick her feet down in the hopes of acquiring enough momentum to return to the station. If the mass of the astronaut with her spacesuit is 105 kg, what must the astronaut's minimum speed relative to the platform be if she is to succeed?

53. ✷✷ A child starts to throw a water balloon at a friend, but the balloon bursts while still in the child's hand. At that instant, the balloon is 90 cm above the ground and moving horizontally at a speed of 2.3 m/s. Where is the centre of mass of the water when all the water has hit the ground? (The balloon is very light.)

54. ✶✶ A 71.0 kg astronaut inside a 2.00 t spacecraft at rest kicks off from one end at a speed of 2.50 m/s, and stops when he reaches the other end having travelled the full length of the 9m-long spacecraft. How far does the spacecraft move?

Section 7-6   Inelastic Collisions

55. ✶ Two huskies collide on frictionless ice. Anyu, with a mass of 27 kg, is heading due east at a speed of 6.5 m/s. Anut, with a mass of 21 kg, is heading due north at a speed of 4.0 m/s. Find the speed of the two dogs after they collide. Assume the collision to be completely inelastic.

56. ✶ Two cars in a completely inelastic head-on collision are at rest immediately after the collision. The mass of one car is 1.7 times the mass of other car. Find the ratio of their initial speeds.

57. ✶✶ An 11 g bullet is fired with speed v into a 68 kg ballistic pendulum. The wires holding the pendulum deviate from the vertical by 27°. The vertical distance from the point of impact to the suspension points for the wires is 92 cm. Find the speed with which the bullet hits the pendulum.

58. ✶✶ A tennis ball hits a wall at an angle of 48° with respect to the wall and loses 25% of its energy in the collision. Find the angle the ball makes with the wall on the way out.

59. ✶✶✶ Derive an expression for the fractional loss of energy for an object that impacts a wall at an angle θ and leaves it an angle θ'.

60. ✶ Two hockey players race for the puck moving approximately in the same direction. Player 1 has a mass of 69 kg and a velocity 5.6 m/s at an angle of 32° with the rink board. Player 2 has a mass of 78 kg and a velocity of 7.1 m/s at an angle of 16° with the same board. The players are locked in a completely inelastic collision. Find the energy lost in the collision.

61. ✶✶ A medicine ball of mass m and speed v is sliding on frictionless ice when it strikes a stationary medicine ball of mass $5m/4$. After the collision, the kinetic energy of the first ball is three-quarters of its initial energy, and this ball moves at an angle of 21° with the incident direction. The second medicine ball moves in a direction that makes an angle of 43° with the final momentum of the first ball. Express the speed of the second ball in terms of v.

62. ✶✶ Two balls collide head-on; one ball (m_1) has twice the speed of the other ball (m_1). The collision is completely inelastic. Derive an expression for the speed of the two balls after the collision.

Section 7-7   Elastic Collisions

63. ✶✶ A 3.0 kg mass is attached to a spring of spring constant $k = 185$ N/m resting on a horizontal frictionless surface. The other end of the spring is fixed to a wall. A 1.6 kg mass is thrown toward the 3.0 kg mass and collides with it in a perfectly elastic collision. The maximum compression of the spring is 80 cm. Find the speed of the incoming mass.

64. ✶✶ A billiard ball of mass m is shot at speed v to strike two adjacent stationary billiard balls with the same mass as the incident ball. The two stationary balls are in contact, and the incoming ball strikes them along the line connecting their centre of mass. The collisions are perfectly elastic. Derive an expression for the final speed of each ball.

65. ✶✶✶ Two balls, one of which is stationary, collide elastically. The speed of one ball after the collision is half the speed of the other. Find the ratio of the masses.

66. ✶✶✶ A moving ball (speed v) collides with another at rest. After the collision, the first ball is at rest. Find the ratio of their masses if the collision is perfectly elastic. What is the speed of the second ball?

67. ✶✶✶ Two balls collide elastically head-on; one ball has mass m and is moving at speed v, the other ball is moving at $3v/2$. After the collision, the lighter mass moves at speed $2v$. Find the ratio of the masses and express the final speed of the heavier ball in terms of v.

68. ✶✶ A billiard ball of mass m and speed v strikes another ball of mass M and speed u as shown in Figure 7-21. The collision is perfectly elastic. Derive expressions for the speed of each ball after the collision.

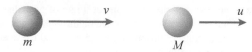

Figure 7-21   Problem 68

Section 7-8   Variable Mass and Rocket Propulsion

69. ✶✶ A rocket is launched vertically. The fuel constitutes half the mass of the rocket. After 21.0 s of flight, half the fuel is burned and the speed of the rocket is 190 m/s. The rocket burns fuel at a rate of 72.0 kg/min. Find the thrust of the rocket and the relative speed of the exhaust with respect to the rocket. Ignore air resistance.

70. ✶✶ A light rocket's engine burns fuel at a rate of 65 kg/s. The rocket is launched in deep space. Twenty seconds later, only half the mass of the fuel is left, and the rocket has reached a speed of 9300 km/h. Ignore the mass of the rocket.
   (a) Find the original mass of the rocket.
   (b) Find the average acceleration of the rocket over the first 10 s.
   (c) Find the rocket's thrust.

71. ✶✶ A giant squid of mass 230 kg propels itself horizontally by swallowing 56 kg of water and then expelling it in 6.0 s. The squid is initially at rest with zero buoyancy. Assuming that the water resistance is negligible and that the squid expels the water at a constant rate through a circular opening 7.0 cm in diameter, calculate estimates for
   (a) the acceleration of the squid when it begins expelling the water
   (b) the speed of the squid 4.0 s after it begins expelling the water
   (c) the squid's average acceleration as it expels the last sixth of the water

72. ✶✶ A 72 kg man exhales deeply trying to blow a wasp away. The air leaves his mouth at rate of 0.70 L/min at a speed of 9.0 m/s. What force must his neck muscles exert on his head to prevent it from moving?

COMPREHENSIVE PROBLEMS

73. ✶✶ A planet orbits its star in an elliptical orbit, with a major-to-minor axis ratio of 1.4. The mass of the planet is 1/100 the mass of the star. The major axis is 7 billion kilometres. Predict the length of the long axes for the orbit of the star. Draw a scale diagram of the two orbits. Assume that there are no other objects in the vicinity.

74. ✳✳✳ A 2.0 m-long bar is held vertical on a frictionless surface. The top is tipped very gently, and the bar begins to swing. What is the speed of the bar when it is fully horizontal? Assume all potential energy becomes kinetic energy of the centre of mass, and the bar moves on the surface after falling down.

75. ✳ A bar of length 1.16 m is tilted 43.0° from the vertical. Find the horizontal displacement of the bar's centre of mass.

76. ✳✳ A child of mass 34 kg is standing on a parade wagon (empty mass $M = 120$ kg), which is loaded with 600 gift baskets, each with a mass of 500 g. She throws the baskets off the side one at a time, such that they leave in a horizontal direction, making an angle of 57° with the length of the wagon. There are no frictional losses between the wagon and the road, and the wagon can move only forward and backward. The wagon is originally at rest. By the time the 70th basket is thrown, the speed of the wagon is 2.1 m/s. Find the magnitude of the relative *velocity* of the baskets with respect to the wagon. Assume this velocity is constant.

77. ✳✳ A 5 g bullet is fired at a speed of 265 m/s into a 9.2 kg watermelon sitting on a thin vertical rod. The bullet shatters the watermelon and leaves with a speed of 235 m/s. Find the speed of the centre of mass of the watermelon immediately after the collision. What is the acceleration of the centre of mass of water melon after the collision?

78. ✳✳✳ A child mounts a toy spring cannon on a sled, and takes this toy out to play on the icy surface of a nearby lake. The mass of the cannon plus sled is M. The projectile that the cannon shoots has a mass m and leaves with a muzzle speed v. The cannon faces forward at an angle θ with respect to the horizontal. The sled-mounted cannon is moving at speed v_c when the projectile is launched.
   (a) Derive an expression for the speed of the cannon immediately after it launches the projectile.
   (b) Find the speed of the cannon, given that $M = 915$ g, $m = 32.0$ g, $v = 17.0$ m/s, $v_c = 5.00$ m/s and $\theta = 41°$.

79. ✳✳✳ Starting from rest, a 2.4 t elephant in a 7.4 t boxcar runs at a speed of 3.0 m/s with respect to the boxcar and then throws herself against a spring with $k = 17.5$ kN/m mounted on the other end of the boxcar, thus compressing the spring. The boxcar is mounted on smooth rails, so losses due to friction as the car moves are minimal. Initially the spring is neither stretched nor compressed and the boxcar is at rest, and the spring is unstretched.
   (a) Find the maximum compression of the spring.
   (b) What is the speed of the boxcar when the spring is at maximum compression?

80. ✳✳✳ An engineer working with an expedition in the Arctic needs to move a heavy rectangular package across the frozen ground. Initially, the package is standing on its edge. He puts two pegs in the ground at one edge of the package and lets the package drop by tilting it gently until it begins to fall. The package has a mass of 120 kg, and travels a total distance of 25 m on the ground, where the coefficient of friction is 0.05. How tall is the package?

81. ✳ A child of mass 19 kg lands on a 34 kg wagon. The wagon then moves with a speed of 2.11 m/s. Find the horizontal speed with which the child lands.

82. ✳✳ Two balls, one twice the mass of the other, are thrown straight up in the air at a speed of 20 m/s, 0.70 s apart. The heavier ball is thrown second.

(a) Where is the centre of mass of the two-ball system 0.20 s after the second ball is thrown?
(b) What is the acceleration of the centre of mass at the following times: (i) before the second ball is thrown, (ii) when the two balls are moving up, (iii) when the lighter ball reaches its maximum height, and (iv) 0.10 s after the lighter ball lands?

83. ✳✳ You are sitting on one end of a canoe ($m = 19$ kg) in calm water. You then throw your 15 kg backpack to your friend at the other end of the canoe, a distance of 3.0 m. Your friend's mass is 65 kg, and your mass is 72 kg. How much would the canoe move if the water resistance were negligible?

84. ✳✳✳ A research team in Antarctica is launching a 32 kg weather probe into the air above the Ross Ice Shelf. They use a special cannon of mass 1600 kg, which launches the probe with a muzzle speed of 710 m/s. The cannon points at an angle of 57° above the horizontal. The team forgets to secure the cannon, which slides backward on the ice after the probe is fired. Find the angle that the probe makes with respect to the ground when it lands. Ignore air resistance, and assume that the probe lands at the same altitude from which it was launched.

85. ✳✳ A low Earth orbit observation station, made of 11 modules each of mass 13 000 kg, moves at a speed of 4.0 km/s. It is equipped with an emergency collision avoidance system whereby an explosive charge can be detonated to release one of the modules. The collision alarm sounds, and the commander detonates a charge, releasing one of the modules at a speed of 1400 km/h with respect to the station, directed opposite to the original direction of travel of the station. Find the resulting change in the speed of the station.

86. ✳✳ During the Iditarod race in Alaska, your sled gets detached from your snow dogs. The sled ($M = 200$ kg) moves at 4.1 m/s, headed for thin ice. You ($m = 70$ kg) jump backward off the sled with a speed of 2.7 m/s with respect to the sled. What is your horizontal speed with respect to the ground while you are in the air?

87. ✳ An astronaut floating in space holds one end of a 20 m-long bar with one 40 kg package on each end. If she rotates the bar by 180°, by how much does she move?

88. ✳✳✳ A man stands on a relatively long sled resting on snow. The sled, of mass M, is tethered with a bungee cord, which can be treated like a spring of spring constant k. The man then moves on the sled toward the bungee cord end, at speed v with respect to the sled. The maximum extension of the cord is L. Derive an expression for the coefficient of friction between the sled and the snow.

89. ✳ A child launches a fireworks rocket at an angle θ from the vertical. When the rocket is at its highest point, its two parts separate by detonating a charge such that the lighter part falls straight down. The speed of the rocket before the separation is v, and the lighter part is one-third the mass of the heavier part. Derive an expression for the speed of the heavier part after the separation.

90. ✳✳ A hockey puck on a frictionless ice strikes the rink board at an angle of 65° with respect to the normal to the board. The collision is partly elastic, and the puck loses 17% of its kinetic energy. Find the angle that the puck makes with the normal to the board after the collision.

91. ✳✳✳ A performance-testing device is made of a long cart of length L and mass M, connected to a wall by a spring of spring constant k. Athletes stand on the end of the cart farthest from the spring and run at maximum speed toward the end where the spring is attached. The maximum extension of the spring is X. Derive an expression for the maximum speed of an athlete of mass m.

92. ✳✳ A polar bear cub is sliding toward his sister on frictionless ice in the Arctic. The sister is stationary before the collision. After the collision, the brother is stationary. Find the ratio of their masses.

93. ✳✳ An archer shoots an arrow of mass 110 g through a vertical plank of cork that is firmly fixed in place on top of a stationary cart. The plank is parallel to the length of the cart, and the cart is free to move on wheels in the forward direction with frictionless bearings. The incident arrow is horizontal when it strikes the vertical plank and makes an angle of 40° with the plane of the plank. When the arrow emerges, it makes an angle of 43° with respect to the now moving plank. The combined mass of the plank and the cart is 3.78 kg, and the incident speed of the arrow is 90 m/s.
    (a) Find the energy lost during the collision.
    (b) Find the speed of the cart immediately after the collision.

94. ✳✳ A pellet gun is used to shoot 0.58 g pellets at a speed of 280 m/s into a piece of wood of mass M, resting on a horizontal frictionless surface. The surface then curves upward into the shape of a hemispherical bowl of radius $R = 70$ cm. The block then moves up the bowl-shaped surface until its angular position measured from the vertical is 37°. The pellet gets completely lodged in the piece of wood. Find M.

95. ✳✳ You are standing on top of Mount Everest with a solid ball made out of neutrons (density 4×10^{17} kg/m³). The ball is 20 cm in radius. You then let the neutron ball roll down to the foothills, 7000 m below.
    (a) How far would Earth have moved along the line connecting you to the centre of Earth? Take Earth's mass to be 6×10^{24} kg.
    (b) Assume there were no frictional losses, and determine the change in Earth's speed.
    (c) If the collision is completely elastic, find the rebound speed of Earth from the point of view of an observer stationary with respect to the centre of mass of Earth–neutron ball system.

96. ✳✳✳ A solid disk of radius R and mass M has a hollow part of radius r, centred at distance x from the centre of the otherwise solid disk, as shown in Figure 7-22. Find the centre of mass of the disk.

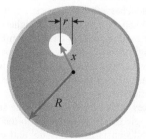

Figure 7-22 Problem 96

97. ✳✳✳ A bar has a linearly increasing density, such that the density at one end is five times the density at the other end. Find the location of the centre of mass as measured from the lighter end.

98. ✳✳✳ A figure skater of mass $3/2\ m$ is holding on to the middle of a bar that connects two other figure skaters of masses m and M. The figure skater in the middle then spins the bar at a constant angular speed. Derive an expression for the radius of the circular trajectory that the centre figure skater will undergo.

99. ✳✳ A spring ($k = 430$ N/m) is compressed 9.0 cm between two masses, $m = 910$ g and $M = 1.4$ kg. While the system moves along a frictionless surface at a constant speed v, the spring is released. What is the speed of the lighter mass when the spring is stretched 7.0 cm?

100. ✳✳✳ A child has a toy made of a spring (spring constant k) connected to a base of mass M to a bobbing monkey head of mass m. When the spring is fully cocked, it is compressed 9.0 cm from equilibrium. The child sits her toy on its base such that the spring is vertical and the bobbing monkey head is attached to the other end. The child then presses a release knob that releases the compressed spring. Find the maximum height reached by the base after the spring is released using the following values: $M = 87$ g, $m = 29$ g, and $k = 92$ N/m.

101. ✳✳ A fire breaks out on the fourth floor of a building. Firefighters bring in a platform of mass 300 kg propped up on four springs, each with a spring constant of 410 N/m. A 67 kg man jumps from a height of 7.0 m above the platform and lands on the platform in such a way that he stays with the platform. Find the maximum compression of the springs.

102. ✳✳✳ You drop a small ball with a mass of 32 g from the roof of a building 40 m above the ground. One second later, you throw a volleyball with a mass of 226 g down with an initial speed such that it hits the small ball 5.0 m above the ground. All the collisions are elastic.
    (a) Find the speed of the small ball immediately after the collision.
    (b) Find the speed of the small ball after it collides with the ground.

103. ✳✳✳ Mass m_1 moves with speed v_1 toward stationary mass m_2. One end of a spring of spring constant k and unstretched length L is attached to the front of mass m_1. Mass m_2 sticks to the other end of the spring when they make contact.
    (a) Derive an expression for the maximum distance between the two masses after the collision.
    (b) Calculate this distance using the values $m_1 = 100$ g, $m_2 = 300$ g, $v_1 = 6.00$ m/s, $L = 28.5$ cm and $k = 50.0$ N/m.

104. ✳✳✳ Where is the centre of mass of the Earth-Moon system relative to the surface of Earth?

105. ✳✳✳ A block of mass m is fitted with a horizontal spring (spring constant k) at on end. The mass is made to move at a speed v, when it strikes a another block of mass M that was moving toward it at speed u. The collision is cushioned by the spring. Derive expressions for
    (a) the maximum compression of the spring;

(b) the speed of m when the spring is at its maximum compression;

(c) the final speed of each mass;

(d) the speed of each mass at maximum spring compression.

106. ✱✱✱ Estimate how far the crew of a submerged submarine could move the vessel by all gathering at one end of it.

107. ✱✱ A child places a spring between two of his toy cars. He holds the two cars together on a frictionless surface in such a way that they compress the spring. He causes the two cars with the spring in-between to move at speed v, keeping the spring compressed between the cars; then he lets the two cars go. Find the final speed of the car in the front. The spring constant is k, and the mass of each car is m.

108. ✱✱ A 7.45 g bullet moving at a speed of 1100 m/s strikes a 27.8 kg block of wood and emerges from the other end to become lodged in 2 kg plate. The plate rests on a horizontal frictionless surface and is attached to a horizontal spring ($k = 1340$ N/m) that is compressed 11 cm as a result of the collision. Find the speed of the block immediately after the bullet emerges from it.

109. ✱✱ Your cat, Boots ($m = 3.15$ kg), is standing in the middle of a cart with frictionless wheels. The cart is at rest on a frictionless surface. Boots then jumps to one end of the cart such that the cart is moving at 0.30 m/s while the cat is in the air. The cat is in the air for 0.8 s and the cart is 1 m long.

(a) Find the mass of the cart.

(b) What is the speed of the cart after the cat lands?

110. ✱✱✱ Two blocks with masses m_1 and m_2 are thrown toward each other on a horizontal frictionless surface with respective speeds v_1 and v_2, respectively. The collision between them is cushioned by a spring of spring constant k that is attached to one of the masses.

(a) Derive an expression for the maximum compression of the spring during the collision.

(b) Calculate the maximum compression in part (a) for $m_1 = 3.20$ kg, $m_2 = 5.24$ kg, $v_1 = 2.09$ m/s, $v_2 = 3.76$ m/s, and $k - 615$ N/m.

111. ✱✱✱ A 32 kg child runs at a speed of 3.00 m/s and jumps onto a wagon of mass $m = 167$ kg sitting on frictionless wheels. The child slides for a distance d on the wagon before coming to a stop relative to the wagon. If the coefficient of kinetic friction between the child and the wagon is 0.7, find a) the final speed of the cart, b) the distance the child slides on the wagon (relative to the wagon).

See the text online resources at
**www.physics1e.nelson.com for Open Problems
and Data-Rich Problems related to this chapter.**

Learning Objectives

When you have completed this chapter you should be able to:

1 Define angular quantities, and relate angular variables to linear variables.

2 Solve kinematics equations for rotation.

3 Define torque, and compare different representations of torque.

4 Explain the relationship between torque, moment of inertia, and angular motion.

5 Calculate moments of inertia for rigid objects, and solve equations of motion for rotation about a fixed axis using Newton's laws.

6 Use the work–mechanical energy approach to solve rotational dynamics problems, and compare that approach to the force–torque approach for solving rotational dynamics problems.

7 Define angular momentum, conservation of angular momentum, and change of angular momentum in terms of Newton's second law.

8 Solve problems using conservation of angular momentum.

For additional Making Connections, Examples, and Checkpoints, as well as activities and experiments to help increase your understanding of the chapter's concepts, please go to the text's online resources at www.physics1e.nelson.com.

Chapter 8
Rotational Dynamics

The Crab pulsar, a pulsating neutron star within the Crab Nebula, is the result of a type II supernova, that was recorded by both Chinese and Arab astronomers in 1054 CE. The light from the supernova explosion was visible to the naked eye for close to 22 months in the night sky before fading from view. The bulk of the exploding star's substance was ejected during the supernova creating the crab nebula, the rest of the stellar mass was crushed together to form the neutron star. The original star was spinning when it exploded, and the mass that became the neutron star continues to spin.

The radius of this neutron star is only 10 km, a tiny fraction (likely less than 1/100 000) of the radius of the progenitor star. As a result, the Crab pulsar spins very fast: the rotational period is only 33.1 ms. The conservation of angular momentum, the principle behind the incredibly high rotational speed of the pulsar (also often observed during figure skating competitions) is covered in this chapter. A spinning neutron star emits electromagnetic radiation along the axis of its magnetic field. If the rotation axis and the magnetic axis do not coincide, the radiation beam is visible to a distant observer once during each revolution, much like the beam of light in a lighthouse. The pulsed nature of the beam and the radio frequency component of the radiation result in the name pulsating radio star, or pulsar.

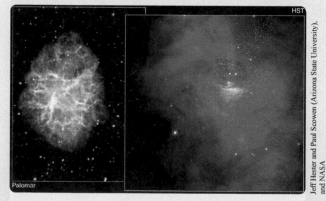

Figure 8-1 The Crab Nebula (left) and the shock waves that bombard it as a result of the radiation from the pulsar within (right)

Jeff Hester and Paul Scowen (Arizona State University), and NASA

8-1 Angular Variables

Consider a particle moving along the circumference of a circle of radius R. When the particle moves from point A to point B (Figure 8-2), the length of the arc travelled by the particle is

KEY EQUATION
$$\Delta s = R\Delta \theta \qquad (8\text{-}1)$$

where $\Delta \theta$ is the **angular displacement**.

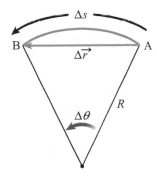

Figure 8-2 (a) Linear displacement: the displacement vector $\Delta \vec{r}$ is a straight line connecting A and B. (b) Tangential displacement: Δs is the full arc length covered by the particle.

In Equation (8-1), R is measured in metres, and θ is measured in radians, which are dimensionless. Hence, Δs has dimensions of length. However, R is a simple length (a line segment), and Δs is an arc length (a segment of a circle).

The differential, or infinitesimal, displacement along the circumference can be written in terms of the infinitesimal angular displacement:

$$ds = Rd\theta \qquad (8\text{-}2)$$

It follows that the instantaneous rate of change with respect to time of the position of the particle along the circumference is related to the instantaneous rate of change of the angular position:

$$\frac{ds}{dt} = R\frac{d\theta}{dt} \qquad (8\text{-}3a)$$

The derivative $\dfrac{ds}{dt}$ is the instantaneous linear speed, v, of the particle along the circumference, or the particle's instantaneous tangential speed (Figure 8-3).

The derivative $\dfrac{d\theta}{dt}$ is the instantaneous rate of change of the angular position with respect to time, or, more

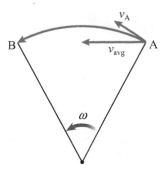

Figure 8-3 The instantaneous linear velocity is tangent to the circle, and the average velocity points directly along the straight line from A to B.

simply, the **angular speed**. The conventional symbol for angular speed is ω:

$$\omega = \frac{d\theta}{dt} \qquad (8\text{-}3b)$$

The SI units for ω are radians per second. Like linear speed, angular speed can be a function of time. Equation (8-3a), therefore, can be rewritten as follows:

KEY EQUATION
$$v(t) = \omega(t)R \qquad (8\text{-}3c)$$

Taking the derivative of Equation (8-3a) with respect to time, we have

$$\frac{d^2s}{dt^2} = R\frac{d^2\theta}{dt^2} \qquad (8\text{-}4a)$$

The derivative on the left-hand side of Equation (8-4a) is the tangential acceleration of the object, not to be confused with the radial acceleration (see Figure 8-4). The derivative on the right-hand side is the **angular acceleration**. The standard symbol for angular acceleration is α:

$$\alpha = \frac{d^2\theta}{dt^2} = \frac{d\omega}{dt} \qquad (8\text{-}4b)$$

Since a and α are not necessarily constant, they are shown as functions of time, t. Thus, Equation (8-4a) can be rewritten as follows:

KEY EQUATION
$$a_t(t) = \alpha(t)R \qquad (8\text{-}4c)$$

PEER TO PEER

An object moving with a constant angular speed has an angular acceleration of zero; therefore, the tangential acceleration is also zero. The radial, or centripetal, acceleration is not zero, however, and is given by $a_r = \dfrac{v^2}{r} = \omega^2 r$.

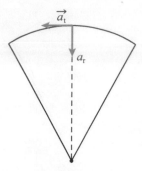

Figure 8-4   An object travelling at constant speed along the circumference has radial, or centripetal, acceleration but no tangential acceleration. If the speed also changes then the object will also have a tangential acceleration

ONLINE ACTIVITY

Ladybug Revolution

The e-resource that accompanies every new copy of this textbook contains an Online Activity using the PhET simulation "Ladybug Revolution." Work through the simulation and accompanying questions to gain an understanding of the relation between angular speed and the acceleration vector for an object moving in a circle.

EXAMPLE 8-1

Rotation of Earth

(a) Determine the values for the linear speed, angular velocity, tangential acceleration, and radial acceleration of a point at the equator. Assume that Earth is a sphere of radius 6340 km.
(b) Calculate how far the rotation of Earth moves a person sitting on an equatorial beach during a 2.00 h sunbath.

SOLUTION

(a) Earth completes a full revolution (2π rad) in about 24 h, so the angular speed is

$$\omega = \frac{2\pi\,\text{rad}}{(24\,\text{h})(3600\,\text{s/h})} = 7.2722 \times 10^{-5}\,\text{rad/s} = 7.27 \times 10^{-5}\,\text{rad/s}$$

to 3 significant figures

The linear speed for a point on the equator is given by Equation (8-3):

$$v = \omega R = (7.2722 \times 10^{-5}\,\text{rad/s})(6340\,\text{km})$$

$$= 461.1\,\text{m/s} = 461\,\text{m/s}$$

to 3 significant figures

Earth's angular velocity is nearly constant; the tangential acceleration as given by Equation (8-4c) is approximately zero.

The radial, or centripetal, acceleration is given by Equation (4-11):

$$a_r = \frac{v^2}{r} = \omega^2 r = (7.2722 \times 10^{-5}\,\text{rad/s})^2\,(6340{,}000\,\text{m})$$

$$= 3.35 \times 10^{-2}\,\text{m/s}^2$$

(b) We can use the linear speed calculated in part (a) to find the distance travelled by a point on the equator in 2 h:

$$s = vt = (461.1\,\text{m/s})(2)(3600\,\text{s}) = 3.32 \times 10^6\,\text{m}$$

Making sense of the result:

The distance covered by a point on Earth's equator during one day is Earth's circumference given by $2\pi r = 2\pi(6380\,\text{km})$ which is approximately 40,080 km, this is covered in 24 hours. if we divide that by 12 we get 3340 km, the distance covered in 2 hours.

☑ CHECKPOINT

C-8-1 Geostationary Orbits

Satellites in a geostationary orbit are directly above Earth's equator and appear motionless in the sky. Such orbits are particularly useful for communication and television satellites. Which of the following statements are true for a satellite in a geostationary orbit?

(a) The linear velocity of the satellite, with respect to Earth's centre, is the same as the linear velocity of a stationary object on Earth's surface.

(b) The angular velocity of the satellite, with respect to Earth's centre, is the same as the angular velocity of a stationary object on Earth's surface at equator.
(c) The tangential acceleration of the satellite is nonzero.
(d) The radial acceleration of the satellite is zero.

C-8-1 (b) all points along line connecting satellite to Earth centre move at same angular speed.

LO 2

8-2 Kinematics Equations for Rotation

Following a treatment similar to the one we used in Section 3-4 to derive the kinematics equations for constant acceleration, we can derive kinematics equations for rotation with a constant angular acceleration.

We can define angular acceleration as

$$\frac{d^2\theta(t)}{dt^2} = \frac{d\omega(t)}{dt} = \alpha \qquad (8\text{-}5)$$

Integrating Equation (8-5) gives an expression for angular velocity as a function of time:

$$\omega(t) = \alpha t + \omega_0 \qquad (8\text{-}6)$$

Equation (8-6) shows that the angular speed at any time t is determined by the initial value at $t = 0$ plus the change incurred as a result of the acceleration.

When we integrate Equation (8-6), we can show that the angular position as a function of time is given by

$$\theta(t) = \frac{1}{2}\alpha t^2 + \omega_0 t + \theta_0 \qquad (8\text{-}7)$$

Combining Equations (8-6) and (8-7) gives us the rotational version of the time-independent kinematics equation, identical in form to Equation (3-20):

$$\omega^2 = \omega_0^2 + 2\alpha\Delta\theta \qquad (8\text{-}8)$$

It is not surprising that Equations (8-6) and (8-7) are almost identical in form to Equations (3-15) and (3-17)

because they are derived in the same fashion and hold for the same condition of constant acceleration. Note that the product of α and $\Delta\theta$ is positive when the two quantities are parallel (i.e., the object speeds up) and negative when the quantities are antiparallel (i.e., the object slows down).

✓ CHECKPOINT

C-8-2 Cranking Your Bike

On a bicycle, a large sprocket is driven by a pedal on crank arms and is connected to the small sprocket on the rear wheel by a chain. Assume that the radius of the large sprocket is R, the radius of the small sprocket is $R/2$, the radius of the wheel is $3R$, and the length of each pedal crank arm is $2R$. Which of the following statements is true?

(a) The rear wheel has the same angular acceleration as the crank arms.
(b) The small sprocket has the same angular speed as the large sprocket.
(c) The radial acceleration of a point on the circumference of the small sprocket is half the acceleration of a point on the circumference of the large sprocket.
(d) The linear speed of a point on the circumference of the small sprocket is the same as the linear speed of a point on the circumference of the large sprocket.

C-8-2 (d) the speed of the chain is the same as that of the tangential speed of each sprocket.

EXAMPLE 8-2

Spinning Flywheel

A flywheel of radius 0.720 m is rotating clockwise at 3.60 rad/s.

(a) Find the constant acceleration needed to slow the wheel to an angular speed of 2.10 rad/s over 1.80 s.
(b) Find the number of revolutions the wheel will go through as it slows down.

SOLUTION

(a) Equation (8-6) relates angular speed, acceleration, and time:

$$\omega = \alpha t + \omega_0$$

We choose the direction of rotation (clockwise) as positive, so ω and ω_0 will be positive. Since the wheel is slowing down, α is directed counterclockwise and is negative according to our sign convention:

$$\alpha = \frac{\omega - \omega_0}{t} = \frac{2.1 - 3.6}{1.8} = \frac{-1.5}{1.8} = -0.833 \text{ rad/s}^2$$

(b) Now we can use either Equation (8-7) or Equation (8-8) to determine the angular displacement of the wheel as it slows down. Here, we use Equation (8-8):

$$\omega^2 = \omega_0^2 + 2\alpha\Delta\theta$$

$$\Delta\theta = \frac{\omega^2 - \omega_0^2}{2\alpha} = \frac{2.1^2 - 3.6^2}{2(-1.5/1.8)} = 5.13 \text{ rad} = 0.816 \text{ rev}$$

Making sense of the result:

The units for θ are $(\text{rad}^2/\text{s}^2)/(\text{rad}/\text{s}^2) = \text{rad}$. We divide by 2π to get the number of revolutions.

8-3 Torque

When you tighten a bolt using a wrench, as in Figure 8-5, three factors determine the effectiveness of the force you apply:

- the strength of the applied force
- the angle θ between the applied force and the wrench
- the distance $|\vec{r}|$ between the bolt and the point where you apply the force

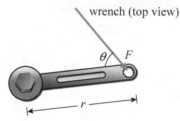

wrench (top view)

Figure 8-5 Torque depends on three parameters.

These three factors determining the **torque** of a force are elegantly summarized using the vector cross product introduced in Chapter 2. The torque of a force about a point, or pivot, is defined as the cross product of the position vector of the point of application of the force (as measured from the point) and the applied force:

KEY EQUATION
$$\vec{\tau} = \vec{r} \times \vec{F} \qquad (8\text{-}9)$$

The magnitude of the cross product in Equation (8-9) is given by

$$|\vec{\tau}| = |\vec{r}||\vec{F}|\sin\theta \qquad (8\text{-}10)$$

where θ is the angle measured counterclockwise from $\vec{r}$ to $\vec{F}$.

 ## ONLINE ACTIVITY

Torque

The e-resource that accompanies every new copy of this textbook contains an Online Activity using the PhET simulation "Torque." Work through the simulation and accompanying questions to gain an understanding of angular momentum.

Supersonic Blades

The propellers on turboprop planes turn at high rotational speeds. The linear speed of the tip of the propeller blades depends on the rotational speed and the distance from the tip to the propeller centre. The outer part of the blades can reach supersonic speeds. However, aircraft designers usually keep blade speeds below the speed of sound to avoid supersonic shockwaves and the increased drag they cause. The Tupolev Tu 114 (Figure 8-6) holds the record for the fastest turboprop passenger plane with a maximum recorded airspeed of 877.21 km/h, and remains one of the fastest passenger planes in use. It has special propellers designed to sustain supersonic blade–tip speeds and cope with the resulting shockwave drag.

Richard Seaman

Figure 8-6 The Tupolev Tu-114

Direction of a Torque Vector

The direction of a torque vector can be determined using the right-hand rule (Figure 8-7 a, b). When the palm of your hand is curled in the direction of rotation, with the thumb aligned with the rotation axis, the thumb points in the direction of the torque vector (Figure 8-7). The direction of the torque vector can also be determined using the Cartesian cross product rules outlined in Chapter 2. It should be mentioned that the angular velocity can be regarded as a vector, whose direction is also determined using the right-hand rule as shown in Figure 8-7(c).

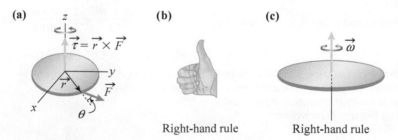

(a) $\vec{\tau} = \vec{r} \times \vec{F}$

(b) Right-hand rule

(c) $\vec{\omega}$ Right-hand rule

Figure 8-7 Demonstrating the right-hand rule for cross products

EXAMPLE 8-3

Torque Without Contact

A forest ranger is using a mallet to drive a peg into the ground to set up her tent in a level area. She drives the peg straight into the ground with an average force of 420 N. Another forest ranger is standing 2.00 m away. Can we say that the force she applies to the peg has a torque about the foot of the other ranger? If so, find the magnitude of the torque.

SOLUTION

The magnitude of the torque is given by $\tau = rF\sin\theta$. Since the peg is being driven straight into the ground, we can say that $\theta = \pi/2$. Therefore, $\sin\theta = 1$ and

$$\tau = rF = 2.0 \text{ m} \times 420 \text{ N} = 840 \text{ N} \cdot \text{m}$$

Making sense of the result:

For the force to exert a torque about a pivot or an axis, it is not necessary that there be a physical *moment arm* between the force and the pivot. (The component of the distance to the pivot that is perpendicular to the force is called the **moment arm** of the force.) So, despite the fact that there is no physical object connecting the mallet to the second ranger, the force used by the first ranger still exerts a torque about the second ranger. Here we are looking at the mathematical definition of torque.

EXAMPLE 8-4

Calculating Torque

Calculate the torque of the force shown in Figure 8-8 about the pivot at point O, given that $r = 1.70$ m, $F = 16.3$ N, and $\theta = 37.0°$.

(a)

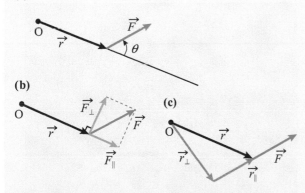

Figure 8-8 Calculating torque (a) as a vector cross product, (b) as a product between r and the component of F perpendicular to r, and (c) as a product between F and the component of r perpendicular to F

SOLUTION

We will carry out this calculation using three equivalent methods. These methods are essentially different perspectives of the way we view the definition of torque as given in Equations (8-9) and (8-10).

METHOD A: USING A CROSS PRODUCT

From Equation (8-10), the magnitude of the torque is

$$\vec{\tau} = \vec{r} \times \vec{F} = |\vec{\tau}| = |\vec{r}||\vec{F}|\sin\theta \qquad (1)$$

Here, we use the definition of torque and simply substitute the given values of $|r|$, $|F|$ and θ into Equation (1) (see Figure 8-8(a)):

$$|\vec{\tau}| = |\vec{r}||\vec{F}|\sin\theta = (1.7)(16.3)\sin(37°) = 16.7 \text{ N} \cdot \text{m}$$

METHOD B: USING THE PERPENDICULAR COMPONENT OF THE FORCE

We can consider the product in Equation (8-10) as a product of $|r|$ and $|F|\sin\theta$ (Figure 8-8(b)). The term $F\sin\theta$ gives the component of F that is perpendicular to the position vector of the force as measured from point O.

$$|\vec{\tau}| = |\vec{r}|\{|\vec{F}|\sin\theta\} = rF_\perp \qquad (8\text{-}11)$$

Alternatively, recognizing that the force can be expressed as the sum of these two components, the expression for the cross product can then be rewritten as

$$\vec{\tau} = \vec{r} \times (\vec{F}_{//} + \vec{F}_\perp) = \vec{r} \times \vec{F}_{//} + \vec{r} \times \vec{F}_\perp \qquad (8\text{-}12)$$

The first term in the expression is identically zero because the angle between $F_{//}$ and r is 0, and only the second term survives:

$$|\vec{\tau}| = |\vec{r} \times \vec{F}| = \sin(90°) = rF_\perp \qquad (8\text{-}13)$$

Both Equations (8-11) and (8-13) tell us that the force component perpendicular to the displacement vector contributes to the cross product for the torque, and the parallel component of F does not. Note that we are loosely using r and F in Equations (8-11) and (8-13) to represent the magnitudes of the position vector and the force vector.

METHOD C: USING THE PERPENDICULAR COMPONENT OF THE DISPLACEMENT

A third way of looking at the cross product is to consider the magnitude of the torque as the product of $r\sin\theta$ and F.

(continued)

203

As shown in Figure 8-8(c), $r\sin\theta$ gives the component of r that is perpendicular to the force F, and the component of r that is parallel to the force does not contribute to the torque:

$$|\vec{\tau}| = |\vec{r}||\vec{F}|\sin\theta = |\vec{r}|\sin\theta|\vec{F}| = r_\perp F \qquad (8\text{-}14)$$

Another way to derive this relationship is to rewrite the expression for the torque as

$$\vec{\tau} = (\vec{r}_{//} + \vec{r}_\perp)\times\vec{F} = \vec{r}_{//}\times\vec{F} + \vec{r}_\perp\times\vec{F} \qquad (8\text{-}15)$$

Here, again, the first term vanishes, so

$$|\vec{\tau}| = |\vec{r}\times\vec{F}| = |\vec{r}||\vec{F}|\sin(90°) = r_\perp F \qquad (8\text{-}16)$$

You can verify that Equations (8-14) and (8-16) yield the same value of $16.7\ \text{N}\cdot\text{m}$ for the required torque.

Making sense of the result:

Note that in this example we have treated the components of a vector as vectors themselves to facilitate a more elegant treatment of the vector product. Only perpendicular components matter.

 ## CHECKPOINT

C-8-3 The Greatest Torque

The forces in Figure 8-9 have the same magnitude and direction. Which force exerts the greatest torque on P?

(a) $\vec{F}_1$
(b) $\vec{F}_2$
(c) $\vec{F}_3$
(d) They are all the same.

Figure 8-9 C-8-3

C-8-3 (d) They all have the same perpendicular distance to P.

 ## MAKING CONNECTIONS

Gears and Transmission Systems

A simple transmission consists of two gears coupled together (Figure 8-10). Typically, one gear (the drive gear) is connected to a power source, such as the engine of a car, and couples to the second gear (the driven gear). If the two gears are not the same size, they will rotate at different speeds. An automotive variable transmission has a mechanism that couples the driven gear to different-sized gears driven by the engine. When the smallest engine gear is coupled to the wheel gear, a relatively high engine rpm results in a relatively low wheel rotation speed. This gear ratio is useful when the car encounters resistance, such as when starting from rest or travelling uphill. Putting the car in low gear effectively assigns a smaller moment arm to the external resisting force, thus reducing the resistive torque. The engine can then turn the wheels with less torque. Once the car is moving, it is more efficient to couple to a larger engine gear. Coupling gears by a chain as shown is only one type of transmission mechanism, in cars this is seen for example in the timing chain (or belt). Other common mechanisms involve direct coupling between gears of different size.

Figure 8-10 A simple transmission with two gears coupled by a chain

LO 4

8-4 Moment of Inertia of a Point Mass

Now, we examine how torque relates to angular motion. We begin by considering the rotating point mass in Figure 8-11.

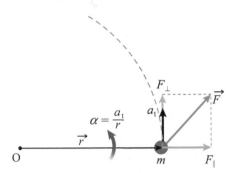

Figure 8-11 A force is applied to a point mass at the end of a rigid massless rod. The radial acceleration is not shown.

The mass in Figure 8-11 is rigidly attached by a massless rod to the pivot at point O. The point mass and the rod are originally at rest and are free to rotate in the plane of the page about the pivot, or, more precisely, about an axis perpendicular to the page running through the pivot. We now apply a force F to the mass as shown. The mass will move upward initially, and, because of the rigid rod, it will also move in a circular trajectory about the axis through O. We will keep the magnitude of the force and its direction relative to the massless rod constant. Note that we do not focus on radial acceleration in this chapter. Since our focus is rotational motion we only consider the tangential direction as tangential acceleration relates directly to angular acceleration as per Equation (8-4).

According to Newton's second law, $F_\perp = ma_t$. Note that the parallel component of the force does not contribute to the motion of the mass. The acceleration a in this case is the tangential acceleration of the point mass m. From Equation (8-4), the angular acceleration of m is given by $\alpha = a_t/r$, where r is the length of the rod. The force F exerts a torque about the pivot such that $|\tau| = |r \times F| = |r|\,|F_\perp|$, which reduces to $|r|\,|F_\perp| = |r|\,|ma_t|$. Since $a_t = \alpha r$,

KEY EQUATION
$$\vec{\tau}_{net} = mr^2\vec{\alpha} \qquad (8\text{-}17a)$$

Moment of Inertia

Equation (8-17a) is the rotational equivalent of Newton's second law for a point mass, $F_{net} = ma$. The torque takes the place of the force, and the angular acceleration replaces the linear acceleration. The mass, m, is replaced by the quantity mr^2.

An object's mass is a measure of its inertia, or resistance to acceleration in a straight line. Newton's second

law tells us that the net force required for a given acceleration of an object, such as the box of Figure 8-12(a) depends solely on the mass of the object. However, if we hold the object at a constant distance from an axis and rotate it about the axis with a given angular acceleration as in Figure 8-12, Equation (8-17) tells us that the torque needed depends not only on the mass of the object, but also on the square of the distance from the rotation axis.

The product mr^2 is called the rotational inertia, or the moment of inertia, of a point mass. The **moment of inertia** is a measure of an object's resistance to rotational acceleration, similar to the way mass is a measure of an object's resistance to linear acceleration. The symbol for moment of inertia is I. For a point mass, Equation (8-17) can be written as follows:

KEY EQUATION
$$\vec{\tau}_{net} = I\vec{\alpha} \qquad (8\text{-}17b)$$

(a)

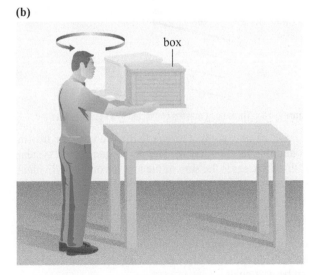

(b)

box

Figure 8-12 (a) In linear motion, resistance to acceleration depends only on the mass of an object. (b) In rotational motion, resistance to acceleration also depends on distance: the farther the object is from the rotation axis, the greater the resistance to rotational acceleration.

C-8-4 Who's Faster?

Two identical point masses are each connected to a pivot by massless rods. The rods are free to rotate about the pivots and are held in a horizontal position. One of the rods is twice as long as the other. At the instant when the rods are released,
(a) the linear acceleration of both masses is the same;
(b) the mass connected to the longer rod will have a smaller linear acceleration;
(c) the bars will have the same radial acceleration;
(d) the longer bar will have the greater angular acceleration.

C-8-4 (a) since the rods are massless, the masses will accelerate down at $a = g$.

LO 5

8-5 Moments of Inertia of Rigid Bodies

Thus far, we have dealt with objects as single point masses. In Chapter 7, we touched upon the idea of a system of particles. We now make the transition from treating point masses and systems of particles to treating rigid bodies. The treatment of a rigid body takes into account the geometry of the body and the mass distribution within it.

Systems of Point Masses

Consider the system of three particles shown in Figure 8-13. The particles are attached to a massless rod that pivots around a perpendicular axis through point O.

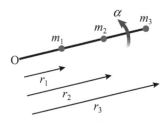

Figure 8-13 Three point masses connected to a rigid massless rod

Let us now apply a torque to the rod to rotate it with an angular acceleration α. The net torque on the system of three masses is the sum of the torques on the individual masses using $\vec{\tau}_{net} = mr^2\vec{\alpha}$ from Equation 8-17a for each of the masses:

$$\vec{\tau}_{net} = \vec{\tau}_1 + \vec{\tau}_2 + \vec{\tau}_3 = m_1 r_1^2 \vec{\alpha} + m_2 r_2^2 \vec{\alpha} + m_3 r_3^2 \vec{\alpha} \quad (8\text{-}18a)$$

we note that the angular accelerations of the 3 masses are the same, $\alpha_1 = \alpha_2 = \alpha_3 = \alpha$. Therefore, Equation 8-18a can be written as

$$\vec{\tau} = m_1 r_1^2 \vec{\alpha} + m_2 r_2^2 \vec{\alpha} + m_3 r_3^2 \vec{\alpha}$$

$$= (m_1 r_1^2 + m_2 r_2^2 + m_3 r_3^2)\vec{\alpha} \quad (8\text{-}18b)$$

This equation can now be written in the following compact form:

$$\vec{\tau}_{net} = I_{tot}\vec{\alpha} \quad (8\text{-}19)$$

where $I_{tot} = m_1 r_1^2 + m_2 r_2^2 + m_3 r_3^2$ is the total moment of inertia of the system of rigidly connected point masses.

In general, the moment of inertia of a two-dimensional system of particles (or point masses) about an axis perpendicular to the plane of the particles is equal to the sum of the individual moments of inertia of each of the particles about that axis.

KEY EQUATION
$$I = \sum_{i=1}^{N} m_i r_i^2 \quad (8\text{-}20)$$

where N is the number of particles.

The summation notation in Equation (8-20) holds as long as we are dealing with a system of discrete (noncontinuous) point masses. We can use this property to derive moments of inertia of some simple objects. For a continuous distribution of mass, the sum is replaced by an integral of infinitesimally small moments of inertia taken over the volume of the object. Then the expression for the total moment of inertia becomes

KEY EQUATION
$$I = \int r^2 dm \quad (8\text{-}21)$$

where r is the distance between each infinitesimal amount of mass dm and the axis of rotation.

C-8-5 Inertia

Consider a thin ring and a solid disk both of mass M and radius R. Both objects are homogeneous (mass is uniformly distributed), and both rotate freely at the same angular speed about fixed frictionless axes running through their respective centres of mass. Which object is more difficult to stop?

C-8-5 The ring has a higher moment of inertia and is therefore more difficult to stop.

ONLINE ACTIVITY

The e-resource that accompanies every new copy of this textbook contains an Online Activity using the PhET simulation "Torque." Work through the simulation and accompanying questions to gain an understanding of the relation between torque and moment of inertia T.

Rigid Bodies

A **rigid body** is a body that does not deform when a force is applied to it. In our study of rigid bodies, we begin by considering a uniform ring with mass M, radius R, and negligible width. We can consider the ring to be composed of identical point masses distributed uniformly along the rim. Let us assign a mass m_i to each of these point masses (Figure 8-14). The moment of inertia of the entire ring about a given rotation axis is simply the sum of all the moments of inertia of the individual point masses about that axis. For the axis that passes through the centre of mass of the ring and is perpendicular to the plane of the ring, the moment of inertia, I_{cm}, can found using Equation (8-20):

$$I_{cm} = \sum m_i r_i^2 = m_1 r_1^2 + m_2 r_2^2 + m_3 r_3^2 + \cdots + m_n r_n^2$$

Since all the point masses are located on the circumference at a distance R from the centre of the ring,

$$I_{cm} = m_1 R^2 + m_2 R^2 + m_3 R^2 + \cdots + m_n R^2$$
$$= (m_1 + m + m_3 + \cdots + m_n)R^2$$

The sum of all the point masses making up the ring adds to the total mass of the ring:

$$I_{cm} = MR^2 \qquad (8\text{-}22)$$

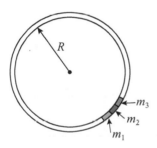

Figure 8-14 Mass distribution in a uniform ring

The distribution of mass in the ring is continuous, so the moment of inertia of the ring is more appropriately calculated using integration. As shown in Figure 8-15, the moment of inertia about the centre of the ring for an infinitesimally small differential element of mass, dm, is

$$dI_{cm} = R^2 dm \qquad (8\text{-}23)$$

Since the ring is uniform, it has uniform linear mass density, λ (mass per unit length). So, $dm = \lambda dl$, where dl is the infinitesimally small length of the mass dm on the ring, and

$$dI = R^2 \, dm = R^2 \, (\lambda \, dl) \qquad (8\text{-}24a)$$

Now, dl can be written as $Rd\theta$, where $d\theta$ is the infinitesimally small angle subtending the infinitesimal ring segment of length dl. Substituting in Equation (8-24a) gives

$$dI = R^2 \, dm$$
$$= R^2 \, (\lambda \, dl)$$
$$= R^2 \lambda R d\theta = \lambda R^3 d\theta \qquad (8\text{-}24b)$$

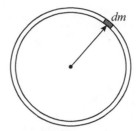

Figure 8-15 Differential elements for the calculation of the moment of inertia of a ring about its centre of mass

To obtain the moment of inertia of the entire ring, we integrate dI over the circumference of the ring:

$$I_{cm} = \int R^2 dm = \int_0^{2\pi} R^3 \lambda d\theta = R^3 \lambda \int_0^{2\pi} d\theta = R^3 \lambda 2\pi \quad (8\text{-}25)$$

Since the total mass of the ring is $M = \lambda(2\pi R)$,

$$I_{cm} = MR^2$$

This expression is the same as Equation (8-22).

We now consider the moment of inertia of a thin uniform disk shown in contrast to that of a ring in Figure 8-16 about an axis perpendicular to its plane, through its centre. The mass per unit area of this disk is a constant, σ, with units of kilogram per square metre.

Figure 8-16 Ring and disk with the same radius

The differential element we begin with is a ring of radius r and infinitesimal width dr (Figure 8-17). The area of this ring is $dA = 2\pi r dr$, and the mass of the ring is $dm = \sigma dA = \sigma 2\pi r dr$. As shown in Equation (8-21), the moment of inertia of an infinitesimally thin ring is mr^2. Therefore, for the ring of width dr,

$$dI = r^2 dm = r^2(2\pi r\sigma dr) = 2\pi r^3 \sigma dr \qquad (8\text{-}26)$$

We integrate this differential element over the range of radii of the rings to obtain the moment of inertia of the disk about its centre of mass:

$$I_{cm} = \int r^2 dm = \int_0^R 2\pi r^3 \sigma dr = 2\pi\sigma \int_0^R r^3 dr = 2\pi\sigma \frac{R^4}{4}$$
$$(8\text{-}27a)$$

Since the total mass of the disk is $\pi R^2 \sigma$, the expression of the total moment of inertia simplifies to

$$I_{cm} = \frac{MR^2}{2} \qquad (8\text{-}27b)$$

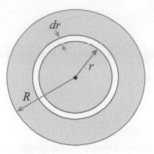

Figure 8-17 The differential element for the calculation of the moment of inertia of a uniform disk

If we think of a disk as being made up of point masses, the point masses located at the edge of the disk will each have a moment of inertia of $m_i R^2$, the same as the point masses making up a ring. However, the rest of the point masses making up the disk are closer to the centre of the disk and will have correspondingly smaller moments of inertia. For example, the point mass located at the centre of the disc has zero moment of inertia. Thus, the moment of inertia of a disk about its centre of mass is less than the moment of inertia of a ring of the same mass and radius.

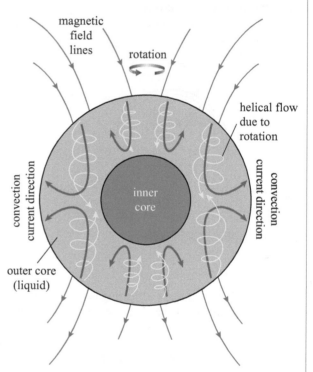

EXAMPLE 8-5

Clever Monkey

The solid drum in Figure 8-19(a) is free to rotate about an axle through its centre. A 23.0 kg monkey decides to ride the rope on the drum down to the ground 12.0 m below. The rope is at rest when the monkey grabs onto it. The drum is a uniform cylinder of radius 0.180 m and mass 190 kg. Find the acceleration of the monkey and the tension in the rope. Assume that the mass of the rope is negligible.

SOLUTION

Let us begin by drawing the FBD for the monkey (Figure 8-19(b)). The forces acting on the monkey are its

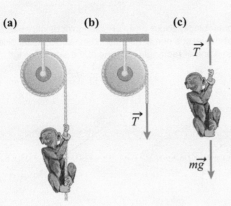

Figure 8-19 (a) Example 8-5. (b) The FBD of the pulley. (c) The FBD of the monkey

weight (mg) and the tension in the rope. Therefore, summing the forces in the y-direction, we have

$$\sum F_y = T_y + mg_y = ma_y$$

We choose up as the positive direction, so

$$T - mg = -ma \qquad (1)$$

where we introduce a as a variable to indicate the magnitude of the acceleration. The same holds for T and g in Equation (1) above.

We have two unknowns so far: T and a. Note that $T = mg$ only when the acceleration of the mass is zero.

The only force acting on the drum is the tension in the rope (Figure 8-19(c)). The tension is applied at the circumference of the drum, so the resulting torque rotates the drum. Next, we sum torques about centre of drum. Using Equations (8-10) and (8-19), and noting that the only force exerting a torque on the drum is the tension in the rope, we have

$$\tau = rT \sin\theta = I\alpha$$

The solid drum has the same basic shape as a solid disk, so its moment of inertia is $\frac{Mr^2}{2}$. The angle between the tension

and its moment arm to the centre is 90°. So, the expression for the torque reduces to

$$\tau = rT = I\alpha = \frac{Mr^2}{2}\alpha \qquad (2)$$

Here, T and α are unknown, so we have a total of three unknowns in two equations. We therefore have to find another relationship between the variables. From Figure 8-19(a) we can see that the linear acceleration of the monkey and the rotational acceleration of the drum are related by Equation (8-4c):

$$a = \alpha r \qquad (3)$$

You can now solve the system of three equations with three unknowns by whatever method you prefer. We will simply substitute the value of T obtained from Equation (1), and the value of a from Equation (3) into Equation (2):

$$r(mg - ma) = \frac{Ia}{r}$$

Solving for a gives

$$a = \frac{mg}{m + I/r^2} = \frac{mg}{m + \frac{1}{2}M}$$

$$= \frac{23 \text{ kg }(9.81 \text{ m/s}^2)}{23 \text{ kg} + \frac{1}{2}(190 \text{ kg})} = 1.91 \text{ m/s}^2$$

for the magnitude of the tension, substituting this value for the acceleration into Equation (1), we get

$$T = mg - ma = m(g - a) = 23 \text{ kg}(9.81 - 1.91) \text{ m/s}^2 = 182 \text{ N}$$

Making sense of the result:

The acceleration has the units of m/s². The acceleration is about one-quarter of g, which seems reasonable because the monkey is not in free fall. The moment of inertia of the drum slows the acceleration of the monkey.

✓ CHECKPOINT

C-8-6 *F = ma*

Suppose we replace the drum in Example 8-5 with a hoop. What is the expression for a now?

C-8-6 The expression for a is now $a = \frac{Mg}{m + M_{ring}}$ because all the mass in the ring would have the same linear acceleration.

Next, we will determine moments of inertia about axes that do not run through the centre of mass. Table 8-1 lists the moments of inertia about the axes of some common geometrical objects.

The Parallel-Axis Theorem

Hold the middle of a horizontal metre stick between your thumb and index finger, and rotate the metre stick back and forth horizontally about an axis perpendicular to its

length. Keep the angle by which you rotate the metre stick relatively small. Gauge the amount of resistance the metre stick presents as you accelerate it. Next, hold the metre stick horizontal with the one end between your thumb and index finger. Swing the metre stick back and forth through approximately the same angular displacement as when you rotated it horizontally. You will notice that you encounter more resistance and need to use more force when you hold the stick at its end. This difference indicates that the moment of inertia of the metre stick is greater about an axis running through its end than about an axis through its centre of mass.

To understand how the choice of axis affects the moment of inertia, we begin by calculating the moment of inertia of a uniform bar about an axis running through its centre of mass perpendicular to its length. The moment of inertia of the differential element of mass dm (Figure 8-20) about the centre-of-mass axis running perpendicular to the length of the bar is

Table 8-1 Moments of Inertia About the Axes of Some Common Objects

Thin rod, mass M and length L; centre-of-mass axis perpendicular to bar

$$I_{cm}^{bar} = \frac{1}{12} ML^2$$

Thin rod, mass M and length L; axis running through edge perpendicular to bar

$$I_{a}^{bar} = \frac{1}{3} ML^2$$

Rectangular block, about centre-of-mass axis perpendicular to plane of rectangle

$$I = \frac{1}{12} M(a^2 + b^2)$$

Rectangular block, mass M, side lengths a and b; about axis along one of the sides of length b.

$$I = \frac{1}{12} Ma^2$$

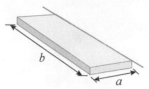

Cylinder, mass M inner radius R_1, outer radius R_2; axis of symmetry

$$I_{cm}^{cyl} = \frac{1}{2} MR_2^2 - \frac{1}{2} MR_1^2$$

Solid cylinder, mass M, radius r, and length L;

$$I_{cm}^{cyl} = \frac{1}{2} MR^2$$

Ring, or hollow cylinder, thin shell, mass M and radius R,

$$I_{cm}^{ring} = MR^2$$

Sphere, mass M and radius R

$$I_{cm}^{sphere} = \frac{2}{5} MR^2$$

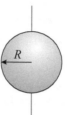

Thin spherical shell, mass M and radius R

$$I_{cm}^{ss} = \frac{2}{3} MR^2$$

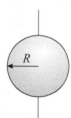

r^2dm. Using λ as the linear density of the rod, we can express dm as λdr. The moment of inertia of the differential element is then $dI = r^2 \lambda dr$. We obtain the moment of inertia of the bar by integrating over the length of the bar:

$$I_{cm}^{bar} = \int_{length} r^2 dm = \int_{-L/2}^{L/2} r^2 \lambda \, dr = \frac{\lambda L^3}{12} \quad (8\text{-}28)$$

The total mass of the rod is given by $M = \lambda L$, so

$$I_{cm}^{bar} = \frac{1}{12} ML^2 \quad (8\text{-}29)$$

We next consider the moment of inertia of the bar about an axis, a, perpendicular to the bar and running

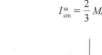

(a)

(b)

Figure 8-20 The moment of inertia of a differential element of a bar about an axis (a) through the centre of mass and (b) through the end of the bar and parallel to the axis through the centre of mass

through one end of it. The axis is parallel to the centre-of-mass axis. The moment of inertia of the differential element of mass dm about the axis a is given by r'^2dm, where r' is the distance from dm to axis a.

The distance r' can be written as $r' = r + d$, where d is the distance between the centre-of-mass axis and the axis a. Then, the moment of inertia of the differential element of mass dm is $(r + d)^2dm$. We integrate this differential over the length of the bar to obtain the moment of inertia of the bar about axis a:

$$I_{a}^{bar} = \int_{length} r'^2 dm = \int_{bar} (r+d)^2 dm = \int_{bar} (r^2 + d^2 + 2rd)dm \quad (8\text{-}30a)$$

$$I_{a}^{bar} = \int_{bar} r^2 dm + \int_{bar} d^2 dm + \int_{bar} 2(d)(r)dm \quad (8\text{-}30b)$$

The first part of the expression is the moment of inertia of the bar about the centre-of-mass axis, as in Equation (8-29). The integral of dm over the length of the bar gives the total mass of the bar, M; therefore, the second term in Equation (8-30b) equals Md^2. The third term can be written as $2d \int_{bar} rdm$. The integral $\int_{bar} rdm$ is Mr_{cm}, as we saw in Chapter 7 Equation (7-9c). Since we have chosen our coordinate system so that the centre of mass of the bar is at the origin, the third term in Equation (8-30b) reduces to zero. So, the expression of the moment of inertia of the bar about axis a can be written as

$$I_{a}^{bar} = I_{cm}^{bar} + Md^2 \quad (8\text{-}30c)$$

Equation (8-30c) is an expression of the **parallel-axis theorem**. This theorem applies for any object and for any parallel axis at any arbitrary distance. The theorem can be stated as follows:

$$I_a^{object} = I_{cm}^{object} + Md^2 \qquad (8\text{-}31a)$$

Equation (8-31a) states that the moment of inertia of an object about an arbitrary axis is equal to the moment of inertia of that object about a parallel axis running through its centre of mass, plus the product of the total mass of the object and the square of the distance between the centre-of-mass axis and the arbitrary axis.

For the above bar, $d = L/2$, and

$$I_a^{bar} = \frac{1}{12}ML^2 + M\left(\frac{L}{2}\right)^2 = \frac{1}{12}ML^2 + M\frac{L^2}{4} = \frac{1}{3}ML^2$$

$$(8\text{-}31b)$$

Note that this moment of inertia is four times greater than the moment of inertia of the bar about the centre-of-mass axis.

✓ CHECKPOINT

C-8-7 Reference Axis

Given that the moment of inertia of an object about an arbitrary axis (not the centre-of-mass axis) is I_a, can we say that the moment of inertia about another axis, axis b, parallel to axis a, is $I_b = I_a + md^2$, where d is the distance between the two parallel axes?

C-8-7 No. $I_b = I_{cm} + md^2$, where r is the distance between the centre-of-mass axis and axis b.

LO 6

8-6 Rotational Kinetic Energy and Work

We next develop an expression for **rotational kinetic energy**, which is the kinetic energy of a rotating object. We start with a point mass connected to a massless rod, rotating freely about the pivot at point O (Figure 8-21). The linear speed of the point mass at the position shown is v. This mass is under the influence of an external force directed *perpendicular* to the massless rod. (The mass also experiences a centripetal force exerted by the rod, but this radial force does no work on the mass.)

The kinetic energy of the mass is given by $KE = \frac{1}{2}mv^2$.

Since the particle is restricted to move in a circle, it has a rotational speed of $v = \omega r$. Substituting for v in the expression for kinetic energy gives an expression for the rotational kinetic energy of the point mass:

$$K_{rot} = \frac{1}{2}m\omega^2 r^2 = \frac{1}{2}I_o\omega^2 \qquad (8\text{-}32)$$

where I_o is the moment of inertia of the point mass about the pivot at O.

We can follow the same reasoning as we did in Section 8-5 to show that Equation (8-32) applies for the kinetic energy of any rigid rotating object.

✓ CHECKPOINT

C-8-8 Wheel and Disk

When a wheel and a disk with the same radius spin such that they have the same kinetic energy,
(a) they have the same mass;
(b) they have the same angular speed;
(c) the mass of the disk must be twice as large as the mass of the ring;
(d) all the above are possible.

C-8-8 (d) all the above are possible, we are demanding the same kinetic energy which depends on a few variables.

The work done by the force $\vec{F}$ in Figure 8-22 can be expressed in terms of angular variables. Since the magnitude of the force is constant in this example, the work done is

$$W_F = \vec{F} \cdot \Delta\vec{s} = |\vec{F}||\Delta\vec{s}|\cos\theta \qquad (8\text{-}33)$$

where θ is the angle between the force and the displacement.

For simplicity, we consider a force that remains perpendicular to the massless rod (Figure 8-22). Then, θ in

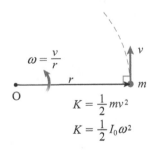

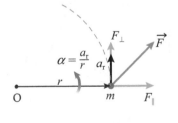

Figure 8-21 The point mass has a linear speed of v and an angular speed of $\omega = v/r$

MECHANICS

Equation 8-33 is zero and $\cos\theta = 1$. Since $\Delta s = r\Delta\theta$, the work done by force F over distance Δs is

$$W_F = F\Delta s = Fr\Delta\theta \qquad (8\text{-}33')$$

The magnitude of the torque is rF in this case, so the work done by F can be expressed as

$$W_F = F\Delta s = Fr\Delta\theta = \tau\Delta\theta \qquad (8\text{-}34a)$$

$$W_\tau = \tau\Delta\theta \qquad (8\text{-}34b)$$

In the case of a nonconstant force F, the work done by the varying torque is

$$W_F = \int_{\Delta s} F ds = \int_{\theta_1}^{\theta_2} \tau d\theta \qquad (8\text{-}34c)$$

If the force were not perpendicular to the rod, only the perpendicular component of the force would contribute to the torque, and the parallel component would not contribute to the work done. We can consider the force F or the torque τ as doing work on the rotating point mass.

The universal relation between the total work done on an object and the change in its kinetic energy is

$$W_T = \Delta K \qquad (8\text{-}35)$$

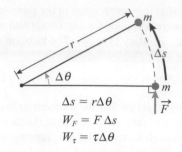

$$\Delta s = r\Delta\theta$$
$$W_F = F\,\Delta s$$
$$W_\tau = \tau\Delta\theta$$

Figure 8-22 Work done on a rotating point mass

Applying this relationship to rotational motion with constant torque gives the following:

KEY EQUATION for constant torque:

$$W_T = \tau\Delta\theta = \frac{1}{2}I\omega_f^2 - \frac{1}{2}I\omega_i^2 \qquad (8\text{-}36a)$$

and

for varying torque: $W_T = \int \tau d\theta = \frac{1}{2}I\omega_f^2 - \frac{1}{2}I\omega_i^2$

$$(8\text{-}36b)$$

EXAMPLE 8-6

Energy and Angular Acceleration

Two masses ($m_1 = 31.0$ kg and $m_2 = 23.0$ kg) are connected with a massless rope across a pulley as shown in Figure 8-23(a). The pulley is free to rotate about an axis through its centre. There is no friction between mass m_1 and the incline. The angle of the incline is $\theta = 32.0°$. The pulley has a radius of 0.300 m and a moment of inertia of 5.00 kg·m². Determine the acceleration of the blocks.

SOLUTION

METHOD A: NEWTON'S SECOND LAW APPLIED TO ROTATION

A quick check shows that $m_2 g$ is greater than $m_1 g \sin\theta$. Therefore, the pulley rotates clockwise.

Since $\sum F_y = ma_y$ for the forces acting on m_2,

$$T_{2y} + m_2 g_y = m_2 a_y$$

Choosing up as our positive direction, we get

$$T_2 - m_2 g = -m_2 a \qquad (1)$$

where we use a as a variable to denote the magnitude of the acceleration. The same holds for T_2 and g. Equation (1) has two unknowns, T_2 and a. For the forces acting on m_1, $\sum F_x = m_1 a_x$, so

$$T_{1x} + m_1 g_x = m_1 a_x$$

Choosing up the incline as our positive direction, we get

$$T_1 - m_1 g \sin\theta = m_1 a \qquad (2)$$

(a) **(b)** **(c)** **(d)**

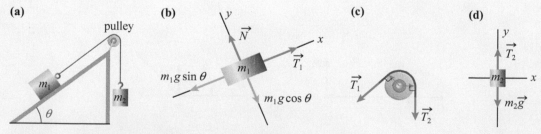

Figure 8-23 (a) Energy and angular acceleration. (b) to (d) The FBDs for Example 8-6

Note that we considered the tension in the rope to be different on each side of the pulley.

In Equations (1) and (2) we have three unknowns. We therefore need at least one more equation. For the pulley, we use the rotational equivalent of Newton's second law. Summing the torques about the centre of mass of the pulley, which we designate as o, we get

$$\sum \vec{\tau}_o = I_o \vec{\alpha}$$

$$\vec{\tau}_{o1} + \vec{\tau}_{o2} = I_o \vec{\alpha}$$

Choosing counterclockwise as our positive direction, we get

$$-\tau_{o1} + \tau_{o2} = I_o \alpha$$

where the variables used in the line above for torque and angular acceleration denote magnitudes. Since the rope applies force tangentially to the circumference of the pulley, the angle between the tension forces and the distance to the centre of rotation (the pivot point) is $\pi/2$ for both forces. Therefore,

$$-rT_1 + rT_2 = I_o \alpha \qquad (3)$$

We now have a total of four unknowns: T_1, T_2, α, and a. From the diagram, we can see another relationship: the acceleration of the rope is the same as the tangential acceleration along the circumference of the pulley:

$$a = \alpha r \qquad (4)$$

Here, again, you have a choice of several methods for solving the system of equations. We will substitute the expressions for T_1 from Equation (1), T_2 from Equation (2), and α from Equation (4) into Equation (3).

$$r(m_2 g - m_2 a) - r(m_1 g \sin\theta + m_1 a) = I_o (a/r)$$

Solving for a gives

$$a = \frac{m_2 g - m_1 g \sin\theta}{m_1 + m_2 + I_o/r^2}$$

$$= \frac{23\,\text{kg}\,(9.81)\,\text{m/s}^2 - 31\,\text{kg}\,(9.81\,\text{m/s}^2)(\sin(32°))}{23\,\text{kg} + 31\,\text{kg} + (5\,\text{kgm}^2/0.3^2\,\text{m}^2)} = 0.583\,\text{m/s}^2$$

Let us examine in detail why the two tension forces need to be different. According to Equation (3), the two tensions can be equal in one of two cases: Either the angular acceleration of the pulley is zero (i.e., the system is in equilibrium) or the moment of inertia of the pulley is zero (i.e., the mass of the pulley is zero). You may recall that in all the problems involving pulleys in previous chapters we assumed that the pulley was frictionless and massless. This assumption guarantees that $T_1 = T_2$. Note that friction at the axle of the pulley would add an extra torque to the equation, increasing the difference between the tension forces on either side of the pulley.

METHOD B: THE WORK-MECHANICAL ENERGY APPROACH

When m_2 descends a distance h, the gravitational potential energy of interaction between m_2 and Earth potential energy

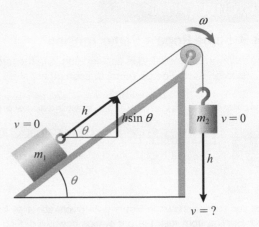

Figure 8-23(e) As m_2 descends and m_1 goes up the incline, some potential energy of the Earth-m_2 system is converted into potential energy for the Earth-m_1 system and into kinetic energy of the blocks and pulley.

decreases by $m_2 g h$. This loss in potential energy goes into increasing the potential energy of the Earth-m_1 system by $m_1 g \sin\theta h$ and increasing the kinetic energies of the pulley and the two masses as shown in Figure 8-23(e). If the two masses are initially at rest,

$$m_2 g h = m_1 g \sin\theta h + \frac{1}{2} m_1 v^2 + \frac{1}{2} m_2 v^2 + \frac{1}{2} I \omega^2 \qquad (5)$$

When we apply the chain rule to take the derivative of Equation (5) with respect to time, noting that h is a function of time, we get

$$m_2 g v = m_1 g \sin\theta v + m_1 v a + m_2 v a + I \omega \alpha$$

Here, we have used the fact that $\dfrac{dv^2}{dt} = \dfrac{dv^2}{dv}\dfrac{dv}{dt} = 2v\dfrac{dv}{dt} = 2va$ as given by the chain rule. Substituting a/r for α and v/r for ω yields

$$m_2 g v = m_1 g \sin\theta v + m_1 v a + m_2 v a + I v a/r^2$$

Dividing this equation by v and solving for a, we have

$$a = \frac{m_2 g - m_1 g \sin\theta}{m_1 + m_2 + I/r^2}$$

This is the same expression as the one derived using Method A.

Making sense of the result:

As a quick check, we can verify the units for the acceleration, a. Every term in the numerator has units of $\text{kg} \cdot \text{m/s}^2$, and the units of I/r^2 are $\text{kg} \cdot \text{m}^2/\text{m}^2 = \text{kg}$, so the units for the denominator are kg. Therefore, the units for the acceleration a are m/s^2, as expected. We were able to use the principle of conservation of energy here because there are no nonconservative forces involved.

For a pulley with negligible mass, I is zero. If we were to substitute this zero value into the expression for a above, we would obtain the same expression that we had for the same problem in Chapter 5, where the pulley was assumed to be massless and frictionless.

CHECKPOINT

C-8-9 Same Rope = Same Tension?

If the pulley in Example 8-6 is accelerating, can the forces of tension on each side be equal to each other?

C-8-9 No. According to Equation (3), $-rT_1 + rT_2 = I_0\alpha$, if the tensions are equal, then either the pulley has a zero moment of inertia or is not accelerating.

EXAMPLE 8-7

Falling Bar

The vertical bar shown in Figure 8-24 pivots about its lower end. Starting from rest, the bar swings downward. Find the speed of the free end when the bar has rotated through angle θ.

SOLUTION

Assuming that friction in the system is negligible, all the gravitational potential energy lost as the bar falls the bar will be converted into kinetic energy. From Chapter 6, we have

$$\Delta K = -\Delta U \qquad (1)$$

Since the bar starts from rest, the change in kinetic energy is

$$\Delta K = K_f - K_i = \frac{1}{2} I\omega^2 - 0 = \frac{1}{2} I\omega^2 \qquad (2)$$

To calculate the change in potential energy, we need to find the distance through which the centre of the mass of the bar falls, Δh_{cm}. We can see from Figure 8-24 that this distance is

$$\Delta h_{cm} = \frac{L}{2} - \frac{L}{2} \cos\theta \qquad (3)$$

Therefore,

$$\frac{1}{2} I\omega^2 = Mg\left(\frac{L}{2} - \frac{L}{2}\cos\theta\right) = Mg\frac{L}{2}(1 - \cos\theta) \qquad (4)$$

The moment of inertia of the bar about the pivot at its end is given by Equation (8-31b):

$$I_{cm} = \frac{1}{3} ML^2$$

Substituting into Equation (4) gives

$$\frac{1}{2}\left(\frac{1}{3} ML^2\right)\omega^2 = Mg\frac{L}{2}(1 - \cos\theta)$$

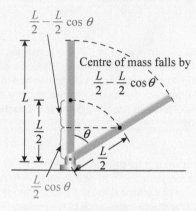

Figure 8-24 The potential energy from the fall of the bar is converted into kinetic energy.

$$L^2\omega^2 = 3gL(1 - \cos\theta)$$

Since the linear speed of the free end of the bar is $v = \omega L$,

$$v^2 = 3gL(1 - \cos\theta)$$

Making sense of the result:

For $\theta = 90°$, the top of the bar would have fallen by a full bar length, giving us $v^2 = 3gL$. If, instead of a swinging bar we had a swinging pendulum with a massless rod and a point mass at its end, such that the mass falls by a height L, the result, as we saw in Chapter 6, would be $v^2 = 2gL$. Since the bar has a moment of inertia that is less than the moment of inertia of a point mass of the same mass, we expect the tip of the bar to be easier to accelerate.

 MAKING CONNECTIONS

Mechanical Hybrids and Continuously Variable Transmissions

In 1900, Ferdinand Porsche developed the first electric hybrid vehicle, the Lohner-Porsche Semper Vivus (Figure 8-25). Most automotive brakes use friction to slow a car, converting most of the kinetic energy into thermal energy. Some hybrid cars use magnetic braking, where kinetic energy is converted first into electrical energy and then into electrochemical energy, which is stored in the car's batteries. However, as in any energy conversion process, there is significant energy loss in the conversion processes of this system. A flywheel with a relatively high moment of inertia can, in principle, be used to store the car's mechanical energy. Such systems generally require a continuously variable transmission for the precise control of energy between the car and the flywheel. Although no fully mechanical hybrids are in production at the time of writing, continuously variable transmissions are already available on a few cars. Fittingly perhaps, the Porsche 911 GT3 R is

the first hybrid vehicle to use mechanical flywheel energy storage instead of the conventional electrochemical battery storage. However, this car still uses magnetic braking to transfer energy to the flywheel which acts as a generator when called upon.

Figure 8-25 (a) The Lohner-Porsche Semper Vivus, the first electrical hybrid, on display. (b) The Porsche 911 GT3 R, the first semi mechanical hybrid, on a racetrack.

LO 7, LO 8

8-7 Angular Momentum

The Angular Momentum of a Point Mass

Let us again consider a point mass attached with a massless rod to a pivot at point O and rotating with speed v as shown in Figure 8-26. In the absence of an external force, the mass will continue to move at a constant speed. The tension in the rod acts perpendicular to the direction of motion of the mass and does not change its speed.

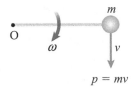

Figure 8-26 A point mass rotating about a pivot

As shown in Figure 8-27, the **angular momentum** of the point mass about an axis through point O, perpendicular to the page, is defined as

$$\vec{L}_o = \vec{r} \times \vec{p} \tag{8-37}$$

where $\vec{p}$ is the linear momentum of the point mass, and $\vec{r}$ is the position vector of the object with respect to the pivot or axis.

Any particle that has linear momentum also has angular momentum about any specified axis. The angular momentum is zero when $\vec{r}$ and $\vec{p}$ are collinear. Note that a particle (or an object) can have angular momentum with respect to an axis even when the particle does not rotate about that axis.

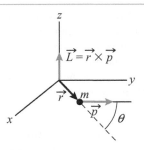

Figure 8-27 The angular momentum for a point mass

✓ CHECKPOINT

C-8-10 Angular Momentum on a Straight Line?

Two cars of the same mass M are moving in straight line in opposite directions on a highway at the same speed v. When the cars pass each other, the distance between them is d. What is the magnitude of the angular momentum of each about the other car as they pass each other?

(a) Mvd
(b) $2Mvd$
(c) cannot be defined as both are moving
(d) none of the above

C-8-10 (a) $L = rp\sin\theta$

✓ CHECKPOINT

C-8-11 Higher Angular Momentum

The mass in Figure 8-28 is travelling at a constant speed v. Compare the angular momentum about point A for the mass at the three locations shown at the three different times, t_1, t_2 and t_3.

C-8-11 The angular momentum is the same because only the perpendicular component of the position vector (as measured from the pivot) matters.

CHAPTER 8 | ROTATIONAL DYNAMICS   215

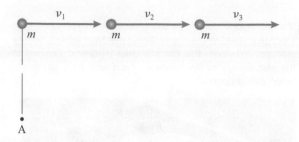

EXAMPLE 8-8

Figure 8-28   A point mass moving in a straight line at a constant speed

For the point mass in Figure 8-27, r and p are perpendicular, so

$$|\vec{L}_o| = |\vec{r} \times \vec{p}| = rp \sin \theta = rp = rmv$$
$$= rm\omega r = mr^2 \omega = I_o \omega \qquad (8\text{-}38a)$$

Stated in vector form:

$$\vec{L}_o = I_o \vec{\omega} \qquad (8\text{-}38b)$$

The subscript O on the symbol for angular momentum indicates that the angular momentum is taken about the point or pivot O (as with torques and moments of inertia).

Since I is a scalar, the angular momentum vector has the same direction as the angular velocity vector. The direction of the angular velocity vector is determined using the right-hand rule as illustrated earlier in Section 8-3.

The Angular Momentum of a Rotating Rigid Body

We can demonstrate that Equation (8-38) applies for any rigid body rotating about a point or an axis. Figure 8-29 shows a disk spinning about its axis of symmetry at angular velocity $\vec{\omega}$. The right-hand rule gives the direction of the angular momentum vector, which, according to Equation 8-38b, is parallel to the angular velocity vector.

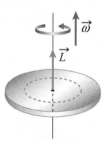

Figure 8-29   The angular momentum of a rigid object

The Rate of Change of Angular Momentum

In Chapter 7, we derived the relationship between the net force on a point mass and the change in its linear momentum:

$$\vec{F}_{net} = \frac{d\vec{p}}{dt} \quad \text{and} \quad \vec{F}_{net}^{avg} = \frac{\Delta \vec{p}}{\Delta t} \qquad (8\text{-}39)$$

When we take the cross product of r with both sides of Equation (8-39), we get

The Angular Momentum of a Spherical Shell

A child spins the brass globe on her parent's desk at 3.00 rev/s. The globe is a thin spherical shell with a mass of 1.10 kg and a radius of 23.0 cm. Find the angular momentum of the spinning globe.

SOLUTION

The angular momentum of a rigid body about an axis of rotation is given by Equation (8-38):

$$L = I\omega$$

From Table 8-1, the moment of inertia of a spherical shell about an axis through its centre of mass is 2/3 MR^2. There are 2π radians in one revolution, so the angular speed is

$$\omega = (2\pi \text{ rad/rev})(3 \text{ rev/s}) = 6.00 \, \pi \text{ rad/s}$$

Substituting into the expression for angular momentum, we get

$$L = I\omega = \frac{2}{3}(1.1 \text{ kg})(0.23 \text{ m})^2(6\pi \text{ rad/s}) = 0.731 \text{ kg} \cdot \text{m}^2/\text{s}$$

Making sense of the result:

The units for L are as expected.

$$\vec{\tau}_{net} = \vec{r} \times \vec{F}_{net} = \vec{r} \times \frac{d\vec{p}}{dt} = \frac{d}{dt}(\vec{r} \times \vec{p})^* = \frac{d\vec{L}}{dt} \quad (8\text{-}40a)$$

and

$$\vec{\tau}_{avg}^{net} = \frac{\Delta \vec{L}_{net}}{\Delta t} \qquad (8\text{-}40b)$$

Thus, a net torque acting on an object changes the angular momentum of the object. Rearranging Equation (8-40) and integrating gives

$$d\vec{L}_o = \vec{\tau}_{net} \, dt$$
$$\Delta \vec{L} = \int_{interval} \vec{\tau}_{net} \, dt = \vec{\tau}_{net} \Delta t \qquad (8\text{-}41)$$

The change in angular momentum, $\Delta \vec{L}$, is called the **angular impulse**. The symbol J is often used for this quantity.

Conservation of Angular Momentum

Equation (8-40) states that the change in the angular momentum of an object depends on the net torque experienced by that object. In the absence of a net torque on an object, the angular momentum of the object does not change and, therefore, is conserved. This principle can be extended to any system of objects.

This principle is demonstrated by the increase in the rate at which a figure skater spins when she reduces her moment of inertia by bringing her arms in close to her torso (the product $I_o \omega$ remains constant).

*Using the product rule, the first term for this derivative is the cross product between the velocity and the linear momentum. Since these are parallel, this term is identically zero; only the second term survives.

EXAMPLE 8-9

Shuttle Payload

A space shuttle orbiter is carrying a 21 000 kg satellite, which has a cylindrical shape with a radius of 1.90 m (Figure 8-30). The satellite needs to be spinning about its axis before it is released into space, so just before the satellite is released, a motor in the cargo bay spins the satellite until the satellite has an angular speed of 45.0 rad/s. The moment of inertia of the orbiter about the axis of rotation of the satellite is 4.20×10^6 kg·m².

(a) Find the angular speed of the orbiter just before the spinning satellite is released.
(b) What effect does releasing the satellite have on the angular speed of the orbiter?
(c) Does the rate at which the satellite is accelerated have an effect on the final angular speed of the orbiter?

SOLUTION

(a) We apply the concept of conservation of angular momentum to the system consisting of the orbiter and the satellite. The force (or torque) exerted by the orbiter on the satellite is internal to the satellite–orbiter system. When the motor in the obiter exerts a force (or torque) on the satellite, the satellite exerts an equal but opposite force (or torque) on the orbiter, in accordance with Newton's third law. Since the action–reaction pair is internal to the system, the total angular momentum is conserved. Therefore, $L_i = L_f$ for the system.

No part of the system was spinning before the motor was turned on, so the angular momentum of the system about the axis of rotation of the satellite is zero, and

$$\vec{L_i} = \vec{L_f} = 0$$

The final angular momentum is the sum of the angular momentum of the satellite and the angular momentum of the orbiter:

$$\vec{L_f} = \vec{L}_{sat} + \vec{L}_{orb} = 0$$

Therefore, the angular momenta of the satellite and the orbiter have equal magnitudes but opposite directions in magnitude we can say:

$$L_{orb} = I_{orb}\omega_{orb} = L_{sat} = I\omega_{sat} = \frac{1}{2} M_{sat} R_{sat}^2 \omega_{sat}$$

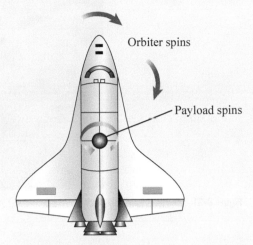

Orbiter spins

Payload spins

Figure 8-30 Example 8-9

Solving for ω_{orb} and then substituting the given values, we get

$$\omega_{orb} = \frac{\frac{1}{2} M_{sat} R_{sat}^2 \omega_{sat}}{I_{obr}}$$

$$\omega_{orb} = \frac{\frac{1}{2}(21\ 000\ \text{kg})(1.9\ \text{m})^2(45\ \text{rad/s})}{4.2 \times 10^6\ \text{kg·m}^2}$$

$$= 0.406\ \text{rad/s} = 23.3°/\text{s}$$

(b) Releasing the satellite has no effect on the angular speed of the orbiter because the satellite was already spinning when it was released. The effect of the satellite's rotation on the orbiter was solved in part (a).
(c) The rate at which the satellite is accelerated has no effect on the final angular speed of the orbiter. All that matters are the initial and final states.

Making sense of the result:

The heavier the orbiter the slower it will spin, according to the answer from (a) an infinitely heavy orbiter will not spin at all and a relatively light payload will not significantly affect the orbiter's rotation.

MAKING CONNECTIONS

Tides and Precise Time

Due to tidal friction, the Earth's rotation loses about 2.3 ms every century compared to atomic clocks. In 2010, NIST researchers introduced a quantum logic clock (Figure 8-31), which uses the vibrations of laser-cooled aluminum and magnesium ions. The new quantum clock is expected to be accurate to within 1 s over 37 billion years. Pulsars can also be used to keep time. Even though pulsar periods are about 100 times less accurate than the quantum clock, pulsars are useful for long-term time measurements because they will likely last much longer that the service life of a clock on Earth. Sudden changes in pulsar periods may indicate gravitational waves originating from black holes.

(continued)

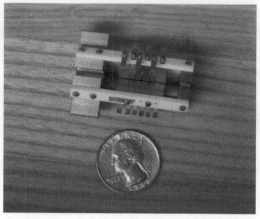

Figure 8-31 The ion trap is a key component of the NIST (National Institute of Standards and Technology) aluminum ion clock.

J. Koelemeij/NIST

ONLINE ACTIVITY

Angular Momentum

The e-resource that accompanies every new copy of this textbook contains an Online Activity using the PhET simulation "Torque." Work through the simulation and accompanying questions to gain an understanding of the conservation of angular momentum.

 CHECKPOINT

C-8-12 Conservation of Angular Momentum

A child steps onto a stationary merry-go-round and starts walking along its edge. Which of the following statements is true? Assume that the friction in the mechanism of the merry-go-round is negligible.

(a) The total angular momentum of the merry-go-round–child system increases as a result of the child walking.

(b) The child has no angular momentum because she walks in a tangential direction around the merry-go-round.

(c) The kinetic energy of the merry-go-round–child system does not change as a result of the child walking.

(d) The angular momentum of the merry-go-round alone does not change.

(e) The child's motion on the merry-go-round has no effect on the total angular momentum.

C-8-12 (e) The force or torque exerted by the child on the merry-go-round is internal to the merry-go-round–child system. The total angular momentum is conserved.

✓ **CHECKPOINT**

C-8-13 Rotational Stabilization

You are standing on a platform that is free to rotate while holding a bicycle wheel that is spinning about its axle, which is horizontal. You then turn the axle vertical (Figure 8-32).

(a) Why do you need to exert a torque to tilt the axle?

(b) When the wheel slows down, what happens to the torque you have to exert to tilt the spinning axle?

(c) Why does the platform you are standing on begin to spin as you tilt the wheel axle?

(a)

The student stands on a stationary platform holding a bicycle wheel that spins counterclockwise; the bicycle wheel's angular momentum points horizontally.

(b)

She flips the moving wheel, reversing its angular momentum. The total angular momentum is conserved, so the wheel and the student must rotate the other way.

Figure 8-32 (a) Initially, the total angular momentum points horizontally. (b) When you turn the axis of rotation of the wheel to the vertical, the direction of the total angular momentum changes

C-8-13 a) Any change to the angular momentum requires a torque. b) the torque you apply will be less if you wish to tilt the axle at the same rate. c) Angular momentum is conserved; because the vertical component of angular momentum was zero initially, it must be zero in the final situation.

MECHANICS

EXAMPLE 8-10

Magnetic Boots

Two astronauts are installing equipment on an orbiting platform that is rotating freely at a rate of 1.10 rad/s about its centre of mass, as shown in Figure 8-33. Astronaut 1 is standing at the centre of the platform, and astronaut 2 is at the edge. The platform has a radius of 11.0 m. The moment of inertia of the platform without the astronauts is $I_p = 2400$ kg·m². Astronaut 1 walks to the edge and hands a container of mass $m_c = 24.0$ kg to astronaut 2 and then returns to the centre. Both astronauts are wearing magnetic boots that keep them on the surface of the platform. Compared to the dimensions of the platform, you can consider the astronauts to be thin cylinders or point masses of mass $m_a = 81.0$ kg each. The container can also be considered a point mass.

(a) Find the resulting angular speed of the platform after astronaut 1 has returned to the centre.
(b) Find the total work done by the astronaut(s) in the process.
(c) If astronaut 1 keeps walking back and forth between the edge of the platform and its centre, will he eventually do enough work on the platform to lose all its kinetic energy and come to a stop? If so, how many times does astronaut 1 have to walk back and forth?

Figure 8-33 Example 8-10

SOLUTION

Any forces that astronaut 1 exerts on the platform to walk to the edge, deliver the container, and then return, along with any force the astronauts exert on the container during transport and delivery, are all internal to the astronauts–platform–container system. There is no net external torque on the system, so the angular momentum of the system does not change. Therefore,

$$\vec{L_i} = \vec{L_f} \quad \text{and} \quad I_i \vec{\omega_i} = I_f \vec{\omega_f}$$

The initial moment of inertia of the system is the total of the moments of inertia of the platform, astronaut 1 and the container at the centre, and astronaut 2 at the edge. Since we can treat the astronauts and the container as point masses, their moments of inertia about the axis are simply mr^2, and r is zero for the astronaut and the container when at the centre. Therefore,

$$I_i = m_a(0)^2 + m_c(0)^2 + I_p + m_a r^2$$
$$= 0 + 0 + 2400 + 81(11)^2 = 12\ 201 \text{ kg·m}^2$$

The final moment of inertia is equal to the initial moment of inertia plus the moment of inertia of the container when at the edge of the platform:

$$I_f = I_p + m_a r^2 + m_c r^2 = 2400 + 81(11)^2 + 24(11)^2$$
$$= 15\ 105 \text{ kg·m}^2$$

So, our conservation of angular momentum equation becomes

$$(12\ 201 \text{ kg·m}^2)(1.1 \text{ rad/s}) = (15\ 105 \text{ kg·m}^2)\omega_f$$

and $\quad \omega_f = 0.889$ rad/s

(b) The work done by astronaut 1 is the only work done on the system. Therefore,

$$W_T = \Delta K_E = \frac{1}{2} I_f \omega_f^2 - \frac{1}{2} I_i \omega_i^2$$
$$= \frac{1}{2}(15\ 105 \text{ kg·m}^2)(0.889 \text{ rad/s})^2$$
$$- \frac{1}{2}(12\ 201 \text{ kg·m}^2)(1.1 \text{ rad/s})^2$$
$$= -1420 \text{ J}$$

(c) Since there are no mechanisms for losing energy, such as an external force of friction, the energy of the whole system is conserved. The work done by astronaut 1 as he walks back and forth on the platform alternates between being positive and negative, such that in each return trip the total work done is zero.

Making sense of the result:

Since moving the container to the edge of the platform results in an increase in the total moment of inertia of the isolated system, we would expect the angular speed of the system to decrease to keep the total angular momentum constant.

 CHECKPOINT

C-8-14 Negative Work?

In Example 8-10, How would the result have been different if the astronaut on the inside chose to slide the container on the surface of the platform to his comrade instead of delivering it himself? How would the result have changed if the astronaut inside threw the package to the other astronaut who caught it in the air?

C-8-14 The final result would be the same in each case, because the initial and final states are the same.

CHECKPOINT

C-8-15 Constant Force?

If the astronauts in Example 8-10 were to walk from the centre of the platform at a constant speed, would the force they exert on the platform surface be constant?

C-8-15 No, the centripetal force they experience increases as they move away from the centre.

FUNDAMENTAL CONCEPTS AND RELATIONSHIPS

Table 8-2 summarizes the main concepts and quantities of rotational dynamics.

Table 8-2   Comparison of Angular and Linear Quantities and Concepts

Linear quantity	Symbol/Expression	Angular quantity	Symbol/Expression
Linear position	x	Angular position	θ
Linear velocity	v	Angular velocity	ω
Linear acceleration	a	Angular acceleration	α
Kinematics equations for constant a	$x(t) = \frac{1}{2}at^2 + v_0 t + x_0$ $v(t) = at + v_0$ $v^2 = v_0^2 + 2a \cdot \Delta x$	Kinematics equations for constant α	$\theta(t) = \frac{1}{2}\alpha t^2 + \omega_0 t + \theta_0$ $\omega(t) = \alpha t + \omega_0$ $\omega^2 = \omega_0^2 + 2\alpha \cdot \Delta\theta$
Mass	M	Moment of inertia	I
Force, average force, and net force	$\vec{F}, \vec{F}_{avg}, \vec{F}_{net}$	Torque, average torque, and net torque	$\vec{\tau}, \vec{\tau}_{avg}, \vec{\tau}_{net}$
Work of a force	constant force $W = \vec{F} \cdot \Delta\vec{s}$ variable force: $W = \int \vec{F} \cdot d\vec{s}$	Work of a torque	constant torque $W = \tau\Delta\theta$ variable torque: $W = \int \tau \, d\theta$
Work–kinetic energy theorem	$W_{Total} = \Delta K$	Work–kinetic energy theorem	$W_{Total} = \Delta K$
Linear momentum	$\vec{p} = m\vec{v}$	Angular momentum	$\vec{L} = \vec{r} \times \vec{p}, \; \vec{L} = I\vec{\omega}$
Linear impulse	$\vec{I} = \Delta\vec{p}$	Angular impulse	$\vec{J} = \Delta\vec{L}$
Conservation of linear momentum	if $\vec{F}_{ext} = 0, \vec{p}_i = \vec{p}_f, \Delta\vec{p} = 0$ (isolated system)	Conservation of angular momentum	If $\vec{\tau}_{net} = 0, \vec{L}_i = \vec{L}_f, \Delta\vec{L} = 0$ (isolated system)
Net force: change in linear momentum	average $\vec{\tau}_{ext} = \dfrac{\Delta\vec{L}}{\Delta t}$ instantaneous $\vec{F}_{ext} = \dfrac{d\vec{p}}{dt}$	Net torque: change in angular momentum	average $\vec{F}_{ext} = \dfrac{\Delta\vec{p}}{\Delta t}$ instantaneous $\vec{\tau}_{ext} = \dfrac{d\vec{L}}{dt}$

Key Applications: navigation, GPS, flywheels, mechanical hybrids, variable transmissions, engines and propellers, tidal friction, Earth's rotation, microgravity, biophysical systems, nuclear rotation, atomic angular momentum, spin angular momentum, atomic and molecular spectra, laser cooling of atoms, nuclear magnetic resonance and MRI, neutron stars, rotating black holes, black Saturns, gravity waves

Key Terms: angular acceleration, angular displacement, angular impulse, angular momentum, angular speed, moment arm, moment of inertia, parallel-axis theorem, rigid body, rotational kinetic energy, torque

QUESTIONS

1. Two spheres of equal mass and radius are spinning at the same angular speed about axes through their centres. One of the spheres is solid, and the other is a spherical shell. Which sphere has the higher kinetic energy?

2. Can one talk about an object's moment of inertia without specifying an axis? Explain.

3. A bicycle wheel mounted on a vertical axis is being accelerated from its initial angular speed at a constant acceleration. Compare the tangential acceleration of a point on the circumference to a point midway between the circumference and the centre. Also compare the speed and the radial acceleration of the two points.

4. Two globes of identical mass and radius are spinning about axes through their centres of symmetry at the same speed. One of the globes is a solid sphere, and the other has a hollow space at its centre. Which globe is easier to stop? Explain.

5. Two disks of different radii are connected by a belt running along their circumference, and driven by a motor that spins the larger disk at a constant angular speed. Can we say that the disks have the same angular speed? Is the radial acceleration of points on the circumference of the two disks the same? What do the two disks have as a common property?

6. Consider the heavy pulley setup in Figure 8-34. Can we say that the tension in the belt is the same everywhere? Which of the two masses would you expect to have the higher acceleration?

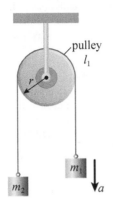

Figure 8-34   Question 6

7. You are standing on a freely spinning disk while holding on to a medicine ball, with your arms extended over the edge of the disk. You then throw the medicine ball such that its speed has only a vertical component when it lands on the ground. Did your and the disk's angular speed increase or decrease after throwing the ball?

8. Two thin rods of the same length are pivoted about a free hinge at their bottom edge while being held vertically. One rod is heavier than the other. The two rods are let go simultaneously, and they begin to fall. Which rod will be rotating faster at the bottom of its trajectory?

9. Two thin rods of the same mass are pivoted about a free hinge at their bottom edge while being held vertically. One rod is longer than the other. The two rods are let go simultaneously, and they begin to fall. Which rod will be rotating faster at the bottom of its trajectory?

10. Two pendulums are held in a horizontal position and then let go. Both pendulums are bars made of the same material, but one is thinner and has a sphere at the end. The masses and overall lengths of the pendulums are the same. Which pendulum will have a higher angular speed when it reaches a vertical position?

11. For the kinematics Equation (8-7), do you have to use radians, or can you use degrees or other units of measuring angles? Explain.

12. The propeller blades of single-engine plane spin at a constant speed. Compare the angular speed in rad/s of a point at the tip of one of the blades to the angular speed of a point at the middle of the blade. Does either of the two points accelerate? If so, compare their accelerations.

13. Your bike is upside down on the floor, and you are fixing the chain. The gear setting is such that the chain winds around the small gear at the rear wheel and around the large gear attached to the pedals (Figure 8-35).

(a) When you spin the pedals at a constant speed, will the wheel necessarily be turning at the same rotational speed as the pedals?

(b) You mark a point with a white marker on the rear tire, and you watch that point as you spin the wheel at a constant speed. Can you say that the white mark accelerates while the wheel spins at a constant speed?

Figure 8-35   Question 13

14. How could you tell a raw egg from a boiled egg without cracking them open?

15. A figure skater is executing a high-speed spin routine with her arms tucked close to her body (Figure 8-36). When she spreads her arms out, is she doing positive or negative work? When she brings her arms in closer to her body again, is she doing positive or negative work? Assuming zero friction, how would her initial angular speed compare to her angular speed after extending her arms out and then bringing then back in?

Figure 8-36   Question 15

16. Would the expression for the moment of inertia of a thin rod about an axis through its centre hold if the diameter of the rod were no longer negligible?

17. Use the moment of inertia for a ring to derive the moment of inertia for a thin cylindrical shell.

221

18. You are sitting on a freely spinning pedestal holding a long bar upright close to your body. You then tilt the bar until it becomes horizontal.
(a) What happens to your angular speed?
(b) Is the work you do as you rotate the bar positive or negative?

19. An insect ($m = 11$ g) is standing on the rim of a horizontal ring of mass $M = 0.90$ kg and radius $R = 9.0$ cm. The ring is free to spin about its centre-of-mass axis. The insect then moves around the ring and reaches a maximum speed of 15 cm/s with respect to the ring. The insect completes two full cycles around the ring and comes to a stop with respect to the ring. Find the final speed of the insect.

20. Why is the moment of inertia of a solid sphere only an approximate model for calculating the moment of inertia of Earth?

21. Using the vector model for angular momentum, explain why torque is required to tilt the axis of a spinning bicycle wheel.

22. Two cars are driving toward each other at the same speed on the highway. The lanes they are travelling in are each a distance d away from the median that divides the highway. Does the angular momentum of the two-car system about a point midway between them change as the cars pass each other?

23. A rope over a pulley with a moment of inertia I has a heavy load m_1 attached to one end and a counterweight m_2 tied to the other end, as in Figure 8-37. Is the tension ever the same in the rope on both sides of the pulley?

24. In the system shown in Figure 8-37, m_2 is greater than m_1, and the pulley has a moment of inertia I. Mass m_1 is given an initial speed downward, and the system slows due to gravity. When the system momentarily stops, which part of the rope (on either side of the pulley) has the higher tension? Does the pulley have an acceleration at that moment?

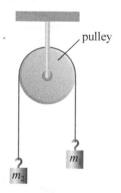

Figure 8-37   Question 24

25. Can a wheel have an angular acceleration if it is not moving? Explain.

26. Can a rotating object have an angular acceleration in one direction while rotating in the other? Explain.

27. Three people are pushing a stalled car toward a gas station. Can they create a torque about a tree a few hundred metres away? Explain.

28. A particle is flying in a straight line at a constant speed. Does the particle have angular momentum? Can we make conclusions regarding the direction of its angular momentum, if it exists? Explain.

29. Can we define the torque of a force without referring to a pivot point or rotation axis? Can we speak of force without referring to a frame of reference coordinate system? Explain.

PROBLEMS BY SECTION

For problems, star ratings will be used, (✷, ✷✷, or ✷✷✷), with more stars meaning more challenging problems.

Section 8-1   Angular Variables

30. ✷   What is the angular speed, in rad/s, of (a) a point on the equator and (b) a point in Antarctica close to the South Pole?

31. ✷   Taking Earth to be a perfect sphere, find the linear speed of a point located on the 32nd parallel, as a result of Earth's rotation.

32. A car is undergoing an emissions test. The car is stationary on rollers while the gas pedal is depressed until the speedometer reads 60 km/h. The wheels are 54 cm in diameter; find their angular speed.

33. ✷   Blades of a plane propeller need to be designed such that the top speed of the blade tip is just under the speed of sound. When the maximum engine speed is 7000 rev/min, determine what length the blade should be.

34. ✷   A Ferris wheel has a radius of 37.0 m. When a point on its circumference has moved 21.0 m, how many degrees did the Ferris wheel rotate?

35. ✷   Find the angular speed of the hour hand of an analog clock.

36. ✷   Astronauts often have to train in high-g environments before they embark on space missions. One way to simulate a high-g environment is to place astronauts in rotating capsules, such as the one in Figure 8-38, used by NASA. What angular speed is required when the radius of the track is 72 m and the acceleration target is $7.0g$?

Figure 8-38   Problem 36

Section 8-2   Kinematics Equations for Rotation

37. ✷   You have your bicycle upside down while you adjust the chain. The chain is wound around the rear wheel's small gear, which has a radius of 4.0 cm. At the pedals, the chain is wound around the large gear, which has a radius of 11.0 cm. The radius of the rear wheel is 35.0 cm. You spin the pedals initially in such a way that they finish two revolutions in 1.0 s, starting from rest. Assume constant acceleration.
(a) Find the angular acceleration of the rear wheel in rad/s².

(b) Find the total acceleration of a point at the edge of the small gear at the end of 1.5 s.

(c) What is the tangential acceleration for that point during the first second?

(d) Calculate the linear acceleration of the chain.

38. ✷ A space-training module consists of a capsule that rotates inside a horizontal circular track of radius 110 m. The rotational speed increases at a constant rate from rest such that the capsule finishes its first cycle in 9.00 s.

(a) How long does it take from the start of the motion for the radial acceleration to be equal to six times the acceleration due to gravity?

(b) How long does it take for the radial acceleration to be equal in magnitude to the linear acceleration?

39. ✷✷ The propeller blades of an airplane are 2.1 m long. The plane is getting ready for takeoff, and the propeller starts turning from rest at a constant angular acceleration. The propeller blades go through two revolutions between the fifth and the seventh second of the rotation. Find the angular speed at the end of 8.0 s.

40. ✷✷ A wheel is spinning about a fixed axis at an angular frequency of 29 rad/s. A braking mechanism reduces the speed by half over the first five revolutions. Assume that the braking force is constant.

(a) Find the angular speed 1.6 s into the braking phase.

(b) Calculate the angular displacement during that time.

(c) Find the angular speed at the end of 4.0 s.

41. ✷✷ A spinning globe is slowed at a constant acceleration of 1.3 rad/s² until it stops. One of the points on the equator moves 23° in the first 0.70 s of the slowing phase.

(a) Find the total angular displacement of the globe during the acceleration phase.

(b) Find the initial angular speed of the globe.

42. ✷ A bicycle wheel initially spinning at 7.00 rad/s is slowed at a constant acceleration such that it reaches a speed of 2.50 rad/s over an angular displacement of 11.2 rad. Find the angular acceleration.

43. ✷ A horizontally mounted bicycle wheel 42 cm in radius is accelerated from rest at 3.5 rad/s² for 7.0 s, after which time it rotates at a constant speed.

(a) Find the radial acceleration of a point on its circumference 3.0 s into the motion.

(b) Find the radial acceleration of the same point 9.0 s into the motion.

Section 8-3 Torque

44. ✷ A carpenter is driving a nail straight into the northern wall of house, with an average force of 1200 N. Find the magnitude of the torque of this force about a point on the wall 3.0 m to the right of the nail.

45. ✷ Find the cross product of $\vec{r}(2\hat{i}, 4\hat{j}, 0)$ m and $\vec{F}(4, 133°, 90°)$ N.

46. ✷ A force $\vec{F} = (12.0\hat{i} + 4.0\hat{j} - 16.0\hat{k})$ N is applied at the point A (−3.0 m, 5.0 m, 2.0 m). Find the torque of $\vec{F}$ about point B (−3.0 m, 2.0 m, −6.0 m).

47. ✷ Calculate the torque about the origin of force $\vec{F} = (11.0\hat{i} + 6.0\hat{j} - 14.0\hat{k})$ N applied at point A (2.0 m, −5.0 m, 7.0 m).

48. ✷ Determine the force F needed to prevent a beam ($m = 1100$ kg) from swinging down under the effect of its own weight (Figure 8-39).

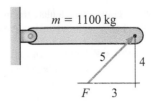

Figure 8-39 Problem 48

49. ✷ Calculate the magnitude of the torque about the origin of force $\vec{F}$ (670 N, 37°) applied in the xy-plane at a point on the x-axis 2.00 m to the right of the origin.

50. ✷ Find the cross product $\vec{r} \times \vec{F}$ where $\vec{r} = (17.0$ m, 33.0°, 90°) m and $\vec{F} = (430$ N, 81.0°, 90°) N. The vectors are given in spherical coordinate notation. Report the result in both spherical and Cartesian notation.

51. ✷ Calculate the vector product of $\vec{r} = (2\hat{i} - 3\hat{j} - 5\hat{k})$ m and $\vec{F} = (-3\hat{i} - 5\hat{j} + \hat{k})$ N. Give the answer in both Cartesian and magnitude angle notation.

Section 8-4 Moment of Inertia of a Point Mass

52. ✷ Three masses, each $m = 12$ kg, are located at the corners of an equilateral triangle. Calculate the moment of inertia of the system about an axis running through one of the point masses perpendicular to the plane of the triangle. The length of one side of the triangle is 0.94 m.

53. ✷ Four 400 g masses are located at the corners of a square with sides 1.1 m long. Find the moment of inertia about one of the diagonals.

Section 8-5 Moment of Inertia of Rigid Bodies

54. ✷✷✷ Using integration, calculate the moment of inertia of a solid disk of mass M and radius R about an axis through its centre of mass when the axis lies in the plane of the disk.

55. ✷ A ceiling fan is made from a cylindrical plate with a mass of 900 g and a radius of 11 cm. Three rectangular blades, 1.1 m in length and 17 cm in width, are attached to the circumference of the cylindrical plate. The total mass of the fan is 1.32 kg. Find the moment of inertia of the fan about an axis through its centre.

56. ✷✷✷ A solid, uniform disk of radius R and mass M has a smaller disk of radius $R/8$ removed from it (Figure 8-40). Find the moment of inertia of the resulting partially hollow disk about an axis running perpendicular to the plane of the disk through (a) the geometrical centre of the large disk and (b) the centre of the void section, in terms of the moment of inertia of the large solid disk. The distance between the centre of the large disk and the centre of the hollow part is $R/4$.

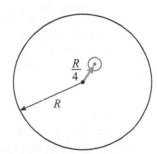

Figure 8-40 Problem 56

57. ★★★ Use integration to calculate the moment of inertia of a sphere about an axis through its centre of mass.

Section 8-6 Rotational Kinetic Energy and Work

58. ★★ A cylindrical bar with a sphere firmly attached to one end is originally at rest in a horizontal position. The bar is pivoted about a free hinge at one end (Figure 8-41). The bar is 4 m long and has a mass of 231 kg. The sphere has a radius of 60 cm and a mass of 53 kg. A variable torque is applied to lift the bar by rotating it 37° about the pivot at a constant angular speed. Calculate the work done by the variable torque.

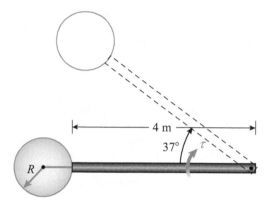

Figure 8-41 Problem 58

59. ★★ Use energy considerations to derive an expression for the angular acceleration of the pulley of the moment of inertia I in Figure 8-42. Assume the friction between the pulley and its axle are negligible. Also assume that the friction between the masses and the surfaces is negligible.

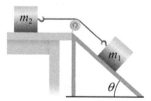

Figure 8-42 Problem 59

60. ★★ You are trying to get a better feel for the effect of geometry and mass distribution on the moment of inertia. You have a solid disk and a thin ring, each of radius $r = 1.3$ m and mass $m = 73$ kg. You mount both on fixed, horizontal frictionless axes about which they can spin freely. Then you spin them both.
(a) How much work do you need to do to get each object to spin at 3.00 rad/s?
(b) Let us assume that you have been causing them to spin by using a constant force applied tangentially to their circumferences. If the above speed is to be reached within 0.700 s, what is the magnitude of the force you need to apply to each object?

(c) You next attempt to stop each object by pressing your finger on the side of each object, right at its outer edge. The coefficient of kinetic friction between your finger and the surface of each object is 0.30. Find the minimum force you have to apply to stop each object within 1.0 min?

Section 8-7 Angular Momentum

61. ★ A car is travelling in the middle lane of a highway at 133 km/h. A police officer has a speed trap set up 210 m down the road. The police officer is standing on the side of the road, 11.0 m from the centre of the middle lane. The car weighs 1250 kg. Calculate the angular momentum of the car about the officer. What would the angular momentum of the car about the officer be at the moment the car passes the officer?

62. ★★ A pellet gun is fired at a solid disk that is free to rotate about a fixed horizontal axis as shown in Figure 8-43. The disk is 21 cm in diameter and has a mass of 370 g. The pellet ($m = 30$ g) gets lodged in the disk and causes it to spin at rate of 11 rad/s. Find the speed of the pellet just before it hits the wheel.

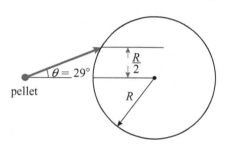

Figure 8-43 Problem 62

63. ★★★ A paint ball gun fires a ball of putty at the pendulum shown in Figure 8-44. The putty has a mass of 53 g and strikes the pendulum at a speed of 14 m/s. The pendulum is made of a thin bar that is 51 cm in length and has a mass of 310 g. The sphere fixed to the end of the pendulum is 17 cm in radius and has a mass of 190 g. The pendulum is originally at rest in the vertical position, and pivots about a free hinge at its top. The putty sticks to the pendulum at the point of impact shown. Find the maximum angle that the pendulum makes with the vertical after the collision. Consider the putty to be a point mass.

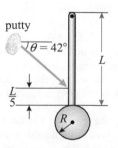

Figure 8-44 Problem 63

COMPREHENSIVE PROBLEMS

64. ✳✳✳ For a particle moving with uniform circular motion, show that $\vec{a}_r = \vec{\omega} \times \vec{v}$. Demonstrate this relationship for a bead moving in a circular path of radius 0.42 m and speed of 3.7 m/s.

65. ✳✳✳ Consider a bicycle wheel that consists of a thin ring of mass 200 g and radius 45 cm along with has 12 thin spokes of mass 21 g each. The wheel is made to spin freely about its axis at an angular speed of 0.30 rad/s. An insect with a mass of 31 g sits at the centre of the ring. The insect then walks radially outward (with respect to the ring) and stops at the rim of the wheel. The insect does not slip.
   (a) Find the final angular speed of the wheel when the insect reaches the rim.
   (b) How much work did the insect do to reach the edge? The coefficient of friction between the insect and the wheel is 0.23.

66. ✳✳✳ A solid disk of radius $R = 89$ cm is spinning about a vertical axis through its centre of mass. The disk is driven by a motor that maintains its angular speed at 4.2 rad/s. The moment of inertia of the disk about the axis is 0.30 kg/m².
   (a) How much work does a 12 g cockroach have to do to walk from the centre of the disk to its edge?
   (b) Is the work in part (a) positive or negative?
   (c) Determine the minimum coefficient of friction needed for the cockroach to make trip in part (a). Consider the cockroach to be a point mass.

67. ✳✳ Calculate the cross product V of the following two vectors: $\vec{V}_1 = 2\hat{i} + 4\hat{j} + 5\hat{k}$ and $\vec{V}_2 = 7\hat{i} + 4\hat{j} - 2\hat{k}$. Verify that $\vec{V} = \vec{V}_1 \times \vec{V}_2$ is perpendicular to the plane containing $\vec{V}_1$ and $\vec{V}_2$. Show that $\vec{V}' = \vec{V}_2 \times \vec{V}_1 = -\vec{V}$.

68. ✳✳ Show that the magnitude of the cross product between two vectors is equal to twice the area of the triangle they define. Assume that the two vectors have the same origin.

69. ✳✳ A worker is felling a 21.0 m-high tree by cutting it at a point 1.1 m above the ground, as shown in Figure 8-45. The mass of the tree is 4200 kg. Approximate the tree as a cylinder of radius 0.27 m (the tree can be treated as a thin rod).
   (a) Find the speed of the tree's centre of mass when the tree trunk is parallel to the ground on its way down. Assume that the tree rotates about the apex of the cut.
   (b) How much kinetic energy does the tree have at that point? How fast would a 700 kg car have to be moving to have that kinetic energy?

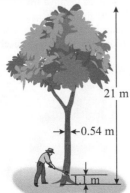

Figure 8-45 Problem 69

70. ✳✳ Find the moment of inertia of a ring about an axis tangent to the ring.

71. ✳ In reduced-gravity environments, astronauts suffer from loss of bone mass and muscle mass. One solution is to rotating sections in space stations and deep space vehicles to simulate Earth's gravity. If a cylindrical segment has a radius of 104 m to the inside surface of its outer wall, what rotational speed is required to create a 1.1g environment at this surface?

72. ✳✳ A movie stunt requires a tiny car with a mass of 380 kg moving at a high speed to collide with a powerful electromagnet. The electromagnet is fixed to the end of a vertical steel bar, which is free to rotate about a fixed axis through its top edge. The steel bar is 7.0 m in length and has a mass of 21.3 t. The magnet is cylindrical in shape and has a mass of 210 kg. When the car collides with the magnet, it completely sticks to it. Treat the car and the electromagnet as point masses. Find approximate values for
   (a) the maximum height that the car reaches when its speed before the collision is 69 km/h;
   (b) the maximum angle the bar makes with the vertical.

73. ✳✳ A disk is rotating at an angular speed of 63 rad/s. It is then slowed according to $\alpha = -12t - 2.5t^3$. How long will it take for the disk to reach half its speed? What is the angular displacement during that time?

74. ✳✳ The angular position of a point on a gear in a machine is given by $3t^3 + 5t - 32$. Determine the angular speed and angular acceleration of this point at $t = 5.0$ s.

75. ✳✳✳ Two disks are spinning freely about axes that run through their respective centres (Figure 8-46). The larger disk ($R_1 = 1.42$ m) has a moment of inertia of 1120 kg·m² and an angular speed of 5.0 rad/s. The smaller disk ($R_2 = 0.60$ m) has a moment of inertia of 910 kg·m² and an angular speed of 8.0 rad/s. The smaller disk is rotating in a direction that is opposite to the larger disk. The edges of the two disks are brought into contact with each other while keeping their axes parallel. They initially slip against each other until the friction between the two disks eventually stops the slipping. How much energy is lost to friction?

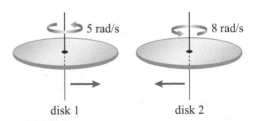

disk 1   disk 2

Figure 8-46 Problem 75

76. ✳✳✳ In Figure 8-47, two pulleys are mounted on fixed axles that have negligible friction. The small pulley has a moment of inertia of 9.0 kg·m²; it is made up of two cylinders welded together, one of radius 7.0 cm and the other of radius 15.0 cm. The large pulley has a radius of 41.0 cm and a moment of inertia of 84 kg·m²; the pulleys are coupled using a light belt. A 7.00 kg mass hangs from the smaller pulley by a rope that is wound around the smaller cylinder. The system is initially at rest. . . The mass is let go and starts to fall.

(a) Find the acceleration of the mass.
(b) Find the tension in the rope.
(c) Is the tension in the belt is the same everywhere? Explain.

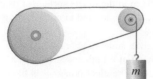

Figure 8-47 Problem 76

77. ✷✷ Calculate the moment of inertia of the object shown in Figure 8-48 about the axis a through one of the five thin bars of mass m and length L each, as shown.
(a) Derive the expression for the moment of inertia of each of the bars parallel to the a axis about that axis.
(b) Can you use the parallel-axis theorem to obtain that expression?
(c) Why is that expression identical in form to the expression for a point mass?
(d) Find the moment of inertia of the whole object about axis a.

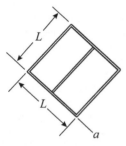

Figure 8-48 Problem 77

78. ✷✷ Using calculus, find the moment of inertia of a solid cylinder about its axis of symmetry. Repeat the same calculation by considering the cylinder to be a stack of disks.
79. ✷✷✷ Calculate the moment of inertia of a thin rectangular plate of length L and width W, about an axis parallel to W, passing through the centre of the plate. How will this change if the plate thickness is T and no longer negligible?
80. ✷✷ Two performers, of masses 62 kg and 84 kg, are standing on two fixed platforms holding on to the ends of a rope wound around the pulley shown in Figure 8-49. This pulley has a moment of inertia of 23 kg·m² and a radius of 37 cm. The platforms are then simultaneously removed, and the pulley starts to rotate.
(a) Find the angular speed of the pulley 3.0 s after the platforms are removed.
(b) After 3.0 s, the lighter performer begins to climb up the rope. What must this performer's acceleration be so that the pulley stops momentarily 7.0 s later?
(c) Find the tension in the rope for parts (a) and (b). Assume no slippage between the rope and pulley.

Figure 8-49 Problem 80

81. ✷✷ Two gears mounted on fixed axes are connected by a belt as shown in Figure 8-50. The large gear has a radius of 52 cm. The smaller gear has a radius of 17 cm. The smaller gear is driven by a motor that causes it to spin at 7.0 rad/s.
(a) Find the linear speed of a point on the circumference of the small gear.
(b) Find the linear speed of a point on the circumference of the larger gear.
(c) Find the linear speed of the belt.
(d) Find the angular speed of the larger gear.
(e) Find the ratio of the acceleration of a point on the circumference of the large gear to the acceleration of a point on the circumference of the small gear.

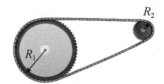

Figure 8-50 Problem 81

82. ✷✷ The pulley in Figure 8-51 has a moment of inertia of 6.4 kg·m² about the axis of rotation and a radius of 0.26 m. The masses are as follows: $m_1 = 31$ kg and $m_2 = 13$ kg. Assume that the tabletop and the pulley axle are frictionless.
(a) Determine the angular speed of the pulley when the system is allowed to move from rest under the effect of gravity 3 s into the motion.
(b) Find the tension force(s) in the rope.
(c) Is the tension force the same on each side of the pulley?

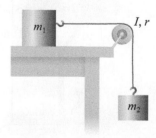

Figure 8-51 Problem 82

83. ★★★ A solid, uniform disk of mass 12 kg and radius 0.91 m is pivoted about an axis through its edge parallel to its axis of symmetry. Initially, the disk is held such that its centre of mass is directly above the axis. The highest point on the disk is marked with a dot. The disk is then let go, and it swings from rest under the effect of gravity.
   (a) Find the speed of the centre of mass when the centre of mass is (i) at the same horizontal level as the axis and (ii) directly below the axis.
   (b) Find the speed of the dot when the centre of mass is 40° below the horizontal.

84. ★★★ A child ($m = 34$ kg) is standing on a merry-go-round midway between the centre and the edge. Both the child and the merry-go-round are initially at rest. The child starts running around the merry-go-round without changing her distance from the centre until she reaches a speed of 3.0 m/s with respect to the merry-go-round surface underneath her feet. The moment of inertia of the merry-go-round is 1600 kg·m², and its radius is 4.0 m. Assume that friction between the merry-go-round and its mounting is negligible.
   (a) Find the final angular speed of the merry-go-round when the child is running.
   (b) What will the speed of the merry-go-round be when the child stops running?
   (c) How much work did the child do to get the merry-go-round to rotate at the angular speed in part (a)?
   (d) Calculate the total work done by the child.

85. ★★ The object shown in Figure 8-52 is made from two solid disks: $m_1 = 23$ kg and $m_2 = 39$ kg, and $r_1 = 17$ cm and $r_2 = 33$ cm. The disks are connected by a thin rod of mass $M = 11$ kg and length $L = 67$ cm. The object is held in a horizontal position and then let go. The object is free to rotate about a pivot through the centre of mass of the larger disk. Find the speed of the centre of mass of the smaller disk when it has swung 90° from its original position.

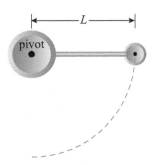

Figure 8-52   Problem 85

86. ★★★ For problem 85 plot the net torque about the pivot as a function of (a) the angle θ that the rod makes with the horizontal and (b) time. How long does it take the bar to reach the vertical position?

87. ★★ You are trying to pry two pieces of wood apart by jamming a 37 cm-long crowbar between them. You drive one end of the crowbar in 0.2 cm in between the pieces of wood, which are held together by six nails. The frictional force from each piece of wood onto a nail is 110 N. With the crowbar perpendicular to the edge of the pieces of wood, what is the force needed if you push at the other end of the crowbar? A clamp on the lower piece of wood keeps it in place.

88. ★★ In Figure 8-53, the block of mass $m_1 = 16$ kg sits on the surface of a bench such that the coefficient of static friction between the block and the surface is 0.20. Find the maximum weight that can hang down from the outer cylinder ($R = 34$ cm) of the pulley without causing m_1 to move. The radius, r, of the inner cylinder is 12 cm.

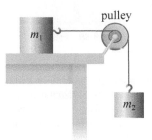

Figure 8-53   Problem 88

89. ★ For the system shown in Figure 8-54, the coefficient of kinetic friction between the mass and the incline is μ_k, and the pulley's moment of inertia is I and radius is r. Derive an expression for the speed of m_1 after m_2 has fallen a distance d from rest along the incline. Assume m_2 is sufficiently larger than m_1 to cause the system to move from rest.

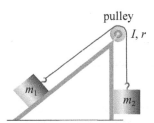

Figure 8-54   Problem 89

90. ★★★ Two mice jump onto the edge of a record turntable, causing it to rotate at the angular speed $\omega = 2.7$ rad/s. After a few seconds, the two mice jump off the turntable along a direction that is tangential to the turntable and opposite to its direction of motion. They jump one after the other at a speed of 2.7 m/s with respect to the turntable. Consider the turntable to be a disk of mass 510 g and radius 18 cm. The mice are 37 g each. Find the final speed of the turntable after the second mouse has jumped off. Consider the mice to be point masses.

91. ★★★ A solid disk with a mass of 32 kg and a radius of 27 cm is spinning counterclockwise at 17 rad/s in the horizontal plane about its axis of symmetry. Another disk with half the mass and half the radius is spinning at 34 rad/s in the opposite direction when it drops onto the first disk, as shown in Figure 8-55.
   (a) Find the final angular speed of the two disks when they eventually spin together.
   (b) How much energy is lost to friction?

227

Figure 8-55 Problem 91

92. ✳✳✳ The beam in Figure 8-56 pivots about one end, swings down from its horizontal position, and strikes a small box, as shown. The collision is perfectly elastic. The beam has a mass of $M = 160$ kg and a length of 1.3 m. Treat the box as a point mass ($m = 32$ kg). The friction at the pivot is negligible.
(a) Find the speed of the box after the collision.
(b) Find the maximum angle the bar makes with the vertical after the collision.

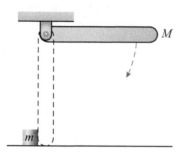

Figure 8-56 Problem 92

93. ✳✳✳ A child of mass $m = 28$ kg is standing at the edge of a playground merry-go-round (moment of inertia $I_{cm} = 2700$ kg·m² and radius $R = 3.0$ m), which is initially turning at 6.0 rev/min. The child walks opposite to the direction of motion of the merry-go-round with a speed of 2.0 m/s with respect to it. He walks along its edge and completes two full revolutions, stopping at his starting point.
(a) Find the speed of the merry-go-round after the child stops.
(b) The child then starts walking along the edge of the merry-go-round until the merry-go-round is spinning with an angular speed of 0.80 rad/s. He then jumps off the merry-go-round by leaping outward in a radial direction with respect to the merry-go-round. Find the angular speed of the merry-go-round just after the child leaps off. The coefficient of static friction between the child and the merry-go-round is 0.50.

94. ✳✳✳ An archaeological research team is attempting to simulate a catapult that might have been used in Europe 12 centuries ago. The catapult arm—a bar of length 3.00 m and mass 170 kg—is pulled from its vertical position against specially mounted springs fixed to the wall as shown in Figure 8-57. At the vertical position, the springs

are unstretched. The catapult arm is then pulled back until it is in the horizontal position. A stone of mass $m = 92.0$ kg is loaded into a light basket at the end of the arm, and the arm is released. At 51.0° above the horizontal, the catapult arm is stopped by a mechanism.
(a) Consider the stone to be a point mass. What should the effective spring constant be to launch the stone at a speed of 190 m/s?
(b) Consider the stone to be a sphere of radius 25.0 cm, and recalculate the required spring constant. Is it reasonable to approximate the stone as a point mass? What is the percent difference between the two results?

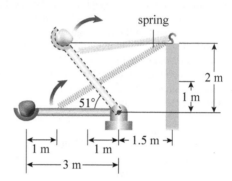

Figure 8-57 Problem 94

95. ✳✳✳ A flywheel is turning at an angular speed of 35 rad/s. It is then coupled with a load that causes it to slow down with a time-dependent acceleration given by $\alpha = -4e^{-2t}$. How many revolutions does the wheel go through before it stops?

96. ✳✳✳ A cylindrical observation station of mass 5200 kg and radius 11 m is spinning freely in space about its central axis. The platform is attended by one astronaut ($m = 72$ kg). She is initially at the centre of the cylinder, but then starts slowly walking toward the edge to look out from one of the windows at the Sun rising behind Earth. The platform is rotating freely at 10 rev/min when the astronaut is at the centre. The cylinder's moment of inertia about the axis is 15 300 kg·m². Find the angular speed of the platform when the astronaut reaches the edge. How much work did the astronaut have to do to reach the edge? What was the net change in the rotational kinetic energy?

97. ✳✳ In Figure 8-58, a rope is wound around a horizontal disk ($M_d = 2.9$ kg and $R_d = 20$ cm), passed across a pulley ($I_{cm} = 48$ kg·cm² and $R_p = 8.0$ cm), and connected to a hanging 4.0 kg mass. Initially the mass is held in place, and the system is at rest. The mass is then released and descends in response to the force of gravity. The rope does not slip on the pulley or the horizontal disk. Both the pulley and disk rotate on bearings that are effectively frictionless.
(a) Find the angular speed of the pulley after the mass has descended 1.2 m.
(b) Using energy considerations, calculate the acceleration of the mass without writing the equations of motion.

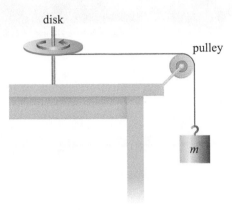

disk

pulley

Figure 8-58 Problem 97

98. ✷✷ You are sitting on a pedestal that is free to rotate about its axis of symmetry; your arms are stretched out fully in front of you. The combined moment of inertia of you and the pedestal is 21 kg·m². You are then handed a wheel of mass $m = 15$ kg and radius $R = 27$ cm, spinning at 21 rad/s in a vertical plane. You tilt the wheel axis from the horizontal to the vertical at an angular speed of 0.15 rad/s.

(a) Find the torque you have to apply as you tilt the wheel.

(b) The wheel axis eventually becomes fully vertical. Find the angular speed of the pedestal at that point. When the axis of the wheel is vertical, the distance between the axis of the wheel and the axis of the pedestal is 1.1 m.

99. ✷✷✷ Two astronauts are standing on a spinning, disk-shaped platform in space. The platform has a radius $R = 17$ m and a mass $M = 15\,000$ kg, and is spinning at an angular speed of 1.3 rad/s. One astronaut ($m_1 = 76$ kg) is standing at the edge of the platform, and the other astronaut ($m_2 = 82$ kg) is standing at the centre. The astronaut at the edge walks toward the centre.

(a) Does the astronaut do positive or negative work while walking toward the centre?

(b) As the astronaut walks in, would you expect the angular speed of the platform to increase or decrease?

(c) Calculate the spinning speed of the platform when the astronaut at the edge reaches the centre.

(d) Next, the two astronauts walk in opposite directions to the edge of the platform. Do you expect the kinetic energy of the system (platform + two astronauts) to increase or decrease?

(e) Find the work done by the two astronauts to reach the edge.

100. ✷✷✷ By how much does Earth's angular momentum change as a result of tidal friction over one year? What is the average torque involved?

See the text online resources at www.physics1e.nelson.com for Open Problems-Ended and Data-Rich Problems related to this chapter.

Learning Objectives

When you have completed this chapter you should be able to:

1 Distinguish between rotation about a fixed axis, and perfect rolling, spinning, and skidding.

2 Relate the translational and rotational motions of rolling objects.

3 Represent rolling as a rotation about a moving axis and as a rotation about a stationary point.

4 Apply Newton's laws to rolling motion and identify forces, torques, and pivot points for a rolling object.

5 Apply work and energy considerations to rolling motion and compare the use of Newton's laws and work–energy approaches for solving rolling motion problems.

6 Examine the relationship between friction and rolling.

7 Explain and quantify rolling friction.

Chapter 9
Rolling Motion

Bear Claw

During the winter in cold-climate countries, roads are often covered with snow, ice, or slush. Ongoing research into skid control, antilock braking systems, and tires aims to improve traction in extreme weather conditions. Winter tires often have more "aggressive" treads with channels that are designed to allow the snow, slush, mud, and water to leave the tires more effectively. However, even winter tires may not have an adequate grip on snowy or icy roads in mountainous terrains. In some parts of the world it is not uncommon to see car tires with studs that bite into the ice. The term "Bear Claw" is sometimes used to describe sharp edged, claw-like projections on tread blocks which enhance winter tire performance. In addition, tires with chains for added traction are required by law on mountain roads in certain areas, such as the Rockies of British Columbia (Figure 9-1).

Blaz Kure/Shutterstock

vesilvio/Shutterstock

Valentin Mosichev/Shutterstock

iStockphoto/Thinkstock

Wayne R Bilenduke/Stone/Getty Images

Figure 9-1 Examples of enhancements used to improve winter tire performance

For additional Making Connections, Examples, and Checkpoints, as well as activities and experiments to help increase your understanding of the chapter's concepts, please go to the text's online resources at www.physics1e.nelson.com.

9-1 Rolling and Slipping

In Chapter 8, we discussed the rotation of an object about a fixed axis. However, many applications involve rotation about a moving axis, such as a ball rolling down hill. Before we begin our discussion however, we need to define a few terms.

Rolling is rotation about an axis that is translating. For example, when you use a baker's rolling pin to flatten dough, you push the axle forward, causing the centre of mass of the rolling pin to translate forward. At the same time, the pin rotates about an axis through its centre of mass.

Spinning occurs when the rotational speed of the surface of an object is too high in comparison to the translational speed of the axis about which the object rotates. For example, when a truck is trying to emerge from a muddy hole, you often see the wheels spinning at relatively high speeds while the truck moves slowly on the muddy surface or does not move at all. Similarly, when a race car driver makes a fast start, the high torque of the powerful engine causes the wheels to spin, even on dry asphalt, and the friction of the asphalt on the spinning tire can burn the surface of the rubber tire (Figure 9-2).

© fastcars/Alamy   © James Warren/Alamy

Figure 9-2   (a) A Porsche Carrera 2 starting rapidly with tire burnout. (b) A Formula 1 car wheel spin

Skidding occurs when the translational speed of a rolling object is too high compared to the rotational speed of its surface. When you are driving a car, it is possible for the wheels to lock when you depress the brake pedal, in which case the wheels stop rotating completely while the car is still moving. The car will also skid without the wheels locking if the tires rotate at a speed slower than is necessary to keep up with the translational speed of the car (Figure 9-3).

Slipping, or the presence of kinetic friction between the tire and the road, refers to both spinning and skidding. A car moving on asphalt such that its wheels are skidding or spinning often leaves skid marks on the road. When a wheel slips, the part of the tire wheel that is touching the asphalt rubs against the asphalt. The asphalt

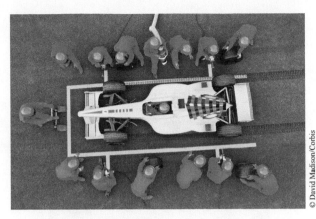

Figure 9-3   Race car skidding into a pit box

actually strips part of the black rubber from the tire in the same way that paper strips away some of a rubber eraser when you erase pencil marks on a page. Slipping often indicates trouble unless, of course, it is done deliberately, such as during a mountain rally or other type of car race (Figure 9-4).

Figure 9-4   French driver Marcel Tarres puts his Citroen Xsara into a controlled skid during the final of the famous car race on ice, *Trophée Andros*, at the Stade de France.

If an object neither spins nor skids as it rolls, its motion is called **rolling without slipping** (also called **ideal rolling** or **perfect rolling**). The relative speed between that part of the wheel in contact with the road and the pavement surface is zero. Figure 9-5(b) shows the motion of a white dot painted on the edge of a tire. Notice that the dot momentarily stops with respect to the ground just as the two come in contact.

In Figure 9-5(a), the car leaves clear tire marks because the part of the tire that touches the sand or dirt is actually stationary with respect to the ground as the car moves. Otherwise, the tire marks would be smeared.

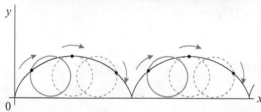

Figure 9-5 (a) Rolling without slipping produces a clear tire track in the Mojave Desert. (b) The trajectory of a point on the edge of a tire

✓ **CHECKPOINT**

C-9-1 Moving or Not?

A car is travelling at a speed of 98 km/h. Its tires have a radius of 31 cm. The speed of a point on the tire that is momentarily touching the ground is

(a) 98 km/h
(b) −98 km/h
(c) 98 km/h + ωr
(d) zero

C-9-1 (d) The speed of the point on the tire in contact with the road is zero for ideal rolling.

✓ **CHECKPOINT**

C-9-2 Why Do Cars Move?

What force moves a car forward when it accelerates from rest without spinning the tires?

(a) static friction
(b) kinetic friction
(c) the torque delivered by the engine
(d) another force

C-9-2 (a) No spinning means static friction.

9-2 Relationships Between Rotation and Translation for a Rolling Object

Consider a wheel of radius r rotating in a clockwise direction about a fixed axis through its centre of symmetry (Figure 9-6). Since the axis is fixed, the velocity of the centre of mass, v_{cm}, is zero. Choosing right as the positive direction, the velocity of a point at the top of the wheel, given by Equation 8-2, is $v_{top} = +\omega r$. The velocity of a point at the bottom of the wheel is then $v_{bottom} = -\omega r$. Since the velocity of the centre of mass is zero, the velocity of a point at the top is ωr greater than the velocity of the centre of mass, and the velocity of a point at the bottom is ωr less than the velocity of the centre of mass:

$$v_{top} = v_{cm} + \omega r$$
$$v_{bottom} = v_{cm} - \omega r$$

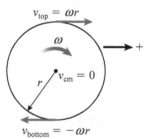

Figure 9-6 Rotation about a fixed axis

Rearranging and combining the above equations give the following relationships:

$$v_{cm} = v_{bottom} + \omega r \qquad (9\text{-}1)$$

$$v_{cm} = v_{top} - \omega r \qquad (9\text{-}2)$$

$$v_{top} = v_{bottom} + 2\omega r \qquad (9\text{-}3)$$

The same reasoning applies for the relation between the acceleration of the centre of mass and the tangential component of the acceleration at the top and the bottom of any rotating object. Taking the derivatives of Equations (9-1), (9-2), and (9-3) with respect to time gives the following:

$$a_{cm} = a_{bottom} + \alpha r \qquad (9\text{-}4)$$

$$a_{cm} = a_{top} - \alpha r \qquad (9\text{-}5)$$

$$a_{top} = a_{bottom} + 2\alpha r \qquad (9\text{-}6)$$

These relationships apply for any rotating object, regardless of whether the rotation axis is stationary or moving.

EXAMPLE 9-1

Stuck Truck

While trying to free a dump truck stuck in the mud, you notice that all four wheels are spinning at approximately 5.00 rev/s, and that the truck is moving forward at 5.00 m/s. The diameter of the wheels is about 1.50 m.

(a) Calculate the velocity with respect to the ground, of the point on the tire that contacts the ground.
(b) Calculate the velocity of the centre of mass of the tire with respect to the ground.
(c) Determine the velocity of a point at the top of a wheel.

SOLUTION

a) From equation (9-1)

$$v_{bottom} = v_{cm} - \omega r$$
$$= 5 \text{ m/s} - 5(2\pi)(0.75)$$
$$= -18.6 \text{ m/s}$$

b) Speed of center of mass of tire is the same as that of the truck = 5 m/s

c) For the top of the wheel we can use Equation 9-2

$$v_{top} = v_{cm} + \omega r$$
$$= 28.6 \text{ m/s}$$

Making sense of the result:

The difference between the velocity at the top and bottom is $2\omega r = 15\pi$.

 CHECKPOINT

C-9-3 Quick Calculation

If you drive a car at the constant speed 10 m/s such that the wheels do not slip, then

(a) the speed of the point at the bottom of the wheel (while in contact with the ground) is −10 m/s;
(b) the velocity of a point on the wheel at the same height as the centre of mass of the wheel is 10 m/s and points either up or down;
(c) the speed of a point on the wheel at the same height as the centre of mass of the wheel is 14.14 m/s;
(d) the acceleration of the point at the top of the wheel is zero.

C-9-3 (c) vertical speed is 10 m/s, horizontal speed is 10 m/s, using Pythagoras we get (c).

LO 3

9-3 Rolling Motion: Two Perspectives

In this section, we will discuss two approaches to analyzing the motion and dynamics of an object rolling without slipping. In the first approach, we consider rolling as a

rotation about a moving centre of mass axis. In the second approach, we use the fact that the point on the rolling body in contact with the surface is momentarily at rest, which allows us to consider rolling as a continuous series of successive rotations of the body about that point.

Rolling as a Rotation About the Moving Centre of Mass

For an object that rolls on a horizontal surface without slipping (Figure 9-7), the velocity of the part of the object in contact with the surface is zero: $v_{bottom} = 0$. Substituting into Equation (9-1) gives

$$v_{cm} = v_{bottom} + \omega r$$
$$v_{cm} = 0 + \omega r \qquad (9\text{-}6)$$

Hence for an object rolling without slipping,

KEY EQUATION $\qquad v_{cm} = \omega r \qquad (9\text{-}7)$

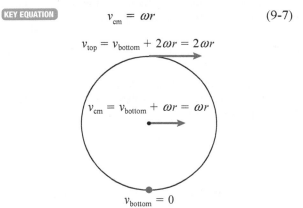

Figure 9-7   Rolling without slipping

Similarly, substituting into Equation (9-3) gives the speed at the top of the object, or more generally, the speed of the point opposite to the point of contact with the surface on which the object is rolling:

$$v_{top} = v_{bottom} + 2\omega r$$
$$= 0 + 2\omega r$$
$$v_{top} = 2\omega r \qquad (9\text{-}8)$$

To determine the acceleration of the centre of mass, we use Equation (9-4) to get

KEY EQUATION $\qquad a_{cm} = \alpha r \qquad (9\text{-}9)$

At the top of the object, the tangential component of the acceleration is

$$a_{top} = 2\alpha r \qquad (9\text{-}10a)$$

For the point on the object that is touching the ground, the tangential acceleration is

$$a_{bottom} = 0 \qquad (9\text{-}10b)$$

In this approach, rolling is regarded as a translation of the centre of mass, combined with rotation of the object

✓ CHECKPOINT

C-9-4

Which of the following statements is *not* true for a wheel rolling without slipping on a flat horizontal surface?

(a) The total acceleration of the point touching the ground is zero.

(b) The radial acceleration of the point touching the ground is zero.

(c) The vertical velocity of the point touching the ground is zero.

(d) The total speed of the point touching the ground is zero.

C-9-4 (a) a is the only false statement.

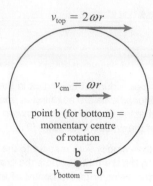

Figure 9-8 Rolling as successive rotations about the contact point b

about an axis through the centre of mass. The equations of motion will have to address both the translational and the rotational components of the motion.

For the translational dynamics of the centre of mass, we can apply Newton's second law:

KEY EQUATION
$$\sum \vec{F} = M\vec{a}_{cm} \qquad (9\text{-}11)$$

For the rotational dynamics, we apply the equivalent of Newton's second law for rotation about the centre of mass (Equation (8-17)):

KEY EQUATION
$$\sum \vec{\tau}_{cm} = I_{cm}\vec{\alpha} \qquad (9\text{-}12)$$

As you will see in the examples, Equations (9-11) and (9-12) often have to be used in conjunction with Equation (9-9).

Rolling as a Rotation about the Point of Contact Between the Object and the Surface

As an object rolls, successive points on its circumference touch the surface on which the object is rolling. If the object does not slip, each of the successive points will be stationary with respect to the surface at the moment that the point touches the surface. Consequently, rolling can be considered as a continuous series of rotations about a contact point, which we label point b (for bottom), as shown in Figure 9-8. This contact point is called a **momentary pivot**.

Since point b is at rest, we can sum the torques about this pivot point and apply the equivalent of Newton's second law for rotation. Although less intuitive than the approach using Equations (9-11) and (9-12), this approach

is an effective method of analysis; it is also often the more convenient approach.

Applying the rotational equivalent of Newton's second law by summing torques about point b gives the following:

KEY EQUATION
$$\sum \vec{\tau}_b = I_b\vec{\alpha} \qquad (9\text{-}13)$$

The object's rotational speed is ω, so the centre of mass located a distance r from the momentary pivot has a speed of $v_{cm} = \omega r$. The top of the object is located a distance $2r$ from the momentary pivot and has the speed $v_{top} = 2\omega r$. These speeds are identical to the speeds obtained in Equations (9-7) and (9-8). The expressions for the tangential acceleration of the centre of mass and the top of the object given in Figure 9-8 and in Equations (9-9) and (9-10) can be obtained using the same reasoning.

Notice how we have labelled the torques and the moment of inertia in Equation (9-13). Since we are summing the torques about point b, we must use I_b, the object's moment of inertia about point b, rather than I_{cm}, the moment of inertia about the object's centre of mass.

LO 4

9-4 Newton's Second Law and Rolling

Next, we study the forces involved in rolling. Let us first analyze the motion of an object rolling freely without slipping down an inclined plane.

EXAMPLE 9-2

Rolling Down an Incline

Consider a solid uniform disk that starts from rest and rolls without slipping down a ramp sloping at an angle θ as

shown in Figure 9-9(a). Derive expressions for the acceleration of the centre of mass of the disk and for the force of friction between the surface of the ramp and the disk.

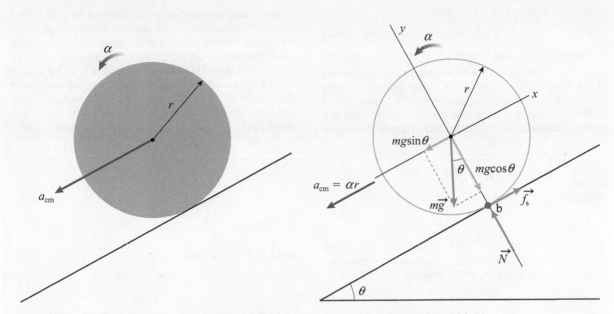

Figure 9-9 (a) A solid uniform disk rolling without slipping down an incline. (b) Free body diagram of the disk in (a)

SOLUTION

Let us begin by drawing the free body diagram of the disk. The forces acting on it are the force of gravity, the normal force from the ramp, and the friction from the surface of the ramp as shown in Figure 9-9(b).

We can analyze this situation using either of the approaches described above. Here, we consider the rolling as a translation of the centre of mass combined with rotation of the object about the centre of mass.

In the free body diagram in Figure 9-9(b), we choose our axes to reflect the direction of the acceleration, which is down the ramp. Applying Newton's second law to translation along the ramp gives

$$\sum F_x = ma_{cm,\,x}$$

$$f_{sx} + mg_x = ma_{cm,\,x}$$

Choosing up the incline as the positive direction, we get

$$f_s - mg\sin\theta = -ma_{cm} \qquad (1)$$

Like f_s and g in Equation (1), a_{cm} is a variable that denotes the magnitude of the acceleration of the centre of mass.

Next, we apply the rotational equivalent of Newton's second law by summing the torques about the centre of mass, noting that only the force of friction exerts a torque about the centre of mass:

$$\sum \vec{\tau}_{cm} = I_{cm}\vec{\alpha}$$

$$rf_s = I_{cm}\alpha \qquad (2)$$

So far, we have two equations and three unknowns. The third equation is the relationship between angular and linear acceleration for an object rolling without slipping (Equation 9-9):

$$a_{cm} = \alpha r \qquad (3)$$

As with most rotation problems, the simplest way to solve a system of three equations is substitution: we substitute expressions for f_s and a into Equation (2), and then solve for a_{cm}, as we did many times in Chapter 8. The key here is to substitute into the torque equation, using the expressions obtained from the other equations.

From Equation (1), $\qquad f_s = mg\sin\theta - ma_{cm}$

From Equation (3), $\qquad \alpha = \dfrac{a_{cm}}{r}$

Substituting these expressions into Equation (2) gives

$$r(mg\sin\theta - ma_{cm}) = I_{cm}\frac{a_{cm}}{r}$$

$$a_{cm} = \frac{mg\sin\theta}{m + \dfrac{I_{cm}}{r^2}} \qquad (4)$$

The moment of inertia of a disk about its centre-of-mass axis is $I_{cm} = \dfrac{mr^2}{2}$. Substituting into the expression for a_{cm} gives

$$a_{cm} = \frac{2}{3}g\sin\theta \qquad (5)$$

To calculate the force of friction, we can substitute this expression for a_{cm} in either Equation (1) or Equation (2) to obtain

$$f_s = \frac{1}{3}mg\sin\theta$$

The same results can be derived by considering rotation about the point of contact between the inclined surface and the disk. This is done in full detail online In Online Example 9-1. It would be worthwhile for you to solve the problem using this approach before you see the online example

(continued)

235

Making sense of the result:

The expression for the acceleration of the centre of mass in Equation (4) has the form of $a = \dfrac{F_{net}}{m}$ and, hence, has the units N/kg, as it should. The driving force is the component of the force of gravity along the ramp. The disk's ability to resist being accelerated comes from its mass, which indicates its ability to resist being accelerated linearly, and from its moment of inertia, which is a measure of its ability to resist being accelerated rotationally. The force of friction was not the maximum force of static friction because we are not told that the object is on the verge of slipping. When the object slides without rotation, we recover $a = g\sin\theta$ which we obtained in Chapter 5, Example 5-13 for a point mass sliding down an incline. This would happen here if the force of friction was zero. If the force of friction is zero, there will be no torque to cause the object to rotate and hence the object will not roll and will simply slide.

EXAMPLE 9-3

The Yo-yo

A yo-yo consists of a disk with a string wound around it as shown in Figure 9-10(a). Derive an expression for the acceleration of the yo-yo as the string unwinds.

SOLUTION

In this example, we consider rolling as a rotation about the point where the string is unwinding from the disk.

Let us begin by drawing the free body diagram of the yo-yo as shown in Figure 9-10(b). The forces acting in the

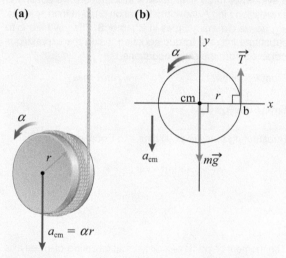

Figure 9-10 (a) A yo-yo. (b) The free body diagram of the forces acting on the yo-yo in part (a)

vertical direction are the weight of the yo-yo and the tension in the string. Since this yo-yo does not slip on the string, the point at which the string leaves the yo-yo is momentarily at rest and can be treated as a temporary, or momentary, pivot, where $\sum \vec{\tau_b} = I_b \vec{\alpha}$. We label this point b but note that this pivot is not at the bottom of the yo-yo.

The only force that exerts a torque about b is the weight of the yo-yo, which acts at the centre of mass, a distance r from point b. Here again we emphasize that in this approach both the torque and the moment of inertia are taken about point b and *not* about the centre of mass. Noting that mg acts at 90° to r, we have

$$rmg = I_b \alpha$$

and $\alpha = \dfrac{rmg}{I_b}$

Since $a_{cm} = \alpha r$, we have

$$a_{cm} = \alpha r = \frac{r^2 mg}{I_b}$$

Using the parallel-axis theorem gives

$$I_b = I_{cm} + md^2 = I_{cm} + mr^2$$

and $a_{cm} = \dfrac{r^2 mg}{I_b} = \dfrac{r^2 mg}{I_{cm} + mr^2} = \dfrac{mg}{\dfrac{I_{cm}}{r^2} + m}$

$$= \frac{mg}{\dfrac{m}{2} + m} = \frac{2g}{3}$$

where a_{cm} is a variable denoting the magnitude of the acceleration of the centre of mass. We have used the fact the moment of inertia of a disk about its centre of mass is $mr^2/2$.

Note that by summing torques about b, we are able to calculate the angular acceleration in a single step instead of solving a system of three equations as we did in Example 9-2.

You are encouraged to solve this problem by considering rolling as a rotation about the moving centre-of-mass axis. This is done fully online in Online Example 9-2. With this approach, you can also show that the tension in the rope is $T = \dfrac{mg}{3}$.

Making sense of the result:

Again, the expression for the acceleration has the correct units of m/s². The terms in the denominator reflect the object's resistance to acceleration: mass (m), for resistance to translational acceleration, and the moment of inertia $\left(\dfrac{I}{r^2}\right)$, for resistance to rotational acceleration. The equation for a_{cm} indicates that acceleration will decrease if either the mass or the moment of inertia of the yo-yo increases, which is reasonable.

C-9-5 Who's Faster?

Compare three yo-yos: one made with a sphere, one made with a cylinder, and one made with a ring, all with the same mass and the same radius. Which of the following is true?

(a) The ring yo-yo will have the smallest downward acceleration.

(b) The sphere yo-yo will have the smallest angular acceleration.

(c) The force for tension will be the highest in the case of the sphere yo-yo.

(d) The tension on the string will be the same for all three yo-yos.

C-9-5 (a) the ring will have the highest moment of inertia.

MAKING CONNECTIONS

Rolling, Friction, and Antilock Braking Systems

When applying the brakes in a car, unless the wheels slip the external force that slows the car is static friction. The direction of the force of static friction exerted by the road on the wheels is opposite to the direction of motion of the car. However, when you accelerate, the force of static friction on the drive wheels is in the direction of motion of the car. If the car is moving at a constant speed without slipping, the friction exerted on the tires by the road is zero.

When a car goes into a skid while trying to stop on an icy surface (Figure 9-11), the braking distance is greatly increased. Antilock braking systems (ABSs) are designed to maximize the force of friction between the road and the tires by keeping the tires on the verge of skidding. Ideally, the ABS will control the car's speed to keep the force of static friction close to its maximum possible value and avoid skidding, which would cause the weaker force of kinetic friction to act between the road and the tires. As described in Section 5-7, the force of static friction can take on values between zero and a maximum value determined by the normal force and the coefficient of friction:

$$0 \le f_s \le \mu_s N \tag{9-14a}$$

When the tires are on the verge of skidding but not actually skidding, the force of friction between the tires and road reaches the maximum value:

$$f_{s,max} = \mu_s N \tag{9-14b}$$

The ABS pumps the brakes automatically, repeatedly reducing the braking force just before the car goes into a skid, and then increasing the braking again. A feedback mechanism regulates the braking to maximize the force of static friction acting on the tires.

Figure 9-11 Extreme driving conditions: a Volvo on an ice-covered road

When slipping does occur, the force of friction is given by

$$f_k = \mu_k N \tag{9-14c}$$

For rubber on ice, the coefficient of kinetic friction can be considerably less than the coefficient for static friction, which makes having ABS quite desirable. Gabriel Voisin is credited with the introduction of the first ABS, used on airplanes in 1929. Figure 9-12 shows a sketch of the Avions Voisin AJ-T.

Figure 9-12 Sketch of the Voisin AJ-T, by Avions Voisin, a Dutch automobile manufacturing firm named in honour of Gabrel Voisin, reputed to be the inventor of the ABS system

✓ CHECKPOINT

C-9-6 Rolling and Friction

Which of the following statements is not correct?

(a) Rolling cannot occur without friction.

(b) Friction is often necessary to cause objects to start rolling initially, but once they do, friction is not always necessary to maintain rolling.

(c) If you throw a marble onto a surface with a horizontal speed v, and give it a forward rotational speed of $\frac{v}{r}$, then the marble will roll with or without friction.

(d) Under ideal conditions, the force of friction on a rigid object, rolling freely on a solid, smooth (not frictionless) horizontal surface, will be zero.

C-9-6 (a) You can cause an object to roll, without friction; see answer (c).

LO 5

9-5 Mechanical Energy and Rolling

How does the kinetic energy of an object rolling without slipping on a smooth horizontal surface differ from the kinetic energy of the same object sliding without rotation on a frictionless surface or that of the same object spinning about a fixed axis? We will first analyze this question using the rotation/translation approach first, and then we will apply the momentary-pivot approach to similar systems.

 EXAMPLE 9-4

Kinetic Energy of a Rolling Wheel

A spoked wheel of mass m and radius r rolls without slipping. Most of the mass of the wheel is distributed along its rim, so the wheel can be treated as a hoop or a ring.

(a) Derive an expression for the total kinetic energy of the wheel when the speed of the centre of mass is v_{cm}.

(b) Starting from rest, the wheel rolls without slipping down a ramp from a height h. Derive an expression for the speed of the centre of mass of the wheel at the bottom of the ramp.

SOLUTION

(a) Since the wheel rolls without slipping, Equation 9-7 gives us $v_{cm} = \omega r$, or $\omega = \frac{v_{cm}}{r}$. We can rewrite the expression for the total kinetic energy (Equation (9-17)) as

$$K = \frac{1}{2} m v_{cm}^2 + \frac{1}{2} I_{cm} \omega^2 \qquad (1)$$

Kinetic Energy of a Rolling Object

The object in Figure 9-13 is rolling without slipping. Since the speed of the centre of mass is v_{cm}, the kinetic energy of translation is

KEY EQUATION $\qquad K_{trans} = \frac{1}{2} m v_{cm}^2 \qquad (9-15)$

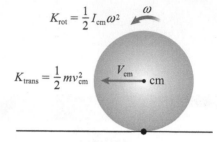

Figure 9-13 Rotation/translation analysis of the kinetic energy of a rolling object

The object is also rotating about the moving centre-of-mass axis at the angular frequency ω. The kinetic energy for this rotation is given by

KEY EQUATION $\qquad K_{rot} = \frac{1}{2} I_{cm} \omega^2 \qquad (9-16)$

The total kinetic energy is therefore the sum of the translational and the rotational kinetic energies:

KEY EQUATION $\qquad K = \frac{1}{2} m v_{cm}^2 + \frac{1}{2} I_{cm} \omega^2 \qquad (9-17)$

The moment of inertia of a ring or a hoop about its centre of mass is $I_{cm} = mr^2$. Substituting this into Equation (1) and using Equation 9-7 we get

$$K = \frac{1}{2} m v_{cm}^2 + \frac{1}{2} \frac{mr^2}{r^2} v_{cm}^2 = \frac{1}{2} m v_{cm}^2 + \frac{1}{2} m v_{cm}^2 = m v_{cm}^2 \qquad (2)$$

(b) The conservation of mechanical energy requires that the gravitational potential energy lost as the wheel rolls down the ramp equal the increase in kinetic energy, from Equation 6-31:

$$\Delta K = -\Delta U \qquad (3)$$

Since gravitational potential energy is lost as the wheel rolls down the ramp,

$$\Delta U = -mgh \qquad (4)$$

Substituting Equations (2) and (4) into Equation (3), we get

$$m v_{cm}^2 = mgh$$

Therefore,

$$v_{cm} = \sqrt{gh}$$

Making sense of the result:

For a ring that rolls without slipping, the rotational kinetic energy is equal to the translational kinetic energy. However, this relationship stems from the expression for the moment of inertia of the ring. We saw in chapter 6, example 6-8, that the expression for an object sliding down a frictionless ramp from height h, the speed was more by a factor of $\sqrt{2}$. This is because part of the potential energy lost as the wheel rolls down the incline goes into the rotational kinetic energy of the object, in addition to the translational kinetic energy. When the object slides down without rolling we recover the result from Example 6-8.

Kinetic Energy Using the Momentary-Pivot Approach

We now apply the momentary-pivot approach to analyze the kinetic energy of rolling objects. As before, the point on the rolling object that contacts the ground or surface is momentarily stationary with respect to the ground, and can be treated as a pivot (Figure 9-14). Using this method, the kinetic energy of a rolling sphere can written as a single term:

KEY EQUATION
$$K = \frac{1}{2}I_b\omega^2 \qquad (9\text{-}18)$$

All the rolling motion of an object is accounted for by a rotation about point b. With this approach, there is no need for a separate term for the translational motion.

Notice that Equations (9-17) and (9-18) give two different expressions for the *same* quantity: the total kinetic energy of a rolling object. Therefore, these two expressions must be equal. (See problem 86.)

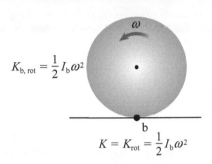

$$K_{b,\,rot} = \frac{1}{2}I_b\omega^2$$

$$K = K_{rot} = \frac{1}{2}I_b\omega^2$$

Figure 9-14 Momentary-pivot analysis of the kinetic energy of a rolling object

✓ CHECKPOINT

C-9-7 Ring or Disk?

When a ring and a disk with the same mass and radius start from rest and roll down a ramp without slipping,
(a) the ring will have the greater angular acceleration;
(b) the ring will have the greater linear acceleration;
(c) the two objects will have the same linear acceleration because they have the same mass;
(d) none of the above.

C-9-7 (d) The ring has the greater moment of inertia, so its angular acceleration is less, which also means that its linear acceleration is less.

EXAMPLE 9-5

The Speed of a Sphere Rolling Downhill

Starting from rest, a solid uniform sphere with mass m rolls without slipping down a ramp with height h (Figure 9-15). Derive an expression for the speed of the sphere at the bottom of the ramp.

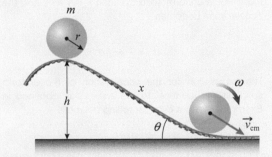

Figure 9-15 Gravitational potential energy is converted to kinetic energy as the sphere rolls down the ramp

SOLUTION

Here, we use the momentary-pivot approach. From Equation (9-18) we have

$$K = \frac{1}{2}I_b\omega^2 \qquad (1)$$

Conservation of mechanical energy gives

$$mgh = \frac{1}{2}I_b\omega^2 \qquad (2)$$

From the parallel-axis theorem,

$$I_b = I_{cm} + md^2 = I_{cm} + mr^2 \qquad (3)$$

$$mgh = \frac{1}{2}(I_{cm} + mr^2)\omega^2 = \frac{1}{2}I_{cm}\omega^2 + \frac{1}{2}mr^2\omega^2 \qquad (4)$$

Substituting $v_{cm} = \omega r$ gives

$$mgh = \frac{1}{2}I_{cm}\omega^2 + \frac{1}{2}mv_{cm}^2 \qquad (5)$$

(continued)

For a sphere, $I_{cm} = \frac{2}{5}mr^2$. Substituting this expression into Equation (5) yields

$$mgh = \left(\frac{1}{2}\right)\frac{2}{5}mv^2_{cm} + \frac{1}{2}mv^2_{cm} = \frac{7}{10}mv^2_{cm}$$

and

$$v_{cm} = \sqrt{\frac{10}{7}gh}$$

To see this same example done by considering rolling as a rotation about the moving centre of mass, go to Online Example 9-4.

Making sense of the result:

While Equation (1) does not explicitly contain an expression for the translational kinetic energy of the sphere, this energy does appear in Equation (5) as a result of applying the parallel-axis theorem in Equation (3). Note: Equation (5) applies to any object of mass m and moment of inertia I starting from rest and rolling downhill without slipping. We will apply this result later in the chapter. Again, we see when the object does not rotate, we recover the scenario in example 6-8.

EXAMPLE 9-6

The Acceleration of a Sphere Rolling Down a Ramp

(a) Using energy considerations, derive an expression for the linear acceleration of an object of mass m, moment of inertia I, and radius r as it rolls without slipping down a ramp with a slope of angle θ.

(b) Apply your results to a solid uniform sphere and to a solid uniform disk.

(c) Compare your results for the disk to the results in Example 9-2, which was done using Newton's second law.

SOLUTION

(a) In Equation (5) in Example 9-5, we applied the conservation of energy and the parallel-axis theorem to show that the increase in kinetic energy for an object rolling down a ramp without slipping is

$$mgh = \frac{1}{2}I_{cm}\omega^2 + \frac{1}{2}mv^2_{cm} \qquad (1)$$

If the object rolls a distance x down the ramp, the change in height is $h = x\sin\theta$. So,

$$\Delta K = -\Delta U \qquad (2)$$

and

$$mgh = mgx\sin\theta = \frac{1}{2}mv^2 + \frac{1}{2}I_{cm}\omega^2 \qquad (3)$$

As we saw in Chapters 6 and 8, taking the derivative of the energy conservation equation with respect to time gives an equation with acceleration as one of the terms. Because we are solving for linear acceleration, it is convenient to replace the angular quantities before taking the derivative. (If we were solving for angular acceleration, it would make more sense to replace the linear quantities.) Since the object is rolling without slipping, the speed of the centre of mass is

$$v_{cm} = \omega r \qquad (4)$$

Substituting this expression into Equation (3) gives

$$mgh = mgx\sin\theta = \frac{1}{2}mv^2_{cm} + \frac{1}{2}\frac{I_{cm}}{r^2}v^2_{cm} \qquad (5)$$

Taking the derivative of both sides with respect to time yields

$$mgv_{cm}\sin\theta = mv_{cm}a_{cm} + \frac{I_{cm}}{r^2}v_{cm}a_{cm} \qquad (6)$$

We have used the fact that the derivative with respect to time of the distance travelled by the object's centre of mass along the ramp is the velocity of the centre of mass, v_{cm}. We have also used the chain rule to obtain the derivative of the square of the velocity with respect to time:

$$\frac{dv^2}{dt} = \frac{dv^2}{dv}\frac{dv}{dt} = 2va \qquad (7)$$

Cancelling v on both sides of Equation (6) leaves us with only one unknown, a_{cm}. Here, the momentary-pivot approach has rather elegantly given us a single equation with the desired variable as the only unknown. Solving for a_{cm}, we get

$$a_{cm} = \frac{mg\sin\theta}{m + \frac{I_{cm}}{r^2}} \qquad (8)$$

(b) For a uniform solid sphere, $I_{cm} = \frac{2}{5}mr^2$. Substituting this expression into Equation (8), we get

$$a_{cm} = \frac{5}{7}g\sin\theta$$

We can consider a solid uniform disk to be a solid cylinder. For a uniform solid cylinder, $I_{cm} = \frac{1}{2}mr^2$, and the acceleration becomes

$$a_{cm} = \frac{2}{3}g\sin\theta$$

(c) The expression for the acceleration of the uniform solid disk is identical to the expression obtained in Example 9-2 for a cylinder rolling downhill.

Making sense of the result:

From Equation 8, we see that if the object was a point mass and had no moment of inertia, a reduces to $g\sin\theta$, as we saw in chapter 5, example 5-13.

C-9-8 Sphere, Cylinder, and Wheel

Starting from rest, a sphere, a solid cylinder, and a wheel with the same mass and radius roll, without slipping, down the same hill from the same height. At the bottom of the hill,

(a) the three objects have the same linear speed;

(b) the ring has the highest angular speed;

(c) the three objects have the same translational kinetic energy;

(d) the three objects have the same rotational kinetic energy;

(e) the three objects have the same total kinetic energy.

C-9-8 (e): As given by conservation of energy.

EXAMPLE 9-7

Acceleration in a Pulley System

Figure 9-16 shows a rope that is wound around a solid cylinder of mass m_c and radius R. The free end of the rope passes over a pulley of radius r and moment of inertia I, and is attached to a hanging mass M. The system is initially at rest. Mass M is then allowed to fall, pulling the rope and causing the pulley to rotate and the cylinder to roll as the rope unwinds from the cylinder. The cylinder rolls without slipping. Derive an expression for the acceleration of the hanging mass.

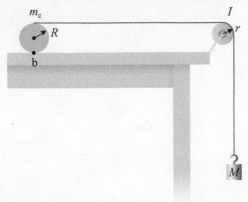

Figure 9-16 Example 9-7

SOLUTION

We begin by drawing free body diagrams for the three objects involved (Figure 9-17).

The forces acting on the hanging mass are its weight and the tension in the rope, T_1. Applying Newton's second law to the hanging mass, in the vertical direction we have

$$\sum F_y = Ma_y$$

$$T_{1y} + Mg_y = Ma_y$$

Choosing upward as the positive direction gives

$$T_1 - Mg = -Ma \qquad (1)$$

where a is a variable that denotes the magnitude of the acceleration. The same is true for T_1 and g in Equation (1).

Next, we sum the torques about the pulley's centre of mass. The torques acting on the pulley are τ_1 exerted by T_1 and τ_2 exerted by T_2:

$$\sum \vec{\tau}_{cm} = I_{cm}\vec{\alpha}$$

$$\vec{\tau}_1 + \vec{\tau}_2 = I\vec{\alpha}$$

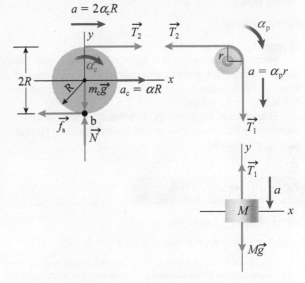

Figure 9-17 Free body diagrams of the (a) cylinder, (b) pulley, and (c) hanging mass

Choosing clockwise as positive,

$$\tau_1 - \tau_2 = rT_1 - rT_2 = I\alpha \qquad (2)$$

the variables in Equation (2) denote magnitudes of the vectors in Equation (1).

For the rolling cylinder, the simplest approach is to consider the torques about the point of contact between the cylinder and the surface, point b:

$$\sum \tau_b = I_b \alpha_c$$

Since the only force that exerts a torque about b is the tension T_2,

$$2RT_2 = I_b \alpha_c \qquad (3)$$

We can use the parallel-axis theorem to determine the moment of inertia of a cylinder about point b:

$$I_b = I_{cm} + md^2$$

The moment of inertia of a cylinder about its symmetry axis is $I_{cm} = \frac{1}{2}mR^2$, and the distance between the centre of mass of the cylinder and the point b is R. Hence,

$$I_b = \frac{m_c R^2}{2} + m_c R^2 = \frac{3}{2}m_c R^2$$

(continued)

Substituting for I_b in Equation (3) gives

$$2RT_2 = \frac{3}{2} m_c R^2 \alpha_c$$

$$2T_2 = \frac{3}{2} m_c R \alpha_c \qquad (3a)$$

So far, we have three equations and five unknowns: the tension forces T_1 and T_2; the angular acceleration of the pulley, α; the angular acceleration of the cylinder α_c; and the linear acceleration of the hanging mass, a. However, we can also write equations relating the linear acceleration of the mass to the angular acceleration of the pulley and that of the cylinder:

For the pulley, $\qquad a = \alpha r \qquad (4)$

For the cylinder, $\qquad a = \alpha_c 2R \qquad (5)$

as given by Equation 9-10a.

Note that the top of the cylinder, the rope, and the hanging mass have the same acceleration a. The centre of mass of the cylinder has acceleration a_c, which according to Equation 9-10 equals $a/2$.

Substituting Equation (5) into Equation (3) and then substituting Equations (1), (3), and (4) into Equation (2) and solving for a gives

$$a = \frac{Mg}{M + \frac{3}{8} m_c + \frac{I}{r^2}}$$

For the force of friction on the cylinder, we can sum the forces acting on the cylinder in the x-direction. We leave it to you to find the tension forces and the force of friction acting on the cylinder.

You can also solve this problem by considering rolling as a rotation about the moving centre of mass, and by using the work–mechanical energy approach as in Online Example 9-5 in the text online resources.

Making sense of the result:

The numerator in the expression for a has units of force, and the denominator has that of mass, as expected. When the pulley is light, and the cylinder does not roll, we recover the expression we obtained for the similar problem in Chapter 5, Example 5-16.

CHECKPOINT

C-9-9 Top vs Centre

In Equation (5) in Example 9-7, why is $a = \alpha_c 2R$?

C-9-9 For an object of radius R rolling without slipping with angular acceleration α, the acceleration of the point at the top of an object (or of the point diametrically opposite to the point of contact between the object and the surface) is $\alpha 2R$. (Refer to Equations (9-3) and (9-10a).)

EXAMPLE 9-8

Ball on a Vertical Circular Track

The metal ball in Figure 9-18 rolls down the ramp as shown, then up the vertical circular track. Find the minimum height h for which the ball will go all the way around the circular part of the track. Assume perfect rolling throughout. The radius of the track is R, and the radius of the ball is r.

Figure 9-18 Example 9-8

SOLUTION

Let us begin by drawing the free body diagram of the ball at the top of the vertical circular track (Figure 9-19).

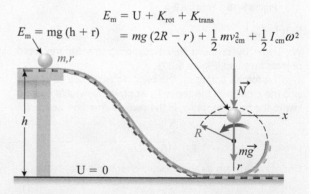

$$E_m = mg(h + r)$$

$$E_m = U + K_{rot} + K_{trans}$$

$$= mg(2R - r) + \frac{1}{2} mv_{cm}^2 + \frac{1}{2} I_{cm} \omega^2$$

Figure 9-19 Free body diagram of the metal ball at the top of the circular track

We apply Newton's second law. In the vertical (r) direction we have

$$\sum F_r = ma_r$$

$$N_r + mg_r = ma_r$$

When travelling around the circular track, the centre of mass of the ball has a trajectory with radius $R - r$. Using the expression for centripetal force for circular motion and choosing down as the positive r-direction, we get

$$N + mg = m\frac{v^2}{R - r}$$

The problem requires that the ball just barely touch the top of the circular track. So, at this point, the track exerts no force on the ball (i.e., the normal force is zero). Hence,

$$mg = m\frac{v_{cm}^2}{R - r}$$

$$g(R - r) - v_{cm}^2 \qquad (1)$$

Conservation of energy requires an increase in the ball's kinetic energy to match the decrease in the Earth-ball's gravitational potential energy. The ball's centre of mass starts at a height of $h + r$ and ends up at the top of the track at a height of $2R - r$. The net change in the ball's height is $2R - 2r - h$; hence, the net change in the ball's potential energy is $-mg(h - 2(R - r))$. Therefore, using Equation 6-31, $\Delta K = -\Delta U$

$$mg(h - 2(R - r)) = \frac{1}{2}mv_{cm}^2 + \frac{1}{2}I_{cm}\omega^2$$

$$= \frac{1}{2}mv_{cm}^2 + \frac{1}{2}\left(\frac{2}{5}\right)mr^2\omega^2 \qquad (2)$$

Since the ball is rolling without slipping, $v_{cm} = \omega r$. Substituting for ωr in Equation (2) gives

$$mg(h - 2(R - r)) = \frac{7}{10}mv_{cm}^2 \qquad (3)$$

Using Equation (1) to substitute for v_{cm}^2 in Equation (3) gives

$$mg(h - 2(R - r)) = \frac{7}{10}mg(R - r)$$

$$h = 2.7(R - r)$$

When $r \ll R$, the expression becomes $h = 2.7R$.

Making sense of the result:

The result we obtained for the related example in Chapter 6, Example 6-15, showed that the minimum height required for an object *sliding* on a frictionless track was $h = 2.5r$. Thus, we can see that it takes more energy for the object to roll (rotational in addition to translational kinetic energy) along the track than to slide without rotating it. For the object to reach the speed necessary, it must start at a larger height when it is rolling as opposed to just sliding. When the object does not roll or when it is a point mass ($r = 0$), we get the same result as in Example 6-15.

 ## EXAMPLE 9-9

Marble on an Inverted Bowl

A bowl of radius R is sitting upside down on a horizontal surface (Figure 9-20). A child holds a marble of radius r at the top of the outside surface of the bowl. The child lets go of the marble, and it rolls without slipping. Derive an expression for the location at which the marble will fly off the surface of the bowl. Comment on the difference between this example and Example 6-16, Sliding on a Spherical Surface, in Chapter 6.

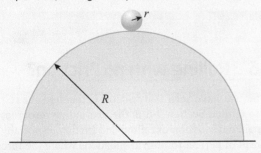

Figure 9-20 Example 9-9

SOLUTION

As the Earth-marble system loses gravitational potential energy, it will gain speed and kinetic energy, which might cause it to lose contact with the surface. We begin by drawing the free body diagram of the marble at the location where it is about to leave the surface of the bowl (Figure 9-21).

Since the marble moves on a semispherical bowl, the acceleration of interest is the radial acceleration. Hence, we need to determine the radial and tangential components of the forces. We apply Newton's second law. In the tangential direction we have

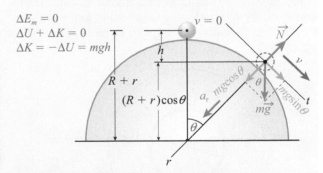

Figure 9-21 Free body diagram of the marble about to leave the surface of the bowl

$$\sum F_t = ma_t$$

$$mg_t = ma_t$$

$$mg\sin\theta = ma_t$$

In the radial direction we have

$$\sum F_r = ma_r$$

$$N_r + mg_r = ma_r$$

Choosing the radial direction toward the centre of the bowl as the positive direction, we get

$$-N + mg\cos\theta = m\frac{v^2}{R + r} \qquad (1)$$

Again, here N is a variable that indicates the magnitude of the normal force.

When the marble is on the verge of leaving the surface, the normal force is zero; hence,

(continued)

$$mg\cos\theta = m\frac{v_{cm}^2}{R+r} \quad \text{or} \quad g\cos\theta = \frac{v_{cm}^2}{R+r} \qquad (2)$$

So far, we have two unknowns and one equation. Since the potential energy lost by the marble is transformed into kinetic energy, using Equation 6-31,

$$mgh = \frac{1}{2}mv_{cm}^2 + \frac{1}{2}I_{cm}\omega^2$$

$$mgh = \frac{1}{2}mv_{cm}^2 + \frac{1}{2}\left(\frac{2}{5}\right)mr^2\omega^2 \qquad (2)$$

Since the marble is rolling without slipping,

$$v_{cm} = \omega r$$

From Figure 9-21, we can see that the height from which the marble drops is

$$h = (R+r)(1-\cos\theta)$$

Substituting for ωr and h in Equation (2), we have

$$mg(R+r)(1-\cos\theta) = \frac{7}{10}mv_{cm}^2 \qquad (3)$$

Using Equation (1) to substitute for v_{cm}^2 in Equation (3) gives

$$mg(R+r)(1-\cos\theta) = \frac{7}{10}mg\cos\theta(R+r)$$

$$\cos\theta = \frac{10}{17}$$

$$\theta = 54.0°$$

In Chapter 6, example 6-16, we obtained a value for θ of 48.2° for an object sliding down a frictionless spherical surface.

Making sense of the result:

If no rotational motion were involved, all the lost potential energy would all be converted to translational kinetic energy, resulting in the object achieving a higher speed as it falls, and hence leaving the surface earlier as we have seen in example 6-16.

✓ CHECKPOINT

C-9-10 Left or Right?

In Figure 9-22, a rope is wound around an axle. The axle is rigidly attached to a large drum, and there is enough friction to prevent slipping. Which way will the object roll?

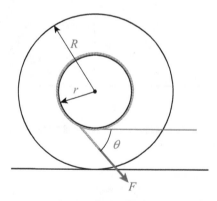

Figure 9-22 Which way will the object roll?

C-9-10 Clockwise. Hint: Sum torques about point of contact between object and surface.

✓ CHECKPOINT

C-9-11 Which Way Should You Push?

The drum shown in Figure 9-23 is machined from a single block such that a narrower cylindrical part protrudes from the centre of one face. A rope is wound around the protruding part. Which of the forces shown will result in the object rolling to the right?

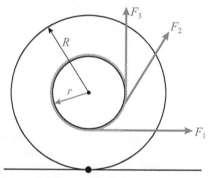

Figure 9-23 Which of the forces would cause the object to move to the right

C-9-11 only F_1. Hint: Sum torques about point of contact between object and surface.

LO 6

9-6 Rolling with no Friction?

So far, we have seen that friction is required for rolling. Go to Online Section 9-6 in the text online resources for a detailed discussion of rolling and friction, especially the concept of perfect rolling, in situations where the force of friction is zero.

LO 7

9-7 Rolling Friction

Go to Online Section 9-7 in the text online resources for a discussion of rolling friction and the effect of surface and object deformation on rolling. This discussion offers a more realistic model of the forces encountered when a rigid object rolls on a smooth surface.

FUNDAMENTAL CONCEPTS AND RELATIONSHIPS

ROLLING

Rolling is a combination of translational and rotational motion. For rolling without slipping,

$$v_{cm} = \omega R$$

$$a_{cm} = \alpha R$$

The part of the object in touch with the surface is momentarily stationary with respect to the surface.

$$v_{bottom} = 0$$

Rolling can also be treated as a series of successive rotations about the point of contact between the object and the surface.

The Dynamics of Rolling

Translation/Rotation Approach

$$\sum \vec{F} = m\vec{a}_{cm} \tag{9-11}$$

$$\sum \vec{\tau}_{cm} = \vec{I}_{cm}\alpha \tag{9-12}$$

Momentary-Pivot Approach (b refers to point of contact between object and surface)

$$\sum \vec{\tau}_b = I_b\vec{\alpha} \tag{9-13}$$

The force of static friction on an object rolling without slipping is usually an unknown:

$$0 \le f_s \le \mu_s N \tag{9-14a}$$

When an object is on the verge of slipping, the force of friction becomes

$$f_{s,max} = \mu_s N \tag{9-14b}$$

If an object slips as it rolls, the force of friction becomes the force of kinetic friction:

$$f_k = \mu_k N \tag{9-14c}$$

WORK AND ENERGY

A. Translation/Rotation Approach

$$K_{translation} = \frac{1}{2}mv_{cm}^2 \tag{9-15}$$

$$K_{rotation} = \frac{1}{2}I_{cm}\omega^2 \tag{9-16}$$

Total kinetic energy:

$$K = \frac{1}{2}mv_{cm}^2 + \frac{1}{2}I_{cm}\omega^2 \tag{9-17}$$

B. Momentary-Pivot Approach

$$K = \frac{1}{2}I_b\omega^2 \tag{9-18}$$

APPLICATIONS

Applications: ABS, traction control, tire design, drive trains, rolling element bearings

Key Terms: ideal rolling, momentary pivot, perfect rolling, rolling without slipping, rolling, skidding, slipping, spinning

QUESTIONS

1. A solid uniform ring and a solid uniform disk of the same radius and mass roll without slipping at the same speed on a horizontal surface, which turns into a ramp sloping upward from the horizontal. Which object will go higher up the incline?
2. A uniform solid sphere and a uniform hollow cylinder roll without slipping at the same speed on a horizontal surface. The surface then becomes frictionless and curves upward into a semicylindrical track of radius r. Will one of these objects go higher up the track? If so, which one? Assume that both objects are in contact with the track at all times and that neither of the objects has enough kinetic energy to make it to a height r above the bottom of the track.
3. A uniform solid cylinder and a uniform solid sphere of the same mass and radius roll without slipping downhill from the same height.

(a) Which object will have the higher rotational kinetic energy at the bottom of the hill?
(b) Which object will have the higher angular speed at the bottom of the hill?

4. Consider a uniform solid sphere rolling without slipping down a ramp of angle θ. How will the force of friction on this object change if you double its mass without changing its radius?
5. What is the main function of an antilock braking system?
6. In the absence of ABS, why is a driver advised to "pump" the brakes when stopping on a slippery surface?
7. A car coasts on the highway at a speed of 130 km/h. Is the friction between the tires and the road kinetic or static?
8. The ball in Figure 9-24 rolls on a flat horizontal surface without slipping. Establish the direction of the total acceleration of points A, B, and C on the wheel at the instant shown.

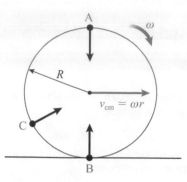

Figure 9-24 Question 8 and Problem 25

9. A solid uniform ring, disk, and sphere with equal masses and equal radii roll without slipping down a smooth ramp from the same height. Which of the three objects will reach the bottom of the ramp first?

10. A ring rolls down a ramp without slipping. At any point, calculate the ratio of the rotational energy to the translational kinetic energy. Repeat the exercise for a uniform solid disk and for a uniform solid sphere.

11. An object rolls down a hill such that $\frac{4}{6}$ of its kinetic energy is rotational. Determine an expression for the object's moment of inertia in terms of its mass and radius.

12. Can an object's rotational and translational kinetic energy ever be equal? Explain.

13. Two uniform solid disks, disk 1 and disk 2, of the same radius and thickness, roll down a ramp from the same height. The mass density of disk 2 is twice the density of disk 1. Will one disk reach the bottom of the hill before the other? If so, which one?

14. Two uniform solid spheres, sphere 1 and sphere 2, are at rest at the top of a hill. The spheres have the same mass but different radii: the radius of sphere 1 is greater than the radius of sphere 2. They roll without slipping downhill from the same height. Will one sphere reach the bottom of the hill before the other? If so, which one?

15. Two cylindrical shells, shell A and shell B, are at rest at the top of a hill. The mass of shell A is 16 times the mass of shell B, and the radius of shell A is one-half the radius of shell B. Both shells roll downhill from the same height. Will one shell reach the bottom of the hill before the other? If so, which one?

16. Two spheres, sphere A and sphere B, are at rest at the same height at the top of a hill. The mass of sphere A is 5 times the mass of sphere B, and the radius of sphere A is $\frac{1}{\sqrt{2}}$ the radius of sphere B. Both spheres roll without slipping downhill. Will one sphere reach the bottom of the hill before the other? If so, which one?

17. Is it possible for a spherical shell and a solid uniform cylinder, originally at rest and at the same height on top of a hill, to roll downhill without slipping and reach the bottom of the hill together? If so, find the ratio of their radii.

18. A cylinder and a sphere of the same mass roll freely down the same slope without slipping. Which of the two objects has a greater force of friction acting on it?

19. A sphere and a ring roll without slipping down a hill. The force of friction acting on the ring is 10 times the force of friction acting on the sphere. Find the mass ratio of the two objects.

20. A cylinder and another object roll without slipping down a slope. The mass of the second object is twice the mass of the cylinder. The force of friction acting on the cylinder is twice the force of friction acting on the other object. Find an expression for the moment of inertia of the second object in terms of its mass and radius.

21. Consider a typical wheel rolling freely on a flat horizontal surface. What could cause the wheel to slow down?

22. A marble rolls on a horizontal, nondeformable, smooth glass surface. Do you expect the marble to slow down because of friction? Explain why or why not.

23. A ball rolls down a hill without slipping. Assume ideal conditions, that is, the object and surface are nondeformable, and the surface is smooth. In which direction does the force of friction act on the ball?

PROBLEMS BY SECTION

For problems, star ratings will be used, (✶, ✶✶, or ✶✶✶), with more stars meaning more challenging problems.

Section 9-1 Rolling and Slipping

24. ✶ A rear-wheel-drive car accelerates from rest, and the wheels turn (roll) without slipping. Determine the direction of the force of friction on the car's front and rear wheels.

Section 9-2 Relationships Between Rotation and Translation for a Rolling Object

25. ✶ The wheel in Figure 9-24 rolls on a horizontal surface without slipping at a constant speed of 15 m/s. The wheel is 74 cm in diameter.
    (a) Determine the speed and acceleration of point A at the top of the wheel.
    (b) Find the magnitude of the total speed and acceleration of the point one-quarter of a period later and one-half of a period later.

26. ✶✶ A tundra buggy, which is a bus fitted with oversized wheels, is stuck in Churchill, Manitoba, on slippery ice. The wheel radius is 0.87 m. The speedometer goes from 0 to 27 km/h while the buggy moves a total distance of 5.0 m in 7.0 s. Find the tangential acceleration and the total acceleration of a point at the bottom of the wheel at the end of 3.0 s.

27. ✶ While driving on the frozen surface of a lake, you notice that your truck is moving at 10 km/h, but the speedometer says 56 km/h. The wheel radius is 46 cm. (Hint: The speedometer reading is obtained from the rotational speed of the tires.)
    (a) Find the speed of a point at the top of a tire.
    (b) Is there a point on the tire where the velocity is zero? If so, where is it located?

28. ✶✶ A cylinder of radius 1.2 m rolls at a constant speed. You apply a brake mechanism that slows the cylinder. At a certain instant, the speed of the cylinder's centre of mass is 11 m/s and the speed of the top of the cylinder is 14 m/s. Calculate the angular speed of the cylinder at that instant.

29. ✶✶ A wheel undergoes acceleration from rest, such that it spins while rolling. The acceleration of the centre of mass is measured to be 2.30 m/s², and the angular acceleration is measured to be 7.10 rad/s². The wheel's radius is 80.0 cm.

(a) Find the magnitude of the tangential acceleration of a point at the top of the wheel.

(b) Find the magnitude and direction of the total acceleration of a point at the bottom of the wheel at 1.70 s into the motion.

Section 9-3 Rolling Motion: Two Perspectives

For the problems in this section, approaching rolling as a rotation about point b, which is the point at which the rolling object is in contact with the surface on which it sits.

30. ✳ Show that it is valid to consider rolling as a rotation about point b.

31. ✳ If we consider rolling to be a rotation about point b, that leaves us with an apparent paradox: If the object were moving at a constant speed, then the centre of mass would be accelerating with respect to b. How do you resolve this apparent paradox?

32. ✳✳ Find the acceleration of the point on a rolling object in contact with the surface on which it rolls.

33. ✳✳ Calculate the maximum angle that the force F can make with the horizontal to cause the object in Figure 9-25 to roll to the right. Assume there is enough friction to ensure rolling without slipping.

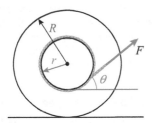

Figure 9-25 Problem 33

34. ✳✳ The object in Figure 9-26 has a moment of inertia about a point on its circumference of 17 kg·m². The object is pulled by a force of 160 N in the direction shown. There is enough friction to prevent slipping. Calculate the angular acceleration of the object, given that $R = 0.91$ m and $r = 0.32$ m.

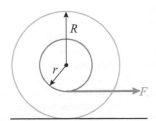

Figure 9-26 Problem 34

35. ✳✳ A form of yo-yo consists of a thin cylindrical shell with a string wound around it. Find the acceleration of this yo-yo when a child lets it go while holding the end of the string.

36. ✳✳ A wheel of radius 72 cm rolls without slipping down a 32° slope starting from rest (Figure 9-27). Point A is shown at $t = 0$. Find the speed and acceleration of point A at 3.0 s into the motion.

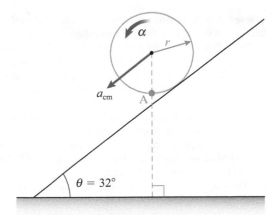

Figure 9-27 Problem 36

37. ✳ Indicate the direction in which the object in Figure 9-28 will roll for each of the given forces.

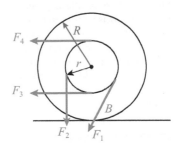

Figure 9-28 Problem 37

38. ✳✳ A car is moving at a constant speed of 29 m/s on a flat horizontal road. The car wheels, 41 cm in radius, roll without slipping. You paint a dot on the edge of one of the tires and monitor the dot's motion. Find the magnitude and direction of the acceleration of the dot when

(a) its speed is zero;

(b) its horizontal speed is zero;

(c) its vertical speed is zero;

(d) the point is at the top of the wheel.

Section 9-4 Newton's Second Law and Rolling

39. ✳✳ A solid uniform spherical shell of mass 1.2 kg rolls from rest without slipping down a hill with a 32° slope. Find the force of friction acting on the spherical shell. What would happen to the force of friction if you tripled the radius of the shell but kept the mass the same?

40. ✳✳ Consider a yo-yo as a uniform solid disk of mass 210 g and radius 6.0 cm. You release the yo-yo while holding the end of the string wound around it.

(a) Find the tension in the string.

(b) Find the acceleration of the centre of mass of the yo-yo.

(c) Find the angular acceleration.

41. ✳✳ Consider a cylindrical shell of radius 21 cm and mass 1.3 kg that rolls without slipping down a 37° hill. Find the force of friction between the surface and the cylindrical shell by taking torques about

(a) the centre of mass;

(b) the stationary point on the shell (the point in contact with the ground).

42. ✷✷✷ A rope is wound around a small, solid uniform ball, which you release from rest and allow to fall vertically while holding the rope (like a yo-yo). The length of the rope is such that the ball comes loose when it hits the ground. The ball is seen to spin for 11 m before coming to a perfect roll. The coefficient of static friction between the ball and the ground is 0.15.
    (a) Determine the height from which the ball was released.
    (b) Find the angular displacement of the ball before it comes to a perfect roll.
    (c) Calculate the displacement of the centre of mass before the ball comes to a perfect roll.
    (d) Does $a_{cm} = \alpha r$ before the ball starts rolling without slipping? Explain.
    (e) How much energy is lost to thermal energy before the ball comes to a perfect roll?
    (f) How long does the ball spin before coming to a perfect roll?
    (g) What is the final speed of the ball?

43. ✷✷✷ The ancient Egyptians may have used the method shown in Figure 9-29 to transfer large stones. The stone in Figure 9-29 of mass M is resting on two logs such that the weight is equally distributed between the logs. The stone is then pushed forward with a constant horizontal force F. Find an expression for the acceleration of the stone. The logs can be considered as ideal cylinders of mass m and radius R. Assume there is enough friction everywhere to prevent slipping.

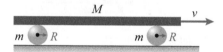

Figure 9-29   Problem 43

44. ✷✷✷ The coefficient of friction between each brake pad and brake disk on the wheels of a sports car is 0.25. The brake pad presses against the disk at a point that is 17 cm from the centre of the disk. The car has a mass of 1630 kg, and the weight is equally distributed on the four wheels. The car is moving at a constant speed on the road where the coefficient of static friction between the tires and the road is 0.56.
    (a) What is the maximum normal force that the brakes can exert on the brake disk so that the wheels do not skid? The wheels have a moment of inertia of 34 kg·m² and a radius of 29 cm.
    (b) Repeat part (a) with the car driving at constant speed up a 23° slope.

45. ✷✷ A uniform solid cylinder sits on a 4 m high hill with a 39° slope. The coefficient of kinetic friction is 0.12, and the coefficient of static friction is 0.17. The mass of the cylinder is 340 g, and its radius is 3 cm.
    (a) Will the cylinder roll without slipping? Explain.
    (b) Find the angular acceleration of the cylinder.
    (c) The cylinder travels from the top to the bottom of the hill. How much energy is lost to friction?

46. ✷✷✷ On an icy road, the coefficient of static friction between the tires of a bicycle and the road is 0.050. The wheels can be considered to be disks of radius 32 cm and mass 3.0 kg. The combined weight of the bicycle and the cyclist is 870 N and is distributed equally between the two wheels. What is the maximum torque that can be delivered to the rear wheel without causing it to spin?

47. ✷✷ You place a sphere on an inclined plane. The coefficient of static friction between the sphere and the plane is 0.30. Determine the maximum angle that the inclined plane can make with the horizontal so that the sphere does not skid. The sphere starts from rest.

48. ✷✷ A car rolls freely down a hill without slipping.
    (a) In which direction must the force of friction act on the tires?
    (b) The driver of the car puts the car in gear and depresses the gas pedal. What must the acceleration of the car be if the force of friction on the wheels is zero? Consider the car wheels to be disks.

49. ✷✷✷ A 52 kg cyclist rides his 7.0 kg bicycle up a 17° hill. The normal force from the road on the bicycle is equally distributed between the two wheels. Each wheel is 27 cm in radius and has a mass of 900 g. Two brake pads act at the rim of each of the wheels. The wheels can be considered as ideal rings. What is the normal force required between the brake pads and the wheels so that the force of friction between the tires and the road is zero?

50. ✷✷✷ A car driver waiting at a traffic light depresses the gas pedal as soon as the light turns green. Find the maximum torque that can be delivered to the wheels of the car without causing them to spin. Assume that the weight of the 2200 kg car is evenly distributed on the four wheels and that the coefficient of static friction between the asphalt and the tires is 0.70. Each wheel has a diameter of 65 cm and a moment of inertia of 25 kg m².

Section 9-5   Mechanical Energy and Rolling

51. ✷ A uniform solid cylinder of mass 1.2 kg rolls without slipping down a hill from a height of 3.4 m.
    (a) Calculate the cylinder's rotational and translational kinetic energies at the bottom of the hill.
    (b) Repeat the problem for a ring of the same mass.

52. ✷ A sphere and a cylinder, both of radius 17 cm and mass 34 kg, roll without slipping down a 4.0 m high. Both objects start from rest.
    (a) Calculate the total kinetic energy of each object at the bottom of the hill.
    (b) Calculate the angular speed of each object at the bottom of the hill.
    (c) Calculate the linear speed of the centre of mass of each object at the bottom of the hill.
    (d) Determine the ratio of the rotational energy to the translational kinetic energy for each object. Explain your results.

53. ✷✷ A solid, uniform bowling ball, 12.7 cm in diameter, starts from rest and rolls down a 7.0 m high ramp with a 37° slope. There is enough friction between the surface of the ramp and the bowling ball to prevent the ball from skidding. At the bottom of the ramp the ball goes up a frictionless ramp with the same slope as the first ramp.
    (a) Determine the maximum height that the ball can reach on the frictionless ramp.
    (b) Is mechanical energy conserved? Explain.
    (c) What is the angular speed of the bowling ball when it reaches its maximum height on the frictionless ramp?

54. ✷✷ A 7.0 kg steel ball of radius 4.0 cm rolls without slipping from rest down a hill of height 3.0 m. The steel ball then rolls on a horizontal surface for a certain distance before striking a spring whose spring constant k is 72 N/m. The spring is originally unstretched. The steel ball

compresses the spring until it comes to a momentary stop. Find the maximum compression of the spring.

55. ⁎⁎ A solid uniform ball of mass m and radius r rolls down a semispherical bowl, starting from a height h above the bottom of the bowl (Figure 9-30). The surface of the left half of the bowl provides enough friction to prevent the ball from slipping on the way down. The surface of the right half of the bowl is frictionless.

(a) Derive an expression for the maximum height the ball will reach on the frictionless side. Is energy not conserved? Explain.

(b) Find the translational kinetic energy of the ball at its highest point on the frictionless side.

(c) What is the rotational kinetic energy at that point?

(d) Explain how would your answer to (a) will change if we replace the ball with an ideal ring.

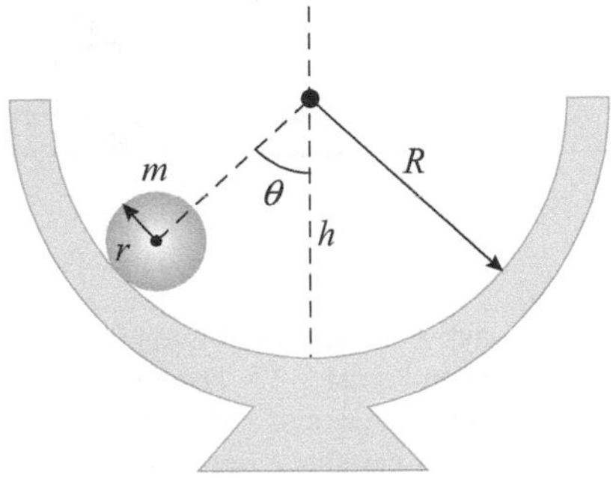

Figure 9-30 Problem 55

56. ⁎⁎ A sphere of mass m starts from rest and rolls without slipping down a ramp from a height h. At the bottom of the ramp, the sphere continues to move on a flat horizontal frictionless surface, and then collides with a horizontal spring whose spring constant is k. The spring is originally unstretched and is fitted with a light, frictionless plate at its end, which the sphere collides with (Figure 9-31).

(a) Derive an expression for the maximum compression of the spring.

(b) Derive an expression for the angular speed of the sphere when it strikes the spring.

(c) What would the result be in part (a) if there were no friction anywhere?

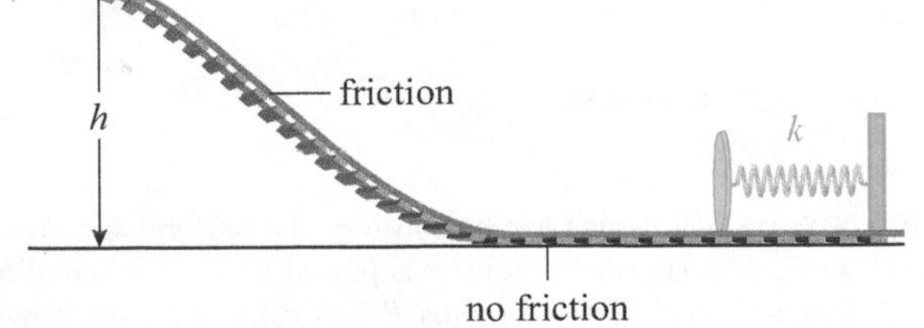

Figure 9-31 Problem 56

57. ⁎⁎ Use the work–mechanical energy approach to solve problem 40.

COMPREHENSIVE PROBLEMS

58. ⁎⁎ A uniform solid sphere of mass m and radius R rolls down an ideally flat ramp with a slope of angle θ with respect to the horizontal, as shown in Figure 9-32. Use energy considerations to derive an expression for the acceleration of the sphere. Do you have to account for friction? Explain. Compare the energy approach to the Newtonian equation of motion approach.

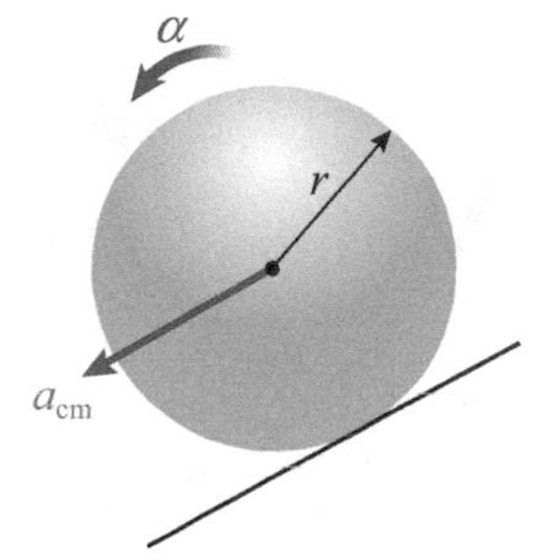

Figure 9-32 Problem 58

59. ⁎⁎⁎ In Figure 9-33, a massless string is wound around a sphere of mass M_s and radius R_s. The string passes over a pulley with moment of inertia I_1 and radius r_1; the string is attached to the mass m on top of the table as shown. The mass m, in turn, is attached to another massless string, which passes over a second pulley with moment of inertia I_2 and radius R_2, and is attached to the centre of a solid uniform cylinder of mass M_c and radius R_c. The system starts from rest, and then the cylinder accelerates downward. Find an expression for the acceleration of the sphere's centre of mass by applying (a) Newton's second law and (b) work–mechanical energy considerations. Assume perfect rolling without slipping anywhere.

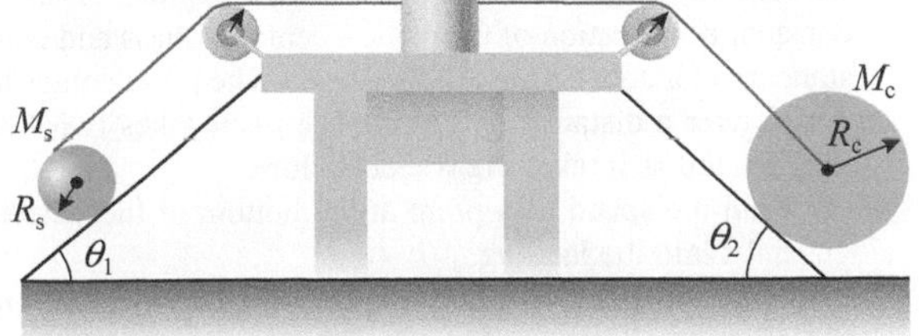

Figure 9-33 Problem 59

MECHANICS

60. ✶✶ The marble of mass m and radius R in Figure 9-34 starts from rest at height h and rolls down the track without slipping. The marble then rolls up the semicircular part of the track, which has a radius r equal to $h/4$. The coefficient of static friction between the marble and the track is μ_s. At the position shown, the normal force from the track on the marble is equal to 2.5 times the marble's weight.

(a) Find an expression for the moment of inertia of the marble.

(b) Determine the force of friction on the marble.

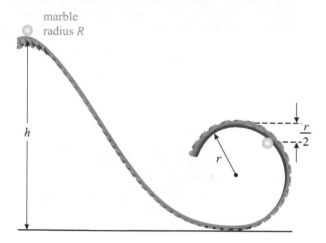

Figure 9-34 Problem 60

61. ✶✶ The object in Figure 9-35 is machined out of a single piece of metal such that its moment of inertia is $\dfrac{mr^2}{3}$. You pull on the rope horizontally as shown.

(a) In which direction will the object roll, assuming there is enough friction between the object and the surface to prevent slipping?

(b) For what value of r will you be able to pull the rope and cause the cylinder to roll such that the force of friction is zero?

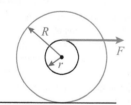

Figure 9-35 Problem 61

62. ✶✶ A pilot applies the wheel brakes upon landing so that his plane skids to a stop on a runway from its landing speed of 130 km/h. The 1.9 m diameter wheels were rolling without slipping before the brakes were applied. Assume constant acceleration of the plane's centre of mass and constant angular acceleration of its wheels. The plane comes to a stop over a distance of 270 m. The plane takes twice as long to stop as it takes the wheels to lock.

(a) Find the speed of a point at the bottom of the wheels 5.4 s into the landing.

(b) Find the speed of a point at the top of a wheel 110 m after the pilot started braking.

(c) Find the total acceleration of a point on the front edge of a wheel at the same height as the centre of the wheel after 3.0 s of braking.

63. ✶✶✶ A uniform solid ball of mass $m = 1.6$ kg and radius $r = 0.11$ m rolls down a 4.3 m hill without slipping. At the bottom of the hill, the surface is horizontal and the ball continues to roll until it strikes a stationary block of mass $M = 9.0$ kg (Figure 9-36). The collision is perfectly elastic; therefore, no energy is lost during the collision. How fast is the ball moving after the collision? Assume the ball never slips.

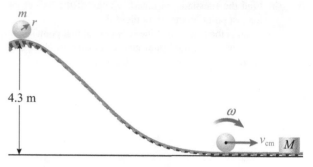

Figure 9-36 Problem 63

64. ✶✶✶ You kick a soccer ball of mass 0.90 kg and radius 14 cm, as shown in Figure 9-37. The kick has a duration of 91 ms and an average force of 161 N. Your foot strikes the ball horizontally 14 cm above the floor. The coefficient of kinetic friction between the ball and the ground is 0.090.

(a) Find the linear speed of the ball immediately after the collision.

(b) What is the angular speed of the ball immediately after the collision?

(c) Determine the linear speed of the ball when it starts rolling.

(d) Find the total amount of energy lost to friction.

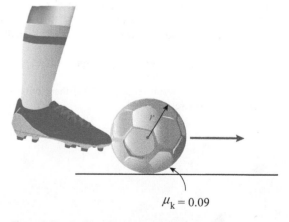

Figure 9-37 Problem 64

65. ✶✶ A ball, which can be considered a uniform spherical shell, of mass m and radius r is placed on the surface of a hemispherical bowl of radius R and rolls down the bowl without slipping (Figure 9-30).

(a) The ball starts from rest at an angle of 72° with the vertical (top of the bowl). What is the normal force on the ball when it is at the bottom of the bowl?

(b) Repeat part (a) if the ball slides without rotating.

(c) Repeat part (a) and (b) with $m = 0.67$ kg, $r = 0.070$ m, and $R = 9.0$ m.

66. ✷✷✷ Starting from rest, a ball of mass 0.63 kg and radius 0.17 m slides from the top to the bottom of a circular track of radius 1.1 m (Figure 9-38). The surface of the track is frictionless. The ball then moves up a ramp that is at an incline of 33° from the horizontal. The coefficient of friction between the ball and the ramp is 0.12.

(a) Find the acceleration of the ball as it skids up the ramp.

(b) What is the angular speed of the ball when it starts rolling perfectly?

(c) What is the acceleration of the ball as it rolls perfectly?

(d) Use energy considerations to find the maximum height the ball will reach above the bottom of the ramp.

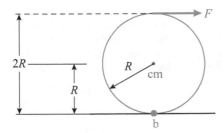

Figure 9-38 Problem 66

67. ✷✷✷ You throw a solid uniform sphere of radius 11 cm and mass 2.1 kg onto a smooth horizontal surface with a horizontal speed of 9.0 m/s. The sphere skids initially and then starts rolling without slipping. The coefficient of static friction between the sphere and the surface is 0.18 and the coefficient of kinetic friction is 0.14.

(a) Find the acceleration of the sphere when it starts rolling without slipping on the horizontal surface.

(b) For how long does the sphere skid?

(c) How much energy is lost to friction?

68. ✷✷ The object shown in Figure 9-39 sits on a horizontal surface. There is enough friction between the surface and the object to prevent slipping. What should the radius of the object be in terms of the moment of inertia and the mass so that the force of friction on the object is zero as it is being pulled? Can you infer what the object is?

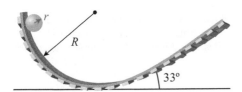

Figure 9-39 Problem 68

69. ✷✷ Redo Example 9-7, this time summing the torques about the point on the cylinder's centre of mass.

70. ✷✷ A 2.0 kg uniform solid bowling ball of radius 0.32 m rolls without slipping down a 43° slope with a smooth surface. At the bottom of the hill, the surface is horizontal.

(a) Find the acceleration of the ball on its way down the hill and when it is rolling on the horizontal surface.

(b) How would the acceleration of the ball on the horizontal surface change if the horizontal surface is covered with glossy ice such that it is almost frictionless?

71. ✷✷✷ Is it possible for an object to roll without slipping on completely frictionless ice? Explain.

72. ✷✷ A solid uniform cylinder is rolling without slipping at a constant speed of 10 m/s on a horizontal surface. The cylinder then rolls 7.0 m down a frictionless 17° slope. Determine the ratio of the rotational kinetic energy to the translational kinetic energy of the cylinder at the bottom of the slope.

73. ✷✷ A marble starts from rest and rolls without slipping down a smooth incline from height h. At the bottom of the incline, the surface is horizontal and frictionless. Will the marble eventually stop rolling on the frictionless surface? If so, what force would be responsible for changing the marble's angular speed?

74. ✷✷✷ You spin a bowling ball of mass m and radius r in your hands until it reaches an angular frequency ω. You then place it gently on a surface where the coefficient of friction between the surface and the ball is μ_k. The ball initially spins on the surface as it picks up linear speed because of kinetic friction (Figure 9-40).

(a) How much time passes before the ball starts rolling without spinning? How many revolutions does the ball go through during that time?

(b) Calculate the distance x that the centre of mass of the ball travels during the time you calculated in part (a).

(c) Does $\Delta x_{cm} = r(\Delta \theta)$ during that time?

(d) Find the final linear speed of the centre of mass of the ball when it starts to roll without slipping.

(e) What is the force of friction on the ball as it rolls on the horizontal surface without slipping?

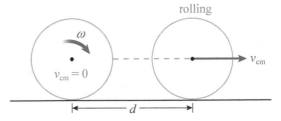

Figure 9-40 Problem 74

75. ✷ Equations (9-17) and (9-18) give two different expressions for the total kinetic energy of a rolling object. Show that these two expressions are always equal.

76. ✷✷✷ Develop parametric equations to describe the motion of a point at the edge of a rolling object.

77. ✷✷ A plank of mass M rests on two solid uniform cylinders as shown in Figure 9-29. The coefficient of static friction between the plank and the cylinders is 0.35, and the coefficient of static friction between the cylinders the horizontal surface they roll on is 0.55. Assume that the weight of the plank is equally distributed on the two cylinders.

(a) Find the maximum force F you can push the plank with so that no slipping occurs anywhere.

(b) What is the acceleration of the plank in part (a)?

251

78. ✳✳✳ A ball of diameter 24 cm is released from rest on a slope that is inclined 42° (Figure 9-41). The surface of the slope is frictionless for the first 7.0 m, and then becomes uniformly rough with some friction.
   (a) Will the ball eventually start rolling without slipping if the rough surface has a coefficient of kinetic friction of 0.13? Explain.
   (b) What is the minimum coefficient of kinetic friction required for the ball to roll without slipping 14 m after it meets the rough surface? (This part will require a graphical solution, i.e. a graphing calculator.)
   (c) How many turns will the ball in part (b) undergo before rolling without slipping? Use the results from your graphing exercise.
   (d) How long will it take the ball in part (b) to start rolling without slipping?
   (e) Find the angular speed of the ball in part (b) when it starts rolling without slipping.

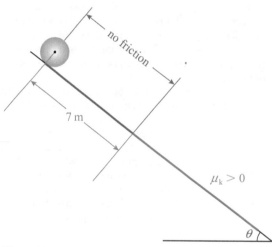

Figure 9-41   Problem 78

79. ✳✳✳ Suppose that the object in Figure 9-42 has a moment of inertia less than mr^2. Show that the force of friction will actually push the object forward as you pull on it. Assume that there is enough friction between the surface and the object to prevent slipping.

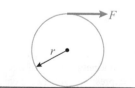

Figure 9-42   Problem 79

80. ✳✳✳ A ball of mass 940 g and radius 5.0 cm rolls down a semicylindrical surface of radius 0.70 m as shown in Figure 9-43. The coefficient of static friction between the surface and the ball is 0.15.
   (a) Determine the location at which the ball begins to skid or slip on the way down.
   (b) What is the normal force exerted by the surface on the ball at that point?

(c) Find the position where the ball leaves the surface. This part will require a graphical solution (i.e. a graphing calculator).
(d) Had the surface been completely frictionless, would the ball have left it at the same position as in part (a)? Explain.

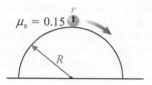

Figure 9-43   Problem 80

81. ✳✳ A rolling marble of radius r enters a hemispherical bowl of radius R through a hole as shown in Figure 9-44. The marble is moving with translational speed v and angular speed $\dfrac{v}{r}$.
   (a) What speed v will cause the marble to maintain a height h above the bottom of the bowl?
   (b) What would happen if the marble entered the bowl at a point slightly above or below the desired height at the same speed as in part (a)?

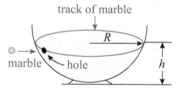

Figure 9-44   Problem 81

82. ✳✳ A hemispherical shell of radius R and mass m is made to spin about a fixed vertical axis as shown in Figure 9-45. As the spinning speed increases very slowly from zero, a small metal ball of radius r at the bottom of the shell starts to move upward. The shell's spinning speed is maintained at a constant value when the ball reaches a height h above the bottom of the bowl.
   (a) Find the spinning speed of the shell in terms of the other variables in the problem.
   (b) Find the work done on the metal ball as the shell accelerates. Assume that the ball rolls without slipping.
   (c) Why does the speed need to be increased slowly? Had the speed been changed quickly, what would you expect the ball to do, assuming it never goes above the rim?

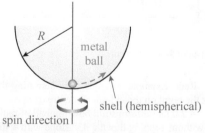

Figure 9-45   Problem 82

83. ✷✷ (a) Use and compare the work–energy and Newton's second law approaches for finding the angular acceleration of the sphere of radius R and mass m as it rolls up the incline in Figure 9-46 without slipping. The pulley has a moment of inertia 0.40 kg m², and radius 11 cm, $m = 1.1$ kg, $R = 23$ cm, and $M = 5.6$ kg. Assume that the hanging mass M accelerates downward.

(b) Find the force of friction between the sphere and the incline.

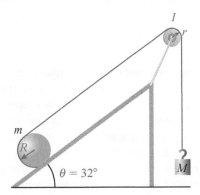

Figure 9-46   Problem 83

84. ✷✷✷ A solid uniform cylinder of mass 0.65 kg and radius 11 cm is held on a horizontal surface where it compresses a spring ($k = 1210$ N/m) by a distance of 41 cm from the equilibrium position. The cylinder is then let go, and it gains kinetic energy from the spring as the spring returns to equilibrium. Initially the surface is frictionless and then it becomes rough suddenly with a coefficient of friction between the surface and the cylinder of 0.12.

(a) What is the speed of the cylinder when it hits the rough part of the surface?

(b) How far does the cylinder go before it stops skidding?

(c) What is the angular displacement of the cylinder before it starts rolling without slipping?

(d) How much energy is lost to friction?

(e) Find the final angular speed of the cylinder.

See the text online resources at
**www.physics1e.nelson.com for Open Problems
and Data-Rich Problems related to this chapter.**

Chapter 10
Equilibrium and Elasticity

Soft Strength

It is desirable to use structural materials that can sag or stretch substantially before fracturing under extreme loading, especially for buildings that may be subject to earthquakes or hurricane-force winds. Generally, the more rigid steel alloys are also more brittle. Most construction-grade steel is designed to sag a significant amount before fracturing. During a severe earthquake, a building supported by this type of steel collapses more slowly than a building with a harder, more rigid steel, thus giving the occupants more time to evacuate.

Researchers at Stanford University have taken earthquake protection a step further by designing special building braces that can be incorporated into new buildings or added to existing buildings to make them more earthquake-resistant. The blue braces in Figure 10-1 are intended to be flexible enough to rock without damage. Strong steel cables hold the brace frames together and help restore them to their original configuration when they rock. As the frame of braces rocks, the bulk of the energy absorbed by the structure from the earthquake goes into deforming arrays of leaf-like steel "fuses" (shown in yellow) at the base of the frame. As the blue central column moves up and down or sideways, it causes the fuses to bend and yield with the motion, thus absorbing kinetic energy from the building. The deformed fuses can be replaced after an earthquake to restore the protection. Go to the text's online resources for a link to a video showing how these structural braces at work.

 For additional Making Connections, Examples, and Checkpoints, as well as activities and experiments to help increase your understanding of the chapter's concepts, please go to the text's online resources at www.physics1e.nelson.com.

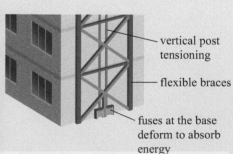

vertical post tensioning

flexible braces

fuses at the base deform to absorb energy

Figure 10-1 During an earthquake, the structure absorbs energy from the earthquake, and the energy deforms the fuses at the base of the frame.

LO 1

10-1 Static and Dynamic Equilibrium

For an object to be in equilibrium two conditions must be met:

1. The vector sum of the external forces acting on the object must be zero:

$$\sum \vec{F} = 0 \qquad (10\text{-}1)$$

2. The vector sum of all the external torques acting on the object about its centre of mass or any other given point must be zero:

$$\sum \vec{\tau} = 0 \qquad (10\text{-}2)$$

These two conditions can be restated in terms of momenta:

1. The linear momentum of the object is constant:

$$\frac{d\vec{p}}{dt} = 0 \qquad (10\text{-}3)$$

2. The angular momentum of the object about a given point is constant:

$$\frac{d\vec{L}}{dt} = 0 \qquad (10\text{-}4)$$

✓ CHECKPOINT

C-10-1 Equilibrium

A marble slides back and forth inside a frictionless bowl, as shown in Figure 10-2. At what point(s) is the marble in equilibrium?

(a) At points A and B
(b) At points A, B, and C
(c) Only point C because the acceleration there is zero
(d) The marble is never in equilibrium because it is always under the effect of a nonzero external force.

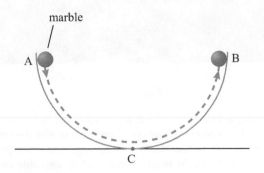

Figure 10-2 C-10-1

When an object is at rest and the above conditions for equilibrium are met, the object is in a state of **static equilibrium**. When the object is moving and the conditions for equilibrium are met, the object is in a state of **dynamic equilibrium**.

LO 2

10-2 Centre of Gravity

The force of gravity is one of the external forces that act on an object. For most applications, we can treat the force of gravity (the object's weight) as acting at a single point on the object. That point is called the **centre of gravity**. The location of the centre of gravity is important when analyzing the equilibrium conditions for a given object, as well as when determining the response of a rigid object to external forces. When the acceleration due to gravity is uniform over the dimensions of the object, the centre of gravity coincides with the centre of mass. The proof is left as an exercise (see problem 47). For most practical applications in this textbook, the centre of gravity coincides with the centre of mass. However, Example 10-1 deals with a situation in which the two centres do *not* coincide.

✓ CHECKPOINT

C-10-2 Tippy Refrigerator

You are trying to push a refrigerator across a room (Figure 10-3). There is friction between the floor and the refrigerator legs. Is the refrigerator more likely to tip if its centre of gravity is higher?

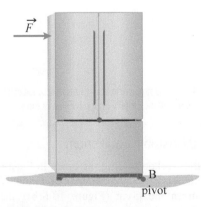

Figure 10-3 C-10-2

C-10-1 (d) there is no point where the net force is non zero.

C-10-2 It makes no difference, the torque exerted by the weight on the point of contact between the refrigerator and the floor does not depend on the vertical position of the centre of mass.

EXAMPLE 10-1

Determining the Centre of Gravity

The force of gravity near a small asteroid changes appreciably over short distances. A point mass $M_2 = m$ is a distance L directly above an identical point mass $M_1 = m$ on the surface of an asteroid (Figure 10-4). The acceleration due to gravity at the elevation of M_2 is 0.80 times the acceleration due to gravity at the elevation of M_1. Find the centre of gravity of the two masses.

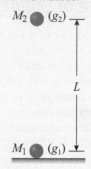

Figure 10-4   Where is the centre of gravity?

SOLUTION

Let Y_g, Y_1, and Y_2 represent the positions of the centre of gravity, M_1, and M_2, respectively. Since M_2 is directly above

M_1 and the line connecting the two masses is parallel to the direction of the force of gravity,

$$Y_g = \frac{Y_1 M_1 g_1 + Y_2 M_2 g_2}{M_1 g_1 + M_2 g_2}$$

Choosing the position of M_1 as our origin and up as the positive direction,

$$Y_g = \frac{(0)M_1 g_1 + LM_2 g_2}{M_1 g_1 + M_2 g_2} = \frac{LM_2 g_2}{M_1 g_1 + M_2 g_2}$$

Since $g_2 = 0.80g_1$ and $M_2 = M_1 = m$,

$$Y_g = \frac{Lm(0.80)g_1}{mg_1 + m(0.80)g_1} = \frac{L(0.80)mg_1}{1.8mg_1} = 0.444\ \text{L}$$

Making sense of the result:

Since the gravitational attraction is stronger at the lower mass, the centre of gravity is closer to the lower mass than to the upper mass, and is below the centre of mass.

For masses that are not positioned such that one is directly above the other, we can find the centre of gravity by summing torques about a given point (see problem 44).

MAKING CONNECTIONS

Balance and Centre of Mass

At first glance, you might think that the fork and spoon in Figure 10-5 should immediately drop off the side of the wineglass. However, the system consisting of the fork, spoon, and toothpick wedged together is actually in equilibrium. If you consider this system carefully, you will realize that its centre of mass is just above the rim of the glass, so the system can remain balanced there indefinitely.

Courtesy of Kris Carstens

Figure 10-5   How is it possible that a fork and spoon can be suspended on a toothpick totally on the outside of the edge of a glass?

Stable and Unstable Equilibrium

Consider the worker holding the beam in Figure 10-6(a). If the worker lets the beam swing down until it is hanging vertically from the pivot (Figure 10-6(b)), the worker has moved the beam to a state of **stable equilibrium**. The centre of mass and the centre of gravity of the beam are now below the pivot. A sideways nudge on the beam will move the beam slightly from its stable equilibrium

position, but the force of gravity will pull the beam back to this stable position. However, if the worker tilts the beam up so that its centre of mass is directly above the pivot (Figure 10-6(c)), the beam is in a state of **unstable equilibrium**. A sideways nudge will cause the beam to swing down. If there were no frictional losses, the beam would continue rotating about the pivot indefinitely.

(a)

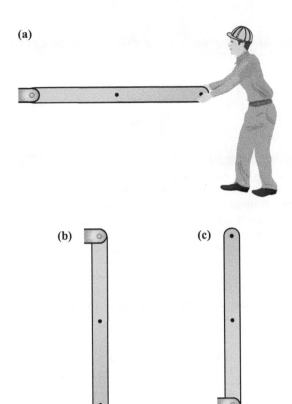

(b)   **(c)**

Figure 10-6 (a) A worker holding a beam (b) Stable equilibrium.
(c) Unstable equilibrium

10-3 Applying the Conditions for Equilibrium

We can use the conditions of equilibrium to determine the forces acting on a mechanical system. We could apply either condition of equilibrium first; however, as we will see throughout the chapter, we can almost always simplify the solution to a given by carefully choosing which condition of equilibrium to consider first. We illustrate this technique in the following example.

A strategy: Note that in the approach we take below, we will be treating the horizontal and vertical reaction forces at a given hinge or support as separate, independent variables rather than as components of a reaction force. This approach is simpler and will prove more efficient, as is used widely in engineering applications. We note that the total reaction force can be obtained from its component using simple vector algebra. We also note that although in the approach we take below the components of forces along a given direction will be treated as scalars, it is also valid to treat them as vectors, as is done elsewhere. We will often sum the forces along a certain direction, choose a positive direction and then use variables to denote the magnitudes of the forces. This approach will prove valuable when we have coupled equations and multiple unknowns.

EXAMPLE 10-2

A Horizontal Beam

The beam in Figure 10-7 has a mass of 120 kg. Assume that the hinge at the left end of the beam is frictionless.

(a) How much force must the worker exert on the beam for it to be in equilibrium?

(b) How much force does the beam then exert on the hinge?

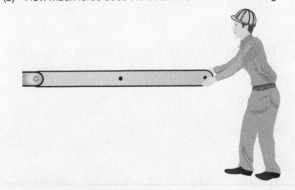

Figure 10-7 Example 10-2

SOLUTION

(a) We start by drawing the FBD of the beam. The FBD in Figure 10-8 shows three forces acting on the beam: the force of gravity at the centre of mass of the beam, the

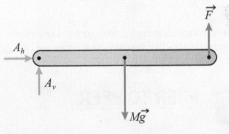

Figure 10-8 The FBD of the hinged horizontal beam

applied vertical force exerted by the worker at the right end, and the reaction force exerted by the hinge at the left end. We will use A_h and A_v to represent the horizontal and vertical components of the reaction force.

When the beam is in equilibrium, the sum of the forces in the x-direction must be zero, so $A_h = 0$. We then have two unknowns: the vertical reaction force, A_v, exerted by the hinge and the force, $\vec{F}$, exerted by the worker. The two conditions for equilibrium give us two equations, so we can solve for the unknown support forces.

Summing the forces in the y-direction, we have

$$\sum F_y = F_y + Mg_y + A_{v,y} = 0 \qquad (1)$$

(continued)

Choosing up as our positive direction, we have

$$F - Mg + A_v = 0 \qquad (2)$$

where the quantities in Equation (2) are variables that denote the magnitudes of the forces in Equation (1).

We have two unknowns in this equation. We have deliberately chosen to begin with the sum of the forces in the y direction to illustrate that although the equation is valid and needed that this is not the most efficient choice for a first step. We would like, whenever possible to begin with an equation that would hand us one of the unknowns. If the worker were to let go of the beam, it would rotate about the pivot at A. The force exerted by the worker prevents the beam from swinging. Summing the torques about point A gives

$$\sum \vec{\tau} = \vec{\tau}_{Mg} + \vec{\tau}_F = 0 \qquad (3)$$

The support forces at the hinge at A exert no torque about A; the moment arm (the perpendicular distance from the pivot to the force) of both supports, A_h and A_v is zero and hence the torque they exert about the point at which they act is zero. Taking clockwise to be positive, we have

$$\tau_{Mg} - \tau_F = 0 \qquad (4)$$

where the quantities in Equation (4) are variable that denote the magnitudes of the torques in Equation (3).

From Equation (8-13), we know that the torque due to a force about a pivot equals the product of the force and the perpendicular distance between the force and the pivot. Therefore,

$$Mg\left(\frac{L}{2}\right) - F(L) = 0$$

$$F = \frac{Mg}{2} = \frac{(120 \text{ kg})(9.81 \text{ m/s}^2)}{2} = 589 \text{ N} \qquad (5)$$

and so considering the torques first is more efficient than starting with the sum of the forces as we did above.

(b) Substituting the value of F from Equation (5) into Equation (2) gives

$$\frac{Mg}{2} - Mg + A_v = 0$$

$$A_v = \frac{Mg}{2} = 589 \text{ N}$$

Alternative Solution

The second condition for equilibrium (Equation (10-2)) states that the sum of the torques about *any* point must be zero. Summing the torques about a pivot or hinge is often the simplest approach; however, we can sum the torques about any other convenient point.

Let us consider the torques about the end of the beam that the worker is holding. The forces exerting torques about this point are the weight of the beam and the reaction force A_v. The worker exerts no torque at this end of the beam because the moment arm is zero. Therefore,

$$\sum \vec{\tau} = \vec{\tau}_{Mg} + \vec{\tau}_{A_v} = 0$$

Taking counterclockwise to be positive, we have

$$\tau_{Mg} - \tau_{A_v} = 0$$

Using the definition for torque (Equation (8-13)), we have

$$Mg\left(\frac{L}{2}\right) - A_v(L) = 0$$

and

$$A_v = \frac{Mg}{2} = 589 \text{ N}$$

Making sense of the result:

Because of the symmetry of the problem, one would expect the weight of the beam to be equally distributed between the hinge and the worker.

PEER TO PEER

To solve a given problem, you need to have at least as many equations as there are unknowns. For equilibrium problems, try to begin with the condition of equilibrium that will give you one of the unknowns immediately.

EXAMPLE 10-3

A Sloping Beam

The beam in Figure 10-9 has a mass of 2200 kg. It is supported at one end by a frictionless hinge, and the other end leans against a wall. Find the force that the wall exerts on the beam, and the reaction forces at the hinge. Assume there is no friction between the wall and the beam.

SOLUTION

We begin by drawing the FBD for the beam, which is shown in Figure 10-10. Since the wall is frictionless, the reaction force from the wall has no vertical component. Since this force is normal to the wall, we label it N. A_h and A_v are the horizontal and vertical components of the support force at the hinge, and the weight of the beam, Mg, acts at the middle of the beam.

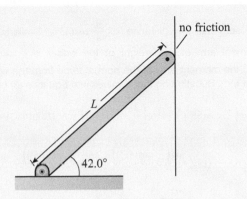

no friction

L

$42.0°$

Figure 10-9 Example 10-3

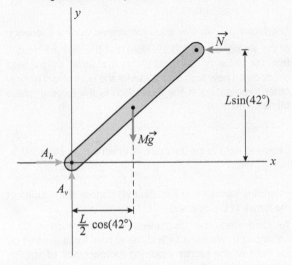

y

$\vec{N}$

$L\sin(42°)$

$M\vec{g}$

A_h

x

A_v

$\dfrac{L}{2}\cos(42°)$

Figure 10-10 The FBD of the forces on the beam

We begin by summing the torques about the hinge at the base of the beam. The forces that exert torques about this point are the weight of the beam and the normal force from the wall, the support forces at the hinge exert no torque about the hinge as their moment arms are zero:

$$\sum \vec{\tau}_A = \vec{\tau}_{Mg} + \vec{\tau}_N = 0 \qquad (1)$$

Taking clockwise to be positive, we have

$$\tau_{Mg} - \tau_N = 0 \qquad (2)$$

where the variables in Equation (2) denote magnitudes of the torques in Equation (1).

From Figure 10-10, we see that the perpendicular distance between the gravitational force and the pivot point of the hinge is $\dfrac{L}{2}\cos(42°)$, and the perpendicular distance between the normal force and the pivot is $L\sin(42°)$. We can express the sum of the torques at the pivot in terms of these moment arms:

$$Mg\left(\frac{L}{2}\cos(42°)\right) - N(L\sin(42°)) = 0 \qquad (3)$$

Solving for N gives

$$N = \frac{Mg}{2\tan(42°)} = 12.0 \text{ kN} \qquad (4)$$

Summing the forces acting on the beam in the y-direction,

$$\sum F_y = 0 \qquad (5)$$

$$A_{v,y} + Mg_y = 0 \qquad (6)$$

Choosing up as our positive direction, we have

$$A_v - Mg = 0 \qquad (7)$$

$$A_v = Mg = 21.6 \text{ kN} \qquad (8)$$

Taking the sum of the forces in the horizontal direction gives

$$\sum F_x = 0 \qquad (9)$$

$$A_{h,x} + N_x = 0 \qquad (10)$$

Choosing right as the positive direction and substituting from Equation (4),

$$A_h - N = 0 \Rightarrow A_h = N = \frac{Mg}{2\tan(42°)} = 12.0 \text{ kN} \qquad (11)$$

we note here that the quantities in Equations (7) and (11) are variables that denote the magnitudes of the forces in Equations (6) and (10) respectively.

Making sense of the result:

The smaller the angle at the base of the beam, the smaller the moment arm of the normal force from the wall, and the greater the normal force has to be to counterbalance the torque from the force of gravity.

EXAMPLE 10-4

Ladder Safety

The worker of mass M in Figure 10-11 wants to climb three-quarters of the way up a 4.00 m–long ladder of mass m. The ladder makes an angle of 65.0° with the floor. Find the minimum coefficient of friction between the ladder and the floor needed to prevent the ladder from sliding on the floor. Assume that the wall is frictionless.

(continued)

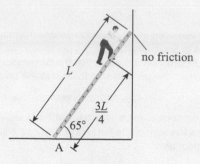

Figure 10-11 Example 10-4

SOLUTION

If the ladder does not slide, the friction between it and the floor is static friction. Equation (5-2) relates the maximum force of static friction to the normal force and the coefficient of static friction:

$$f_s = N_g \mu_s \tag{1}$$

We start by drawing the FBD of the ladder. Five forces are acting on the ladder: the weight of the ladder (mg), the weight of the worker (Mg), the normal support force (N_g) exerted by the floor, the force of static friction (f), and a horizontal support force (N_w) exerted by the wall (Figure 10-12). Since the wall is frictionless, it cannot exert a vertical force on the ladder.

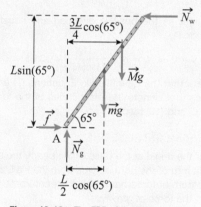

Figure 10-12 The FBD of the ladder

Now consider the torques about A, the point of contact between the ladder and the floor. The force of friction and the normal force from the floor do not exert a torque about point A because their moment arms are zero. Since the ladder is in equilibrium if it does not slide, the sum of the torques of the other three forces must be zero:

$$\sum \vec{\tau}_A = \vec{\tau}_{mg} + \vec{\tau}_{Mg} + \vec{\tau}_{N_w} = 0 \tag{2}$$

Taking clockwise to be positive,

$$\tau_{mg} + \tau_{Mg} - \tau_{N_w} = 0 \tag{3}$$

where the quantities in Equation (3) denote magnitudes of the torques in Equation (2)

The torque exerted by the worker's weight can be written as the product of Mg and the perpendicular distance between the force and the foot of the ladder. From Figure 10-12

we see that the distance is $\dfrac{3L}{4}\cos(65°)$. Similarly, the moment arm of the weight of the ladder is $\dfrac{L}{2}\cos(65°)$, and the moment arm of the normal force from the wall is $L\sin(65°)$. Substituting these values into Equation (3) gives

$$mg\left(\frac{L}{2}\cos(65°)\right) + Mg\left(\frac{3L}{4}\cos(65°)\right) - N_w(L\sin(65°)) = 0$$

and

$$N_w = \frac{mg\left(\dfrac{L}{2}\cos(65°)\right) + Mg\left(\dfrac{3L}{4}\cos(65°)\right)}{L\sin(65°)}$$

$$= \frac{\dfrac{mg}{2} + \dfrac{3}{4}Mg}{\tan(65°)} \tag{4}$$

The factor of $\dfrac{3}{4}$ in this equation comes from the location of the worker on the ladder. Note that this factor indicates that the value of the normal force exerted by the wall depends on how high up the ladder the worker is. The only other force acting in the x-direction is the force of static friction. Therefore,

$$\sum F_x = N_{wx} + f_x = 0 \tag{5}$$

Choosing right to be the positive direction, we have

$$-N_w + f = 0 \quad \Rightarrow \quad f = N_w \tag{6}$$

where the variables in Equation (6) denote magnitudes of the forces in Equation (5)

To determine the coefficient of static friction from Equation (1), we need to find the normal force exerted by the floor on the ladder. Applying the condition of equilibrium for forces in the y-direction, we get

$$\sum F_y = Mg_y + mg_y + N_{gy} = 0 \tag{7}$$

Taking up to be positive,

$$-Mg - mg + N_g = 0 \quad \text{and} \quad N_g = Mg + mg \tag{8}$$

Now that we have what we need, putting the value of f from Equation (6) and that of N_g from Equation (8) into Equation (1), we have

$$\mu_s = \frac{f}{N_g} = \frac{\dfrac{mg}{2} + \dfrac{3}{4}Mg}{(Mg + mg)\tan(65°)} = \frac{2m + 3M}{4(M + m)\tan(65°)}$$

Note here that the quantities in Equations (6) and (8) are variables that denote the magnitudes of the forces in Equations (5) and (7) respectively.

Making sense of the result:

The larger the angle that the ladder makes with the vertical, the less friction there needs to be between the floor and the ladder. When the ladder is vertical, it does not exert a horizontal force on the ground, and no friction is needed. However, this position would be an unstable equilibrium because any outward force will cause the ladder to fall away from the wall. For safety, a ladder is normally positioned such that the distance from the foot of the ladder to a vertical wall is one-quarter of the height of the point of contact of the ladder with the wall.

PEER TO PEER

After writing the condition for equilibrium for torques in vector form, you need to pick a positive direction. Choose either clockwise or counterclockwise as the positive direction. At the point about which you are summing the torques, any torque that pushes the beam or element about the pivot in the direction you have picked as positive you denote by a curved arrow with a plus sign. This indicates that the torque is positive. Any torque pushing the beam or element opposite to the direction of the curved arrow is negative.

Choosing up or down or left or right as a positive direction for forces has no bearing on the sign of the torque. The sign of the torque depends only on whether the torque rotates the object in the rotational direction (clockwise or counterclockwise) that you have picked as positive.

EXAMPLE 10-5

A Beam Supported by a Cable

A 500 kg crate sits 3.00 m from a hinge on a 4.00 m–long horizontal beam, as shown in Figure 10-13. The beam is supported at one end with a cable that is attached to a wall; the cable makes an angle of 42.0° with the horizontal. The beam has a mass of 270 kg. Determine the tension in the cable and the reaction force at the hinge.

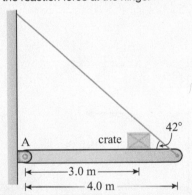

Figure 10-13 Example 10-5

SOLUTION

The weight of the beam, the weight of the crate, and the tension in the cable, as well as the horizontal and vertical components of the reaction force at the hinge are shown in the FBD of Figure 10-14.

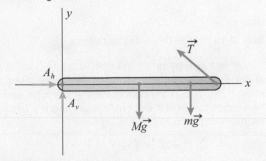

Figure 10-14 The FBD of the beam

We apply the condition for equilibrium for the torques about the hinge at A. The forces that exert torques about A are the weight of the beam (Mg), the weight of the crate (mg), and the tension in the cable (T). A_h and A_v do not exert a torque about A because the moment arm is zero. Therefore,

$$\sum \vec{\tau}_A = \vec{\tau}_{Mg} + \vec{\tau}_{mg} + \vec{\tau}_T = 0$$

Taking clockwise to be positive,

$$\tau_{Mg} + \tau_{mg} - \tau_T = 0$$

the variables in the line above denote magnitude of the torques. From Figure 10-14, using Equation 8-10 for the torque exerted by the tension and Equation 8-16 for the torque exerted by Mg and mg, and taking the weight of the beam to be at its centre, we get

$$Mg(2 \text{ m}) + mg(3 \text{ m}) - T(4 \text{ m})\sin(42°) = 0$$

Solving for T gives

$$T = \frac{(500 \text{ kg})(9.81 \text{ m/s}^2)(3 \text{ m}) + (270 \text{ kg})(9.81 \text{ m/s}^2)(2 \text{ m})}{(4 \text{ m})\sin(42°)}$$

$$= 7477 \text{ N} = 7480 \text{ N}$$

Next, we sum the forces in the y-direction. Here, we have the weight of the beam, the weight of the crate, the vertical component of the tension, and the vertical component (A_v) of the reaction force at the hinge:

$$\sum F_y = Mg_y + mg_y + T_y + A_{v,y} = 0$$

From Figure 10-14, the y-component of the tension force is $T\sin(42°)$.

Choosing up as our positive direction, we have

$$-Mg - mg + T\sin(42°) + A_v = 0$$

and

$$A_v = Mg + mg - T\sin(42°)$$

$$= (500 \text{ kg})(9.81 \text{ m/s}^2) + (270 \text{ kg})(9.81 \text{ m/s}^2) - 7477 \sin(42°) \text{ N}$$

$$= 2550.6 \text{ N} = 2550 \text{ N}$$

to 3 significant figures.

Now, the only unknown is the horizontal component, A_h, of the reaction force at the hinge. We take the sum of the forces acting on the beam in the x-direction, A_h, and the horizontal component of the tension, T:

$$\sum F_x = T_x + A_{h,x} = 0$$

(continued)

From Figure 10-14, the x-component of the tension is $T\cos(42°)$.

Choosing right as our positive direction,

$$-T\cos(42°) + A_h = 0 \quad \text{or} \quad A_h = T\cos(42°)$$

Therefore,

$$A_h = T\cos(42°) = (7477 \text{ N}) \cos(42°) = 5556.5 \text{ N} = 5560 \text{ N}$$

EXAMPLE 10-6

Working with a Force with an Unknown Direction

Assume that the weight of the V-shaped structural element shown in Figure 10-15 is negligible compared to the other forces acting on it. Find the reaction forces acting on this element and the force it exerts on the roller. Assume that the roller is frictionless and that $F_1 = 600.0$ N, $F_2 = 400.0$ N, and $F_3 = 500.0$ N.

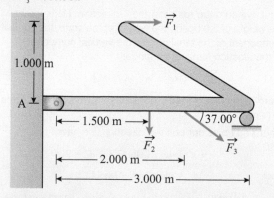

Figure 10-15　Example 10-6

SOLUTION

We begin with the FBD of the structural element (Figure 10-16). At the hinged end, we have the reaction force with horizontal and vertical components, A_h and A_v. At the roller end, we have only a normal force, N, since there is no friction. We do not know the directions of A_h or A_v, we will assume that A_v points down and that A_h points to the left.

Making sense of the result:

The cable supports part of the weight of the beam, but since the cable is on an angle, part of the force it exerts is horizontal, which results in the horizontal component of the reaction force at the hinge. A quick check would be to sum torques about the other end of the beam, which gives $A_v = 2550$ N as well.

LO 4

10-4　Working with Unknown Forces

For most of the problems in this chapter, the directions of unknown forces are fairly obvious. However, for more complex situations, it can be difficult to predict the direction of a given unknown force. For such problems, we can simply assume a direction. If the magnitude of the force then comes out to be positive, our assumption was correct. However, if we get a negative value for the magnitude of the force, then the direction of the force is opposite to what we had assumed.

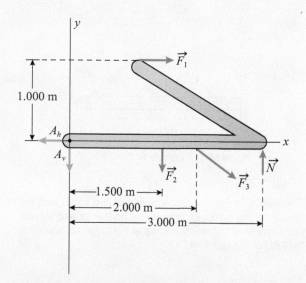

Figure 10-16　The FBD of the structural element

Let us begin by taking the sum of the torques about the hinge. The forces that exert torques about A are the forces F_1, F_2, F_3 and the normal force at the roller end, N. The reaction forces at A exert no torques about A because their moment arm is zero. Thus,

$$\sum \vec{\tau}_A = \vec{\tau}_{F_1} + \vec{\tau}_{F_2} + \vec{\tau}_{F_3} + \vec{\tau}_N = 0 \tag{1}$$

Taking clockwise to be positive,

$$\tau_{F_1} + \tau_{F_2} + \tau_{F_3} - \tau_N = 0 \tag{2}$$

where the variables in Equation (2) denote the magnitudes of the torques in Equation (1)

then

$$F_1(1.00\text{ m}) + F_2(1.50\text{ m}) + F_3(2.00\text{ m})\sin(37°) - N(3.00\text{ m}) = 0$$

Note that for the torques exerted by F_1, F_2 and N, we have used the moment arm concept (perpendicular distance from pivot to force) as given in Equation 8-13 and 8-16. For torque exerted by F_3 we have used the definition of torque as given by Equation 8-10.

Solving for N gives

$$N = \frac{(600\text{ N})(1.00\text{ m}) + (400\text{ N})(1.50\text{ N}) + (500\text{ N})(2.00\text{ m})\sin(37°)}{3.00\text{ m}}$$

$$= 600.6\text{ N} \tag{3}$$

Next, we sum the forces in the y-direction:

$$\sum F_y = A_{v,y} + F_{2y} + F_{3y} + N_y = 0 \tag{4}$$

Choosing up as our positive direction,

$$-A_v - F_2 - F_3\sin(37°) + N = 0$$

in the line above the variables denote the magnitudes of the forces in Equation (4), then,

$$A_v = -F_2 - F_3\sin(37°) + N$$
$$= -400\text{ N} - (500\text{ N})\sin(37°) + 600.6\text{ N} = -100.3\text{ N}$$

The negative value for A_v indicates that our guess for the direction of this component is not correct, and that the force points up, opposite to what we had guessed.

To determine A_h, we apply the condition for equilibrium for forces in the x-directions:

$$\sum F_x = A_{h,x} + F_{1x} + F_{3x} = 0 \tag{5}$$

Choosing right as our positive direction,

$$-A_h + F_1 + F_3\cos(37°) = 0 \tag{6}$$
$$A_h = F_1 + F_3\cos(37°) = 600\text{ N} + (500\text{ N})\cos(37°) = 999.3\text{ N} \tag{7}$$

we note here that the variables in Equations (6) and (7) denote the magnitudes of the quantities in Equation (5).

Making sense of the result:

The normal force has a longer moment arm than all the other forces exerting a torque about the hinge. This explains its relatively low value considering that it is the only force exerting a torque in a direction opposite to those exerted by the other forces.

EXAMPLE 10-7

Analyzing a Multipart Structure

The shorter beam in Figure 10-17 has a mass $m_1 = 1100$ kg and is supported by a free hinge at point A. The longer beam has a mass $m_2 = 2100$ kg and is supported by a free hinge at point B. The two beams are fixed to each other by a free hinge at C. Find the reaction forces at the hinges.

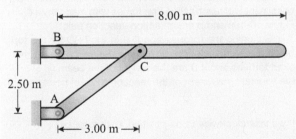

Figure 10-17   Example 10-7

SOLUTION

Let us first look at the FBD of the two-beam structure as a whole, which is shown in Figure 10-18. The forces acting on this system are the reaction forces at points A and B and the weights of the two beams. We do not consider the force acting on the hinge at C because this force is internal to the system. An FBD shows only the external forces acting on an object or a system.

We could try to infer the direction of the components of the reaction forces, but we will simply take a guess. If

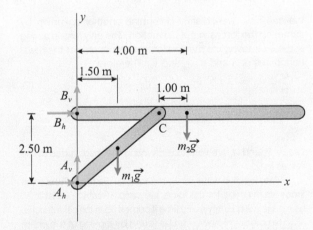

Figure 10-18   The FBDs of both beams

our choice for the direction of these components is correct, then the values that we obtain for the forces will be positive; if the value that we obtain for a component is negative, then we know that our choice for that component is incorrect. An incorrect choice of direction for a component will not affect the magnitude of the force.

If we take the sum of the forces in the x-direction at this stage, we will end up with two unknowns, A_h and B_h, in one equation. Similarly, if we take the sum of the forces in the y-direction, both A_v and B_v will be in the same equation. Although these are valid equations, starting with the sum of the torques will prove to be a simpler and more efficient approach.

(continued)

The forces that exert torques about B in figure 10-18 are the weights of the beams, m_1g and m_2g, as well as A_h, the horizontal component of the reaction force at A. The components of the reaction force at B, B_h and B_v exert no torque about B because the moment arm of these components about B is zero. The vertical component of the reaction force at A exerts no torque about point B because the line of action of this component passes through the hinge at B; therefore, the moment arm for A_v about B is also zero. So, the sum of the torques about point B is

$$\sum \vec{\tau}_B = \vec{\tau}_{m_1g} + \vec{\tau}_{m_2g} + \vec{\tau}_{A_h} = 0$$

Taking clockwise to be positive, the torques of m_1g and m_2g are positive, and the torque of A_h is negative. Thus,

$$\tau_{m_1g} + \tau_{m_2g} - \tau_{A_h} = 0$$

where the variables in the line above denote the magnitudes of the torques.

From Figure 10-18, the perpendicular distances from the pivot at B to m_1g and m_2g are 1.5 m and 4.0 m, respectively. Therefore,

$$m_1g(1.5\text{m}) + m_2g(4.0\text{ m}) - A_h(2.5\text{ m}) = 0$$

$$A_h = \frac{m_1g(1.5\text{ m}) + m_2g(4\text{ m})}{2.5\text{ m}}$$

$$= \frac{1100\text{ kg }(9.81\text{ m/s}^2)(1.5\text{ m}) + 2100\text{ kg }(9.81\text{ m/s}^2)(4\text{ m})}{2.5\text{ m}}$$

$$= 39\ 436\text{ N} = 39.4\text{ kN} \tag{1}$$

We can now immediately determine another unknown by summing the forces in the x-direction. The only forces acting in the x-direction are the horizontal components of the reaction forces at A and B, A_h and B_h. Therefore,

$$\sum F_x = A_{h,x} + B_{h,x} = 0 \tag{2}$$

Choosing right as our positive direction,

$$B_h = -A_h = -39.4\text{ kN} \tag{3}$$

where B_h and A_h are variables denoting the magnitudes of the forces in (2).

The negative value for B_h means that we have chosen the incorrect direction for the force. We simply treat the variable B_h as a negative quantity every time it comes up in the calculations.

No other unknown can be found by looking at the entire two-beam structure as one body, so we now apply the conditions for equilibrium to the components of the structure. We still need to find the vertical reactions at each hinge at A and B, as well as the forces acting on the hinge at C.

We now draw the FBDs for each beam (Figure 10-19). The forces acting on the long beam are its weight, the horizontal and vertical components of reaction force at B, B_h and B_v, and the force acting on the hinge at C. Since we do not know the direction of the force at C, we make the assumptions shown in Figure 10-19.

Examine closely how C_h and C_v, the components of the force at C, act one way on the top beam and the opposite way on the bottom beam. According to our assumption, C_v acts down on the top beam, which means that the bottom beam pulls the top beam down. Consequently, the

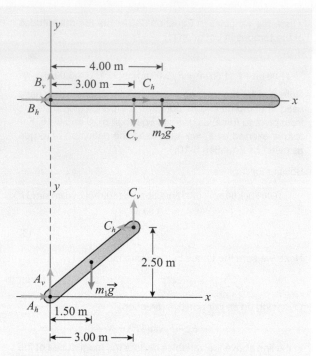

Figure 10-19 (a) The FBD of the top beam (b) The FBD of the bottom beam

top beam must pull the bottom beam up with an equal but opposite force in accordance with Newton's third law.

Note that to illustrate what happens when you make the incorrect assumption about the direction of an unknown force, we have made an absurd assumption that the vertical component of the reaction force at C acts downward on the top beam. In fact, C_v must be directed upward so that it counters the torque due to the weight of the top beam and hence hold the beam up. However, choosing the incorrect direction will not affect the end result of the analysis. Nonetheless, the directions we choose have to be consistent with each other. The forces with magnitude C_v in Figure 10-19a and b are an action–reaction pair.

Summing the torques about B in the FBD for the top beam gives us one of the unknowns. Since the forces that exert torques about B are the weight of the beam (m_2g) and the vertical component of the reaction force at point C,

$$\vec{\tau}_{m_2g} + \vec{\tau}_{C_v} = 0 \tag{4}$$

If we take clockwise to be positive, both torques

$$\tau_{m_2g} + \tau_{C_v} = 0 \tag{5}$$

$$m_2g(4\text{ m}) + C_v(3\text{ m}) = 0 \tag{6}$$

which gives us

$$C_v = \frac{-m_2g(4\text{ m})}{3\text{ m}} = \frac{-2100\text{ kg}(9.81\text{ m/s}^2)(4\text{ m})}{3\text{ m}}$$

$$= -27\ 468\text{ N} = -27.5\text{ kN} \tag{7}$$

where the variables in Equation (5) denote magnitudes of the torques in Equation (4)

The negative value for the variable C_v means that our guess for the direction of C_v is incorrect. We can redraw the FBD to show the correct direction of C_v or we can keep the FBD as is and use the negative value for C_v in all subsequent calculations involving that variable. We will always choose to do the latter.

To find the vertical reaction at the hinge at B, we apply the equilibrium conditions for the vertical forces acting on the top beam. Here, we have the weight of the beam, the vertical reaction at C, and the vertical reaction at B, therefore

$$\sum F_y = m_2 g_y + C_{v,y} + B_{v,y} = 0$$

Choosing up as our positive direction, we have

$$-m_2 g - C_v + B_v = 0$$

$$B_v = m_2 g + C_v$$

where the variables B_v and C_v denote magnitudes.

Keeping in mind that the value for C_v is negative, we have

$$B_v = Mg + C_v = 2100 \text{ kg}(9.81 \text{ m/s}^2) + (-27\,468 \text{ N})$$

$$-6867 \text{ N} = -6870 \text{ N} \qquad (8)$$

The negative value for the variable B_v means that our assumption regarding the direction for B_v in Figure 10-18 was not correct.

Next, we find the horizontal component of the force acting on the hinge at C, C_h, by summing the forces acting on the top beam in the x-direction. Since the only forces here are the reaction forces at B and C,

$$\sum F_x = C_{h,x} + B_{h,x} = 0$$

Choosing right as our positive direction,

$$C_h + B_h = 0 \Rightarrow C_h = -B_h = -(-39.4 \text{ kN}) = 39.4 \text{ kN}$$

where the variables C_h and B_h in the line above denote magnitudes of their respective forces.

Note that we have used the value of B_h as -39.4 kN as obtained in Equation 3 above. The positive value indicates that we picked the correct direction for C_h in the FBD in Figure 10-19. Now we can go back to the FBD for the entire system. The only unknown we have left is the vertical component of the reaction force at A. We take the sum of the vertical forces on the entire structure: the weight of the top beam ($m_2 g$), the weight of the bottom beam ($m_1 g$), and the vertical components of the reaction forces at A and B, A_v and B_v. Thus,

$$\sum F_y = m_1 g_y + m_2 g_y + A_{v,y} + B_{v,y} = 0$$

Choosing up as our positive direction,

$$-m_1 g - m_2 g + A_v + B_v = 0 \qquad (9)$$

$$A_v = m_1 g + m_2 g - B_v$$

$$= 1100 \text{ kg } (9.81 \text{ m/s}^2) + 2100 \text{ N } (9.81 \text{ m/s}^2) - (-6860) \text{ N}$$

$$= 38\,259 \text{ N} = 38.2 \text{ kN}$$

where the variables in Equation (9) denote magnitudes of the forces.

Making sense of the result:

One expects the vertical component at A to be larger than the vertical component at B because the bottom hinge is supporting the weight of the bottom beam, and some of the weight of the top beam is transferred to the bottom beam through the hinge at C.

 CHECKPOINT

C-10-4 Internal Forces

Why do we not include the reaction force at C when summing the forces on the entire system in Example 10-7?

C-10-4 The reaction force at C is internal to the system.

 PEER TO PEER

Sometimes, it helps to look at the FBD for the whole structure before focusing on individual components of the structure.

LO 5

10-5 Deformation and Elasticity

So far, we have treated structural components such as beams and supports as rigid objects that do not deform. However, deformation often occurs when external forces act on objects. If sufficient force is applied, steel and concrete structural elements will deform, break, or buckle.

The three types of forces of primary interest to us are tension, compression, and shear.

When you pull each end of a bar away from the centre of the bar, the bar is under **tension.** This is shown in Figure 10-20(a). When you push each end of a bar toward the centre, you are exerting a **compression force** as in Figure 10-20(b). When you clamp one end of a bar and push on the side of the free section, you are exerting a **shear force** as shown in Figure 10-20(c). The forces exerted by scissors and tin snips when cutting paper and sheet metal are examples of shear forces.

Stress

Consider the cylinder of length L and cross section A shown in Figure 10-21(a). Two forces equal in magnitude (F) and opposite in direction pull the ends of the cylinder away from the centre. This tension on the cylinder causes it to elongate. Similarly, if the forces push the ends of the cylinder toward the centre, the compression causes the cylinder to shorten as in Figure 10-21(b).

The effect of the force exerted on a beam depends on the cross section of the beam. We are interested in the force per cross section of the beam, called **stress**, σ:

$$\sigma = \frac{F}{A} \qquad (10\text{-}5)$$

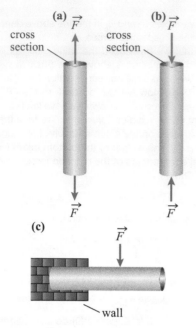

cross section

cross section

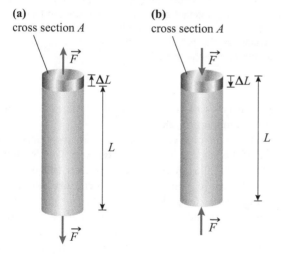

wall

Figure 10-20 (a) Beam is under tension. (b) Beam is under compression. (c) Shear force

(a) cross section A

(b) cross section A

Figure 10-21 (a) Tension causes the cylinder to elongate. (b) Compression causes the cylinder to shorten.

Strain, Elastic Deformation, and the Proportional Limit

The deformation of an object in response to stress is characterized by a quantity called **strain**, which is the fractional change in the length of the object. A symbol commonly used for strain is the Greek letter epsilon:

$$\epsilon = \frac{\Delta L}{L} \qquad (10\text{-}6)$$

An **elastic deformation** is a deformation that is completely reversible: when the stress is removed, the object returns to its original length. It is common for objects to deform elastically in response to applied stress as long as the stress is not too large. For many materials of interest, such as steel alloys, the elastic response region has two parts. The first part is characterized by a linear relation between stress and strain, and the material follows Hooke's law:

$$\sigma = Y\epsilon \qquad (10\text{-}7)$$

or

$$\frac{F}{A} = Y\frac{\Delta L}{L} \qquad (10\text{-}8)$$

The parameter Y is called **Young's modulus**, which is a measure of a material's elasticity. For many materials, especially steel alloys, the value of Young's modulus is the same for compressive and tensile stress. However, the response of a composite material, such as concrete, to tension can differ drastically from its response to compression.

Although tension and compression forces act in a direction parallel to the length of the object in figure 10-20, (perpendicular to the cross section), a shear force acts perpendicular to the length of the object, (parallel to the cross section) as shown in Figure 10-20(c). The deformation of the object along the direction of the shear force is given by

$$\frac{F}{A} = G\frac{\Delta L}{L}$$

The constant of proportionality between the shear stress and the strain is called the **shear modulus**, G.

MAKING CONNECTIONS

Classic Arches

Stone, like concrete, is a lot weaker in tension than it is in compression. Therefore, stone buildings are designed to have the structural elements under compression rather than tension. This is true of the stones in a classic arch. This design has been used since the times of the Etruscans and the ancient Greeks.

The maximum stress for which the response of the material is linear is called the **proportional limit**. This point is labelled P in the typical stress versus strain curve of Figure 10-22. Past the proportional limit the slope of the stress versus strain curve changes: The response of the material is no longer linear, and Hooke's law does not apply. However, the deformation remains elastic past the proportional point until the stress reaches a value called the **yield point**, **yield strength**, or the **elastic limit**, labelled E in Figure 10-22. Beyond this limit, the deformation is not completely reversible.

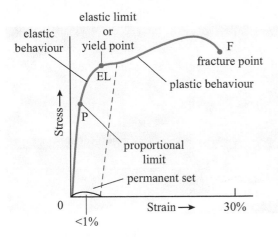

Figure 10-22 Typical stress versus strain curve for a ductile material

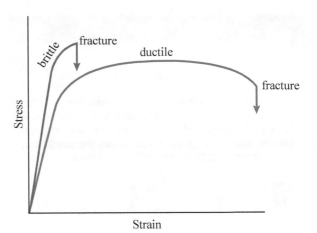

Figure 10-23 Ductile versus brittle behaviour

✓ CHECKPOINT

C-10-5 The Yield Point

The yield point of a given material is
(a) its ultimate tensile stress;
(b) the maximum stress that the material will take such that its strain is directly proportional to the applied stress;
(c) slightly lower than the proportional point;
(d) the maximum stress for which the deformation of the material is completely, or almost completely, reversible.

C-10-5 (d) d is the definition of the yield point.

If the material is strained beyond the yield point, the material does not return to its original length when the stress is removed. Past the yield point, the material undergoes **plastic deformation**, which is not completely reversible. The material has a **permanent set**, meaning that a permanent deformation has occurred. The greatest stress value on the graph of stress versus tensile strain is called the **ultimate tensile strength**. Increasing the strain beyond the ultimate tensile strength point fractures the object.

LO 6

10-6 Ductile and Brittle Materials

If a significant amount of plastic deformation occurs before the material fractures, the material is called a **ductile** material. If fracturing occurs shortly beyond the yield point, the material is called **brittle** (Figure 10-23). For example, the harder steel alloys tend to be more brittle, and the softer alloys tend to be more ductile. The behaviour of a material in compression may differ from its behaviour under tension. Steel alloys are made in various classes: usually the harder (or stronger) the steel, the more brittle it is.

MAKING CONNECTIONS

The *Titanic*

There is evidence that the steel used in the hull of the *Titanic*, while ductile at room temperature, might have become brittle at lower temperatures. This loss of ductility may have substantially increased the size of the fracture in the hull when the ship hit the iceberg in the icy Atlantic waters.

 ## EXAMPLE 10-8

Animal versus Mineral

The yield point of both A36 steel (a grade of structural steel) and spider silk is 250 MPa, and the yield point for human tendon is 28.0 MPa. Compare the maximum mass than can be carried in tension by a steel rod or spider silk, without permanent deformation, with the mass that can be supported by a human tendon of the same size. The rod (and spider silk) is 2.00 m long and has a radius of 2.00 cm.

SOLUTION

According to Equation (10-5), $F = \sigma A = Mg$.
For steel and silk,

$$M = \frac{\sigma A}{g} = \frac{(250 \text{ MPa})(\pi)(0.020 \text{ m})^2}{9.81 \text{ m/s}^2} = 3.20 \times 10^4 \text{ kg}$$

For tendon,

$$M = \frac{\sigma A}{g} = \frac{(28 \text{ MPa})(\pi)(0.020 \text{ m})^2}{9.81 \text{ m/s}^2} = 3.59 \times 10^3 \text{ kg}$$

This is an order of magnitude less than the mass that can be carried by an equivalent piece of silk or steel.

Making sense of the result:

The mean diameter of an Achilles tendon is approximately 5 mm, which makes it capable of carrying approximately 60 kg.

MAKING CONNECTIONS

Is Silk Stronger than Steel?

The ultimate tensile strength (UTS) for many kinds of steel is of the order of 500 MPa, while some spider silk threads have a UTS upward of 1.5 GPa. Thus, a strand of spider silk can support, in tension, three times the weight that can be supported by a strand of steel of the same size.

(c) **(d)**

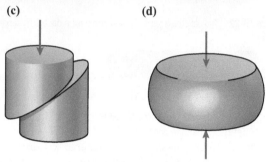

✓ CHECKPOINT

C-10-6 Ductile Metals

A ductile metal
(a) is in general harder than a brittle metal;
(b) is not elastic;
(c) displays more plastic behaviour before fracture than a brittle material.
(d) None of the above.

C-10-6 (c) a ductile metal undergoes a more significant amount of plastic deformation before it fractures in comparison to a brittle metal.

(e)

Failure Modes in Compression and Tension

Whether under compression or tension, concrete and stone structures will fail by fracturing as shown in Figure 10-24(a). Under sufficient tension, a steel beam will eventually snap in two as shown in Figure 10-24(b). Steel under compression has several possible failure modes. Brittle steel will likely crack as shown in Figure 10-24(c). More ductile steel will bulge out as shown in Figure 10-24(d), and plastic deformation will take place, a process called **ductile failure**. Ductile failure is applied in fabricating processes such as extruding and stamping metals as shown in Figure 10-24(e). However, a steel beam is more likely to buckle under compression as shown in Figure 10-24(f) before reaching its compressive strength limit and undergoing ductile failure.

Beyond the ultimate tensile strength, a steel bar, for example, will experience what is called necking before

(f)

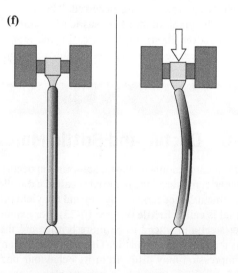

(a)

Figure 10-24 (a) Concrete failure by fracture (b) Failure of steel under tension (c) Failure of brittle steel by cracking (d) Ductile failure (e) Ductile failure used in stamping (f) Buckling of steel

Figure 10-25 A metal strip experiences necking before snapping

Courtesy of Barry Goodno

it eventually snaps (Figure 10-25). During **necking**, the cross section of the material shrinks as it elongates due to plastic deformation past the ultimate tensile strength.

Apparent stress, also called **engineering stress**, is defined as the force per unit area of the original cross section. However, the apparent stress is not the real stress that the material undergoes. **True stress** is defined as the force divided by the actual cross section as the material undergoes deformation. True stress continues to increase until fracture, and apparent stress may drop somewhat, as shown in Figure 10-26. Depending on the application,

MECHANICS

stress versus strain curves may show either the true stress or the apparent stress.

Maximum Tensile and Compressive Strength

For most metals, for example, steel and copper, the stress versus strain curve under compression is similar to the stress versus strain curve under tension. The maximum stress that a material can take under compression, called the **compressive strength** (not the ultimate compressive strength), cannot always be as precisely defined as the ultimate tensile strength. For materials that fail by shattering under compression, such as concrete, the compressive strength is defined as the strain at which the fracture occurs. However, for materials that do not shatter under compression, the definition of the compressive strength has to be defined in terms of the stress required to cause a somewhat arbitrarily chosen amount of permanent deformation. A deformation of 0.2% is used as the standard for many applications. Table 10-1 lists the tensile and compressive properties for a variety of materials.

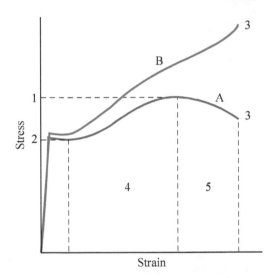

Figure 10-26 Engineering stress (red) versus true (blue) stress

Table 10-1 The Tensile and Compressive Properties for Various Materials

Material	Yield strength	Ultimate tensile strength	Young's modulus	Compressive strength
A36 steel	250 MPa	400–550 MPa	200 GPa	152 MPa
Stainless steel A316L	290 MPa	579 MPa	193 GPa	170 MPa
Aluminum	95 MPa	110 MPa	69 GPa	
Rubber		15 MPa		
Concrete		5 MPa	30 GPa (compression)	14–42 MPa
Wood		~88–150 MPa	6–15 GPa	~14–50 MPa (parallel to grain)
Tendon			0.6–1.8 GPa	
Hair	~200 MPa	380 MPa	~5 GPa	
Spider silk	225 MPa	Up to 1.9 GPa	1.5 GPa	
Sandstone		4–25 MPa	1–20 GPa	20–170 MPa
Granite		7–25 MPa	10–70 GPa	100–250 MPa
Femur bone (along axis)	~100 MPa	135 MPa	17.4 GPa	190–205 MPa

✓ CHECKPOINT

C-10-7 Compressive Strength

Which of the following statements are true?

(a) One cannot define a compressive strength for most steel alloys.

(b) The compressive strength for steel is the point at which steel fractures.

(c) The compressive strength for steel is the stress point for which the steel goes back to within an arbitrary percentage of original length, often 0.2%.

(d) Statement (b) is true for brittle steel, and statement (c) is true for ductile steel.

C-10-7 (c) The accepted arbitrary limit is set to 0.2% for most standards.

EXAMPLE 10-9

The Strength of Human Bone

Young's modulus for the human femur bone for longitudinal stress is approximately 17.0 GPa, and the compressive strength is approximately 190 MPa. The average length of a female femur is 48.0 cm, and the average diameter is 25.4 mm.

(a) Compare the maximum mass that can be carried by the femur to the maximum mass that can be carried by a column of concrete that has the same dimensions and a compressive strength of 42.0 MPa.

(b) Assuming a linear relation between stress and strain until the point of maximum compression, estimate the compressive strain at which the femur will fracture.

SOLUTION

(a) At a given stress, the force is given by Equation (10-5):

$$F = \sigma A$$

The force in this context is the weight carried by the material. So,

$$M = \frac{\sigma A}{g}$$

Since we are looking at the maximum mass, we use the maximum value of the stress, which is the compressive stress.

For the femur:

$$M = \frac{\sigma A}{g} = \frac{(190 \text{ MPa})(\pi)(0.0127 \text{ m})^2}{9.81 \text{ m/s}^2} = 9.81 \times 10^3 \text{ kg}$$

For concrete:

$$M = \frac{\sigma A}{g} = \frac{(42 \text{ MPa})(\pi)(0.0127 \text{ m})^2}{9.81 \text{ m/s}^2} = 2.17 \times 10^3 \text{ kg}$$

(b) Using Equation (10-7) or Equation (10-8),

$$\sigma = Y\epsilon \quad \text{or} \quad \frac{F}{A} = \frac{Y\Delta L}{L}$$

$$\epsilon = \frac{\sigma}{Y} = \frac{190 \text{ MPa}}{17 \text{ GPa}} = \frac{190 \times 10^6 \text{ Pa}}{17 \times 10^9 \text{ Pa}} = 1.12\%$$

Making sense of the result:

The bone will compress by 1.1% of the length at fracture. Since the tensile strain of bone at ultimate tensile strength is slightly over 1%, the value makes sense.

✓ CHECKPOINT

C-10-8 Compressive Strength

Which of the following statements is true?

(a) The compressive strength for concrete is the point at which the concrete returns to within 0.2% of its original length.

(b) Concrete shatters at its compression strength.

(c) One cannot define a compression strength for concrete.

(d) None of the above statements are true.

C-10-8 (b) The compressive strength for concrete and other composite materials is defined as the point at which the material shatters.

Steel Grades

Steel is an alloy composed primarily of iron. Depending on the desired properties, steel contains various amounts of other elements, such as carbon, manganese, chromium, nickel, copper, aluminum, boron, and molybdenum. For example, 18-8 stainless steel, often used for quality cookware, contains 18% chromium and 8% nickel by mass. All types of stainless steel contain at least 10.5% chromium by mass.

The higher the carbon content of a steel alloy, the greater the strength, hardness, rigidity, and brittleness of the alloy. The carbon content of steel alloys is generally between 0.2% and 2.1%. Alloys with a carbon content greater than 2.1% are called cast iron. Such alloys are easy to cast but rather brittle; a hard impact can crack or shatter cast iron. The alloy used for wrought iron contains minimal carbon and is quite malleable, making it easy to work but less strong.

A36 is an ASTM designation for an alloy widely used for construction and structural supports in the United States and Europe (Figure 10-27). The number 36 comes from the alloy's yield point of 36 000 psi, which is equivalent to 250 MPa. A36 was the preferred grade for

Figure 10-27 A36 steel beams

construction steel for a long time, but it has now given way to better-performing alloys. Canada's version of A36 steel is an alloy designated 44W. This alloy has slightly better properties, including a greater yield strength and more elongation before snapping.

Steel will eventually fracture once the ultimate tensile strength is reached. A 5 cm–long piece of A36 steel will elongate by a minimum of 23% of its original length before fracturing, and a 20 cm–long piece will stretch by 20% before fracture. The amount by which an element stretches before snapping is called **elongation at fracture**. The stress at which a steel beam buckles depends on its length and cross section. The buckling stress for a cylindrical A36 beam with a length 10 times its radius is approximately 145 MPa, but an A36 steel beam with a length 100 times its radius buckles as little as 52 MPa under stress.

✓ CHECKPOINT

C-10-9 A36 Steel

A36 steel has
(a) an ultimate tensile strength of 36 000 psi;
(b) a yield point of 36 000 psi;
(c) an ultimate tensile strength of 36 MPa;
(d) a yield point of 36 MPa.

C-10-9 (b) The yield point of A36 steel is 36,000 psi.

FUNDAMENTAL CONCEPTS AND RELATIONSHIPS

THE CONDITIONS OF EQUILIBRIUM

When an object is in equilibrium, the vector sum of all the external forces acting on the object must be zero:

$$\sum \vec{F} = 0 \qquad (10\text{-}1)$$

When an object is in equilibrium, the vector sum of all the external torques acting on the object about its centre of mass or any other given point must be zero:

$$\sum \vec{\tau} = 0 \qquad (10\text{-}2)$$

In terms of momenta:

The linear momentum of an object in equilibrium is constant:

$$\frac{d\vec{p}}{dt} = 0 \qquad (10\text{-}3)$$

The angular momentum of an object in equilibrium about a given point is constant:

$$\frac{d\vec{L}}{dt} = 0 \qquad (10\text{-}4)$$

Elasticity

Stress, σ, is the force per unit area:

$$\sigma = \frac{F}{A} \qquad (10\text{-}5)$$

Strain is the fractional change in the length of an object in response to stress:

$$\epsilon = \frac{\Delta L}{L} \qquad (10\text{-}6)$$

The linear part of an elastic deformation is characterized by a linear relation between stress and strain:

$$\sigma = Y\epsilon \qquad (10\text{-}7)$$

or

$$\frac{F}{A} = Y\frac{\Delta L}{L} \qquad (10\text{-}8)$$

Young's modulus is the same for compressive and tensile stress.

The response of certain materials to tension can differ drastically from the materials' response to compression.

The deformation of an object along the direction of a shear force is given by

$$\frac{F}{A} = G\frac{\Delta L}{L} \qquad (10\text{-}9)$$

Plastic Deformation and Failure Modes

The yield point is the stress beyond which deformation is irreversible. Plastic deformation occurs beyond the yield point.

Ductile materials endure significant plastic deformation before fracture. Brittle materials fracture shortly past the yield point.

Materials can fail by fracture, buckling or ductile failure.

APPLICATIONS

Applications: earthquake-resistant structures; construction materials; cast iron and stainless steel

Key Terms: apparent stress, brittle, centre of gravity, compression force, compressive strength, ductile, ductile failure, dynamic equilibrium, elastic deformation, elastic limit, elongation at facture, necking, permanent set, plastic deformation, proportional limit, shear force, shear modulus, stable equilibrium, static equilibrium, strain, stress, tension, true stress, ultimate tensile strength, unstable equilibrium, yield point, yield strength, Young's modulus

QUESTIONS

1. Will a ball thrown straight up be in equilibrium at any point while in the air? In particular, will it be in equilibrium when it reaches its maximum height?
2. Is a car moving on level ground at a constant speed in equilibrium?
3. When a soccer ball rolls down a hill at a constant speed,
   (a) the ball cannot be in equilibrium because it is going downhill;
   (b) the sum of the forces on the ball is never zero;
   (c) the ball is in equilibrium.
   (d) None of the above are true.

4. A particle's position as a function of time is given by $x = x_0\cos\omega t$. Which of the following statements is true?
   (a) The particle is in equilibrium when $x = x_0$.
   (b) The particle is in equilibrium when $x = 0$.
   (c) The particle is never in equilibrium.
   (d) None of the above are true.
5. When you are standing in an elevator moving straight up at a constant speed, you are
   (a) not in equilibrium because the elevator is moving;
   (b) not in equilibrium because you are moving;
   (c) in equilibrium because your velocity is constant;
   (d) in equilibrium because your acceleration is constant.

6. You step onto a platform resting on top of a spring with a spring constant of k. The spring is fixed to the ground and is then compressed until it comes to a momentary stop. Which of the following statements is true?
   (a) You are in equilibrium throughout.
   (b) You are only in equilibrium when you stop momentarily.
   (c) You are in equilibrium when you are halfway to the point where you stop momentarily.
   (d) You are never in equilibrium during this process.

7. With one hand you hold one end of a string fixed. With the other hand you hold a small mass, which is attached to the other end of the string. You are holding the string such that it is initially horizontal. Then you let the mass go. The mass
   (a) is in equilibrium;
   (b) is in equilibrium at the bottom of its trajectory;
   (c) is in equilibrium when it stops momentarily at either end of its trajectory.
   (d) None of the above are true.

8. A rocket is launched and accelerates vertically, lifted by the thrust of its engine. When the engine runs out of fuel, the rocket begins to slow down under the effect of the force of gravity. Which of the following statements is true?
   (a) After ignition, the rocket is in equilibrium at only one point in its journey upward.
   (b) The rocket is in equilibrium at two points after ignition.
   (c) The rocket is only in equilibrium when it reaches its maximum height and stops momentarily.
   (d) The rocket is in equilibrium as long as it is on the ground, and again at another point between liftoff and maximum height.

9. Is an object that is undergoing uniform circular motion in equilibrium?

10. A disk is spinning about its centre of mass axis at a constant angular speed. Which of the following statements is true?
    (a) The disk is in equilibrium.
    (b) Any given point on the disk is not in equilibrium.
    (c) Both (a) and (b) are true.
    (d) None of the above are true.

11. Can an object's centre of gravity lie outside the physical dimensions of the object? Explain.

12. Two point masses are tied to a massless rod, one at the centre of the rod, the other at its end. The rod is made to spin about the end with no mass on it at a fixed angular speed. The force of gravity where this is done is zero. Since there is an acceleration experienced by each mass, one can define a centre of gravity. To which mass will the centre of gravity be closer?

13. A ball rolls downhill at a constant speed. Aside from the centre of mass of the ball, is any other point on the ball in equilibrium?

14. Calculate the centre of gravity for a uniform bar spinning in a zero gravity environment about a pivot through one end.

15. Two Earth-sized planets orbiting the same star are r and $2r$ from the centre of the star. Find the centre of gravity of the planets when the three objects are collinear with both planets on the same side of the star.

16. Why do you have to lean forward when walking against the current in a river?

17. Does doubling the cross section of a steel rod increase its Young's modulus?

18. Is bone stronger than concrete?

19. Which is tougher, steel or human hair?

20. Which of the following statements is true?
    (a) Steel is mined in steel mines.
    (b) Steel is an alloy that can contain any amount of carbon.
    (c) The amount of carbon in steel alloys ranges between 0.2% and 2.1%; the higher the carbon content, the stronger but more brittle the steel.
    (d) High-carbon steel is more ductile than medium- and low-carbon steel.

21. Does concrete behave better in compression or in tension?

22. The compressive strength for a material that fails by shattering is the strain in which
    (a) you will get the original length back if you remove the load
    (b) the material fractures;
    (c) a permanent set of 0.2% occurs.
    (d) None of the above are true.

23. The yield point for a ductile material is
    (a) the highest stress for which the material will return to its original length after the stress has been removed;
    (b) the stress for which the material will return to within 0.2% of its original length after the stress has been removed.
    (c) Both (a) and (b) can be true.
    (d) Neither (a) nor (b) is true.

24. Compression ductile failure is
    (a) plastic deformation in which a material shrinks in the direction of an applied force and expands along the transverse direction;
    (b) deformation similar to the deformation seen when a train runs over a coin;
    (c) actually "ducktile" failure, because the material makes a distinct "quack" sound like that of a duck.
    (d) Both (a) and (b) are true.

25. Which of the following statements is true?
    (a) Steel goes into plastic behaviour before reaching ultimate tensile strength.
    (b) Steel goes into plastic behaviour after reaching ultimate tensile strength.
    (c) Steel will begin to fail after reaching ultimate tensile strength.
    (d) None of the above are true.

26. Under compression,
    (a) ductile steel fails by cracking;
    (b) ductile steel fails by buckling;
    (c) ductile steel fails by undergoing ductile failure.
    (d) Any of the above could be true.

27. A brittle steel can fail by
    (a) ductile failure;
    (b) fracturing;
    (c) buckling.
    (d) None of the above are true.

28. Which of the following statements is true?
    (a) Steel behaves in a similar way under both compression and tension.
    (b) The compressive yield point of steel is often slightly higher than its tensile yield point.
    (c) Steel will most likely buckle before reaching its compression strength.
    (d) Only (a) and (c) are true.

273

29. The proportional limit is
   (a) the highest point at which the stress versus strain curve is linear;
   (b) the point at which a material goes into plastic deformation;
   (c) outside the elastic limit.
   (d) None of the above are true.

PROBLEMS BY SECTION

For problems, star ratings will be used, (✷, ✷✷, or ✷✷✷), with more stars meaning more challenging problems.

Section 10-1   Static and Dynamic Equilibrium

30. ✷ Consider the simplified model of an arm shown in Figure 10-28. What must be the force T applied by the muscle when the arm is in static equilibrium? Treat the elbow as the pivot, and assume that all forces are perpendicular to the arm as shown. Use the following data: $L = 5.00$ cm, $c = 15.0$ cm, $d = 30.0$ cm, the weight of the forearm alone is 20.0 N, and the weight of the moving mass is 10.0 N.

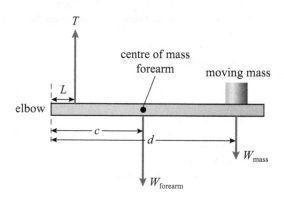

Figure 10-28   Problem 30

31. ✷ A 5.0 kg crate hangs vertically from a rope wound around the pulley as shown in Figure 10-29. The other end of the rope is tied to a 12 kg mass sitting on the horizontal surface. Find the coefficient of static friction between the second mass and the surface it sits on, given that the mass is on the verge of slipping.

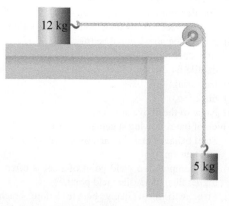

Figure 10-29   Problem 31

32. ✷ A 10 kg mass hangs from a rope as shown in Figure 10-30. The rope passes over a fixed pulley, then under a movable pulley, and is attached to the ceiling as shown. The movable pulley has a mass m attached to it and moves at a constant speed. Find the tension in the rope. Assume that the masses of the pulleys are negligible.

Figure 10-30   Problem 32

33. ✷ A 65 kg climber in a slippery crevice wedges an adjustable bar between the two icy, vertical walls (Figure 10-31). What must the compression force in the bar be to support the climber's weight? The coefficient of friction between the bar and the walls is 0.12.

Figure 10-31   Problem 33

34. ✷ While climbing a chute inside a cave, what force must you apply against the walls to support your weight? The coefficient of static friction between your boots and the wall is 0.56, and that between your gloves and the wall is 0.39. Assume that you exert equal force with your hands and feet.

35. ✷ The bridge in Figure 10-32 is supported at its end by forces F_1 and F_2. Solve for these forces, given that the mass of the bridge is 14.0 t, its length is 5.00 m long, and its mass is distributed uniformly.

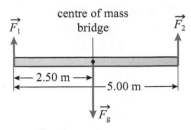

centre of mass
bridge

$\vec{F}_1$ $\vec{F}_2$

← 2.50 m →
← 5.00 m →

$\vec{F}_g$

Figure 10-32 Problem 35

36. ✳ The 62 kg athlete in Figure 10-33 hangs from a 3.0 m–long horizontal bar. The athlete is one-fifth of the length from one end. Find the support forces at each end of the bar, assuming that the mass of the bar is negligible.

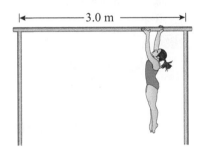

← 3.0 m →

Figure 10-33 Problem 36

37. ✳✳ A child rests two branches against each other as shown in Figure 10-34. There is enough friction between the branches and the ground to prevent them from slipping. The longer of the two branches is 1.6 m in length and has a mass of 4.0 kg; the shorter branch has a mass of 2.7 kg. The shorter branch makes an angle of 67° with the horizontal ground and an angle of 105° with the longer branch. What must the coefficient of friction between the two branches be so that the structure does not collapse?

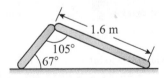

1.6 m
105°
67°

Figure 10-34 Problem 37

38. ✳ The 110 kg sign in Figure 10-35 is suspended by two cables as shown. Find the tension in each cable.

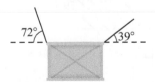

72°   39°

Figure 10-35 Problem 38

39. ✳ The 300 kg horizontal bar in Figure 10-36 is supported by two cables as shown. Find the tension in each cable, given that the bar is 7.0 m long.

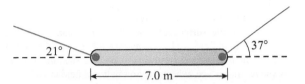

21°   37°

← 7.0 m →

Figure 10-36 Problem 39

40. ✳ A person's leg is suspended to lie horizontally as shown in Figure 10-37. All lengths shown are measured from the left edge of the leg, which is supported by the rest of the body. The mass of the leg alone is 6.50 kg, and its centre of mass is a distance of 0.450 m from the left end. A 2.00 kg mass is attached to the leg 0.750 m from the left end as shown. The leg is supported by the person's body through a vertical force F_p and by a tension T in a vertical cord attached at a distance of 0.650 m.
(a) Find the torque due to the weight of the leg alone.
(b) Find the torque due to the 2.0 kg mass alone.
(c) Find the tension T.
(d) Find the magnitude of the vertical force F_p exerted at the left end of the leg.
(e) If the 2.0 kg mass were moved farther to the right, would the required tension T increase or decrease?

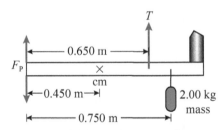

T

← 0.650 m →

F_P

×
cm

← 0.450 m →|

2.00 kg
mass

← 0.750 m →

Figure 10-37 Problem 40

Section 10-2 Centre of Gravity

41. ✳✳ A 75 kg passenger on a cruise ship leans forward into a strong horizontal head wind. When his centre of gravity is 24 cm in front of his ankles, he does not have to exert any effort to keep his balance. His centre of gravity is 110 cm above the deck when he stands upright. What force does the wind exert on this passenger?

42. ✳ Find the distance between the centre of mass and the centre of gravity of a 30 m–long vertical bar on the surface of an asteroid where the gravitational gradient is such that the acceleration due to gravity at the top of the bar is 90% of that at the bottom of the bar.

43. ✳✳ The gravitational field near an anomalous phenomenon in space varies along the z-direction as $1400e^{-(r-r_0)/120}$, where $r_0 = 0$ is the reference point where the acceleration due to gravity is equal to the constant $A = 1400$. Find the distance between the centre of mass and the centre of gravity for a bar of length L that is aligned parallel to the z-axis and has its lower end at $r = r_0$.

44. **✱✱** Show that the centre of mass coincides with the centre of gravity for an object if the acceleration due to gravity is constant over the dimensions of the object. (Hint: Add the torques due to the point masses that make up the object about a point outside the object.)

45. **✱✱** Find the centre of gravity for two equal masses, one located at the surface of a small neutron star, where the gravitational field is 110.0 g, and the other located 100.0 m from the first mass along the surface of the star and 200.0 m above the surface, where the gravitational field is 107 g.

Section 10-3   Applying the Conditions for Equilibrium

46. **✱✱** For the 20 kg sphere shown in Figure 10-38, find the normal forces from each surface. Assume that all the surfaces are frictionless.

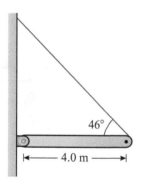

$R = 40$ cm

49°   27°

Figure 10-38 Problem 46

47. **✱✱** Determine the tension in the rope in Figure 10-39. The beam is 4.0 m long, has a mass of 210 kg, and is supported by a free hinge at one end and a rope at the other end, making an angle of 46° with the horizontal.

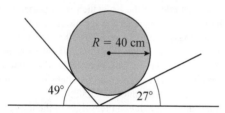

46°

|← 4.0 m →|

Figure 10-39 Problem 47

48. **✱✱** Determine the tension in the horizontal rope supporting the beam in Figure 10-40. The 320 kg beam is 2.4 m long.

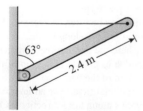

63°

2.4 m

Figure 10-40 Problem 48

49. **✱** If the system in Figure 10-41 is in equilibrium, how are the masses m_1 and m_2 related? Assume that the rope and pulleys have negligible mass.

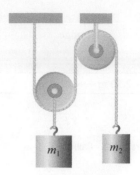

m_1 m_2

Figure 10-41 Problem 49

50. **✱** The 4.0 kg monkey in Figure 10-42 hangs from a rope that is wound around a fixed pulley and around a pulley attached to a banana crate as shown. The crate is on the verge of moving. Find the tension in the rope.

m

Figure 10-42 Problem 50

51. **✱✱** The worker in Figure 10-43 is trying to push the 120 kg wheel over a curb 20 cm high. What force must he apply horizontally if he pushes on the wheel at the point shown in the figure? The wheel has a radius of 40 cm.

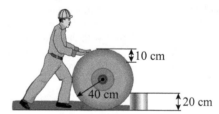

10 cm

40 cm   20 cm

Figure 10-43 Problem 51

52. **✱✱** A winch used to lift a car (mass 1350 kg) is supported by a cable as shown in Figure 10-44. The 450 kg mass of the winch arm is uniformly distributed. Find the tension in the supporting cable and the reaction forces at the hinge.

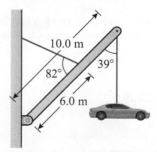

10.0 m

82°   39°

6.0 m

Figure 10-44 Problem 52

Section 10-4 Working with Unknown Forces

53. ✱✱✱ The mass of the upper beam in Figure 10-45 is 2100.0 kg and that of the lower beam is 700.0 kg. Find the horizontal and vertical reaction forces at the hinge at C and at the free hinges at A and B.

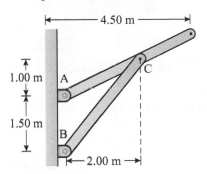

Figure 10-45   Problem 53

54. ✱✱✱ You are designing the crosspiece for the A-frame structure in Figure 10-46. Beams AB and AC are 4.00 m long and have a mass of 300.0 kg each. How much tension must the crosspiece EF withstand? Assume that the mass of the crosspiece and the friction at points B and C are negligible.

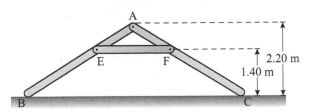

Figure 10-46   Problem 54

55. ✱✱✱ Each beam in Figure 10-47 has a mass of 1100 kg. The length of the shorter beam is 4.0 m. The assembly is in equilibrium. Find the force of friction on each beam, and the horizontal and vertical forces exerted by one beam on the other.

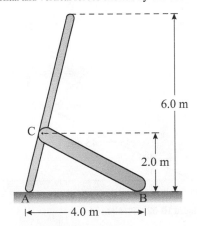

Figure 10-47   Problem 55

Section 10-5 Deformation and Elasticity

56. ✱ Below the proportional point, within the elastic limit, a steel rod can be considered like a spring. Find the effective spring constant of a steel rod of cross section A and length L, given that Young's modulus is Y. How much work would you have to do to stretch the rod by an amount x?

57. ✱ At a construction site, a 4.0 m–long beam supports a winch at its end (Figure 10-48). When the winch lifts a crate up from the ground, the beam deflects by 13 cm. The cross section of the beam is 25 cm², and its shear modulus is 190 MPa. Find the mass of the crate.

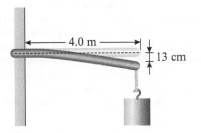

Figure 10-48   Problem 57

58. ✱ Vertebrate muscle typically produces approximately 25 N of tensile force per square centimetre of cross-sectional area. Calculate the stress.

59. ✱✱ The shear strength of limestone is approximately 25 MPa. Find the maximum load that can be supported at the end of a limestone beam that has a square cross section of 100 cm² and extends horizontally 2.3 m from its supporting structure?

60. ✱✱ A 22.0 tonne platform is supported by four vertical beams of concrete with a compressive strength of 42 MPa. What diameter of beams must be used?

61. ✱ A 3.0 m–long cylindrical alloy beam is 21 cm in radius. It is found to stretch by 1.5 cm under a tension of 2.00 kN. Find its Young's modulus.

62. ✱✱ Derive an expression for the elastic potential energy stored in a metal rod in terms of its length (L), cross-sectional area (A), Young's modulus (Y), and elongation (ΔL) within the elastic limit.

Section 10-6 Ductile and Brittle Materials

63. ✱✱ A36 structural steel has a yield point of 250 MPa and an ultimate tensile strength of approximately 450 MPa.
    (a) What is the maximum mass that you can suspend from a 6.0 cm–diameter, 2.0 m–long A36 steel bar before the bar gets a permanent set?
    (b) What is the maximum mass that you can suspend from the same bar such that it does not break?
    (c) What diameter of spider silk thread would you need to suspend the same mass without the thread reaching its ultimate tensile strength of 1000 MPa?

64. ✱ Rubber has an ultimate tensile strength of 15 MPa. Find the maximum force that you can apply to a strip of rubber 10 cm long with a cross section of 5.0 mm². Assume that the cross section of the rubber strip does not change appreciably as you pull on it.

65. ✱✱ What proportion of the yield strength of bone is the stress on the femur of an 82 kg person standing on one leg?

66. ✱✱ If the concrete in a concrete-only building were replaced with enough bone to have the same strength, how much lighter would the building be? What would the change be if the concrete were replaced with steel instead? A typical femur bone has a length of 48 cm, an average diameter of 25 mm, and a mass of 260 g. The density of steel is approximately 8000 kg/m³, and the density of concrete is 1680 to 3000 kg/m³.

67. ✷✷ A shipment of limestone blocks has a density of 2700 kg/m³ and a compressive strength of 250 MPa. The blocks are cubical with a side length of 30.0 cm. How many of these blocks can you stack up before the bottom one cracks?

68. ✷✷ Some types of granite have a Young's modulus of 70 GPa and a compressive strength of 250 MPa. Assume linear behaviour until the compressive strength point.
   (a) Find the amount by which granite compresses before shattering.
   (b) How much weight can a 1 m–long, 20 cm–radius granite beam take in compression?

69. ✷✷✷ A sample of basalt has a compressive strength of 300 MPa and a Young's modulus of 73 GPa.
   (a) Assuming linear stress versus strain behaviour, estimate the amount of work it would take to break a column of basalt 2.0 m in length and 49 cm in diameter.
   (b) What is the minimum mass that you can drop from a height of 20 m to cause the column to shatter?

70. ✷ The ultimate tensile strength of concrete is 3.0 MPa, and the compressive strength is 42 MPa. Find the maximum weight that can be supported by a concrete pillar 50 cm in diameter and 4.0 m long under both tension and compression.

COMPREHENSIVE PROBLEMS

71. ✷✷ The 900 kg, 3.7 m–long beam in Figure 10-49 is pivoted at one end, and the other end rests on a sloped, frictionless incline. A 200 kg mass hangs from the beam 3.0 m from the hinge. Determine the normal force from the surface on the beam and the reaction force at the hinge.

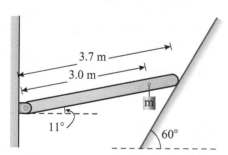

Figure 10-49　Problem 71

72. ✷✷✷ A disk rests on two other disks as shown in Figure 10-50. All three disks are the same size, and each has a mass of 600.0 kg. There is enough friction among the disks and between the disks and the floor to prevent the *m* from rolling apart.
   (a) Find the normal force between the top disk and one of the lower disks.
   (b) Find the magnitudes of the force of friction between each lower disk and the floor, and the normal force from the floor on each disk.

Figure 10-50　Problem 72

73. ✷ In Figure 10-51, a sphere of radius *R* is placed on a frictionless slope. A rope of length 6*R* is attached to the top of the sphere. The other end of the rope is attached to the slope above the sphere. What angle does the rope make with the slope when the sphere reaches a position of stable equilibrium?

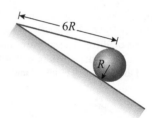

Figure 10-51　Problem 73

74. ✷✷ A 125 kg wheel with a radius of 2.5 m is being pulled up an incline by a 550 kg horse (Figure 10-52). There is a 45 cm–high protrusion in the incline, as shown. What is the tension in the rope when the wheel is just beginning to lift off the incline to go over the protrusion? What is the minimum possible coefficient of friction between the horse and the incline for the horse to be able to pull the wheel over the protrusion?

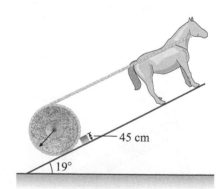

Figure 10-52　Problem 74

75. ✷ The 3.0 kg mass in Figure 10-53 rests on a horizontal surface. The mass is attached to one end of a rope, most of which is wound around a drum of radius *R* on the side of a pulley, as shown. Another rope is wound around the pulley, which has a radius of 2*R*. The coefficient of friction between the mass and the surface is 0.35. What is the largest mass than can hang from the vertical rope without causing the pulley to rotate?

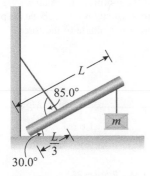

Figure 10-53 Problem 75

76. ✷✷✷ The 80 kg painter in Figure 10-54 is using a large stepladder. The base of the ladder is 3 m wide, and the ladder is 7 m high. The painter is on a step that is 2.0 m above the floor. The horizontal bar of the ladder has a mass of 10 kg, and the entire ladder has a mass of 60 kg. Assume that the floor is frictionless.
  (a) Find the force of tension in the horizontal bar and the normal forces that the floor exerts on the ladder.
  (b) Find the reaction forces at the hinges A, B, and C.

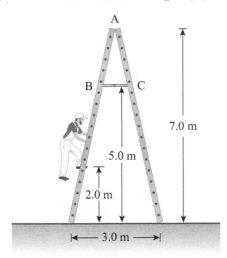

Figure 10-54 Problem 76

77. ✷✷ The beam in Figure 10-55 is 3.4 m long and has a mass of 760 kg. A 220 kg crate is suspended from the beam seven-eighths of its length from the free hinge at the left end of the beam. A rope supports the beam as shown. Find the tension in the rope and the components of the reaction force at the hinge.

78. ✷✷ A stick of mass m and length L is supported at one end by a string, and the other end leans against a wall as shown in Figure 10-56. The string makes an angle θ with the wall, and the stick is horizontal. Derive expressions for the force of friction on the stick, and the minimum coefficient of friction that will stop the stick from sliding down the wall.

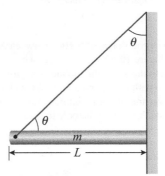

Figure 10-56 Problem 78

79. ✷✷✷ The sphere in Figure 10-57 rests on a friction-less floor. At the point of contact between the sphere and the wall, the coefficient of static friction is μ. Derive an expression for the maximum height d at which a horizontal force F on the sphere will not cause it to spin.

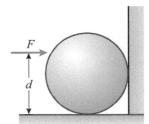

Figure 10-57 Problem 79

80. ✷✷ The drum-and-axle assembly in Figure 10-58 has mass M. The radius of the drum is R, and the radius of the axle is r. You pull on a rope wound around the axle with a force F parallel to the surface of the incline. The drum does not move, but it is about to slide up the incline. Derive an expression for the coefficient of friction between the incline and the drum.

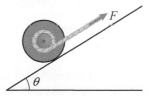

Figure 10-58 Problem 80

81. ✷✷✷ You pull the rope wound around the axle of the drum-and-axle assembly in Figure 10-59 with a force of 120 N directed horizontally as shown. The coefficient of friction between the drum and the inclined surface is 0.20. The mass of the drum is 4.0 kg. The axle radius is 20 cm, and the drum radius is 50 cm. The drum is on the verge of spinning in place. Find the coefficient of friction between the drum and the horizontal surface.

CHAPTER 10 | **EQUILIBRIUM AND ELASTICITY**   279

MECHANICS

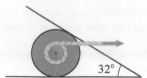

Figure 10-59   Problem 81

82. ✷✷✷   The coefficient of friction between the pole and the incline in Figure 10-60 is 0.45. A guy wire runs from the top of the pole to the top of the incline. You push the 45 kg pole at its centre with a force parallel to the incline and directed down the incline. What is the maximum magnitude of the force that will not cause the pole to slide on the incline?

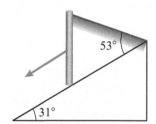

Figure 10-60   Problem 82

83. ✷✷   A 4.0 m–long bar is supported by a string as shown in Figure 10-61. What must be the coefficient of friction between the bar and the floor if the bar is on the verge of slipping?

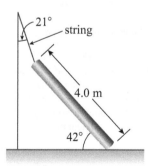

Figure 10-61   Problem 83

84. ✷✷✷   The vertical bar of length L in Figure 10-62 is supported at the top by a string that makes an angle θ with the vertical. The bottom of the bar rests on a surface where the coefficient of friction is μ_s. You have another string that you can move up and down the bar and use it to pull horizontally on the bar.
   (a) Derive an expression for the maximum distance y above the floor at which you can position the string and still cause the bottom of the bar to slide on the floor.
   (b) You now bring the string to the middle of the bar. Derive an expression for the maximum horizontal force exerted on the string without the bar sliding on the floor.

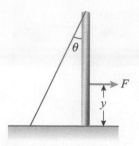

Figure 10-62   Problem 84

85. ✷✷✷   You are pushing a 92 kg refrigerator on a horizontal floor (Figure 10-63). The coefficient of static friction between the refrigerator legs and the floor is 0.34, and the refrigerator's centre of gravity is 80 cm above the floor.
   (a) Find the horizontal force that causes the refrigerator to just move when applied 30 cm above the centre of gravity.
   (b) Will the refrigerator tip or slide first?
   (c) What is the maximum height at which you can apply the horizontal force without tipping the refrigerator?

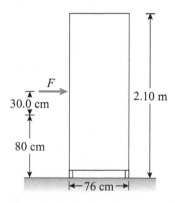

Figure 10-63   Problem 85

86. ✷✷✷   A forklift driver does not notice a truck right behind him and attempts to back up, pushing against the truck with a horizontal force F (Figure 10-64). The wheels of the truck are 1.9 m apart, and its mass is 4800 kg. The friction between the truck's wheels and the pavement prevents the truck from slipping. The point of contact between the truck and the forklift is 1.6 m above the pavement. The truck's centre of gravity is 0.9 m above the pavement. What is the minimum magnitude of the force F that will tip the truck?

Figure 10-64   Problem 86

87. ✳✳✳ A child pulls on an empty 15 kg bookcase as shown in Figure 10-65. She pulls on a shelf 1.4 m above the floor and exerts a force at angle of 30° below the vertical. What is the maximum force the child can exert without causing the bookcase to tip? Assume that the mass of the bookcase is distributed symmetrically around its geometric centre and that there is enough friction to prevent the bookcase from sliding on the floor.

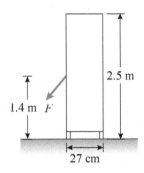

Figure 10-65   Problem 87

88. ✳✳✳ The bar of length L in Figure 10-66 extends by a distance x over the edge of a table.
   (a) What is the maximum distance x for which the bar will not fall?
   (b) You put an identical bar on top of the first bar. What is the maximum distance that the right edge of the second bar can extend over the right edge of the first bar so that the system is stable?
   (c) Find the corresponding position for a third bar.

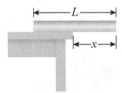

Figure 10-66   Problem 88

89. ✳✳ Figure 10-67 shows a typical stress versus strain curve for spider silk.
   (a) Find Young's modulus for spider silk by looking at the stress between 0 and 200 MPa.
   (b) Draw the force versus elongation curve for a thread diameter of 1.5 μm and an original length of 11 cm.
   (c) How much energy does it take to stretch the thread by 1% of its original length?
   (d) Identify the proportional limit on the curve.
   (e) Estimate the minimum speed that a 50 mg fly needs to break a strand of spider silk 20 cm in length. Is this speed practical?

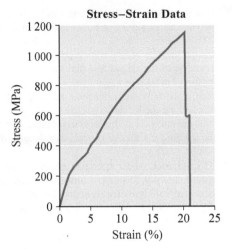

Figure 10-67   Problem 89

90. ✳ For the curve in Figure 10-67, the silk snaps at a stress of approximately 1150 MPa. Find the maximum mass than can be supported by a 3.0 cm–diameter silk thread. What mass can an A36 steel beam of the same diameter support? (A36 has an ultimate tensile strength of approximately 440 MPa.)

91. ✳✳ Figure 10-68 shows the linear region of the stress versus strain curve for a spider silk thread 15 mm long and 1 μm in diameter. Find the spring constant for this material. (Within the elastic limit, the material behaves like a spring.)

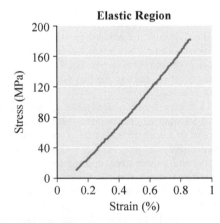

Figure 10-68   Problem 91

92. ✳✳✳ Use Figure 10-67 to estimate the kinetic energy needed to break a 15 mm thread. (You may have to reconstruct the force versus elongation graph for the thread.)

93. ✳✳ Four vertical steel bars, each 2.0 m in length and 10 cm in diameter, support a 120 kg platform with a 4500 kg load of marble slabs on top of it. Find the stress and strain in each bar.

94. ✳✳ Kevlar is used in bulletproof vests. The yield point of Kevlar is approximately 3600 MPa, and its Young's modulus is approximately 100 GPa.

(a) What diameter of a 30 cm–long Kevlar thread would one need to stop a 7.5 g bullet moving at a speed of 420 m/s if the Kevlar is to stay within its elastic limit? Assume that the only stopping power comes from the stretching of the thread and that the bullet strikes the taut thread at its centre.

(b) What is the maximum distance that the bullet moves after striking the Kevlar?

(c) If one did not want the bullet to move more than 5.0 mm after striking the Kevlar thread, what length of thread must be used?

95. ✷✷ The ultimate shear strengths of many steel alloys are approximately 0.75 of their ultimate tensile strength. The shear yield point of steel is approximately 0.58 of the tensile yield point. If you wanted to construct a bullet-proof vest using a square steel plate 4.0 cm on a side, how thick should the plate be to remain within its shear elastic limit while stopping a 7.5 g bullet moving at a speed of 390 m/s?

96. ✷✷ A steel cable 10 cm in diameter secures a ship to a bollard. Wind and wave action combine to push the ship and stretch the 12 m–long cable by 1.0 cm. Find the elastic energy stored in the cable.

97. ✷✷✷ Human leg bone typically has a Young's modulus of 17 GPa, and a compressive strength of 190 MPa.

(a) From what height can a 70 kg athlete jump without fracturing a femur? Assume an impact time of 220 ms and that the femur has a length of 52 cm and a radius of 1.7 cm.

(b) How much can the femur compress before fracturing? Use the average force from the impact time, and assume that all the weight of the athlete is supported by the femur.

98. ✷✷ Some types of bone have a compressive yield strength of 190 MPa and a tensile yield strength of 130 MPa. Estimate the amount by which the femur of a 90 kg person compresses when the person jumps from a height of 2.0 m and lands on one foot. Use the data from the previous question, and assume that the person takes 40 ms to come to a stop.

99. ✷✷ A human femur typically has a longitudinal compressive strength of approximately 200 MPa, a strain at fracture of 0.019, an ultimate tensile strength of 135 MPa, a compressive strength in the transverse direction of 131 MPa, and a shear strength of 65 to 71 MPa. Estimate the maximum longitudinal and transverse forces that a femur can support. Assume a length of 53 cm and a radius of 1.5 cm.

100. ✷✷✷ Figure 10-69 shows the true compressive stress versus true strain curve for AISI 304L stainless steel.

(a) Find Young's modulus for this steel.

(b) What is the maximum elongation shown on the curve?

(c) Find the spring constant for the elastic part of this curve for a 1.7 m–long cylindrical beam that is 3.0 cm in radius.

(d) Plot the force versus elongation for this steel.

(e) Find the energy that it takes to compress this beam by 6% of its original length.

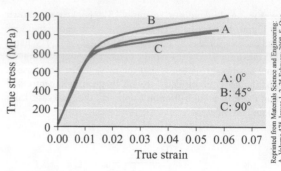

Reprinted from Materials Science and Engineering: A, Volume 475, Issues 1–2, 25 February 2008, S. Qua, C.X. Huanga, Y.L. Gaob, G. Yangb, S.D. Wu a, Q.S. Zanga, Z.F. Zhang, "Tensile and compressive properties of AISI 304L stainless steel subjected to equal channel angular pressing," Pages 207–216, Copyright 2008, with permission from Elsevier.

Figure 10-69 Problem 100

101. ✷✷ Figure 10-70(a) show the stress versus strain curve of stainless steel that has undergone severe plastic deformation in a process called equal-channel angular pressing. The steel is pushed down a vertical channel and undergoes severe deformation while passing through the junction with a horizontal channel (Figure 10-70(b)).

(a) Describe how this process changes the ultimate tensile strength of the steel.

(b) Why would this change be desirable for construction grade steel?

(a)

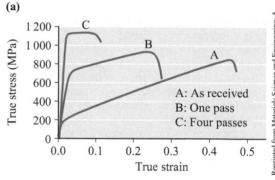

Reprinted from Materials Science and Engineering: A, Volume 475, Issues 1–2, 25 February 2008, S. Qua, C.X. Huanga, Y.L. Gaob, G. Yangb, S.D. Wu a, Q.S. Zanga, Z.F. Zhang, "Tensile and compressive properties of AISI 304L stainless steel subjected to equal channel angular pressing," Pages 207–216, Copyright 2008, with permission from Elsevier.

(b)

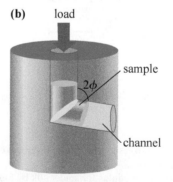

Figure 10-70 Problem 101

See the text online resources at www.physics1e.nelson.com for Open Problems and Data-Rich Problems related to this chapter.

Chapter 11
Gravitation

LOOK AHEAD

There are more than a million asteroids with diameters larger than a kilometre (Figure 11-1). Most asteroids are located in the region between Mars and Jupiter; however, some have orbits that bring them close to Earth. These near-Earth asteroids pose a small, but not insignificant, hazard to life on Earth. It is believed that, 65 million years ago, an asteroid or a comet impact caused the demise of the dinosaurs—and about three-quarters of all other species. Asteroids also have potential benefits. Within your lifetime we may well begin to harness resources from asteroids.

Understanding the physics of gravitation is essential for assessing the risks from asteroids and for exploiting their resources. Gravitational theory enables you to answer questions such as, What is the gravitational acceleration at the surface of an asteroid? How fast would you need to throw an object for it to escape from the gravitational attraction of the asteroid? How long does an asteroid take to make one orbit around the Sun? When will an asteroid be closest to Earth?

As you will learn in this chapter, gravitational theory already has many practical applications here on Earth.

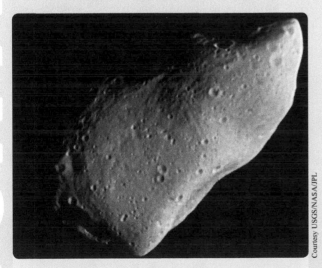

Courtesy USGS/NASA/JPL

Figure 11-1 Asteroid 951 Gaspra is about 18 km long. Would you be able to jump from the surface fast enough to escape from Gaspra's gravitational pull?

Learning Objectives

When you have completed this chapter you should be able to:

1 Solve problems using Newton's law of universal gravitation.

2 Derive and apply an expression for acceleration due to gravity near a surface.

3 Explain what is meant by weightlessness, and describe several ways that weightlessness can be achieved.

4 Use expressions for gravitational potential energy in the general astrophysical case.

5 Calculate force components by taking the derivatives of potential energy functions.

6 Derive and apply an expression for escape from a planet or other object.

7 State Kepler's laws, and solve related problems.

8 Classify orbits as bound or unbound using energy considerations.

9 Calculate the centre of mass of a two-body system, and explain how masses of exoplanets are calculated.

For additional Making Connections, Examples, and Checkpoints, as well as activities and experiments to help increase your understanding of the chapter's concepts, please go to the text's online resources at www.physics1e.nelson.com.

LO 1

11-1 Universal Gravitation

One of Isaac Newton's many contributions to physics is the **law of universal gravitation**: Every mass in the universe attracts every other mass with a force directed along a line joining the centres of the two masses, and with a magnitude given by the following equation:

KEY EQUATION

$$|\vec{F}_G| = \frac{GMm}{r^2} \qquad (11\text{-}1)$$

where M and m are the values of the masses (in kilograms), r is the distance between the centres of mass of the two objects (in metres), and $|\vec{F}_G|$ is the magnitude of the gravitational force (in newtons).

The **centre of mass** is the mean position of the mass in an object or a system. The centre of mass of a homogeneous sphere is at the centre of the sphere. (Newton proved this while developing his theory of universal gravitation.) The value G represents the gravitational constant. In SI units, G has the value 6.672×10^{-11} N m²/kg². You can find the gravitational force for objects of arbitrary shape by integrating over differential mass elements.

As shown in Figure 11-2, the direction of the force is toward the centre of the masses causing the gravitational force. The strength of the gravitational force depends directly on the two masses, M and m, and inversely on the square of the distance between the two masses, r.

Figure 11-2 The gravitational force that mass A exerts on mass B, $\vec{F}_{A \text{ on } B}$, is equal in magnitude but opposite in direction to the force that mass B exerts on mass A, $\vec{F}_{B \text{ on } A}$.

Every pair of objects in the universe exerts gravitational forces on each other, and the magnitudes of these forces can be determined using Equation (11-1). For example, you can use Equation (11-1) to calculate both the force that Earth exerts on your body (i.e., your weight on Earth) and the force that Earth exerts on the Moon. Since Newton's third law states that for every force exerted by one object on another, there is an equal and opposite reaction force, you can similarly calculate the force that you exert on Earth (or on any other object in the universe).

MAKING CONNECTIONS

Isaac Newton

Sir Isaac Newton (1643–1727) developed the key ideas for the law of universal gravitation during a two-year period, shortly after graduating with his first university degree in 1665. However, his work was not published until much later. Newton had planned to go to graduate school immediately, but the University of Cambridge was closed to help prevent the spread of the plague. So, Newton worked largely alone at his home (Figure 11-4). In addition to his work on universal gravitation, he made major contributions to the development of calculus, optics, and mechanics during this time.

Figure 11-3 Sir Isaac Newton at age 46

Figure 11-4 Woolsthorpe Manor, where Sir Isaac Newton developed his most important contributions to physics

EXAMPLE 11-1

The Sun's Gravitational Force

(a) Find the gravitational force exerted by the Sun on a 50.0 kg person on Earth.

(b) Find the gravitational force exerted by Earth on the same person.

SOLUTION

(a) We can use the law of universal gravitation, substituting the person's mass (50.0 kg) for m, the mass of the Sun (1.989×10^{30} kg) for M, and the distance from Earth to the Sun (1.496×10^{11} m) for r. Since r is approximately 10^4 times larger than the diameter of Earth, the particular location of the person on Earth has negligible effect on the magnitude of the force in this calculation:

$$|\vec{F}_G| = \frac{GMm}{r^2}$$

$$= \frac{(6.672 \times 10^{-11}\ \mathrm{N \cdot m^2/kg^2})(1.989 \times 10^{30}\ \mathrm{kg})(50.0\ \mathrm{kg})}{(1.496 \times 10^{11}\mathrm{m})^2}$$

$$= 0.296\ \mathrm{N}$$

(b) We use the same method as in part (a) and substitute Earth's mass and radius (6380 km) for M and r, respectively:

$$|\vec{F}_G| = \frac{GMm}{r^2}$$

$$= \frac{(6.672 \times 10^{-11}\ \mathrm{N \cdot m^2/kg^2})(5.97 \times 10^{24}\ \mathrm{kg})(50.0\,\mathrm{kg})}{(6.38 \times 10^6 \mathrm{m})^2}$$

$$= 489\ \mathrm{N}$$

Making sense of the result:

The gravitational force of the Sun on a person is much less than the gravitational force of Earth on the person (about 0.06%). However, for high-precision calculations the gravitational force exerted by the Sun on the person is not negligible compared to the gravitational force exerted by Earth.

 CHECKPOINT

C-11-1 Forces on a Satellite

A satellite is orbiting Earth. Which of the following statements is true?

(a) The magnitude of the gravitational force exerted by Earth on the satellite is larger than the gravitational force exerted by the satellite on Earth.

(b) The magnitude of the gravitational force exerted by the satellite on Earth is larger than the gravitational force exerted by Earth on the satellite.

(c) Both forces have exactly the same magnitude.

(d) The relationship between the magnitudes of the two forces depends on the distance between Earth and the satellite.

C-11-1 (c) Gravitational forces exist in pairs, with the magnitudes equal but the directions of the forces on the different objects opposite.

You encountered mass earlier when considering Newton's second law. In that context, we had **inertial mass**, which is defined as the amount of force required to accelerate the object divided by the acceleration. In this chapter, we have **gravitational mass**, which is a measure of the gravitational influence on other objects through the law of universal gravitation. While inertial mass and gravitational mass might be different, experiments suggest that they are equal.

In this chapter, we will refer to a number of celestial objects. A **planet** is a solid approximately spherical celestial object that orbits a star. A **satellite** is an object, either natural or artificial, that is in orbit around a planet. Smaller objects in orbits around stars are often called **asteroids**. A **meteoroid** is a smaller object in orbit about a star (typically from meter-sized to dust-sized).

The term "tidal force" refers to the difference in gravitational forces across any body, not just oceans. For example, if a spacecraft were near an extremely dense star, the difference between the gravitational forces on the side facing the star and the side facing away would produce a tidal force strong enough to tear the spacecraft apart. In Chapter 29, you will see how Einstein's general theory of relativity presents a radically different explanation of the nature of gravity.

MAKING CONNECTIONS

Tides

For understanding ocean tides, we can reasonably approximate Earth as a rigid body that responds to gravitational forces as if its mass were concentrated at its centre. Water on the side of Earth facing the Moon is somewhat closer to the Moon than Earth's centre of mass, and therefore experiences a corresponding greater gravitational acceleration. Conversely, water on the side of Earth away from the Moon experiences a lesser gravitational acceleration and therefore gets "left behind." As shown in Figure 11-5, high tides occur at opposite points on Earth. A given point has two high tides each day as Earth rotates.

For a more complete understanding of tides, we have to take into account ocean currents and the gravitational force exerted by the Sun. Tides are higher at times of new and full moon because the gravitational forces of the Sun and the Moon approximately align. The most extreme tides (called spring tides) occur when the Sun, the Moon, and

Earth are in a line. The mean tidal range of the oceans is about 1 m, but the shape and orientation of coastlines affect the local differences between the water level at low and high tide. For example, in the Bay of Fundy a combination of resonance and funnel effects produces a tidal range of up to 16 m (Figure 11-6).

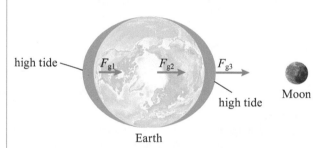

Earth

Figure 11-5 The pattern of two tides each day results from the relative magnitudes of gravitational acceleration due to the gravitational force of the Moon on water on the side of Earth nearer the Moon, the solid Earth, and water on the far side of Earth.

Figure 11-6 High and low tide at Alma, New Brunswick, in the Bay of Fundy

LO 2

11-2   Acceleration Due to Gravity

We designate the **acceleration due to gravity** near the surface of a planet (or a satellite of a planet) as the vector quantity $\vec{g}$. This vector is directed toward the centre of the planet, for example, toward the centre of Earth if we are dealing with acceleration due to gravity near Earth's surface. Without a vector sign, g represents the magnitude of the acceleration due to gravity. On Earth, g is approximately 9.81 m/s² at sea level. We can also write the acceleration due to gravity near Earth's surface as $-g\hat{k}$, where $+\hat{k}$ is a unit vector in the upward direction.

In a region where the acceleration due to gravity is $\vec{g}$, the gravitational force on a mass m is given by the following relationship:

$$\vec{F}_G = m\vec{g} \qquad (11\text{-}2)$$

This force is called the **weight** of the mass. We now have two ways to calculate $\vec{F}_G$: When we know the local acceleration due to gravity, $\vec{g}$, we can use Equation (11-2), and when we know the mass and radius of the large object causing the gravitational force, we can use Equation (11-1). In Example 11-2, we demonstrate how these two methods are related.

When we measure the gravitational force on a small test mass, and then divide by that mass, we have the vector $\vec{g}$. This vector represents two equivalent

EXAMPLE 11-2

Finding the Acceleration Due to Gravity, g

Show that the observed value of approximately 9.81 m/s² for the magnitude of the gravitational acceleration at Earth's surface is consistent with Newton's law of universal gravitation.

SOLUTION

From Newton's law of universal gravitation, we have the following equation for the magnitude of the force due to gravity:

$$F_G = \frac{GMm}{r^2}$$

Equating this expression with $F_G = mg$ and then solving for g gives

$$g = \frac{GM}{r^2}$$

Next, we substitute values for Earth's radius and mass:

$$g = \frac{(6.672 \times 10^{-11}\,\text{N} \cdot \text{m}^2/\text{kg}^2)(5.974 \times 10^{24}\,\text{kg})}{(6.378 \times 10^6\,\text{m})^2} = 9.80\,\text{m/s}^2$$

In fact, this value for g applies only at the equator because Earth is slightly oblate—its polar radius is approximately 6357 km, and its equatorial radius is approximately 6378 km. As a result, the value of g at Earth's surface varies slightly with latitude.

Making sense of the result:

As expected, the value for g calculated using Newton's law of universal gravitation agrees with observed values.

---

quantities: the gravitational field strength (in newtons per kilogram) and the gravitational acceleration (in metres per second squared). A plot of this vector at different points around a mass shows the gravitational field created by the mass. In Chapter 20, we will apply the same concept to describe electrical fields around charged objects.

We can use the method in Example 11-2 to see how g varies at different altitudes above Earth's surface. In fact, the equation below applies for locations on or near any planet, satellite, or other body.

$$g = \frac{GM}{r^2} \qquad (11\text{-}3)$$

It is important to realize that G is a fundamental constant of physics and, as far as we know, does not depend on where or when you measure it. Some physicists are currently working on theories in which G is not strictly

✓ CHECKPOINT

C-11-2 What Do We Call the Force of Gravity?

Which of the following terms describes the force of gravity on an object by a planet?
(a) the gravitational acceleration, g
(b) the universal gravitational constant, G
(c) the weight of the object, W
(d) the mass of the object, m

C-11-2 (c) Weight is a force, depending on both mass and the local acceleration due to gravity.

constant at different times, but the preponderance of evidence suggests that G is constant. On the other hand g—the magnitude of the acceleration due to gravity—is definitely not a universal constant. It is quite different on the Moon and Mars, for example, and even varies somewhat for different locations on Earth's surface.

EXAMPLE 11-3

Finding g on an Asteroid

(a) Find the value of g on the surface of an asteroid that has a density of 3200 kg/m³ and a radius of 5.00 km. Assume that the asteroid is homogeneous and spherical.
(b) How long would it take an object starting from rest to fall through a height of 1.0 m above the surface of the asteroid? Assume that g is constant during the fall.
(c) Show that the assumption in part (b) is reasonable.

SOLUTION

(a) First, find the mass of the asteroid:

$$M = \rho V = \rho \frac{4\pi r^3}{3} = 3200\,\text{kg/m}^3\,\frac{4\pi(5.00 \times 10^3\,\text{m})^3}{3}$$

$$= 1.68 \times 10^{15}\,\text{kg}$$

Now, calculate the acceleration due to gravity at the asteroid's surface:

$$g = \frac{GM}{r^2} = \frac{(6.672 \times 10^{-11}\,\text{N} \cdot \text{m}^2/\text{kg}^2)(1.68 \times 10^{15}\,\text{kg})}{(5.00 \times 10^3\,\text{m})^2}$$

$$= 4.47 \times 10^{-3}\,\text{m/s}^2$$

(continued)

(b) From Section 3-4 we have

$$y - y_0 = v_0 t + \frac{1}{2} a t^2$$

We choose a convention with positive upward, so $y - y_0 = -1.0$ m. Since the object starts from rest, $v_0 = 0$. The acceleration is -4.48×10^{-3} m/s². The acceleration is negative because it is in the downward direction. Therefore,

$$-1.0 \text{ m} = 0 + \frac{1}{2}(-4.48 \times 10^{-3} \text{ m/s}^2)t^2$$

$$t^2 = \frac{2(-1.0 \text{ m})}{-4.48 \times 10^{-3} \text{ m/s}^2}$$

$$t = 21 \text{ s}$$

It would take 21 s for an object to fall 1.0 m on the asteroid.

(c) If we repeat part (a) with $r = 5.001 \times 10^3$ m, we obtain a value for g of 4.47×10^{-3} m/s². Therefore, the assumption that g is approximately constant is justified.

Making sense of the result:

As expected, we found that the gravitational acceleration is also much less than near the surface of Earth. Since 1.0 m is small compared to the radius of the asteroid, it is not surprising that the change in g would not affect calculations that have a precision of three significant figures.

MAKING CONNECTIONS

Mapping with *g*

The acceleration due to gravity, g, varies slightly from place to place due to different subsurface rock densities. Scientists use precise maps of the acceleration due to gravity to study geologic features. Figure 11-7 shows a gravity map of the region around the Chicxulub crater on the Yucatán peninsula in Mexico. This crater was created by the impact of an asteroid or a comet 65 million years ago, an event that is linked to the mass extinction of the dinosaurs and many other species living at that time. Dr. Alan Hildebrand of the University of Calgary discovered the crater, which is not visible on the surface.

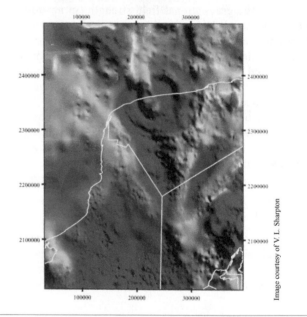

Image courtesy of V. L. Sharpton

Figure 11-7 A gravity map of part of the Yucatán peninsula in Mexico. The curving white line shows the boundary between land and sea. The circular pattern in the measurements of *g* reveals the buried remnants of the Chicxulub crater.

LO 3

11-3 Orbits and Weightlessness

It is easy to see that an object in deep space, far from any other masses, is almost totally weightless. The gravitational forces acting on it are too weak to measure. Figure 11-8 shows the International Space Station (ISS). Inside the ISS objects appear to be weightless. Is the gravitational force actually zero inside the ISS? Example 11-4 answers this question.

Example 11-4 demonstrates that the force of gravity can provide the centripetal acceleration necessary to keep an object in a circular orbit. Figure 11-9 shows the

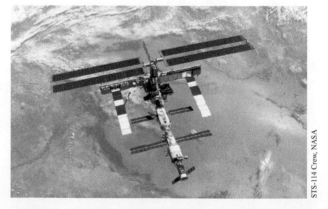

STS-114 Crew, NASA

Figure 11-8 The ISS is an orbiting laboratory built by a consortium of many countries, including Canada.

EXAMPLE 11-4

Gravity at the ISS

(a) Calculate the gravitational force exerted by Earth on a 1.00 kg object at the ISS. Assume that the ISS is in a circular orbit at an altitude of 353 km above Earth's surface.
(b) Calculate the value of g at the ISS.
(c) Calculate the centripetal acceleration required to keep an object in the circular orbit of the ISS, given that the orbital speed of the ISS is 27 740 km/h.

SOLUTION

(a) We can apply the law of universal gravitation with the mass of the object (1.00 kg) for m and the mass of Earth (5.974×10^{24} kg) for M. The distance r is the sum of the radius of Earth (we will use a mean value of 6371 km) and the altitude of the ISS:

$$r = 6371 \text{ km} + 353 \text{ km} = 6724 \text{ km} = 6.724 \times 10^6 \text{ m}$$

$$F_G = \frac{GMm}{r^2}$$

$$= \frac{(6.672 \times 10^{-11} \text{ N} \cdot \text{m}^2/\text{kg}^2)(5.974 \times 10^{24} \text{ kg})(1.00 \text{ kg})}{(6.724 \times 10^6 \text{ m})^2}$$

$$= 8.82 \text{ N}$$

(b) Using Equation (11-3),

$$g = \frac{GM}{r^2} = \frac{(6.672 \times 10^{-11} \text{ N} \cdot \text{m}^2/\text{kg}^2)(5.974 \times 10^{24} \text{ kg})}{(6.724 \times 10^6 \text{ m})^2}$$

$$= 8.816 \text{ m/s}^2$$

(c) We convert the speed to units of metres per second, and then apply the formula for centripetal acceleration:

$$v = 27 \, 740 \, \frac{\text{km}}{\text{h}} \times \frac{1000 \text{ m}}{1 \text{ km}} \times \frac{1 \text{ h}}{3600 \text{ s}} = 7.7056 \times 10^3 \text{ m/s}$$

$$a_c = \frac{v^2}{r} = \frac{(7.7056 \times 10^3 \text{ m/s})^2}{6.724 \times 10^6 \text{ m}} = 8.830 \text{ m/s}^2$$

Making sense of the result:

The gravitational force on the 1.00 kg object is approximately 10% less than the 9.81 N force that the object would experience on Earth's surface. The gravitational force at the ISS is certainly not zero. The centripetal acceleration is (within the precision of data) equal to the gravitational acceleration. Thus, the gravitational force provides just enough acceleration to keep the object in a circular orbit with the ISS. As a result, the object appears weightless in the ISS frame of reference.

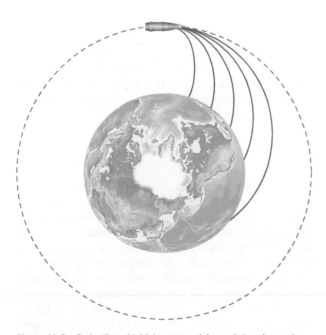

Figure 11-9 Projectiles with higher tangential speeds travel a greater distance before reaching Earth's surface. At just the right speed, the path of the projectile will curve such that the projectile remains at a constant distance from Earth.

trajectories of objects with various initial velocities. The greater the initial tangential velocity, the farther the object travels before it falls to Earth. However, when the object has just the right speed, it will fall toward Earth at the same rate that Earth's surface curves away, and the object will circle Earth at a constant altitude. So, an object in a circular orbit is falling continuously and appears weightless, similar to the way that you would feel weightless if you were in free fall. It is important to realize that it is an apparent weightlessness, and there is still a weight force.

✓ CHECKPOINT

C-11-3 Weightless in Orbit

Which of the following, if any, have a zero value for an object in a circular orbit?
(a) the gravitational force on the object
(b) the weight of the object
(c) the net force on the object
(d) the difference between the gravitational force on the object and its mass times its acceleration

C-11-3 (d)

Achieving Weightlessness

Aircraft like the one in Figure 11-10 produce weightless conditions more readily and at a lower cost than in an orbiting spacecraft, but only for brief periods, typically 15 to 20 s. The aircraft accelerates steeply upward, as shown in Figure 11-11, then the engines are idled and the aircraft follows a parabolic path like a projectile. During the parabolic portion of the flight path, passengers in the aircraft experience weightlessness, although usually not quite complete weightlessness. These flights are called parabolic flights, or sometimes suborbital zero gravity flights. The European Space Agency (ESA), NASA, the Canadian Space Agency (CSA), and others use these flights to achieve weightless conditions for short periods.

Figure 11-10 The NASA parabolic flight aircraft during ascent

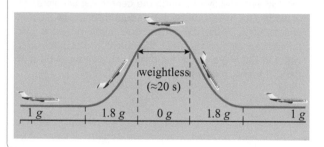

Figure 11-11 This flight path produces a brief period of weightlessness. The acceleration is expressed in units of *g* (equal to 9.81 m/s²).

LO 4

11-4 Gravitational Potential Energy

Because it takes work to move objects in gravitational fields there must be **gravitational potential energy**, which is the work required to move from some zero point to a given configuration of masses. In Chapter 6, we derived the relationship $U = mgh$ for the gravitational potential energy (U) of an object with mass m at a height h above the surface of a much larger mass that causes a gravitational acceleration of g. This derivation assumed that g is constant over the distance h, and set $U = 0$ at the surface of the larger mass. It used the concept that the amount of potential energy increase between two points is equal to the work done in moving between the points. That work is found by integrating the area under the force versus displacement graph.

You can integrate the force of universal gravitation to obtain an expression for the gravitational potential energy difference. The gravitational force is infinite for the case

of zero distance, so we start the integration at an infinite distance (where the gravitational potential energy is defined as zero). That integration (which you complete in problem 60) produces the following more general definition of gravitational potential energy that does not assume that g is constant.

KEY EQUATION
$$U = -\frac{GMm}{r}$$
(11-4)

As in earlier equations, M and m are the masses of the two objects, and r is the distance between their centres of mass. With this definition, the gravitational potential energy of the mass m has its greatest possible value – zero – when the masses are infinitely far apart and becomes increasing negative as r decreases. This choice for the zero point may seem strange, but reflection will show that it is a very reasonable definition for a universal zero point.

Although the two definitions of gravitational potential energy have different zero points, the two techniques should largely agree for calculations of changes in potential energy that occur in a region where g is approximately constant. Example 11-5 compares the results of one such calculation.

PEER TO PEER

At first, I found the idea of gravitational potential energy always being negative strange. It just seemed to me that an energy should be positive. Then I began thinking of it as the amount of energy needed to get away (kind of like being in debt by that amount). Really that is how it works. If at a certain position the gravitational potential energy is −15 J then you will need +15 J of kinetic energy to just barely escape. It is important to remember that potential energies in physics are always relative to some defined zero level.

CHECKPOINT

C-11-4 Gravitational Potential Energy

(a) When you increase the distance between two masses, does the gravitational potential energy increase or decrease?

(b) Do you need to do work to increase the distance between two masses?

C-11-4 (a) When you increase the distance between two masses, the gravitational potential energy will *increase*. The increased distance results in a negative energy with a smaller magnitude, so the energy increases. Since it takes work to move the objects farther apart, the potential energy must increase. (b) Yes, you need to do work to increase the distance between two masses. The force required to separate the masses has the same direction as the increase in displacement, so the work is positive.

EXAMPLE 11-5

Calculating Changes in Gravitational Potential Energy

Calculate the work required to lift a 1.00 kg object from sea level to the top of Mount Everest, which has an elevation of 8848 m. Assume that Earth is spherical with a radius of 6371 km. Perform the calculation in the following two ways.

(a) Use the *mgh* expression for potential energy, and assume that *g* remains constant at its value at sea level, 9.81 m/s².

(b) Use Equation (11-4).

SOLUTION

(a) Since $U = 0$ at sea level,

$$\Delta U = mgh - 0 = 1.00 \text{ kg} \times 9.81 \text{ m/s}^2 \times 8848 \text{ m} = 86.8 \text{ MJ}$$

The gravitational potential energy of the object is greater at the top of the mountain. Therefore, it takes 86.8 MJ of work to lift the object to the top of Mount Everest.

(b) We find the difference between the gravitational potential energy at the top of the mountain and at sea level. The distance from the centre of Earth to the top of Mount Everest is

$$r_{m} = 6.371 \times 10^6 \text{ m} + 8848 \text{ m}$$

$$r_{m} = 6.3798 \times 10^6 \text{ m}$$

Using r_s to represent the distance from the centre of Earth to the surface at sea level, we can express the difference in gravitational potential energy as

$$\Delta U = -\frac{GMm}{r_{m}} - \left(-\frac{GMm}{r_{s}}\right) = GMm\left(\frac{1}{r_{s}} - \frac{1}{r_{m}}\right)$$

Substituting in the known values gives

$$\Delta U = (6.672 \times 10^{-11} \text{ N} \cdot \text{m}^2/\text{kg}^2)(5.974 \times 10^{24} \text{ kg} \times 1.00 \text{ kg})$$

$$\times \left(\frac{1}{6.3710 \times 10^6 \text{ m}} - \frac{1}{6.3798 \times 10^6 \text{ m}}\right)$$

$$= 86.8 \text{ MJ}$$

Therefore, the work required to lift the 1.00 kg mass to the mountaintop is 86.8 MJ.

Making sense of the result:

As expected, the two ways of calculating the work required give similar results. Since *g* decreases with height, the assumption that *g* remains equal to its value at sea level gives a result that slightly overestimates the work required.

EXAMPLE 11-6

Meteor Speeds

Estimate the speed that a meteoroid attains just before reaching Earth's surface and becoming a meteor (see Figure 11-12). Use the following simplifying assumptions in obtaining the estimate:

- The meteoroid started from rest at such a great distance from Earth that the two can be considered to have been infinitely far apart.

- The influence of all other objects (e.g., the Sun and the Moon) is negligible.
- Earth's atmosphere does not significantly slow the meteoroid.

While the last assumption in particular may seem unreasonable, in fact all three are valid to the first order (see the Solution for more details).

(continued)

Courtesy R.L. Hawkes

Figure 11-12 A meteor is the bright trail or streak caused by a meteorite entering Earth's atmosphere at high speed.

SOLUTION

When the meteoroid is infinitely far away, its gravitational potential energy is 0.

At Earth's surface, the meteoroid's gravitational potential energy is $-\dfrac{GMm}{r}$, where r is the radius of Earth, M is the mass of Earth, and m is the unknown mass of the meteoroid. Since we are assuming that there is no loss to aerodynamic drag, the amount of gravitational potential energy converted to kinetic energy as the meteorite falls to Earth is $\dfrac{GMm}{r}$. Using v_s to represent the speed of the meteorite at Earth's surface, we have

$$\frac{1}{2}mv_s^2 = \frac{GMm}{r}$$

$$v_s = \sqrt{\frac{2GM}{r}} = \sqrt{\frac{2(6.672 \times 10^{-11}\,\mathrm{N \cdot m^2/kg^2})(5.974 \times 10^{24}\,\mathrm{kg})}{6.3710 \times 10^6\,\mathrm{m}}}$$

$$= 1.119 \times 10^4\ \mathrm{m/s}$$

Making sense of the result:

A meteoroid that fell without air resistance to Earth's surface would accelerate due to the Earth's gravity to a speed of approximately 11.2 km/s. In fact, this approximation is within a few percent of the observed minimum speed of meteors in the upper atmosphere; most meteors have higher speeds. Astronomers who measure properties of meteors do need to provide a small correction for the aerodynamic drag in the high atmosphere.

LO 5

11-5 Force from Potential Energy

When a potential energy function is defined, we can determine force from potential energy using derivatives, as we will show in this section. Alternatively, we can determine a potential energy function using integration when we know the force everywhere. The concept that derivatives of potential energy functions represent the associated forces is applied in many branches of physics, for example, electromagnetism.

When a potential energy function, U, exists, we can obtain the force components through the following relationships:

$$F_x = -\frac{\partial U}{\partial x} \qquad (11\text{-}5)$$

$$F_y = -\frac{\partial U}{\partial y} \qquad (11\text{-}6)$$

$$F_z = -\frac{\partial U}{\partial z} \qquad (11\text{-}7)$$

Note that these are partial derivatives, which means that the differentiation is only done with respect to an *explicit* appearance of the variable. This means that if, for example, U does not explicitly depend on y, then the partial derivative with respect to y will be zero and there will be no force in the y-direction. If you have only one-dimensional situations, you can think of the partial derivatives just like ordinary derivatives. We show in Example 11-7 how to obtain the gravitational force near Earth's surface from these relationships.

✓ CHECKPOINT

C-11-5 Force from Potential Energy

What are the direction and magnitude of the force when the potential energy function is $U = 3z^2$?

C-11-5 The partial derivatives in the x- and y-directions are 0, so there are no force components in those directions. The z-component is

$$F_z = -\frac{\partial U}{\partial z} = -6z$$

Therefore, when z is positive, the force is in the negative z-direction, and when z is negative, the force is in the positive z-direction.

We can extend the relationship between force components and derivatives to situations in which a potential energy function, $U(r)$, depends only on radial distance. In the following equation, $\hat{r}$ is a unit vector in the direction of increasing r:

$$\vec{F} = -\frac{\partial U(r)}{\partial r}\hat{r} \qquad (11\text{-}8)$$

EXAMPLE 11-7

Gravitational Force Near Earth's Surface

Find the gravitational force near Earth's surface where the potential energy function is given by $U = mgy$.

SOLUTION

The y-component of the force is

$$F_y = -\frac{\partial U}{\partial y} = -mg$$

Since U does not depend on x or z, the partial derivatives yield a 0 result for these force components:

$$F_x = 0 \qquad F_z = 0$$

Making sense of the result:

As expected, we found that the direction of the gravitational force is downward (negative y-direction) and the magnitude of the force is simply the weight of the object at the surface, mg.

We demonstrate in Example 11-8 how the gravitational force can be obtained from the expression for gravitational potential energy.

EXAMPLE 11-8

Deriving Gravitational Force from Gravitational Potential Energy

Determine the gravitational force, given that the function for gravitational potential energy can be written as $U(r) = -\dfrac{GMm}{r}$.

SOLUTION

Substituting $U(r)$ into Equation (11-8), we get

$$\vec{F} = -\frac{\partial U(r)}{\partial r}\hat{r} = -\frac{\partial\left(-\dfrac{GMm}{r}\right)}{\partial r}\hat{r}$$

$$= GMm\frac{\partial\left(\dfrac{1}{r}\right)}{\partial r}\hat{r} = -\frac{GMm}{r^2}\hat{r}$$

Making sense of the result:

The derivative gives the magnitude of the force introduced at the beginning of the chapter for the force of universal gravitation. The force on mass m points in the $-\hat{r}$ direction, back toward the mass M because $\hat{r}$ points from M to m.

LO 6

11-6 Escape Speed

In this section, we will consider the **escape speed**, which is the speed needed for an object to escape the gravitational pull of another object. What do we mean by "escape"? If you throw an object straight up, it will rise for a time, slowing as its kinetic energy decreases and its gravitational potential energy increases. The object will stop momentarily and then reverse direction. If we give the object a somewhat greater initial speed, it will go higher but will still eventually slow to a stop and then return. However, gravitational escape can occur when the object has sufficient initial speed that it does not ever stop completely, and instead travels infinitely far away.

For an object to escape, the total orbital energy must be positive: the object must have enough positive kinetic energy from its orbital motion to overcome its negative gravitational potential energy. Then the object will still have some kinetic energy when it reaches a point infinitely far away, where its gravitational potential energy is zero.

We call the orbit or trajectory "bound" when the total orbital energy is negative and "unbound" when the total is positive. To determine the minimum speed for escape, we set the total orbital energy equal to zero:

$$E_{\text{total}} = T + U = \frac{1}{2}mv_{\text{esc}}^2 - \frac{GMm}{r} = 0 \qquad (11\text{-}9)$$

Solving for the escape speed, we derive the relationship given in Equation (11-10) below, where r is the initial distance of the escaping object from the centre of the larger object, and M is the mass of the larger object. Note that the mass of the escaping object, m, cancels out and does not affect the escape speed.

$$v_{\text{esc}} = \sqrt{\frac{2GM}{r}} \qquad (11\text{-}10)$$

☑ CHECKPOINT

C-11-6 Total Orbital Energy

Is the sum of the kinetic energy and the gravitational potential energy of Earth in its orbit about the Sun positive, zero, or negative?

C-11-6 The total energy is negative. We can show this starting with $F = ma$ and substituting the centripetal acceleration for a and the gravitational force for F; then compare the potential energy and the kinetic energy. We see in the next section that a circular orbit (or an elliptical orbit) has a negative total orbital energy.

EXAMPLE 11-9

Escape from Earth

Find the speed required to escape from Earth's surface. Neglect any work needed to overcome aerodynamic drag in the atmosphere.

SOLUTION

We use Equation (11-10), substituting the mass of Earth for M and the radius of Earth for r:

$$v_{esc} = \sqrt{\frac{2GM}{r}}$$

$$= \sqrt{\frac{2(6.672 \times 10^{-11}\,\text{N} \cdot \text{m}^2/\text{kg}^2)(5.974 \times 10^{24}\,\text{kg})}{6.3710 \times 10^6\,\text{m}}}$$

$$= 1.119 \times 10^4\,\text{m/s}$$

Making sense of the result:

The speed needed to escape from Earth's surface is about 11.2 km/s. Note that this escape speed matches the speed for the falling meteoroid in Example 11-8. This makes sense because the potential energy that an object loses while falling from an infinite distance is equal to the energy needed to take the object to an infinite distance from the surface. Since the aerodynamic drag on an object passing through Earth's atmosphere is not negligible, the speed required to escape from Earth's surface is actually somewhat greater than 11.2 km/s.

Looking at Equation (11-10), we can see that the escape speed for a planet varies with $\sqrt{\frac{M}{r}}$, the mass of the planet divided by its radius. The volume of a sphere increases with the cube of its radius, so a large planet will have a greater escape speed than a smaller planet with the same density. In Example 11-10, we calculate the escape speed of a mid-sized asteroid.

While the calculations in this section have dealt with escape from the surface of an object, the same techniques can be applied for cases of escape starting from an altitude above the surface of an object.

11-7 Kepler's Laws

Historically, the orbits of celestial objects were first understood empirically and only later explained in terms of Newtonian mechanics. The German mathematician and astronomer Johannes Kepler (1572–1630) deduced

EXAMPLE 11-10

Escape from an Asteroid

What speed is needed to escape from the surface of a spherical, homogeneous asteroid with a radius of 5.00 km and a density of 3200 kg/m³? The asteroid has no atmosphere.

SOLUTION

First, find the mass of the asteroid:

$$M = \rho V = \rho\,\frac{4\pi r^3}{3}$$

$$= 3200\,\text{kg/m}^3\,\frac{4\pi(5.00 \times 10^3\,\text{m})^3}{3} = 1.68 \times 10^{15}\,\text{kg}$$

Now, substitute values for M and r in Equation (11-6):

$$v_{esc} = \sqrt{\frac{2GM}{r}}$$

$$= \sqrt{\frac{2(6.672 \times 10^{-11}\,\text{N} \cdot \text{m}^2/\text{kg}^2)(1.68 \times 10^{15}\,\text{kg})}{5.00 \times 10^3\,\text{m}}}$$

$$= 6.69\,\text{m/s}$$

Making sense of the result:

As expected, the escape speed from the asteroid is very small compared to the escape speed from Earth.

three laws of motion that describe the orbital motion of objects with negative total orbital energy. Although Kepler devised his laws to describe the orbits of planets in the solar system, these laws apply to any object in a bound orbit.

ONLINE ACTIVITY

Exploring Orbital Motion

The e-resource that accompanies every new copy of this textbook contains an Online Simulation Activity using the PhET simulation "My Solar System." Work through the simulation and accompanying questions to gain an understanding of planetary orbits.

Kepler's Laws

1. Each planet moves in an elliptical orbit with the Sun at one of the two foci of the ellipse (Figure 11-13).

2. A line drawn from the Sun to a planet will sweep out the area inside the elliptical orbit at a constant rate (Figure 11-14).

3. The square of the period of the orbit is proportional to the cube of the semi-major axis of the orbit. The semi-major axis (a) is one-half of the longest dimension of the elliptical orbit.

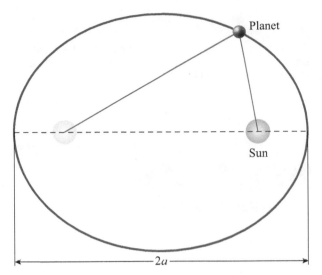

Figure 11-13 Kepler's first law: Planets move in elliptical orbits with the Sun at one focus.

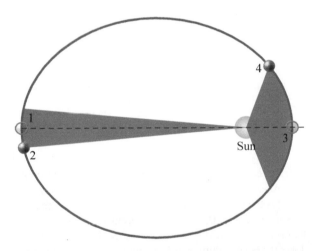

Figure 11-14 Kepler's second law: A line from the Sun to a planet will sweep equal areas in equal time periods. The areas shown in red are equal; therefore, the planet moves from point 1 to point 2 and from point 3 to point 4 in the same length of time.

The motion is confined to a single orbital plane because of the central nature (along the line joining the centres) of the gravitational force. This means that torque out of that plane is not possible. Kepler's second law is a consequence of the law of conservation of angular momentum. In order for angular momentum to remain constant, the planet must move faster when it is closer to the Sun. Conservation of energy also indicates that the kinetic energy must be greater when the planet is closer to the Sun and has less gravitational potential energy. In fact, the speed of an orbiting object a distance r from the Sun (or other relatively large mass) is given by the following relationship, where a is the semi-major axis of the ellipse:

$$v = \sqrt{GM\left(\frac{2}{r} - \frac{1}{a}\right)} \qquad (11\text{-}11)$$

If we take the case of a circular orbit where $r = a$, this can readily be shown to be equivalent to that obtained by using $F = ma$ with the centripetal acceleration for a.

Newton's law of universal gravitation gave astronomers an expression for the constant of proportionality in Kepler's third law. This constant of proportionality depends on the sum of the masses of the two objects (M and m), as shown in Equation (11-12) below.

KEY EQUATION $\qquad T^2 = \dfrac{4\pi^2}{G(M + m)}\, a^3 \qquad (11\text{-}12)$

When one of the masses is much less than the other, we can replace the sum of the masses with the mass of the larger object:

$$T^2 = \frac{4\pi^2}{GM}\, a^3 \quad \text{where } M \gg m \qquad (11\text{-}13)$$

The Sun is much more massive than other objects in the solar system (e.g., the mass of Earth is about one-millionth the mass of the Sun), so this approximation gives accurate results for orbits around the Sun. Example 11-11 shows the derivation of Equation (11-13) for a circular orbit.

 ## EXAMPLE 11-11

Kepler's Third Law

Derive Kepler's third law for a planet moving in a circular orbit around the Sun. Assume that the mass of the planet is negligible compared to the mass of the Sun.

SOLUTION

The gravitational force provides the centripetal acceleration to maintain the circular orbit. Since both the force and the acceleration are in the same direction,

$$F_{\text{G}} = \frac{GMm}{r^2} = \frac{mv^2}{r}$$

Simplifying gives

$$\frac{GM}{r} = v^2$$

The period is the time for one orbit, and the circumference of a circle is $2\pi r$, so

$$v = \frac{2\pi r}{T}$$

Substituting for v in the above simplified equation gives

$$\frac{GM}{r} = \frac{4\pi^2 r^2}{T^2}$$

(continued)

$$T^2 = \frac{4\pi^2 r^3}{GM}$$

For a circular orbit, the radius r is the semi-major axis a, and

$$T^2 = \frac{4\pi^2}{GM} a^3$$

Kepler's third law is a universal relationship and can be applied to any two masses in a bound orbit, including asteroids, comets, and meteoroids in elliptical orbits around the Sun. Kepler's laws also apply to the orbit of a satellite around a planet, and we can use measurements of the satellite's motion to determine the mass of the planet (see Example 11-12).

Any orbit can be completely specified with six parameters. The most important of these are the semi-major

CHECKPOINT

C-11-7 Exoplanet Orbits

An exoplanet around some star is at radius R and has a circular orbit of period T. The same star has a second planet in a circular orbit of radius $4R$. What is the period of the planet?

(a) $\dfrac{T}{8}$

(b) $\dfrac{T}{4}$

(c) $4T$

(d) $8T$

C-11-7 (d) From Kepler's third law, the ratio of the square of the period over the cube of the semi-major axis (equal to radius for a circular orbit) is constant, so we have the following relationship,

$$\frac{T^2}{R^3} = \frac{T'^2}{(4R)^3}$$

which yields $8T$ for the unknown period.

EXAMPLE 11-12

Mass of Saturn

Titan, the largest natural satellite of Saturn, takes 15.945 Earth days for one complete orbit. The semi-major axis of Titan's orbit around Saturn is 1.22×10^6 km. Find the mass of Saturn. Assume that the mass of Titan is much less than the mass of Saturn.

SOLUTION

First, convert the period into SI units of seconds:

$$T = 15.945 \text{ days} \times \frac{24 \text{ h}}{1 \text{ day}} \times \frac{60 \text{ m}}{1 \text{ h}} \times \frac{60 \text{ s}}{1 \text{ m}}$$

$$= 1.3776 \times 10^6 \text{ s}$$

When one mass is considerably smaller than the other, Kepler's third law becomes

$$T^2 = \frac{4\pi^2}{GM} a^3$$

Rearranging to solve for M and substituting the given data, we have

$$M = \frac{4\pi^2}{GT^2} a^3 = \frac{4\pi^2 (1.22 \times 10^9 \text{ m})^3}{(6.672 \times 10^{-11} \text{ N} \cdot \text{m}^2/\text{kg}^2)(1.3776 \times 10^6 \text{ s})^2}$$

$$= 5.66 \times 10^{26} \text{ kg}$$

Making sense of the result:

This result agrees with published values for the mass of Saturn. In fact, the published values were obtained using Kepler's third law with measurements of various natural and artificial satellites.

axis, the eccentricity, and the inclination, represented by the symbols a, e, and i, respectively. The **semi-major axis** specifies the size of the orbit, and the eccentricity specifies the shape. If $e = 0$, the orbit is circular; for $0 < e < 1$, the orbit is elliptical; if $e = 1$, the orbit is parabolic. The closer the eccentricity is to 1, the more elongated the shape of the orbit.

The inclination is the angle of the orbit relative to some reference plane. For motion around the Sun, the reference plane is the plane of Earth's orbit, which is called the ecliptic.

The **perihelion**, or closest point to the Sun, of an orbit and the **aphelion**, or most distant point from the Sun, are readily calculated from the following relationships:

$$\text{Perihelion:} \quad r_{\text{per}} = a(1 - e) \quad \quad (11\text{-}14)$$

$$\text{Aphelion:} \quad r_{\text{aph}} = a(1 + e) \quad \quad (11\text{-}15)$$

If $e = 0$, the orbit is circular, and the perihelion and aphelion are equal.

We frequently express solar system distances in terms of the astronomical unit (AU). One astronomical unit is defined as the mean distance between the Sun and Earth, which is approximately 1.50×10^{11} m.

LO 8

11-8 Types of Orbits

The total orbital energy (kinetic plus gravitational potential) is used to classify the type of orbit. Gravitational orbits are classified as one of the types of conic sections: ellipse, parabola, or hyperbola, with the circle being a special case of the ellipse. As the name implies, a **conic section** is a planar geometric shape corresponding to one of the ways that a cone can be cut (Figure 11-15). Cutting parallel to the base of a cone produces a circle. Cutting at an angle less than the angle of a cone's edge produces an ellipse. Cutting parallel to a cone's edge produces a parabola, and cutting at an even greater angle produces a hyperbola. The shape of each conic section can be defined by a relatively simple equation.

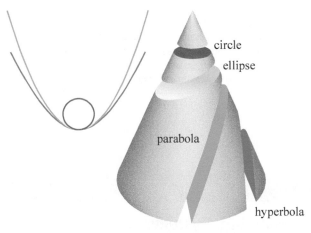

circle
ellipse
parabola
hyperbola

Figure 11-15 A cone can be cut at different angles to produce four types of conic sections. These sections correspond to the possible types of orbits.

 ONLINE ACTIVITY

Types of Orbits

The e-resource that accompanies every new copy of this textbook contains an Online Simulation Activity using the PhET simulation "Types of Orbits." Work through the simulation and accompanying questions to gain an understanding of different types of orbits.

The type of gravitational orbit is determined by the total orbital energy:

- When the total orbital energy is negative, the orbit is an ellipse.
- When the total orbital energy is zero, the orbit is a parabola.
- When the total orbital energy is positive, the orbit is a hyperbola.

A **hyperbolic orbit** is an unbound orbit, and an **elliptical orbit** is a bound orbit; a **parabolic orbit** is on the borderline between bound and unbound orbits. When an object is in a hyperbolic or parabolic orbit with respect to the Sun, the object will sweep past the Sun once and never return. Later chapters will describe how electromagnetic and nuclear forces can create similar types of motion for subatomic particles. In Example 11-13 we demonstrate how to determine the type of orbit.

☑ CHECKPOINT

C-11-8 Total Orbital Energy

Classify the orbit of each of the following:
(a) A meteoroid approaches Earth with a geocentric (Earth-relative) speed of 25 km/s.
(b) The altitude of a satellite orbiting Earth's varies from 1000 km to 3000 km.

C-11-8 (a) The meteoroid is approaching Earth with a speed higher than the escape velocity for Earth, so it is clear that the orbit must be hyperbolic (relative to the Earth). (b) The satellite orbit is elliptical from the description given.

EXAMPLE 11-13

Interstellar Visitor

A spacecraft is observed to be at a distance of 2.00 AU from the Sun and is and moving at a speed of 38.0 km/s.

(a) What type of orbit (relative to the gravitational field of the Sun) is the spacecraft following?
(b) Could the spacecraft have come from interstellar space?

SOLUTION

(a) To determine the type of orbit, we need to know if the total orbital energy of the spacecraft is negative, zero, or positive. Although the mass of the spacecraft is unknown, we can calculate the total orbital energy per unit mass:

$$\frac{E_{\text{total}}}{m} = \frac{T}{m} + \frac{U}{m} = \frac{1}{2}v^2 - \frac{GM}{r}$$

(continued)

where M is the mass of the Sun, $v = 3.80 \times 10^4$ m/s^2, and $r = 2.00 \times 1.496 \times 10^{11}$ m $= 2.99 \times 10^{11}$ m. Substituting into the above equation gives

$$\frac{E_{total}}{m} = \frac{1}{2}(3.8 \times 10^4 \text{ m/s})^2$$

$$- \frac{(6.672 \times 10^{-11} \text{ N} \cdot \text{m}^2/\text{kg}^2)(1.989 \times 10^{30} \text{ kg})}{2.992 \times 10^{11} \text{ m}}$$

$$= +2.78 \times 10^8 \text{ J/kg}$$

The total orbital energy is positive, so the object must be in a hyperbolic orbit.

(b) Since the orbit is hyperbolic, the spacecraft would have come from interstellar space unless its engines (or booster rocket) provided enough kinetic energy to move it into its current orbit.

Making sense of the result:

Earth moves at a speed of approximately 30 km/s in its near-circular orbit. Since the spacecraft is moving faster at twice the distance from the Sun, it is not unreasonable that the orbit is hyperbolic. Note that we did not need to know the direction that the spacecraft was moving to determine the type of orbit.

MAKING CONNECTIONS

Space Debris and Meteoroids

Both meteoroids and space debris pose small, but not insignificant, threats to space operations. The impact of even a tiny object can be disastrous for a satellite. Space debris impacts other objects in orbit with a lower relative speed than meteoroids, but the probability of impact is greater for space debris. The term "space debris" refers to artificial material that has ended up in Earth orbit as a result of our use of space. Such debris varies from specks of paint to much larger pieces of spacecraft. In 2008, an astronaut on a space walk accidentally released a tool kit that ended up in Earth orbit. NASA and other space agencies track thousands of pieces of space debris larger than a centimetre across, but the majority of space debris is much smaller. Space debris eventually spirals down toward Earth. Even at orbital altitudes, there is a tenuous atmosphere that produces a tiny drag force that ultimately results in the debris falling from orbit. Relative to Earth, space debris has negative total orbital energy, and meteoroids have positive total orbital energy.

MAKING CONNECTIONS

Finding Debris from Other Planetary Systems

Planets form along with their parent stars from a cloud of gas and dust called a nebula. The formation process likely also produces large numbers of small meteoroids, some of which escape the gravitational pull of the star (through collisions, near collisions, and radiation processes). We can detect escaped meteoroids that reach our solar system by finding meteoroids not gravitationally bound to the Sun. Canada is a world leader in the study of meteoroids, including the search for debris from other planetary systems. Researchers Peter Brown, Margaret Campbell-Brown, and Rob Weryk of the University of Western Ontario use the Canadian Meteor Orbit Radar (CMOR) and an automated super-sensitive digital camera system to compute meteor orbits and to isolate those on hyperbolic orbits. Figure 11-16 shows a map of meteor origins from a single day of observations with this facility. Each year, close to one million meteoroids are detected by this facility. A tiny

fraction of these are not in orbits within our solar system, but are material ejected from another planetary system.

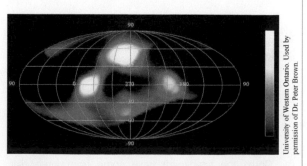

Figure 11-16 A plot of meteor origin directions from the CMOR meteor radar: the brighter colour indicates more meteors from that direction.

LO 9

11-9 Detection of Exoplanets

The distinguished Russian American astronomer Otto Struve (1897–1963) made numerous advances, one of which is his technique for detecting exoplanets; **exoplanets** are planets that orbit stars other than the Sun. Twenty-five years after his death, a Canadian research team used his technique to provide the first evidence that exoplanets do indeed exist. There are now more than 500 known exoplanets, and more are being discovered every year, many of them by NASA's Kepler space telescope.

While at least seven techniques have been used to detect exoplanets, the most successful so far has been Doppler spectroscopy (also called radial velocity spectroscopy). When a planet and a star are in orbit, we often consider the position of the star to be fixed because the star is so much more massive than the planet. However, it is the centre of mass of the system

ONLINE ACTIVITY

Star and Planet Orbital Motion

The e-resource that accompanies every new copy of this textbook contains an Online Activity using the PhET simulation "My Solar System." Work through the simulation and accompanying questions to gain an understanding of the orbital motion of stars and planets.

MAKING CONNECTIONS

The Discovery of Exoplanets

In 1988, Canadian scientists published a paper in the prestigious *Astrophysical Journal*, which presented clear evidence that seven of the stars they studied had objects in orbit with masses that could be planets. Led by Bruce Campbell, then of the University of Victoria and the Herzberg Institute of Astrophysics, the discovery team included Gordon Walker of the University of British Columbia and Stephenson Yang, who was then working at the Canada-France-Hawaii Telescope (CFHT) but is now at the University of Victoria. They found very small but regular shifts in the wavelengths of light from the stars. These shifts showed that the stars were in orbital motion, most likely with unseen planets.

In 1992, Dale Frail, a graduate of Acadia University and the University of Toronto working at the National Radio Astronomy Observatory (NRAO) in New Mexico, conclusively detected two planets in orbit around a pulsar (a rotating neutron star). Some astronomers consider 1992 to be the date of the first detection of an exoplanet because Frail made a more definitive claim of planet detection.

that is fixed, and both the star and the planet move in elliptical orbits about this centre. The star's orbit is very small compared to the planet's orbit. In Example 11-14, we examine the orbital motion of the Sun caused by its interaction with Jupiter.

We cannot directly see a planet orbiting a distant star, but we can infer that the planet is there from the slight

EXAMPLE 11-14

The Sun–Jupiter System

The mass of the Sun, M_s, is 1.99×10^{30} kg; the mass of Jupiter, M_J, is 1.90×10^{27} kg; and the mean distance between them, centre to centre, is 7.80×10^{11} m.

(a) Find the location of the centre of mass of the Sun–Jupiter system.
(b) Find the orbital velocity of the Sun about this centre of mass.

SOLUTION

We use x to represent the distance from the centre of the Sun to the centre of mass of the Sun–Jupiter system, and d to represent the distance from the centre of the Sun to the centre of Jupiter, as shown in Figure 11-17.

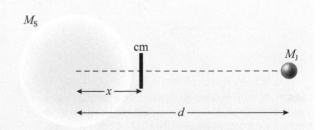

Figure 11-17 Example 11-14: The Sun–Jupiter system

(continued)

(a) For a system of just two masses, the product of one mass times its distance from the centre of mass must equal the corresponding product for the other mass. Therefore,

$$xM_s = (d - x)M_J$$

and

$$x = \frac{dM_J}{M_s + M_J} = \frac{(7.80 \times 10^{11} \text{ m})(1.90 \times 10^{27} \text{ kg})}{1.99 \times 10^{30} \text{ kg} + 1.90 \times 10^{27} \text{ kg}}$$

$$= 7.44 \times 10^8 \text{ m}$$

(b) The Sun makes a complete circuit of its orbit in one orbital period. The period of the Sun and Jupiter are the same because the two masses are in a joint orbit about a common centre. This period can be determined from Kepler's third law, or we can simply use the observed value, which is 11.86 y.

$$T = 11.86 \text{ yr} \times \frac{365.25 \times 24 \times 60 \times 60 \text{ s}}{1 \text{ yr}} = 3.74 \times 10^8 \text{ s}$$

Assuming that the Sun's orbit is circular with radius x, we have

$$v = \frac{2\pi x}{T} = \frac{2\pi (7.44 \times 10^8 \text{ m})}{3.74 \times 10^8 \text{ s}} = 12.5 \text{ m/s}$$

Making sense of the result:

(a) The radius of the Sun is 6.96×10^8 m, so the centre of mass is close to the surface of the Sun. This location is reasonable because the mass of the Sun is so much larger than the mass of Jupiter. (b) The orbital speed of the Sun is just 12.5 m/s. The speed of light is about 30 million times greater, so the shift caused by the presence of Jupiter is correspondingly small.

motion of the star—as the star orbits the centre of mass of the system, the star periodically moves toward us and away from us (Figure 11-18).

The light from a star moving away from us shifts slightly to a lower frequency (and a correspondingly longer wavelength); we call this effect redshift. (This is similar to a sound having a lower frequency when the source is moving away from you.) When the star moves toward us, its light shifts to shorter wavelengths, a blueshift (Figure 11-19). Note that the term "redshift" does not mean that the light becomes red; it means the light will have a longer wavelength (more toward the red end of the spectrum). We can use redshift and blueshift to determine star motions, and from that infer the presence of orbiting exoplanets. You can learn how to calculate the mass of an exoplanet with the online explanation and example in the accompanying eBook.

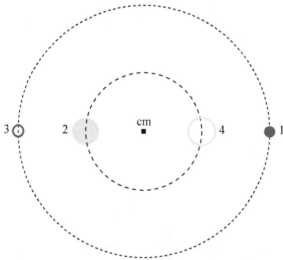

Figure 11-18 The motion of an exoplanet and a star about their centre of mass. When the planet (orange) is at point 1 on the right of the centre of mass, the star (yellow) will be on the left of the centre of mass at point 2. Similarly, when the planet is on the left at point 3, the star will be at point 4 on the right of the centre of mass. Note that the diagram is not to scale: the relative size of the orbit of the star will be much smaller than shown here.

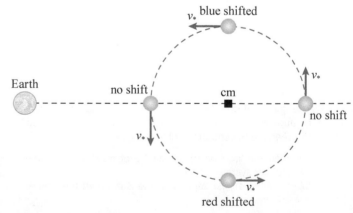

Figure 11-19 A star–planet system observed from Earth. The light from the star is blueshifted when the star moves toward us and redshifted when the star moves away from us.

According to the law of universal gravitation, every object in the universe attracts every other object with a force that depends directly on the two masses and inversely on the square of the distance between the centres of the objects. A corresponding universal form of gravitational potential energy can be derived from the law of universal gravitation, and has a negative value for all finite distances. The orbit of an object can be classified by the total of the object's kinetic energy and gravitational potential energy. Gravitational escape requires a positive total orbital energy.

Universal Gravitational Force

The gravitational force between any pair of masses in the universe is $|\vec{F}_G| = \dfrac{GMm}{r^2}$,

where M and m are the masses, r is the distance between their centres, and G is the universal gravitational constant. The force is directed toward the centre of the attracting mass.

The acceleration due to gravity near a point a distance r from the centre of a spherically symmetric mass can be readily shown from Newton's second law and the universal gravitational force to have the following magnitude: $g = \dfrac{GM}{r^2}$.

Force from Potential Energy

The derivatives of a potential energy function U give the components of the associated force, for example, $F_x = -\dfrac{\partial U}{\partial x}$, $F_y = -\dfrac{\partial U}{\partial y}$, $F_z = -\dfrac{\partial U}{\partial z}$, and $F_r = -\dfrac{\partial U}{\partial r}$.

Gravitational Potential Energy

The general expression for gravitational potential energy between spherically symmetric masses M and m, with centres separated by a distance r, is given by $U = -\dfrac{GMm}{r}$.

The gravitational potential energy is zero when the masses are infinitely far apart and negative everywhere else.

Total Orbital Energy

By calculating the sum of an object's kinetic energy and gravitational potential energy, we can determine the type of orbit. If the total is negative, the orbit is elliptical; if the total is zero, the orbit is parabolic; if the total is positive, the orbit is hyperbolic. A positive total also corresponds to gravitational escape.

Kepler's Laws

1. Planetary orbits are ellipses with the Sun at one focus.

2. A line joining a planet to the Sun sweeps out equal areas in equal times.

3. The square of the orbital period (T) is proportional to the cube of the semi-major axis (a) of the orbit. For a pair of objects with masses M_1 and M_2 orbiting around their centre of mass, this law is expressed as

$$T^2 = \frac{4\pi^2}{G(M_1 + M_2)}a^3 \qquad (11\text{-}12)$$

Orbital Speed

The speed of an object in an elliptical orbit with semi-major axis a about a mass M varies with distance r according to the following relationship:

$$v = \sqrt{GM\left(\frac{2}{r} - \frac{1}{a}\right)} \qquad (11\text{-}11)$$

Key Applications: gravitational field mapping in geophysics, satellite orbits, interplanetary travel, predicting tides, detecting exoplanets, predicting the paths of comets and asteroids, producing weightlessness

Key Terms: acceleration due to gravity, aphelion, asteroid, centre of mass, conic sections, elliptical orbit, escape speed, exoplanet, gravitational mass, gravitational potential energy, hyperbolic orbit, inertial mass, Kepler's laws, law of universal gravitation, meteoroid, parabolic orbit, perihelion, planet, satellite, semi-major axis, weight

QUESTIONS

1. When Earth exerts a force of 550 N on a person, what is the magnitude of the force that the person exerts on Earth?
   (a) exactly 0
   (b) 550 N
   (c) 550 N times the ratio of the mass of the person to Earth's mass
   (d) 550 N times the ratio of the mass of Earth to the mass of the person

2. Consider a planet that has the same mass as Earth, but eight times the average mass density. What would your weight on this planet be if your weight on Earth is W?
   (a) W
   (b) $2W$
   (c) $4W$
   (d) $8W$

3. Weigh scales are often marked in units of kilograms. Why is this marking incorrect?

4. What is your weight at the centre of Earth?

PUTTING IT ALL TOGETHER

5. Consider two spherical homogeneous planets, both with exactly the same density. The radius of one planet is twice as large as the radius of the other planet. Which of the following statements is true?
   (a) The two planets have exactly the same gravitational acceleration at their surfaces.
   (b) The larger planet has a greater gravitational acceleration at its surface.
   (c) The smaller planet has a greater gravitational acceleration at its surface.
   (d) More information is needed to compare the gravitational accelerations.

6. On a planet that has the same mass as Earth but a smaller radius, the escape speed from the surface is
   (a) less than the escape speed on Earth;
   (b) more than the escape speed on Earth;
   (c) the same as on Earth;
   (d) impossible to compare without more information.

7. At the surface of the planet in the previous question, the gravitational acceleration is
   (a) < 9.81 m/s^2;
   (b) ≈ 9.81 m/s^2;
   (c) > 9.81 m/s^2;
   (d) impossible to compare without more information.

8. Will it require less energy to launch a satellite to orbit from east to west or from west to east? Explain.

9. Which of the following statements is true?
   (a) G is constant, but g varies with location.
   (b) Both G and g vary with location.
   (c) Both G and g are constant everywhere in the universe.
   (d) G varies with location, and g is constant.

10. Three objects fall to Earth from a location some distance away. Object A starts from rest; object B has an initial velocity directed toward Earth; and object C has the same initial speed but directed away from Earth. Which of the following statements correctly describes the speeds of the objects when they reach Earth's surface?
    (a) All three objects will have exactly the same speed.
    (b) Object A will have the greatest speed.
    (c) Objects B and C will have exactly the same speed, which will be greater than the speed of object A.
    (d) Object C will have the greatest speed, and object B will have the lowest speed.

11. Does the speed of a satellite increase or decrease when atmospheric drag causes it to lose altitude? Explain your reasoning.

12. (a) If you want to produce a straight-line graph by plotting a set of data for orbital periods and semi-major axes, what function of the periods would you plot on the y-axis? What function of the semi-major axes would you plot on the x-axis?
    (b) If you want the slope of the graph from part (a) to be 1, what units should be used for periods and semi-major axes?

13. A planet is in a circular orbit around the Sun. Its distance from the Sun is four times the average distance between Earth and the Sun. The period of this planet, in Earth years, is
    (a) 4;
    (b) 8;
    (c) 16;
    (d) 64.

14. When we are observing a planetary system (assume a single planet and star) at a certain instant in time, the star is observed to be in motion toward us. What is the direction of motion of the exoplanet at this time?
    (a) toward us
    (b) away from us
    (c) stationary relative to us
    (d) cannot be determined from the information given

15. Consider a star and an exoplanet that orbit in a plane perpendicular to the direction from the star to Earth. What shifts would we observe in the light from the star?
    (a) The light would be redshifted at all positions in the star's orbit.
    (b) The light would be blueshifted at all positions of the orbit.
    (c) There would be no observable shift from the orbital motion.
    (d) The light would be redshifted at one side of the orbit and blueshifted at the opposite side.

16. In 2005, a satellite was discovered in orbit around the distant dwarf planet Eris. What key information about Eris can we learn by observing this satellite?

17. Suppose that a powerful rocket engine is set up on an asteroid in an elliptical orbit around the Sun. When the asteroid is at its perihelion, the engine is fired for a moment to increase the speed of the asteroid. Which of the following statements correctly describes the subsequent motion of the asteroid, assuming that no other forces act to alter its orbit. Justify the answer you select.
    (a) The perihelion, aphelion, and semi-major axis of the orbit will change.
    (b) The perihelion and semi-major axis, but not the aphelion, will change.
    (c) The aphelion and semi-major axis, but not the perihelion, will change.
    (d) The perihelion, aphelion, and semi-major axis are unchanged.

18. In 2006, the International Astronomical Union redefined "planet." As a result, Pluto is no longer considered to be one of the major planets.
    (a) Use print and/or Internet resources to research and summarize the criteria for an object to be considered a planet.
    (b) One of these criteria is that the object pull itself into an approximately spherical shape. Using concepts from this chapter, explain why a molten body in space will assume a spherical shape.

PROBLEMS BY SECTION

For problems, star ratings will be used, (✶, ✶✶, or ✶✶✶), with more stars meaning more challenging problems.

Section 11-1 Universal Gravitation

19. ✶✶ What force does a person of mass 50 kg exert on the Andromeda Galaxy, M31? The Andromeda Galaxy is believed to have a total mass of approximately one trillion times the mass of the Sun and is about 2.3 million light years away. (A light year is the distance that lights travels in one year.)

20. ✷✷ Assume that Earth's orbit is circular with a radius of 1.50×10^{11} m. From first principles, calculate the approximate speed of Earth in its orbit.

21. ✷ Calculate the difference in force per kilogram for water on the side of Earth nearest the Moon compared to water on the side farthest from the Moon. Use 6378 km as Earth's equatorial radius, and assume that the centre of the Moon is 3.84×10^8 m from the centre of Earth.

22. ✷✷ Assume that you are 25 km from the centre of a neutron star that has a radius of 10.0 km and a mass of 4.4×10^{30} kg. Your feet are pointed at the neutron star, and your head is 1.5 m above your feet. What is the *difference* in gravitational force per kilogram on your head compared to your feet?

Section 11-2 Acceleration Due to Gravity

23. ✷ Calculate the value of the acceleration due to gravity on the surface of Mars, given that the mass of Mars is 6.42×10^{23} kg. Assume that Mars is a homogeneous sphere with a radius of 3.40×10^6 m.

24. ✷ Find the radius of a spherical asteroid that has a density of 2900 kg/m^3 and an acceleration due to gravity of 0.10 m/s^2 at the surface.

25. ✷✷✷ Assume that the normal value of g in a region is 9.800 m/s^2. However, one part of the region has an ore deposit directly below the surface. This deposit has a density of 7800 kg/m^3, and the surrounding material has a mean density of 3000 kg/m^3. Assume that the ore deposit is spherical with a radius of 500 m. Calculate the value of g above the ore deposit.

26. ✷✷ The acceleration due to gravity at the surface of a fictional planet is 3.11 m/s^2, and the mean density of the planet is 880 kg/m^3. Find the radius and mass of the planet. Assume that it is spherical.

Section 11-3 Orbits and Weightlessness

27. ✷✷ A satellite is in a circular orbit 3000 km above the surface of a fictional planet. The satellite has a speed of 8000 m/s and is 8000 km from the centre of the planet.
(a) Find g at the altitude of the satellite.
(b) Find g at the surface of the planet.
(c) Find the mass of the planet.

28. ✷✷ If a parabolic flight aircraft starts the parabolic portion of a flight at an angle of 40° above the horizontal, and this portion of the flight lasts 20 s, what was the speed of the aircraft when it started the parabolic portion? Assume that the air resistance is negligible.

Section 11-4 Gravitational Potential Energy

29. ✷✷ Olympus Mons is an extinct volcano on Mars that is approximately 22 km high (almost three times the height of Mount Everest). Calculate the work required to raise a 25 kg object from the surface of Mars to the top of Olympus Mons. Assume that Mars is effectively spherical with a radius of 3400 km. The mass of Mars is 6.42×10^{23} kg.

30. ✷✷ An object is fired vertically from the surface of the Moon with a speed of 1000 m/s. What is the maximum height that it will reach?

Section 11-5 Force from Potential Energy

31. ✷ Find the components of the force in a region where the potential energy function is $U = ky^2$.

32. ✷✷ If gravitational potential energy had the form $U(r) = -\dfrac{GMm}{\sqrt{r}}$, what would the gravitational force be?

Section 11-6 Escape Speed

33. ✷ Calculate the escape speed from the surface of the Moon.

34. ✷✷ The dwarf planet Ceres is approximately spherical with a radius of 488 km. The density of Ceres is 2078 kg/m^3.
(a) What is the escape velocity from the surface of Ceres?
(b) If an object was launched vertically from the surface of Ceres with a speed that is exactly one-half the escape speed, what height would it reach above the surface of Ceres?

Section 11-7 Kepler's Laws

35. ✷✷ It is believed that there is a black hole with a mass of approximately four million times the mass of the Sun near the centre of the Milky Way galaxy. Some stars are relatively close to this black hole. Assume that the semi-major axis of the orbit of one of these stars is 3.5 light days (i.e., the distance that light would travel in 3.5 days).
(a) Find the orbital period of the star.
(b) Find the orbital speed of the star, assuming that it is in a circular orbit.

36. ✷✷ An asteroid has a perihelion distance of 0.55 AU (astronomical unit) and an aphelion distance of 1.65 AU. Predict the orbital period of the asteroid.

37. ✷ Since the Moon has virtually no atmosphere, an object could orbit just above the Moon's surface without experiencing any aerodynamic drag. What speed would be needed to launch an object so that it would orbit just slightly above the Moon's surface? Treat the Moon as a smooth sphere for this question.

Section 11-8 Types of Orbits

38. ✷✷✷ A sensitive Doppler radar instrument determines that a small object at a height of 1800 km above Earth's surface has a radial velocity component of 1.3 km/s directed away from Earth's centre. Night-vision images indicate that the object has a transverse velocity component of 6.7 km/s. Assuming that the transverse and radial components are perpendicular, determine whether the object is in a gravitationally bound orbit with Earth or in a hyperbolic orbit that will cause it to escape Earth's gravitational field. Assume no upper atmosphere drag, and neglect the influence of any other masses such as the Moon and the Sun.

39. ✷✷ (a) What is the centre of mass of the Sun-Earth system?
(b) Express your answer from the first part as a fraction of the radius of the Sun.

40. ✷✷✷ Spectroscopy of a star produces radial velocity plots suggesting a peak stellar speed of 34 m/s. The period of the cycle is 25 days. Other measurements give an estimate of 4.0×10^{30} kg for the mass of the star. Calculate the mass of the exoplanet that would account for the radial velocity of the star. Consult the eBook section on exoplanet mass determination for help in doing this problem.

COMPREHENSIVE PROBLEMS

41. ✷✷✷ Mars has two small satellites called Phobos and Deimos. Assume that these satellites are spherical with a mean density of 1880 kg/m³ and that the diameter of Phobos is 21 km.
    (a) If you dropped an object from a height of 1.5 m above the surface of Phobos, how long would it take the object to fall to the surface?
    (b) Phobos orbits Mars with a period of 0.319 days at a mean distance of 9400 km from the centre of Mars. Calculate the mass of Mars.

42. ✷ For a museum exhibit, you want to make a set of scales that show the weight a person would have on different planets. If you start with scales that measure 0 to 1000 N, how should you alter the markings on the scales for each of the following celestial objects?
    (a) Mars
    (b) Venus
    (c) Jupiter (if it had a surface you could stand on)
    (d) Earth's Moon
    (e) Pluto
    (f) Mercury
    Use reference works to find the information needed for each planet.

43. ✷✷✷ This problem involves calculations related to the Cavendish experiment. Refer to the online example for help with these calculations.
    (a) In the original Cavendish experiment, two small spheres with a mass, m, of about 0.73 kg each, were attached to the end of a balance arm 1.8 m in length. What is the moment of inertia, I, of the two balls as they rotate about the centre of the balance? Ignore the mass of the balance arm.
    (b) The period of a torsion balance is given by the following relationship:

    $$T = 2\pi\sqrt{\frac{I}{\kappa}}$$

    where T is the period, I is the moment of inertia, and κ is the torsion coefficient of the suspending wire. The observed period in the Cavendish experiment was about 420 s. Calculate the torsion coefficient for the apparatus.
    (c) The large masses, M, each had a mass of 158 kg each. Calculate the gravitational force between a small sphere and the adjacent large sphere when the distance between their centres is 230 mm.
    (d) The moment arm for each force is half the length of the balance arm, 0.90 m. Calculate the torque associated with one of the masses, using the force from part (c). The total torque will be double this value (since there is an equal torque at each end of the arm).
    (e) The rotational analog to Hooke's law is given by $\tau = -\kappa\theta$, where τ is the torque, κ is the torsion coefficient, and θ is the deflection angle. Predict the deflection angle for the apparatus described above. (Remember that the total torque acting on the balance arm is double the value calculated in part (d).)

44. ✷✷ On Earth, a person has a weight of 520 N. What weight would the person have on Mars? The radius of Mars is 53% of the radius of Earth, and its mass is 11% of the mass of Earth.

45. ✷✷✷ Consider a planet that has a gravitational acceleration of 1.8 m/s² at its surface. Assume that the planet is spherical with a radius of 4300 km.
    (a) Calculate the mass of the planet.
    (b) Find the escape speed from the surface of this planet, assuming no air resistance.

46. ✷✷✷ Consider a geosynchronous satellite with a mass of 140 kg. Geosynchronous satellites orbit Earth with a period of approximately 24 h. The mass of Earth is 5.97×10^{24} kg. Assume that Earth is exactly spherical with a radius of 6378 km.
    (a) At what height above Earth's surface will the geosynchronous satellite orbit?
    (b) Calculate the angular velocity for the orbit.
    (c) Calculate the angular momentum of the satellite in orbit.
    (d) How much additional gravitational potential energy does the satellite have when it is in orbit, compared to when it is on Earth's surface?
    (e) The orbital period of a geosynchronous satellite is not precisely 24 h. Explain why.

47. ✷✷ What is the fastest speed at which an object 1.0 AU from the Sun can move and still be gravitationally bound to the Sun? The object is positioned such that the gravitational forces exerted by Earth and other planets are negligible.

48. ✷✷ Europa is one of the large "moons" of Jupiter and is believed to have a major ocean under a thick frozen surface. Europa has a radius of 1570 km and a mass of 4.87×10^{22} kg. Assume that Europa is completely spherical.
    (a) Find the escape speed for an object at the surface of Europa.
    (b) Find the acceleration due to gravity, g, at the surface of Europa.
    (c) What would the weight of a 64 kg person be on the surface of Europa?

49. ✷✷✷ The Sun subtends an angle of about 0.53° when photographed from Earth. (This angle varies slightly because Earth follows a slightly elliptical orbit.) The Sun is a distance of 1.50×10^{11} m from Earth. Calculate the mean density of the Sun, assuming that it is spherical. (Hint: The period of Earth's orbit is one year.)

50. ✷✷ Pluto has at least three satellites, the largest of which is Charon. Charon orbits Pluto once every 6.387 days, and the semi-major axis of the orbit is 19 570 km. Find the sum of the masses of Pluto and Charon.

51. ✷✷ Derive the relationship $M = \dfrac{rv^2}{G}$ for a small mass, m, in a circular orbit at a distance r from the centre of a much larger mass, M. The speed of mass m is v.

52. ✷✷✷ The solar system orbits the centre of our Milky Way galaxy with a speed of approximately 230 km/s at a distance from the centre of the galaxy of about 26 000 light years.
    (a) Use the relationship from problem 51 to estimate the mass within the orbit of the solar system around the centre of our galaxy.
    (b) Express the mass calculated in part (a) in units of solar mass, that is, as a multiple of the mass of the Sun.

(c) How long will it take our solar system to make one orbit around the galaxy? Express your answer in years.

53. ✸✸ Imagine that a space colony living in a large vessel is in a circular orbit at a distance of 55 km from the centre of a neutron star. The neutron star has a radius of 10.0 km and a mass of 4.4×10^{30} kg.
(a) Find the velocity of the vessel in its orbit.
(b) Find the period of the orbit.

54. ✸✸ A comet discovered in an orbit around the Sun has a semi-major axis of 36 AU (astronomical units).
(a) Use Kepler's third law to estimate the period for the orbit of this comet. Express your answer in years.
(b) If the same comet were in orbit around an exoplanet of mass 8.6×10^{29} kg with the same semi-major axis of 36 AU, what would be the period of the orbit? Convert the final answer to years.
(c) Would the orbit about the exoplanet be circular? Justify your answer.

55. ✸✸ Go to the PhET site (http://phet.colorado.edu/), and download the "My Solar System" simulation. Select Trojan asteroids from the pulldown menu. Start the simulation, and describe the subsequent motion. Describe as precisely as possible the location of the Trojans relative to the planet as viewed from the Sun. Read the online material on Lagrange points and write an explanation of how the positions of the Trojans might relate to Lagrange points.

56. ✸✸ Stars lose mass from gradual processes, such as the solar wind, and from the ejection of material in eruptions and explosions. Consider a star with several planets in circular orbits.
(a) Describe qualitatively what would happen to the orbits of the planets when the star loses a significant part of its mass (say 20%).
(b) How much of its original mass would a star have to lose to make the orbits of the planets hyperbolic?

57. ✸✸ Consult the e-book online material on exoplanet mass determination for background in doing this problem. Consider an exoplanet system with one star and two planets. Doppler spectroscopy shows that one planet causes motion of the star with a peak radial speed of 40 m/s and a period of 10 days, and the other planet causes motion with a peak radial speed of 20 m/s and a period of 50 days.
(a) Which planet is closer to the star?
(b) Which planet is more massive?

58. ✸✸✸ Consider an exoplanet system with a star of mass 2.5×10^{30} kg and a single planet of mass 1.5×10^{28} kg in a circular orbit around their centre of mass. The orbital period is 25 days.
(a) Calculate the semi-major axis of the planet's orbit.
(b) Where is the centre of mass of the system located?
(c) Find the approximate orbital speed of the planet.
(d) Find the approximate orbital speed of the star.
(e) What would be the range of observed wavelengths for a spectral line with a wavelength of 656.280 nm emitted by the star in this system?

59. ✸✸ Some meteorites come from the Moon. Calculate the minimum ejection speed needed to move a stone from the surface of the Moon to Earth, ignoring the orbital motion of the Moon and Earth. (Hint: First find the point on the line from the Moon to Earth where the gravitational forces balance.)

60. ✸✸✸ Using work–energy principles, derive the expression for gravitational potential energy at distance r from a symmetric spherical mass M.

61. ✸✸ Suppose that the potential energy between two masses, M and m, had the form $U(r) = -\dfrac{kMm}{r^3}$, where k is a constant. Use calculus to find an expression for the magnitude of the related force.

62. ✸✸✸ A visible star with an unseen companion star often emits X-rays. The visible star in one such pair has spectral characteristics that indicate it has a mass of about 7.6 solar masses (i.e., 7.6 times the mass of the Sun). After careful observations you have determined that the semi-major axis of the orbit of the visible star is 26 million km. The orbital period of that star is 8.5 Earth days.
(a) Use the general version of Kepler's third law to determine the mass of the unseen star.
(b) Theory suggests that a stellar remnant will not form a black hole with a mass less than 3.0 solar masses. Is the unseen object a black hole?

63. ✸✸ Assume that the Sun started as a thin, spherical shell of material with the same mass that it has now and with a radius of one light year.
(a) If we take three-quarters of the current radius of the Sun as the average distance of its mass from the centre of mass, how much gravitational potential energy would be released as the Sun contracts from the large shell to its current size?
(b) For what period of time could the released energy power the Sun at its current output of about 3.85×10^{26} W?

64. ✸✸ Assume that an object is perfectly spherical and has a uniform density equal to the density of water (1000 kg/m^3).
(a) Find the radius and mass of the object that would make the escape speed at the surface equal to the speed of light.
(b) What does the answer to part (a) tell you about black holes?

65. ✸✸✸ Access the e-book online resources on exoplanet mass determination for background in doing this problem. Consider an exoplanet system that consists of a star of mass 1.8×10^{30} kg and a single planet of mass 4.5×10^{26} kg in a circular orbit around their centre of mass. The orbital period is 12 days. Find
(a) the maximum orbital speed of the star;
(b) the range of observed values for a spectral line that has an unshifted wavelength of 396.781 00 nm on Earth.

DATA RICH PROBLEM

66. ✸✸✸ A space probe to a fictitious planet accurately measures the radius of the (assumed perfectly spherical) planet as 5038 km. The space probe measures the g value at a number of heights above the surface of the planet as given in Table 11-1. From these data determine
(a) the mass of the planet;
(b) the mean density of the planet;
(c) the value of g at the surface of the planet.

305

Table 11-1 Problem 66

Height (m)	g (m/s^2)
10 000	3.92583
15 000	3.91807
20 000	3.91033
25 000	3.90261
30 000	3.89491
35 000	3.88724
40 000	3.87959
45 000	3.87196
50 000	3.86435
55 000	3.85677
60 000	3.84921
65 000	3.84167
70 000	3.83415
75 000	3.82666

OPEN PROBLEM

67. ✱✱✱ At the beginning of the chapter, it is suggested that the event 65 million years ago that resulted in the extinction of the dinosaurs and perhaps 75% of all species was probably caused by a comet or an asteroid impact.

(a) Under a worst-case scenario, what would be the approximate maximum speed that a comet could impact Earth? Show calculations to support your answer, and assume that the comet is gravitationally bound to the solar system.

(b) Consult sources to suggest a reasonable radius and density for a comet, and use this information to calculate the comet's mass.

(c) Using your results from parts (a) and (b), find the kinetic energy of the comet just before impact. To put this in perspective, convert the result into the energy equivalent of a number of tonnes of TNT.

See the text online resources at www.physics1e.nelson.com for Open Problems and Data-Rich Problems related to this chapter.

Chapter 12
Fluids

The human cardiovascular system consists of a pump (the heart), a fluid (blood), and blood vessels of various cross-sectional areas and structure (Figure 12-1). Oxygen-rich blood leaves the heart at a pressure greater than atmospheric pressure. As blood flows through smaller arteries, frictional forces result in a loss of pressure. By the time the blood enters the capillaries, it is almost at atmospheric pressure. Clogged arteries restrict blood from flowing freely, and the heart has to pump harder to maintain circulation, resulting in higher blood pressure. Thus, high blood pressure is an indicator of clogged arteries. To understand how the pressure and flow of fluids are related, we will analyze some basic properties of static and moving fluids.

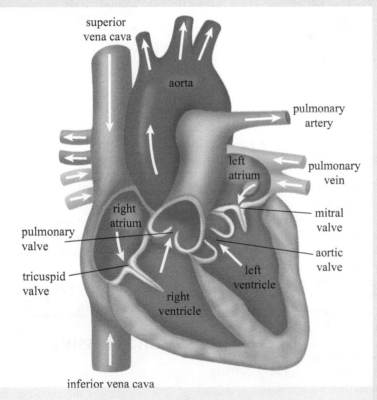

Figure 12-1 The human heart is a powerful muscle that controls the blood flow through the body.

Learning Objectives

When you have completed this chapter you should be able to:

1 Describe three phases of matter.

2 Calculate pressure at a given depth inside a liquid.

3 Explain Pascal's principle and the operation of a hydraulic lift.

4 Calculate buoyant forces on floating and immersed objects.

5 Calculate the average density of an object by determining its weight in air and in a fluid of known density.

6 Derive the equation of continuity for fluids.

7 Demonstrate how conservation of energy in a moving fluid leads to the derivation of Bernoulli's equation.

8 Use Bernoulli's equation to calculate flow speed and pressure of a moving fluid for simple situations.

9 Explain the consequence of conservation of momentum for moving fluids.

10 Explain the effect of viscosity on volume flow rate through a pipe.

 For additional Making Connections, Examples, and Checkpoints, as well as activities and experiments to help increase your understanding of the chapter's concepts, please go to the text's online resources at www.physics1e.nelson.com.

12-1 Phases of Matter

Most matter at everyday temperatures and pressures exists in three **phases** (or **states**), classified as gases, liquids, and solids (Figure 12-2). The physical properties of a given phase are determined by the average distance between its constituents (atoms and molecules) and the strength of the electromagnetic forces between them.

solid keeps its shape and does not deform easily. However, solid molecules are not completely stationary: they vibrate rapidly about their equilibrium positions.

There is a fourth phase of matter, called **plasma**. Plasmas form at very high temperatures and in strong electric fields. In this phase, collisions between molecules are so strong that the electrons can be knocked from the outer atomic orbits. Consequently, plasmas are a combination of ions and electrons. This state of matter exists in lightning, fluorescent lamps, neon signs, the solar wind, and stars.

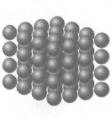

solid   liquid   gas

Figure 12-2 Solid, liquid, and gas states of matter

In gases, the average distance between the molecules is large compared to the size of the atoms; hence, the electromagnetic forces between the molecules are so weak that the molecules do not bind with each other and are able to move freely. Only when two molecules are very close to each other do they experience an appreciable force, and we say that "a collision has occurred between the two molecules." Such collisions obey Newton's laws of motion. At 20°C and sea level, the average distance between air molecules is 33×10^{-10} m, which is approximately 10 times larger than the diameter of a nitrogen molecule. Because of the large distances between gas molecules, gases are easily compressed.

In liquids, the molecules are very close to each other, and the electromagnetic forces between them are sufficiently strong to weakly bind the molecules but not strong enough to hold them at fixed positions. The molecules are still free to slide around each other without leaving the body of the liquid, enabling a liquid to flow under the influence of an external force. The average distance between water molecules is approximately 3×10^{-10} m, which is about the same as the diameter of a water molecule. Because the molecules in liquids are very close to each other, liquids are difficult to compress.

In solids, the electromagnetic forces hold the molecules at fixed positions relative to each other. Therefore, a

Gases and liquids are collectively called **fluids** (from the Latin word *fluidus*, meaning "to flow") because of their ability to flow. Fluids do not retain a specific shape, so we will define several new variables to describe their physical properties. Although the mathematical equations describing the behaviour and motion of fluids look different from what we have studied so far, the underlying physics is still based on Newton's laws and principles of conservation of energy and momentum.

12-2 Density and Pressure

Mass Density

The mass density, ρ, of a substance is the mass per unit volume of the substance. For an object of mass M that occupies a volume V,

$$\rho = \frac{M}{V} \tag{12-1}$$

Mass density is a scalar quantity. Note that Equation (12-1) gives the **average mass density** of an object. When we use the word "density" by itself, we are referring to the mass density. (The concept of density can be applied to other physical quantities, such as charge and current.) The SI units for density are kilograms per cubic metre (kg/m³). Densities are also often expressed in grams per cubic centimetre (g/cm³):

$$1 \frac{kg}{m^3} = 1 \frac{10^3 g}{(10^2 cm)^3} = 10^{-3} \frac{g}{cm^3} \quad or \quad 1 \frac{g}{cm^3} = 1000 \frac{kg}{m^3}$$

The density of a substance depends on the mass of its constituents (atoms) and the spacing between them. Table 12-1 lists the densities of various substances. The densities of gases are typically thousands of times less than the densities of liquids because the molecules are much farther apart in a gas than in a liquid. By definition, the density of a pure substance is independent of its volume. A steel nail has exactly the same density as a large block of the same material.

The density of an unconfined gas depends on its temperature. As the temperature of a gas increases, its volume increases and, therefore, its density decreases. When quoting the density of a gas, we will assume that the density is given at 0°C. The temperature dependence of density for liquids and solids is much less than that for gases.

Table 12-1 Mass densities of common substances

Substance		Mass density, ρ (kg/m³)
Solids		
Aluminum		2 700
Copper		8 890
Diamond		3 520
Gold		19 300
Ice		917
Lead		11 300
Silver		10 500
Wood (various types)	Balsa	160
	Maple	600–750
Liquids		
Alcohol		806
Blood (at 37°C)		1 060
Mercury		13 600
Water (at 4°C)		1 000
Gases		
Air		1.29
Helium		0.179
Hydrogen		0.0899
Nitrogen		1.25
Oxygen		1.43

EXAMPLE 12-1

The Density of the Sun

The mass of the Sun can be accurately determined from the length of the year on Earth, the distance between Earth and the Sun, and the value of the gravitational constant, G. The mass of the Sun is 1.99×10^{30} kg, and its radius is approximately 6.96×10^8 m. What is the average density of the Sun? Assume that the Sun is spherical.

SOLUTION

The Sun is assumed to be a sphere of radius 6.96×10^8 m. Therefore, the volume of the Sun is

$$V_{sun} = \frac{4\pi}{3} (6.96 \times 10^8 \text{ m})^3 = 1.41 \times 10^{27} \text{m}^3$$

The mass density of the Sun, ρ_{sun}, is now calculated from Equation (12-1):

$$\rho_{sun} = \frac{M_{sun}}{V_{sun}} = \frac{1.99 \times 10^{30} \text{ kg}}{1.41 \times 10^{27} \text{ m}^3} \cong 1410 \text{ kg/m}^3$$

Making sense of the result:

The *average* density of the Sun is approximately 40% greater than the density of water. The inner core of the Sun has a density of approximately 1.5×10^5 kg/m³; the outer corona consists of hot gases with an average density of approximately 10^{-11} kg/m³. For intermediate layers, the density varies between these values. The value of 1410 kg/m³ therefore represents an average density for the entire Sun.

✓ CHECKPOINT

C-12-1 The Density of a Wooden Block

Three blocks of wood are cut from the same rectangular plank of wood, as shown in Figure 12-3. Which of the following statements about the densities of the blocks is correct?
(a) Block A has the highest density.
(b) Block B has the highest density.
(c) Block C has the highest density.
(d) All three blocks have the same density.

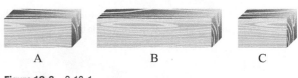

Figure 12-3 C-12-1

C-12-1 (d)

Specific gravity

The **specific gravity** (s.g.) of a material is the ratio of the mass of the material to the mass of an equal volume of water. Since mass = density × volume, the specific gravity of a material can also be defined as the ratio of its density to the density of pure water at the same temperature:

$$\text{s.g.} = \frac{\rho_{material}}{\rho_{water}} \qquad (12\text{-}2)$$

As you can see, specific gravity is a ratio of two quantities that have the same units, so it is a dimensionless quantity.

In Table 12-1, the density of blood is listed as 1060 kg/m^3. Therefore, its specific gravity is

$$\text{s.g.} = \frac{\rho_{material}}{\rho_{water}} = \frac{1060 \text{ kg/m}^3}{1000 \text{ kg/m}^3} = 1.06$$

Pressure

When you push against a wall with your hand, the force exerted by your hand on the wall is distributed over the area of your hand that is touching the wall. **Pressure**, P, is defined as the magnitude of the component of the force perpendicular to a surface divided by the area of the surface on which the force acts.

Consider a force $\vec{F}$ acting on a surface of area A. Let us denote by F_n the component of $\vec{F}$ that is perpendicular to the surface. Then, the pressure exerted by the force on the surface is given by

$$P = \frac{F_n}{A} \qquad (12\text{-}3)$$

Since the component of a force and the area are both scalars, pressure is a scalar quantity. The SI unit of pressure is the **pascal** (symbol Pa), which is equal to one newton per square metre (N/m^2):

$$1 \text{ Pa} = 1 \text{ N/m}^2$$

The name "pascal" honours the French mathematician, philosopher, and physicist Blaise Pascal (1623–62), who, among many other things in his short life, did some of the pioneering work on the properties of fluids.

For a given magnitude of force, pressure can be increased by applying the force over a smaller surface area. When you stand on one foot, the pressure that your foot exerts on the ground is twice the pressure that it exerts when you stand on both feet.

EXAMPLE 12-2

Calculating Pressure

A 0.500 kg book is resting on a table. Its cover has a surface area of 0.060 m^2. You push down on the book with your hand with a force of magnitude 50.0 N, with your arm inclined at an angle of 40.0° with respect to the normal to the book's surface (Figure 12-4). The area of your hand in contact with the book, A_{hand}, is 0.012 m^2.

(a) Find the pressure exerted by your hand on the book.
(b) What is the pressure exerted on the table by the book when your hand is pressing on the book?

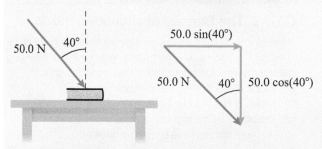

Figure 12-4 The force exerted by the hand on the book is represented by an arrow that makes an angle of 40° with respect to the normal to the book.

SOLUTION

(a) The pressure exerted by your hand on the book is equal to the normal component of your hand's force on the book, $\vec{F}_{hand\text{-}book}$, divided by the area of the hand in contact with the book.
From Figure 12-4, the normal component of the force exerted by the hand on the book $F_{n,hand\text{-}book}$, is given by

$$F_{n,hand\text{-}book} = 50.0 \text{ N} \times \cos(40°) = 38.3 \text{ N}$$

Therefore, the pressure exerted by the hand on the book is

$$P_{hand\text{-}book} = \frac{F_{n,hand\text{-}book}}{A_{hand}} = \frac{38.3 \text{ N}}{0.012 \text{ m}^2} = 3.19 \text{ kN/m}^2 = 3.19 \text{ kPa}$$

(b) The net force exerted on the table by the book is the vector sum of the forces exerted by the hand on the book and the weight of the book:

$$\vec{F}_{book\text{-}table} = \vec{F}_{hand\text{-}book} + \vec{W}_{book}$$

The normal component of $\vec{F}_{book\text{-}table}$ is the sum of the normal components of the two forces on the right-hand side of the above equation:

$$F_{n,book\text{-}table} = F_{n,hand\text{-}book} + m_{book}g$$
$$= 50.0 \text{ N} \cos(40°) + 0.5 \text{ kg} \times 9.81 \text{ m/s}^2$$
$$F_{n,book\text{-}table} = 43.2 \text{ N}$$

The pressure exerted by the book on the table is equal to the normal component of the net force exerted by the book on the table, divided by the cross-sectional area of the book:

$$P_{book\text{-}table} = \frac{F_{n,book\text{-}table}}{A_{book}} = \frac{43.2 \text{ N}}{0.060 \text{ m}^2} = 720 \text{ Pa}$$

Making sense of the results:

The pressure in part (b) is considerably smaller than the pressure calculated in part (a), even though more force is being exerted than in part (a). This result is reasonable because the contact area of the book on the table is greater than the contact area between the hand and the book.

C-12-2 The Pressure on a Surface

Figure 12-5 shows four different situations in which a force of magnitude 45.0 N is exerted on a surface of area 0.10 m². The direction of the force is represented by the direction of the arrow. Which of the following statements correctly describes the relationship between the pressures P_A, P_B, P_C, and P_D being exerted on the surface?

(a) $P_A < P_B < P_C < P_D$
(b) $P_B < P_A < P_C < P_D$
(c) $P_C < P_B < P_A < P_D$
(d) $P_A = P_B = P_C = P_D$

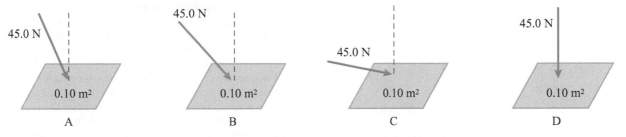

A   B   C   D

Figure 12-5 A force of magnitude 45.0 N is exerted at different angles on a surface.

C-12-2 (c)

LO 3

12-3 Pressure in Fluids

A fluid exerts pressure at every point within it and at every point where it contacts a surface, such as the wall of a container. This pressure is called **fluid pressure**. Fluid pressure results from the weight of the fluid and from the constant collisions of the molecules with each other and with the walls of the containing vessel. For liquids,

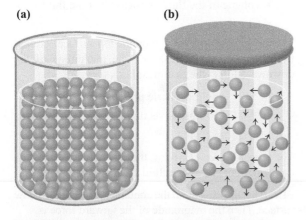

Figure 12-6 (a) In a liquid, the major source of pressure is the weight of the liquid. (b) In a gas, where densities are approximately 1000 times smaller than the densities of the liquids, pressure exists because of collisions between gas particles.

the major contribution to fluid pressure is the weight of the liquid. In contrast, the pressure exerted by a gas usually arises mainly from molecular collisions because the molecules of a gas are free to move and the densities of gases are much less than the densities of liquids (Figure 12-6). However, due to the enormous size of the atmosphere, the weight of air is the dominant contributor to atmospheric pressure.

Atmospheric Pressure

Earth's atmosphere consists mostly of nitrogen (78.1%), oxygen (20.9%), and argon (0.9%). Although there is no well-defined boundary between the atmosphere and space, it is generally agreed that Earth's atmosphere extends to about 120 km above sea level. The weight of the air molecules exerts a pressure at every point in the atmosphere and on Earth's surface. This pressure is called **atmospheric pressure**. The atmospheric pressure at sea level at 20°C is experimentally determined to be 1.013×10^5 Pa (≈ 101 kPa).

Objects on Earth are subject to atmospheric pressure. When you hold your hand flat, the air above the hand exerts a downward force on top of your hand. For a hand with an area of 130 cm², the magnitude of this force is approximately 1.3×10^3 N, which is equivalent to holding a 130 kg object on your palm. Why are you able to hold your hand flat without feeling this force? It is because the air below the palm exerts an upward force on the other side of your hand, almost exactly balancing the downward force from the top (the difference is minuscule).

Moving higher up the mass (hence the weight) of air above a surface decreases, causing a decrease in atmospheric pressure with increasing height. Table 12-2 shows how atmospheric pressure varies with height above sea level.

Table 12-2 Variation of Atmospheric Pressure with Altitude

Height above sea level (m)	Atmospheric pressure* (kPa)
0	101.3
10 000	31.7
20 000	9.9
30 000	3.1
40 000	0.98
50 000	0.30
60 000	0.095

*These values were calculated assuming an exponential decrease in air density with increasing altitude.

EXAMPLE 12-3

Measuring the Mass of Earth's Atmosphere

Given that atmospheric pressure at Earth's surface is 1.01×10^5 Pa and the mean radius of Earth is about 6370×10^3 m, calculate the approximate mass of the air surrounding Earth. What assumptions do you need to make to calculate the mass of the air?

SOLUTION

The atmospheric pressure at Earth's surface results from the weight of the air. If we assume that Earth is spherical, we can calculate Earth's surface area from its radius. The total downward force exerted by the air, which is equal to the weight of the entire atmosphere, is the product of Earth's surface area and the atmospheric pressure at the sea level.

Assuming that the acceleration due to gravity from sea level to the top of the atmosphere remains constant, the magnitude of the downward force on Earth's surface due to the weight of the air is

$$F_{air,down} = M_{air} g \qquad (12\text{-}4)$$

This force is equal to the product of the atmospheric pressure at sea level and Earth's surface area:

$$F_{air,down} = P_0 \times 4\pi R_{earth}^2 \qquad (12\text{-}5)$$

From Equations (12-4) and (12-5), we get

$$M_{air} = \frac{P_0 \times 4\pi R_{earth}^2}{g} = \frac{1.01 \times 10^5 \dfrac{N}{m^2} \times 4\pi \times (6.370 \times 10^6 m)^2}{9.81 \text{ m/s}^2}$$

$$\cong 5.2 \times 10^{18} \text{ kg} \qquad (12\text{-}6)$$

MAKING CONNECTIONS

Why Do Your Ears Pop?

In the human ear, the middle ear and the inner ear are separated from the outer ear by the eardrum, which is a flexible membrane. While taking off or landing in a plane, you often feel your ears pop. At ground level, the air pressure is the same in the outer ear and the middle ear. When taking off, the cabin pressure gradually decreases, so the pressure in the outer ear decreases and the high-pressure air in the middle ear pushes outward on the ear-drum, making it pop (flex outward). Since the difference in pressure is often uncomfortable, many people move their jaws or hold their noses and swallow to try to equalize the pressure immediately. During the flight, air seeps out of the inner ear until the pressure in the ear matches the pressure in the cabin, which is usually kept at approximately 80 kPa. During landing, the pressure in the outer ear increases, but the pressure in the middle ear remains low. The high-pressure air from outside pushes inward on your eardrum, making it pop inward. Similar popping can occur when you drive from a lower altitude to a higher altitude and vice versa.

Hydrostatic Pressure

Consider a container that is open to the atmosphere and filled with a fluid of density ρ_f (Figure 12-7). Imagine a cylindrical element of this fluid with cross-sectional area A, with its top end at the surface and the bottom end at a depth d below the surface. The following external forces are acting on this element in the vertical direction:

■ The weight ($\vec{W}$) of the fluid element acting vertically downward. The magnitude of the weight is

W = mass of the fluid element $\times g$

= volume of the fluid $\times$ density of the fluid $\times g$

= $(Ad)\rho_f g$

■ The downward force ($\vec{F}_{down}$) on the top surface, exerted by the atmospheric pressure. The magnitude of the downward force due to atmospheric pressure is

F_{down} = atmospheric pressure

$\times$ cross-sectional area of the cylinder

= $P_0 A$

■ The upward force ($\vec{F}_{up}$) that the rest of the fluid exerts on the bottom end of the fluid element.

Let P be the pressure in the container at a depth d below its surface. Then the magnitude of the upward force is

F_{up} = Pressure at the bottom of fluid element

$\times$ cross-sectional area of the fluid element

= PA

(a)

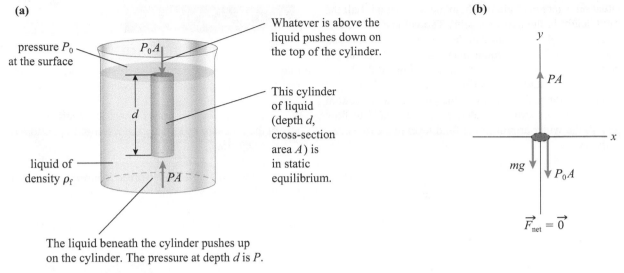

pressure P_0 at the surface

P_0A

Whatever is above the liquid pushes down on the top of the cylinder.

d

This cylinder of liquid (depth d, cross-section area A) is in static equilibrium.

liquid of density ρ_f

PA

The liquid beneath the cylinder pushes up on the cylinder. The pressure at depth d is P.

(b)

y

PA

x

mg

P_0A

$\vec{F}_{net} = \vec{0}$

MECHANICS

Figure 12-7   (a) Forces acting on a cylindrical element of a fluid in a container (b) The FBD of the column of liquid

There is no net horizontal flow of fluid because the *horizontal* force exerted on any point of the cylinder is exactly balanced by an equal horizontal force on the opposite side of the point. Taking the y-axis along the vertical direction with the positive y-axis upward, the net vertical force on the cylindrical volume is given by

$$\sum F_y = F_{up} - F_{down} - W$$
$$\sum F_y = PA - P_0A - \rho_f g\,(A\ d) \qquad (12\text{-}7)$$

Since the fluid element is stationary, the downward forces due to the weight of the fluid and the atmospheric pressure must be balanced by an upward force exerted by the rest of the fluid. Therefore, $\sum F_y = 0$, and we obtain the following relationship for the fluid pressure at a depth d below the surface of a fluid that is open to the atmosphere:

KEY EQUATION $$P = P_0 + \rho_f gd \qquad (12\text{-}8)$$

The second term on the right-hand side of Equation (12-8) arises due to the weight of the fluid. Notice two important consequences of this simple equation:

■ The pressure within a fluid increases with depth.

■ All points at a given depth within a connected body of fluid are at the same pressure.

The second point means that the pressure along a horizontal line through a connected body of a fluid is the same, no matter how the container is shaped. The pressure depends only on the depth below the surface. The pressure at a depth of 5 m inside a narrow tube filled with water is exactly the same as the pressure at a depth of 5 m in a large swimming pool, for example.

Equation (12-7) assumes that the fluid density does not change with depth. Therefore, it is not applicable for

the atmosphere, where the density of air changes with elevation.

Water Keeps Its Level

Water level in two arms of a U-shaped tube always rises to the same level (Figure 12-8). Why? Suppose that the water level is higher in the left arm. According to Equation (12-8), the pressure at the bottom of the left arm is higher than the pressure on the right side. The higher pressure at the left forces the water to flow from left to right until the level is the same in the two arms.

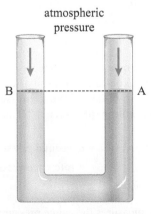

atmospheric pressure

B

A

Figure 12-8   The pressure at the bottom of the two arms of the tube is the same, so the height of the fluid in the two arms must be equal.

Suppose that the cross-sectional area of one arm of the tube is twice the cross-sectional area of the other (Figure 12-9). Is it possible to have an equilibrium situation where the fluid height in the wider arm is half the fluid height in the narrower arm? The volume and, therefore, the weight of the fluid in the two arms are then equal. Therefore, a force of equal magnitude is exerted at the bottom of the two arms of the tube. You might think that the fluid is in static equilibrium. However, with unequal levels of fluid in the two arms, the pressure at the bottom of the narrower arm is greater, causing the fluid to flow toward the wider arm until the fluid level in the two arms is the same.

Figure 12-9 Is it possible for a fluid at rest in a tube with two arms of unequal cross-sectional area to be at two different levels?

Equation (12-8) can be used to find the pressure difference between two points within a fluid. Let P_1 be the pressure at a point that is located at a depth d_1 within a fluid and P_2 be the pressure at a depth d_2. The pressure difference between the two points is

$$P_2 - P_1 = (P_0 + \rho_f g d_2) - (P_0 + \rho_f g d_1) = \rho_f g (d_2 - d_1)$$
(12-9)

In a room that is 5 m high, the increase in the air pressure at the floor level, compared to the pressure at the ceiling, is $(1.3 \text{ kg/m}^3) \times (9.81 \text{ m/s}^2) \times (5.0 \text{ m}) = 63.8 \text{ Pa}$. This increase is a tiny fraction of the average atmospheric pressure of 1.01×10^5 Pa.

Gauge Pressure

Pressure is usually measured with respect to a reference pressure, usually taken to be atmospheric pressure. The absolute pressure (P), the reference pressure (P_0), and the measured or **gauge pressure** (P_G) are related as follows:

absolute pressure = measured pressure
+ reference pressure

$$P = P_G + P_0$$

A tire gauge measures the gauge pressure inside a tire with respect to the atmospheric pressure. So, a reading of zero on a tire gauge indicates that the pressure inside the tire is equal to the atmospheric pressure, indicating that the tire is flat. Similarly, blood pressure measured by a sphygmomanometer is the gauge pressure measured relative to atmospheric pressure.

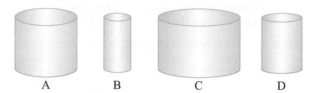

☑ **CHECKPOINT**

C-12-3 Pressure and Force

Four cylinders of different cross-sectional areas but the same height are completely filled with water, as shown in Figure 12-10.

(a) Order the cylinders, from highest to lowest, in terms of the pressure exerted by the water at the bottom of each cylinder.

Highest _____ _____ _____ _____ Lowest

(b) Order the cylinders, from highest to lowest, in terms of the force exerted by the water at the bottom of each cylinder.

Highest _____ _____ _____ _____ Lowest

A   B   C   D

Figure 12-10 Four cylinders of equal height and different cross-sectional areas

C-12-3 (a) A = B = C = D; (b) CADB

Measuring Pressure

A common instrument used to measure gauge pressure is a manometer. A manometer consists of a U-shaped glass tube of uniform cross-sectional area, partially filled with a liquid of known density (Figure 12-11). If the two arms of the manometer tube are at different pressures, then the fluid in the arm at the higher pressure is pressed downward, and the fluid level in the other arm rises.

Figure 12-11 A U-shaped manometer. The height difference in the column of fluid between the two arms is proportional to the pressure difference between the two sides.

Consider a manometer of uniform cross-sectional area partially filled with a liquid of density ρ_f (Figure 12-12). One arm (A) of the manometer is connected to a gas-filled cylinder that is maintained at a pressure P. The other arm (B) is open to the atmosphere and under atmospheric pressure P_0, which is assumed to be less than P. Let h be the difference in the heights of columns of fluids in the two arms. The pressure difference $P - P_0$ between the two arms of the manometer is proportional to h. To see this, draw a straight horizontal line that passes through a point (a) at the gas–fluid interface in arm A and a corresponding point (b) in arm B. Since the pressure in a static fluid is the same at all points at given level,

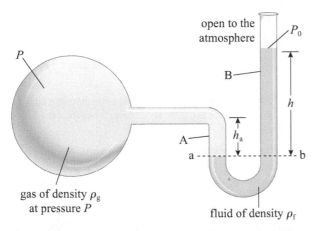

Figure 12-12 A U-shaped manometer connected to a pressurized device

$$\text{pressure at a } (P_a) = \text{pressure at b } (P_b) \quad (12\text{-}10)$$

The pressure at the top of arm A is P; therefore,

$$P_a = P + \text{pressure because of the weight}$$
$$\text{of the gas in the left arm}$$

$$P_a - P + \rho_g g h_a$$

where h_a is the height of the column of gas in arm A, and ρ_g is the gas density. The pressure at point b is the sum of the atmospheric pressure and the weight of the column of liquid above the point b:

$$P_b = P_0 + \rho_f g h$$

Since $P_a = P_b$, we get

$$P + \rho_g g h_a = P_0 + \rho_f g h$$

Therefore,

$$P - P_0 = \rho_f g h - \rho_g g h_a \quad (12\text{-}11)$$

The quantity on the left-hand side is the gauge pressure. Since liquid densities are generally approximately 1000 times greater than gas densities, the second term on the right-hand side of Equation (12-11) can be ignored. With this approximation,

$$P - P_0 = \rho_f g h \quad (12\text{-}12)$$

This simple equation can be used to determine the pressure in a gas or a liquid pipeline by using a manometer.

EXAMPLE 12-4

Pipeline Pressure

The total pressure in a gas pipeline is to be maintained at three times atmospheric pressure. Figure 12-12 shows the cross-sectional area of the pipeline that is connected to a manometer filled with mercury. An inspector measures a difference of 1.25 m between the two columns of mercury. Is the gas pressure being maintained at the desired value? The density of mercury is 13 600 kg/m³.

SOLUTION

The open end (B) of the manometer is at atmospheric pressure. We do not know the height of the gas column in arm A, but we can ignore the pressure from the weight of the gas because the gas is much less dense than mercury.

Consider a horizontal line that passes through the gas–fluid interface in arm A at point a and through point b on arm B. The pressures at these two points are equal. If the pipeline is at the desired pressure then,

$$3P_0 = P_0 + \rho_f g h$$

Solving for h, we get

$$h = \frac{2P_0}{\rho_f g} = \frac{2 \times 1.01 \times 10^5 \text{ N/m}^2}{(13.6 \times 10^3 \text{ kg/m}^3)(9.81 \text{ m/s}^2)} = 1.51 \text{ m}$$

The measured height is only 1.25 m, so the gas pressure is below the desired value of $3P_0$.

ONLINE ACTIVITY

Variation of pressure with depth

The e-resource that accompanies every new copy of this textbook contains an Online Activity using the PhET simulation "Studying Variation of Fluid Pressure with Depth." Work through the simulation and accompanying questions to gain an understanding of the variation of pressure within the body of a fluid.

MAKING CONNECTIONS

Bourdon Pressure Gauges

Many mechanical pressure gauges use a Bourdon tube, a coiled metal tube that straightens when the pressure in it increases (patented in France by Eugene Bourdon in 1849). The end of the tube is connected to a gear mechanism that rotates a pointer on a calibrated dial (Figure 12-13). Alternatively, a Bourdon tube can be connected to a transducer to give an electronic readout. Bourdon gauges are used to measure pressure in bicycle tires, boilers, and tanks of compressed gases.

Figure 12-13   A Bourdon-type pressure meter

✓ CHECKPOINT

C-12-4 Pressure in a Closed Tube

In Figure 12-14, the tube is open at one end, closed at the other end, and filled with water. The pressure at the closed end is

(a) greater than the atmospheric pressure;
(b) equal to the atmospheric pressure;
(c) less than the atmospheric pressure but nonzero;
(d) zero.

Figure 12-14   C-12-4

C-12-4 (a)

LO 4

12-4   Pascal's Principle

Consider a container that is fitted with a movable piston and filled with a liquid (Figure 12-15).

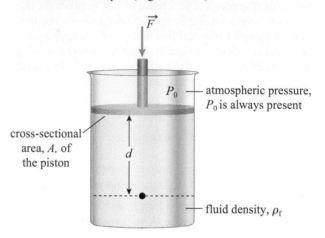

Figure 12-15   The total pressure at any point within the fluid is equal to the sum of the pressures due to the atmosphere, the weight of the fluid, and the external force.

When the piston is pushed downward, the layer of liquid in contact with it experiences an increase in pressure. The molecules of a liquid interact with each other and are free to move, so the increase in the pressure is quickly transferred to every part of the liquid and to the walls of the container. This simple fact was first realized and stated by Blaise Pascal, and is called **Pascal's principle**:

A change in pressure at any point of an enclosed fluid that is at rest is transmitted, without any loss, to every point of the fluid, including the walls of the container.

Notice that for Pascal's principle to hold true, the fluid must be confined. We encounter Pascal's principle in action all around us. When you squeeze the end of a toothpaste tube, the resulting increase in pressure is transmitted throughout the body of the tube, forcing the toothpaste through the opening (Figure 12-16).

Hydraulic Systems

A hydraulic system, based on Pascal's principle (Figure 12-17), is an essential component of many modern-day machines, for example, excavators, pumps, turbines, automobile braking systems (Figure 12-18), and hydraulic elevators. A simple hydraulic system consists of a sealed tube filled with a liquid and fitted with two pistons, usually called an input piston and an output piston. A force exerted on the input piston results in an increase in pressure within the fluid and at the output piston.

Consider two cylindrical containers of cross-sectional areas A_1 and A_2, connected by a tube as shown in Figure 12-19. The cylinders are fitted with movable pistons and are partially filled with a liquid of density ρ_f.

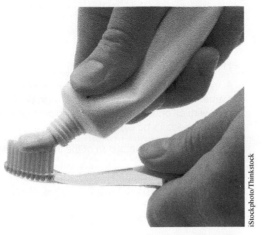

Figure 12-16 Pressure applied at one end of the tube is transmitted to all parts of the tube, forcing the toothpaste out at the open end.

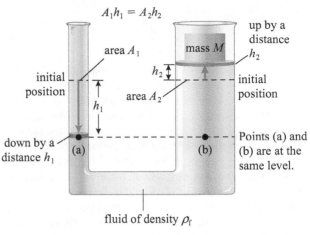

$$A_1 h_1 = A_2 h_2$$

Figure 12-19 Applying a downward force to the input piston raises the mass on the output arm of a hydraulic lift.

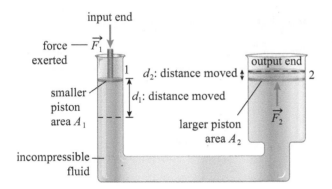

Figure 12-17 A simple hydraulic lift

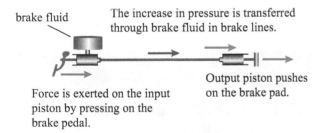

brake fluid

The increase in pressure is transferred through brake fluid in brake lines.

Force is exerted on the input piston by pressing on the brake pedal.

Output piston pushes on the brake pad.

Figure 12-18 Hydraulic brakes in a car. When the brake pedal is depressed, it pushes on the input piston in the hydraulic cylinder, causing the output pistons to move pads at each wheel.

This system is a **hydraulic lift** in its simplest form. To analyze this system, we will assume that the pistons are massless and that the frictional forces are negligible.

First, consider the case when the only force exerted on the two pistons is due to atmospheric pressure. The total pressure (atmospheric plus pressure due to the weight of the fluid) is the same for all points in the cylinder that are on the same horizontal line. Therefore, the height of the liquid in the two cylinders is the same, and the bottom surfaces of the two pistons are level with each other. Now, assume that the input piston is pushed down by applying a force of magnitude F_1. The increase in the pressure, ΔP, on the input piston is

$$\Delta P = \frac{F_1}{A_1} \qquad (12\text{-}13)$$

By Pascal's principle, the pressure within the fluid increases by the same amount. This increase in the pressure results in an upward force exerted on the output piston by the fluid. The magnitude of the upward force, F_2, is

$$F_2 = \Delta P \times A_2$$
$$= \left(\frac{A_2}{A_1}\right) \times F_1 \qquad (12\text{-}14)$$

Note the amplification factor A_2/A_1 on the right-hand side of (12-14). By making the cross-sectional area of the output piston larger than the cross-sectional area of the input piston, a small amount of force can be amplified into a large force. This amplification is one of the most useful features of hydraulic systems.

Does the amplification of the applied force violate the law of conservation of energy? Consider the situation shown in Figure 12-19. An object (e.g., an automobile) of mass M is placed on the output piston. To hold the two pistons level, a force of magnitude $F_1 = \left(\frac{A_1}{A_2}\right) Mg$ must be exerted on the input piston. Suppose we need to elevate the object by a distance h_2. For this, the input piston needs to be pushed downward to force the fluid into the output end. Since the fluid is neither leaving nor entering the system, the decrease in the volume of the fluid in the input cylinder equals an increase in the volume in the output cylinder.

Let h_1 be the distance by which the input piston needs to be moved downward to raise the output piston by a distance h_2. Then,

$$A_1 \times h_1 = A_2 \times h_2$$

and

$$h_1 = \left(\frac{A_2}{A_1}\right)h_2 \qquad (12\text{-}15)$$

If $A_2 > A_1$, then $h_1 > h_2$. For example, if A_2 is 20 times larger than A_1 and the output piston is to be raised 10 cm, then the input piston must be pushed down 200 cm. Equation (12-15) is a consequence of the law of conservation of energy. The work done by the input piston on the liquid is equal to the work done by the liquid on the output piston.

As the elevation of the output piston increases, the force at the input end must also increase. Let F_1' be the magnitude of the force needed to hold the mass M at height h_2 with respect to the initial equilibrium position. In this configuration, the difference between the levels of the two pistons is $h_1 + h_2$. Consider points a and b on a horizontal line that passes through the fluid–piston interface in the input cylinder, as shown in Figure 12-19. The two points are at the same pressure; therefore,

$$P_0 + \frac{F_1'}{A_1} = P_0 + \frac{Mg}{A_2} + \rho_f g(h_1 + h_2) \qquad (12\text{-}16)$$

and

$$F_1' = \left(\frac{A_1}{A_2}\right)Mg + \rho_f g A_1(h_1 + h_2) \qquad (12\text{-}17)$$

The first term on the right-hand side of Equation (12-17) is the force needed to keep the two pistons at the same level. The second term is the additional force needed to balance the weight of the volume $A_1(h_1 + h_2)$ of the fluid above the horizontal line that is level with the input arm. The force exerted at the input arm must balance the combined weight of the object and the lifted fluid.

✅ CHECKPOINT

C-12-5 Hydrostatic Pressure

A cylindrical U tube, with both arms open to the atmosphere, is partially filled with water and oil (Figure 12-20). The density of oil is less than the density of water. Consider the pressure within the two fluids at points A, B, C, D, E, F, G, and H. Rank the points, from highest to lowest, in terms of the total pressure at these points. Use an equality sign where needed.

Highest _____ _____ _____ _____ Lowest

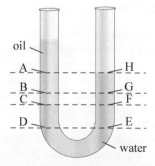

Figure 12-20 A U-shaped manometer

C-12-5 D = E, C = F, B, G, A, H

EXAMPLE 12-5

Hydraulic Car Jack

A hydraulic car jack has an output cylinder 30.0 cm in diameter and an input cylinder 5.00 cm in diameter, and is filled with oil of density 900 kg/m³. An automobile of mass 1100 kg needs to be elevated 40.0 cm.

(a) How far does the piston of the input cylinder need to move?

(b) What is the magnitude of the force on the input piston required to keep the car elevated?

SOLUTION

From Equation (12-15),

$$h_1 = \left(\frac{A_2}{A_1}\right)h_2$$

$$h_1 = \frac{\pi(0.15 \text{ m})^2}{\pi(0.025 \text{ m})^2} \times 0.4 \text{ m} = 14.4 \text{ m}$$

The magnitude of the force that needs to be applied on the input piston to keep the car elevated is given by Equation (12-17):

$$F_1' = \left(\frac{A_1}{A_2}\right)Mg + \rho_f g A_1 (h_1 + h_2)$$

$$F_1' = (1/36) \times (1100 \text{ kg}) \times (9.81 \text{ m/s}^2) + (900 \text{ kg/m}^3)$$
$$\times (9.81 \text{ m/s}^2) \times \pi(0.025 \text{ m})^2 \times (14.8 \text{ m})$$

$$= 299.7 \text{ N} + 256.6 \text{ N} = 556.3 \text{ N}$$

The second term in Equation (12-17) is usually negligible if the fluid is a compressed gas.

LO 5

12-5 Buoyancy and Archimedes' Principle

If you hold a table tennis ball underwater and then release it, the ball accelerates upward and shoots above the surface. Clearly, the surrounding water must exert an upward force on the ball.

What causes this upward force? The answer lies in the facts that pressure exists at all points within a fluid and that this pressure increases with depth. Therefore, a fluid exerts forces that are normal to every point on the surface of an immersed object. The forces exerted by the water at various points on the surface of a fully immersed ball are shown in Figure 12-21. Since the lower half of the ball is at a greater depth, upward forces on the lower half of the ball are stronger than the downward forces on the upper half. The vector sum of all these forces is a single force that acts upward on the ball.

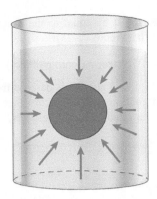

Figure 12-21 The forces due to fluid pressure on a sphere that is fully immersed in a liquid

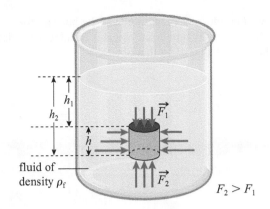

$F_2 > F_1$

Figure 12-22 The forces exerted by the fluid on a fully submerged cylinder

We can conclude that the horizontal components of forces all cancel out because there is no horizontal motion when the ball is released.

The upward force exerted on an object that is fully or partially immersed in a fluid is called the **buoyant force**. We will denote the buoyant force by the symbol $\vec{F}_B$ and its magnitude by the symbol F_B. The direction and the magnitude of the buoyant force are stated by **Archimedes' principle**:

When an object is fully or partially immersed in a fluid, the fluid exerts an upward force (called the buoyant force) on the object. The magnitude of this upward force is equal to the weight of the fluid displaced by the object.

The genius of Archimedes was in realizing that the magnitude of the buoyant force is equal to the weight of the fluid displaced by the object. The more you immerse a table tennis ball into the water, the greater the volume of water displaced by the ball and the greater the upward force that the water exerts on it. Once the ball is completely submerged, it has displaced the maximum amount of water. If the ball is then moved anywhere within the body of the water, while remaining below the surface, it experiences exactly the same buoyant force (assuming that the density of the water does not change with depth and the ball does not deform).

Imagine a vertical cylinder of height h and cross-sectional area A that is fully immersed in an open container of fluid of density ρ_f (Figure 12-22). The top of the cylinder is at a depth h_1 below the fluid surface, and the bottom of the cylinder is at a depth $h_2 = h_1 + h$ below the surface.

The pressure at the top end of the cylinder, P_1, is

$$P_1 = P_0 + \rho_f g h_1 \qquad (12\text{-}18)$$

Therefore, the magnitude of the downward force, F_1, on the top of the cylinder is

$$F_1 = (P_0 + \rho_f g h_1)A \qquad (12\text{-}19)$$

The pressure at the bottom end of the cylinder, P_2, is

$$P_2 = P_0 + \rho_f g h_2 \qquad (12\text{-}20)$$

The magnitude of the upward force, F_2, on the bottom end of the cylinder is

$$F_2 = (P_0 + \rho_f g h_2)A \qquad (12\text{-}21)$$

The horizontal force at any given point on the side of the cylinder is cancelled by the corresponding force on the diametrically opposite point. Since $h_2 > h_1$, the magnitude of the upward force is greater than the magnitude of the downward force. The upward force exerted on the cylinder by the fluid, F_B, is given by

$$\begin{aligned} F_B &= F_2 - F_1 = (P_0 + \rho_f g h_2)A - (P_0 + \rho_f g h_1)A \\ &= (\rho_f g h_2 - \rho_f g h_1)A \qquad (12\text{-}22) \\ &= \rho_f g (h_2 - h_1)A = \rho_f g V_o \end{aligned}$$

The factor $V_o = (h_2 - h_1)A$ is the volume of the cylinder and is equal to the volume of the fluid displaced by the cylinder. The quantity $\rho_f V_o$ is the mass of the fluid displaced by the cylinder, and $\rho_f V_o g$ is the weight of the fluid displaced by the cylinder. Therefore, the magnitude of the buoyant force exerted by the fluid on the cylinder is equal to the weight of the fluid displaced by the cylinder.

KEY EQUATION

$$\begin{aligned} F_B &= \text{weight of the displaced fluid} \\ &= (\rho_f V_o)g \qquad (12\text{-}23) \end{aligned}$$

Although we have derived this result for a vertically immersed cylinder, the general result holds for any orientation or shape of an object. Consider a container filled with a fluid, and imagine that a three-dimensional segment of this fluid is suspended within it (Figure 12-23). The fluid is stationary, so this segment is in static equilibrium with the rest of the fluid. Therefore, the buoyant force exerted on this segment by the rest of the fluid must be equal to the weight of the fluid within the segment. Now, suppose that we replace this segment by a solid object of exactly the same shape. The fluid around the object would exert exactly the same buoyant force on the object. Therefore, the magnitude of the buoyant force on a fully or partially immersed object is equal to the weight of the fluid displaced by that object.

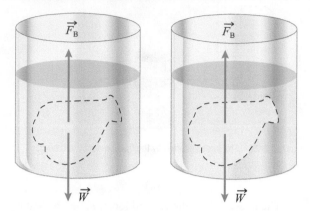

Figure 12-23 The buoyant force on an object immersed in a fluid is equal to the weight of the fluid displaced by the object.

✓ CHECKPOINT

C-12-6 The Buoyant Force on Spheres

Three identical spheres of different materials are held fully immersed in a container of water, as shown in Figure 12-24. Which of the following statements concerning the magnitude of buoyant force on each sphere is correct?

(a) $F_{B,cork} < F_{B,aluminum} < F_{B,lead}$
(b) $F_{B,cork} = F_{B,aluminum} = F_{B,lead}$
(c) $F_{B,cork} < F_{B,aluminum} = F_{B,lead}$
(d) $F_{B,cork} > F_{B,aluminum} > F_{B,lead}$

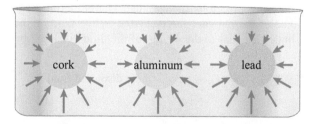

Figure 12-24 C-12-6

C-12-6 (b)

Flotation

The net vertical force on an object floating in a fluid is zero:

$$\vec{F}_{net} = \vec{F}_B + \vec{W} = 0 \quad \text{(for a floating object)} \quad (12\text{-}24)$$

Since the buoyant force and the weight act in opposite directions, the magnitude of the buoyant force on a floating object must be equal to the magnitude of the object's weight, W (Figure 12-25):

$$F_B = W \quad \text{(for a floating object)} \quad (12\text{-}25)$$

Note that Equation (12-24) is derived from Newton's second law of motion, and confirms that Archimedes' principle can be applied to floating objects.

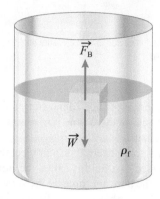

Figure 12-25 For a floating object, the magnitude of the upward buoyant force exerted on the object by the fluid is equal to the magnitude of the weight of the object.

Consider an object of volume V_o and density ρ_o that is floating in a fluid of density ρ_f. Assume that a volume V_{sub} of the object is submerged in the fluid. Then

W = weight of the object = $(\rho_o V_o) g$

F_B = bouyant force on the object

= weight of the fluid displaced by the object

= (mass of fluid displaced by the object) g

= $(\rho_f V_{sub})g$

Substituting into Equation (12-25), we get

$$\rho_o V_o g = \rho_f V_{sub} g$$

$$\text{and} \quad \frac{V_{sub}}{V_o} = \frac{\rho_o}{\rho_f} \quad (12\text{-}26)$$

Therefore, the fraction of the volume of a floating object below a fluid's surface is equal to the ratio of the density of the object to the density of the fluid.

The density of ocean water is approximately 1020 kg/m³, and the density of sea ice is 920 kg/m³. Therefore,

$$\frac{V_{sub}}{V_o} = \frac{920 \text{ kg/m}^3}{1020 \text{ kg/m}^3} = 0.902$$

approximately 90% of an iceberg is beneath the ocean's surface (Figure 12-26).

Figure 12-26 Approximately 90% of the volume of an iceberg is beneath the ocean's surface.

Apparent Weight in a Fluid

If the average density of an object is greater than the density of a fluid, the object will submerge in that fluid. The weight of the object is greater than the weight of the displaced fluid. The **apparent weight**, $W_{apparent}$, of the fully submerged object is defined as the difference between the weight of the object when not immersed and the buoyant force exerted on the object when it is immersed in the fluid. The magnitude of the apparent weight is given by

$$W_{apparent} = W - F_B \qquad (12\text{-}30)$$

A convenient method to determine the buoyant force on an object is to first weigh the object outside the fluid and then when it is completely immersed in a fluid. The difference between these two weights equals the magnitude of the buoyant force on the object. This method can also be used to determine the average density of an object.

EXAMPLE 12-6

How Large Is That Helium Balloon?

A balloon is carrying two people in a basket. The total mass of the people, the basket, and the material of the balloon, but excluding the mass of helium gas, is 280 kg. What volume of helium is needed to lift the balloon (Figure 12-27)? The density of the helium is 0.18 kg/m³, and the density of the air is 1.30 kg/m³.

Figure 12-27 A helium (or hot air) filled balloon floats when the buoyant force exerted on the balloon by the surrounding air is equal to the weight of the balloon plus the enclosed gas.

SOLUTION

When a balloon is filled with helium gas, the surrounding air exerts a buoyant force on the balloon and the gas inside it. As more helium is added, the volume of the balloon

increases, so more air is displaced and the buoyant force on the balloon increases. When the buoyant force equals the total weight of the balloon (including the helium), the balloon will begin to float; adding any more helium will cause the balloon to accelerate upward.

The total weight of the balloon is

$W =$ weight of balloon's contents
   $+$ weight of the helium gas filling the balloon

$$W = 280\,g + m_{He}\,g = 280\,g + (V_{He}\,\rho_{He})\,g \qquad (12\text{-}27)$$

In Equation (12-27), V_{He} is the volume of the helium filled-balloon, and ρ_{He} is the density of helium gas. The buoyant force exerted by the air on the balloon is

$F_B =$ weight of the air displaced by the balloon

$$= m_{air}\,g = (V_{He}\,\rho_{air})\,g \qquad (12\text{-}28)$$

At liftoff, the buoyant force equals the total weight of the balloon, so

$$280\,g + V_{He}\,\rho_{He}\,g = (V_{He}\,\rho_{air})\,g$$

and $$V_{He} = \frac{280\text{ kg}}{(\rho_{air} - \rho_{He})} = \frac{280\text{ kg}}{(1.30 - 0.18)\text{ kg/m}^3} = 250\text{ m}^3$$

$$(12\text{-}29)$$

With 250 m³ of helium, the balloon is ready to lift off. To initiate an upward velocity, the amount of helium pumped into the balloon must be greater than the above calculated value.

EXAMPLE 12-7

Is the Ring Made of Pure Gold?

Your friend asks you to determine whether the ring that she recently bought is made of pure gold. (The density of gold is 19.3×10^3 kg/m³.) Using a thread of negligible mass and volume, you tie the ring to a very accurate spring balance and determine that in air it weighs 6.40×10^{-3} N. You then fully immerse the ring in a beaker full of water and find that its apparent weight is 6.00×10^{-3} N. Is the ring made of pure gold?

SOLUTION

To solve this problem, we apply the key points of this section as follows:

1. The difference between the weight of the ring in air and the weight of the ring in water is equal to the magnitude of the buoyant force on the ring.
2. The magnitude of the buoyant force is equal to the weight of the water displaced by the ring.
3. Knowing the density of the water and the value of g, the volume of water displaced by the ring can be calculated.
4. The volume of the ring is equal to the volume of the displaced water.

5. The weight of the ring is known; therefore, we can calculate the density of the ring.

Let us now do these calculations step by step:

1. $F_B = W_{ring,air} - W_{ring,water}$

   $= 6.40 \times 10^{-3} \, \text{N} - 6.00 \times 10^{-3} \, \text{N} = 0.40 \times 10^{-3} \, \text{N}$

2. $W_{displaced} = $ Weight of displaced water

   $= \rho_{water} V_{water} \, g = 0.40 \times 10^{-3} \, \text{N}$

   $V_{water} = \dfrac{0.40 \times 10^{-3} \, \text{N}}{\rho_{water} g} = \dfrac{0.40 \times 10^{-3} \, \text{N}}{1000 \, \text{kg/m}^3 \times 9.81 \, \text{m/s}^2}$

   $= 4.08 \times 10^{-8} \, \text{m}^3$

3. $V_{ring} = V_{water} = 4.08 \times 10^{-8} \, \text{m}^3$

4. $W_{ring,air} = m_{ring,air} \, g = (V_{ring} \, \rho_{ring}) g$

5. $\rho_{ring} = \dfrac{W_{ring,air}}{V_{ring} \, g} = \dfrac{6.4 \times 10^{-3} \, \text{N}}{(4.08 \times 10^{-8} \, \text{m}^3) \times 9.81 \, \text{m/s}^2}$

   $= 16.0 \times 10^3 \, \text{kg/m}^3$

Since the density of the ring is considerably less than the density of pure gold (19.3×10^3 kg/m³), the ring is not made of pure gold.

✓ CHECKPOINT

C-12-7 Pressure on the Scale

A container, partially filled with water, is resting on a scale that measures its weight. If you immerse your hand in the water, without touching any part of the container and not spilling any water, what happens to the reading on the scale? Explain your reasoning.

(a) It stays the same.
(b) It increases.
(c) It decreases.
(d) Insufficient information is provided to answer this question.

C-12-7 (b)

LO 6

12-6 Fluids in Motion

There is an obvious difference in the energy of water in a large wave and stationary water in a cup. The water in a wave has significant kinetic energy. Water in a cup is at rest and has no net kinetic energy. Any physical description of a moving fluid must include its kinetic energy. We start our discussion of moving fluids by first introducing the concepts of kinetic energy per unit volume and potential energy per unit volume for a fluid.

Kinetic and Potential Energy Per Unit Volume

Consider a fluid of uniform density ρ moving with a velocity $\vec{v}$. The mass of a small volume ΔV of this fluid is $\Delta m = \rho \Delta V$, and its kinetic energy is

$$\Delta K = \frac{1}{2}(\Delta m)v^2 = \frac{1}{2}\rho v^2 (\Delta V) \qquad (12\text{-}31)$$

The kinetic energy per unit volume of the fluid is

$$\frac{\Delta K}{\Delta V} = \frac{1}{2}\rho v^2 \qquad (12\text{-}32)$$

The SI unit for kinetic energy per unit volume is the pascal (Pa), the same as pressure:

$$\frac{\text{kg}}{\text{m}^3} \times \frac{\text{m}^2}{\text{s}^2} = \frac{\text{kg}}{\text{ms}^2} = \text{Pa}$$

We can similarly define potential energy per unit volume for a fluid. The gravitational potential energy of a small element of fluid that has volume ΔV and is located at a height h from the ground is

$$\Delta U = (\Delta m) gh = \rho gh(\Delta V) \qquad (12\text{-}33)$$

The potential energy per unit volume of the fluid is then

$$\frac{\Delta U}{\Delta V} = \rho gh \qquad (12\text{-}34)$$

The SI unit for potential energy per unit volume is also the pascal.

Ideal Fluids

A general description of moving fluids is complicated because fluids exhibit a wide variety of behaviour that depends on the type of fluid and its speed. The motion of water smoothly emerging from a tap is very different from the chaotic flow of water over Niagara Falls. Similarly, it is easier to suck water through a straw then it is to suck honey through the same straw. However, we can develop a fairly reasonable description of moving fluids by making some key assumptions about the physical properties of the fluids and how the fluids flow. These assumptions are as follows:

A fluid is incompressible

We assume that the density of a fluid remains constant and does not change with pressure. This assumption closely approximates the behaviour of liquids. For example, the density of water increases by approximately 1% at a depth of 1000 m below the ocean's surface, where the absolute pressure is 100 times greater than the atmospheric pressure. The assumption that a gas is incompressible is only valid if the change in the gas's pressure is small.

A fluid is inviscid

Viscosity is the measure of a fluid's resistance to flow and to any deformation of its shape. For example, water flows much more readily than honey. The assumption that a fluid is inviscid implies that frictional forces within the fluid, as well as between the fluid and the walls of the containing vessel, are negligible. A frictionless flow is an idealization because all real fluids have nonzero friction. We will discuss viscous flow later in the chapter. A fluid that is incompressible and inviscid is called an **ideal fluid**.

The flow of the fluid is steady

A **steady flow** is smooth and uniform. In a steady flow, the velocity of the fluid at any *given* point within the region of flow has a fixed magnitude and direction and remains constant with time. The velocities at *different* points in the region of flow may differ from each other. Water slowly flowing from a tap (Figure 12-28) is an example of a steady flow. In this case, the velocity of water at a fixed distance below the tap is well-defined and does not change with time as long as the flow rate through the tap remains the same.

A flow that is not steady is called **turbulent**. In a turbulent flow, the velocity at a given point in the region of flow varies with time. The flow of gases and ash from the eruption of Mount St. Helens is an example of turbulent flow (Figure 12-29). The mathematical description of turbulent flows is quite complex.

The flow of the fluid is irrotational

In an irrotational flow, a fluid element does not move in a circular path. An example of irrotational flow is water moving in a straight pipe or falling straight downward under the influence of gravity.

Figure 12-28   Water gently flowing from a tap is an example of a steady flow.

Chaotic flow from the mouth of Mount St. Helens (Washington, U.S.A.). Compare this flow to the flow of water slowly emerging from a tap.

Figure 12-29   The eruption of Mount St. Helens. The flow from the volcano is chaotic and turbulent.

Streamlines and Flow Tubes

When the flow is steady, fluid particles move along well-defined paths called streamlines. A **streamline** is a curve drawn in the body of a moving fluid such that the velocity of a particle at any point along the streamline is tangent to the streamline at that point. Figure 12-30 shows velocity vectors at various points along a streamline. The instantaneous velocity of a particle must be unique, and for steady flow the velocity at any point in the region of the flow does not change with time. *Therefore, in a steady flow, streamlines do not cross each other.* A steady flow is also called a **streamline**, or **laminar flow**.

Imagine a set of neighbouring streamlines that form a closed path in a given region of a moving fluid. The particles of the fluid that are either on the boundary of this closed path or are enclosed within it would follow their respective streamlines as they move along the fluid. These neighbouring streamlines form a tube, called a **flow tube** (Figure 12-30). The fluid within a flow tube must remain within the tube. There is no mixing of fluid from different flow tubes. The cross-sectional area of a given flow tube may change along its length, but the amount of the fluid that enters at one end of a flow tube must be equal to the amount that leaves through another end down the stream. You can envision a flow tube as an invisible pipe that keeps the fluid within it from mixing with the fluid outside the pipe.

TAKESHI TAKAHARA/Photo Researchers/Getty Images

Figure 12-31   Streamlines of smoke passing over an automobile in a wind tunnel

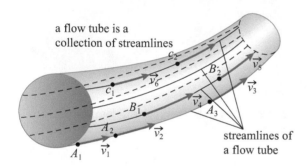

a flow tube is a collection of streamlines

streamlines of a flow tube

Figure 12-30   A flow tube. Solid lines represent streamlines. The magnitude and direction of the flow velocity at any point along the flow line is given by the length and direction of the arrow at that point. The direction of the arrow is tangential to the streamlines. The group of streamlines forms a stream tube (or flow tube).

There are several ways to visualize streamlines and flow tubes. Smoke in gases and dyes in liquids are often used by engineers and scientists to study flow patterns of fluids around objects (Figure 12-31). With computers, scientists can calculate the pattern of streamlines around objects. A study of such patterns provides information about the resistance (or drag) experienced by an object as it moves through a fluid. Smooth and continuous streamlines indicate a steady flow with minimum drag. Closely spaced streamlines indicate high speed.

Keep in mind that a description of an object moving through a stationary fluid is the same as that of a fluid that is moving past a stationary object. That is why streamlines created by moving air around a stationary automobile in a wind tunnel are exactly the same as the streamlines of an automobile moving through stationary air.

Flow Rate

The **flow rate** (Q) is the volume of fluid that flows through a cross-sectional area per unit time:

$$Q = \frac{\text{volume of fluid that flows through a surface}}{\text{time taken}}$$

$$(12\text{-}35)$$

The SI units for flow rate are cubic metres per second (m^3/s). Other convenient units are litres per second (L/s) and cubic centimetres per second (cm^3/s):

$$1 \text{ m}^3/\text{s} = 10^6 \text{ cm}^3/\text{s} = 1000 \text{ L/s}$$

For a fluid moving with a speed v through a cross-sectional area A that is perpendicular to the direction of flow, the volume flow rate is the product of A and v:

$$Q = Av \qquad (12\text{-}36)$$

LO 7

12-7   The Continuity Equation: Conservation of Fluid Mass

Consider water flowing through a pipe of variable cross-sectional area (Figure 12-32). If there are no other sources of water and no holes in the pipe, then the amount of water that enters the pipe at one end in some time Δt must be equal to the amount that leaves at the other end in the same time. In other words, what comes in must go out. The mathematical equation that describes this "law of conservation of fluid mass" is called the **equation of continuity**.

Consider a section of a pipe (or a flow tube) where the cross-sectional area is A_1, the speed of the fluid is v_1, and the density of the fluid is ρ_1. In time Δt, a length $\Delta x_1 = v_1 \, \Delta t$ of the fluid passes through A_1. The volume, ΔV_1, of the

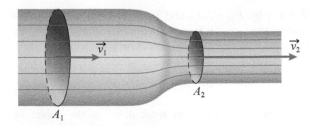

Figure 12-32 Fluid flowing through a horizontal pipe of variable cross-sectional area

fluid passing through A_1 is $\Delta V_1 = A_1(v_1 \Delta t)$. The mass of this volume of the fluid is

$$\Delta M_1 = \rho_1 \Delta V_1 = (\rho_1 A_1 v_1)\Delta t \qquad (12\text{-}37)$$

In the same time interval, a mass ΔM_2 of the fluid must pass through another section of the pipe where the cross-sectional area is A_2, the fluid speed is v_2, and the fluid density is ρ_2:

$$\Delta M_2 = \rho_2 \Delta V_2 = (\rho_2 A_2 v_2)\Delta t \qquad (12\text{-}38)$$

Since the mass of the fluid entering through A_1 must be equal to the mass leaving through A_2, we must have

$$\rho_1 A_1 v_1 = \rho_2 A_2 v_2 \qquad (12\text{-}39)$$

The quantity ρAv has the dimensions of kg/s and is called the **mass flow rate**. For incompressible fluids $(\rho_1 = \rho_2)$ the above equation reduces to a simpler relationship:

KEY EQUATION $A_1 v_1 = A_2 v_2$ (for incompressible fluids)

$$(12\text{-}40)$$

The quantity Av is the volume flow rate, Q. The continuity equation states that for incompressible fluids, the volume flow rate remains constant through all sections of a flow tube. An important consequence of the equation of continuity is that the fluid moves faster through a narrower section of a flow tube and slows when moving through a wider section (Figure 12-33).

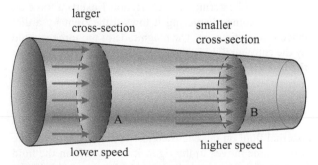

Figure 12-33 When a fluid moves through a horizontal pipe that gradually narrows, the fluid speed increases as the cross-sectional area of the pipe decreases.

MAKING CONNECTIONS

Water Flowing Through a Tap

When you turn on a water tap, keeping the flow smooth and slow, you may notice that as the stream of water flows downward, it continuously narrows (Figure 12-28). Why does the stream narrow as it accelerates downward? The water flows out of the tap with an initial flow rate that is determined by the cross-sectional area of the mouth of the tap and the speed with which the water leaves the tap. Water accelerates due to gravity, so its speed increases. Since the flow rate must remain constant, the cross-sectional area of the stream must decrease with increasing speed of water.

The same phenomenon occurs when you gently tilt a spoonful of honey and watch the stream of honey as it accelerates downward.

✓ CHECKPOINT

C-12-8 Flow Rate Through Pipes

Water flows continuously in and out of an arrangement of pipes as shown in Figure 12-34. What is the flow rate from the pipe A? Is the water flowing into or out of this pipe?

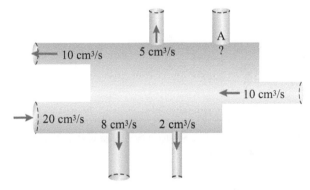

Figure 12-34 Water flow through an arrangement of pipes

C-12-8 5 cm³/s; the water is flowing out of the pipe.

LO 8

12-8 Conservation of Energy for Moving Fluids

According to the equation of continuity, as a fluid passes through a narrower section of a pipe (or a flow tube), its speed increases. For an element of fluid to speed up, the fluid behind it must exert a force on it. Therefore, the pressure in the wider section of the pipe

325

must be higher than the pressure in the narrower section. We thus reach an interesting and nonintuitive conclusion: *In the regions of flow where the fluid moves faster, the pressure within the fluid is lower than the pressure in the regions where the fluid speed is slower.* In this section, we will show that this relationship is a consequence of the law of conservation of energy as applied to moving fluids.

Consider an ideal fluid flowing through a flow tube of variable cross-sectional area and variable height (measured with respect to an arbitrary ground level). The fluid is moving from left to right (Figure 12-35). Let us focus on the body of the fluid that is contained between two sections (1 and 2) of the flow tube. We will calculate the work done on this "test volume" of fluid by the surrounding fluid (Figure 12-36). At section 1, the pressure in the fluid is P_1, the speed of the fluid is v_1, the cross-sectional area of the tube is A_1, and the height of the tube is y_1. Corresponding quantities at section 2 are

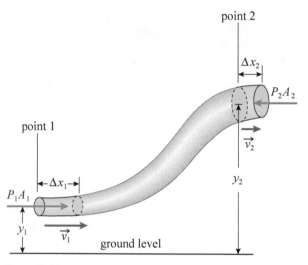

Figure 12-35 An ideal fluid flowing through a pipe of variable height and cross-sectional area

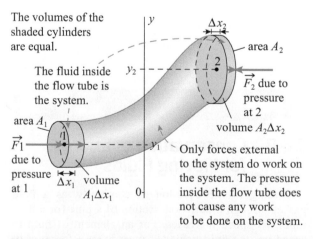

Figure 12-36 Fluid flow through a flow tube of variable cross-sectional area and height

denoted by P_2, v_2, A_2, and y_2. According to the work–energy theorem, in the absence of frictional forces, the net work done by external forces on an object is equal to the change in the object's total energy:

$$W_{net} = \Delta K + \Delta U \qquad (12\text{-}41)$$

Here, W_{net} is the net work done on the object by external forces, ΔK is the change in the kinetic energy of the object, and ΔU is the change in the object's potential energy. Note that only external forces are considered when calculating the work done on the object because internal forces always come in action–reaction pairs and cancel out. Also, the total energy of the object is the sum of its kinetic and potential energies. Although potential energy could be of several forms, in dealing with fluids we will only consider gravitational potential energy.

In time Δt, a length $\Delta x_1 = v_1 \Delta t$ of the fluid enters through section 1. The volume of the entering fluid is

$$\Delta V = A_1 (v_1 \Delta t) \qquad (12\text{-}42)$$

In the same time, the volume of fluid that leaves through section 2 is $A_2(v_2\Delta t)$. By the equation of continuity,

$$\Delta V = A_1 v_1 \Delta t = A_2 v_2 \Delta t \qquad (12\text{-}43)$$

The entering volume has speed v_1, and the speed of the leaving volume is v_2. Therefore, the change in the kinetic energy (ΔK) of the test volume is

$\Delta K = $ (kinetic energy of leaving fluid)

$\qquad$ − (kinetic energy of entering fluid)

$$\Delta K = \frac{1}{2}(\rho \Delta V)v_2^2 - \frac{1}{2}(\rho \Delta V)v_1^2 \qquad (12\text{-}44)$$

The fluid enters at a height y_1 and leaves at a height y_2. Therefore, in the time Δt, the change in the potential energy (ΔU) of the test volume is

$\Delta U = $ (potential energy of leaving fluid)

$\qquad$ − (potential energy of entering fluid)

$$\Delta U = (\rho \Delta V) g y_2 - (\rho \Delta V) g y_1 \qquad (12\text{-}45)$$

The fluid entering through section 1 exerts a force on the test volume, displacing it toward the right by a distance $v_1\Delta t$ in time Δt. The magnitude of the force exerted by the entering fluid is $P_1 A_1$. Therefore, the **work done on the test volume** by the entering fluid, W_1, is

$$W_1 = P_1 A_1 (v_1 \Delta t) \qquad (12\text{-}46)$$

At section 2, the magnitude of the force exerted by the external fluid on the enclosed test volume is $P_2 A_2$. As the test volume moves to the right, it does work on the fluid in front of it. The **work done by the leaving fluid**, W_2, is

$$W_2 = -P_2 A_2(v_2 \Delta t) \qquad (12\text{-}47)$$

The minus sign in W_2 implies that the force exerted on the enclosed test volume is in a direction opposite to that of fluid flow. The net work done on the test volume by the surrounding fluid in time Δt is therefore

$$W_{net} = W_1 + W_2 = P_1 A_1(v_1 \Delta t) - P_2 A_2(v_2 \Delta t) \quad (12\text{-}48)$$

Using Equation (12-43), we can write Equation (12-48) as

$$W_{net} = (P_1 - P_2)\Delta V \quad (12\text{-}49)$$

Inserting Equations (12-44), (12-45), and (12-49) into Equation (12-41), we obtain

$$(P_1 - P_2)\Delta V = \frac{1}{2}(\rho\Delta V)v_2^2 - \frac{1}{2}(\rho\Delta V)v_1^2$$
$$+ (\rho\Delta V)\,gy_2 - (\rho\Delta V)\,gy_1 \quad (12\text{-}50)$$

Dividing by ΔV gives

$$P_1 - P_2 = \frac{1}{2}\rho v_2^2 - \frac{1}{2}\rho v_1^2 + \rho gy_2 - \rho gy_1 \quad (12\text{-}51)$$

Rearranging so that all terms corresponding to a given section are on the same side,

KEY EQUATION $\quad P_1 + \frac{1}{2}\rho v_1^2 + \rho gy_1 = P_2 + \frac{1}{2}\rho v_2^2 + \rho gy_2$

$$(12\text{-}52)$$

Equation (12-52) is called **Bernoulli's equation**. Bernoulli's equation is a statement of the law of conservation of energy for ideal fluids in the absence of frictional forces. It states that during the steady flow of an ideal fluid, the sum of the fluid pressure, the kinetic energy per unit volume, and the potential energy per unit volume remains constant throughout the body of the fluid. Since points 1 and 2 can be located anywhere along the flow tube, we can also write Bernoulli's equation in the following form:

$$P + \frac{1}{2}\rho v^2 + \rho gy = \text{constant}$$

(for every point along a streamline in a flow tube)

$$(12\text{-}53)$$

Equations (12-40) and (12-52) are the two fundamental equations describing the flow of ideal fluids. In many situations a description of the flow requires using both of these equations. Let us examine two special cases.

Fluid flow through a horizontal pipe Consider an ideal fluid flowing through a horizontal pipe of variable cross-sectional area, as shown in Figure 12-37. Imagine a streamline that passes through the centre of the pipe with point 1 located in the wider section and point 2 in the narrower section of the pipe. Since the pipe is horizontal, $y_1 = y_2$, and the gravitational potential energy of

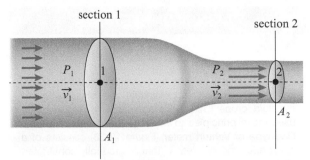

Since $v_1 < v_2$, P_1 must be greater than P_2. Bernoulli's equation predicts that the pressure is lower in the section of a tube in which the fluid moves faster.

Figure 12-37 Fluid flow along a horizontal pipe of variable cross-sectional area

the fluid does not change. Bernoulli's equation simplifies to the following:

$$P_1 + \frac{1}{2}\rho v_1^2 = P_2 + \frac{1}{2}\rho v_2^2 \quad (12\text{-}54)$$

Equation (12-54) implies that the sum of pressure and kinetic energy per unit volume remains constant in a horizontal flow. When passing through the narrow section of the pipe, the fluid speeds up (because of the continuity equation) and, hence, the pressure within the fluid must decrease. Conversely, when a fluid slows down, the pressure within the fluid must increase. You might think that as a fluid speeds up, the pressure within the fluid should also increase, but the opposite is true.

Applying the equation of continuity at points 1 and 2, we get $v_2 = (A_1/A_2)\,v_1$, and the pressure difference between the two sections of the pipe is

$$P_1 - P_2 = \frac{1}{2}\rho v_1^2\left(\frac{A_1^2}{A_2^2} - 1\right) \quad (12\text{-}55)$$

Thus, $P_1 > P_2$ if $A_1 > A_2$. Bernoulli's equation explains many commonly observed phenomena and has many practical applications. Some of these are discussed below.

MAKING CONNECTIONS

Blowing Over a Paper Strip

Cut a small strip of paper (about 2 cm × 16 cm), and place it close to the lower lip of your mouth, holding it with your fingers. The strip will hang downward because of its weight. Now blow hard such that the air moves horizontal to the top surface of the strip. The strip will rise. As you blow, the air above the top surface moves faster, causing the pressure above the strip to drop below atmospheric pressure. Since pressure beneath the strip remains equal to atmospheric pressure, an upward force acts on the strip, lifting it up.

MAKING CONNECTIONS

The Venturi Meter

As a fluid passes through a constricted region, its speed increases (equation of continuity) and pressure within the fluid drops (Bernoulli's equation). A Venturi meter uses these principles to measure the speed of a fluid. One type of Venturi meter (Figure 12-38) consists of a horizontal pipe of area A_1 that is gradually constricted at the centre where the area is A_2. A fluid of density ρ_0, such as air, enters the horizontal tube at one end and leaves through the other. A U-tube of uniform area and partially filled with a liquid of density ρ is attached to the horizontal tube.

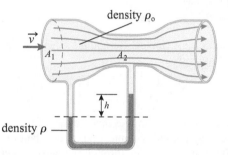

density ρ_0

$\vec{v}$

A_1 A_2

h

density ρ

Figure 12-38 A Venturi meter

When air in the horizontal tube is stationary, the height of the liquid in the two arms of the U-tube is the same. Suppose the air flows through A_1 with speed v. Its speed will increase as it passes through the constricted region, resulting in a pressure drop at the constriction, and the liquid in that arm of the U-tube will rise. The following equation relates the air speed, v, and the difference between the levels of the liquid in the two arms, h:

$$v = \sqrt{2g\left(\frac{\rho - \rho_0}{\rho_0}\right)\left(\frac{A_2^2}{A_1^2 - A_2^2}\right)}\ \sqrt{h}$$

For a given Venturi meter, all quantities in the square root of the first factor are known, so the air speed is determined by measuring h.

 CHECKPOINT

C-12-9 Blood Pressure in a Clogged Artery

Blood passes through a section of human artery that is partially blocked due to cholesterol buildup. As the blood enters from the unblocked (healthy) part of the artery to the blocked part, the blood pressure in the blocked artery

(a) decreases;
(b) remains the same;
(c) increases;
(d) may increase or decrease, depending on the pressure in the unblocked section.

C-12-9 (a)

Constant velocity flow For flow through a pipe of uniform cross section, the equation of continuity requires that the fluid speed remain constant. For this case, we obtain a simplified form of Bernoulli's equation:

$$P_1 + \rho g y_1 = P_2 + \rho g y_2 \qquad (12\text{-}56)$$

Here, the sum of pressure and the gravitational potential energy per unit volume remains constant along a streamline. As the height of the fluid increases, the pressure within the fluid decreases. Conversely, a decrease in the height results in an increase in fluid pressure. Water pressure is lower on higher floors of a building compared to the ground floor.

 ## EXAMPLE 12-8

Water Pressure in a Home

Water flows with a constant speed through a pipe of constant cross-sectional area in the ground floor of a house. The pipe rises 5.00 m to reach a showerhead on the second floor (Figure 12-39). The gauge pressure at ground level is 140 kPa.

(a) What is the gauge pressure at the shower head?
(b) What is the maximum height that water from ground level could reach?

SOLUTION

(a) Since the water is moving, its behaviour is described (assuming a steady flow) by both the continuity equation and Bernoulli's equation. The area of the pipe does not change, so the flow speed remains constant.

We designate the segment of the pipe at ground level as 1 and the segment at the height of the showerhead as 2.

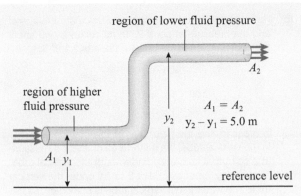

region of lower fluid pressure

region of higher fluid pressure

A_2

$A_1 = A_2$
y_2 $y_2 - y_1 = 5.0$ m

A_1 y_1

reference level

Figure 12-39 Water flow to a showerhead

gauge pressure at ground level, $P_1 = 140.0 \times 10^3$ Pa

height of the section at ground level $= y_1$

height of the section at the showerhead level $= y_2$

gauge pressure at the showerhead level, P_2 is to be determined

For a pipe of constant cross section, Bernoulli's equation reduces to the form

$$P_1 + \rho g y_1 = P_2 + \rho g y_2$$
$$P_2 = P_1 + \rho g y_1 - \rho g y_2 \tag{12-57}$$

Inserting the given values of the quantities on the right-hand side yields

$$P_2 = 140 \times 10^3 + (1000 \text{ kg/m}^3)(9.81 \text{ m/s}^2)(-5 \text{ m})$$
$$P_2 = 91.0 \times 10^3 \text{ Pa}$$

(b) At the maximum height, the gauge pressure is zero. Let us denote this height by y_{max}. Then, $P_2 = 0$ at $y_2 = y_{max}$. From Equation (12-57) we get

$$0 = P_1 - \rho g (y_{max} - y_1)$$

and $y_{max} - y_1 = \dfrac{P_1}{\rho g} = \dfrac{140.0 \times 10^3 \text{ Pa}}{(1000 \text{ kg/m}^3)(9.81 \text{ m/s}^2)} = 14.3$ m

Making sense of the result:

The potential energy per unit volume of the fluid increases as the fluid moves upward. There is a corresponding decrease in fluid's pressure and that is why P_2 is less than P_1.

EXAMPLE 12-9

Water Flow Through a Pressurized Tank

Water flows from a large enclosed tank through a horizontal pipe of variable cross-sectional area, as shown in Figure 12-40. The upper end of the tank is maintained at an absolute pressure of 1.20×10^5 N/m^2 by pumping compressed air into it. The horizontal outlet of the pipe has a cross-sectional area of 4.00×10^{-2} m^2 in the larger section and 2.00×10^{-2} m^2 in the smaller section. A vertical tube of cross-sectional area 4.00×10^{-2} m^2 is connected to the horizontal tube and open to the atmosphere.

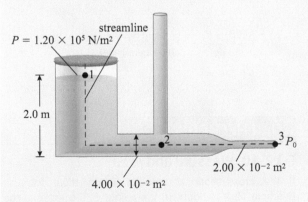

streamline
$P = 1.20 \times 10^5$ N/m^2

2.0 m

1

3 P_0

2

2.00×10^{-2} m^2

4.00×10^{-2} m^2

Figure 12-40 Example 12-9

(a) What is the speed of the fluid leaving the outlet when the height of the water in the large tank is 2.00 m?

(b) What is the speed of the water in the horizontal pipe of cross-sectional area 4.00×10^{-2} m^2?

(c) To what height does the water stand in the vertical tube when the water height in the large tank is 2.00 m?

SOLUTION

To determine the unknown quantities for parts (a), (b), and (c), we consider a streamline that starts at the top surface of the water in the large tank, passes through the horizontal pipe, and leaves through the outlet. We label three points on this streamline as follows: Point 1 is at the top of the water surface in the large tank because we know the height and the pressure at this point. Point 2 is at the centre of that portion of the vertical tube that is at the bottom of the vertical tube, because we need to determine the height of the column of water in the vertical tube. Point 3 is at the outlet where the water leaves the horizontal pipe, because we know that the pressure at the outlet must be equal to the atmospheric pressure.

The variables in Bernoulli's equation and the continuity equation are pressure, speed, height, and the cross-sectional area of the surface perpendicular to the fluid flow. The values of these quantities at the points of interest are:

point 1:

$$P_1 = 1.20 \times 10^5 \text{ Pa}, \quad v_1 = \text{unknown},$$
$$y_1 = 2.00 \text{ m}, \quad A_1 = \text{unknown}$$

point 2:

$$P_2 = \text{unknown}, \quad v_2 = \text{unknown},$$
$$y_2 = 0.00 \text{ m}, \quad A_2 = 4.00 \times 10^{-2} \text{ m}^2$$

(continued)

point 3:

$$P_3 = P_0, \quad v_3 = \text{unknown},$$

$$y_3 = 0.00 \text{ m}, \quad A_3 = 2.00 \times 10^{-2} \text{ m}^2$$

Here, P_0 is atmospheric pressure, and we have chosen to measure all heights with respect to the streamline that passes through points 2 and 3. There are five unknowns: v_1, A_1, P_2, v_2, and v_3. We can use Bernoulli's equation between points 1 and 2 (or 3) and between points 2 and 3, with two corresponding continuity equations, giving a total of four equations so there is not enough information to solve for five unknown quantities and we have to make an approximation.

Since the tank is large, the rate at which the level of the water in the tank decreases is negligible. Therefore, we can make the approximation that the speed at which the water is falling in the tank is negligible:

$$v_1 \approx 0 \text{ (large cross-sectional area approximation)}$$

Notice that once we assume that $v_1 \approx 0$, we cannot use the continuity equation between points 1 and 2 (or between points 1 and 3) to determine v_2 or A_2 as both v_1 and A_1 have dropped out of the problem.

Since all the variables except v_3 are known at point 3, we apply Bernoulli's equation between points 1 and 3:

$$P_1 + \frac{1}{2}\rho v_1^2 + \rho g y_1 = P_3 + \frac{1}{2}\rho v_3^2 + \rho g y_3 \qquad (12\text{-}58)$$

Inserting the given values and the approximation $v_1 \approx 0$, and solving for v_3 we get,

$$v_3 = \sqrt{\frac{2}{\rho}(P_1 - P_0 + \rho g y_1)} \qquad (12\text{-}59)$$

Note that the speed v_3 depends on both the pressure difference $(P_1 - P_0)$ and the height difference $(y_1 - 0)$. Inserting the given values, we get

$$v_3 = \sqrt{\frac{2}{1000 \text{ kg/m}^3}(1.20 \times 10^5 \text{ Pa} - 1.01 \times 10^5 \text{ Pa} + (1000 \text{ kg/m}^3) \times (9.81 \text{ m/s}^2) \times (2.00 \text{ m}))}$$

$$v_3 = 8.80 \text{ m/s} \qquad (12\text{-}60)$$

(b) Having calculated v_3, we can now calculate the speed and the pressure at point 2. To calculate v_2, we apply the continuity equation between points 2 and 3:

$$A_2 v_2 = A_3 v_3$$

$$v_2 = \frac{A_3}{A_2}v_3 = \frac{2.00 \times 10^{-2} \text{ m}^2}{4.00 \times 10^{-2} \text{ m}^2} \times 8.80 \text{ m/s} = 4.40 \text{ m/s}$$

(c) In order to find the height, h, of the column of water in the vertical tube, we need to calculate the pressure at a point that is directly underneath the vertical tube. That is why we chose point 2 to be under the vertical tube. The water in the vertical tube is stationary, so the pressure P_2 at the bottom of the tube and the height h are related by

$$P_2 = P_0 + \rho g h$$

Pressure P_2 can be calculated by applying Bernoulli's equation between points 2 and 3:

$$P_2 + \frac{1}{2}\rho v_2^2 + \rho g y_2 = P_3 + \frac{1}{2}\rho v_3^2 + \rho g y_3$$

$$P_2 = P_3 + \left(\frac{1}{2}\rho v_3^2 - \frac{1}{2}\rho v_2^2\right) + (\rho g y_3 - \rho g y_2)$$

Inserting $P_3 = P_0$, $y_2 = y_3 = 0$, $v_2 = 4.39$ m/s, $v_3 = 8.80$ m/s, we get

$$P_2 = 1.01 \times 10^5 \text{ Pa} + \frac{1}{2} \times (1000 \text{ kg/m}^3)$$

$$\times ((8.80 \text{ m/s})^2 - (4.40 \text{ m/s})^2)$$

$$P_2 = 1.30 \times 10^5 \text{ Pa}$$

The height, h, of the water in the vertical tube can now be calculated:

$$h = \frac{P_2 - P_0}{\rho g} = \frac{1.30 \times 10^5 \text{ Pa} - 1.01 \times 10^5 \text{ Pa}}{(1000 \text{ kg/m}^3) \times (9.81 \text{ m/s}^2)}$$

$$h = 2.95 \text{ m}$$

ONLINE ACTIVITY

Non-viscous flow through a horizontal pipe

The e-resource that accompanies every new copy of this textbook contains an Online Activity using the PhET simulation "Non-Viscous Flow Through a Horizontal Pipe of Variable Cross-Sectional Area." Work through the simulation and accompanying questions to gain an understanding of the flow of a non-viscous fluid through a horizontal pipe of variable cross-sectional area.

 CHECKPOINT

C-12-10 Pressure and Area

If the cross-sectional area of the vertical tube in Example 12-9 is doubled, the height of the water in the tube would

(a) increase by a factor of 4;
(b) increase by a factor of 2;
(c) remain the same;
(d) decrease by a factor of 2.

C-12-10 (c)

12-9 Conservation of Fluid Momentum

Consider a fluid moving through a straight horizontal tube that gradually narrows (Figure 12-37). As the cross-sectional area of the tube decreases, the speed of the fluid increases. According to Newton's second law of motion, any change in a fluid's velocity indicates that an external force is acting on the fluid. Similarly, there must be a net force on a fluid that passes through a bend in a tube because its direction changes. To study situations in which a fluid's momentum changes, we need to understand how Newton's second law of motion is applied to moving fluids.

Consider an element of fluid flowing between two sections of a tube of variable cross-sectional area (Figure 12-37). At section 1, the fluid pressure is P_1, the velocity is $\vec{v}_1$, and the cross-sectional area of the tube is A_1. In time Δt, the mass of the fluid that enters section 1 from the left is

$$\Delta M_1 = \rho(A_1\, v_1\, \Delta t)$$

The momentum of the entering fluid is

$$\Delta \vec{p}_1 = \Delta M_1\, \vec{v}_1 = (\rho A_1\, v_1 \Delta t)\, \vec{v}_1$$

In the same time, the mass of fluid leaving the tube at section 2 is

$$\Delta M_2 = \rho(A_2\, v_2\, \Delta t)$$

and the momentum of the leaving fluid is

$$\Delta \vec{p}_2 = \Delta M_2\, \vec{v}_2 = (\rho A_2\, v_2 \Delta t)\, \vec{v}_2$$

The rate of change of momentum of the fluid between the two sections of the tube is, therefore,

$$\frac{\Delta \vec{p}_2 - \Delta \vec{p}_1}{\Delta t} = (\rho A_2\, v_2)\vec{v}_2 - (\rho A_1\, v_1)\, \vec{v}_1 \quad (12\text{-}61)$$

By the equation of continuity, $A_1 v_1 = A_2 v_2 \equiv Q$, and the rate of change of the fluid's momentum is

$$\frac{\Delta \vec{p}_2 - \Delta \vec{p}_1}{\Delta t} = \rho Q\,(\vec{v}_2 - \vec{v}_1) \quad (12\text{-}62)$$

The rate of change of momentum of a physical object is equal to the sum of all external forces acting on that object. We can then write Equation (12-62) in the form of Newton's second law of motion:

$$\sum \vec{F} = \rho Q\,(\vec{v}_2 - \vec{v}_1) \quad (12\text{-}63)$$

Here, $\sum \vec{F}$ denotes the vector sum of all external forces acting on the fluid element. Equation (12-63) is Newton's second law of motion as applied to moving fluids. This equation indicates that a fluid element accelerates or decelerates in response to external forces. By Newton's third law of motion, the fluid exerts an equal and opposite force on the surrounding fluid or the containing vessel.

EXAMPLE 12-10

Keeping a Fire Hose Stationary

Consider a fire hose that has an inner diameter of 4.40 cm. The diameter of the nozzle connected to the end of the hose tapers from 4.40 cm to 2.00 cm. Water is flowing out of the nozzle at a rate of 10 L/s, and the gauge pressure at the wide end of the nozzle is 6.80×10^5 Pa. What force must a firefighter exert to keep the nozzle stationary? (Ignore the weight of the nozzle.)

SOLUTION

To hold the nozzle steady, the firefighter has to exert a force to counter the force that the water exerts on the nozzle. Imagine a flow tube that consists of the fluid inside the nozzle. The forces acting on the element of water within the flow tube are as follows (See Figure 12-41):

■ $\vec{F}_1$ is the force exerted from the left by the water entering the nozzle;

■ $\vec{F}_2$ is the force exerted from the right by the atmospheric pressure P_0 on the water that is leaving the nozzle; and

■ $\vec{F}_w$ is the force that the nozzle exerts on the water.

For this situation, Equation (12-63) gives

$$\vec{F}_1 + \vec{F}_2 + \vec{F}_w = \rho Q(\vec{v}_2 - \vec{v}_1) \quad (1)$$

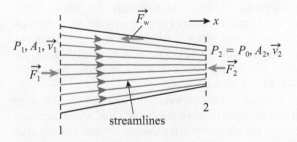

Figure 12-41 Example 12-10

Taking the flow direction as the positive x-axis,

$$\vec{F}_1 = P_1 A_1 \hat{x}$$
$$\vec{F}_2 = -P_2 A_2 \hat{x} = -P_0 A_2 \hat{x}$$
$$\vec{F}_w = F_w \hat{x}$$
$$\vec{v}_1 = v_1 \hat{x}, \quad \vec{v}_2 = v_2 \hat{x}$$

Substituting into Equation (1) gives

$$P_1 A_1 - P_0 A_2 + F_w = \rho Q(v_2 - v_1)$$

and the magnitude of the force exerted by the nozzle on the water is

$$F_w = -P_1 A_1 + P_0 A_2 + \rho Q(v_2 - v_1) \quad (2)$$

(continued)

All the quantities on the right-hand side of Equation (2) are known:

$$P_0 = 1.01 \times 10^5 \text{ Pa}$$

$$P_1 = P_0 + 6.80 \times 10^5 \text{ Pa} = 7.81 \times 10^5 \text{ Pa}$$

$$A_1 = \frac{\pi D_1^2}{4} = \frac{\pi (4.40 \times 10^{-2} \text{ m})^2}{4} = 1.52 \times 10^{-3} \text{ m}^2$$

$$A_2 = \frac{\pi D_2^2}{4} = \frac{\pi (2.00 \times 10^{-2} \text{ m})^2}{4} = 3.14 \times 10^{-4} \text{ m}^2$$

$$Q = 10.0 \times 10^{-3} \text{ m}^3/\text{s}$$

$$\rho = 1000 \text{ kg/m}^3$$

$$v_1 = \frac{Q}{A_1} = \frac{10^{-2} \text{ m}^3/\text{s}}{1.52 \times 10^{-3} \text{ m}^2} = 6.58 \text{ m/s}$$

$$v_2 = \frac{Q}{A_2} = \frac{10^{-2} \text{ m}^3/\text{s}}{3.14 \times 10^{-4} \text{ m}^2} = 31.8 \text{ m/s}$$

Substituting these quantities into Equation (2) gives

$$F_w = -(7.81 \times 10^5 \text{ Pa}) \times (1.52 \times 10^{-3} \text{ m}^2)$$
$$+ (1.01 \times 10^5 \text{ Pa}) \times (3.14 \times 10^{-4} \text{ m}^2)$$
$$+ (1000 \text{ kg/m}^3) \times (10^{-2} \text{ m}^3/\text{s}) \times (31.8 \text{ m/s} - 6.58 \text{ m/s})$$

$$F_w = -1187 \text{ N} + 32 \text{ N} + 252 \text{ N} = -903 \text{ N}$$

Making sense of the result:

The sign of F_w is negative; therefore, the force exerted by the nozzle on the water is toward the negative x-axis, opposite to the direction of water flow, as it should be. How large is this force? It is equivalent to lifting a 92 kg weight. The firefighters need to be strong and fit.

LO 10

12-10   Viscous Flow

It takes less force to stir a spoon in a cupful of water than in a cupful of honey. We say that the honey is more *viscous* than water. Recall from Section 12-6 that viscosity is a measure of the resistance of a fluid to flow or change in its shape. A fluid with nonzero viscosity is called a **viscous fluid**. Frictional forces are always present (except for a few liquids at very low temperatures), so ignoring the viscosity of a moving fluid can give inaccurate results in many situations.

Viscosity can be measured in different ways. Consider the following scenario. Two horizontal flat plates, each of cross-sectional area A, are a distance d apart, and the region between the two plates is filled with a fluid of unknown viscosity (Figure 12-42). The bottom plate is held stationary, and the top plate is moved with a constant velocity by exerting a force on it.

Imagine that the enclosed fluid consists of a very large number of layers of small thickness. Fluid speed decreases from the top plate, where the fluid in contact with the plate moves with speed v, to zero for the fluid that is in contact with the stationary bottom plate. This variation in the fluid

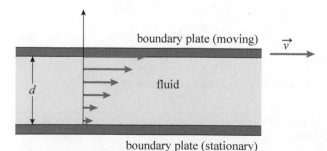

Figure 12-42   Measuring the viscosity of a viscous fluid between two parallel plates

speed is a consequence of the frictional forces between the fluid and the plates and between various layers of the fluid.

The speed of a given layer increases with its vertical distance from the bottom plate. The magnitude of the force F required to move the top plate with a speed v is

- proportional to the cross-sectional area A of the plates;
- inversely proportional to the distance between the plates; and
- proportional to the viscosity of the fluid between the plates.

Therefore,

$$F = \mu \left(\frac{A}{d}\right) v \qquad (12\text{-}64)$$

The constant of proportionality, μ, is called the viscosity (or the coefficient of viscosity) of the fluid. From Equation (12-64), we see that the units of viscosity are as follows:

$$\frac{\text{N} \cdot \text{m}}{\text{m}^2 \cdot (\text{m/s})} = \left(\frac{\text{N}}{\text{m}^2}\right) \text{s} = \text{Pa} \cdot \text{s} = \text{kg} \cdot \text{m}^{-1} \cdot \text{s}^{-1}$$

If a fluid with a viscosity of 1 Pa·s is placed between the two plates, and the top plate is pushed by applying a stress of 1 Pa, then in 1 s the plate will move a distance equal to the separation between the two plates.

A commonly used unit for viscosity is the poise (P), named after French physicist Jean Poiseuille:

$$1 \text{ P} = 1 \text{ g} \cdot \text{cm}^{-1} \cdot \text{s}^{-1} = 0.1 \text{ kg} \cdot \text{m}^{-1} \cdot \text{s}^{-1} = 0.1 \text{ Pa} \cdot \text{s}$$

Therefore,

$$10 \text{ P} = 1 \text{ Pa} \cdot \text{s}$$

Another commonly used measurement unit is the centipoise (cP):

$$1 \text{ cP} = 10^{-2} \text{ P} = 10^{-3} \text{ Pa} \cdot \text{s}$$

In general, viscosities of liquids are greater than viscosities of gases. In addition, the viscosity of liquids decreases with increasing temperature. In gases, viscosity arises from the diffusion of gas molecules within the gas.

Table 12-3 Table of Viscosities

Material	Viscosity (Pa·s)	Material	Viscosity (Pa·s)
Air (0°C)	17.4×10^{-6}	Honey	2–10
Hydrogen (0°C)	8.4×10^{-6}	Corn syrup	1.38
Xenon (0°C)	2.12×10^{-5}	Olive oil	0.081
Water (20°C)	8.94×10^{-4}	Peanut butter	~250
Blood	3 to 4×10^{-3}	Ketchup	50–100

CRC handbook of chemistry and physics: a ready-reference book of chemical and physical data by LIDE, DAVID R., JR Copyright 1992 Reproduced with permission of TAYLOR & FRANCIS GROUP LLC - BOOKS in the format Textbook via Copyright Clearance Center.

Therefore, the viscosity of a gas increases with temperature. Table 12-3 lists the viscosities of several fluids.

Poiseuille's Law for Viscous Flow

Consider streamlined flow of a viscous fluid flow through a pipe of length L and radius R. As discussed above, the layer in contact with the walls of the pipe is almost at rest. The layer adjacent to it moves with a slightly higher speed. The farther a layer is from the walls, the faster it moves, with the fluid at the centre of the pipe moving with the highest speed. The speed profile of a fluid is shown in Figure 12-43.

Because of viscosity, the kinetic energy and, hence, the speed of the fluid, decreases as the fluid moves through the pipe. Therefore, work must be done on the fluid by applying an external pressure to maintain a constant flow rate. Bernoulli's equation contains no information about frictional forces, so it is not applicable in this case. Poiseuille showed that the volume flow rate Q of a viscous fluid through a horizontal and cylindrical pipe is related to the pressure difference between the two ends of the pipe by the following equation:

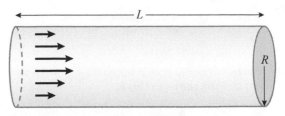

Figure 12-43 The speed profile of a fluid moving through a horizontal tube. The length of the arrows is proportional to the speed of the fluid. The layers of the fluid that are close to the walls of the pipe move slowly. The farther a layer is from the walls, the faster its speed. Fluid moving along the central axis of the pipe has the highest speed.

$$P_1 - P_2 = 8\mu\frac{LQ}{\pi R^4} \qquad (12\text{-}65)$$

Here, $P_1 - P_2$ is the pressure difference between the two ends of the pipe. The quantity $\frac{8}{\pi}\left(\frac{\mu L}{R^4}\right)$ can be interpreted as the resistance offered to fluid flow by a pipe of length L and radius R. This quantity is called the total peripheral resistance (TPR). Equation (12-65) shows that in order to maintain a constant volume flow rate, a nonzero pressure difference needs to be maintained between the ends of a pipe. The required pressure difference is proportional to the length of the pipe and inversely proportional to its radius. The longer the pipe, the greater the energy loss; hence, a greater pressure difference is required to maintain a constant flow rate. For smaller radii, a large fraction of the fluid is close to the inner surface of the pipe, where fluid speed is small. Poiseuille's equation can be derived from Newton's laws of motion.

Note that the pressure difference required to maintain a constant flow rate is inversely proportional to the fourth power of the radius of the pipe. As radius increases, the required pressure difference decreases rapidly. Consequently, Bernoulli's equation can be used for flow through pipes of large cross-sectional areas with negligible error.

EXAMPLE 12-11

Drawing Blood Through a Needle

A needle of inner diameter 0.50 mm is used to draw blood from a patient's forearm for a medical test. The average blood pressure in the patient's arm is 85 mm Hg, and the viscosity of the blood is 3.5×10^{-4} Pa·s. The length of the needle is 6.0 cm, and the blood needs to be drawn at a constant flow rate of 0.10 cm³/s. What pressure must be applied to the plunger?

SOLUTION

The diameter of the needle is small, so the viscosity of the blood cannot be ignored. We must use Poiseuille's equation to determine the pressure difference needed to maintain the required flow rate. Let P_{arm} be the pressure at the end of the needle that is inserted in the patient's arm and

P_{pull} the pressure that needs to be maintained to pull the plunger outward. We will assume that the pressure at the end of the needle that is closer to the plunger is P_{pull}.

Known quantities:

radius of the needle, $R = 0.50$ mm/2 $= 0.25 \times 10^{-3}$ m

length of the needle, $L = 6.0$ cm $= 6.0 \times 10^{-2}$ m

volume flow rate of the blood through the needle, $Q = 0.1$ cm³/s $= 0.1 \times (10^{-2}$ m$)^3$/s $= 10^{-7}$ m³/s

blood viscosity, $\mu = 3.5 \times 10^{-4}$ Pa·s

pressure at the end of the needle that is inserted in the arm, $P_{arm} = 85$ mm Hg

(continued)

Since 760 mm Hg = 1.0 atmospheric pressure
= 1.01×10^5 Pa,

85 mm Hg = 11 300 Pa

P_{pull} can be calculated from Equation (12-65), which for the present situation can be written as

$$P_{pull} = P_{arm} + 8\mu \frac{LQ}{\pi R^4}$$

$P_{pull} = 11\ 300$ Pa

$$+ \frac{8 \times (3.5 \times 10^{-4}\ \text{Pa·s}) \times (6.0 \times 10^{-2}\ \text{m}) \times (10^{-7}\ \text{m}^3/\text{s})}{\pi (0.25 \times 10^{-3}\ \text{m})^4}$$

$P_{pull} = 11\ 300$ Pa + 1 370 Pa = 12 670 Pa

Making sense of the result:

As the radius of the needle is very small the viscosity of the blood cannot be ignored. Therefore, the pressure at the plunger side needs to higher than the blood pressure in the forearm to maintain a flow rate.

 CHECKPOINT

C-12-11 Viscous Flow Through a Pipe

Figure 12-44 shows viscous fluid in a horizontal pipe and two connected vertical tubes that are open to the atmosphere. Which of the following statements correctly describes the motion of the fluid in the pipe? Explain your reasoning.

(a) The fluid is not moving.
(b) The fluid is moving from left to right.
(c) The fluid is moving from right to left.
(d) The motion cannot be determined from the diagram.

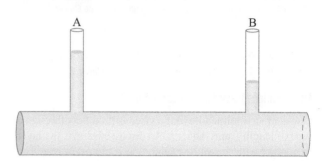

Figure 12-44 C-12-11

C-12-11 (b)

 MAKING CONNECTIONS

Blood Pressure and Blood Flow

The laws of fluid dynamics apply to the flow of blood through the human body. The human heart has two ventricles, which are chambers that contract to pump blood to the rest of the body (Figure 12-45). With each contraction (heartbeat), the pressure in the arteries—which carry blood from the heart—cycles between a minimum *diastolic* pressure (when the ventricles relax) and a maximum *systolic* pressure (when the ventricles contract). In a healthy adult, the systolic pressure is approximately 16 kPa (120 mm Hg), and the diastolic pressure is approximately 10 kPa (80 mm Hg).

The heart of a healthy adult at rest pumps blood at a rate of approximately 5 L/min, giving an average flow rate of approximately 80 cm³/s. During strenuous physical activity, the flow rate can increase by a factor of 5. Since the cross-sectional area of the major arteries does not increase significantly during exercise, the blood speed in the major arteries must increase by approximately the same factor.

When a person is standing, the weight of the blood has a substantial effect on the blood pressure throughout the body. Let h_{head} be the vertical distance between the head and the heart, and let h_{feet} be the distance between the feet and the heart. From Equation (12-8), the blood pressure at the head and the feet level is related to the pressure at the heart level as follows:

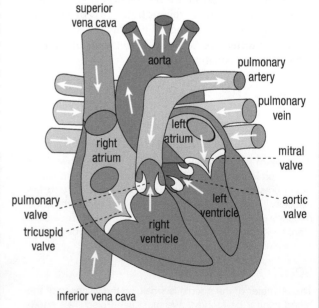

Figure 12-45 An interior view of the human heart showing major blood vessels

(continued)

$$P_{\text{head}} = P_{\text{heart}} - \rho_{\text{blood}}\, g h_{\text{head}}$$

$$P_{\text{feet}} = P_{\text{heart}} + \rho_{\text{blood}}\, g h_{\text{feet}}$$

Taking typical values of $P_{\text{heart}} = 13$ kPa, $\rho_{\text{blood}} = 1060$ kg/m³, $h_{\text{head}} = 50$ cm, and $h_{\text{feet}} = 130$ cm, we find that $P_{\text{head}} = 7.8$ kPa and $P_{\text{feet}} = 26.5$ kPa. This large difference in relative pressure explains why lying down increases blood flow to the brain and can help relieve swelling in the feet.

Because of viscosity, the blood pressure drops as the blood moves through the arteries. For example, taking the viscosity of blood to be 4×10^{-3} Pa·s and using a flow rate of 80 cm³/s, the pressure drop per unit length in an artery of radius 1.0 cm is

$$\frac{P_1 - P_2}{L} = \frac{8\mu}{\pi}\frac{Q}{R^4} = \frac{8 \times 4.0 \times 10^{-3}\ \text{Pa·s}}{\pi}$$

$$\times \frac{80 \times 10^{-6}\ \text{m}^3/\text{s}}{(10^{-2}\ \text{m})^4} \approx 80\ \text{Pa/m}$$

By the time the blood reaches the small capillaries, it is at atmospheric pressure. Since changing R greatly affects the total peripheral resistance, a buildup of fats in and around the walls of arteries greatly reduces the blood flow, and the heart has to work harder to maintain the needed flow rate.

Fluids are substances that deform and flow under an applied stress. Liquids and gases are collectively called fluids. Physical laws describing fluid motion are derived using Newton's laws of motion along with the concepts of momentum and energy conservation. The concepts of density and pressure are useful for describing the properties of fluids.

Density

The density of a material is defined as its mass per unit volume. The average density is given by

$$\rho = \frac{M}{V}$$

The units of density are kg/m³.

Pressure

The pressure exerted by a force on a surface element of area A is defined as the component of the force perpendicular to the surface (F_n) divided by the area of the surface:

$$P = \frac{F_n}{A}$$

The SI unit of pressure is the pascal (1 Pa = 1 N/m²).

Pressure in Fluids

A fluid exerts pressure within itself and on the walls of the containing vessel. This pressure arises due to the weight of the fluid and due to collisions between the constituents of the fluids. In liquids, the primary source of pressure is the weight of the liquid. Gas pressure arises mainly due to collisions between gas molecules and between the gas and the walls of the containing vessel.

Atmospheric Pressure

Atmospheric pressure is due to the weight of the air above Earth's surface.

Gauge Pressure

Gauge pressure is the difference between the actual pressure and a reference pressure, usually taken to be atmospheric pressure.

Pressure Variation with Depth

Pressure within the body of a fluid increases with depth. If P_a denotes the pressure at a point a within a fluid of density ρ, and P_b is the pressure at a point b that is at a depth d below a, then

$$P_b = P_a + \rho g d$$

When point a is at the surface of the fluid where the pressure is equal to atmospheric pressure, P_0, then

$$P_b = P_0 + \rho g d$$

Pascal's Principle

Pascal's principle states that any change in the pressure of an enclosed fluid is transmitted undiminished to every part of the fluid and to the walls of the containing vessel.

Archimedes' Principle

Archimedes' principle states that, when an object is fully or partially immersed in a fluid, the fluid exerts a net upward force on the object (called the buoyant force); the magnitude of the buoyant force is equal to the weight of the fluid displaced by the object.

Flotation

An object will float in a fluid when the average density of the object is less than or equal to the density of the fluid.

Ideal Fluids

An ideal fluid is an incompressible fluid that has no viscosity.

Ideal Fluid Flow

A flow that is steady and nonturbulent is called an ideal flow.

Volume Flow Rate (Q)

The volume flow rate of a fluid through a surface that is normal to the direction of flow is equal to the volume of the fluid that passes through the surface per unit time:

$$Q = Av$$

Here, A is the cross-sectional area of the surface perpendicular to the direction of flow, and v is the speed of the fluid.

Streamlines

A streamline is the trajectory taken by an infinitesimal element of fluid in the region of flow. In a steady flow, streamlines never cross.

Continuity Equation

The amount of fluid that passes through a flow tube remains constant:

$$A_1 v_1 = A_2 v_2$$

Bernoulli's Equation

For the flow of an ideal fluid, at any point along a streamline through the fluid, the sum of pressure, kinetic energy per unit volume, and potential energy per unit volume remains constant:

$$P + \frac{1}{2}\rho v^2 + \rho g y = \text{constant}$$

Poiseuille's Law for Viscous Flow

When viscosity is taken into account, the volume flow rate Q of a fluid of viscosity μ from a pipe of length L and radius R is related to the pressure difference between the two ends of the pipe by the following relationship:

$$P_1 - P_2 = 8\mu\frac{LQ}{\pi R^4}$$

Applications: airplanes and hot-air balloons, spray bottles, hydrometers, hydraulic lifts and brakes, water distribution and irrigation systems, sphygmomanometers, Venturi flow meters, submarines

Key Terms: apparent weight, Archimedes' principle, atmospheric pressure, average mass density, Bernoulli's equation, buoyant force, equation of continuity, flow rate, flow tube, fluid pressure, fluids, gauge pressure, hydraulic lift, ideal fluid, mass flow rate, pascal, Pascal's principle, phase (state), plasma, pressure, specific gravity, steady flow, streamline (laminar flow), viscosity, viscous fluid, work done by the leaving fluid, work done on a test volume

QUESTIONS

1. Helium gas contracts when cooled. A rubber balloon is filled with helium gas at room temperature and closed tightly so that the gas cannot escape. When the balloon is cooled, the density of helium inside the balloon
   (a) decreases;
   (b) remains the same;
   (c) increases.
2. A cylindrical glass is filled with water. The gauge pressure at the bottom of the glass is P. A second cylindrical glass with twice the diameter is filled with water to the same height. The gauge pressure at the bottom of the second glass is
   (a) $P/2$
   (b) P
   (c) $2P$
   (d) $4P$
3. A cylindrical glass is filled with water. The magnitude of the net force at the bottom of the glass due to the weight of the water is F. A second cylindrical glass with twice the diameter is filled with water to the same height. The magnitude of the net force at the bottom of the second glass due to the weight of the water is
   (a) $F/2$
   (b) F
   (c) $2F$
   (d) $4F$
4. Is the following statement true or false? Explain your reasoning: The magnitude of the buoyant force on an object that is fully immersed in a fluid can never be greater than the magnitude of the weight of the object.
5. Figure 12-46 shows mass versus volume plots for three fluids, A, B, and C. Order the fluids from highest to lowest densities.

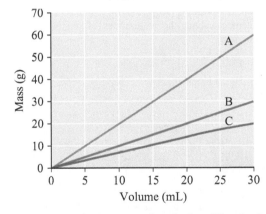

Figure 12-46 Plots of mass versus volume for three different materials

6. Four objects with identical volumes are submerged in a fluid, as shown in Figure 12-47. Each object is in a state of equilibrium. Which of the following statements is true about their densities?
   (a) $\rho_1 < \rho_2 < \rho_3 < \rho_4$
   (b) $\rho_1 = \rho_2 = \rho_3 = \rho_4$
   (c) $\rho_1 > \rho_2 > \rho_3 > \rho_4$
   (d) $\rho_1 < \rho_2 = \rho_3 < \rho_4$

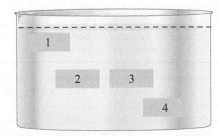

Figure 12-47 Question 6

7. A plastic toy boat carrying a small iron cube is floating in a glass container. The cube is then gently dropped into the water. The level of the water in the container
   (a) goes down;
   (b) goes up;
   (c) stays the same.

   Explain your reasoning. Now perform this experiment to determine whether or not your prediction agrees with your observations.

8. A rubber ball with density 500 kg/m³ is floating in a container of water. Oil with a density less than the density of water but greater than the ball's density is slowly poured in the container until it completely covers the ball. When completely immersed in the oil, what fraction of the volume of the ball is submerged in the water? Explain your reasoning.
   (a) half (the same as before the oil was poured)
   (b) less than half
   (c) more than half

9. An ice cube is floating in a partially filled glass of water. After the ice cube melts, the level of the water in the glass
   (a) goes up;
   (b) goes down;
   (c) remains the same.

10. Three identical rectangular pieces of metal are held submerged under water in different orientations, as shown in Figure 12-48. The magnitudes of the buoyant forces on these pieces are $F_{B,1}$, $F_{B,2}$, and $F_{B,3}$. Which of the following statements about the magnitudes of the buoyant forces is correct?
    (a) $F_{B,1} = F_{B,2} = F_{B,3}$
    (b) $F_{B,1} < F_{B,2} = F_{B,3}$
    (c) $F_{B,1} = F_{B,2} < F_{B,3}$
    (d) $F_{B,1} < F_{B,2} < F_{B,3}$

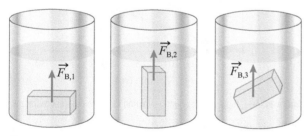

Figure 12-48   Question 10

11. A container, partially filled with water, is resting on a scale that measures its initial weight. A solid iron ball is placed in the container, without spilling any water. What happens to the reading on the scale after the iron ball is placed in the container?
    (a) The reading does not change because the weight of the iron ball is exactly balanced by the upward buoyant force exerted by the water on the ball.
    (b) The reading increases by an amount exactly equal to the weight of the ball.
    (c) The reading increases by an amount that is less than the weight of the ball because the buoyant force cancels a fraction of the weight of the ball.
    (d) The reading increases by an amount that is greater than the weight of the ball because the water on top of the ball further pushes the ball down, thus increasing its weight.

    (e) The reading decreases because the buoyant force on the ball is greater than its weight.

12. A gas-filled balloon of mass 2.25 kg (including the mass of the gas) and volume 5.00 m³ is released from a tall building. The density of air is 1.30 kg/m³. Which of the following statements is correct? Explain your reasoning.
    (a) The balloon will remain stationary.
    (b) The balloon will fall.
    (c) The balloon will rise.

13. Three containers of different shapes but equal base area are filled with water to the same height (Figure 12-49).

    Is the hydrostatic pressure at the base of each container the same? Explain your reasoning.

Figure 12-49   Question 13

14. A container filled with a liquid has a hole in its side through which the liquid is coming out. If the container (with the liquid in it) is dropped from a height and allowed to fall freely, what will happen to the liquid emerging from the container?
    (a) It will keep coming out following a similar trajectory as before.
    (b) It will start flowing upward.
    (c) It will stop flowing.

15. A piece of solid metal of mass M has a density four times the density of water. It is attached to a string and suspended in water, fully immersed. The tension in the string is
    (a) $\frac{3}{4}Mg$
    (b) Mg
    (c) $3Mg$
    (d) $4Mg$

16. Scuba divers control buoyancy by wearing a vest that can be inflated or deflated by adding or removing air. The density of water slightly increases with depth. A scuba diver is in a state of neutral buoyancy (neither sinking or rising) at a certain depth. To establish neutral buoyancy at a greater depth the diver should
    (a) inflate its vest;
    (b) deflate its vest;
    (c) do nothing as he/she is already in a state of neutral buoyancy

PROBLEMS BY SECTION

For problems, star ratings will be used, (✷, ✷✷, or ✷✷✷), with more stars meaning more challenging problems.

Section 12-1   Phases of Matter

17. ✷✷   Water has a density of 1.0 g/cm³, and 18 g of water contains 6.02×10^{23} molecules. From this information, estimate the average distance between water molecules.

18. ✳✳ Given that the molar mass of dry air is 29 g/mol and at STP the density of air is 1.30 kg/m³, estimate the average separation between air molecules.

Section 12-2 Density and Pressure

19. ✳ Your mass is 70 kg. The area of the parts of your shoes that are in contact with the ground is 0.010 m². While standing stationary on your two feet, what is the pressure that you exert on the ground?

20. ✳✳ Two rectangular blocks of masses M_1 and M_2 and cross-sectional areas A_1 and A_2, respectively, are lying on a smooth surface as shown in Figure 12-50. What is the pressure exerted by the blocks on the surface?

(a) $\dfrac{M_1 g}{A_1} + \dfrac{M_2 g}{A_2}$

(b) $\dfrac{M_1 g}{A_1} + \dfrac{M_2 g}{A_1}$

(c) $\dfrac{M_1 g}{A_2} + \dfrac{M_2 g}{A_2}$

(d) $\dfrac{M_1 g}{A_1} - \dfrac{M_2 g}{A_2}$

Figure 12-50 Problem 20

21. ✳ A neutron star results when a star in its final stages collapses due to gravitational pressure, forcing the electrons to combine with the protons in the nucleus and converting them into neutrons.
    (a) Assuming that a neutron star has a mass of 3.00×10^{30} kg and a radius of 1.20×10^3 m, determine the density of a neutron star.
    (b) How much would a 1.0 cm³ (the size of a sugar cube) of this material weigh at Earth's surface? Do you think you would be able to lift it?

22. ✳ Wind blowing at 25.0 m/s hits a glass window in a high-rise office building. The cross-sectional area of the window is 4.0 m². Calculate the magnitude of the force exerted on the window by the wind.

23. ✳✳ A spherical metal shell has an outer radius of 15.0 cm and an inner radius of 14.0 cm. The density of the metal is 6000 kg/m³. What is the average density of the shell? Will this shell float when immersed in water? If yes, what fraction of the volume of the shell will be beneath the water's surface?

Section 12-3 Pressure in Fluids

24. ✳ The air provided by a diver's scuba equipment is at the same pressure as the surrounding water. At absolute pressures greater than approximately 1.0×10^6 Pa, the nitrogen in the air becomes dangerously poisonous. At what depth in water does nitrogen become dangerous? The density of the ocean water is 1030 kg/m³.

25. ✳ A submarine is at a depth of 45.0 m under the ocean surface. The inside of the submarine is kept at atmospheric pressure. What is the net force being exerted on a circular window of the submarine that has a radius of 20 cm? The density of the ocean water is 1030 kg/m³.

26. ✳✳ A glass tube of cross-sectional area 10^{-4} m² is partially filled with water. An oil with a density of 800 kg/m³ is slowly poured into the tube so that it does not mix with the water and floats on top of it. The height of the oil above the water surface is 8 cm.
    (a) What is the change in pressure at a depth of 10 cm below the water surface?
    (b) Would the pressure at this depth increase or decrease if you were to add more oil to the tube? Why?

27. ✳ A cylindrical tube closed at one end and open at the other is filled with water, as shown in Figure 12-51. Determine the pressure exerted by the water on the closed end of the tube.

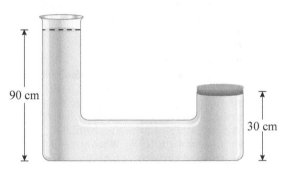

Figure 12-51 Problem 27

28. ✳✳ A container of fluid is accelerated vertically downward with acceleration a. Show that the pressure in the fluid at a depth h below the surface is given by

$$P = P_0 + \rho h(g - a)$$

where P_0 is the atmospheric pressure and ρ is the density. If the container is allowed to fall freely ($a = g$), why does the pressure not change with depth?

Section 12-4 Pascal's Principle

29. Two pistons of a hydraulic lift have radii of 2.67 cm and 20.0 cm. A mass of 2.00 t is placed on the larger piston. Calculate the minimum downward force needed to be exerted on the smaller piston to hold the larger piston at level with the smaller piston.

30. Consider the two hydraulic presses shown in Figure 12-52 (page 340). A force of 1000 N needs to be maintained at the output arm of each press. What is the relationship between the magnitudes of the forces that need to be exerted on the two input arms? Explain your result.
    (a) $F_1 = F_2$
    (b) $F_1 < F_2$
    (c) $F_1 > F_2$

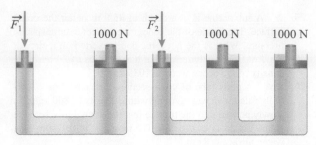

Figure 12-52　Problem 30

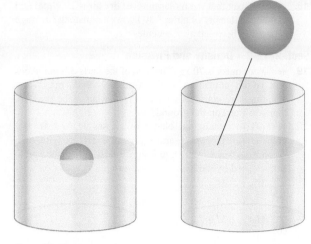

Figure 12-53　Problem 36

Section 12-5　Buoyancy and Archimedes' Principle

31. ✹✹　A rubber ball of uniform density 0.7 g/cm³ is floating in a glass of water. Oil, with a density less than the density of the water, is slowly poured into the glass so that the oil floats on top of the water, and the ball is completely submerged under the oil. What fraction of the volume of the ball is now immersed in the water?
    (a) less than 0.7
    (b) 0.7
    (c) between 0.7 and 1.0
    (d) 1.0 (i.e., the ball is completely immersed in the water)

32. ✹✹　An object with specific gravity 0.85 is immersed in a liquid with specific gravity 1.2. What is the fraction of the volume of the object that submerges in the liquid?

33. ✹✹　A typical supertanker has a mass of 2.0×10^6 kg and carries oil of mass 4.0×10^6 kg. When empty, 9.0 m of the tanker is submerged in water. What is the minimum water depth needed for it to float when full of oil? Assume the sides of the supertanker are vertical and its bottom is flat.

34. ✹✹　An object weighs 500.0 N in air. A force of 30.0 N is needed to push the object down to completely submerge it under water. What is the average density of the object?

35. ✹✹　A 0.48 kg piece of wood floats in water but is found to sink in alcohol. When in the alcohol it has an apparent weight of 0.46 N. What is the density of the wood? The density of alcohol is 790 kg/m³.

36. ✹✹　A solid sphere of mass m floats in a container of water, half submerged, as shown in Figure 12-53. A second solid sphere of the same material but twice the mass is placed in a second container of water.
    (a) Sketch the final position of the second sphere.
    (b) What is the magnitude of the buoyant force on the first sphere?
    (c) What is the magnitude of the buoyant force on the second sphere?
    (d) Is the buoyant force acting on the second sphere larger, smaller, or equal to the buoyant force on the first sphere?
    (e) Is the volume of water displaced by the second sphere larger, smaller, or equal to the volume displaced by the first sphere?

37. A piece of cork with a mass of 10 g is held in place under water by a string tied to the bottom of the container. What is the tension in the string? The density of cork is 0.33 g/cm³.

38. ✹✹✹　A hydrometer consists of a spherical bulb with a cylindrical stem (Figure 12-54). The cross-sectional area of the stem is 0.40 cm². The total volume of bulb and stem is 13.2 cm³. When immersed in water, the hydrometer floats with 8.0 cm of the stem above the water surface. In alcohol, 1.0 cm of the stem is above the water surface. Find the density of alcohol.

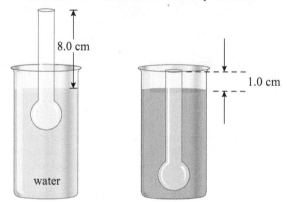

Figure 12-54　Problem 38

39. ✹✹✹　A glass beaker is placed in a large container of water. The beaker itself has a mass of 390.0 g and an interior volume of 500.0 cm³. A student starts filling the beaker with water and finds that when the beaker is less than half full, it floats (Figure 12-55). But when it is more than half full, it sinks to the bottom. What is the density of the glass making up the beaker?

Figure 12-55　Problem 39

40. ✷✷ You need to construct a hollow cubical box of aluminum that has an outside length of 20 cm and an average density of 1000 kg/m³. What must be the thickness of the aluminum sheet used to construct this box? The density of aluminum is 2700 kg/m³. Did you have to make any approximations to solve this problem?

41. ✷✷✷ A wooden boat has a density of 700 kg/m³ and a volume of 5.5 m³.
    (a) What fraction of volume of the boat is submerged in water when empty?
    (b) The boat needs to carry a number of survivors from a desert island. The average mass of each survivor is 65 kg. How many survivors can the boat carry without sinking?
    (c) If some of the survivors sit at the edge of the boat with their legs partially submerged in the water, would the boat be able to carry more people? Why?

42. ✷✷ An evacuated hollow spherical shell of aluminum floats almost completely submerged under water. The outer diameter of the shell is 100.0 cm, and the density of aluminum is 2.7 g/cm³. What is the inner diameter of the shell? Would your answer increase or decrease if you were to assume that the inside of the shell is filled with air?

43 ✷✷ A solid sphere of aluminum (density 2.7 g/cm³) is gently dropped into a deep ocean. (The density of ocean water is approximately 1.03 g/cm³.) Calculate the sphere's acceleration at the point where it is completely submerged in the ocean. As the sphere drops deeper into the ocean, does the acceleration of the sphere increase or decrease compared to its acceleration just beneath the surface? Explain your reasoning.

44. ✷✷✷ A rectangular block of wood weighs 4.0 N in air and has a density of 600 kg/m³. The block is immersed in water.
    (a) What fraction of the volume of the block is immersed in the water?
    (b) A piece of iron (density 6000 kg/m³) is placed on top of the block so that 85% of the block is now immersed in the water. What is the weight of the iron placed on the block?
    (c) The same piece of iron is now attached to the bottom of wood that is immersed in the water. What fraction of volume of the block is now immersed in the water?

45. ✷✷ Most balloons use hot air to fly. At 50°C the density of hot air is approximately 0.85 kg/m³. What would be the volume of the balloon if hot air is used in the Example 12-6? Assume the density of the ambient air to be 1.28 kg/m³.

Section 12-7   The Continuity Equation: Conservation of Fluid Mass

46. ✷ Water is travelling at a speed of 0.1 m/s in a portion of pipe that has a radius of 2.0 cm. The water enters another section of the pipe that has a radius of 1.0 cm. What is the speed of the water in this section of the pipe?

47. ✷✷ The cross-sectional area of the lake behind a dam is approximately 1.00×10^8 m². Water due to heavy rains enters the lake at a rate of 2.5×10^4 m³/s and leaves through the dam at a rate of 1.9×10^4 m³/s. By how much will the water level increase in the lake in a period of 6.0 h?

48. ✷✷ A large spherical balloon made of flexible rubber is filled with water at the rate of 0.1 L/s. What is the rate of increase of the radius of the balloon at an instant when its radius is 20.0 cm?

49. ✷✷ A water faucet has an inner area of 3.0 cm². The flow of water through the faucet is such that it fills a 500 mL container in 15 s.
    (a) What is the flow rate of the water as it comes out of the faucet?
    (b) What is the velocity with which the water emerges from the faucet?
    (c) What is the velocity of the water 20 cm below the faucet?
    (d) What is the area of the water stream at 20 cm below the faucet?

Section 12-8   Conservation of Energy for Moving Fluids

50. ✷✷ Water in a main pipe at street level is at a gauge pressure of 170 kPa and is moving with negligible speed. A pipe connected to the main pipe is used to deliver water to a kitchen located on the third floor of a building, at a 15.0 m height above street level (Figure 12-56). What is the maximum possible speed with which the water can emerge from an open kitchen faucet?

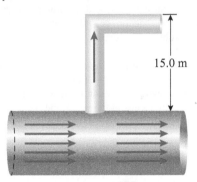

15.0 m

Figure 12-56   Problem 50

51. ✷ Water flows through a horizontal pipe that gradually narrows so that the final inside diameter of the pipe is one-half of the original diameter. In the wide section, the flow speed is 2.0 m/s, and the water pressure is two times the atmospheric pressure. Determine
    (a) the flow speed of the water in the narrower section;
    (b) the water pressure in the narrower section.

52. ✷ A horizontal pipe, which carries water with a speed of 5.0 cm/s, is smoothly connected to a pipe with a smaller cross-sectional area. The pressure in the small pipe is 7.0 kPa lower than the pressure in the large pipe. What is the velocity of the water in the small pipe?

53. ✷ Normal blood pressure in a human artery is 13.6 kPa; the blood flows with a speed of 0.12 m/s. Suppose that part of the artery is clogged such that only 20% of the artery's area is open (assume the artery is horizontal).
    (a) What is the blood pressure in the clogged part of the artery?
    (b) What is the percentage change in the blood pressure in the clogged part?

54. ✷✷ A garden hose has an inside cross-sectional area of 4.00 cm², and the opening in the nozzle has a cross-sectional area of 0.50 cm². The water speed is 0.50 m/s in a segment of the hose that lies on the ground.
    (a) With what speed does the water leave the nozzle when it is held 1.00 m above the ground?
    (b) What is the water pressure in the hose on the ground?

55. ✳✳✳ A 12.0 m long outlet pipe with a diameter of 4.0 cm stands vertically in a water tank that has a diameter of 4.0 m and is open to the atmosphere. The section of the pipe above the water surface is 5.0 m long. Water is pumped from the tank by lowering the pressure at the top end of the pipe. The desired flow rate is 4.0 L/s.
   (a) By how much should the pressure at the top end of the pipe differ from atmospheric pressure?
   (b) Assuming that the pressure at the top of the outlet pipe remains constant, what is the lowest level from which the water can be pumped from the tank?

56. ✳✳ Water flows through a horizontal pipe of radius 40.0 cm that is located 20.0 m below the surface of a large lake next to a dam. The end of the pipe is connected to a nozzle of radius 10.0 cm, and the outlet end of the nozzle is open to the atmosphere. Assuming that the water behaves as an ideal fluid, find
   (a) the speed with which the water leaves the nozzle;
   (b) the water pressure inside the pipe.

57. ✳✳ In problem 56, if the nozzle is closed so that there is no water flow through it, would the pressure in the pipe be less than, greater than, or equal to the pressure calculated in part (b) of problem 53? Ignore viscosity.

58. ✳✳ Air is blown through a circular and horizontal Venturi tube, as shown in Figure 12-57. The diameters of the narrow and wider sections of the tube are given, and the height h of the mercury column is measured to be 1.0 mm. What are the air speeds in the wider and the narrower sections of the tube? The density of mercury is 13 600 kg/m³.

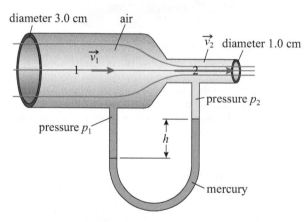

Figure 12-57   Problem 58

59. ✳✳ A tank of cross-sectional area A_2 is filled to a height y_2 with a fluid of density ρ. The tank has a small hole of cross-sectional area A_1 in its side at y_1 from the base. The pressure at the top of the fluid is kept at P_t. Determine an expression for the speed with which the fluid leaves the hole.

60. ✳✳ A very large tank, closed from above and not open to the atmosphere, is partially filled with water. The pressure at the top surface of the water is kept at three times atmospheric pressure by constantly injecting pressurized air from above. Water emerges from the tank through a circular nozzle that is located close to the base of the tank and is open to the atmosphere. At the instant when the water stands 3.0 m high above the nozzle, what is the water flow rate from the nozzle? The cross-sectional area of the nozzle is 2.0×10^{-4} m².

Section 12-9   Conservation of Fluid Momentum

61. ✳✳✳ A fluid of density ρ is placed in a cylindrical container which is rotated about its symmetry axis with a constant angular velocity ω.
   (a) Show that in the radial direction, the variation of pressure within the fluid is given by
   $$\frac{dp}{dr} = \rho\, \omega^2 r$$
   where r is the radial distance from the axis of rotation.
   (b) Suppose the pressure along the axis of rotation is p_0. Show that the pressure at a radial distance R from the central axis is
   $$p_R = p_0 + \frac{1}{2}\rho\omega^2 R^2$$
   (c) Show that the surface of the rotating fluid forms a paraboloid.

62. ✳✳✳ Water passes through a horizontal pipe of uniform cross-sectional area that has a 90° bend. The diameter of the pipe is 10.0 cm, and the volume flow rate is 50.0 L/s. Find the magnitude and direction of the net radial force exerted on the bend as the water passes through it.

Section 12-10   Viscous Flow

63. ✳ A cylindrical pipe of length 5.0 m and cross-sectional area 1.0×10^{-4} m² needs to deliver oil at a rate of 5.0×10^{-4} m³/s. What must be the pressure difference between the two ends of the pipe if the viscosity of the oil is 1.00×10^{-3} Pa·s?

64. ✳✳ An ostrich's neck is approximately 1.2 m long. Assume that the artery supplying blood to the ostrich's head has a radius of 2.0 mm and that the blood flows through it with an average speed of 2.0 cm/s. If the blood arrives at the ostrich's head (when upright) at one atmospheric pressure, with what pressure should the ostrich's heart pump the blood? Assume that the density of the ostrich's blood is 1060 kg/m³ and that the blood's viscosity is 4.0×10^{-3} Pa·s.

COMPREHENSIVE PROBLEMS

65. A pitot-static tube is a device that is used to measure the speed of gases. Its uses include measuring the speed of air flow in a wind tunnel and the speed of an airplane in air. In its simplest form, a pitot-static tube consists of a U-shaped tube that is open at both ends and is partially filled with a liquid, as shown in Figure 12-58. One end of the tube (A) is perpendicular to the incoming flow, and the other end (B) is horizontal to the direction of flow. The tube is closed in the middle (it contains liquid), so the air impinging on end A is brought to rest. The pressure at this end is then the total pressure (also called dynamic pressure) of the flowing air. Air is not flowing right at the edge of the end B, and the pressure at this end is equal to the atmospheric pressure (also called static pressure). Because of the pressure difference at the two ends, the liquid rises in the arm that is at the lower pressure. The height difference, h, between the two arms of the tube is related to the pressure difference between the two ends.
   (a) Consider the pitot-static tube in Figure 12-58, which is attached to a pipe through which air is flowing with speed v. The tube is partially filled with a liquid of

density ρ_f. Show that, if the difference between the heights of the fluid column in the two arms of the tube is h, then the air speed is related to h by the following relationship:

$$v = \sqrt{2gh\left(\frac{\rho_f}{\rho_{air}}\right)}$$

Thus, by measuring h and knowing the densities, flow speed can be calculated. When pitot-static tubes are used to measure the speed of an airplane, ρ_{air} is the density of the air at the flight altitude and needs to be measured or calculated.

(b) In modern pitot-static tubes, instead of using a liquid, an electronic pressure-measuring instrument is placed between the two ends of the tube. Show that the speed, v, of airflow is related to the measured pressure difference ΔP by the following relationship:

$$v = \sqrt{\frac{2(\Delta P)}{\rho_{air}}}$$

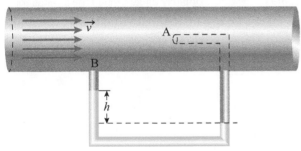

Figure 12-58 A simple pitot-static tube. The height difference of the column of liquid in the two arms of the tube is proportional to the speed of the air flowing through the tube.

66. The density of body fat in humans is approximately 900 kg/m³, and the density of lean muscle is 1100 kg/m³. Suppose a person weighs 790.0 N in air and 50.0 N when completely immersed in water (density 1000 kg/m³).
   (a) Find the person's volume.
   (b) What is the person's density?
   (c) Show that the fraction x_F of a person's body mass that is fat is given by

$$x_F = \frac{\rho_F}{\rho_P}\left(\frac{\rho_L - \rho_P}{\rho_L - \rho_F}\right)$$

   where ρ_F is the density of body fat, ρ_L is the density of the lean muscle, and ρ_P is the average density of a person.
   (d) What approximations did you have to make to derive the above result?
   (e) What is x_F for the given data?

67. A rotating garden sprinkler consists of a pipe of diameter 2.0 cm that is open and slightly bent at both ends so that the water rushes out from the ends. The pipe is free to rotate around its centre (Figure 12-59). The flow rate from each opening of the pipe is 2.0 L/s. What is the net torque generated by the sprinkler?

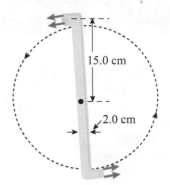

Figure 12-59 Problem 67

68. A wooden cube is floating at rest in a glass of water with a certain fraction of the cube's volume immersed. In this equilibrium position, the upward buoyant force exerted on the cube by the water is balanced by the weight of the cube's weight. The cube is now gently pressed farther into the water, held stationary, and then released. It bobs up and down before coming to rest in its original configuration.
   (a) Show that when the cube is gently pressed into the water and released, the net force exerted on the cube is proportional to the height of the cube immersed in the water.
   (b) Show, ignoring the damping effect of the water, that the resulting motion of the cube is simple harmonic motion.
   (c) Find the period of oscillation of the cube in terms of the height of the cube pressed down and the acceleration due to gravity.

69. A container of water has a block of wood floating in it (Figure 12-60). The block is half immersed in the water. The container is now placed in an elevator and accelerated upward with a constant acceleration. The fraction of the block that is immersed in the water
   (a) remains the same as before;
   (b) increases (i.e., more than half is immersed);
   (c) decreases (i.e., less than half is immersed).
   (d) There is not enough information to answer this question.

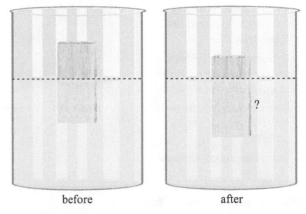

Figure 12-60 Problem 69

 See the text online resources at www.physics1e.nelson.com for Open Problems and Data-Rich Problems related to this chapter.

Learning Objectives

When you have completed this chapter you should be able to:

1 Interpret oscillations as periodic motion of a single particle.

2 Describe simple harmonic motion mathematically, and calculate displacement, velocity, and acceleration of a simple harmonic oscillator.

3 Describe uniform circular motion, and compare it to simple harmonic motion.

4 Explain what a restoring force is and why it causes a system to undergo simple harmonic motion.

5 Describe the relationship between the spring constant, k, and the natural frequency of a mass–spring oscillator.

6 Calculate the frequency and the period for simple systems that undergo simple harmonic motion.

7 Apply conservation of energy to calculate potential and kinetic energy in simple harmonic motion.

8 Describe the relationship between the length of a pendulum and its period.

9 Describe a physical pendulum, and calculate the period for a physical pendulum.

10 Interpret displacement versus time and velocity versus time graphs for simple harmonic motion.

11 Compare ideal simple harmonic motion to damped and driven oscillations.

12 Qualitatively explain resonance.

For additional Making Connections, Examples, and Checkpoints, as well as activities and experiments to help increase your understanding of the chapter's concepts, please go to the text's online resources at www.physics1e.nelson.com.

Chapter 13
Oscillations

Located between New Brunswick and Nova Scotia, the Bay of Fundy is home to some of the highest tides in the world. The average tide height worldwide is about 1 m, and Bay of Fundy tides rise as high as 16 m.

The high tides at the Bay of Fundy result from the shape of the bay (Figure 13-1). The mouth of the bay is 100 km wide and between 120 m and 215 m deep. At its end near Hopewell Rocks, the bay is about 2.5 km wide and 14 m deep, at low tide. This shape funnels the incoming tidal water to create spectacular high tides at the end of the bay. The time it takes for the tidal water to travel the 290 km length of the bay is approximately equal to the time it takes for a new tide to come from the ocean. As a result, the previous tide reinforces the incoming tide in a phenomenon called tidal resonance.

The rise and fall of the water height at inlets like the Bay of Fundy follow a pattern called periodic motion. In this chapter, we will examine the nature of this type of motion.

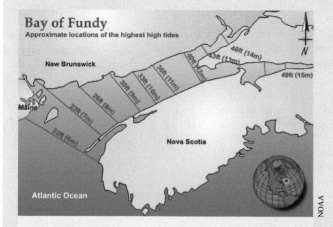

Figure 13-1 The width of the Bay of Fundy at various locations

13-1 Periodic Motion

A motion that repeats after a finite amount of time is called **periodic motion**, or **harmonic motion**. The amount of time after which the motion repeats is called the **period** of the motion, and is denoted by the symbol T. Thus, if a motion is repeated after 20.0 s, then $T = 20.0$ s.

From very small to extremely large scales, examples of periodic motions are all around us. The motion of an electron around an atom, of atoms in molecules, of a mass attached to a spring, of Earth around the Sun, and of the Sun around the centre of Milk Way are all examples of periodic motion. Although the forces underlying these periodic motions may be different, the mathematical equations describing them are similar. The mathematical descriptions developed in this chapter apply to many types of periodic motion. In later chapters, we will use the descriptions to understand sound, electromagnetic waves, and atomic structure.

We will investigate periodic motion in one dimension. Although motion in one dimension can be discussed using vector notation, the underlying physics is more transparent by avoiding vector notation and distinguishing motion in the positive and negative directions by using plus and minus signs. We will use vector notation where necessary.

The statement that an object repeats its motion periodically means that both the displacement and the velocity of the object will be the same at the beginning and at the end of any interval of T seconds. Let $x(t)$ denote the position of the object, as measured from an equilibrium point, and $v(t)$ its velocity. Then, for any integer n,

$$x(t + nT) = x(t) \tag{13-1}$$

$$v(t + nT) = v(t) \tag{13-2}$$

One complete back-and-forth motion is called an **oscillation**. The number of oscillations completed in one second is called the **frequency**, f, of the periodic motion. The frequency is related to the period of the motion:

$$\text{frequency} = \frac{1}{\text{period}}$$

KEY EQUATION
$$f = \frac{1}{T} \tag{13-3}$$

The SI unit for frequency is the hertz (Hz):

1 Hz = 1 s⁻¹ = one oscillation per second

For periodic motion, a plot of the displacement $x(t)$ versus time t shows a curve whose shape repeats after every T seconds. Thus, the period and the frequency of a periodic motion can be determined if we are given a graph of its displacement versus time. Two examples of periodic

EXAMPLE 13-1

Periodic Motion of Earth

Find the period and frequency of the motion of Earth around the Sun.

SOLUTION

There are approximately 365.25 days in a year. Therefore,

$T = 365.25$ days = 365.25 days $\times$ 24 h/day $\times$ 3600 s/h

$\quad = 3.1557 \times 10^7$ s

$f = 1/T = 1/(3.1557 \times 10^7 \text{ s}) = 3.1688 \times 10^{-8}$ Hz

motion are shown in Figures 13-2 and 13-3: Figure 13-2 shows an electrocardiogram (ECG) as a function of time, and Figure 13-3 shows a graph of the displacement of a mass that is attached to a vertical spring and is oscillating about an equilibrium position (the other end of the spring is attached to a rigid support). The graph in Figure 13-3 ignores frictional forces.

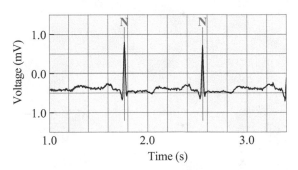

Figure 13-2 An ECG. A plot of voltage as a function of time

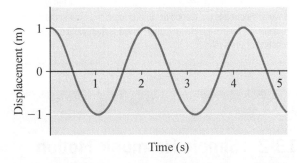

Figure 13-3 A plot of the displacement of a mass, attached to a spring, as a function of time.

Notice two things from these plots. First, both motions are periodic, so we can determine their periods from the graphs. Second, the ECG graph is more complicated than the mass–spring graph. It would be difficult to write an equation that describes the shape of the ECG graph as a function of time, whereas the displacement of the mass–spring system can be described by a cosine or a sine function. In this chapter, we consider a particular type of periodic motion that can be described by a sine or a cosine function, called simple harmonic motion.

What Is an Electrocardiogram?

The heart generates rhythmic electrical signals that cause the heart's muscle fibres to contract, pumping blood through the body. An electrocardiogram (ECG or EKG) is a plot of these electric signals (voltage versus time) made while the heart is beating (Figure 13-4). Many heart problems can be diagnosed by comparing the shape of a patient's electric pulses on an ECG to an ECG of a healthy heart.

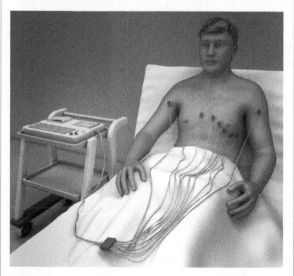

Figure 13-4 An electrocardiogram test being administered. Several electrodes positioned at various locations around the heart monitor the heart's electrical signal, which is then plotted as a function of time.

LO 2

13-2 Simple Harmonic Motion

When the displacement of an oscillating object from an equilibrium position varies with time as a cosine (or a sine) function, the periodic motion is called **simple harmonic motion**. Therefore, for simple harmonic motion,

KEY EQUATION
$$x(t) = A \cos(\omega t + \phi) \tag{13-4}$$

Note that we have chosen to describe the simple harmonic motion by a cosine function. A description using a sine function is equivalent since the two functions

are related by $\cos\left(\theta - \dfrac{\pi}{2}\right) = \sin(\theta)$. The cosine and the sine functions are periodic and have a period of 2π rad. An object that is undergoing simple harmonic motion is called a **simple harmonic oscillator**. The argument of a cosine function is a dimensionless quantity and is measured in radians. The quantities in Equation (13-4) are defined as follows.

A is called the **amplitude**. It is a positive number and describes the maximum displacement of the oscillator from its equilibrium position. As the object oscillates, its displacement from the equilibrium position continuously varies between the values $-A$ and $+A$ because the cosine function oscillates between the values -1 and $+1$. The SI unit for amplitude is the metre (m).

ω is called the **angular frequency**. Angular frequency is related to the frequency, f, of the simple harmonic motion as follows:

$$\omega = 2\pi f = \frac{2\pi}{T} \tag{13-5}$$

Angular frequency is measured in radians per second (rad/s); since radians are dimensionless, we can simply write per second (s^{-1}).

ϕ is called the **phase constant**. The phase constant is an angle, measured in radians, and its value is determined by the position and the velocity of the motion at a given time, usually taken to be $t = 0$.

We will now explore the role of the amplitude, angular frequency, and the phase constant in describing simple harmonic motion. Figure 13-5 shows the displacements of two oscillators described by $x_1(t) = (1.0 \text{ m})\cos t$ and $x_2(t) = (2.0 \text{ m})\cos t$. For these motions, we have chosen $\omega = 1$ rad/s and $\phi = 0$ rad. Comparing with Equation (13-4), the first oscillator has amplitude $A = 1.0$ m, and the second oscillator has amplitude $A = 2.0$ m. The displacement of the first oscillates between the value $x = +1.0$ m and $x = -1.0$ m, and the displacement of the second oscillates between $x = +2.0$ m and $x = -2.0$ m. Both oscillators reach the maximum displacement, pass through

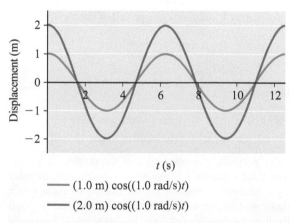

t (s)
—— (1.0 m) cos((1.0 rad/s)t)
—— (2.0 m) cos((1.0 rad/s)t)

Figure 13-5 Plots of (1.0 m) cos$t$ and (2.0 m) cos$t$ as a function of time.

the equilibrium position, and then reach the maximum displacement at the opposite side of the equilibrium position at the same time.

For a given simple harmonic motion, the amplitude is a *constant* and does not change with time. We will see later that the amplitude of a simple harmonic oscillator is related to the energy of the oscillator: a particle with greater amplitude has more energy.

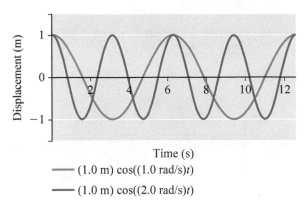

Figure 13-6 Plots of (1.0 m) cos$t$ and (1.0 m) cos$2t$ as a function of time

motions, with $\omega = 1.0$ rad/s and $\omega = 2.0$ rad/s, $A = 1.0$ m, and $\phi = 0$ rad for both. For $\omega = 1.0$ rad/s, $T = 6.28$ s; for $\omega = 2.0$ rad/s, $T = 3.14$ s. The values of the periods can be read from the graphs.

 CHECKPOINT

C-13-1 Amplitude of a Simple Harmonic Oscillator

Can the amplitude of a simple harmonic motion be negative?

C-13-1 No.

Now, we derive the relationship (13-5) between the angular frequency, ω, and the frequency, f. By the definition of the period,

$$x(t + T) = x(t)$$

Inserting Equation (13-4) in the above relationship, we get

$$A \cos(\omega(t + T) + \phi) = A \cos(\omega t + \phi) \qquad (13\text{-}6)$$

$$\cos(\omega t + \omega T + \phi) = \cos(\omega t + \phi) \qquad (13\text{-}7)$$

The cosine function repeats after 2π rad, so Equation (13-7) can only be satisfied when

$$\omega T = 2\pi \text{ rad} \qquad (13\text{-}8)$$

This condition implies that

$$\omega = \frac{2\pi}{T} = 2\pi f \text{ rad/s} \qquad (13\text{-}9)$$

The angular frequency is 2π times the frequency. By adjusting ω, we can describe fast and slow oscillations. Figure 13-6 shows graphs of two simple harmonic

CHECKPOINT

C-13-2 Period of a Simple Harmonic Oscillator

What is the period of a simple harmonic motion described by $x(t) = (3.5 \text{ m})\cos(2\pi t)$? Is the period different at $t = 2.0$ s?

C-13-2 1 s. No, the period is not different at $t = 2.0$ s because the period does not change with time.

We now examine the meaning of the phase constant, ϕ. Setting $t = 0$ in Equation (13-4) we get

$$x(0) = A \cos(\phi)$$

Therefore, the phase constant is related to the position of the oscillatory motion at $t = 0$. Different values of ϕ describe different possible initial locations of the oscillator, as shown in the following example.

EXAMPLE 13-2

Calculating the Phase Constant

A simple harmonic oscillator has $T = 2$ s and $A = 0.5$ m. Calculate the values of the phase constant, ϕ, for the following starting ($t = 0$) positions of the oscillatory motion:

(a) $x(0) = 0.5$ m;
(b) $x(0) = -0.5$ m;
(c) $x(0) = 0.3$ m;
(d) $x(0) = 0.0$ m.

SOLUTION

For the given values, $T = 2$ s, $\omega = \pi$ rad/s, and $A = 0.5$ m, the equation for simple harmonic motion is given by

$$x(t) = (0.5 \text{ m}) \cos(\pi t + \phi)$$

At $t = 0$,

$$x(0) = (0.5 \text{ m}) \cos \phi$$

For each starting position, the corresponding value of the phase constant can be determined by setting the above equation to the given value.

(a) $x(0) = 0.5$ m, then

$$0.5 \text{ m} = (0.5 \text{ m}) \cos \phi$$

$$\cos \phi = 1 \Rightarrow \phi = \cos^{-1}(1) = 0 \text{ rad or } 2\pi \text{ rad}$$

The oscillatory motion of the particle is described by the equation

$$x(t) = (0.5 \text{ m}) \cos(\pi t)$$

The displacement versus time graph for the above equation is shown in Figure 13-7.

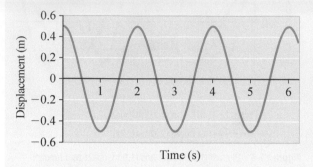

Figure 13-7 A plot of $(0.5 \text{ m}) \cos(\pi t)$ as a function of time

(b)
$$x(0) = -0.5 \text{ m}$$
$$-0.5 \text{ m} = (0.5 \text{ m})\cos\phi$$
$$\cos\phi = -1 \Rightarrow \phi = \cos^{-1}(-1) = \pi \text{ rad}$$

The oscillatory motion of the particle is described by the equation

$$x(t) = (0.5 \text{ m}) \cos(\pi t + \pi) \qquad (13\text{-}10)$$

A graph of this equation is shown in Figure 13-8. Note that the displacements described by (a) and (b) are equal and opposite for all times.

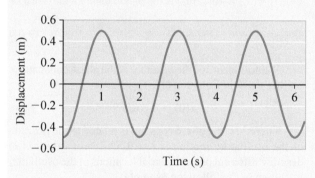

Figure 13-8 A plot of $(0.5 \text{ m}) \cos(\pi t + \pi)$ as a function of time

(c)
$$x(0) = +0.3 \text{ m}$$
$$+0.3 \text{ m} = (0.5 \text{ m})\cos\phi$$
$$\cos\phi = \frac{0.3}{0.5} = 0.6$$
$$\phi = \cos^{-1}(0.6) = 0.93 \text{ rad or } (2\pi - 0.93) \text{ rad}$$

Since $\cos\theta = \cos(2\pi - \theta)$ therefore, there are two possible values for the phase constant. To distinguish between the two values, additional information is needed, for example, the oscillator's velocity at $t = 0$. In general, a unique determination of the phase constant requires information about the oscillator's position and velocity at $t = 0$.

(d)
$$x(0) = 0.0 \text{ m}$$
$$0.0 \text{ m} = (0.5 \text{ m})\cos\phi$$
$$\cos\phi = \frac{0.0}{0.5} = 0$$
$$\phi = \cos^{-1}(0) = \frac{\pi}{2} \text{ or } \left(2\pi - \frac{\pi}{2}\right) = \frac{3\pi}{2} \text{ rad}$$

Again, there are two possible values for ϕ, and we need additional information to determine a unique value. Figure 13-9 shows a plot of $x(t)$ versus t for $\phi = \pi/2$ and $\phi = 3\pi/2$.

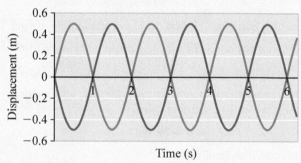

Figure 13-9 A plot of $x(t) = (0.5 \text{ m}) \cos(\pi t + \phi)$ for $\phi = \pi/2$ (blue) and $\phi = 3\pi/2$ (red) as a function of time

---

✓ **CHECKPOINT**

C-13-3 Displacement Versus Time Graph

What are the amplitude (A) and angular frequency (ω) of the displacement versus time graph shown in Figure 13-10?

(a) $A = 8$ m, $\omega = 4$ rad/s
(b) $A = 4$ m, $\omega = 2\pi/3$ rad/s
(c) $A = 0$ m, $\omega = 2\pi$ rad/s
(d) $A = -4$ m, $\omega = 2\pi/3$ rad/s

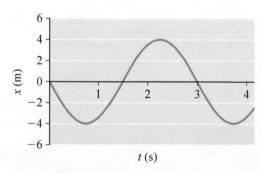

C-13-3 (b).

Figure 13-10 C-13-3

The argument of the cosine function, $\omega t + \phi$, is called the *phase* of the simple harmonic motion. Do not confuse the *phase constant* ϕ, which does not change with time, and the phase, $\omega t + \phi$, which does change with time.

The Velocity of a Simple Harmonic Oscillator

The velocity $v(t)$ of a simple harmonic oscillator is obtained by taking the derivative of the displacement $x(t)$ with respect to time, t:

$$v(t) = \frac{dx(t)}{dt} = \frac{d}{dt}[A \cos(\omega t + \phi)]$$

$$= A \frac{d}{dt}[\cos(\omega t + \phi)] = A[-\omega \sin(\omega t + \phi)]$$

$$= -(\omega A) \sin(\omega t + \phi) \quad (13\text{-}11)$$

KEY EQUATION $v(t) = -(\omega A) \sin(\omega t + \phi) \quad (13\text{-}12)$

The sine function oscillates between -1 and $+1$, so the velocity of the oscillator varies between the values $+\omega A$ and $-\omega A$; its maximum value (v_{max}) is ωA.

$$v_{max} = \omega A \quad (13\text{-}13)$$

In a displacement versus time graph, the direction of the velocity is given by the slope of the graph at that time.

The Phase Difference between an Oscillator's Velocity and Position The velocity of a simple harmonic oscillator varies with time as a sine function and the displacement varies as a cosine function. Therefore the position and velocity of the oscillator are $\pi/2$ rad (90°) out of phase. We can derive this result by using the trigonometric identity.

$$-\sin\theta = \cos\left(\theta + \frac{\pi}{2}\right)$$

Equation (13-12) for $v(t)$ can now be written as follows:

$$v(t) = \omega A \cos\left(\omega t + \phi + \frac{\pi}{2}\right) \quad (13\text{-}14)$$

Both $x(t)$ and $v(t)$ are now written in terms of the cosine function, so the phase difference between $x(t)$ and $v(t)$ is

phase of $v(t)$ − phase of $x(t)$

$$= \left(\omega t + \phi + \frac{\pi}{2}\right) - (\omega t + \phi) = \frac{\pi}{2} \text{ rad} \quad (13\text{-}15)$$

Graphs of $x(t)$ and $v(t)$ for an oscillator with $A = 1.0$ m, $\phi = 0$ rad, and $\omega = 2.0$ rad/s are shown in Figure 13-11.

Note that when an oscillator's displacement from equilibrium is maximum ($x = \pm A$), its velocity is zero, and when an oscillator passes through the equilibrium position ($x = 0$), its velocity is maximum ($v = \pm \omega A$). We will see later that this relationship between the position and the velocity is a result of conservation of the oscillator's energy.

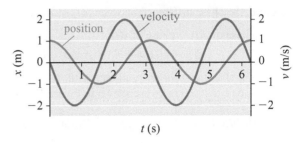

Figure 13-11   Position (blue) and velocity (red) plots for a harmonic oscillator with $A = 1.0$ m, $\phi = 0$ rad, and $\omega = 2.0$ rad/s.

The Acceleration of a Simple Harmonic Oscillator

The acceleration $a(t)$ of a simple harmonic oscillator is obtained by taking the derivative of the velocity, $v(t)$, with respect to time, t:

$$a(t) = \frac{dv(t)}{dt} = \frac{d}{dt}[-\omega A \sin(\omega t + \phi)]$$

$$= -\omega A \frac{d}{dt}[\sin(\omega t + \phi)] = -\omega A[\omega \cos(\omega t + \phi)]$$

$$= -(\omega^2 A) \cos(\omega t + \phi) \quad (13\text{-}16)$$

KEY EQUATION $a(t) = -(\omega^2 A) \cos(\omega t + \phi) \quad (13\text{-}17)$

The cosine function oscillates between -1 and $+1$, so the acceleration of the oscillator varies between the values $-\omega^2 A$ and $+\omega^2 A$; its maximum value (a_{max}) is $\omega^2 A$.

Using Equation (13-17), we can write the acceleration of a simple harmonic oscillator in terms of its displacement from equilibrium as

KEY EQUATION
$$a(t) = -\omega^2(A \cos(\omega t + \phi))$$
$$= -\omega^2 x(t) \quad (13\text{-}18)$$

This relationship between the acceleration and the displacement is a fundamental property of all simple harmonic oscillators: *For a motion to be simple harmonic motion, the acceleration of the object must be proportional to its displacement from the equilibrium position and opposite in direction, at all times.*

The constant of proportionality between the acceleration and the displacement of a simple harmonic oscillator is equal to the square of its angular frequency.

The Phase Difference Between the Acceleration and Position of an Oscillator

The minus sign in Equation (13-17) means that the acceleration and the displacement are π rad (180°) out of phase. Using the trigonometric identity

$$-\cos(\theta) = \cos(\theta + \pi)$$

we can write

$$a(t) = \omega^2 A \cos(\omega t + \phi + \pi) \qquad (13\text{-}19)$$

Therefore, the phase difference between $a(t)$ and $x(t)$ is

$$\text{phase of } a(t) - \text{phase of } x(t)$$
$$= (\omega t + \phi + \pi) - (\omega t + \phi) = \pi \text{ rad} \qquad (13\text{-}20)$$

Graphs of $x(t)$ and $a(t)$ for an oscillator with $A = 1.0$ m, $\phi = 0$ rad, and $\omega = 2.0$ rad/s are shown in Figure 13-14.

The Restoring Force and Simple Harmonic Motion

As an object executes simple harmonic motion, its acceleration changes with time, so there must be a time-dependent force acting on the object. What is the nature of this force? Using equation (13-18) for the acceleration of a simple harmonic oscillator, a mass m that executes simple harmonic motion experiences a force,

$$F(t) = ma(t) = -m\omega^2 x(t) \qquad (13\text{-}21)$$

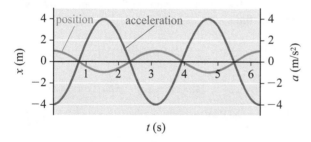

Figure 13-12 Position (blue) and acceleration (red) plots for a harmonic oscillator with $A = 1.0$ m, $\phi = 0$ rad, and $\omega = 2.0$ rad/s.

Since both the mass, m, and the angular frequency, ω, of the motion are fixed, the quantity $m\omega^2$ is constant for a given oscillator. Therefore,

$$F(t) \propto -x(t) \qquad (13\text{-}22)$$

Figure 13-13 shows the relationship between the direction of the force on the oscillator and its displacement from equilibrium. When the displacement is positive (x along the positive x-axis), the force is negative, pointing toward the origin. When the displacement is negative, the force is positive and again points toward the origin. The relationship described by Equation (13-22) is called **Hooke's law**. Elastic objects (e.g., springs) obey Hooke's law when they are deformed, as long as the deformation is small. A force that causes an object to return to its equilibrium position is called a **restoring force**. *Simple harmonic motion is motion by an object that is subjected to a restoring force.*

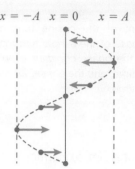

Figure 13-13 The direction of the restoring force for various displacements of a simple harmonic oscillator. The restoring force is always directed toward the equilibrium position.

To determine whether an object will undergo simple harmonic motion, we can use the following procedure:

1. Displace the object from the equilibrium position by a small amount, x.

2. Calculate the net force, F, on the object at the displaced position.

3. Determine whether F can be written as

$$F = -\text{constant} \times x \qquad (13\text{-}23)$$

If the displacement from the equilibrium position of the object is proportional to the net force exerted on the object but opposite in sign, the object will undergo simple harmonic motion. The angular frequency of oscillation is determined by comparing Equations (13-21) and (13-23):

$$\text{constant} = m\omega^2$$

$$\omega = \sqrt{\frac{\text{constant}}{m}} \qquad (13\text{-}24)$$

This procedure also applies for oscillations in two and three dimensions.

✓ CHECKPOINT

C-13-4 A Ball Bouncing on a Floor

Consider an ideal situation where a ball is bouncing elastically on a perfectly elastic floor without losing any energy. The motion of the ball is an example of

(a) periodic motion that is also simple harmonic motion;

(b) periodic motion that is not simple harmonic motion;

(c) simple harmonic motion that is not periodic motion;

(d) motion that is neither simple harmonic motion nor periodic motion;

Explain your reasoning.

C-13-4 (b)

13-3 Uniform Circular Motion and Simple Harmonic Motion

A point moving around a circle with uniform angular speed undergoes periodic motion. When this motion is projected onto the diameter of a circle, the point appears to be moving back and forth in periodic motion. As the point completes one revolution around the circle, its projection onto the diameter completes one oscillation. This analogy between the two motions provides a geometric meaning of the phase constant, ϕ, and the angular frequency, ω.

Consider a point P moving in a circle of radius R with an angular speed of ω rad/s. Let Q be the projection of P onto the x-axis. The time taken by P to complete one revolution (2π rad) is equal to $\dfrac{2\pi}{\omega}$ s, which is equal to the period of a simple harmonic oscillator of angular frequency ω rad/s. Therefore, the angular frequency of a simple harmonic oscillator is the same as the angular speed of a point moving around a circle.

Suppose at $t = 0$, P makes an angle ϕ with respect to the x-axis, as shown in Figure 13-14. A time t later, it undergoes an angular displacement of ωt and makes an angle of $(\omega t + \phi)$ with respect to the x-axis. The x-coordinate of P at time t is

$$x(t) = OQ = R\cos(\omega t + \phi) \qquad (13\text{-}25)$$

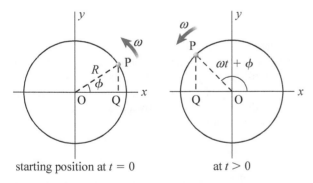

starting position at $t = 0$ at $t > 0$

Figure 13-14 A particle moving with an angular speed ω in a circle of radius R. At $t = 0$, the particle makes an angle ϕ with respect to the x-axis.

Equation (13-25) states that as P moves in a circle, its projection onto the x-axis undergoes simple harmonic motion. The phase constant, ϕ, for simple harmonic motion is therefore analogous to the starting angle of a circular motion.

Table 13-1 compares uniform circular motion and simple harmonic motion.

Table 13-1 Comparison Between Uniform Circular Motion and Simple Harmonic Motion

Uniform circular motion	Simple harmonic motion
Radius: R	Amplitude: A
Angular speed: ω	Angular frequency: ω
One complete revolution	One complete oscillation
Time taken for one complete revolution $= 2\pi/\omega$	Time period for one oscillation $= 2\pi/\omega$
Starting angle: ϕ	Phase constant: ϕ

✓ CHECKPOINT

C-13-5 Circular Motion and the Phase Constant

Starting positions ($t = 0$) for four particles moving around a circle of radius R are shown in Figure 13-15. Rank the starting positions in order of increasing phase constant of the corresponding simple harmonic motion.

Lowest 1. _____   2. _____   3. _____

4. _____ Highest

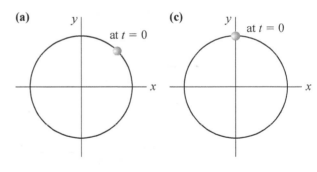

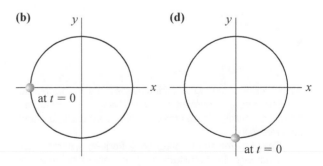

Figure 13-15 C-13-5. Starting positions for four particles moving around a circle of radius R

C-13-5 (a), (c), (b), (d)

351

13-4 Mass-Spring Systems

A Horizontal Mass-Spring System

Figure 13-16 shows a mass, m, attached to one end of a spring of unstretched length L_0 and resting on a horizontal, frictionless surface. The other end of the spring is attached to a rigid support. The mass is restricted to move along the horizontal direction, so gravity does not play any role in its motion. We choose a coordinate system with the origin located at the equilibrium position of the mass and the positive x-axis toward the right. Suppose a force is exerted on the mass (by pulling on the mass) to displace it toward the positive x-axis by an amount x that does not exceed the elastic limit of the spring. At the displaced position, the change in the length of the spring is x, and the spring exerts an elastic restoring force on the mass, trying to return the mass to the equilibrium position. The elastic restoring force of a spring follows Hooke's law:

$$F_{spring, x} = -kx \qquad (13\text{-}26)$$

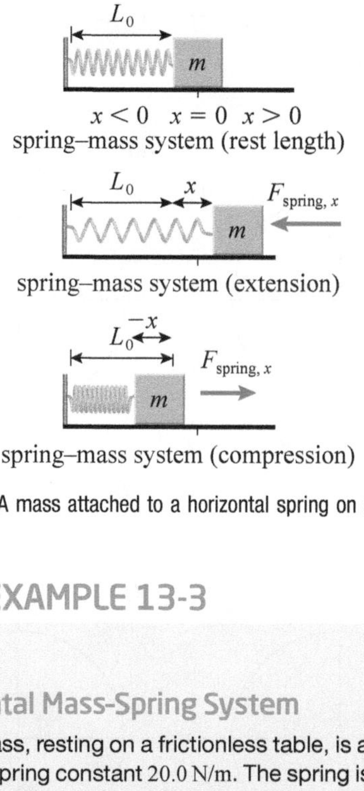

spring–mass system (rest length)

spring–mass system (extension)

spring–mass system (compression)

Figure 13-16 A mass attached to a horizontal spring on a frictionless surface

The minus sign in Equation (13-26) indicates that the direction of the force exerted by the spring on the mass is opposite to the direction of the displacement of the mass. As long as the mass is held stationary by the external force, the net force on the mass is zero:

$$F_{net, x} = F_{spring, x} + F_{external, x} = 0$$

When the external force is removed, the only force acting on the mass is the elastic restoring force of the spring. Applying Newton's second law of motion,

$$F_{net, x} = F_{spring, x} = -kx = ma_x \qquad (13\text{-}27)$$

$$a_x = -\left(\frac{k}{m}\right)x \qquad (13\text{-}28)$$

The acceleration of the mass is proportional to the displacement of the mass from the equilibrium position and opposite in sign. The mass–spring system will undergo simple harmonic motion, with the displacement of the mass given by the standard equation:

$$x(t) = A\cos(\omega t + \phi) \qquad (13\text{-}4)$$

The square of the angular frequency of the motion is obtained by comparing Equations (13-18) and (13-28):

$$\omega^2 = \frac{k}{m}$$

Therefore, for a horizontal mass–spring system,

KEY EQUATION
$$\omega = \sqrt{\frac{k}{m}} \qquad (13\text{-}29)$$

KEY EQUATION
$$T = 2\pi\sqrt{\frac{m}{k}} \qquad f = \frac{1}{2\pi}\sqrt{\frac{k}{m}} \qquad (13\text{-}30)$$

Equation (13-30) tells us that the frequency of oscillation depends on the stiffness of the spring (k) and the mass of the oscillating object. A light mass attached to a stiff spring has a large frequency and, hence, a small period. Note that the frequency of oscillation is independent of the amplitude of motion.

EXAMPLE 13-3

A Horizontal Mass-Spring System

A 2.00 kg mass, resting on a frictionless table, is attached to a spring of spring constant 20.0 N/m. The spring is stretched 20.0 cm by pulling on the mass, which is held stationary and then released.
(a) Calculate the angular frequency, frequency, and period.
(b) What are the amplitude and the phase constant of oscillations?
(c) What are the maximum speed and maximum acceleration of the mass?
(d) What is the displacement of the mass from equilibrium as a function of time?

SOLUTION

The mass–spring system forms a simple harmonic oscillator, and when the mass is displaced from its equilibrium, it will undergo simple harmonic motion.

We choose a coordinate system with the origin at the equilibrium position of the mass and the positive x-axis toward the right. The initial conditions for this problem are as follows:

$$x(0) = 0.200\text{ m} \quad \text{and} \quad v(0) = 0\text{ m/s}$$

(a) The angular frequency, frequency, and period of the system are as follows:

$$\omega = \sqrt{\frac{20.0 \text{ N/m}}{2.00 \text{ kg}}} = 3.16 \text{ rad/s}$$

$$f = \frac{\omega}{2\pi} = \frac{3.16 \text{ rad/s}}{2\pi \text{ rad}} = 0.503 \text{ Hz}$$

$$T = \frac{1}{f} = 1.99 \text{ s}$$

(b) As the system oscillates, the position and velocity of the mass at any time t are given by

$$x(t) = A \cos(\omega t + \phi)$$

$$v(t) = -\omega A \sin(\omega t + \phi)$$

To determine A and ϕ, we use the given initial position and velocity of the mass.
Evaluating $x(t)$ and $v(t)$ at $t = 0$, we obtain

$$x(0) = A \cos\phi \tag{1}$$

$$v(0) = -\omega A \sin\phi \tag{2}$$

Squaring and adding Equations (1) and (2) and using the trigonometric identity $\sin^2\phi + \cos^2\phi = 1$, we can determine A:

$$A = \sqrt{x^2(0) + \frac{v^2(0)}{\omega^2}} \tag{3}$$

To determine ϕ, divide Equation (2) by Equation (1):

$$\tan\phi = \frac{\sin\phi}{\cos\phi} = -\frac{v(0)}{\omega x(0)} \tag{4}$$

$$\phi = \tan^{-1}\left(-\frac{v(0)}{\omega x(0)}\right) \tag{5}$$

Equations (3) and (5) tell us that A and ϕ can be calculated when we know the position and the velocity of an oscillator at $t = 0$.

Inserting the values for $x(0)$ and $v(0)$ into Equations (3) and (5), we get

$$A = \sqrt{(0.200 \text{ m})^2 + \frac{(0 \text{ m/s})^2}{(3.16 \text{ rad/s})^2}} = 0.200 \text{ m}$$

$$\phi = \tan^{-1}\left(-\frac{0 \text{ m/s}}{(3.16 \text{ rad/s}) \times (0.20 \text{ m})}\right) = \tan^{-1}(0) = 0 \text{ rad}$$

(c) The maximum speed of the mass is

$$v_{\max} = \omega A = 3.16 \text{ rad/s} \times 0.200 \text{ m} = 0.632 \text{ m/s} \tag{6}$$

and its maximum acceleration is

$$a_{\max} = \omega^2 A = (3.16 \text{ rad/s})^2 \times 0.200 \text{ m} = 2.00 \text{ m/s}^2$$

(d) The displacement of the mass from the equilibrium position at any time t is given by

$$x(t) = (0.200 \text{ m}) \cos(3.16\,t) \tag{7}$$

Notice that a unique determination of the phase constant requires information about the oscillator's displacement and velocity at the start. A plot of the function $\tan^{-1}(x)$ is shown in Figure 13-17. For any x there is a single value between $-\pi/2$ and $\pi/2$.

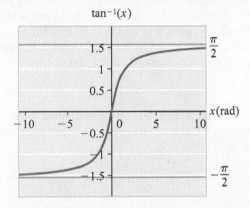

Figure 13-17 A plot of the function $\tan^{-1}(x)$ versus x, where x is a real number.

C-13-6 A Horizontal Mass-Spring System

The amplitude of a simple harmonic oscillator is A. In one period,
(a) What is the total distance travelled by the oscillator?
(b) What is the net displacement of the oscillator from the starting position?

C-13-6 (a) 4A; (b) 0

A Vertical Mass-Spring System

Consider a mass, m, suspended vertically from a spring of unstretched length L_0. The hanging mass experiences the elastic restoring forces of the spring and gravity. We take the positive y-axis vertically upward along the spring with $y = 0$ at the equilibrium position of the mass, as shown in Figure 13-18. At the equilibrium position, the spring stretches by the length ΔL such that the elastic restoring force on the mass acting upward balances the weight of the mass acting downward:

$$k\Delta L - mg = 0 \tag{13-31}$$

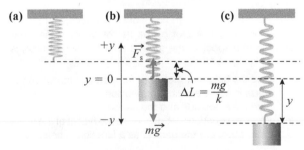

Figure 13-18 A mass attached to a vertically hanging spring (a) An unstretched spring (b) A mass, m, hanging free and motionless with the spring, with $y = 0$ taken as the equilibrium position of the mass. (c) The mass displaced from the equilibrium position by an external force.

Therefore, at the equilibrium position,

$$\Delta L = \frac{mg}{k} \qquad (13\text{-}32)$$

Suppose we displace the mass from the equilibrium position and release it. Would the mass–spring system undergo simple harmonic motion? Let the position of the displaced mass relative to the equilibrium position be y. Compared to the unstretched length L_0, the change in the length of the spring is $\Delta L - y$. Note that y is positive when the mass is above the equilibrium position, that is, when the spring has compressed. When y is negative, the spring has stretched. The net force on the displaced mass is the sum of the weight and elastic restoring force that could be acting either upward or downward:

$$F_{\text{net},y} = k(\Delta L - y) - mg = k\Delta L - ky - mg$$

Using Equation (13-31) to substitute for ΔL, we get

$$F_{\text{net},y} = -ky \qquad (13\text{-}33)$$

When displaced from the equilibrium position, the net force on the mass is proportional to its displacement and in a direction that is opposite to the displacement. The motion of the mass is therefore simple harmonic motion. Comparing Equation (13-33) to Equation (13-21), we see that

$$m\omega^2 = k$$

and that the angular frequency of oscillation is therefore the same as the angular frequency of a horizontal mass–spring oscillator:

$$\omega = \sqrt{\frac{k}{m}} \qquad (13\text{-}34)$$

$$T = 2\pi\sqrt{\frac{m}{k}} \qquad f = \frac{1}{2\pi}\sqrt{\frac{k}{m}} \qquad (13\text{-}35)$$

EXAMPLE 13-4

A Vertical Mass-Spring System

A mass attached to a vertical spring is undergoing simple harmonic motion. A plot of the height of the mass measured with respect to the ground as a function of time, is shown in Figure 13-19.

(a) Find the height of the mass from the ground when the mass is at the equilibrium position.
(b) Find the amplitude and the period.
(c) Write an equation that describes the height of the mass with respect to the ground as a function of time.

SOLUTION

(a) This situation is slightly different in that the height, $h(t)$, of the mass is measured with respect to the ground

The displacement of the mass from its equilibrium is given by the standard equation for a simple harmonic oscillator:

$$y(t) = A\cos(\omega t + \phi) \qquad (13\text{-}36)$$

Notice that by choosing to measure the change in the length of the spring with respect to the equilibrium position, we have removed the effect of gravity from this problem. While solving a problem, always think whether a particular choice of coordinate system will make the problem easier to solve.

 ## ONLINE ACTIVITY

The Spring Constant and the Period

The e-resource that accompanies every new copy of this textbook contains an Online Activity using the PhET simulation "Determining the spring constant and the period of a mass spring oscillator." Work through the simulation and accompanying questions to gain an understanding of the motion of mass spring oscillators.

 ## MAKING CONNECTIONS

A Vertical Mass-Spring System

Equation (13-32) provides a convenient way of measuring the spring constant. Attach one end of a spring to a rigid support, and hang a known mass to its free end. Let the freely hanging mass come to a rest, and measure the change in the length (ΔL) of the spring in this equilibrium position. The spring constant is equal to the weight of the mass divided by ΔL.

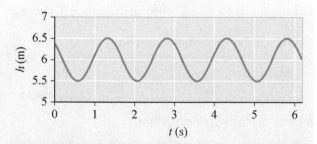

Figure 13-19 Position versus time plot of an oscillating vertical mass–spring system

and not with respect to the equilibrium position of the mass. From the graph, $h(t)$ varies between 6.5 m and 5.5 m. The equilibrium position of the mass is halfway between these two heights. Therefore, the height of the mass from the ground when the mass is in its equilibrium position is 6.0 m.

(b) From the plot, the period and the amplitude of the oscillations are

$$T = 1.5 \text{ s} \quad \text{and} \quad A = 0.50 \text{ m}$$

(c) To write a general equation for $h(t)$, we need to determine the phase constant, ϕ, from the graph. The equation for $h(t)$ can be written as follows:

$$h(t) = (0.50 \text{ m}) \cos\left(\frac{2\pi}{1.5}t + \phi\right) + 6.0 \text{ m} \qquad (1)$$

Note that we have added the equilibrium height of the mass to the standard equation for a simple harmonic oscillator. Equation (1) describes a **displaced harmonic oscillator**, where the origin of the coordinate system is not located at the equilibrium position of the oscillator.

To determine ϕ, we note from the graph that $h(0)$ is approximately 6.3 m. Inserting this value and $t = 0$ into the equation for $h(t)$ gives

$$h(0) = (0.50 \text{ m})\cos\phi + 6.0 \text{ m} = 6.3 \text{ m}$$

Therefore,

$$0.50 \cos\phi = 0.30$$

$$\cos\phi = 0.60$$

$$\phi = \cos^{-1}(0.60) \approx +0.93 \text{ rad}$$

$$\text{or } (2\pi - 0.93) \text{ rad}$$

The correct value of the phase constant can be determined from Figure 13-19. Choose a point, for example, at $t = 0.5$ s. From the plot, we note that $h(0.50) = 5.5$ m. Now, evaluate $h(0.50)$ from Equation (1) using both values of ϕ:

For $\phi = +0.93$ rad:

$$h(0.50) = (0.50 \text{ m}) \cos\left(\frac{2\pi}{1.5} \times 0.50 + 0.93\right)$$
$$+ 6.0 \text{ m} = 5.5 \text{ m}$$

For $\phi = (2\pi - 0.93)$ rad:

$$h(0.50) = (0.50 \text{ m}) \cos\left(\frac{2\pi}{1.5} \times 0.50 + 2\pi - 0.93\right)$$
$$+ 6.0 \text{ m} = 6.2 \text{ m}$$

Thus, $\phi = 0.93$ rad is the correct value for the phase constant because it reproduces the given value for $h(0.50)$. The general equation for $h(t)$ is

$$h(t) = (0.50 \text{ m}) \cos\left(\frac{2\pi}{1.5}t + 0.93\right) + 6.0 \text{ m}$$

LO 5

13-5 Energy Conservation in Simple Harmonic Motion

In the absence of frictional forces, the total energy (E) of an oscillating horizontal mass–spring system is the sum of the kinetic energy (K) of the mass and the elastic potential energy (U) of the spring (Figure 13-20).

Suppose that the mass is pulled from its equilibrium position to $x = A$ and held stationary. In doing so, work has been done to stretch the spring by a length A. This work is stored as the elastic potential energy of the spring. The total energy of the mass–spring system at $x = A$ is

$$E = K + U = 0 + \frac{1}{2}kA^2 \quad \text{(Total energy at } x = A) \quad (13\text{-}37)$$

where we have used the fact that the work done in stretching (or compressing) a spring by a length x is equal to $\frac{1}{2}kx^2$. When the mass is released, it accelerates toward the equilibrium position due to the elastic restoring force of the spring. Therefore, its velocity increases, and the length of the spring decreases. Let $x(t)$ be the position of the mass and $v(t)$ its velocity at some time t. At this instant, the kinetic energy of the mass is $\frac{1}{2}mv^2(t)$, and the elastic potential energy of the spring is $\frac{1}{2}kx^2(t)$. The total energy of the mass–spring system is

$$E = \frac{1}{2}mv^2(t) + \frac{1}{2}kx^2(t) \qquad (13\text{-}38)$$

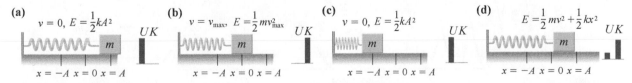

Figure 13-20 A mass–spring system oscillating in a straight line on a horizontal, frictionless surface (a) The spring is fully stretched, and the mass is at rest. (b) The spring is unstretched, and the mass is passing through the equilibrium position with maximum speed. (c) The spring is fully compressed, and the mass is at rest. (d) The spring is not fully stretched, and the mass has a nonzero speed.

Equating Equations (13-38) and (13-37), we get

$$\frac{1}{2}kA^2 = \frac{1}{2}mv^2(t) + \frac{1}{2}kx^2(t) \qquad (13\text{-}39)$$

At the equilibrium position, the spring is unstretched and the total energy of the system is stored as the kinetic energy of the mass. So the mass must move with maximum speed as it passes through the equilibrium position. Let v_m denote the maximum speed of the mass. Then, at equilibrium,

$$E = \frac{1}{2}mv_m^2 \quad \text{(Total energy at } x = 0) \qquad (13\text{-}40)$$

We can express v_m in terms of the maximum displacement A by using Equations (13-37) and (13-40):

$$E = \frac{1}{2}kA^2 = \frac{1}{2}mv_m^2$$

$$v_m = \sqrt{\frac{k}{m}}\,A = \omega A \qquad (13\text{-}41)$$

This expression for maximum speed is the same as Equation (13-13), which was derived previously. We see that the relationship between the amplitude and the maximum speed of a simple harmonic oscillator is a consequence of the conservation of energy.

Because of inertia, the mass overshoots the equilibrium position and continues to move toward the left, compressing the spring. The resulting restoring force slows the mass until it comes to rest at $x = -A$, where all the energy is again stored as elastic potential energy in the spring:

$$E = K + U = 0 + \frac{1}{2}k(-A)^2 = \frac{1}{2}kA^2$$

$$\text{(Total energy at } x = -A) \qquad (13\text{-}42)$$

This continuous transformation of energy between different modes (kinetic energy and elastic potential energy) is a fundamental property of all harmonic motions. In the absence of frictional forces, a harmonic oscillator, once set in motion, will oscillate forever. However, in reality, frictional forces are always present and oscillations eventually stop.

Figure 13-21 shows a plot of the potential, kinetic, and total energy of the mass–spring oscillator. For any

position between $-A$ and $+A$, the values of U and K can be read from the corresponding curves. Note that the sum of U and K is always equal to $\frac{1}{2}kA^2$. Using Equation (13-39), we can find the velocity of the mass for a given location x:

$$v = \pm\sqrt{\frac{k}{m}(A^2 - x^2)} \qquad (13\text{-}43)$$

Similarly, for a given value of velocity, the location of the mass is given by

$$x = \pm\sqrt{A^2 - \frac{m}{k}v^2} \qquad (13\text{-}44)$$

The variation of kinetic and potential energy with time can be determined by using

$$x(t) = A\cos(\omega t + \phi)$$

$$v(t) = -\omega A\sin(\omega t + \phi)$$

Then,

$$U = \frac{1}{2}kx^2(t) = \frac{1}{2}kA^2\cos^2(\omega t + \phi)$$

$$K = \frac{1}{2}mv^2(t) = \frac{1}{2}m\omega^2 A^2\sin^2(\omega t + \phi)$$

$$= \frac{1}{2}kA^2\sin^2(\omega t + \phi) \qquad (13\text{-}45)$$

Note that, individually, both U and K vary with time. However, the sum of the two is always equal to the total energy, E:

$$U + K = \frac{1}{2}kA^2\cos^2(\omega t + \phi) + \frac{1}{2}kA^2\sin^2(\omega t + \phi)$$

$$= \frac{1}{2}kA^2(\cos^2(\omega t + \phi) + \sin^2(\omega t + \phi))$$

$$= \frac{1}{2}kA^2 \qquad (13\text{-}46)$$

The total energy of a mass–spring oscillator is proportional to the square of the amplitude. Doubling the amplitude requires increasing the energy of the oscillator by a factor of 4.

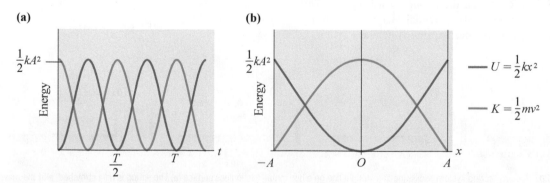

Figure 13-21 (a) A plot of kinetic energy (K) and elastic potential energy (U) of a simple harmonic oscillator as a function of time. (b) A plot of K and U as a function of displacement

EXAMPLE 13-5

Energy of a Mass-Spring System

A mass–spring oscillator consists of a 1.00 kg mass and a spring with a spring constant of 200 N/m. The oscillator is set in motion by stretching the spring 0.300 m and releasing the mass from rest.

(a) What is the total energy of the mass–spring system?
(b) What is the maximum speed of the mass as it oscillates?
(c) What is the speed of the mass when it is 0.100 m from the equilibrium position?
(d) For which displacements of the mass is the kinetic energy half its total energy?

SOLUTION

We choose a coordinate system with the equilibrium position at the origin and the positive x-axis toward the right. The mass is released from rest at $t = 0$. Therefore,

$$x(0) = 0.300 \text{ m} \quad \text{and} \quad v(0) = 0.00 \text{ m/s}$$

(a) The total energy (E) is the sum of the kinetic energy (K) of the mass and the elastic potential energy (U) of the spring. Using the values of displacement and velocity at $t = 0$,

$$K = \frac{1}{2}mv(0)^2 = 0.00 \text{ J} \quad \text{and}$$

$$U = \frac{1}{2}kx(0)^2 = \frac{1}{2} \times (200 \text{ N/m}) \times (0.300 \text{ m})^2 = 9.00 \text{ J}$$

Therefore, $E = K + U = 9.00$ J.

(b) When the mass is moving with its maximum speed, the total energy of the oscillator is in the form of the kinetic energy of the mass. From Equation (13-41):

$$v_{\text{m}} = \sqrt{\frac{2E}{m}} = \sqrt{\frac{2 \times 9.00 \text{ J}}{1.00 \text{ kg}}} = 4.24 \text{ m/s}$$

(c) The total energy of the mass–spring oscillator is given by Equation (13-38). Rearranging this equation to solve for v and then evaluating for $x = 0.100$ m, we get

$$E = \frac{1}{2}mv^2 + \frac{1}{2}kx^2$$

$$v = \sqrt{\frac{2E}{m} - \frac{k}{m}x^2}$$

$$v = \sqrt{\frac{2 \times 9.00 \text{ J}}{1.00 \text{ kg}} - \frac{200 \text{ N/m}}{1.00 \text{ kg}}(0.100 \text{ m})^2}$$

$$= 4.00 \text{ m/s}$$

(d) When the kinetic energy of the mass is 4.50 J, the potential energy of the spring must also equal 4.50 J, since the sum of the two must equal 9.00 J. Therefore,

$$4.50 \text{ J} = \frac{1}{2}kx^2$$

$$x = \pm\sqrt{\frac{2 \times 4.50 \text{ J}}{200 \text{ N/m}}} = \pm 0.212 \text{ m}$$

Making sense of the results:

The speed calculated in part (c) when the mass is not at the equilibrium position is less than the maximum speed of 4.24 m/s, and the displacement calculated in part (d) is less than the amplitude of oscillation.

 CHECKPOINT

C-13-7 Energy of a Mass-Spring System

A mass attached to a spring is undergoing simple harmonic motion with an amplitude of 20 cm. When the mass is 10 cm from the equilibrium position, what fraction of its total energy is the kinetic energy of the mass?

(a) 1/4
(b) 1/2
(c) 3/4
(d) 3/8

C-13-7 (c)

LO 6

13-6 The Simple Pendulum

A **simple pendulum** consists of a point mass attached to a string of negligible mass that does not stretch. The other end of the string is tied to a rigid and frictionless support. When displaced from the vertical equilibrium position and released, the pendulum swings back and forth in a plane under the influence of gravity, which acts as the restoring force. The motion of the pendulum is periodic. Is the motion simple harmonic?

Figure 13-22 shows a mass, m, hanging from a string of length L and making an angle θ with respect to the equilibrium position ($\theta = 0$). Angular displacement is taken to be positive to the right of the equilibrium position

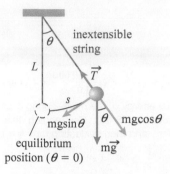

inextensible
string

L

$\overrightarrow{T}$

s

$mg\sin\theta$ θ $mg\cos\theta$

equilibrium
position ($\theta = 0$) $\overrightarrow{mg}$

Figure 13-22 A mass attached to an inextensible string of length L. The equilibrium position is the vertical position at which $\theta = 0$.

and negative to the left. The displacement, s, along the arc of the circle is related to the angular displacement as:

$$s = L\theta \qquad (13\text{-}47)$$

The forces acting on the mass are $\overrightarrow{mg}$, its weight, acting vertically downward, and $\overrightarrow{T}$, the tension in the string. To analyze the pendulum's motion, we choose a coordinate system with one axis, called the radial axis, along the length of the string. The other axis, called the tangential axis, is tangent to the circular motion of the mass. Note that the radial and the tangential axes are perpendicular to each other and their directions change as the mass oscillates. This choice of coordinate system simplifies the analysis because the mass does not move along the radial axis.

The Radial Axis The weight, $\overrightarrow{mg}$, makes an angle θ with the radial axis. Therefore, the component of the weight along the radial axis is $mg\cos\theta$, and it points away from the suspension point. The tension, $\overrightarrow{T}$, is directed toward the suspension point. There is no motion along the radial axis, so the radial components of the weight and the tension are equal. Therefore,

$$T - mg\cos\theta = 0$$

$$T = mg\cos\theta \qquad (13\text{-}48)$$

Note that Equation (13-48) gives the correct result for $\theta = 0$ rad ($T = mg$) and $\theta = \pi/2$ rad ($T = 0$)

The Tangential Axis The tangential component of the weight is $mg\sin\theta$, and it provides the restoring force that tends to pull the mass toward the equilibrium position. Newton's second law of motion for the tangential component is given by

$$F_{net,t} = -mg\sin\theta = ma_t \qquad (13\text{-}49)$$

where a_t denotes the tangential component of the acceleration. Since $\theta = s/L$,

$$a_t = -g\sin\theta = -g\sin\left(\frac{s}{L}\right) \qquad (13\text{-}50)$$

Note that the acceleration is not proportional to the displacement, s, but to $\sin(s/L)$. In general, pendulum motion is not simple harmonic motion. However, if the angle is small, we can use the small-angle approximation for the sine function:

$$\sin x = x \quad \text{(small-angle approximation)} \quad (13\text{-}51)$$

Figure 13-23 shows a plot of $\sin x$ versus x. We see that the small-angle approximation is reasonable for $x \cong 0.2$ rad (approximately 12°).

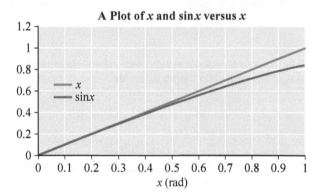

A Plot of x and $\sin x$ versus x

— x
— $\sin x$

x (rad)

Figure 13-23 A plot of $\sin x$ and x versus x (in rad)

Using the small-angle approximation,

$$\sin(s/L) = s/L \qquad (13\text{-}52)$$

Therefore,

$$a_t = -\left(\frac{g}{L}\right)s \quad \text{(small-angle approximation)} \quad (13\text{-}53)$$

For small angular displacements, the acceleration of the mass is proportional to the displacement and opposite in sign. Therefore, the motion of the pendulum is simple harmonic motion. The angular frequency is obtained by comparing Equation (13-53) with the standard equation for simple harmonic motion:

KEY EQUATION

$$\omega = \sqrt{\frac{g}{L}} \qquad (13\text{-}54)$$

The period and the frequency of oscillation of a pendulum are given by

$$T = \frac{2\pi}{\omega} = 2\pi\sqrt{\frac{L}{g}}$$

and $f = \dfrac{\omega}{2\pi} = \dfrac{1}{2\pi}\sqrt{\dfrac{g}{L}} \qquad (13\text{-}55)$

Notice that the period of a simple pendulum depends on the length of the pendulum and the acceleration due to gravity. The value of g varies slightly across Earth's surface and with height from sea level, so the period of a pendulum of fixed length will vary slightly with the value of g at a given location. Time can be measured fairly accurately with a simple stopwatch, so we can use a pendulum to measure the variation in the value of g across Earth's surface.

CHECKPOINT

C-13-8 Tangential Acceleration

What is the tangential acceleration of the above mass when $\theta = 0$ rad and $\theta = \pi/2$ rad? Is your answer sensible?

C-13-8 0, −g

ONLINE ACTIVITY

Exploring a Simple Pendulum

The e-resource that accompanies every new copy of this textbook contains an Online Activity using the PhET simulation "Exploring a Simple Pendulum." Work through the simulation and accompanying questions to gain an understanding of the motion of a simple pendulum.

EXAMPLE 13-6

Period and Length of a Simple Pendulum

You need to design a simple pendulum that has a period of 1.00 s. The acceleration due to gravity in your lab is 9.81 m/s². What should the length of the string be?

SOLUTION

The period of a simple pendulum is given by

$$T = 2\pi\sqrt{\frac{L}{g}}$$

Solving for L, we get

$$L = \frac{gT^2}{4\pi^2}$$

$$L = \frac{9.81 \text{ m/s}^2 \times (1.00 \text{ s})^2}{4 \times (3.14)^2} = 2.48 \times 10^{-2}\text{m}$$

CHECKPOINT

C-13-9 Decreasing a Pendulum's Length

A pendulum of length L is undergoing simple harmonic motion. When the length of the string is halved, the period of the pendulum
(a) increases by a factor of $\sqrt{2}$;
(b) increases by a factor of 2;
(c) decreases by a factor of $\sqrt{2}$;
(d) decreases by a factor of 2.

C-13-9 (c)

LO 7

13-7 The Physical Pendulum

An extended object that, when displaced from its equilibrium position, oscillates about an axis is called a **physical pendulum**. This includes objects that oscillate about a pivot point, such as a sphere attached to the end of a rod or a string. Swings and baseball bats can also be considered as physical pendulums.

Consider a sphere of uniform density, radius R, and mass M attached to a rod of length L and negligible mass. The system is free to rotate about an axis that is perpendicular to the plane of the paper and passes through the other end of the rod at the pivot point P (Figure 13-24). Since the rod is massless, the centre of mass, C, of the rod–sphere system is located at the centre of the sphere. In the equilibrium position, C is directly below the pivot point P, and there is no torque on the sphere due to its weight.

The sphere is then displaced by an angle θ. In the displaced position (Figure 13-24(b)), the gravitational force exerts a torque $\vec{\tau}$ on the sphere about the pivot point, P.

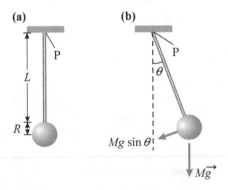

Figure 13-24 A sphere of mass M and radius R is attached to a massless rod of length L, which is hanging from a fixed, frictionless pivot point P. The rod is free to rotate about P. (a) The equilibrium position (b) The rod is displaced by an angle θ from the equilibrium position.

The direction of the torque is perpendicular to the plane of the paper, and the magnitude of the torque is given by

$$\tau = -(Mg)(L_{cm}) \sin\theta \qquad (13\text{-}56)$$

Here, $L_{cm} = L + R$ is the distance between the pivot point and the centre of mass of the rod–sphere pendulum. The minus sign indicates that the direction of the torque is opposite to the direction of the angular displacement, θ. Therefore, the torque tends to rotate the sphere toward the equilibrium position. According to Newton's second law of motion for rotational motion, the magnitude of the torque on the sphere is equal to the product of the angular acceleration, α, of the sphere and its moment of inertia, I_p, about the axis passing through the pivot point, P. Therefore,

$$I_p \alpha = -MgL_{cm}\sin\theta$$

$$\alpha = -\left(\frac{MgL_{cm}}{I_p}\right)\sin\theta \qquad (13\text{-}57)$$

By the parallel-axis theorem, the sphere's moment of inertia about the axis through P is

$$I_p = I_{cm} + M(L_{cm})^2$$

where $I_{cm} = \frac{2}{5}MR^2$ is the moment of inertia of the sphere about its centre of mass. When the angular displacement is small, we can use the small-angle approximation $\sin\theta = \theta$, in which case Equation (13-57) becomes

$$\alpha = -\left(\frac{MgL_{cm}}{I_p}\right)\theta \qquad (13\text{-}58)$$

This relationship between the angular acceleration, α, and the angular displacement, θ, is similar to the relationship between the acceleration, a, and displacement, x, of a simple harmonic oscillator. The angular acceleration is proportional to the angular displacement and opposite in sign. Therefore, for small angular displacement, the sphere will undergo simple harmonic motion about the pivot point, P. The angular frequency of oscillation is the square root of the constant of proportionality between α and θ:

KEY EQUATION
$$\omega = \sqrt{\frac{MgL_{cm}}{I_p}} \qquad (13\text{-}59)$$

and the period is

$$T = 2\pi\sqrt{\frac{I_p}{MgL_{cm}}} \qquad (13\text{-}60)$$

Inserting the expressions for I_p and L_{cm}, the period for a rod–sphere physical pendulum is

$$T = 2\pi\sqrt{\frac{\frac{2}{5}MR^2 + M(L+R)^2}{Mg(L+R)}}$$

Notice that for $R = 0$, the above expression reduces to $T = 2\pi\sqrt{\dfrac{L}{g}}$, the period for a simple pendulum. The angular displacement of the oscillating sphere is described by an equation analogous to that for linear oscillation:

$$\theta(t) = \theta_m\cos(\omega t + \phi) \qquad (13\text{-}61)$$

Here, θ_m is the maximum angular displacement of the sphere and is the analog of the amplitude A for linear oscillations.

The moment of inertia I_p in Equation (13-60) is calculated with respect to the pivot point, P, and a particular axis of rotation. If either or both of these change, I_p and hence ω and T will change. By comparing Equations (13-60) and (13-55), we notice that a physical pendulum and a simple pendulum have the same time period when

$$L = \frac{I_p}{ML_{cm}} \qquad (13\text{-}62)$$

EXAMPLE 13-7

A Cardboard Sheet as a Physical Pendulum

Consider a square sheet of cardboard with a mass of 20.0 g and a length of 28.0 cm. The sheet is free to rotate about a pivot point, P, that is located directly above the centre of mass of the sheet and 1.00 cm from the top edge. Calculate the period for small angular displacement about the pivot point P.

SOLUTION

A sheet of cardboard that freely rotates about a frictionless pivot point forms a physical pendulum. The period of the sheet is given by Equation (13-60). The sheet's moment of inertia about the point P is

$$I_p = I_{cm} + ML_{cm}^2$$

The moment of inertia of a square sheet of mass M and length L about its centre of mass is

$$I_{cm} = \frac{1}{12}ML^2$$

The distance L_{cm} is the distance from the centre of the mass of the sheet to the pivot point, so

$$L_{cm} = 14.0\ \text{cm} - 1.00\ \text{cm} = 13.0\ \text{cm}$$

Using the given values for the mass and the length of the sheet, we get

$$I_p = \frac{1}{12} \times (0.0200\ \text{kg}) \times (0.280\ \text{m})^2 + (0.0200\ \text{kg}) \times (0.130\ \text{m})^2$$

$$= 4.69 \times 10^{-4}\ \text{kg m}^2$$

Substituting into Equation (13-60) gives

$$T = 2\pi\sqrt{\frac{4.69 \times 10^{-4}\ \text{kg m}^2}{(0.0200\ \text{kg}) \times (9.81\ \text{m/s}^2) \times (0.130\ \text{m})}} = 0.852\ \text{s}$$

Walking Motion and the Physical Pendulum

The natural walking motion of animals can be modelled as the free swinging motion of legs pivoting about the hip joints, as shown in Figure 13-25. A leg can be approximated as a physical pendulum of length L and moment of inertia I. During the walking motion, one foot remains in contact with the ground, and the other foot swings through from behind ($\theta = -\theta_0$) to the front of the planted foot ($\theta = +\theta_0$). This motion constitutes one-half the period of the pendulum. For most animals, the maximum deflection angle, θ_0, is typically 20°.

(a)

(b)

Figure 13-25 (a) Free walking motion of a person (b) A schematic representation of free walking motion, where a physical pendulum of length L oscillates between $\theta = -\theta_0$ and $\theta = +\theta_0$.

The period of a physical pendulum of length L, mass M, and moment of inertia I is

$$T = 2\pi\sqrt{\frac{I}{MgL_{cm}}}$$

where L_{cm} is the distance from the centre of mass of the pendulum from the pivot point, and I is calculated about an axis that is perpendicular to the plane of rotation and passes through the pivot point. Let us consider a simple model, where a leg is approximated as a rod, and the mass of the leg is uniformly distributed along the length of the rod. This model is reasonable for animals with legs that do not taper much, for example, humans, elephants, and ostriches. The moment of inertia for a rod of mass M and length L that is free to rotate about an axis passing through an end point of the rod is $\frac{1}{3}ML^2$. The centre of mass of the rod is at its centre, so $L_{cm} = L/2$. Therefore, the period of the rod is

$$T = 2\pi\sqrt{\frac{ML^2/3}{Mg(L/2)}} = \sqrt{\frac{2}{3}} \times 2\pi\sqrt{\frac{L}{g}} \qquad (13\text{-}63)$$

The amplitude of oscillation of the rod is given by

$$A = L\sin\theta_0 \qquad (13\text{-}64)$$

where θ_0 is the maximum angle of deflection. The maximum walking speed is, therefore,

$$v_{walk} = \omega A = \frac{2\pi}{T} \times L\sin\theta_0 = \sqrt{\frac{3}{2}}\sin\theta_0 \times \sqrt{gL} \qquad (13\text{-}65)$$

Taking $L = 1.0$ m as the average length of an adult human leg and $\theta_0 = 20°$, we obtain $v_{walk} = 1.3$ m/s $\cong 4.7$ km/h, which is a reasonable result for the maximum walking speed of an adult.

Several interesting points emerge from this simple model:

- The walking speed is proportional to $\sqrt{g}$. On the Moon, $g = 1.6$ m/s². Using the numbers for leg length and displacement angle as above, we obtain the value $v_{walk} = 0.5$ m/s $\cong 1.8$ km/h, which is approximately 2.6 times slower than the walking speed on Earth. Have you ever watched a video of astronauts on the Moon and wondered why they were walking so slowly?

- The walking speed depends on the moment of inertia of the leg. Fast animals (e.g., horses and lions) have legs that are thick at the top and slender at the bottom. A greater fraction of the leg's mass for these animals is located closer to the hip joint, thus decreasing the moment of inertia of the leg about the hip joint. This distribution decreases the period of the leg's oscillations, resulting in quicker steps and a faster walk.

- As a leopard chases its prey, it hunches down and pounds hard on the ground with its feet (Figure 13-26). By hunching down, the leopard decreases its moment of inertia, and by pounding hard on the ground, the leopard increases its effective g as it accelerates upward. Both of these effects decrease the period of the legs, thus helping the leopard to run faster.

- The walking speed is proportional to the square root of the length of the leg. Animals with longer legs have a faster natural walking speed. The walking speed of a child with legs half the length of an adult is reduced by a factor of $\sqrt{1/2}$, or approximately 30% compared to that of the adult.

Figure 13-26 A leaping leopard

C-13-10 Walking Speed

Two animals have similarly shaped legs of the same length. One animal's legs are twice as thick as the other animal's legs. Which animal has the faster walking speed?

(a) the animal with thicker legs;

(b) the animal with thinner legs;

(c) They both have the same walking speed;

(d) This question cannot be answered without knowing the leg mass for each animal.

C-13-10 (b)

LO 8

13-8 Time Plots for Simple Harmonic Motion

The period, amplitude, and phase constant of a simple harmonic motion can also be determined from the graph of the displacement, the velocity, or the acceleration as a function of time. In this section, we show how this is done through two examples.

EXAMPLE 13-8

Determining an Equation for Simple Harmonic Motion from a Displacement Graph

Figure 13-27 shows a graph of displacement $x(t)$ versus time t for a particle. From this graph, determine an equation for $x(t)$.

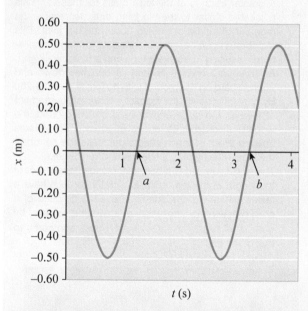

Figure 13-27 The displacement versus time graph for a particle

SOLUTION

We can see from Figure 13-27 that the graph is a sinusoidal function. Therefore, the motion is simple harmonic motion. We can also see that the maximum displacement of the particle is ±0.50 m. Therefore, the amplitude of the oscillation is 0.50 m.

We can determine the period by calculating the time taken to complete one oscillation. While any point on the curve can be taken as the starting point, using points where

the displacement is zero is often convenient. Using t_a and t_b as the start and end times of an oscillation, the period is

$$T = t_b - t_a = 3.2 \text{ s} - 1.2 \text{ s} = 2.0 \text{ s}$$

The angular frequency is

$$\omega = \frac{2\pi}{T} = \frac{2\pi \text{ rad}}{2.0 \text{ s}} = 3.1 \text{ rad/s}$$

To determine the phase constant, ϕ, we write the standard equation for $x(t)$ with $A = 0.5$ m and $\omega = 3.1$ rad/s:

$$x(t) = (0.50 \text{ m}) \cos(3.1t + \phi) \qquad (1)$$

From the graph, $x(0)$ is approximately 0.35 m. Substituting this value into Equation (1):

$$x(0) = 0.35 \text{ m} = (0.50 \text{ m})\cos\phi$$

$$\cos\phi = \frac{0.35 \text{ m}}{0.50 \text{ m}} = 0.70$$

$$\phi = \cos^{-1}(0.70) = 0.79 \text{ rad} \quad \text{or} \quad (2\pi - 0.79) \text{ rad}$$

To determine the correct sign of ϕ, we need another piece of information, such as the velocity of the particle at $t = 0$ or its position at a time near $t = 0$. Let us choose $t = 0.1$ s. From the graph, $x(0.1) = 0.2$ m, approximately. Calculating $x(0.1)$ from Equation (1) for $\phi = 0.79$ rad and $\phi = (2\pi - 0.79)$ rad:

$$\phi = +0.79 \text{ rad}$$

$$x(0.1) = (0.50 \text{ m}) \cos(3.1 \times 0.10 + 0.79) = 0.23 \text{ m}$$

$$\phi = (2\pi - 0.79) \text{ rad}$$

$$x(0.1) = (0.50 \text{ m}) \cos(3.1 \times 0.10 + 2\pi - 0.79) = 0.44 \text{ m}$$

Comparing these two values of $x(0.1)$ with the value from the graph, we see that +0.79 rad is the correct value for the phase constant. Thus, the equation for the displacement shown in Figure 13-29 is

$$x(t) = (0.50 \text{ m}) \cos(3.1t + 0.79) \qquad (2)$$

EXAMPLE 13-9

Determining an Equation for Simple Harmonic Motion from Position and Velocity Graphs

Position and velocity graphs of a simple harmonic oscillator are shown in Figure 13-28. Determine the equation of motion for the oscillator from these graphs.

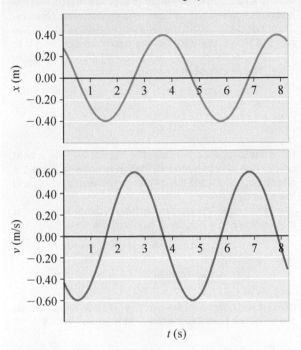

Figure 13-28 Position (blue) and velocity (red) plots for a harmonic oscillator

SOLUTION

From the position graph, the amplitude of the oscillations is $A = 0.40$ m, and the period is approximately $T = 4.2$ s, which gives the following value for angular frequency:

$$\omega = \frac{2\pi}{4.2 \text{ s}} = 1.5 \text{ rad/s}$$

The equations for $x(t)$ and $v(t)$ are as follows:

$$x(t) = (0.40 \text{ m}) \cos(1.5t + \phi)$$

$$v(t) = -(1.5 \times 0.40 \text{ m/s}) \sin(1.5t + \phi) \quad (1)$$

We can use Equation (1) together with the graphs to determine ϕ. From the plots, approximately, $x(0) = 0.28$ m and $v(0) = -0.42$ m/s. Evaluating $x(t)$ and $v(t)$ for $t = 0$, we get

$$x(0) = (0.40 \text{ m}) \cos\phi = 0.28 \text{ m}$$

$$v(0) = -(0.60 \text{ m/s}) \sin\phi = -0.42 \text{ m/s}$$

Thus

$$\cos\phi = 0.70$$
$$\sin\phi = 0.70$$

and

$$\tan\phi = \frac{\sin\phi}{\cos\phi} = 1.0 \Rightarrow \phi = \tan^{-1}(1.0) = 0.79 \text{ rad}$$

Thus, the equation of motion for the oscillator is

$$x(t) = (0.40 \text{ m}) \cos(1.5t + 0.79)$$

WAVES AND OSCILLATIONS

✓ CHECKPOINT

C-13-11 Determining the Direction of Motion from a Displacement Graph

A graph of the displacement $x(t)$ versus time for a simple harmonic motion is given in Figure 13-29. Which of the following statements is true for the velocity of the particle at $t = 1$ s?

(a) The velocity is positive at $t = 1$ s.
(b) The velocity is zero at $t = 1$ s.
(c) The velocity is negative at $t = 1$ s.
(d) A displacement versus time graph has no information about the velocity of the oscillator.

C-13-11 (c)

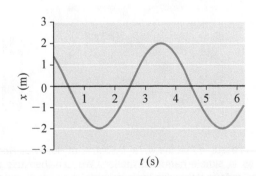

Figure 13-29 A displacement versus time plot for a simple harmonic motion

EXAMPLE 13-10

A Boat Stuck at Hopewell Rocks During Low Tide

At 09:00, you find that your sailboat is stuck in 0.3 m-deep water during low tide at Hopewell Rocks in the Bay of Fundy (Figure 13-30). You look at the tide table and verify that the next high tide is at 15:15, when the water height will be 13.1 m. Your boat needs a water depth of at least 3.0 m to float. Assuming that the tides cause the water height at Hopewell Rocks to change in harmonic motion, estimate the earliest time when your boat will be able to float. Use the tide data given in Table 13-2.

Courtesy of Michael Taylor

Figure 13-30 Hopewell Rocks, Bay of Fundy, Nova Scotia, at low tide

Table 13-2 Tides Table at Hopewell Rocks for April 9, 2008

April 9, 2008 (Wednesday)	
Time (ADT)	Height (m)
02:48	13.7
09:23	0.3
15:15	13.1
21:44	0.8

Fisheries and Oceans Canada, http://www.lau.chs-shc.gc.ca/english/Canada.shtml. Reproduced with the permission of the Canadian Hydrographic Service. Not to be used for navigation.

SOLUTION

Let us assume that the water height, h, at Hopewell Rocks varies in simple harmonic motion. We can then use the given data to determine the amplitude, the period, and the phase constant for this motion. From the equation for $h(t)$ we can determine the time when the height will be 3.0 m. The period of the tides is the time between two successive high tides or two successive low tides. Taking the time between two successive high tides, we get the following:

$$T = 15:15 - 02:48 = 12 \text{ h} \quad \text{and} \quad 27 \text{ min} = 44\,820 \text{ s}$$

Let us determine the amplitude of the tides. Note that there is a variation in the height of successive high and low tides—the generation of tides is a complicated phenomenon that depends on the relative positioning of the Sun and the Moon. We will take the maximum tide height to be 13.1 m, corresponding to 15:15 ADT (Atlantic Daylight Time). The amplitude of the tides is, therefore,

$$A = \frac{\text{water height at high tide } - \text{ water height at low tide}}{2}$$

$$= \frac{13.1 \text{ m} - 0.30 \text{ m}}{2} = 6.4 \text{ m}$$

Since the difference between the highest and the lowest positions in simple harmonic motion is equal to twice the amplitude, we divided the difference between the high and low tides by 2.

In this case, the equilibrium height, denoted by $h(0)$, is the average of the heights at high and low tides.

$$h_0 = \frac{\text{height at high tide + height at low tide}}{2}$$

$$= \frac{13.1 \text{ m} + 0.30 \text{ m}}{2} = 6.7 \text{ m}$$

Given these quantities, we can now write an equation that describes the height of the water at Hopewell Rocks:

$$h(t) = 6.7 \text{ m} + (6.4 \text{ m}) \cos\left(\frac{2\pi}{T} t + \phi\right) \quad (1)$$

The first term in Equation (1) is the equilibrium height, h_0. *The variation in the water height is measured with respect to h_0.* This term is absent when $x = 0$ is taken as the equilibrium point.

From the previous examples, we know that we can determine ϕ if we know $h(t)$ at a given time. For ease of calculating, we take the morning low tide time, 09:23, as the start time $t = 0$. At this time, the sea is 0.30 m high. Inserting this information in the above equation yields

$$h(0) = 6.7 \text{ m} + (6.4 \text{ m}) \cos\phi = 0.30 \text{ m}$$

Therefore,

$$\cos\phi = \frac{0.30 \text{ m} - 6.7 \text{ m}}{6.4 \text{ m}} = -1.0$$

$$\phi = \cos^{-1}(-1.0) = 3.1 \text{ rad}$$

Now we can write Equation (1) for the water height:

$$h(t) = 6.7 \text{ m} + (6.4 \text{ m}) \cos\left(\frac{2\pi}{T} t + 3.1\right) \quad (2)$$

The last task is to determine the time when the water height is 3.0 m. Setting $h(t) = 3.0$ m in Equation (2) and using the trigonometric identity $\cos(\theta + \pi) = -\cos(\theta)$, we get

$$3.0 \text{ m} = 6.7 \text{ m} + (6.4 \text{ m}) \cos\left(\frac{2\pi}{T} t + 3.1\right)$$

$$= 6.7 \text{ m} - (6.4 \text{ m}) \cos\left(\frac{2\pi}{T} t\right)$$

WAVES AND OSCILLATIONS

Solving for t gives

$$\frac{2\pi}{T} t = \cos^{-1}\left(\frac{6.7 \text{ m} - 3.0 \text{ m}}{6.4 \text{ m}}\right) = 0.95$$

$$t = \frac{T}{2\pi} \times 0.95 = \frac{44\,820 \text{ s} \times 0.95}{2\pi} \approx 6.8 \times 10^3 \text{ s} = 1 \text{ h } 53 \text{ min}$$

Since $t = 0$ was at 09:23, your boat should be ready to float again at about 11:16. Happy sailing, and don't leave your tide table at home!

13-9 Damped Oscillations

So far in our study of simple harmonic motion, we have ignored frictional forces. As a result, our model predicts that once a simple harmonic oscillator is set in motion, it will oscillate indefinitely. However, frictional forces are always present. These forces transform a fraction of an oscillator's kinetic energy into thermal energy, thus decreasing the amplitude of the oscillation. Eventually, all the energy of the oscillator is transformed into thermal energy, and it comes to rest.

Frictional forces can be quite complicated. For a vertical mass–spring oscillator in air, the frictional force between the mass and the surrounding air depends on the shape of the mass and the speed of oscillation. There is also friction at the contact points where the mass is attached to the spring and where the spring is connected to the rigid support. Furthermore, there are internal frictional forces between the molecules of the spring as it stretches and compresses. A spring that is continuously stretched and compressed gets hot. (Try stretching and compressing a metal spring.) Although an exact description of frictional forces is complicated, two features are common for frictional forces experienced by objects executing simple harmonic motion:

1. The magnitude of the frictional force is proportional to the speed of the object.

2. The direction of the frictional force is opposite to the direction of motion of the object.

Oscillations in the presence of frictional forces are called **damped oscillations**. In the study of fluids, such a frictional force is called a **drag force**, or **fluid resistance**. The simplest form of the drag force, $\vec{F}_D$, that an object experiences when moving with a velocity $\vec{v}$ in a medium can be written as

$$\vec{F}_D = -b\,\vec{v} \qquad (13\text{-}66)$$

The minus sign indicates that the direction of the drag force is opposite to the direction of the velocity of the object. The proportionality constant, b, is a measure of the strength of the drag force, and for oscillating objects it is called the **damping constant**, or **drag constant**. Its value is positive and depends on the properties of the medium in which the object is moving, as well as the shape of the object (Figure 13-31).

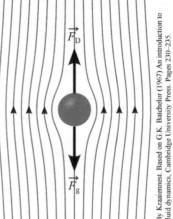

Figure 13-31 The drag force ($\vec{F}_D$) on a sphere falling through the air. Note that the direction of the drag force is opposite to the direction of the gravitational force, $\vec{F}_g$.

By Kraaiennest. Based on G.K. Batchelor (1967) An introduction to fluid dynamics, Cambridge University Press. Pages 230–235.

Since the dimensions on both sides of an equation must be same, we can use Equation (13-66) and dimensional analysis to find the dimension of b, denoted $[b]$:

$$\text{kg m/s}^2 = [b] \times \text{m/s}$$

$$[b] = \frac{\text{kg}}{\text{s}}$$

Equation (13-66) is only one of many forms for drag forces, and is valid only for special conditions, such as the air resistance on small, slowly moving objects. A fast-moving baseball experiences air resistance that is proportional to the square of the speed of the ball.

Let us analyze the motion of a mass–spring system, oscillating in the air, in the presence of the drag force of Equation (13-66). The net force, $\vec{F}_{net}$, on the oscillating mass is the sum of the elastic restoring force exerted by the spring and the drag force exerted by the air:

$$\vec{F}_{net} = \vec{F}_{spring} + \vec{F}_D \qquad (13\text{-}67)$$

Taking the x-axis along the direction of motion,

$$F_{net,x} = F_{spring,x} + F_{D,x} = -kx + (-bv_x) \qquad (13\text{-}68)$$

Using Newton's second law of motion, the acceleration of the mass is

$$ma_x = -kx - bv_x$$

$$a_x = -\frac{k}{m}x - \frac{b}{m}v_x \qquad (13\text{-}69)$$

Will the mass–spring system undergo simple harmonic motion in the presence of the damping force? Comparing Equation (13-69) with Equation (13-18), we see that in this case the acceleration has an additional term that depends on the velocity of the oscillator. So, in general, the motion is not simple harmonic motion. However, if the velocity-dependent term in Equation (13-69) is small compared to the displacement term, then the acceleration of the mass is "approximately" proportional to the displacement, and the motion will be nearly simple harmonic motion. Using calculus, it can be shown (see Online Section 13-1) that the damping force in Equation (13-69) modifies the motion of a simple harmonic oscillator as follows:

1. The amplitude of a damped oscillator decreases exponentially with time.
2. The oscillation frequency of a damped oscillator is lower than the oscillation frequency of an undamped oscillator.

A decrease with time in the amplitude of a damped oscillator is expected. The energy of a simple harmonic oscillator is proportional to the square of its amplitude. In the presence of a drag force, the oscillator loses its energy by doing work against the drag force. As its energy decreases, so does the amplitude of oscillation. The decrease in the oscillation frequency is due to the fact that as it loses energy, a damped oscillator slows down and therefore takes more time to complete a cycle.

As shown in Online Section 13-1, the displacement $x(t)$ of the damped oscillator described by Equation (13-69) is given by

KEY EQUATION $\quad x(t) = A e^{-bt/2m} \cos(\omega_D t + \phi) \qquad (13\text{-}70)$

where

$$\omega_D = \sqrt{\frac{k}{m} - \frac{b^2}{4m^2}} = \sqrt{\omega_0^2 - \left(\frac{b}{2m}\right)^2} \qquad (13\text{-}71)$$

and $\omega_0 = \sqrt{k/m}$

In Equation (13-71), ω_D is the oscillation frequency in the presence of the drag force, and ω_0 is the oscillation frequency when $b = 0$. The angular frequency ω_0 is called the **natural frequency** of the oscillation. The constant e is the base of the natural logarithm, and $e^{-bt/2m}$ denotes an exponential function that decreases with time. Note that $\omega_D < \omega_0$, so a damped oscillator oscillates at a lower frequency than a corresponding undamped oscillator. We can combine the amplitude A and the exponential term to define a time-dependent amplitude $A(t)$ for a damped oscillator and write Equation (13-70) in a form that is similar to the form for an undamped oscillator:

KEY EQUATION $\qquad A(t) = A e^{-bt/2m} \qquad (13\text{-}72)$

KEY EQUATION $\qquad x(t) = A(t) \cos(\omega_D t + \phi) \qquad (13\text{-}73)$

We now examine the motion of a damped oscillator over the range of possible values for the damping constant, b.

An Underdamped Oscillator $\left(\omega_0 > \dfrac{b}{2m}\right)$

In this case ω_D is positive, and the argument of the cosine function is a real number. The system oscillates with the frequency ω_D, and the oscillations are called **underdamped**. For example, when $k = 4.0$ N/m, $m = 0.20$ kg, and $b = 0.20$ kg/s,

$\omega_0 = 4.47$ rad/s, $\ b/2m = 0.50$ rad/s, and $\ \omega_D = 4.44$ rad/s

A plot of $x(t)$ for $A = 0.2$ m and $\phi = 0$ rad is shown in Figure 13-32.

For underdamped oscillations, the main effect of the drag force is to decrease the amplitude exponentially with time. The oscillation frequency, ω_D, is close to the natural frequency, ω_0. The dashed curve in Figure 13-32 is called the **envelope** of the damped oscillations. As the damping constant is increased, the amplitude decreases more rapidly with time, and the difference between ω_D and ω_0 increases. Figure 13-33 shows a plot of $x(t)$ with $b = 1.50$ kg/s, with other parameters kept the same as above. Increasing the value of b by a factor of 3 causes only a small decrease in ω_D (4.21 rad/s), but the effect on the amplitude is quite pronounced. The displacement of the oscillator is unnoticeable after two oscillations.

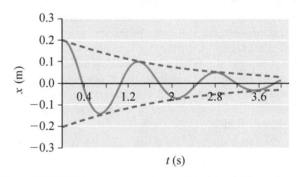

Figure 13-32 Displacement versus time graph for a damped harmonic oscillator, with $k = 4.0$ N/m, $m = 0.20$ kg, and $b = 0.20$ kg/s. The dashed lines are called the "envelope" of the oscillation and are shown to emphasize the damping.

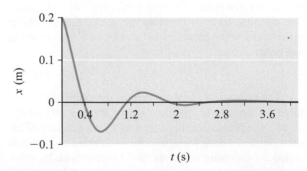

Figure 13-33 A displacement versus time graph for a damped harmonic oscillator, with $k = 4.0$ N/m, $m = 0.20$ kg, and $b = 0.60$ kg/s.

Most everyday oscillations that you observe—for example, a swinging pendulum and a child on a swing—are situations with low damping, where $\omega_0 \ll b/2m$. These systems lose a fraction of their energy with each oscillation, until finally all the energy has been lost. The greater the damping, the quicker the system comes to a stop.

C-13-12 Damped Mass-Spring Oscillator

Four identical mass–spring systems are subjected to damping forces of the form $F_x = -bv$, with a different value of b for each system. Each system is displaced by 0.10 m from equilibrium and released. The amplitude, $A(t)$, for each is shown in Figure 13-34. Rank in order, from smallest to largest, the damping constants for these systems.

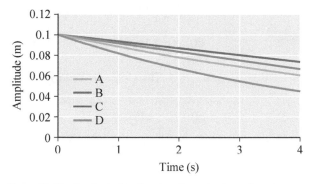

Figure 13-34 Amplitude as a function of time for four different values of damping constant b.

Oscillators from smallest to largest b:
Smallest ---------- ---------- ----------- ----------- Largest

C-13-12 B, C, A, D

A Critically Damped Oscillator $\left(\omega_0 = \dfrac{b}{2m} \right)$

As b increases, the frequency of the damped oscillator decreases, and the oscillator takes more and more time to complete an oscillation. The resulting motion is called **critically damped** when $b = 2\sqrt{mk}$ (i.e., $b/2m = \omega_0$). In this case, $\omega_D = 0$; therefore, the argument of the cosine function does not change with time, and there is no back-and-forth motion of the oscillator. Using calculus, it can be shown that for a critically damped oscillator the displacement from the equilibrium position as a function of time is given by

$$x(t) = (a_1 + a_2 t)e^{-\omega_0 t} \qquad (13\text{-}74)$$

Constants a_1 and a_2 are determined by the initial conditions, $x(0)$ and $v(0)$. When a critically damped oscillator is displaced, it returns to its equilibrium position without completing an oscillation. The time taken for the oscillator to return to the equilibrium position is related to the natural frequency of oscillations, ω_0. *A critically damped oscillator reaches its equilibrium position, without oscillating, faster than for any other value of the damping constant, b.* Figure 13-35 shows a displacement graph for a critically damped mass–spring system. For the graph, we chose $x(0) = 1.0$ m, $v(0) = 0.0$ m/s, $m = 0.20$ kg, and $k = 4.0$ N/m. With these values, $\omega_0 = \sqrt{20}$ rad/s, and critical damping occurs for $b = 2\sqrt{0.20 \times 4.0} = 1.8$ kg/s.

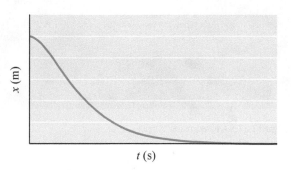

Figure 13-35 A displacement versus time graph for a critically damped oscillator with the initial conditions $x(0) = 1.0$ m, $v(0) = 0.0$ m/s, $\omega_0 = \sqrt{20}$ rad/s, and $b = 1.8$ kg/s.

Shock absorbers in cars are a good example of a critically damped system (Figure 13-36). When a car goes over a bump in the road, properly damped shock absorbers should settle the car back to its normal state in a minimal amount of time.

Figure 13-36 A typical spring-based shock absorber. The inside hydraulic piston contains oil that acts as a damping medium, which absorbs the vibrational energy of the spring and dissipates vibrations.

Another example of a critically damped oscillator is a **tuned-mass damper** (Figure 13-37), which is used in skyscrapers. Tuned-mass dampers are huge blocks of concrete or metal, suspended inside buildings to damp strong vibrations of the building caused by the wind and earthquakes. A detailed explanation is complicated, but basically, the damper absorbs the vibrational energy from the structure when the oscillation frequency of the damper matches the oscillation frequency of the structure.

Figure 13-37 A tuned-mass damper

An Overdamped Oscillator $\left(\omega_0 < \dfrac{b}{2m}\right)$

When the damping constant increases beyond the critically damped value, the oscillator is called an **overdamped** oscillator. There is still no oscillatory motion, but as b increases beyond the critical value, the oscillator takes more and more time to return to its equilibrium position. Heavy doors in public places have hydraulic devices that are overdamped. This keeps the doors from slamming quickly.

Energy in a Damped Harmonic Oscillator

The energy of an oscillator is proportional to the square of its amplitude, so the energy of a damped harmonic oscillator also decays exponentially with time. Consider an

ONLINE ACTIVITY

Determining the Damping Constant of a Damped Oscillator

The e-resource that accompanies every new copy of this textbook contains an Online Activity using the PhET simulation "Determining the Damping Constant of a Damped Oscillator." Work through the simulation and accompanying questions to gain an understanding of a damped harmonic oscillator.

underdamped mass–spring oscillator with an amplitude given by Equation (13-72). The energy of the oscillator at a given instant is

$$E(t) = \frac{1}{2} kA^2(t)$$

$$= \frac{1}{2} kA^2 (e^{-bt/2m})^2$$

$$E(t) = \frac{1}{2} kA^2 e^{-bt/m} \qquad (13\text{-}75)$$

At $t = 0$, $A(0) = Ae^{-0} = A$. Therefore, the energy at time $t = 0$ is

$$E(0) = \frac{1}{2} kA^2 \qquad (13\text{-}76)$$

After one period, $T_D = 2\pi/\omega_D$, the energy of the oscillator is

$$E(T_D) = \frac{1}{2} kA^2 e^{-bT_D/m} = E(0)e^{-bT_D/m} \qquad (13\text{-}77)$$

The **fractional loss of energy** in one oscillation is therefore

$$\frac{E(0) - E(T_D)}{E(0)} \equiv \frac{\Delta E}{E(0)} = 1 - e^{-bT_D/m} \qquad (13\text{-}78)$$

EXAMPLE 13-11

A Damped Harmonic Oscillator

A damped harmonic oscillator consisting of a block of mass 2.00 kg and a spring of spring constant 10.0 N/m has an amplitude of 25.0 cm at $t = 0$ s. The amplitude falls to 75.0% of its initial value after four oscillations. Assuming that the damping force is of the form $F_{Dx} = -bv_x$, calculate
(a) the value of the damping constant, b;
(b) the amount of energy lost during four oscillations.

SOLUTION

(a) We are given that the oscillations are damped. Since we know the spring constant and the amplitude at

$t = 0$, we can calculate the initial total energy of the oscillator:

$$E(0) = \frac{1}{2} kA^2 = \frac{1}{2} \times (10.0 \text{ N/m}) \times (0.250 \text{ m})^2 = 0.312 \text{ J}$$

We need to determine the period. Since we do not know the damping constant, ω_D is not known. Let us assume that ω_D is approximately equal to ω_0. This is a reasonable assumption because the oscillator loses only 25% of its initial energy in four oscillations so the damping must be weak. Therefore, the time period of the damped oscillator can be approximated by using the natural frequency, ω_0. Once we have determined the damping constant, we will check the validity of this approximation. Calculating the period, we get

$$T = \frac{2\pi}{\omega_0} = 2\pi\sqrt{\frac{m}{k}} = 2\pi\sqrt{\frac{2.00 \text{ kg}}{10.0 \text{ N/m}}} = 2.81 \text{ s}$$

To determine the damping constant, we use the information that the amplitude falls to 75% of its initial value in four oscillations. Substituting into Equation (13-72) gives

$$A(4T) = Ae^{-4bt/2m} = \frac{3}{4}A$$

$$e^{-2bt/m} = \frac{3}{4}$$

Taking the natural logarithm (ln) of both sides of the above equation, we get

$$-\frac{2bT}{m} = \ln\left(\frac{3}{4}\right)$$

$$b = -\frac{m}{2T} \times \ln\left(\frac{3}{4}\right)$$

$$= -\frac{2.00 \text{ kg}}{2 \times 2.81 \text{ s}} \times \ln\left(\frac{3}{4}\right) = 0.102 \text{ kg/s}$$

Let us check whether our original assumption that ω_D is approximately equal to ω_0 was valid. Using Equation (13-71),

$$\omega_D = \sqrt{\frac{k}{m} - \frac{b^2}{4m^2}}$$

$$= \sqrt{\frac{10.0 \text{ N/m}}{2.00 \text{ kg}} - \frac{(0.102 \text{ kg/s})^2}{4 \times (2.00 \text{ kg})^2}} = 2.23 \text{ rad/s}$$

This value is very close to the undamped value, $\omega_0 = \sqrt{5}$ rad/s, so our approximation was excellent!

(b) Using Equation (13-77), the energy of the oscillator after four cycles is related to its energy at $t = 0$:

$$E(4T) = E(0)e^{-4bt/m}$$

Using values of b, T, and m:

$$E(4T) = (0.312 \text{ J})e^{-4 \times 0.100 \text{ kg/s} \times 2.81 \text{ s}/2.00 \text{ kg}} = 0.178 \text{ J}$$

This is the energy of the oscillator after four cycles. Therefore, the energy lost in four cycles is 0.312 J − 0.178 J = 0.134 J.

Note two limiting cases of Equation (13-78):

1. When $b = 0$, $\frac{\Delta E}{E(0)} = 1 - 1 = 0$, indicating that there is no loss of energy as the system oscillates.

2. At the critical damping limit, $\omega_D = 0$ and therefore $T_D \to \infty$. Since $e^{-\infty} \to 0$, we obtain the result $\Delta E = E(0)$, indicating that a critically damped oscillator loses all its energy in one oscillation.

PEER TO PEER

If you make an approximation to solve a problem, after solving the problem, evaluate your initial approximation to ensure that it is valid.

LO 10

13-10 Resonance and Driven Harmonic Oscillators

Consider a simple mechanical system, for example, a swing, with a natural frequency of oscillation, f_0. Left on its own, an oscillating swing loses energy due to frictional forces and comes to a rest after a while. The motion can be sustained if energy is transferred to the swing by an external mechanism, for example, by pushing the swing periodically. If the swing is pushed after each oscillation, then the time between successive pushes (t_{push}) is the same as the period of the swing ($t_{push} = T = 1/f_0$). In this case, the frequency of the externally applied force, called the **driving frequency**, is the same as the natural frequency of the swing.

The swing can also absorb energy at other driving frequencies. If the swing is pushed after every other oscillation, the period of successive pushes is twice the period of the swing, and the driving frequency is $f_0/2$. Other possible driving frequencies are $f_0/3$, $f_0/4$, $f_0/5$, ...

A mechanical system (such as a swing, a wine glass, a bridge, or a building) is said to be in **resonance** with an externally applied force when the frequency of the externally applied force ($f_{driving}$) matches the natural frequency (f_0) of the system.

$$f_{driving} = f_0 \text{ (resonance condition)} \quad (13\text{-}79)$$

When the driving and the resonance frequencies match, there is an efficient transfer of energy from the external force to the mechanical system.

There are numerous examples of mechanical resonances: resonating strings in a string instrument such as a violin, resonating membranes of a drum, resonance of the basilar membrane in the ear, and breaking a wine glass using sound waves. Resonance phenomena also play a key role in electromagnetism: when the input circuit of a radio receiver is tuned to match the frequency broadcast by a particular radio station, the amplitude of the received signal is greatly increased.

Consider the damped mass–spring system of Section 13-9 that is subjected to an external harmonic force. Such an oscillator is called a **driven oscillator**, and the resulting oscillations are called **forced oscillations**, or **driven oscillations**. The harmonic driving force has the form

$$F_{driven,x} = F_0 \cos(\omega_{driven} t) \qquad (13\text{-}80)$$

Here, F_0 is the magnitude of the applied force, and ω_{driven} is its angular frequency. The net force on the mass is the sum of the elastic restoring force exerted by the spring, the damping force exerted by the surface, and the external driving force:

$$\Sigma F_x = F_{spring,x} + F_{damping,x} + F_{driven,x}$$

$$\Sigma F_x = -kx + (-bv_x) + F_0 \cos(\omega_{driven} t) \qquad (13\text{-}81)$$

The acceleration of the mass is therefore given by

$$a_x = -\frac{k}{m}x - \frac{b}{m}v_x + \frac{F_0}{m}\cos(\omega_{driving} t) \qquad (13\text{-}82)$$

where $\omega_0 = \sqrt{k/m}$ is the natural frequency of the oscillator.

What is the equation for the displacement of the oscillator as a function of time? Using calculus, it can be shown that the displacement, $x(t)$, of the driven oscillator described by Equation (13-82) is given by

$$x(t) = A_{driven}\cos(\omega_{driven} t + \phi_{driven}) \qquad (13\text{-}83)$$

where

$$A_{driven} = \frac{F_0}{\sqrt{m^2(\omega_{driven}^2 - \omega_0^2)^2 + b^2 \omega_{driven}^2}}$$

$$\phi_{driven} = \tan^{-1}\left(\frac{b\omega_{driven}}{m(\omega_{driven}^2 - \omega_0^2)}\right) \qquad (13\text{-}84)$$

A_{driven} and ϕ_{driven} are, respectively, the amplitude and the phase constant of the driven oscillator. Equation (13-83) is valid after transient effects have died down. (Transient effects are temporary and depend on how the initial motion was started.) Note several properties of driven oscillations:

1. An oscillator that is subjected to a sinusoidal driving force oscillates at the frequency of the driving force, *not* at the system's natural frequency. For example, an adult human ear resonates at approximately 3000 Hz. However, when an ear is subjected to a sound, for example, at a frequency of 500 Hz, the eardrum vibrates at 500 Hz rather than at its natural frequency.

2. The amplitude and the phase constant of the driven oscillations are determined not by the initial conditions, but by the parameters of the system: the natural frequency of the oscillator, the driving frequency of the external force, the magnitude of the driving force, the damping constant, and the mass of the oscillator. This situation is different from a simple harmonic oscillator, where the amplitude and the phase constant depend on the initial position and velocity of the oscillator.

3. As ω_{driven} approaches ω_0, the amplitude increases and reaches a maximum at $\omega_{driven} = \omega_0$. As ω_{driven} is increased past the value ω_0, the amplitude decreases again. If the damping constant, b, is small, the amplitude becomes quite large as ω_{driven} approaches ω_0. For $b = 0$, A_{driven} goes to infinity as $\omega_{driven} \to \omega_0$. In real life, damping is never zero, and the amplitude of driven oscillations can become large but not infinite. This dependence of A_{driven} on ω_{driven} forms the basis of resonance phenomena. For example, a wine glass can shatter when subjected to sound waves whose frequency is very close to the natural frequency of oscillation of the wine glass. Constructing earthquake-resistant buildings requires incorporating enough damping in the structures so that when a building is subjected to oscillations of frequency close to its natural oscillation frequency, the amplitude never becomes large. A plot of the amplitude as a function of b clearly shows the effect of resonance near the natural frequency of the oscillator. For the plot in Figure 13-38, we chose $m = 1.0$ kg, $\omega_0 = 1.0$ rad/s, and $F_0 = 1.0$ N and then plotted A_{driven} for $b = 0.01, 0.1, 0.5$, and 0.8 kg/s. Near the resonance frequency, the amplitude is strongly dependent on the damping constant. Notice that the position of the peak amplitude shifts slightly away from the natural frequency as b increases. From Equation (13-84) it can be shown (by finding the maximum of A_{driven} as a function of ω_{driven}) that the amplitude is maximum at

$$\omega_{driven} = \sqrt{\omega_0^2 - \frac{b^2}{2m^2}} \qquad (13\text{-}85)$$

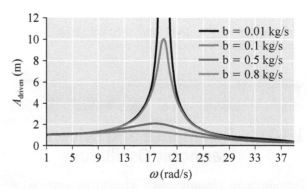

Figure 13-38 The amplitude for a forced harmonic oscillator for various values of damping constant, b, and $\omega_0 = 20$ rad/s.

Simple Harmonic Motion

A motion that repeats after a finite amount of time is called periodic motion. The time that it takes for the motion to repeat is called the period (T). One complete back-and-forth movement is called an oscillation. The number of oscillations per second completed by an object is the oscillation frequency (f) of the object and is related to the period:

$$f = \frac{1}{T} \qquad (13\text{-}3)$$

Periodic motion is called simple harmonic motion when, as a function of time, the displacement $x(t)$ of an object from the equilibrium position is given by a cosine (or a sine) function:

$$x(t) = A \cos(\omega t + \phi) \qquad (13\text{-}4)$$

The amplitude, A, and the phase constant, ϕ, are determined by the position and the velocity of the oscillator at $t = 0$. The angular velocity, ω, depends on the physical properties of the oscillating object and is related to the period:

$$\omega = 2\pi/T \qquad (13\text{-}5)$$

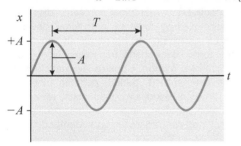

Figure 13-39   Simple harmonic motion; displacement as a function of time for simple harmonic motion.

The velocity and acceleration of a simple harmonic oscillator are determined by taking first and second derivatives, respectively, of Equation (13-4) as a function of time:

$$v(t) = -\omega A \sin(\omega t + \phi) \qquad (13\text{-}12)$$

$$a(t) = -\omega^2 A \cos(\omega t + \phi) \qquad (13\text{-}17)$$

An object undergoes simple harmonic motion when the net force acting on the object is proportional to its displacement from the equilibrium position and is opposite in sign of the displacement. Such a force is called a restoring force.

For horizontal and vertical mass–spring systems,

$$\omega = \sqrt{\frac{k}{m}}; \quad T = 2\pi\sqrt{\frac{m}{k}} \qquad (13\text{-}34), (13\text{-}35)$$

For a pendulum,

$$\omega = \sqrt{\frac{g}{L}}; \quad T = 2\pi\sqrt{\frac{L}{g}} \qquad (13\text{-}54), (13\text{-}55)$$

SHM and Uniform Circular Motion

The projection of a motion with a constant angular velocity along the circumference of a circle is mathematically equivalent to simple harmonic motion. The radius of the circle

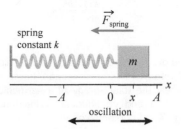

Figure 13-40   A horizontal mass–spring simple harmonic oscillator

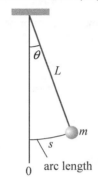

Figure 13-41   A simple pendulum

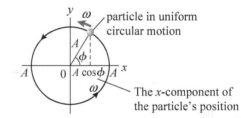

Figure 13-42   Comparing simple harmonic motion and circular motion

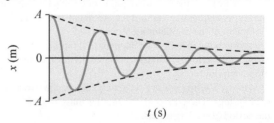

Figure 13-43   Displacement versus time plot for a damped harmonic oscillator

corresponds to the amplitude of the simple harmonic motion, and the angular velocity along the circle corresponds to the angular frequency of simple harmonic motion.

Damped Harmonic Oscillator

When frictional forces are present, the oscillations are damped and the oscillator is called a damped harmonic oscillator. For a velocity-dependent damping force ($\vec{F}_D = -b\vec{v}$), the displacement of a damped harmonic oscillator is given by

$$x(t) = Ae^{-bt/2m} \cos(\omega_D t + \phi) \qquad (13\text{-}70)$$

(continued)

where b is the damping constant, and the oscillation frequency ω_D is given by

$$\omega_D = \sqrt{\frac{k}{m} - \frac{b^2}{4m^2}} \qquad (13\text{-}71)$$

Energy

The total energy, E, of a simple harmonic oscillator is proportional to the square of its amplitude. In the absence of damping forces, the total energy of a harmonic oscillator remains constant and continuously transforms between kinetic energy, K, and potential energy, U:

$$E = K + U = \frac{1}{2}kA^2 \qquad (13\text{-}37)$$

Energy for a damped harmonic oscillator decreases exponentially with time and is given by

$$E(t) = \frac{1}{2}kA^2 e^{-bt/m} \qquad (13\text{-}75)$$

where A is the amplitude of the damped oscillator at $t = 0$.

Applications: pendulum clocks, shock absorbers, tune-mass dampers, seismometers

Resonance

A harmonic oscillator that is subjected to a sinusoidal external force oscillates with the frequency of the force. A resonance occurs when the frequency of the external force equals the natural frequency of the oscillator:

$$f_{driving} = f_0 \qquad (13\text{-}79)$$

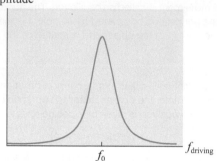

Figure 13-44   The amplitude of a driven oscillator as a function of driving frequency

Key Terms: amplitude, angular frequency, critically damped, damped oscillations, damping constant, displaced harmonic oscillator, drag constant, drag force, driven oscillations, driven oscillator, driving frequency, envelope, fluid resistance, forced oscillations, fractional loss of energy, frequency, harmonic motion, Hooke's law, natural frequency, oscillation, overdamped, period, periodic motion, phase constant, physical pendulum, resonance, restoring force, simple harmonic motion, simple harmonic oscillator, simple pendulum, tuned-mass damper, underdamped

QUESTIONS

1. What is the total distance travelled in one cycle by a simple harmonic oscillator of amplitude A? What is the net displacement of the oscillator after one cycle?
2. If a spring is cut in half, what happens to its spring constant?
3. Why is the period of a mass–spring oscillator independent of the amplitude?
4. Two masses, m and $4m$, are separately attached to two identical springs. The period of the smaller mass is T. What is the period of the larger mass?
   (a) $T/2$ (b) T (c) $2T$ (d) $4T$
5. What would happen to the period of a mass–spring oscillator if the mass of the spring is not ignored?
6. As a pendulum oscillates, is there a point along its path where both its velocity and tangential acceleration are zero? Is there a point where both its displacement and tangential acceleration are zero?
7. What would happen to the period of a simple pendulum if the mass of the string is not ignored?
8. If the maximum angular displacement of a pendulum is increased, what happens to its speed as it passes through the equilibrium?
9. A pendulum is suspended from the ceiling of an elevator. When the elevator is at rest, the period of the pendulum is T. When the elevator accelerates upward, the period of the pendulum
   (a) decreases;   (b) remains T;   (c) increases.

10. You displace the mass of a mass–spring oscillator from equilibrium by 0.10 m, hold it stationary, and then release it. If you were to give the mass a kick as you released it, would the resulting amplitude be less than, equal to, or greater than 0.1 m? Why? Does it matter if the kick is toward the equilibrium or away from it? Explain.
11. A child is swinging on a playground swing while sitting. When the child stands up, what happens to the period of the swing?
12. A child is swinging on a playground swing while sitting. Another child sits on the same swing. Assuming that the children and the swing form a simple pendulum, what happens to the period of the swing? What happens to the speed of the swing as it passes through the equilibrium position?
13. The period of a pendulum is 2.00 s in a vacuum. Its period in air is
   (a) less than 2.00 s;
   (b) 2.00 s;
   (c) greater than 2.00 s.
14. In the following equations, x is displacement and t is time. Which equation describes simple harmonic motion?
   (a) $t(x) = (2.0)\cos(3.0x + \pi/4)$
   (b) $x(t) = 1.0/\sin(2t + \pi)$
   (c) $x(t) = 1.0\cos^2(4t)$
   (d) $x(t) = 3.0 + 5.0\cos(2t)$
15. Which of the following forces, acting on a particle of mass m, will *not* result in simple harmonic motion? The displacement of the particle from the equilibrium position is denoted by $x(t)$.

(a) $F(t) = -2x^2(t)$
(b) $F(t) = +2x^2(t)$
(c) $F(t) = +2x(t)$
(d) $F(t) = -2x(t)$

16. A mass attached to a spring is undergoing simple harmonic motion with an amplitude of 10.0 cm. When the mass is 5.0 cm from the equilibrium position, what fraction of the total energy is the *kinetic energy* of the mass?
(a) 0.25   (b) 0.50   (c) 0.75   (d) 1.00

17. When the amplitude of a simple harmonic oscillator is doubled, the total energy of the oscillator
(a) remains the same
(b) increases by a factor of 2;
(c) decreases by a factor of 2;
(d) increases by a factor of 4;
(e) decreases by a factor of 4.

18. The graph in Figure 13-45 shows the displacement versus time plot of an object that is undergoing simple harmonic motion. For the displacement at $t = 3$ s, which of the following statements is correct?
(a) the velocity of the particle is positive, and its acceleration is positive;
(b) the velocity of the particle is positive, and its acceleration is negative;
(c) the velocity of the particle is negative, and its acceleration is positive;
(d) the velocity of the particle is negative, and its acceleration is negative.

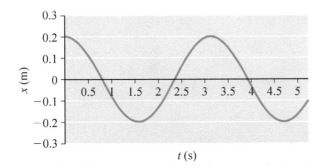

Figure 13-45   Question 18

PROBLEMS BY SECTION

For problems, star ratings will be used, (✶, ✶✶, or ✶✶✶), with more stars meaning more challenging problems.

Section 13-2   Simple Harmonic Motion

19. ✶ A mass of 2.0 kg is connected to a spring with a spring constant of 5.0 N/m. The mass is oscillating on a horizontal, frictionless surface. At time $t = 0$, the mass is 0.30 m from the equilibrium position and has zero velocity.
(a) What is the amplitude?
(b) What is the maximum speed of the mass?
(c) What is the maximum acceleration of the mass?
(d) Write an equation that describes the displacement of the mass from the equilibrium position as a function of time.

20. ✶ A mass–spring system has $k = 18.0$ N/m and $m = 0.71$ kg. The system oscillates with an amplitude of 0.54 m.
(a) Determine the angular frequency of the oscillations.
(b) What is the speed of the mass at $x = 0.034$ m?
(c) What is the displacement of the mass from the equilibrium position when its speed is 0.18 m/s?

21. ✶✶✶ A mass attached to a horizontal spring is pulled to the left and released at $t = 0$ s. The mass passes through $x = +10.0$ cm at $t = 0.3$ s with a velocity of 85.7 cm/s.
(a) What is the lowest frequency of the system?
(b) What is the amplitude?
(c) What is the phase constant?
(d) Write an equation that describes the position of the mass from the equilibrium position as a function of time.

22. ✶✶✶ Two identical mass–spring oscillators are $\pi/3$ rad out of phase and are undergoing simple harmonic motion with amplitude A about the same origin. One of the oscillators is at $x = +A$ at $t = 2.0$ s. When will the second oscillator next reach $x = +A$?

23. ✶✶ A mass–spring oscillator undergoing simple harmonic motion about $x = 0$ has a period of 2.0 s. At $t = 0$ s, the mass is at $x = 0$; at $t = 0.5$ s, the mass is at $x = -0.40$ m.
(a) Write an equation that describes the position of the mass as a function of time.
(b) What are the maximum velocity and acceleration of the mass?
(c) The mass is 0.10 kg. What is the total energy of the oscillator?

24. ✶✶✶ Two particles oscillate in simple harmonic motion with amplitude A, about the centre of a common straight line of length $2A$. Each particle has a period of 1.5 s, and their phase constants differ by $\pi/6$ rad.
(a) How far apart are the particles from each other (in terms of A) 0.50 s after the lagging particle leaves one end of the path?
(b) What are their velocities at that time?

25. ✶✶✶ Can we determine the amplitude, A, and the phase constant, ϕ, for a simple harmonic motion given the values of displacement at $t = 0$ and $t = T/2$, where T is the period? Explain your reasoning.

26. ✶✶ An object, oscillating with simple harmonic motion, has a period of 4.0 s and an amplitude of 0.20 m.
(a) How long does the object take to move from $x = 0.0$ m to $x = 0.07$ m?
(b) Would the object take the same time to move from $x = 0.07$ m to $x = 0.14$ m? Explain your answer without any calculation.

27. ✶✶ Equations for five simple harmonic oscillators are given below:

$$x_1(t) = 0.2 \cos(5t + \pi/4)$$
$$x_2(t) = 0.1 \cos(2t + \pi)$$
$$x_3(t) = 0.3 \cos(t/5)$$
$$x_4(t) = 0.2 \cos(7t/2 + \pi/2)$$
$$x_5(t) = 0.6 \cos(15t)$$

(a) Rank the oscillators, from high to low, in order of period:
High ---------- ---------- ---------- ---------- ---------- Low

(b) Rank the oscillators, from high to low, in order of maximum velocity:
High ---------- ---------- ---------- ---------- ---------- Low
(c) Rank the oscillators, from large to small, in order of the phase constant:
Large ---------- ---------- ---------- ---------- ---------- Small
(d) Rank the oscillators, from high to low, in order of maximum acceleration:
High ---------- ---------- ---------- ---------- ---------- Low

28. ✱✱✱ Two particles are executing simple harmonic motion with same amplitude and frequency about the same equilibrium position. They pass each other moving in opposite direction, each time their displacement is half their amplitude. What is the difference between their phase constants?

Section 13-3 Uniform Circular Motion and Simple Harmonic Motion

29. ✱ Consider a point located on the equator of Earth. Find the angular speed (in rad/s) and the period of this point.

30. ✱ A particle initially at rest on the x-axis, at $x = 3.0$ m, starts moving with a constant angular velocity of 2.0 rad/s in a counterclockwise direction in a circle of radius 3.0 m around the origin. The x-component of the particle's position at any time t is given by which of the following equations?
(a) $x(t) = 2.0 \sin(3.0t)$
(b) $x(t) = 3.0 \cos(2.0t)$
(c) $x(t) = 3.0 \sin(2.0t)$
(e) $x(t) = 3.0 + \cos(2.0t)$

Section 13-4 Mass–Spring Systems

31. ✱ The spring constants and masses for six mass–spring oscillators are given in Table 13-3. Rank the oscillators in the order of their periods, from small to large.
Small ---------- ---------- ---------- ---------- ----------Large

Table 13-3 Spring Constants and Masses for Six Mass–Spring Oscillators

Oscillator	A	B	C	D	E	F
k (N/m)	2.0	4.5	1.0	7.0	11.6	4.6
m (kg)	0.5	1.2	2.0	0.3	3.2	3.1

32. ✱ The displacement from equilibrium position of an oscillator is given by the equation $x(t) = (11.0 \text{ cm}) \cos(3.0 t + 5.0)$
(a) What are the period and the phase constant of the oscillation?
(b) What is the phase of the oscillator at $t = 0$ s and $t = 2.0$ s?
(c) What are the position and the velocity of the oscillator at $t = 0$ s?
(d) At what time after $t = 0$ s is the oscillator at $x = 0$ cm?
(e) What is the total distance covered by the oscillator between $t = 1.3$ s and $t = 1.5$ s?

33. ✱✱ A simple harmonic oscillator consists of a 0.10 kg mass attached to a spring with a spring constant of 100 N/m. The mass is displaced 0.20 m from the equilibrium position, held motionless, and then released. Calculate the following quantities for this system:
(a) the angular frequency and the period;
(b) the maximum speed and the maximum acceleration;
(c) the total energy of the mass–spring system.

34. ✱✱ In problem 33, if the mass is released at time $t = 0$ s, at what time will the kinetic energy of the mass be equal to the potential energy of the spring?

35. ✱✱ A 2.0 kg mass rests on a smooth, horizontal surface and is attached to a spring, with the other end of the spring attached to a rigid support. An applied force of 10.0 N causes a displacement of 5.0 cm from the equilibrium.
(a) Find the spring constant of the spring.
(b) What is the frequency of the mass–spring system?
(c) When oscillating, the mass is at a maximum displacement of 5.0 cm at $t = 0.50$ s. Write an equation that describes the position of the mass from the equilibrium position as a function of time.
(d) What is the total distance travelled by the mass in the interval $t = 0.5$ s to $t = 2.0$ s?

36. ✱✱✱ Figure 13-46 shows a 0.20 kg mass resting on top of a 2.0 kg mass. The lower mass is attached to a spring with a spring constant of 100 N/m. The two masses are oscillating on a frictionless surface. The coefficient of static friction between the two masses is 0.60. What is the maximum amplitude for which the upper block does not slip?

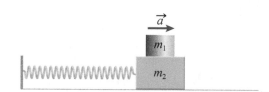

Figure 13-46 Problem 36

37. ✱✱✱ A 500 g block slides along a frictionless surface at a speed of 0.35 m/s. It runs into a horizontal massless spring that extends outward from a wall; the spring constant is 50 N/m. The block compresses the spring and is then pushed back in the opposite direction by the spring, eventually losing contact with the spring.
(a) How long does the block remain in contact with the spring?
(b) How would your answer to part (a) change if the block's initial speed were doubled?

38. ✱✱✱ A 0.20 kg block attached to a horizontal spring is oscillating with an amplitude of 5.0 cm and a frequency of 2.0 Hz. Just as it passes through the equilibrium point, moving to the right, a sharp blow directed to the left exerts a 10 N force for 1.0 ms. Calculate the new frequency and amplitude of oscillations.

39. ✱✱ An empty compact car has a mass of 1200 kg. Assume that the car has one spring on each wheel, the springs are identical, and the mass is equally distributed over the four springs.

(a) What is the spring constant of each spring if the car bounces up and down 2.0 times per second?

(b) What will be the car's oscillation frequency while carrying four passengers each of mass 70 kg?

40. ✹✹✹ Two masses, m_1 and m_2, are connected to two ends of a spring with a spring constant k. The masses are free to slide on a horizontal, frictionless surface. The spring is compressed and then released. Find an expression for the frequency of the system.

41. ✹✹✹ A block of mass m, resting on a frictionless surface, is connected to two springs, 1 and 2, with spring constants k_1 and k_2, respectively, as shown in Figure 13-47. The block is displaced from the equilibrium position and then released.

(a) Show that the resulting motion of the block is simple harmonic motion.

(b) Show that the frequency of oscillations of the block is given by

$$f = \sqrt{\frac{f_1^2 f_2^2}{f_1^2 + f_2^2}}$$

where f_1 is the frequency of the mass when only spring 1 is attached to the mass, and f_2 is the frequency of the mass when only spring 2 is attached to the mass.

(c) Let U_1 and U_2 denote the elastic potential energies stored in the two springs as the mass oscillates. Show that

$$\frac{U_1}{U_2} = \frac{k_2}{k_1}$$

(d) As the mass oscillates, is there a position of the mass where only one of the springs is stretched (or compressed) and the other spring is in its equilibrium position? Explain your reasoning.

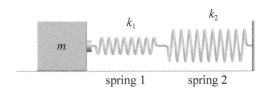

spring 1   spring 2

Figure 13-47 Problem 41

42. ✹✹✹ A block of mass m, resting on a frictionless surface, is connected to two springs, 1 and 2, with spring constants k_1 and k_2, respectively, as shown in Figure 13-48. The block is displaced from the equilibrium position and then released.

(a) Show that the resulting motion of the block is simple harmonic motion.

(b) Show that the frequency of the block is given by

$$f = \sqrt{f_1^2 + f_2^2}$$

where f_1 is the frequency of the mass when only spring 1 is attached to the mass, and f_2 is the frequency when only spring 2 is attached to the mass.

(c) Let U_1 and U_2 denote the elastic potential energies stored in the two springs as the mass oscillates. Show that

$$\frac{U_1}{U_2} = \frac{k_1}{k_2}$$

(d) As the mass oscillates, is there a position of the mass where only one of the springs is stretched (or compressed) and the other spring is in its equilibrium position? Explain your reasoning.

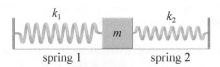

spring 1   spring 2

Figure 13-48 Problem 42

43. ✹✹ A spring hangs vertically from a fixed support. A mass is then attached to the lower end of a spring. When this system undergoes simple harmonic motion, it has a period of 0.50 s. By how much is the spring stretched from its initial length when the mass and spring are hanging motionless?

44. ✹✹✹ A mass, m, is attached to a spring with a spring constant k. The other end of the spring is connected to a fixed support. Initially, the mass is supported from underneath so that the spring remains in its unstretched configuration. The support is then removed, and the mass falls vertically downward under the influence of gravity.

(a) Show, using conservation of energy, that the maximum distance, ΔL, that the mass falls from its initial position is given by $\Delta L = \dfrac{2mg}{k}$.

(b) Once dropped, the mass will undergo simple harmonic motion about an equilibrium position. Show that the equilibrium position is at a distance $\dfrac{mg}{k}$ below the unstretched position of the spring, the amplitude of oscillations is $\dfrac{mg}{k}$, and period is $T = 2\pi\sqrt{\dfrac{\Delta L/2}{g}}$.

45. ✹✹ A 1.0 kg mass is hanging motionless from a spring that is attached to a fixed support. The mass is pulled down 5.0 cm, held motionless, and then released. The mass–spring system begins to oscillate with a period of 2.0 s.

(a) What is the spring constant of the spring?

(b) What is the total energy of the mass–spring system?

(c) What fraction of the system's total energy is the kinetic energy of the mass at $x = 3.0$ cm from the equilibrium position?

(d) Write an equation that describes the position of the mass from the equilibrium position as a function of time.

Section 13-5 Energy Conservation in Simple Harmonic Motion

46. ✹ A 2.0 kg object, attached to a spring, oscillates with simple harmonic motion. The spring constant is 200 N/m, and the maximum force acting on the object is measured to be 3.5 N.

(a) What is the amplitude of the oscillation?

(b) What is the total energy of the mass–spring system as it oscillates?

(c) What is the maximum speed of the mass?

47. ★★ An archer pulls her bow string back by 0.50 m before releasing the arrow. The bow–string system has an effective spring constant of 500 N/m.
 (a) What is the elastic potential energy of the drawn bow?
 (b) The arrow has a mass of 30.0 g. What is the speed of the arrow as it leaves the bow?

48. ★★ A particle undergoes simple harmonic motion with amplitude A. At what position (in terms of A) is the kinetic energy of the particle equal to its potential energy?

49. ★★ A mass–spring system is oscillating with an amplitude of 10.0 cm. What is the speed of the mass at a location where the kinetic energy and the potential energy of the mass are equal?

50. ★★ A student observes that when a mass of 2.0 kg is attached to one end of a vertical spring, the other end of which is connected to a fixed support, the spring stretches 10.0 cm. The student then pulls the mass farther down by 5.0 cm from the equilibrium position and lets it go. As expected, he observes that the mass–spring system undergoes simple harmonic motion.
 (a) What is the amplitude and frequency of oscillations?
 (b) What is the kinetic energy of the mass when it is 3.00 cm below the equilibrium position?
 (c) Assuming that the upward direction is positive, write an equation for the displacement, y, from the equilibrium position, assuming that the mass was released at $t = 0$. Evaluate all parameters.
 (d) If the student had connected two springs identical to the spring above, and then hung a mass of 2.0 kg on them, by how much would the combined springs stretch?

Section 13-6 The Simple Pendulum

51. ★★ The lengths and masses for six pendulums are listed in Table 13-4. Rank the pendulums in order of their periods, from small to large.
 Small ---------- ---------- ---------- ---------- ---------- Large

Table 13-4 Lengths and Masses for Six Pendulums

Pendulum	A	B	C	D	E	F
L (m)	0.8	1.1	1.7	0.9	2.1	3.2
m (kg)	0.5	1.2	2.0	0.3	3.2	3.1

52. ★★ A simple pendulum of length 1.00 m is measured to have a period of 2.00 s.
 (a) What is the value of g at the location of the pendulum?
 (b) The pendulum is now taken to the top of Mount Everest. Does the period increase or decrease? Justify your answer.
 (c) What would be the period of this pendulum on the Moon?
 (d) What would be the period of this pendulum on the space shuttle at the International Space Station?

53. ★★ A pendulum with a length of precisely 1.000 m can be used to measure the acceleration due to gravity, g. Such a device is called a *gravimeter*.
 (a) How long do 100 oscillations take at sea level, where $g = 9.81$ m/s²?

(b) Suppose you take a gravimeter to the top of Mount Everest, where the value of g is 0.28% less than the value at sea level. How long do 100 oscillations take at the top of Mount Everest? Would you be able to measure the difference with a stopwatch that can measure time differences of 0.1 s?

54. ★★ A simple pendulum, consisting of a 0.30 kg mass attached to a string of length 1.0 m, is undergoing simple harmonic motion with a maximum angular displacement of 15°.
 (a) What is the total energy of the pendulum?
 (b) What is the speed of the mass as it passes through the equilibrium position?

55. ★★ (a) Consider a simple pendulum of length L. Show that a change in the length by an amount dL changes the pendulum's period by an amount dT where

$$dT = \left(\frac{T}{2L}\right)dL$$

(b) If the pendulum clock runs faster by 5 s/h, how should the length be adjusted?

Section 13-7 The Physical Pendulum

56. ★★★ A thin circular ring of mass m and radius R hangs freely on a small peg at its rim. The ring is gently displaced from its equilibrium position and released. Derive an expression for the frequency of small oscillations of the ring.

57. ★★★ A physical pendulum consists of a sphere of mass m and radius R that is attached to a massless string. The other end of the string is attached to a fixed support, and the distance between the centre of the bob and the fixed support is L. Show that the period of small oscillations for this pendulum is given by

$$T = T_0 \sqrt{1 + \frac{2R^2}{5L^2}}$$

where $T_0 = 2\pi\sqrt{L/g}$ is the period of a simple pendulum of length L.

58. ★★★ A physical pendulum consists of a sphere of mass m_s and radius R that is attached to a uniform rod of mass m_r and length L. The other end of the rod is attached to a fixed support, and the rod is free to rotate in a plane about an axis that is perpendicular to the plane of rotation and passes through the fixed support.
 (a) Show that the period of small oscillations for this physical pendulum is given by

$$T = 2\pi \sqrt{\frac{\frac{2}{5}m_s R^2 + m_r\frac{L^2}{3} + m_s(L+R)^2}{g\left(m_s(L+R) + m_r\frac{L}{2}\right)}}$$

(b) Calculate the period for the following parameters: $m_s = 1.00$ kg, $m_r = 0.20$ kg, $R = 5.0$ cm, and $L = 1.0$ m. Compare this period to the period of a simple pendulum with $L = 1.05$ m. What is the percentage difference between the two periods? Why is the physical pendulum faster than an equivalent simple pendulum?

Section 13-8 Time Plots for Simple Harmonic Motion

59. ✶ The position versus time graph for a simple harmonic oscillator is shown in Figure 13-49.
(a) What are the amplitude and the period?
(b) What is the phase constant?
(c) What is the phase at $t = 1$ s and $t = 2$ s?
(d) Write an equation for the oscillator that describes its position $x(t)$ as a function of time.
(e) What is the velocity of the oscillator at $t = 1.0$ s?
(f) What is the acceleration of the oscillator at $t = 2$ s?

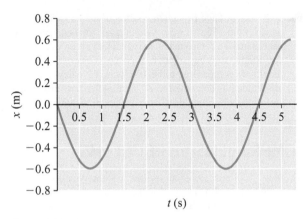

Figure 13-49 Problem 59

60. ✶✶ The velocity versus time graph for a simple harmonic oscillator is shown in Figure 13-50.
(a) What are the maximum speed and the period?
(b) What is the amplitude?
(c) What is the phase constant?
(d) Write an equation for the oscillator that describes its velocity $v(t)$ as a function of time.
(e) What is the velocity of the oscillator at $t = 1.0$ s?
(f) What is the position of the oscillator at $t = 3.0$ s?

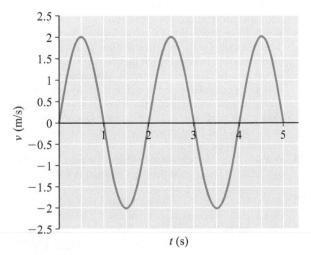

Figure 13-50 Problem 60

61. ✶✶ The position versus time graph for a simple harmonic oscillator is shown in Figure 13-51.

(a) What are the amplitude and the period?
(b) What is the equilibrium position?
(c) What is the phase constant?
(d) Write an equation for the oscillator that describes its position $x(t)$ as a function of time with respect to the point $x = 0$.
(e) Write an equation for a simple harmonic oscillator that is $\pi/2$ rad out of phase with this oscillator.
(f) What is the acceleration of the oscillator at $t = 1.0$ s?

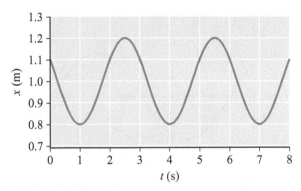

Figure 13-51 Problem 61

62. ✶✶ The position versus time plots of two oscillators are shown in Figure 13-52.
(a) What is the phase difference between the two oscillators at $t = 1.0$ s?
(b) At what times, closest to $t = 0$ s, do the two oscillators have zero velocities?
(c) At what times, closest to $t = 0$ s, are the two oscillators moving with maximum velocity toward the negative x-axis?

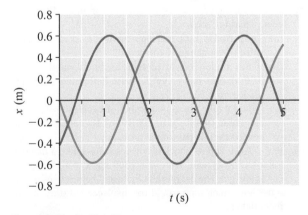

Figure 13-52 Problem 62

63. ✶ Figure 13-53 shows the displacement, $x(t)$, curves for three experiments involving three different masses attached to identical springs oscillating with simple harmonic motion. Rank the curves as indicated.

(a) the system's period
Greatest ----------- ---------------- -------------- Lowest
(b) the total energy of the mass–spring system
Greatest ----------- ---------------- -------------- Lowest
(c) the potential energy of the spring at $t = 0$
Greatest ----------- ---------------- -------------- Lowest
(d) the kinetic energy of the spring at $t = 0$
Greatest ----------- ---------------- -------------- Lowest

Table 13-5 For Problem 67

Damped harmonic oscillator	Amplitude (cm)	b/m (s^{-1})
A	10	0.01
B	20	0.02
C	30	0.03

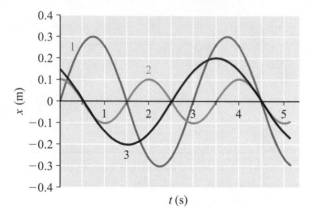

Figure 13-53 Problem 63

Section 13-10   Damped Oscillations

64. ✹✹   A damped harmonic oscillator consists of a block of mass 2.0 kg and a spring with a spring constant $k = 10.0$ N/m. Initially, the system oscillates with an amplitude of 25 cm. Because of the damping, the amplitude decreases by 75% of its initial value at the end of four oscillations.
(a) What is the value of the damping constant, b?
(b) What percentage of initial energy has been lost during these four oscillations?

65. ✹✹   A damped mass–spring oscillator loses 5% of its energy during each cycle.
(a) The period is 2.0 s, and the mass is 0.25 kg. What is the value of the damping constant?
(b) If the motion is started with an initial amplitude of 0.30 m, how many cycles elapse before the amplitude reduces to 0.20 m?
(c) Would it take the same number of cycles to reduce the amplitude from 0.20 m to 0.10 m? No calculations are needed for this part of the problem.

66. ✹✹   A car and its shock absorbers behave like a damped mass–spring system, with $m = 1500$ kg, $k = 60$ N/m, and $b = 1500$ kg/s. The car hits a pothole and begins to oscillate. After how much time would the displacement drop to half its initial value?

67. ✹✹   The initial amplitudes and b/m values for three damped oscillators are given in Table 13-5. Rank their amplitudes after 100 s, from smallest to largest, indicating any equivalencies by using the equality sign (=).

Smallest ---Largest

68. ✹✹   (a) A large pendulum consists of a 50.0 kg mass suspended from a 15.0 m wire. Assuming no damping, how many oscillations are completed by the pendulum in one week?
(b) Slight damping, of course, exists. At a certain time, the amplitude is measured to be 100.0 cm. Five days later, the amplitude has decreased to 80.0 cm. After how many days would the amplitude drop from 100.0 cm to 20.0 cm?

COMPREHENSIVE PROBLEMS

69. ✹✹   The position of an oscillator is known at two times, $t = 0$ and $t = T/4$, where T is the period. Use this information to determine the amplitude and the phase constant of the oscillator.

70. ✹✹✹   A spherical object of mass m and radius r is placed in a large, hemispherical bowl of radius R. The mass rests in the equilibrium position at the bottom of the bowl. The mass is then displaced from its equilibrium position by a small angle, still touching the inner surface of the bowl, and released from rest. The mass rolls down the inner surface of the bowl without slipping. The mass of the bowl is large enough that the bowl remains stationary as the mass moves back to the equilibrium position.
(a) Show that, when the friction between the mass and the bowl is ignored, the mass will undergo simple harmonic motion about its equilibrium position.
(b) What is the frequency of the oscillation, in terms of the variables m, r, and R?

71. ✹✹✹   The gravitational force acting on a particle that is located inside a solid sphere of uniform density is directly proportional to the distance of the particle from the centre of the sphere. Imagine a narrow, evacuated cylindrical hole along the polar diameter of Earth, passing through its centre. A small object is dropped into this hole at one pole. How much time would the object take to reach the other pole? Does this time depend on the mass of the object? Explain why or why not.

72. ✹✹✹   A simple pendulum consists of a mass m attached to a string of linear mass density μ and length L. Determine the period for the pendulum.

73. ✹✹   An object of uniform cross-sectional area A floats in a liquid. When at rest, a volume V_0 of the object is immersed in the liquid. Show that when pushed slightly farther into the liquid and released, the object will undergo simple harmonic motion with the period $2\pi\sqrt{\dfrac{V_0}{gA}}$, where g is the acceleration due to gravity.

74. ✷✷ A mass attached to a spring has a period of T seconds. The mass is displaced from its equilibrium position and then released from rest. Determine the time (in terms of T) when
   (a) the kinetic energy of the mass equals the potential energy of the spring;
   (b) the kinetic energy of the mass is twice the potential energy of the spring.

75. ✷✷ A block of mass M, resting on a frictionless surface, is attached to a spring with a spring constant k (Figure 13-54). The other end of the spring is tied to a rigid support. The mass is initially at rest. A bullet of mass m, moving horizontally with speed v, strikes the block and bounces off it, straight back. Determine expressions for
   (a) the speed of the block immediately after the collision;
   (b) the amplitude of the block–spring oscillations as a result of the collision.

Figure 13-54   Problem 75

76. ✷✷ (a) Show that when the bob of a simple pendulum is displaced by an angle θ from the equilibrium position, the potential energy of the pendulum increases by $mgL(1 - \cos\theta)$.
   (b) If the bob is displaced by an angle θ and released from rest, show that the speed of the bob when it passes through the equilibrium position is given by

$$v = \sqrt{2gL(1 - \cos\theta)}$$

See the text online resources at
www.physics1e.nelson.com for Open Problems and Data-Rich Problems related to this chapter.

Chapter 14

Waves

Learning Objectives

When you have completed this chapter you should be able to:

1 Describe the nature and classification of waves.

2 Describe how a mechanical wave travels through a medium.

3 Write equations for a travelling waveform moving toward the left or the right from a given algebraic form at a fixed time.

4 Distinguish between the wave speed and the instantaneous speed of a segment of a medium, calculate the instantaneous speed of a segment of a medium, and explain how the wave speed in a string depends on the inertial and elastic properties of the string.

5 Determine the amplitude, wavelength, frequency, and phase constant for a travelling harmonic wave from a given mathematical form.

6 Describe what is meant by position plots and time plots of a wave. Determine the mathematical equation for a wave from its position and time plots.

7 Distinguish between the phase and the phase constant for a harmonic wave. Calculate the phase difference between two points of a medium in the presence of a wave.

8 Explain how energy is carried along a string by a travelling wave and how the kinetic energy and the potential energy of an element of string vary with time.

9 Describe the principle of superposition for waves. Explain the consequence of this principle when two or more waves are simultaneously present in a medium.

10 Determine the waveform of the resultant wave when two or more waves are simultaneously present in a medium.

11 Describe the behaviour of reflected and transmitted waves at the boundary between two media.

12 Explain what standing waves are, and determine the location of nodes and antinodes of a standing wave from its mathematical equation.

13 Calculate the resonant frequencies of a vibrating string that is fixed at both ends.

For additional Making Connections, Examples, and Checkpoints, as well as activities and experiments to help increase your understanding of the chapter's concepts, please go to the text's online resources at www.physics1e.nelson.com.

Earthquakes release an enormous amount of energy (Figure 14-1). This energy is carried from the origin of the earthquake, through Earth, by longitudinal and transverse waves over large distances. What are waves? How do waves transport energy? How does a longitudinal wave differ from a transverse wave? What factors determine the speed with which waves travel? Is the motion of a wave different from the motion of the constituents of the medium through which it moves?

© Maciej Dakowicz/Alamy

Figure 14-1 Damage caused by the 2005 earthquake in Islamabad, Pakistan

WAVES AND OSCILLATIONS

14-1 The Nature, Properties, and Classification of Waves

Wave motion is very different from the motion of objects. Some waves, such as waves in an ocean, need a medium in which to travel, but other waves, such as light, do not. A wave passes through a medium unaffected by the presence of other waves in that medium. In this chapter, we will explore the laws governing wave motion and derive mathematical relationships for some properties of waves.

What is a wave? Consider a long rope stretched taut between two rigid supports. If you pluck the rope near one end and then release it, a disturbance originates from where you plucked the rope and travels along the length of the rope. The disturbance moves along the rope, but it does not take the rope with it. As the disturbance passes through a given section of the rope, that section is displaced from its equilibrium position and then returns to it after the disturbance has passed through. Therefore, we can define **wave** as the motion of a disturbance. Waves are classified into two major categories: mechanical waves and electromagnetic waves.

Mechanical Waves A **mechanical wave** travels through a physical material, or **medium**, and the speed of the wave depends on the properties of the medium. The particles of the medium must interact such that a disturbance of particles at any given point is transmitted to adjacent particles. The vibration of a violin string, ripples on the surface of water (Figure 14-2), and seismic waves in Earth's crust are all examples of mechanical waves.

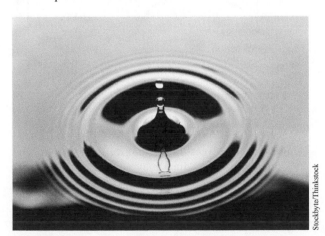

Figure 14-2 Concentric waves on a water surface

Electromagnetic Waves Light, microwaves, radio waves, and X-rays are all manifestations of **electromagnetic waves**, or **electromagnetic radiation**. Electromagnetic waves are generated by charged particles that accelerate; for example, electrons moving back and forth in the antenna of a radio transmitter. Unlike mechanical waves, electromagnetic waves do not need a medium in which to travel. The light that we receive from the Sun travels through the vacuum between the Sun and Earth. This chapter deals with mechanical waves; electromagnetic waves are discussed in Chapter 26.

Waveform The shape or pattern of a wave is called its **waveform**. Waveforms can be represented by plotting some property of the medium as the wave moves through it. For example, a sound wave in air creates regions of higher and lower pressure relative to undisturbed air. If we graph the pressure as a function of time at a given location, the resulting pressure versus time graph would show the shape of the sound wave at that location. A waveform can also be displayed by plotting the shape of the disturbance at a fixed time as a function of position in the medium through which the wave is travelling. Figure 14-3 shows the waveform of a sinusoidal wave on a long string, and Figure 14-4 shows the more complex waveform of a note played on a violin string. Figure 14-5 shows an image of a complex waveform for waves on the surface of an ocean at a given instant.

The waveform for a sound wave in the air can also be obtained by graphing either the variation in the air pressure or the displacement from the equilibrium position of a small element of air as a function of time. These are all different ways of visualizing the same wave.

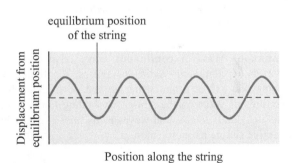

Figure 14-3 The waveform (at a fixed time) generated on a long string by shaking one end of the string in simple harmonic motion

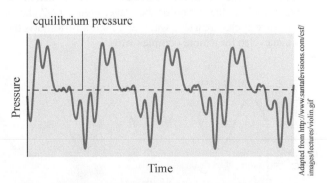

Figure 14-4 A waveform of sound waves generated when playing a note on a violin. The plot shows the change in air pressure with respect to the equilibrium pressure (dashed line, corresponding to no sound) at a given distance from the violin.

Figure 14-5   Complex waveforms on the ocean surface

Pulse A **pulse** is a disturbance of short duration. The disturbance generated on a stretched string by a quick flip of the hand, up and back to the original position, is a pulse (Figure 14-6). Similarly, a single handclap generates a sound pulse. The shape of a pulse depends on the motion of the source that generates the pulse.

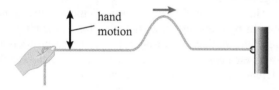

hand
motion

Figure 14-6   A quick flick on one end of a stretched string creates a pulse that travels along the string.

Continuous Wave A **continuous wave** (also called a wave) is a continuous series of pulses. Continuously flipping one end of a taut string up and down generates a continuous wave on the string. A vibrating tuning fork generates continuous sound waves as its tines oscillate in repetitive simple harmonic motion.

Wave Cycle A continuous wave has a waveform that repeats over and over again. The fundamental unit of the repeating pattern is called a **wave cycle**. The duration of the wave cycle is the period of the wave. Figure 14-7 shows the wave cycles for a sine wave, and Figure 14-8 shows the wave cycles for a more complex wave.

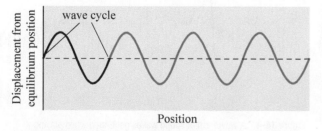

wave cycle

Position

Figure 14-7   The black portion of the wave represents a wave cycle for a sinusoidal wave.

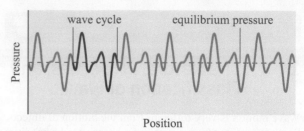

wave cycle   equilibrium pressure

Position

Figure 14-8   The black portion of the wave represents a wave cycle for a more complex wave.

Wave Speed The speed at which a disturbance travels through a medium is called the **wave speed**, usually denoted by the letter v. For example, the speed of sound in dry air at sea level at 20°C is approximately 343 m/s.

Transverse Waves As a wave passes through a medium, the constituents (atoms and molecules) of the medium are temporarily displaced from their equilibrium positions in response to the forces exerted by the wave. In a **transverse wave**, the constituents of the medium move *perpendicular* to the direction of propagation of the wave (Figure 14-9). For example, a pulse generated by moving one end of a taut horizontal string up and down travels horizontally along the length of the string, and the segment of the string that the pulse is passing through is displaced vertically from the rest position.

Transverse Waves

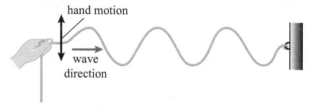

hand motion

wave
direction

Figure 14-9   A transverse wave on a string

Longitudinal Waves In a **longitudinal wave**, the constituents of the medium move *parallel* to the direction of motion of the wave. For example, alternately pushing and pulling on one end of a long spring generates a wave that moves along the length of the spring (Figure 14-10). As the wave passes through a given section of the spring, the coils of that section move parallel to the direction of motion, being alternately compressed and stretched. Sound waves are examples of three-dimensional longitudinal waves. As a sound wave passes through air, the air molecules oscillate in the direction of wave motion, alternately creating regions of higher and lower pressure.

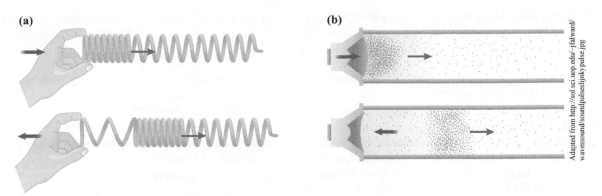

(a)

(b)

Figure 14-10 (a) A longitudinal wave moving along a stretched string. The movement of the coils is in the direction of wave motion, creating regions of higher and lower pressure (compressions and rarefactions, respectively). (b) A sound wave traveling along the length of an air-filled tube. As the diaphragm of the speaker oscillates, it creates periodic pattern of compression and expansion of air.

 MAKING CONNECTIONS

Earthquake Waves

During an earthquake, both transverse and longitudinal waves are generated in the Earth's mantle. The longitudinal waves, called P waves (primary waves), alternately compress and stretch the ground in the direction of wave motion. Transverse waves, called S waves (secondary waves), displace the ground up and down, perpendicular to the direction of wave motion (Figure 14-11). In solids, P waves move approximately twice as fast as S waves. Therefore, during an earthquake, we usually feel two jolts because the P waves arrive ahead of the S waves. The major damage in an earthquake is caused by the up-and-down motion of the ground, that is, by the S waves. Thus, the P waves can give some warning that the more destructive S waves are coming.

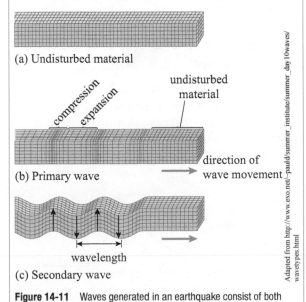

(a) Undisturbed material

(b) Primary wave

(c) Secondary wave

Figure 14-11 Waves generated in an earthquake consist of both longitudinal (P) and transverse (S) waves. The horizontal arrows show the direction of wave propagation.

 LO 2

14-2 The Motion of a Disturbance in a String

 ONLINE ACTIVITY

A Travelling Pulse

The e-resource that accompanies every new copy of this textbook contains an Online Activity using the PhET simulation "A Travelling Pulse." Work through the simulation and accompanying questions to gain an understanding of the motion of a travelling pulse.

Consider a long, stretched string with one end tied to a rigid support and the other pulled taut by your hand. If you quickly flip your hand up and down, you will generate a pulse. The pulse travels along the length of string with a constant speed. This example demonstrates two important properties of waves:

- **Waves carry energy.** In moving the string up and down, your hand did work on the string, thus imparting energy to the string. The pulse carries this energy as it travels along the string. It takes energy to generate a pulse, and the pulse carries this energy as it propagates through the medium. A wave can therefore be described as the motion of energy through a medium.

- **The medium does not travel with the wave.** A given segment of the string is displaced from its equilibrium position while the pulse passes through it and then returns to the equilibrium position once the pulse has gone by. There is no net movement of the string in the direction of motion of the pulse. Similarly, a sound wave does not carry air along with it; the air molecules only oscillate back and forth in the direction of the motion of the sound wave. Do not confuse the motion of the wave with the motion of the particles of the medium.

What causes a pulse to move through a string? In Figure 14-12, a pulse is moving from left to right along a string that is held under a constant tension of magnitude T. Let us analyze the motion of a segment of the string located at point P as the pulse passes through it. Point E is at the leading edge of the pulse, and point A at the trailing edge.

(a) Before the pulse arrives, point P is at rest and is being pulled with tension T by the string to its left and its right. When point E of the pulse reaches P, a slightly increased tension is exerted on P from the string to its left, causing it to accelerate upward.

(b) As point P accelerates upward, it experiences a downward force due to the tension from the string to its right. At the same time, the increased tension from the left keeps it moving upward.

(c) As P moves farther up, the increased pull from its right starts to decelerate it, until it momentarily comes to a stop at C, the point of maximum displacement on the pulse. At this displacement, P is at rest and is now being pulled downward by the string from both sides. Therefore, it has maximum downward acceleration at its farthest displacement.

(d) As P accelerates toward the equilibrium position, it is being pulled upward by the string to its right and downward by the string to its left.

(e) The point P passes through point A on the pulse, finally coming to rest after the pulse has passed through.

So, a pulse (or a wave) moves through a medium due to the forces exerted on a given segment of the medium in the presence of the pulse, by the elements of the medium around that segment.

Figure 14-13 shows a plot of displacement versus time for point P. Time $t = t_1$ corresponds to the arrival of the leading edge (E) of the pulse, and $t = t_2$ corresponds to the departure of the trailing edge (A).

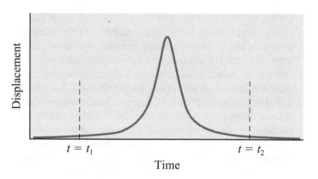

Figure 14-13 A displacement versus time plot for point P (Figure 14-12) as a pulse passes through it

✓ **CHECKPOINT**

C-14-1 Displacement Graph

The pulse in Figure 14-14 is about to pass through point P. Which of the graphs in Figure 14-15 correctly describes the displacement of point P, as a function of time, when the pulse passes through P?

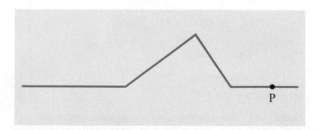

Figure 14-14 C-14-1

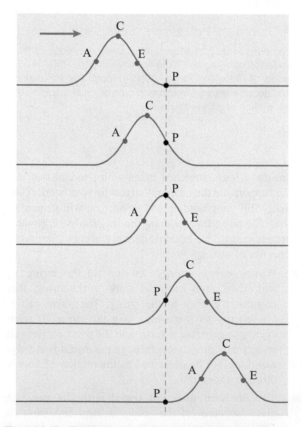

Figure 14-12 The motion of a point (P) on a string as a pulse passes through it. The pulse exerts a net force at P, causing it to be displaced from its equilibrium position.

(a)

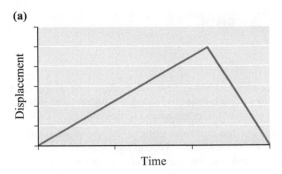

(b)

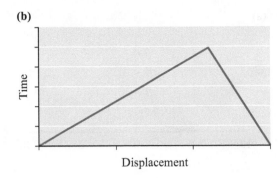

(c)

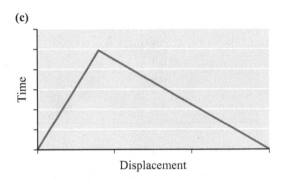

(d)

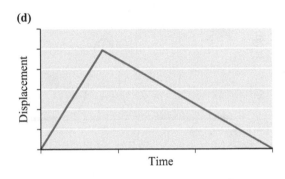

Figure 14-15   C-14-1

LO 3

14-3 Equation for a Pulse Moving in One Dimension

Consider a pulse travelling along a stretched string along the x-axis. The transverse displacement of an element of the string from its equilibrium position is denoted by $D(x, t)$ and is plotted along the y-axis. The displacement is a function of position along the string (x) and time (t). For example, $D(1, 0)$ represents the displacement from the equilibrium position at time $t = 0$ of an element of the string that is positioned at $x = 1$ m. In the absence of any disturbance, every element of the string is at its equilibrium position, and

$$D(x, t) = 0 \quad \text{for all } x \text{ and } t \quad (14\text{-}1)$$

A waveform can be described by giving the displacement of each element of the medium at a fixed time. Figure 14-16(a) shows a pulse with a waveform given by the equation (x and D are in meters and t in seconds).

$$D(x, 0) = \frac{4}{x^4 + 2}$$

Let us assume that the pulse is moving in the positive x-direction with a speed of 3.0 m/s. After 1.0 s, the pulse

has moved 3.0 m toward the right, and the peak of the pulse is at $x = 3.0$ m, as shown in Figure 14-16(b). The waveform of the pulse at $t = 1.0$ s is given by the equation

$$D(x, 1.0) = \frac{4}{(x - 3.0)^4 + 2}$$

After 2.0 s, the pulse has moved 3.0 m farther and it's peak is centred 6.0 m from the origin. Its waveform at $t = 2.0$ s is described by the equation

$$D(x, 2.0) = \frac{4}{(x - 6.0)^4 + 2}$$

For any given time t, the peak of the pulse is at $x = 3.0t$, and its waveform is given by

$$D(x, t) = \frac{4}{(x - 3.0t)^4 + 2}$$

If the pulse is moving to the right at an arbitrary speed v (instead of the speed of 3.0 m/s assumed above), we can describe the waveform of the pulse by substituting v for the speed in the preceding equation:

$$D(x, t) = \frac{4}{(x - vt)^4 + 2}$$

The above analysis can be generalized to write an equation for the waveform of any travelling pulse, given

(a)

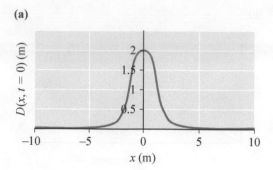

(b)

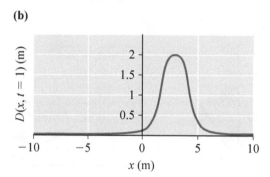

(c)

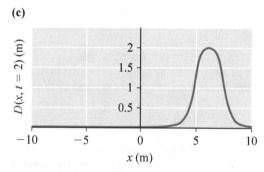

Figure 14-16 Displacement versus position plots of a pulse travelling in the positive x-direction with a speed of 3.0 m/s: (a) at $t = 0$, (b) at $t = 1.0$ s, and (c) at $t = 2.0$ s.

its waveform at $t = 0$. If at $t = 0$ the waveform of a pulse is described by a function $D(x)$ and the pulse is moving in the direction of increasing x with speed v, then the waveform at any time t is obtained by the substitution $x \rightarrow x - vt$ in $D(x)$. Similarly, if the pulse is moving in the direction of decreasing x with speed v, the waveform at any time t is obtained by substituting $x \rightarrow x + vt$ in $D(x)$. A function of the form $D(x \mp vt)$ is called a **wave function** or a **displacement function** because it describes a travelling wave by giving the displacement of each point of the medium at any given time. In a wave function for a travelling wave, the variables x and t always occur in the combination $x \mp vt$.

 EXAMPLE 14-1

Pulse on a String

A pulse is travelling from left to right with a speed of 2.0 m/s along a stretched string. Figure 14-17 shows the waveform of the pulse when $t = 1.0$ s.

(a) Find the displacements of the string elements at $x = 0.0$ m and $x = 5.0$ m at this time.

(b) Where was the peak of the pulse at $t = 0.0$ s?

(c) Where will the peak of the pulse be 5.0 s later?

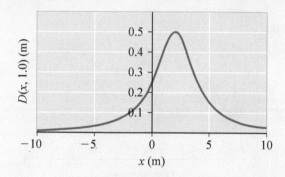

Figure 14-17 A displacement versus position plot of a pulse travelling in the positive x-direction

SOLUTION

(a) From the graph, $D(0.0 \text{ m}, 1.0 \text{ s}) = 0.25$ m, and $D(5.0 \text{ m}, 1.0 \text{ s}) = 0.15$ m.

(b) At $t = 1.0$ s the peak is located at $x = 2.0$ m. Since the pulse speed is 2.0 m/s, one second earlier the waveform was shifted towards the left by 2.0 m. The peak was therefore located at $x = 0.0$ m.

(c) In 5.0 s, the pulse will travel a distance of 10.0 m farther right from its location at $t = 1.0$ s, so the peak will be at $x = 12.0$ m.

✓ CHECKPOINT

C-14-2 A Travelling Pulse

Which of the following equations does *not* describe a pulse moving toward the direction of increasing x?

(a) $D(x, t) = \dfrac{2}{(5.0t - x)^2 + 5}$

(b) $D(x, t) = \dfrac{2}{(x + 0.5t)^2 + 5}$

(c) $D(x, t) = \dfrac{2}{(x - 5.0t)^2 + 5}$

(d) $D(x, t) = \dfrac{3}{(2.0x - 4.0t)^4 + 6}$

EXAMPLE 14-2

Using Waveform Equations

The equation for a pulse traveling through a medium is given by

$$D(x, t) = \frac{3.0}{(x - 2.0t)^2 + 5.0}$$

where x and D are in metres, and t is in seconds.

(a) What is the speed of the pulse, and in which direction is it travelling?
(b) What is the displacement of a point of the medium located at $x = 0.30$ m at $t = 1.0$ s?
(c) What is the maximum displacement of a point as the pulse passes through it? Plot the displacement as a function of time of the point at $x = 15.0$ m as the pulse passes through it.
(d) What is the equation of a pulse that has the same speed and shape as the above waveform but is inverted relative to it?

SOLUTION

(a) Since the x and t variables appear in the form $x - 2.0t$, the pulse speed is 2.0 m/s. The minus sign between x and t indicates that the pulse is moving in the direction of increasing x.
(b) Inserting $x = 0.30$ m and $t = 1.0$ s in the given waveform, we get

$D(0.30 \text{ m}, 1.0 \text{ s}) = 0.38$ m

(c) The numerator of the given waveform is a constant. Therefore, the displacement is maximum when the denominator is minimum. The term $(x - 2.0 t)^2$ has a

minimum value of zero. Therefore, the minimum value of the denominator is 5.0. The maximum displacement of a point of the medium is therefore 3.0 m/5.0 m = 0.60 m. A plot of $D(15, t)$ as function of t is shown in Figure 14-18.

(d) The displacement of the inverted pulse is the negative of the displacement of the given pulse. We therefore multiply the numerator of the given wave function by -1 to obtain the wave function of the inverted pulse. The wave function of the inverted pulse is

$$D(x, t) = \frac{-3.0}{(x - 2.0t)^2 + 5.0}$$

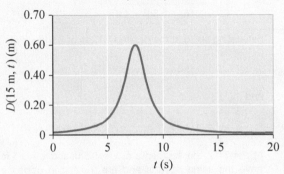

Figure 14-18 The displacement versus time graph at $x = 15.0$ m for the wave function of Example 14-2

Making sense of the result:

Since the wave speed is 2.0 m/s, the peak of the pulse reaches the point at $x = 15.0$ m at $t = 7.5$ s.

LO 4

14-4 Transverse Speed and Wave Speed

In the presence of a wave, the instantaneous speed, $u(x, t)$, of a string element is equal to the rate of change of its displacement, $D(x, t)$, with time. The displacement function depends on both position and time, so the instantaneous speed of an element at x_0 is obtained by taking the derivative of $D(x, t)$ with respect to time while holding the position fixed at $x = x_0$:

$$u(x_0, t) = \left.\frac{\partial D(x, t)}{\partial t}\right|_{x=x_0} \tag{14-2}$$

The symbol ∂ indicates a partial derivative, and the term $\partial/\partial t$ indicates a partial derivative with respect to time. The instantaneous speed can also be obtained by graphing $D(x, t)$ as a function of time for a fixed x. The slope of the graph at a given time is equal to the transverse speed of the segment at that time, and the sign of the slope gives the direction of motion.

EXAMPLE 14-3

Transverse Displacement

A pulse travelling through a string is described by the following wave function:

$$D(x, t) = \frac{2.0}{(x - 0.5t)^2 + 4.0}$$

where x and D are in metres, and t is in seconds.

(a) Find the velocity of the pulse.
(b) At $t = 2.0$ s, what are the speeds of the string elements located at $x = 0.5$ m and $x = 1.5$ m?
(c) Graph the speed of the particle located at $x = 10.0$ m between $t = 0.0$ s and $t = 40.0$ s.

(continued)

SOLUTION

(a) The coefficient of the t term is 0.5, so the pulse speed is 0.5 m/s. The sign between the position x and the time t is negative, so the pulse is travelling in the direction of increasing x.

(b) To determine the speed of a string element, we take the derivative of the wave function with respect to t, while holding x constant:

$$u(x, t) = \frac{\partial}{\partial t} \left(\frac{2.0}{(x - 0.5t)^2 + 4.0} \right)$$

$$= 2.0 \times (-1) \frac{2.0 \times (x - 0.5t) \times (-0.5)}{\left((x - 0.5t)^2 + 4.0\right)^2}$$

$$= \frac{2.0(x - 0.5t)}{\left((x - 0.5t)^2 + 4.0\right)^2}$$

Evaluating $u(x, t)$ at the desired values of x and t gives

$$u(0.5 \text{ m}, 2.0 \text{ s}) = \frac{-1}{(4.25)^2} = -0.055 \text{ m/s}$$

and

$$u(1.5 \text{ m}, 2.0 \text{ s}) = \frac{1}{(4.25)^2} = +0.055 \text{ m/s}$$

At $t = 2.0$ s, the segments at $x = 0.5$ m and $x = 1.5$ m have the same speeds and are moving in opposite directions.

(c) A plot of $u(10.0 \text{ m}, t)$ as a function of time is shown in Figure 14-19.

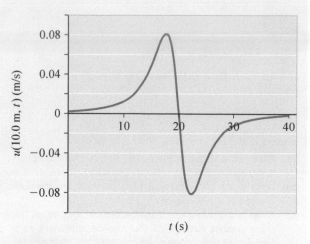

Figure 14-19 The velocity versus time plot of a segment located at $x = 10.0$ m as the pulse described in Example 14-3 passes through the segment.

Making sense of the result:

From the plot, we see that the string element at $x = 10.0$ m begins to move up (positive speed) slowly, but increases its speed with time until it reaches a maximum speed. The element then starts to slow down and momentarily comes to a stop. The instant where the velocity changes sign corresponds to the centre of the pulse passing through the element. The element then moves downward, back to the equilibrium position, first gaining speed and then slowing down until it returns to its equilibrium position.

 CHECKPOINT

C-14-3 Transverse Speed of a String

A waveform is passing through a given element of a string with speed v. What is the speed of the element as a function of time?

(a) v

(b) $-v$

(c) zero

(d) changes with time

C-14-3 (d)

Wave Speed on a String

The speed of a mechanical wave in a medium depends on the elastic and inertial properties of the medium. For a string, the elastic property can be determined from the tension in the string, and the inertial property depends on the linear mass density (mass per unit length), denoted by the Greek letter mu, μ.

If M is the total mass of a string of length L, then

$$\mu = \frac{M}{L} \tag{14-3}$$

The units of linear mass density are kilograms per metre.

As we discussed earlier, in the presence of a wave, a given element of a string is displaced from its equilibrium position by the forces exerted on it by the neighbouring string elements. The greater the tension (the restoring force), the faster a string element moves in response. However, a greater linear mass density means greater inertia and, hence, slower movement. The wave speed is determined by the response time of the medium as a wave passes through it. We should therefore expect the wave speed in a string to be greater for greater tension in the string and less for greater linear mass density.

Consider a pulse moving from left to right with a speed v along a string that has a linear mass density μ and is under a tension of magnitude T_S (Figure 14-20). The pulse speed is measured with respect to a reference frame in which the string is at rest, which we will call the string frame. We assume that the height of the pulse is small compared to the length of the string so that the tension and the linear mass density do not change appreciably in the presence of the pulse. To analyze the motion of the pulse, we choose a frame of reference that is moving with the pulse (call it the pulse frame). In this frame, the pulse is stationary and the string is moving with a speed v from right to left. The string and pulse frames are moving with a constant speed relative to each other, so Newton's laws of motion are identical in both frames. Let us focus on a very small element of string at the top of the pulse at some instant, as shown in Figure 14-21. The string element has length ΔL; hence mass $\Delta m = \mu \Delta L$. Since ΔL is very small, this element can be considered to form an arc of a circle of radius R subtending an angle θ with respect to the centre of a circle such that $\Delta L = R\theta$. In the pulse frame, the mass Δm is moving in a circle of radius R with speed v, so it has a centripetal acceleration of magnitude $a_c = \dfrac{v^2}{R}$. The centripetal acceleration is provided by the tension force $\vec{T}_s$ in the string, acting tangentially at each end of the element.

Figure 14-20 A pulse moving with speed v toward increasing x in a string that is under a tension T_s.

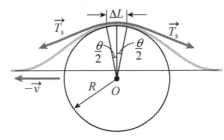

Figure 14-21 The same pulse, viewed from a reference frame in which the pulse is at rest and the string is moving toward the left with speed v.

The two horizontal components of the tension force each have magnitude $T_s \cos(\theta/2)$ and cancel because they are pointing in opposite directions. Each radial component has a magnitude $T_s \sin(\theta/2)$ pointing toward the centre of the circle. Therefore, the net restoring force on the string element is $2T_s \sin(\theta/2)$. According to Newton's second

law, the net restoring force on this element is equal to its mass times the radial acceleration:

$$\Delta m \frac{v^2}{R} = 2T_s \sin\left(\frac{\theta}{2}\right) \qquad (14\text{-}4)$$

Since ΔL and, hence, angle θ are small, we can use the small-angle approximation, $\sin(\theta) \approx \theta$:

$$\Delta m \frac{v^2}{R} \approx 2T_s\left(\frac{\theta}{2}\right) = T_s\theta = T_s\left(\frac{\Delta L}{R}\right)$$

Solving for the wave speed yields

$$v^2 = T_s\left(\frac{\Delta L}{\Delta m}\right) = \frac{T_s}{\mu}$$

$$v = \sqrt{\frac{T_s}{\mu}} \qquad (14\text{-}5)$$

This analysis confirms our assumption that the wave speed increases with greater tension in the string and decreases with greater linear mass density. Since we did not specify a particular waveform, Equation (14-5) is valid for a disturbance of any shape.

✓ **CHECKPOINT**

C-14-4   Wave Speed in a String

A long string of nonzero linear mass density is hanging from a ceiling. A pulse is generated by shaking the string at the bottom end. As this pulse moves upward toward the ceiling, what happens to its speed? Explain your answer.
(a) The speed remains the same.
(b) The speed increases.
(c) The speed decreases.
(d) The speed increases at first, but then decreases closer to the ceiling.

 EXAMPLE 14-4

Pulse Speed on a Hanging String

A string of linear mass density 20.0 g/m and length 10.0 m is hanging vertically from a high ceiling. A mass of 2.00 kg hangs freely from the lower end of the string. A pulse is generated at the lower end of the string and travels upward. What is the speed of the pulse at the lower end of the string, at the middle of the string, and at the top of the string?

SOLUTION

Known quantities:
linear mass density of the string, μ

$$20.0 \text{ g/m} = 20.0 \times 10^{-3} \text{ kg/m}$$

mass of the block, M 2.00 kg
length of the string, L 10.0 m

The tension at any given point along the string is due to the weight of the hanging mass and the weight of the portion of the string below the point. Let x denote the height of a segment of the string, measured from the lower end. The total weight hanging below the point x is $Mg + (\mu x)g$, which is equal to the tension in the string at a height x. Therefore, the wave speed, $v(x)$, at a height x from the lower end of the string is

$$v(x) = \sqrt{\frac{Mg + \mu x g}{\mu}}$$

(continued)

Evaluating the wave speed at $x = 0$, $x = 5.0$ m, and $x = 10.0$ m, we get

$$v(0) = \sqrt{\frac{Mg}{\mu}} = \sqrt{\frac{2.00 \text{ kg} \times 9.81 \text{ m/s}^2}{20.0 \times 10^{-3} \text{ kg/m}}} = 31.3 \text{ m/s}$$

$$v(5.0 \text{ m}) = \sqrt{\frac{Mg + 5.0 \, \mu g}{\mu}}$$

$$= \sqrt{\frac{(2.00 \text{ kg} + 5.0 \text{ m} \times 20.0 \times 10^{-3} \text{ kg/m}) \times 9.81 \text{ m/s}^2}{20.0 \times 10^{-3} \text{ kg/m}}}$$

$$= 32.1 \text{ m/s}$$

$$v(10.0 \text{ m}) = \sqrt{\frac{Mg + 10.0 \, \mu g}{\mu}}$$

$$= \sqrt{\frac{(2.00 \text{ kg} + 10.0 \text{ m} \times 20.0 \times 10^{-3} \text{ kg/m}) \times 9.81 \text{ m/s}^2}{20.0 \times 10^{-3} \text{ kg/m}}}$$

$$= 32.8 \text{ m/s}$$

Making sense of the result:

The tension in the string increases with height, so the pulse speeds up.

LO 5

14-5 Harmonic Waves

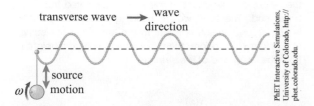

ONLINE ACTIVITY

Harmonic Waves

The e-resource that accompanies every new copy of this textbook contains an Online Activity using the PhET simulation "Harmonic Waves." Work through the simulation and accompanying questions to gain an understanding of harmonic waves.

A wave generated by a source that is undergoing simple harmonic motion is called a **harmonic wave** or a **sinusoidal wave**. Imagine a long string with one end attached to a source. As the source oscillates, it generates a continuous wave that travels along the string in simple harmonic motion (Figure 14-22).

A snapshot of the string at, say, $t = 0$ corresponds to a sine (or a cosine) function (Figure 14-22) and can be represented in the following form:

$$D(x) = A \sin(kx) \quad \text{at } t = 0 \quad \quad (14\text{-}6)$$

Here, $D(x)$ denotes the transverse displacement of the element of the string located at x. The parameters A and k are described below.

Amplitude As the sine function oscillates between $+1$ and -1, the displacement $D(x)$ oscillates between the values $\pm A$. A is called the **amplitude** of the wave and corresponds to the maximum displacement from the equilibrium position of any element of the string. Amplitude is always a positive quantity. For waves on a string, amplitude has the dimension of length. Figure 14-23 shows two harmonic waves with different amplitudes.

The point of maximum displacement ($D(x) = +A$) in a cycle is called a **crest**, and the point of minimum displacement ($D(x) = -A$) is called a **trough**.

Figure 14-22 A source executing a simple harmonic motion generates a simple harmonic wave.

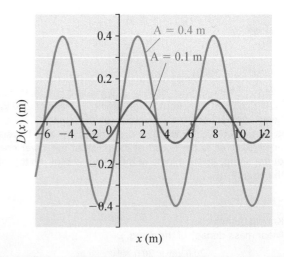

Figure 14-23 A snapshot at a fixed time of two harmonic waves with amplitudes 0.1 m and 0.4 m

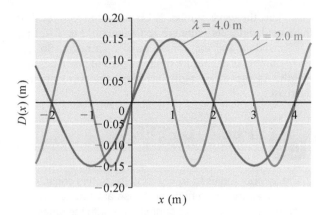

Figure 14-26 Harmonic waves with wavelengths of 2.0 m and 4.0 m

✓ CHECKPOINT

C-14-5 Amplitude of a Harmonic Wave

What is the amplitude of the harmonic wave shown in Figure 14-24?

(a) $+0.15$ m
(b) -0.15 m
(c) $+0.30$ m
(d) zero

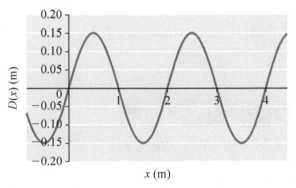

Figure 14-24 C-14-5

C-14-5 (a)

Wavelength The shortest distance over which a wave shape repeats is called the **wavelength**. The usual symbol for wavelength is the Greek letter lambda, λ. Wavelength has dimensions of length. In Figure 14-25, the points at x_1 and x_2 are one wavelength apart and points x_1 and x_3 are two wavelengths apart. Figure 14-26 shows two waves with wavelengths of 2.0 m and 4.0 m.

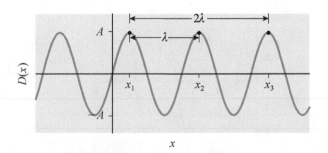

Figure 14-25 The points at $x = x_1$ and $x = x_2$ are one wavelength apart. The points at $x = x_1$ and $x = x_3$ are two wavelengths apart.

✓ CHECKPOINT

C-14-6 Wavelengths

Figure 14-27 shows a plot of three waves with wavelengths λ_1, λ_2, and λ_3. Which of the following wavelength comparisons is correct?

(a) $\lambda_1 < \lambda_2 < \lambda_3$
(b) $\lambda_1 > \lambda_2 > \lambda_3$
(c) $\lambda_1 < \lambda_3 < \lambda_2$
(d) $\lambda_1 > \lambda_2 < \lambda_3$

C-14-6 (c)

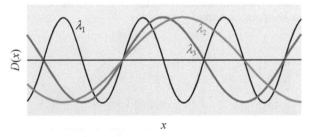

Figure 14-27 C-14-6

Wave Number The parameter k in Equation (14-6) is called the **wave number**. To determine the relationship between k and λ, we use the fact that a waveform repeats over a length λ. Therefore,

$$D(x) = D(x + \lambda) \quad \text{for any given location } x \quad (14\text{-}7)$$

Applying this relationship to the wave function for harmonic waves (Equation (14-6)) gives

$$A \sin(kx) = A \sin(k(x + \lambda)) \quad (14\text{-}8)$$

$$\sin(kx) = \sin(kx + k\lambda) \quad (14\text{-}9)$$

The sine function repeats after 2π rad, so Equation (14-9) is satisfied when $k\lambda = 2\pi$ rad. Therefore,

$$k = \frac{2\pi}{\lambda} \text{ rad/m}$$

Thus, the wave number is 2π times the reciprocal of the wavelength. Its units are radian/metre. For $\lambda = 1$ m, $k = 2\pi$ rad/m, and for $\lambda = 0.5$ m, $k = 4\pi$ rad/m. So, the wave number is a measure of the change of phase per unit length. As the wavelength increases, the wave number decreases.

PEER TO PEER

The symbol k is used to describe the spring constant as well as the wave number. These are two very different quantities; do not confuse one with the other.

Period and Wave Frequency A continuous wave repeats itself in space and time (Figure 14-28). If the period of the source generating a continuous wave is

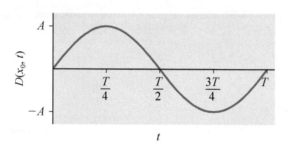

Figure 14-28 The displacement as a function of time as one wave cycle passes through a given point x_0 of the string.

T seconds, one wave cycle is produced in T seconds. As the wave moves through the medium, it takes T seconds for a wave cycle to pass a given point in the medium. The frequency (f) of a mechanical wave is equal to the number of wave cycles passing a fixed point of the medium in one second. Therefore, in one second, an element of the medium will undergo f oscillations.

✓ CHECKPOINT

C-14-7 Wave Numbers

Which of the following relations applies for the waveforms in Figure 14-27?
(a) $k_1 < k_2 < k_3$
(b) $k_1 > k_3 > k_2$
(c) $k_1 < k_2 > k_3$
(d) $k_1 > k_2 < k_3$

C-14-7 (b)

✓ CHECKPOINT

C-14-8 Wave Frequency

Figure 14-29 shows four plots of displacement versus time for four waves passing through a string element at x_0. Order the frequencies of these waves from lowest to highest.

Lowest _____ _____ _____ _____ Highest

C-14-8 (b), (a), (c), (d)

(a)

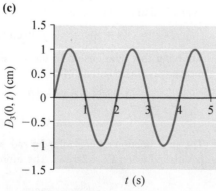

t (s)

(b)

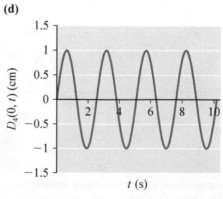

t (s)

(c)

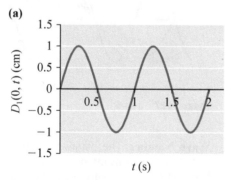

t (s)

(d)

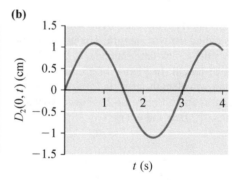

t (s)

Figure 14-29 C-14-8

Wave Speed Relationships

Imagine a train moving with a constant speed with respect to a stationary observer standing beside the tracks. Each car of the train has a length L, and N cars pass by the observer in one second. Therefore, the speed, v, of the train with respect to the observer is $v = LN$. The same reasoning applies to the speed of a mechanical wave with respect to a medium's frame of reference. Consider a harmonic wave of wavelength λ and frequency f. In one second, f wave cycles, each of length λ, pass through a fixed point of the medium. Therefore, the wave speed with respect to the medium's frame of reference is

 KEY EQUATION $$v = \lambda f \qquad (14\text{-}10)$$

Equation (14-10) is valid for all periodic waves. Using $k = 2\pi/\lambda$ and $f = \omega/2\pi$, we can write the wave speed in terms of the wave number and the angular frequency of a harmonic wave:

$$v = \lambda f = \left(\frac{2\pi}{k}\right)\left(\frac{\omega}{2\pi}\right) = \frac{\omega}{k} \qquad (14\text{-}11)$$

EXAMPLE 14-5

Wave Speed and Wavelength

The speed of sound waves in air is approximately 340 m/s and 1400 m/s in fresh water. What is the change in the wavelength of a sound wave of frequency 400 Hz when it crosses from air into water?

SOLUTION

The frequency of a wave is determined by the frequency of the source and does not change when the wave moves from one medium into another. Using Equation (14-10), the wavelengths of the sound wave in air and in water are as follows:

$$\lambda_{air} = \frac{v_{air}}{f} = \frac{340 \text{ m/s}}{400 \text{ Hz}} = 0.85 \text{ m};$$

$$\lambda_{water} = \frac{v_{water}}{f} = \frac{1400 \text{ m/s}}{400 \text{ Hz}} = 3.5 \text{ m}$$

Making sense of the result:

Since the speed of sound in water is about four times greater than its speed in air, for the same number of wave cycles to pass through a given point in water the wavelength of the sound wave in water must be four times longer than the wavelength in air.

Travelling Harmonic Waves

The displacement of any element of a medium from its equilibrium position changes with time as a wave passes through it. Figure 14-30 shows the displacement of two

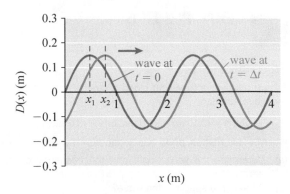

Figure 14-30 The graphs of displacement versus position at time $t = 0$ and $t = \Delta t$ for a sinusoidal wave travelling from left to right

string elements as a wave passes through the string from left to right. At $t = 0$, the point at x_1 is at the crest of the wave and is displaced by 0.15 m, and the point at x_2 is displaced by +0.10 m. A short time later, the wave has moved to the right, and the point at x_1 is displaced by +0.10 m and the point at x_2 by +0.15 m. We can write an equation describing a travelling harmonic wave using the same method that we used to describe a moving pulse. If the equation describing the shape of a wave with a speed of v is known for some fixed time, say $t = 0$, then the wave equation for any arbitrary time t is obtained by

- replacing x by $x - vt$ for a wave travelling in the direction of increasing x
- replacing x by $x + vt$ for a wave travelling in the direction of decreasing x

For a harmonic wave travelling in the direction of increasing x, replacing x by $x - vt$ in Equation (14-6) gives

$$D(x, t) = A \sin(k(x - vt)) \qquad (14\text{-}12)$$

Since $\qquad kv = \left(\frac{2\pi}{\lambda}\right)v = 2\pi\left(\frac{v}{\lambda}\right) = 2\pi f = \omega,$

Equation (14-12) can be written as

$$D(x, t) = A \sin(kx - \omega t) \qquad (14\text{-}13)$$

Similarly, the wave function for a harmonic wave travelling in the direction of decreasing x is given by

$$D(x, t) = A \sin(kx + \omega t) \qquad (14\text{-}14)$$

We can also write Equation (14-13) in terms of the wavelength and the time period by using the relationships $k = 2\pi/\lambda$ and $\omega = 2\pi f = 2\pi/T$:

$$D(x, t) = A \sin\left(\frac{2\pi}{\lambda}x - \frac{2\pi}{T}t\right) \qquad (14\text{-}15)$$

Since a harmonic wave is always described by a sine (or a cosine) function, its properties are determined by its amplitude (A), its wavelength(λ), and its period (T).

EXAMPLE 14-6

Travelling Harmonic Wave

A harmonic wave travelling along a string is described by the wave function

$$D(x, t) = (0.10 \text{ m}) \sin[(2.0 \text{ rad/m})x - (3.0 \text{ rad/s})t]$$

where x is in metres, and t is in seconds.

(a) Determine the wavelength and the frequency of the wave.
(b) What is the speed of the wave?
(c) What is the displacement of a segment of the string located at $x = 0.30$ m at $t = 1.0$ s and at $t = 1.2$ s? Determine the average speed of the segment during this time, and compare it to the wave speed.
(d) Plot the displacement of the segment at $x = 0.3$ m between $t = 1.0$ s and $t = 5.0$ s.
(e) Plot the shape of the section of the string between $x = 0.50$ m and $x = 3.0$ m at 0.50 s and 0.60 s.

SOLUTION

(a) By comparing the given equation with the standard equation for a travelling harmonic wave, Equation (14-13), we find that

$$A = 0.10 \text{ m}, k = 2.0 \text{ rad/m, and } \omega = 3.0 \text{ rad/s}$$

Therefore, the wavelength and frequency of the wave are

$$\lambda = \frac{2\pi}{k} = \frac{2\pi \text{ rad}}{2.0 \text{ rad/m}} = 3.1 \text{ m}$$

$$f = \frac{\omega}{2\pi} = \frac{3.0 \text{ rad/s}}{2\pi \text{ rad}} = \frac{1.5}{\pi} \text{ Hz} = 0.48 \text{ Hz}$$

(b) The wave speed can be calculated from Equation (14-11):

$$v = \frac{\omega}{k} = \frac{3.0 \text{ rad/s}}{2.0 \text{ rad/m}} = 1.5 \text{ m/s}$$

Since the argument of the sine function is of the form $kx - \omega t$, the wave is travelling in the direction of increasing x.

(c) Substituting the given values for x and t, we get

$$D(0.30 \text{ m}, 1.0 \text{ s}) = (0.10 \text{ m}) \sin(2.0 \times 0.3 - 3.0 \times 1.0)$$
$$= (0.10 \text{ m}) \times (-0.68) = -0.068 \text{ m}$$

$$D(0.30 \text{ m}, 1.2 \text{ s}) = (0.10 \text{ m}) \sin(2.0 \times 0.3 - 3.0 \times 1.2)$$
$$= (0.10 \text{ m}) \times (-0.14) = -0.014 \text{ m}$$

In 0.20 s, this segment moved a distance of 0.054 m, so the average speed was

$$0.054 \text{ m}/0.20 \text{ s} = 0.27 \text{ m/s}.$$

This speed is not the same as the wave speed.

(d) A plot of the displacement of the segment located at $x = 0.30$ m between $t = 1.0$ s and $t = 5.0$ s is obtained by setting $x = 0.30$ m in the wave equation and plotting

the resulting equation between $t = 1.0$ s and 5.0. We thus need to plot

$$D(0.3 \text{ m}, t) = (0.10 \text{ m}) \sin(0.60 - 3.0t) \quad 1.0 \leq t \leq 5.0$$

The resulting plot is shown in Figure 14-31. Note that the above equation is similar to the equation for a simple harmonic oscillator (described by a cosine function) with an angular frequency 3.0 rad/s, amplitude 0.10 m, and phase constant of $\left(\frac{\pi}{2} - 0.60\right)$ rad.

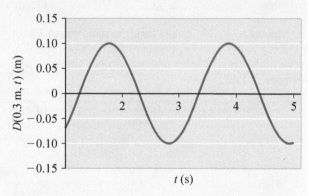

Figure 14-31 The graph of the wave function for Example 14-6 at $x = 0.3$ m for 1.0 s $\leq t \leq 5.0$ s

(e) A plot of the shape of the segment of the string between $x = 0.50$ m and $x = 3.0$ m at 0.50 s and 0.60 s is obtained by setting $t = 0.50$ s and $t = 0.60$ s in the wave equation and plotting the resulting equations between $x = 0.50$ m and $x = 3.0$ m:

$$D(x, 0.50 \text{ s}) = (0.10 \text{ m}) \sin(2.0x - 1.5) \quad 0.5 \leq x \leq 3.0$$

$$D(x, 0.60 \text{ s}) = (0.10 \text{ m}) \sin(2.0x - 1.8) \quad 0.5 \leq x \leq 3.0$$

The resulting graphs are shown in Figure 14-32.

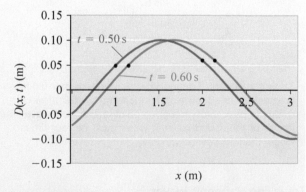

Figure 14-32 Graphs of the wave function for Example 14-6 at $t = 0.50$ s and $t = 0.60$ s

C-14-9   Harmonic Wave Functions

Which of the following wave functions does *not* describe a travelling wave? Explain your reasoning.

(a) $D_1(x, t) = A \sin(x - 4t - 3)$
(b) $D_2(x, t) = A \cos(x - 4t - 3)$
(c) $D_3(x, t) = A \sin(x^2 - 4t)$
(d) $D_4(x, t) = A \sin(x + 4t)$

C-14-9 (c) For a travelling wave, the position x and time t must always occur in the combination $x \mp vt$, where v is the wave speed.

The Phase Constant (ϕ)

Equation (14-13) is not the most general equation for a travelling wave. It restricts the point at $x = 0$ to have zero displacement at $t = 0$. To generalize Equation (14-13), we add a phase constant, ϕ, to the argument of the sine function, as we did for simple harmonic oscillators in Chapter 13. The most general form of the wave function for a harmonic wave travelling in the direction of *increasing* x is

$$D(x, t) = A \sin(kx - \omega t + \phi)$$

$$= A \sin\left(\frac{2\pi}{\lambda} x - \frac{2\pi}{T} t + \phi\right) \qquad (14\text{-}16)$$

Similarly, a wave travelling in the direction of *decreasing* x is described by the wave function

$$D(x, t) = A \sin(kx + \omega t + \phi)$$

$$= A \sin\left(\frac{2\pi}{\lambda} x + \frac{2\pi}{T} t + \phi\right) \qquad (14\text{-}17)$$

The phase constant ϕ can be positive or negative. The sine function is periodic with a period of 2π rad, so it is sufficient to specify ϕ within a range of 0 to 2π rad. The phase constant is determined by the initial conditions. If the segment at $x = 0$ at $t = 0$ has a displacement D_0, then from Equation (14-16)

$$D_0 = D(0 \text{ m}, 0 \text{ s}) = A \sin(\phi) \quad \text{and} \quad \phi = \sin^{-1}\left(\frac{D_0}{A}\right)$$

We will later see that the phase constant can be uniquely determined from the position or time plot of a harmonic wave.

EXAMPLE 14-7

Phase Constant of a Harmonic Wave

A sinusoidal wave travelling in the positive x-direction has an amplitude of 0.15 m, a wavelength of 0.40 m and a frequency of 8.0 Hz. The displacement at $t = 0$ of a point located at $x = 0$ is 0.15 m. What is the phase constant of the wave? Write an equation for the wave function of this wave.

SOLUTION

The given wave is travelling along the positive x-axis. For this wave,

$$A = 0.15 \text{ m}, \quad \lambda = 0.40 \text{ m}, \quad \text{and} \quad f = 8.0 \text{ Hz}$$

Inserting these values into Equation (14-17) we obtain

$$D(x, t) = (0.15 \text{ m}) \sin\left(\frac{2\pi}{0.40 \text{ m}} x - 2\pi \times 8.0t + \phi\right)$$

The phase constant is determined by the initial conditions. We are given that the segment at $x = 0$ is displaced by 0.15 m from equilibrium at $t = 0$. Therefore,

$$D(0 \text{ m}, 0 \text{ s}) = 0.15 \text{ m} = (0.15 \text{ m}) \sin\phi$$

$$\sin\phi = 1$$

$$\phi = \frac{\pi}{2} \text{ rad}$$

Therefore, the equation for the wave function is

$$D(x, t)$$
$$= (0.15 \text{ m}) \sin\left((5.0\pi \text{ rad/m})x - (16.0\pi \text{ rad/s})t + \frac{\pi}{2} \text{ rad}\right)$$

In general, ϕ cannot be uniquely determined from a single initial condition because sine is a multi-valued function.

C-14-10   Harmonic Wave Functions

Figure 14-33 shows waveforms (Equation (14-16)) of four waves with the same amplitude and wavelength at $t = 0$. Rank these waveforms in the order of their phase constants, from small to large. Assume that the phase constant is between 0 and 2π rad.

C-14-10 (d), (a), (c), (b)

(a)

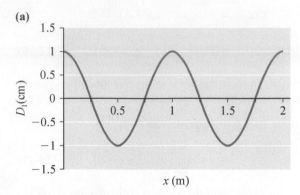

(b)

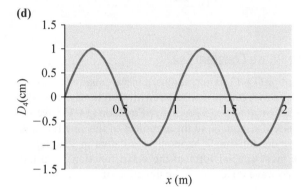

(c)

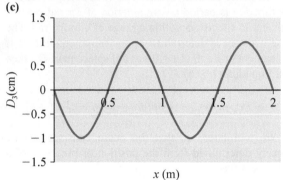

(d)

Figure 14-33   C-14-10

Transverse Velocity and Acceleration for Harmonic Waves

As a harmonic wave travels through a medium, each segment of the medium undergoes periodic motion about its equilibrium position. The instantaneous speed of a segment located at $x = x_0$ is equal to the rate of change of the displacement of the segment. Inserting Equation (14-13) for a harmonic wave into Equation (14-2) and setting $x = x_0$, we obtain the following:

$$u(x_0, t) = \frac{\partial}{\partial t} (A \sin(kx - \omega t + \phi))\big|_{x=x_0}$$

$$= (-\omega A) \cos(kx_0 - \omega t + \phi) \qquad (14\text{-}18)$$

Equation (14-18) shows that the velocity of the segment changes with time, oscillating in the range of $\pm \omega A$. This result is the same as the result for the velocity of a simple harmonic oscillator. The wave speed and the velocity of a segment of the medium are quite different: a wave moves through a medium with a constant speed, while each segment of the medium oscillates about its equilibrium position with a continuously changing velocity.

The instantaneous acceleration of a segment of a string, $a(x, t)$, is equal to the rate of change of the instantaneous velocity of the segment:

$$a(x_0, t) = \frac{\partial u(x, t)}{\partial t}\bigg|_{x=x_0} = \frac{\partial^2 D(x_0, t)}{\partial t^2}$$

$$= -\omega^2 A \sin(kx_0 - \omega t + \phi) = -\omega^2 D(x_0, t) \quad (14\text{-}19)$$

The acceleration of a segment is proportional to its displacement and is opposite in sign, a result that also applies to the motion of simple harmonic oscillators. Notice that each segment of the medium has instantaneous acceleration, but the wave itself has no acceleration.

EXAMPLE 14-8

A Harmonic Wave

A harmonic wave on a string is described by the wave function $D(x, t) = (0.20 \text{ m}) \sin(2.0x - 3.0t)$

(a) What is the velocity at $t = 0.0$ s of a segment of the string located at $x = 0.50$ m?

(b) What is the maximum positive velocity of this segment? When is the first time after $t = 0.0$ that the segment attains this velocity?

(c) Plot the segment's velocity for $0.0 \text{ s} \leq t \leq 10.0 \text{ s}$.

SOLUTION

(a) We take the partial derivative of the displacement to find an expression for the velocity:

$$u(x, t) = \frac{\partial}{\partial t}[(0.20 \text{ m}) \sin(2.0x - 3.0t)]$$

$$= (-0.60 \text{ m/s}) \cos(2.0x - 3.0t)$$

Substituting $x = 0.50$ m and $t = 0.0$ s into this expression, we get

$$u(0.50 \text{ m}, 0.0 \text{ s}) = (-0.60 \text{ m/s}) \cos(2.0 \times 0.5 - 3.0 \times 0)$$

$$= -0.32 \text{ m/s}$$

(b) The maximum positive velocity of the segment is $+0.60$ m/s (when the cosine function is -1). To find the time when the segment first attains this velocity, we substitute $+0.60$ m/s in the velocity equation and solve for t:

$$u(0.50 \text{ m}, t) = (-0.60 \text{ m/s}) \cos(1.0 - 3.0t) = +0.60 \text{ m/s}$$

$$(-0.60 \text{ m/s}) \cos(1.0 - 3.0t) = +0.60 \text{ m/s}$$

$$\cos(1.0 - 3.0t) = -1.0$$

$$1.0 - 3.0t = \cos^{-1}(-1.0) = \pm\pi$$

Here, we have retained both signs for π since $\cos(\pm\pi) = -1$. Solving for t, we get

$$t = \frac{1.0 \mp \pi}{3.0}$$

Time must be positive, so

$$t = \frac{1.0 + \pi}{3.0} = 1.4 \text{ s}$$

(c) A plot of the velocity of the segment is shown in Figure 14-34.

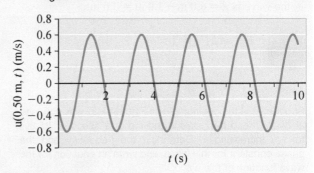

Figure 14-34 A graph of the velocity of the string segment in Example 14-8 at $x = 0.50$ m for 0.0 s $\leq t \leq 10.0$ s

✓ CHECKPOINT

C-14-11 Transverse Velocity

A displacement versus time graph for a particular segment of a medium is shown in Figure 14-35. List the labelled points on the displacement curve where the segment's velocity is
(a) positive;
(b) negative;
(c) zero.

C-14-11 (a) A, F, K; (b) C, D, I; (c) B, E, G, J

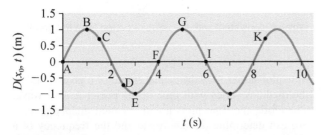

Figure 14-35 C-14-11

LO 6

14-6 Position Plots and Time Plots

A plot of $D(x, t)$ as a function of position x and time t results in a three-dimensional graph. Figure 14-36 shows a plot of $D(x, t) = (1.0 \text{ m}) \sin(2.0\pi x - \pi t)$, with position along the x-axis, time along the y-axis, and the displacement $D(x, t)$ along the z-axis. Although such a plot is useful in visualizing wave motion, it can be difficult to determine the parameters of a wave (wavelength, frequency, amplitude, and phase constant) from a three-dimensional plot. It is easier to use plots in which one of the variables is kept constant and the wave function is plotted against the other variable.

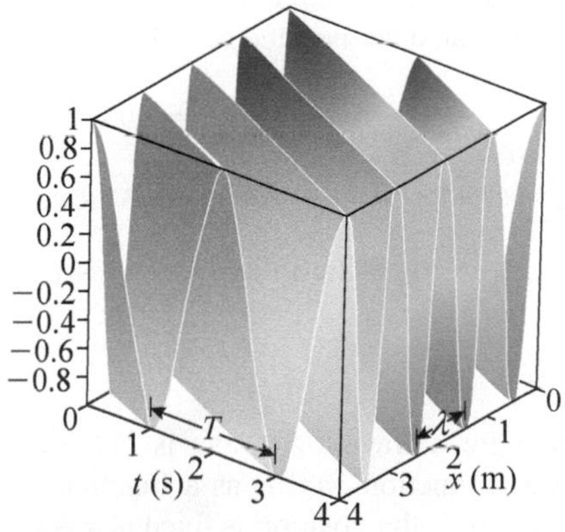

Figure 14-36 A three-dimensional plot of the displacement function $D(x, t) = (1.0 \text{ m}) \sin(2.0\pi x - \pi t)$

Position Plots

A **position plot** for a travelling wave is obtained by keeping the time fixed and plotting the wave function as a function of position (x). If the time is fixed at $t = t_0$, then the position plot is a graph of $D(x, t = t_0)$ as a function of x. A position plot shows the displacement of every section of the medium at a fixed time. It is like taking a picture or a snapshot of the medium, and for this reason, position plots are also called **snapshot graphs**. We can determine the amplitude and the wavelength of the wave from its position plot. Figure 14-37 shows a displacement versus position plot of a harmonic wave. From the graph, notice the following:

- The maximum displacement of a point is ±1.0 m. Therefore $A = 1.0$ m.

- The points at $x = 3.0$ m and $x = 6.0$ m are located at two successive crests. Therefore, the wavelength of the wave is $\lambda = 6.0$ m $- 3.0$ m $= 3.0$ m.

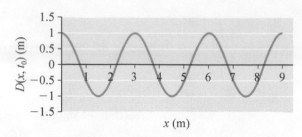

Figure 14-37 The displacement versus position plot of a harmonic wave at a fixed time

EXAMPLE 14-9

The Position Plot of a Harmonic Wave

Suppose that the position plot in Figure 14-37 was obtained at $t = 1.0$ s and the wave is moving at a speed of 1.5 m/s toward increasing x. Determine the frequency and the phase constant for this wave, and write an equation for the wave function of this wave.

SOLUTION

The general equation for a sinusoidal wave travelling toward increasing x is given by

$$D(x, t) = A \sin\left(\frac{2\pi}{\lambda} x - \frac{2\pi}{T} t + \phi\right)$$

Substituting $A = 1.0$ m, $\lambda = 3.0$ m, and $t = 1.0$ s, we get

$$D(x, 1.0\,\text{s}) = (1.0\,\text{m}) \sin\left(\frac{2\pi}{3.0} x - \frac{2\pi}{T} + \phi\right)$$

The wave speed is 1.5 m/s, and the wavelength is known; therefore, we can determine the period:

$$T = \frac{1}{f} = \frac{\lambda}{v} = \frac{3.0\,\text{m}}{1.5\,\text{m/s}} = 2.0\,\text{s}$$

Insert this value for T into above equation:

$$D(x, 1.0\,\text{s}) = (1.0\,\text{m}) \sin\left(\frac{2\pi}{3.0} x - \pi + \phi\right)$$

We can determine ϕ by determining the displacement for a given x from the graph and then inserting that information in the wave equation. At $x = 0.0$ m, $D(0.0\,\text{m}, 1.0\,\text{s}) = 1.0$ m. Inserting these values into the equation yields

$$D(0.0\,\text{m}, 1.0\,\text{s}) = (1.0\,\text{m}) \sin(-\pi + \phi) = 1.0\,\text{m}$$

$$\sin(-\pi + \phi) = 1$$

$$-\pi + \phi = \sin^{-1}(1) = \frac{\pi}{2}$$

Therefore, $$\phi = \frac{3\pi}{2}\,\text{rad}$$

Therefore, the wave function for the wave is

$$D(x, t) = (1.0\,\text{m}) \sin\left(\frac{2\pi}{3.0} x - \pi t + \frac{3\pi}{2}\right) \quad (14\text{-}20)$$

Making sense of the result:

Evaluating (14-20) for $x = 0.0$ m and $t = 1.0$ s, we get

$$D(0.0\,\text{m}, 1.0\,\text{s}) = (1.0\,\text{m}) \sin\left(-\pi + \frac{3\pi}{2}\right)$$

$$= (1.0\,\text{m}) \sin\left(\frac{\pi}{2}\right) = 1.0\,\text{m}$$

which is consistent with the value obtained from the graph.

Time Plots

A time plot for a travelling wave is obtained by plotting the wave function $D(x, t)$ as a function of time for a fixed position. If the position is fixed at $x = x_0$, then the time plot is a graph of $D(x = x_0, t)$ as a function of t. A time plot shows how the displacement of a given point of the medium varies with time in the presence of a wave. Therefore, the time plots are also called history graphs. We can determine the amplitude and the frequency of a wave from its time plot.

EXAMPLE 14-10

The Time Plot of a Harmonic Wave

A sinusoidal wave is travelling with a speed of 20.0 m/s along the positive x-axis on a string. The time plot of the displacement of the section of the string located at $x = 2.0$ m is shown in Figure 14-38. Using the given data and the plot,

(a) determine the wavelength and angular frequency of the wave;

(b) determine the phase constant of the wave;

(c) write an equation for the wave function $D(x, t)$ for the wave.

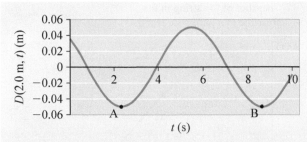

Figure 14-38 The displacement versus time plot of a segment of string located at $x = 2.0$ m as a harmonic wave passes through the point.

SOLUTION

(a) The general equation for a harmonic wave travelling toward increasing x is

$$D(x, t) = A \sin(kx - \omega t + \phi)$$

Inserting $x = 2.0$ m in the above equation we get

$$D(2.0 \text{ m}, t) = A \sin(2.0k - \omega t + \phi)$$

From the graph, $A = 0.05$ m, corresponding to the maximum positive displacement. The period can be calculated from the graph by finding the time difference between two consecutive troughs, for example, points A and B on the graph. Approximately, the period T is:

$$T = t_{\text{B}} - t_{\text{A}} = 6.4 \text{ s}$$

We can now determine the wavelength:

$$\lambda = \frac{v}{f} = vT = (20.0 \text{ m/s})(6.4 \text{ s}) = 128.0 \text{ m}$$

The angular frequency is

$$\omega = \frac{2\pi}{T} = \frac{2\pi \text{ rad}}{6.4 \text{ s}} = 0.98 \text{ rad/s}$$

(b) Since the wave speed is given and we now know the angular frequency, we can determine k:

$$k = \frac{\omega}{v} = \frac{0.98 \text{ rad/s}}{20.0 \text{ m/s}} = 0.05 \text{ rad/m}$$

Inserting the values of A, ω, and k into the wave equation, we get

$$D(2.0 \text{ m}, t) = (0.05 \text{ m}) \sin(2.0 \times 0.05 - 0.98t + \phi)$$

$$= (0.05 \text{ m}) \sin(0.10 - 0.98t + \phi)$$

We can read the displacement at any time on the graph, and then solve the wave equation to determine ϕ. At $t = 0.0$ s, $D(2.0, 0.0) = 0.035$ m, approximately. Therefore,

$$D(2.0, 0.0) = (0.05 \text{ m}) \sin(0.10 + \phi) = 0.035 \text{ m}$$

$$\sin(0.10 + \phi) = \frac{0.035 \text{ m}}{0.05 \text{ m}} = 0.70$$

$$0.10 + \phi = \sin^{-1}(0.70) = 0.77 \text{ rad} \text{ or } (\pi - 0.77) \text{ rad}$$

$$\phi = 0.67 \text{ rad} \text{ or } 2.27 \text{ rad}$$

(c) The value $\pi = 0.67$ rad reproduces the graph shown in figure 14-38. We can now write the general equation for the travelling wave:

$$D(x, t) = (0.05 \text{ m}) \sin(0.05x - 0.98t + 0.67)$$

A common error while doing problems is to omit the phase constant ϕ in the wave function.

LO 7

14-7 Phase and Phase Difference

The argument of the sine (or the cosine) function of a harmonic wave is called the **phase** of the wave. It is measured in radians and is usually denoted by the Greek letter phi, Φ. For the harmonic waves of Equations (14-16) and (14-17), the phase is

$$\Phi(x, t) = kx \mp \omega t + \phi = \left(\frac{2\pi}{\lambda}\right)x \mp 2\pi ft + \phi \quad (14\text{-}21)$$

Do not confuse the phase $\Phi(x, t)$ with the phase constant ϕ. The phase constant is a constant determined by the initial conditions, whereas the phase of a wave is a variable that depends on position and time. Consider two points, x_1 and x_2, that are a distance Δx apart. The difference between the phase at two points, at the same time, is called the **phase difference** and is denoted as $\Delta\Phi$. For harmonic waves,

$\Delta\Phi$ = phase of the wave at x_2 − phase of the wave at x_1

$$= (kx_2 - \omega t + \phi) - (kx_1 - \omega t + \phi)$$

$$= k(x_2 - x_1)$$

$$= k\Delta x$$

$$\Delta\Phi = 2\pi\left(\frac{\Delta x}{\lambda}\right) \quad (14\text{-}22)$$

Table 14-1 shows the phase differences for various distances between two points on a wave.

Two points on a wave that are an integer multiple of wavelengths apart have a phase difference of 2π rad (or an even multiple of π rad). Such points are said to be **in phase** with each other. These points have equal displacements at all times.

Two points that are an odd half-integer multiple (1/2, 3/2, 5/2, ...) of a wavelength apart are π rad (or an odd multiple of π rad) **out of phase** with each other. These points have equal but opposite displacements from the

Table 14-1 Phase Differences for a Periodic Wave

Distance between points, Δx (in multiples of the wavelength)	Phase difference, $\Delta\Phi$ (rad)
0	0
$\lambda/4$	$\pi/2$
$\lambda/2$	π
$3\lambda/4$	$3\pi/2$
λ	2π

equilibrium position at all times. These relationships follow from Equation (14-22). For example, let $\Delta\Phi = \pi$ rad. Then Equation (14-22) can be rearranged as

$$(kx_2 - \omega t + \phi) = (kx_1 - \omega t + \phi) + \pi$$

and

$$\begin{aligned} D(x_2, t) &= A\sin(kx_2 - \omega t + \phi) \\ &= A\sin(kx_1 - \omega t + \phi + \pi) \\ &= -A\sin(kx_1 - \omega t + \phi) \\ &= -D(x_1, t) \end{aligned} \quad (14\text{-}23)$$

Here, we have used the trigonometric identity $\sin(\theta + \pi) = -\sin(\theta)$.

 CHECKPOINT

C-14-12 Phase Difference

A harmonic wave is travelling along a string. Two points that were 0.30 m apart when the string was at rest oscillate such that their displacements are always equal and opposite. What is the longest possible wavelength of the wave?

(a) 0.3 m
(b) 0.6 m
(c) 0.9 m
(d) 1.2 m

C-14-12 (b)

LO 8

14-8 Energy and Power in a Travelling Wave

It takes energy to generate a wave, and the wave carries that energy with it as it travels along its medium. Consider a harmonic wave passing through a string of linear mass density μ. A segment of this string has a mass of $\Delta m = \mu\Delta x$, where Δx is the undisturbed length of the segment (i.e., the length of the segment when the string is at rest). In the presence of the wave, the instantaneous kinetic energy of the oscillating segment is

$$\Delta K = \frac{1}{2}\Delta m\big(u(x, t)\big)^2 \quad (14\text{-}24)$$

where $u(x, t)$ is the instantaneous velocity of the segment.

Using Equation (14-18) for the velocity of the string segment, we can write ΔK as

$$\begin{aligned} \Delta K &= \frac{1}{2}(\mu\Delta x)\big(A^2\omega^2\cos^2(kx - \omega t)\big) \\ &= \frac{1}{2}(\mu\Delta x)\big(A^2\omega^2(1 - \sin^2(kx - \omega t))\big) \\ &= \frac{1}{2}(\mu\Delta x)\omega^2\big(A^2 - D^2(x, t)\big) \quad (14\text{-}25) \end{aligned}$$

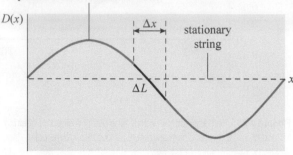

snapshot of the string in the presence of a harmonic wave

Figure 14-39 The change in the length of an infinitesimal segment of a string in the presence of a harmonic wave. The change is maximum when the segment passes through the origin and is zero at the maximum displacement.

Notice that the vibrating segment has zero kinetic energy at $D(x, t) = \pm A$, the maximum displacement from the equilibrium position where the segment momentarily comes to rest. The segment has a maximum kinetic energy as it passes through its equilibrium position where $D(x, t) = 0$.

As the wave passes through the segment, the length of the segment changes as it is continuously stretched and compressed by the wave. To determine how the change in the length of a segment depends upon the shape of the wave consider a curve in one dimension that is described by a continuous function $f(x)$. The length L of the curve between $x = a$ and $x = b$ is given by the integral

$$L = \int_a^b \sqrt{1 + \left(\frac{df}{dx}\right)^2}\, dx$$

If $\dfrac{df(x)}{dx}$ does not change appreciably in the interval $[a, b]$ then L can be approximated as

$$L = \sqrt{1 + \left(\frac{df}{dx}\right)^2}\,(b - a)$$

Now consider a wave passing through a string. Consider an infinitesimal segment of the string of undisturbed length Δx. In the presence of the wave the string is deformed and let ΔL be the length of the deformed segment (Figure 14-39). Using the previous equation, ΔL can be approximated as,

$$\Delta L = \left(\sqrt{1 + \left(\frac{\partial D(x, t)}{\partial x}\right)^2}\right)\Delta x$$

where we have assumed that the derivative of $D(x, t)$ does not change appreciably within the segment Δx. A general description of wave propagation is quite complicated and we will assume that the amplitude of the wave is small compared to its wavelength. In this approximation the length of the segment in the presence of a wave is not very different from its undisturbed length. Therefore, the second term under the square root is much smaller than 1 and we can expand the square root to write,

$$\Delta L = \left(1 + \frac{1}{2}\left(\frac{\partial D(x, t)}{\partial x}\right)^2\right)\Delta x$$

If Δl is the change in the length of the segment due to the presence of the wave, then $\Delta L = \Delta x + \Delta l$ and Δl is given by,

$$\Delta l = \Delta L - \Delta x = \frac{1}{2}\left(\frac{\partial D(x, t)}{\partial x}\right)^2 \Delta x \qquad (14\text{-}26)$$

For a harmonic wave, $D(x, t) = A\sin(kx - \omega t)$. Therefore,

$$\frac{1}{2}\left(\frac{\partial D(x, t)}{\partial x}\right)^2 = \frac{1}{2}\left(\frac{2\pi A}{\lambda}\right)^2 \cos^2(kx - \omega t)$$

and we can see that in the approximation $2\pi A \ll \lambda$, $\frac{1}{2}\left(\frac{\partial D(x, t)}{\partial x}\right)^2 \ll 1$.

In the absence of a wave a string segment is under a tension T (the tension can vary from segment to segment as is the case for a string of non-negligible linear mass density hanging from a support). As the wave passes through the segment the tension in the segment varies continuously. We now assume that for waves with $A \ll \lambda$ the change in the tension in the presence of a wave is negligible. In this approximation the work done by the wave in changing the length of the segment, W, is given by:

W = Tension in the string × change in the length of the segment

$$= T(\Delta l) = T\left(\frac{1}{2}\left(\frac{\partial D(x, t)}{\partial x}\right)^2 \Delta x\right)$$

It is this work that is stored as the instantaneous elastic potential energy, ΔU of the segment. Therefore,

$$\Delta U = \frac{1}{2}T\left(\frac{\partial D(x, t)}{\partial x}\right)^2 \Delta x \qquad (14\text{-}27)$$

For a harmonic wave, $\dfrac{\partial(A\sin(kx - \omega t))}{\partial x} = Ak\cos(kx - \omega t)$; therefore, the stored potential energy is

$$\Delta U = \frac{1}{2}\left(T\Delta x\right)\left(A^2 k^2 \cos^2(kx - \omega t)\right)$$

$$= \frac{1}{2}\left(Tk^2 \Delta x\right)\left(A^2 - A^2 \sin^2(kx - \omega t)\right)$$

$$= \frac{1}{2}\left(Tk^2 \Delta x\right)\left(A^2 - D^2(x, t)\right) \qquad (14\text{-}28)$$

Using $v = \sqrt{\dfrac{T}{\mu}}$ for the wave speed along a string and $v = \dfrac{\omega}{k}$, we get $Tk^2 = \mu\omega^2$, and Equation (14-28) can then be written as

$$\Delta U = \frac{1}{2}(\mu\Delta x)\omega^2\left(A^2 - D^2(x, t)\right) \qquad (14\text{-}29)$$

Notice that the equation for stored instantaneous potential energy is exactly the same as the equation for the instantaneous kinetic energy of the segment (Equation (14-25)). As a harmonic wave travels through a string, the instantaneous kinetic energy and the potential energy of any segment of the string are *equal*. When the segment is at its maximum displacement ($D = \pm A$), both the kinetic energy and the potential energy of the segment are zero, so the segment has zero total energy. When the segment is passing through the equilibrium position ($D = 0$), it has maximum potential energy as well as maximum kinetic energy. Thus, the behaviour of a string element in the presence of a travelling wave is different from the behaviour of a simple harmonic oscillator. The kinetic and potential energies of a single oscillator continually transform back and forth into each other. When the kinetic energy of a simple harmonic oscillator is maximum, its potential energy is minimum, and vice versa.

The total energy of the segment (ΔE) is equal to the sum of its instantaneous kinetic and potential energies:

$$\Delta E = \Delta K + \Delta U = (\mu\Delta x)\omega^2 A^2 \cos^2(kx - \omega t) \qquad (14\text{-}30)$$

This energy flows in the direction of the wave motion. The segment receives energy at one end, stores it briefly, and then passes it to the neighbouring segment at the other end. If energy ΔE is transmitted through a segment of length Δx in time Δt, then the energy transmitted across the segment per unit time is

$$\frac{\Delta E}{\Delta t} = \mu\left(\frac{\Delta x}{\Delta t}\right)\omega^2 A^2 \cos^2(kx - \omega t)$$

The ratio $\Delta x/\Delta t$ is the wave speed. Therefore, the instantaneous power, P, delivered by the wave is

$$P = \frac{\Delta E}{\Delta t} = \mu v\omega^2 A^2 \cos^2(kx - \omega t) \qquad (14\text{-}31)$$

The average rate at which the power is transmitted by the wave is obtained by integrating Equation (14-31) over a complete wave cycle. The average value of the square of a cosine (or a sine) function over an integer number of cycles is ½. Thus, the average power transmitted by a harmonic wave, P_{avg}, is

$$P_{avg} = \frac{1}{2}\mu v\omega^2 A^2 = 2\pi^2 \mu v f^2 A^2 \qquad (14\text{-}32)$$

The average power delivered by a wave is proportional to the linear mass density of the string, the wave speed, the square of the amplitude of the wave, and the square of its frequency. The dependence of the average power of a wave on the square of its amplitude and frequency is valid for other types of waves as well, including ocean waves, seismic waves, and electromagnetic waves. In real-life situations frictional forces cause a wave to lose its energy as it moves in a medium. As a wave gradually loses its energy it finally disappears. We will neglect effects of frictional forces on wave motion. This means that a wave travels through a uniform medium without a change in its energy.

Note that Equations (14-24) and (14-27) are strictly valid only in the limit when $\Delta x \rightarrow 0$. For a segment of finite length, we must integrate these equations over the length of the segment.

☑ CHECKPOINT

C-14-13   Wave Power

Figure 14-40 shows time plots of four waves passing through four identical strings held under the same tension. Rank these waves, from highest to lowest, on the basis of the power carried by the waves.

Highest _____   _____   _____   _____   Lowest

C-14-13 (b), (d), (c), (a)

(a)

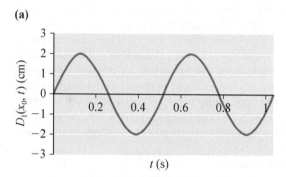

(b)

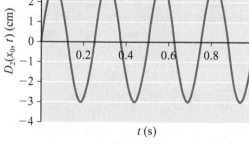
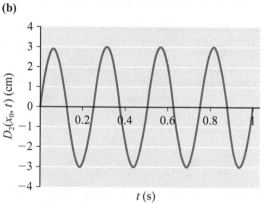

(c)

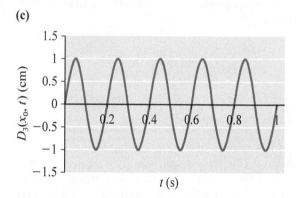

(d)

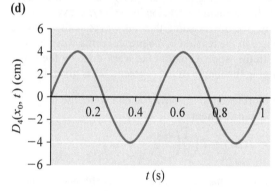

Figure 14-40   C-14-13

MAKING CONNECTIONS

Energy from Ocean Waves

Waves in an ocean are generated by the interaction of the wind with the water's surface. If the wind speed just above the water surface is greater than the speed of the water waves, then energy is transferred from the wind to the water, causing the amplitude of the waves to increase. When the following three conditions are satisfied (this happens during ocean storms), the waves are transformed into a series of repetitive swells:

■ The wind speed is large (several tens of kilometres per hour).

■ The water surface over which the wind is blowing is large (thousands of square kilometres).

■ The wind blows in one direction for a sustained period of time (several hours).

A swell is a long, thick wave that has a height of several metres. Swells are characterized by the wave height, H, which is the crest-to-trough height of the swell, and the period T between successive swells. A phrase such as, "a 3 metre swell at 12 seconds" means 3 metre-high swells that are 12 seconds apart.

Swells carry a large amount of energy and travel over large distances from the point of origin. An analysis similar to the analysis that led to the expression for the average power for one-dimensional waves (Equation (14-32)) shows that the average power per unit length carried by swells is given by

$$P_{avg} = \left(\frac{1}{32\pi}\right)\rho_{ocean}g^2TH^2$$

Here, g is the acceleration due to gravity, ρ_{ocean} is the density of ocean water, T is the period of the swells, and H is the wave height. Note that just as for one-dimensional waves, the power carried by swells is proportional to the square of the swell height and the density of the medium.

Daily swell heights for all oceans are published by the NOAA (National Oceanic and Atmospheric Administration). The period between the swells can range from 12 s to 25 s. Let us assume the following data for swells in the deep ocean:

$$\rho_{ocean} = 1030 \text{ kg/m}^3, H = 3.0 \text{ m, and } T = 14.0 \text{ s}$$

The average power per unit length carried by these swells is

$$P_{avg} = \left(\frac{1}{32\pi}\right)(1030 \text{ kg/m}^3)(9.81 \text{ m/s}^2)^2(14.0 \text{ s})(3.0 \text{ m})^2$$

$$\approx 124\,000 \text{ W/m}$$

This is a large amount of freely available, environment-friendly and renewable power, which if efficiently harnessed could greatly help in solving the world's energy problems.

There are a number of efforts underway to harvest this power. One such method is called the oscillating water column (OWC) generator (Figure 14-41). It consists of an air-filled enclosure that has two openings, one underneath and the other at the top. An air turbine (a device that generates electricity when rotated) is connected to the top opening. The enclosure is partially submerged in the water such that the bottom opening is below the waterline and the top opening is above the maximum swell height. When a swell reaches the enclosure, the column of water in the enclosure rises. The air inside the enclosure is compressed and is forced out from the top opening past the turbine. This causes the turbine to rotate and generate electricity. As the swell recedes, the air pressure within the enclosure decreases. This forces the outside air into the enclosure past the turbine, again rotating the turbine and generating electricity.

Located on the island of Islay, off Scotland's west coast, LIMPET (Land Installed Marine Powered Energy Transformer) is the first commercial OWC generator and produces 500 kW of power (Figure 14-42). Scotland is the world leader in producing energy from renewable resources.

(a)

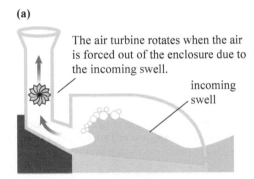

The air turbine rotates when the air is forced out of the enclosure due to the incoming swell.

incoming swell

(b)

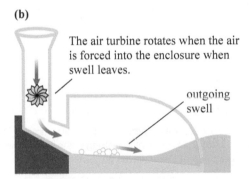

The air turbine rotates when the air is forced into the enclosure when swell leaves.

outgoing swell

Figure 14-41 An oscillating water column generator. The incoming swell forces the air out of the enclosure, which rotates the air turbine and generates electricity. The outgoing swell forces the air into the enclosure, again generating electricity.

Figure 14-42 This depiction of LIMPET, located off the west coast of Scotland, is the first commercial OWC generator.

LO 9

14-9 Superposition of Waves

The space around us is filled with electromagnetic waves of various frequencies from radio and TV stations, cellphones, satellites, and other sources. The electric current induced by this multitude of waves in the antennas of a radio receiver is quite complicated. However, if you tune a radio to a particular frequency, say 106.9 MHz, you clearly hear the broadcast at that frequency. The electromagnetic waves of a particular frequency are completely unaffected by the presence of other waves. The same is true of sound waves in air. While listening to a concert, you can often identify the sound generated by a particular instrument or singer, even though the sound waves from a number of sources reach your ears simultaneously.

Principle of Superposition of Waves When more than one wave is present in a medium at the same time, the resultant wave at any point in the medium is equal to the algebraic sum of the individual waves at that point.

For mechanical waves, the superposition principle states that a medium responds to the effects of each wave individually. The presence of other waves does not change in any way the effect of a particular wave on a medium. When multiple waves are travelling along a string, the net displacement of a segment of the string at any time is equal to the sum of the displacements produced by the individual waves.

For a one-dimensional medium carrying a set of waves with individual waveforms given by $D_1(x, t), D_2(x, t), D_3(x, t), \ldots,$ the resultant waveform, $D(x, t)$, of the medium (i.e., the shape of the resultant wave) is equal to the sum of the individual waveforms:

$$D(x, t) = D_1(x, t) + D_2(x, t) + D_3(x, t) + \cdots \quad (14\text{-}33)$$

The wave obtained by adding the component waves is called the **resultant wave**. The process of combining the waves to produce a resultant wave is called the **superposition** of waves. The physical phenomenon of two or more waves combining to produce a resultant wave is called the **interference** of waves.

EXAMPLE 14-11

Superposition of Two Harmonic Waves

Two harmonic waves with different frequencies, moving in opposite directions, are present in a string at the same time. The wave functions of these waves are given by

$$D_1(x, t) = (0.10 \text{ m}) \sin(2.0x - 10.0t)$$

$$D_2(x, t) = (0.15 \text{ m}) \sin(4.0x + 20.0t)$$

Find the displacement of a segment of the string located at $x = 1.5$ m at $t = 2.0$ s. Plot the displacement of the string between $x = 0.0$ and $x = 2.0$ m at $t = 1.0$ s.

SOLUTION

We apply the principle of superposition to obtain the resultant displacement at $x = 1.5$ m and $t = 2.0$ s.

$$D(1.5, 2.0) = D_1(1.5, 2.0) + D_2(1.5, 2.0)$$
$$= (0.10 \text{ m}) \sin(2.0 \times 1.5 - 10.0 \times 2.0)$$
$$\quad + (0.15 \text{ m}) \sin(4.0 \times 1.5 + 20.0 \times 2.0)$$
$$= 9.6 \text{ cm} + 13.5 \text{ cm}$$
$$= 23.1 \text{ cm}$$

Figure 14-43 shows the position plot of the two waves and the resultant wave. Note that displacement of each point on the resultant wave is equal to the sum of the displacements due to the component waves.

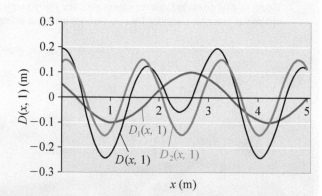

Figure 14-43 Position plots of $D_1(x, 1.0 \text{ s})$ (red), $D_2(x, 1.0 \text{ s})$ (blue), and the resultant wave (black)

The principle of superposition of waves is easy to understand for mechanical waves. A wave exerts a tension force at the points of the medium through which it is passing. When two or more waves are present, the net force at a point is equal to the sum of all the forces. Therefore, the resultant displacement of a point is equal to the sum of the displacements due to the individual waves. The fact that two or more waves can simultaneously exist in a medium without changing the shape of each other is quite remarkable.

When particles collide, they scatter off each other and their velocities change in accordance with Newton's laws of motion. Waves, on the other hand, pass through each other unaffected.

Figure 14-44 shows a series of position plots of two pulses moving in opposite direction along a stretched string. When the pulses are far apart, the segment of the string where each pulse happens to be conforms to the shape of the pulse. When the two pulses overlap, the shape of the string at the overlapping positions corresponds to the combined effect of the two pulses, in accordance with the principle of superposition.

Now consider two pulses moving in opposite directions with identical shapes but opposite displacements.

Figure 14-45 shows a series of position plots of the two pulses at different times. At the instant their centres overlap, the individual displacements cancel each other completely and the string is fully straight. Then each pulse reappears, continuing to travel in its original direction of motion with its original shape.

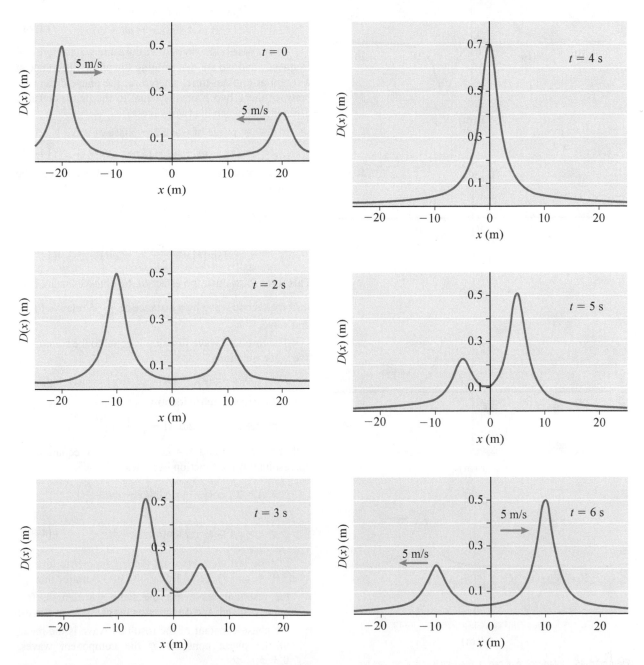

Figure 14-44   Interference of two pulses travelling in opposite direction

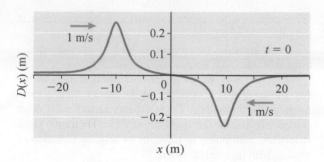

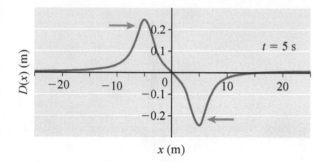

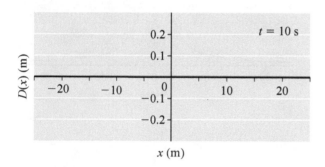

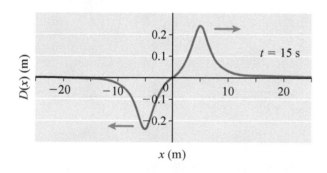

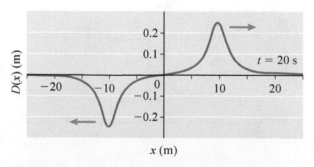

Figure 14-45 Interference of two pulses with identical shapes but opposite amplitudes travelling in opposite directions

14-10 Interference of Waves Travelling in the Same Direction

We now derive a mathematical expression for the resultant wave from the superposition of two waves with the same wavelength, frequency, amplitude, and direction but different phase constants. To simplify the algebra, we will assume that the phase constant of one of the waves is zero. Then, the wave functions of the two waves are

$$D_1(x, t) = A \sin(kx - \omega t)$$
$$D_2(x, t) = A \sin(kx - \omega t + \phi) \qquad (14\text{-}34)$$

In this case both waves have the same wavelength and frequency and we are determining the phase difference at a point at a given time. Therefore, the phase difference between these two waves is equal to the phase constant difference between the two waves:

$$\Delta\Phi = \text{phase of wave 2} - \text{phase of wave 1}$$
$$= (kx - \omega t + \phi) - (kx - \omega t) = \phi \qquad (14\text{-}35)$$

The wave function of the second wave can be rewritten as

$$D_2(x, t) = A \sin\left(k\left(x + \frac{\phi}{k}\right) - \omega t\right)$$

$$= A \sin\left(k\left(x + \frac{\phi}{2\pi}\lambda\right) - \omega t\right) \qquad (14\text{-}36)$$

This form shows that the effect of the phase constant is to shift the second wave by a distance of $\dfrac{\lambda}{2\pi}\phi$ relative to the first wave.

The resultant wave function is equal to the sum of the two wave functions:

$$D(x, t) = A \sin(kx - \omega t) + A \sin(kx - \omega t + \phi)$$

Using the trigonometric identity

$$\sin(a) + \sin(b) = 2 \cos\left(\frac{a - b}{2}\right) \sin\left(\frac{a + b}{2}\right)$$

with $a = kx - \omega t$ and $b = kx - \omega t + \phi$, the equation for the resultant wave function becomes

$$D(x, t) = 2A \cos\left(-\frac{\phi}{2}\right) \sin\left(kx - \omega t + \frac{\phi}{2}\right)$$

$$= 2A \cos\left(\frac{\phi}{2}\right) \sin\left(kx - \omega t + \frac{\phi}{2}\right) \qquad (14\text{-}37)$$

For the last step, we used the trigonometric identity $\cos(-\theta) = \cos(\theta)$. From Equation (14-37) notice that

- the resultant wave has the same wavelength, frequency, speed, and direction as the component waves;

- the phase constant of the resultant wave is the mean of the phase constant of the component waves, $\dfrac{0 + \phi}{2} = \dfrac{\phi}{2}$;

- the amplitude of the resultant wave, called the **resultant amplitude**, is $2A \cos\left(\dfrac{\phi}{2}\right)$.

The resultant amplitude depends on the amplitude of the component waves and on the difference between their phase constants. Figure 14-46 shows a plot of the resultant amplitude as a function of ϕ.

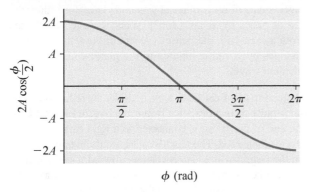

Figure 14-46 The amplitude of the resultant wave as a function of the difference between the phase constants of two waves that have the same wavelength, frequency, amplitude, and direction

When the phase difference between the two waves is zero or an integer multiple of 2π radians, the waves are said to be in phase. In this case the waves reinforce each other, with the crest of one coinciding with the crest of the other, as shown in Figure 14-47. The amplitude of the resultant wave is $2A$. Such an alignment of the waves is called **constructive interference** of waves.

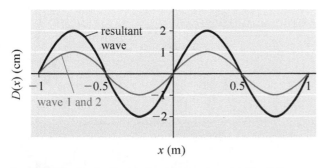

Figure 14-47 Constructive interference of two waves moving in the same direction with equal amplitudes and wavelengths, and a phase constant difference of 0 rad

When the phase difference between the two waves is an odd integer multiple of π radians the waves are said to be out of phase. In this case the two waves completely cancel each other, as shown in Figure 14-48. Now the displacements of the two component waves are exactly equal and opposite at every point, with the crest of one coinciding with the trough of the other, resulting in **destructive interference** of the waves.

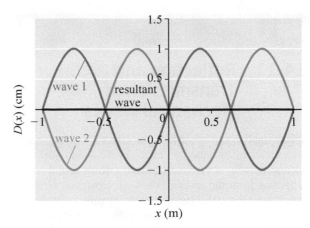

Figure 14-48 Destructive interference of two waves moving in the same direction with equal amplitudes and wavelengths, and a phase constant difference of π rad

For the phase difference that is other than an integer multiple of π rad, the resultant amplitude is intermediate between the constructive ($2A$) and destructive (0) values. As an example, Figure 14-49 shows the interference at time $t = 0$ of two waves with amplitudes of 1.0 m, wavelengths of 1.0 m, and a phase constant difference of $\pi/3$ rad. From Equation (14-37), the resultant wave has an amplitude of $2 \times (1.0 \text{ m}) \times \cos(\pi/6) = 1.73$ m. From the plot, we can see that the crests of the component waves are displaced with respect to each other by

$$\frac{\phi}{2\pi}\lambda = \frac{\pi/3 \text{ rad}}{2\pi \text{ rad}} \times 1.0 \text{ m} = 0.16 \text{ m}.$$

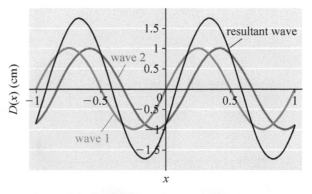

Figure 14-49 The interference of two waves of equal amplitudes and wavelengths and a phase constant difference of $\pi/3$ rad

✓ CHECKPOINT

C-14-14   Wave Interference

Two waves with the same amplitude and frequency are $\pi/2$ rad out of phase when they arrive at point P. What is the amplitude of the resultant wave at P?

(a) $2A$

(b) $1.7A$

(c) $1.4A$

(d) $0.7A$

C-14-14 (c)

14-11 Reflection and Transmission of Mechanical Waves

When you listen to music in a concert hall, you may notice reverberation, an important factor in the sound quality of a concert hall. Reverberation results from sound waves reflecting from the walls of the hall. However, someone standing in the lobby can also hear the music, albeit considerably muffled. So, a portion of the sound waves have gone through the walls. In this section, we introduce the concepts of reflection and transmission of mechanical waves at boundaries.

Reflection at a Fixed End Consider a transverse pulse travelling along a string toward an end that is tied to a rigid support, as shown in Figure 14-50. The transverse displacement of the **incident pulse** shown is upward, so the string exerts an upward force on the support when the pulse reaches the fixed end. The rigid support exerts an equal and opposite reaction force on the string, in accordance with Newton's third law of motion. The reaction force generates an inverted pulse that moves in the opposite direction. A reflection from a rigid boundary is called a **hard reflection**.

When a wave reflects from a fixed end, the reflected wave is inverted (compared to the incident wave) and moves in the opposite direction. Since the rigid end has zero displacement, the incident and the reflected waves always cancel each other at this end. Therefore, they are out of phase by π rad (180°). When a wave is reflected from a *fixed* end, the phase constant of the reflected wave changes by π rad compared to the phase constant of the incident wave.

Reflection at a Free End Now consider the reflection of the same pulse from an end that is free to move. In Figure 14-51, the end of the string is connected to a light ring that slides freely along a rod. The incident pulse exerts an upward force on the ring, causing it to accelerate upward. Due to inertia, the ring keeps moving upward, overshooting the maximum height of the pulse and pulling the string with it. The reaction force exerted on the string by the ring generates a *backward-moving pulse that is not inverted*. Such a reflection is called

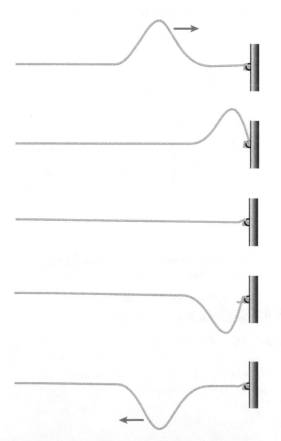

Figure 14-50 Hard reflection: A pulse reflected from a fixed end is inverted. The phase of reflected pulse is changed by π radians.

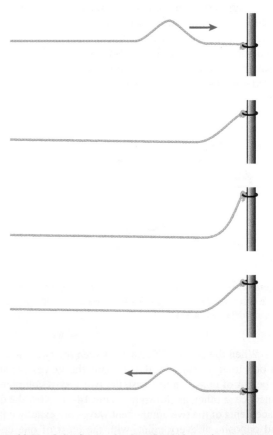

Figure 14-51 Soft reflection: A pulse reflected from a free end is not inverted. The phase of the reflected pulse does not change.

a **soft reflection**. If the amplitude of the pulse is A, the maximum displacement of the ring from its equilibrium position is $2A$. Therefore, for a soft reflection, the incident and the reflected pulses are in phase with each other.

The same is true for waves. When a wave reflects from a free end, the reflected wave is not inverted (compared to the incident wave) and moves in the opposite direction. In this case, the incident and the reflected waves are in phase with each other.

Often, an end is neither completely rigid nor fully free. Consider a string of linear mass density μ_1 attached to a string of linear mass density μ_2, with $\mu_1 < \mu_2$. When a pulse travelling in the lighter string reaches the junction between the two strings, part of the incident pulse is reflected and part is transmitted into the thicker string. The reflected pulse is inverted, but the transmitted pulse is not (Figure 14-52). For the reverse situation, when an incident pulse travelling in the thicker string reaches the boundary with the lighter string, neither the reflected pulse nor the transmitted pulse is inverted (Figure 14-53).

The total energy of the incident wave is equal to the sum of the energies of the reflected and transmitted waves.

Although our discussion here has dealt only with one-dimensional waves on strings, two- and three-dimensional mechanical waves and electromagnetic waves show the same behaviour at boundaries between media.

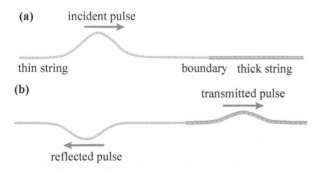

(a) incident pulse

thin string   boundary   thick string

(b) transmitted pulse

reflected pulse

Figure 14-52 When a pulse travels from a less dense medium to a denser medium ($\mu_1 < \mu_2$), the reflected pulse is inverted, but the transmitted pulse is not.

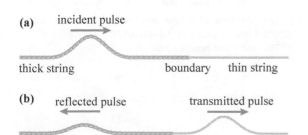

(a) incident pulse

thick string   boundary   thin string

(b) reflected pulse   transmitted pulse

Figure 14-53 When a pulse travels from a dense medium to a less dense medium ($\mu_1 > \mu_2$), neither the reflected pulse nor the transmitted pulse is inverted.

ONLINE ACTIVITY

Reflection of Waves

The e-resource that accompanies every new copy of this textbook contains an Online Activity using the PhET simulation "Reflection of Waves." Work through the simulation and accompanying questions to gain an understanding of how waves reflect from boundaries.

LO 12

14-12 Standing Waves

In Section 14-10, we discussed the interference of two waves that are moving in the same direction. Now we consider two harmonic waves of equal amplitude, wavelength, and frequency that are moving in *opposite* directions. For mathematical simplicity, we assume that the phase constants of both waves are zero; however, in general, they may be unequal. The wave functions for the two waves are

$$D_1(x, t) = A \sin(kx - \omega t)$$

(wave moving in direction of increasing x)

$$D_2(x, t) = A \sin(kx + \omega t)$$

(wave moving in direction of decreasing x) (14-38)

According to the principle of superposition, the resultant wave function is

$$\begin{aligned} D(x, t) &= D_1(x, t) + D_2(x, t) \\ &= A \sin(kx - \omega t) + A \sin(kx + \omega t) \\ &= A[\sin(kx - \omega t) + \sin(kx + \omega t)] \end{aligned} \quad (14\text{-}39)$$

Using the trigonometric identity
$$\sin(a - b) + \sin(a + b) = 2\sin(a) \cos(b),$$
with $a = kx$ and $b = \omega t$, we obtain

$$D(x, t) = 2A \sin(kx) \cos(\omega t) \quad (14\text{-}40)$$

What kind of wave does this wave function describe? A travelling wave must have the position x and time t in the combination $x \mp vt$, where v is the wave speed. In Equation (14-40), the x and t variables appear separately, with x in the argument of the sine function and t in the cosine function. Therefore, the wave described by Equation (14-40) is not a travelling wave. We call it a **standing wave**. Let us define a position-dependent amplitude, $A(x)$, as

$$A(x) = 2A \sin(kx) = 2A \sin\left(2\pi \frac{x}{\lambda}\right) \quad (14\text{-}41)$$

Using this definition, we can rewrite Equation (14-40) as

$$D(x, t) = A(x) \cos(\omega t) \quad (14\text{-}42)$$

When two waves of equal wavelength, frequency, and amplitude but moving in opposite directions combine, each segment of the string oscillates in simple harmonic motion with the frequency of the individual waves and an amplitude that depends on the location of the segment along the string.

Figure 14-54 shows a plot of $A(x)$ as a function of position. Amplitude is a sine function, so certain points on the string have zero amplitude and remain at rest at all times. These points are called **nodes**. The points that move with the maximum possible amplitude of $2A$ are called **antinodes**. All the other points have amplitudes between zero and $2A$. Any two points that are one wavelength apart have the same amplitude because

$$A(x_0 + \lambda) = 2A \sin\left(2\pi \frac{x_0 + \lambda}{\lambda}\right) = 2A \sin\left(2\pi \frac{x_0}{\lambda} + 2\pi\right)$$

$$= 2A \sin\left(2\pi \frac{x_0}{\lambda}\right) = A(x_0) \qquad (14\text{-}43)$$

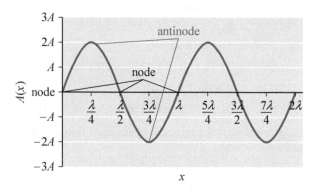

Figure 14-54 The amplitude $A(x)$ of a standing wave as a function of position along a string

Since nodes do not move, no energy flows along a standing wave. As each segment of the medium between two consecutive nodes oscillates in simple harmonic motion its energy continuously transforms between kinetic energy and elastic potential energy, just as for a single harmonic oscillator.

Location of Nodes and Antinodes At the nodes of a standing wave, $A(x) = 0$. Therefore, nodes occur when

$$\sin\left(\frac{2\pi}{\lambda} x\right) = 0$$

$$\frac{2\pi}{\lambda} x = m\pi \qquad m = 0, \pm 1, \pm 2, \ldots$$

$$x = m \frac{\lambda}{2} \qquad m = 0, \pm 1, \pm 2, \ldots$$

$$x = 0, \pm \frac{\lambda}{2}, \pm \lambda, \pm \frac{3\lambda}{2}, \pm 2\lambda, \ldots \quad (14\text{-}44)$$

Thus, the distance between two consecutive nodes is half a wavelength.

At the antinodes of a standing wave, $A(x) = \pm 2A$, which occurs when

$$\sin\left(\frac{2\pi}{\lambda} x\right) = \pm 1$$

$$\frac{2\pi}{\lambda} x = \left(m + \frac{1}{2}\right)\pi \qquad m = 0, \pm 1, \pm 2, \ldots$$

$$x = \left(m + \frac{1}{2}\right)\frac{\lambda}{2} \qquad m = 0, \pm 1, \pm 2, \ldots$$

$$x = \pm \frac{\lambda}{4}, \pm \frac{3\lambda}{4}, \pm \frac{5\lambda}{4}, \ldots \qquad (14\text{-}45)$$

EXAMPLE 14-12

Standing Wave

The amplitude of a standing wave is given by

$$A(x) = (25.0 \text{ cm}) \sin(2.00x)$$

where x is in meters. Determine

(a) the amplitude and the wavelength of the constituent travelling waves;

(b) the location of the first three nodes and the first three antinodes along the positive x-axis;

(c) the location of the first point from the origin where the amplitude is 0.20 m.

SOLUTION

(a) Comparing the given amplitude to Equation (14-41), we find that

$$2A = 25.0 \text{ cm} \quad \text{and} \quad \frac{2\pi}{\lambda} = 2.00 \text{ rad/m}$$

Therefore,

$$A = 12.5 \text{ cm} \quad \text{and} \quad \lambda = \pi \text{ m} = 3.14 \text{ m}$$

(b) The positions of nodes and antinodes can be calculated from Equations (14-44) and (14-45).
The first three nodes occur at

$$x = 0.00 \text{ m}, \quad x = \frac{\lambda}{2} = \frac{3.14 \text{ m}}{2} = 1.57 \text{ m}, \quad x = \lambda = 3.14 \text{ m}$$

The first three antinodes occur at

$$x = \frac{\lambda}{4} = \frac{3.14 \text{ m}}{4} = 0.78 \text{ m}, \quad x = \frac{3\lambda}{4} = 2.36 \text{ m},$$

$$x = \frac{5\lambda}{4} = 3.92 \text{ m}$$

(c) To determine the location of the point that has amplitude of 0.20 m, we substitute this value in the amplitude equation and solve for x:

$$0.20 \text{ m} = (0.25 \text{ m}) \sin(2.00x)$$

$$\sin(2.00x) = 0.80 \Rightarrow 2.00x = \sin^{-1}(0.80) \Rightarrow x = 0.46 \text{ m}$$

Making sense of the result:

The amplitude is zero at $x = 0.00$ m, and the first antinode at $x = 0.78$ m has a displacement of 0.25 m. Therefore, the first occurrence of an amplitude of 0.20 m must be somewhere between 0.00 m and 0.78 m from the origin, in agreement with the location calculated.

The distance between consecutive antinodes is also half a wavelength. An adjacent node and an antinode are a quarter of a wavelength apart.

To see how a string oscillates in a standing wave pattern, we write the wave function in terms of the time period ($T = 1/f$):

$$D(x, t) = 2A \sin\left(\frac{2\pi}{\lambda}x\right) \cos\left(\frac{2\pi}{T}t\right) \qquad (14\text{-}46)$$

Figure 14-55 shows the displacement of a section of the oscillating string at intervals of $T/8$ from $T = 0$ to $t = T/2$. Table 14-2 compares the displacements of the section of the string between the first two nodes during the first half of a cycle. The motion for the next half of the period is in the opposite direction. This oscillatory motion repeats every cycle.

All points between two consecutive nodes oscillate in phase with each other. The mean speed is greatest for the antinode because it has to cover the longest distance ($8A$) in one period. The speed decreases away from the antinode and is zero at the nodes. Note that the motion of the section between the next two nodes is π rad out of phase with the first section. The sine term in the wave function accounts for this property:

$$A\left(x_0 + \frac{\lambda}{2}\right) = 2A \sin\left(\frac{2\pi}{\lambda}\left(x_0 + \frac{\lambda}{2}\right)\right) = 2A \sin\left(\frac{2\pi}{\lambda}x_0 + \pi\right)$$

$$= -2A \sin\left(2\pi \frac{x_0}{\lambda}\right) = -A(x_0) \qquad (14\text{-}47)$$

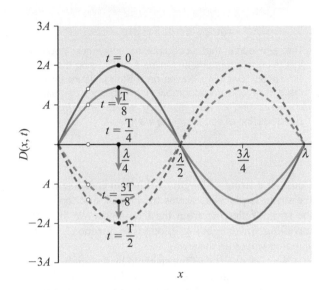

Figure 14-55 Displacement versus position for a standing wave on a string during the first half of a cycle. The position x is expressed in multiples of the wavelength λ.

Table 14-2 Displacements of string elements between two nodes of a standing wave during half a time period

Time	Wave function	Displacement
0	$D(x, 0) = 2A \sin\left(\frac{2\pi}{\lambda}x\right) \cos(0) = 2A \sin\left(\frac{2\pi}{\lambda}x\right)$	each segment is at its maximum displacement
$T/8$	$D\left(x, \frac{T}{8}\right) = 2A \sin\left(\frac{2\pi}{\lambda}x\right) \cos\left(\frac{2\pi}{T}\frac{T}{8}\right) = \sqrt{2}A \sin\left(\frac{2\pi}{\lambda}x\right)$	displacement of each segment decreases by factor of $\sqrt{2}$
$T/4$	$D(x, T/4) = 2A \sin\left(\frac{2\pi}{\lambda}x\right) \cos\left(\frac{2\pi}{T}\frac{T}{4}\right) = 0$	each segment reaches equilibrium position
$3T/8$	$D\left(x, \frac{3T}{8}\right) = 2A \sin\left(\frac{2\pi}{\lambda}x\right) \cos\left(\frac{2\pi}{T}\frac{3T}{8}\right) = -\sqrt{2}A \sin\left(\frac{2\pi}{\lambda}x\right)$	each segment reaches $1/\sqrt{2}$ of maximum displacement at the opposite side of the equilibrium position
$T/2$	$D(x, T/2) = 2A \sin\left(\frac{2\pi}{\lambda}x\right) \cos\left(\frac{2\pi}{T}\frac{T}{2}\right) = -2A \sin\left(\frac{2\pi}{\lambda}x\right)$	each segment is at its maximum displacement at the opposite of the equilibrium position

14-13 Standing Waves on Strings

ONLINE ACTIVITY

Standing Waves on a String

The e-resource that accompanies every new copy of this textbook contains an Online Activity using the PhET simulation "Standing Waves on a String." Work through the simulation and accompanying questions to gain an understanding of how standing waves arise from the superposition of two travelling waves.

Consider a string of length L with both ends fixed. When the string is plucked, waves travel back and forth along the string, reflecting from the fixed ends. We then have travelling waves moving in opposite directions, creating standing waves on the string.

We take one end of the string to be the origin, and the other end to be at $x = L$. Because the string is clamped at both ends the amplitude must be zero at $x = 0$ and at $x = L$. The amplitude in 14-41 is zero at $x = 0$ and for it to be zero at $x = L$ we must have,

$$\sin\left(\frac{2\pi}{\lambda}L\right) = 0 \qquad (14\text{-}48)$$

and

$$\frac{2\pi}{\lambda}L = m\pi \qquad (14\text{-}49)$$

where m is a positive, nonzero integer $(1, 2, 3, \ldots)$.

Rearranging Equation (14-49), we find that a string with both ends fixed can oscillate in a standing wave pattern only with following wavelengths:

$$\lambda_m = \frac{2L}{m} \quad m = 1, 2, 3, 4, \ldots \qquad (14\text{-}50)$$

The longest of these wavelengths is $\lambda_1 = 2L$. The series of wavelengths continues with $\lambda_2 = L$, $\lambda_3 = \frac{2L}{3}$, $\lambda_4 = \frac{L}{2}$, and so on. Notice that each of these wavelengths is such that an integral number of half wavelengths Figures 14-56 and 14-57. These standing waves are called the **normal modes** of vibration of the string.

The frequencies corresponding to the normal modes of vibration are

$$f_m = \frac{v}{\lambda_m} = \frac{m}{2L}v = \frac{m}{2L}\sqrt{\frac{T}{\mu}} \qquad (14\text{-}51)$$

The lowest frequency corresponds to the longest wavelength, $\lambda_1 = 2L$, and is called the **fundamental frequency** or the **first harmonic**:

Fundamental frequency:

$$f_1 = \frac{v}{\lambda_1} = \frac{1}{2L}v = \frac{1}{2L}\sqrt{\frac{T}{\mu}} \qquad (14\text{-}52)$$

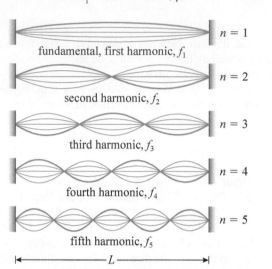

Figure 14-56 Standing waves on a string: the first five harmonics

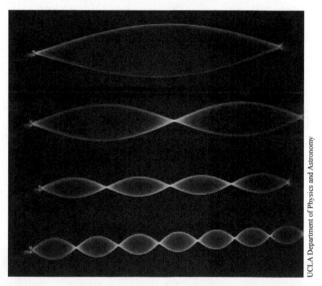

Figure 14-57 The envelopes of a fluorescent string vibrating in various normal modes. The images were taken by shining a strobe light on the string.

Notice that the fundamental frequency of a string is proportional to the square root of the tension in the string and is inversely proportional to its length and to the square root of the linear mass density. Higher frequencies of vibration correspond to higher values of m and are integer multiples of the fundamental frequency, since:

$$f_m = mf_1 \quad m = 1, 2, 3, 4, \ldots \qquad (14\text{-}53)$$

The allowed frequencies are called **harmonics** or **resonant frequencies**. The first few harmonics are

$$f_1 = \frac{1}{2L}\sqrt{\frac{T}{\mu}}$$ fundamental or first harmonic

$$f_2 = 2f_1$$ second harmonic

$$f_3 = 3f_1$$ third harmonic

$$f_4 = 4f_1$$ fourth harmonic   (14-54)

When plucked, a string can vibrate in a single normal mode or a combination of several modes. Notice that m is equal to the number of antinodes between the fixed ends of the string.

As we will see in the next chapter, sound waves in a wind instrument also form standing wave patterns.

EXAMPLE 14-13

String Vibrations

A violin string is 0.32 m long and has a linear mass density of 3.83×10^{-4} kg/m. The tension in the string is kept at 70.0 N.

(a) What is the wave speed in the string?

(b) If the string is plucked and allowed to vibrate, approximately how many times would a wave reflect from one end of the string in one second?

(c) What are the wavelengths and frequencies of the first three normal modes of vibration of the string? What is the spacing between two consecutive nodes for each mode? Sketch the vibration pattern of these modes.

(d) By what amount should the tension be changed to decrease the fundamental frequency by 7 Hz?

SOLUTION

When plucked, the string will vibrate in a standing wave pattern with the wavelengths and frequencies of normal modes given by Equations (14-50) and (14-53).

(a) The wave speed in the string is determined by the tension and linear mass density of the string:

$$v = \sqrt{\frac{T}{\mu}} = \sqrt{\frac{70.0\ \text{N}}{3.83 \times 10^{-4}\ \text{kg/m}}} = 427\ \text{m/s}$$

(b) The time that the waves take to travel from one end of the string to the other is

$$\Delta t = \frac{\text{length of the string}}{\text{wave speed}} = \frac{0.32\ \text{m}}{427\ \text{m/s}} = 7.5 \times 10^{-4}\ \text{s}$$

Each reflection from a given end requires a back-and-forth trip. Therefore, the number of times a wave reflects from a given fixed end in one second is

$$\frac{1.0\ \text{s}}{2 \times \Delta t} = \frac{1.0\ \text{s}}{2 \times 7.5 \times 10^{-4}\ \text{s}} = 668$$

(c) For the first harmonic:

    wavelength: $\lambda_1 = 2L = 2 \times 0.32\ \text{m} = 0.64\ \text{m}$

    frequency: $f_1 = \dfrac{v}{\lambda_1} = \dfrac{427\ \text{m/s}}{0.64\ \text{m}} = 668\ \text{Hz}$

The distance between consecutive nodes: $= \lambda_1/2 = 0.32$ m

For the second harmonic:

    wavelength: $\lambda_2 = 2L/2 = 0.32$ m

    frequency: $f_2 = 2f_1 = 1336$ Hz

The distance between consecutive nodes: $= \lambda_2/2 = 0.16$ m

For the third harmonic:

    wavelength: $\lambda_3 = 2L/3 = 0.21$ m

    frequency: $f_3 = 3f_1 = 2004$ Hz

The distance between consecutive nodes: $= \lambda_3/2 = 0.11$ m

The vibration patterns of these modes are shown in Figure 14-58.

(d) The new fundamental frequency will be $f_1' = 668\ \text{Hz} - 7\ \text{Hz} = 661\ \text{Hz}$. To decrease the fundamental frequency, the tension in the string must be decreased. If T' is the new tension, then

$$f_1' = \frac{1}{2L}\sqrt{\frac{T'}{\mu}}$$

Squaring the above equation, we can write

$$T' = 4L^2\mu(f_1')^2$$

The fractional decrease in tension needed to lower the frequency by 7.000 Hz is therefore,

$$\frac{T - T'}{T} = \frac{(f_1)^2 - (f_1')^2}{(f_1)^2} = \frac{(668\ \text{Hz})^2 - (661\ \text{Hz})^2}{(668\ \text{Hz})^2} = 0.0208$$

Therefore, the tension in the string needs to be decreased by $0.0208 \times 70.0\ \text{N} = 1.46\ \text{N}$.

Making sense of the result:

The fundamental frequency calculated in part (c) is the same as the number of times the wave reflects from a fixed end of the string (calculated in part (b)).

Figure 14-58 Example 14-13. Envelopes of the displacements for (a) the first harmonic, (b) the second harmonic, and (c) the third harmonic.

C-14-15 Standing Wave Nodes

How many nodes are present between the fixed ends of a string vibrating at its seventh harmonic?

(a) 5
(b) 6
(c) 7
(d) 8

C-14-15 (b)

 MAKING CONNECTIONS

Lasers

Standing waves are also formed by electromagnetic waves. For example, there are standing waves in lasers. A gas laser consists of a cylindrical tube that is filled with gases, such as helium and neon. The cavity has a reflecting mirror at one end and a partially reflecting mirror at the other (Figure 14-59). An electric voltage pumps energy into the gases, putting many of the gas atoms into an excited state. The atoms produce light when they release the absorbed energy. This light reflects back and forth between the two mirrors, thus creating standing waves within the cavity. Each pass of the light between the mirrors stimulates more of the excited atoms to emit light. The partially reflecting mirror allows some of the light to pass through. Because of the stimulated emission process, laser light has spatial coherence, allowing it to be very tightly focused.

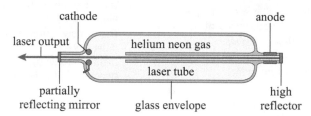

Figure 14-59 The inner workings of a laser

WAVES AND OSCILLATIONS

FUNDAMENTAL CONCEPTS AND RELATIONSHIPS

A mechanical wave is the propagation of a disturbance through a medium. It takes energy to generate a wave, and the wave carries this energy with it as it passes through a medium. In the presence of a wave, the particles of the medium oscillate about their equilibrium positions but are not carried along with the wave. The wave speed in a medium depends on the properties of the medium.

In longitudinal waves, the motion of the particles of the medium is in the direction of propagation of the wave. In transverse waves, the motion of the particles of the medium is perpendicular to the direction of wave propagation.

The shape of a wave is a function of position (x) and time (t). If the shape of a wave is described by a function $D(x)$ at $t = 0$, then its shape at time x is obtained by the substitution $x \rightarrow x \mp vt$, where v is the speed of the wave. $D(x, t)$ is called the wave function.

A sinusoidal wave is generated by a source that is undergoing simple harmonic motion. The wave function of a sinusoidal wave is given by

$$D(x, t) = A \sin(kx - \omega t + \phi)$$

$$= A \sin\left(\frac{2\pi}{\lambda} x - 2\pi ft + \phi\right)$$

where A is the amplitude, λ is the wavelength, f is the frequency, $k = \dfrac{2\pi}{\lambda}$ is the wave number, $\omega = 2\pi f = \dfrac{2\pi}{T}$ is the angular frequency, and ϕ is the phase constant.

For a continuous wave, the wave speed is related to the wavelength and frequency by $v = \lambda f$.

A position plot displays the shape of the wave as a function of position for a fixed time. A time plot displays the displacement of a point of the medium as a function of time.

The value of the phase constant can be determined from a position or a time plot.

The average power transported by a wave is

$$P_{avg} = 2\pi^2 \mu v f^2 A^2$$

When two or more waves are simultaneously present in a medium, the resultant wave function is equal to the sum of the individual wave functions.

When two waves with the same wavelength and frequency and travelling in the same direction combine, the amplitude of the resultant wave is maximum when the phase difference between the two waves is an integer multiple of 2π rad, and zero when the difference is an odd multiple of π rad.

Two waves with the same frequency and amplitude travelling in opposite directions produce standing waves. The wave function of a standing wave is

$$D(x, t) = 2A \sin(kx) \cos(\omega t)$$

where A is the amplitude of each wave. In a standing wave, energy is not transported through the medium.

The harmonics for a string that is fixed at both ends are

$$\lambda_m = \frac{2L}{m}; \quad f_m = m \frac{v}{2L}$$

where L is the length of the string, v is the wave speed, and $m = 1, 2, 3, \ldots$.

Key Terms: amplitude, antinodes, constructive interference, continuous wave, crest, destructive interference, displacement function, electromagnetic radiation, electromagnetic wave, first harmonic, fundamental frequency, hard reflection, harmonic wave, harmonics, in phase, incident pulse, interference, longitudinal wave, mechanical wave, medium, nodes, normal modes, out of phase, phase difference, phase, position plot, principle of superposition, pulse, resonant frequencies, resultant amplitude, resultant wave, sinusoidal wave, snapshot graphs, soft reflection, standing wave, superposition, transverse wave, trough, wave cycle, wave function, wave number, wave speed, wave, waveform, wavelength

QUESTIONS

1. Indicate whether each of the following statements is true (T) or false (F). Explain your answer.
   (a) As a wave passes through a medium, it carries the particles of the medium with it.
   (b) The wave speed of a mechanical wave depends on the frequency of the source.
   (c) Waves with different wavelengths, travelling in the same medium, can have different speeds.
   (d) Mechanical waves do not require a medium to travel.
   (e) A wave, travelling in a uniform medium, can have a nonzero acceleration.
   (f) A travelling wave can have zero energy.
   (g) When two waves of amplitudes 1.0 m and 2.0 m interfere, the amplitude of the resultant wave can be greater than 3.0 m.
   (h) Standing waves can be produced by combining two waves that are moving in the same direction.
2. Waves carry energy. Do they carry momentum?
3. The speed of sound waves in air is 340 m/s, and the speed of light is 3×10^8 m/s. Consider a sound wave and a light wave of 20 000 Hz. Which wave has the longer wavelength?

4. In the presence of a travelling wave, are there any particles of the medium that are always at rest?

5. Suppose a longitudinal harmonic wave is moving through a spring with a speed of 10.0 m/s. Does each coil of the spring oscillate about its equilibrium position with a speed of 10.0 m/s?

6. A pulse with total energy E_1 is travelling along a string that is connected to another string. At the boundary of the two strings, the pulse is partly transmitted into the second string and partly reflected into the first string. The energy of the reflected pulse is E_2, and the energy of the transmitted pulse is E_3. How are E_1, E_2, and E_3 related? What assumptions did you make to reach your conclusion?

7. When you throw a stone on the surface of a pond, the wave travels away from the stone and its amplitude decreases. Why?

8. Two wires, stretched under the same tension, have linear mass densities that differ by a factor of 4. The wave speeds in the wires differ by a factor of
   (a) ½ (b) 1 (c) 2 (d) 4 (e) 16

9. When the frequency of a wave is doubled, the energy required to generate the wave
   (a) remains the same;
   (b) increases by a factor of 2;
   (c) increases by a factor of 4;
   (d) decreases by a factor of 2;
   (e) decreases by a factor of 4.

10. A source is generating waves at a certain frequency. When the power of the source is doubled, the amplitude of the wave
    (a) remains the same;
    (b) increases by a factor of $\sqrt{2}$;
    (c) increases by a factor of 2;
    (d) decreases by a factor of $\sqrt{2}$;
    (e) decreases by a factor of 2.

11. Why do two travelling waves (moving in opposite directions) with the same amplitude and frequency generate a standing wave? Can we produce a standing wave by combining two travelling waves of (a) the same amplitude but different frequencies, and (b) different amplitudes but the same frequency?

12. A travelling wave transports energy between two points. Can a standing wave transport energy between two points?

13. Standing waves with a frequency of 440 Hz are generated on two strings of the same length. Must the two strings have exactly the same linear mass density?

14. Two waves of amplitude A are travelling on a string. The waves interfere to produce a resultant wave of amplitude $A/2$. Is the total energy of the resultant wave equal to the sum of the energies of the constituent waves? Explain your reasoning.

PROBLEMS BY SECTION

For problems, star ratings will be used, (✷, ✷✷, or ✷✷✷), with more stars meaning more challenging problems.

Section 14-1 The Nature, Properties, and Classification of Waves

15. ✷ A wave has a wavelength of 2.00 m and a frequency of 10 Hz. Determine
    (a) the wave number;
    (b) the angular frequency;
    (c) the wave speed.

16. ✷ The speed of sound in air is 340.0 m/s. The angular frequency of a sound wave is 600.0 rad/s. Determine the wave's
    (a) frequency;
    (b) period;
    (c) wavelength.

17. ✷ A wave is travelling along a stretched string with a speed of 150.0 m/s. The wave number of the wave is 6.00 rad/m. Determine the wave's
    (a) frequency;
    (b) period;
    (c) wavelength.

18. ✷ Electromagnetic waves in a microwave oven have a frequency of 4.0×10^9 Hz. What is the wavelength of the waves?

19. ✷ Human ears can detect sounds with frequencies from 20 Hz to 20 000 Hz. The speed of sound in air is approximately 340 m/s. What is the corresponding range of wavelengths that the human ear can detect?

20. ✷ A transverse wave takes 2.0 s to travel the full length of a 3.0 m string. What is the frequency of the wave when its wavelength is 0.50 m?

21. ✷ The speed of sound waves in ocean water is approximately 1500 m/s. Dolphins produce sound waves with frequencies in the range of 250 Hz to 150 kHz. What is the range of wavelengths of the waves in ocean water and in air?

22. ✷ The speed of sound in air is 340 m/s, and the speed of light is 3.0×10^8 m/s. What frequency of light has the same wavelength as a 200.0 Hz sound wave?

Sections 14-2 and 14-3 The Motion of a Disturbance in a String and Equation for a Pulse Moving in One Dimension

23. ✷✷ At time $t = 0$, the transverse displacement of a string due to a pulse travelling through it is given by the equation

$$D(x, 0) = \frac{0.3}{x^2 + 1.2}$$

where x and D are measured in metres. When the pulse is moving at 4.0 m/s toward the positive x-axis, what is the displacement of a segment of the string that is located at $x = 1.5$ m at $t = 2.0$ s? What are the maximum and minimum displacements of this segment ?

24. ✷ Rank the following pulses in order of wave speed and amplitude, from lowest to highest, using an equality sign where needed. Also, indicate the direction of motion of each pulse.

(a) $D(x, t) = \dfrac{3}{4 + (x - 0.5t)^2}$

(b) $D(x, t) = \dfrac{2}{3 + (x + t)^2}$

(c) $D(x, t) = \dfrac{11}{15 + (x - 5t)^2}$

(d) $D(x, t) = \dfrac{7}{5 + (x + 4t)^2}$

25. ✶✶ The displacement function for a pulse passing through a string is given by

$$D(x, t) = \frac{-3.0}{6.0 + (x + 3.0t)^2}$$

where x and D are in metres, and t is in seconds.
(a) What is the speed of the pulse, and in which direction is it travelling?
(b) What is the displacement of a point located at $x = 2.0$ m at $t = 3.0$ s?
(c) What is the maximum displacement of a point on the string?
(d) What is the transverse speed of a point that is located at $x = 2.0$ m at $t = 3.0$ s?
(e) Draw a displacement versus position (x) graph for $t = 0, 1, 2,$ and 3 s for this waveform.

26. ✶ At time $t = 0$, the transverse displacement of a string due to a pulse travelling at a speed of 2.0 m/s is given by

$$D(x, 0) = \frac{3.0}{x^4 + 10.0}$$

where x and D are in metres.
(a) What is the amplitude of the pulse?
(b) What is the displacement of a point located at $x = 1.0$ m at $t = 0.0$ s?
(c) The pulse is moving toward increasing x. Write an equation that describes the displacement of the string as a function of x and t.
(d) What is the transverse speed of a point located at $x = 2.0$ m at $t = 1.0$ s?

27. ✶✶ The displacement function for a travelling pulse is given by the equation

$$D(x, t) = \begin{cases} +2 \text{ m} & \text{if } |x - t| \le 1 \\ -2 \text{ m} & \text{if } |x - t| > 1 \end{cases}$$

where x is in metres, and t is in seconds.
(a) Draw a displacement versus position graph for the pulse for $t = 1, 2,$ and 3 s.
(b) Determine the wave speed from the graph. In which direction is the disturbance travelling?
(c) What is the displacement of a point located at $x = 0.5$ m at $t = 1.0$ s?
(d) Write the equation for a pulse that has the same shape but is moving in the opposite direction.

28. ✶ The displacement function for a pulse travelling along a string is given by the equation

$$D(x, t) = \frac{4.0}{(x - 2.0t - 10.0)^2 + 6.0}$$

where x is in metres, and t is in seconds.
(a) What is the speed of the pulse, and in which direction is it travelling?
(b) What is the displacement of a point located at $x = 4.0$ m at $t = 1.0$ s?
(c) What is the speed of a point that is located at $x = 5.0$ m at $t = 1.0$ s?
(d) What is the maximum displacement of a point as the pulse passes through it?

29. ✶✶ Two pulses travelling on a string are described by the displacement functions

$$D_1(x, t) = \frac{2}{(x - 5t + 10)^2 + 4}$$

and

$$D_2(x, t) = \frac{-2}{(x + 5t - 10)^2 + 4}$$

(a) Sketch the shape of the pulses at $t = 0$.
(b) Where is the peak of each pulse located when $t = 0$?
(c) At what time do the two pulses cancel each other?
(d) Is there a point along the string that has zero displacement for all t? If yes, where is it located?

Section 14-4 Transverse Speed and Wave Speed

30. ✶ The E string on a violin has a diameter of 0.0039 cm and is made of steel of density (ρ) 7.86 g/cm^3. What is the linear mass density of the string?
31. ✶ The density of steel is 7.86 g/cm^3. It is observed that transverse waves in a 0.40 mm diameter steel wire propagate at 160 m/s. What is the tension in the wire?
32. ✶✶ A steel wire of cross-sectional area 1.00×10^{-4} m^2 is under a tension of 1.00×10^3 N. At what speed does a transverse wave move along the wire? The density of steel is 7860 kg/m^3.
33. ✶ A 50.0 m string of uniform density has a mass of 0.10 kg. When the tension in the string is 100.0 N, what is the speed of a pulse travelling along the length of the string?
34. ✶✶ The string in problem 33 is hanging from a tall ceiling, and a weight is tied to its lower end to produce a tension of 10.0 N. A pulse generated at the lower end travels upward toward the ceiling. The pulse passes through the midpoint of the string.
(a) What is the speed of the pulse at the midpoint if the mass of the string is ignored?
(b) What is the speed of the pulse at the midpoint if the mass of the string is *not* ignored?
35. ✶ String A is 10.0 m long and has a linear mass density of 2.0 g/m. String B is 20.0 m long and has a linear mass density of 5.0 g/m. The two strings are tied together and kept under a constant tension of 50.0 N. Two identical pulses are generated at the two opposite ends of the strings. Where will the two pulses first meet along the string?
36. ✶✶✶ A thick rope of uniform density hangs vertically from a fixed support.
(a) Show that the wave speed at a distance y from the lower end is given by $v = \sqrt{yg}$.
(b) Show that the time taken by a pulse to travel the entire length L of the rope is given by $t = 2\sqrt{L/g}$.
37. ✶✶ The speed of waves on the surface of a deep ocean is given by $v = \sqrt{\dfrac{g\lambda}{2\pi}}$, where λ is the wavelength of the waves, and g is the acceleration due to gravity.
(a) Show that the expression on the right side of this equation has the dimensions of speed.
(b) What is the period of the waves?
(c) Successive crests pass a stationary oil platform every 12 s. What is the speed of the waves?

Section 14-5　Harmonic Waves

38. ✶　The displacement versus position graphs at a fixed time for four waves, travelling in the same medium, are shown in Figure 14-60. Rank these waves in order of frequency, from highest to lowest.

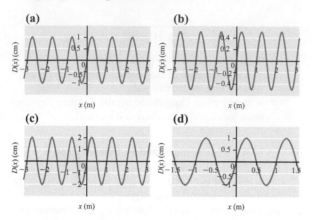

(a)

(b)

(c)

(d)

Figure 14-60　Problem 38

39. ✶　Rank the following waves in order of wavelength and frequency, from lowest to highest. Also, indicate the direction of motion, toward increasing or decreasing x, for each wave. Use an equality sign where needed. Here x is in meters and t in seconds.
(a) $D(x, t) = (0.25 \text{ m}) \sin(x - t)$
(b) $D(x, t) = (1.50 \text{ m}) \sin(2x - t)$
(c) $D(x, t) = (1.25 \text{ m}) \sin(x - 2t)$
(d) $D(x, t) = (0.75 \text{ m}) \sin\left(\dfrac{x}{2} - t\right)$

40. ✶　Wave functions for five waves are given below. Here, x is in metres, and t is in seconds.
(a) $D(x, t) = (0.3 \text{ m}) \sin(x - 3t)$
(b) $D(x, t) = (0.5 \text{ m}) \sin(2x - 6t)$
(c) $D(x, t) = (0.6 \text{ m}) \sin(2x - 3t)$
(d) $D(x, t) = (0.1 \text{ m}) \sin(4x - 7t + \pi)$
(e) $D(x, t) = (0.5 \text{ m}) \sin(7x - 13t + \pi/2)$
Using equality signs where needed, rank the waves from lowest to highest in terms of
(a) amplitude;
(b) wavelength;
(c) frequency;
(d) wave speed.

41. ✶✶　Write a displacement equation for each wave shown in Figure 14-61. The wavelength of D_2 is 1.0 m. Assume that both waves are moving in the same direction in the same medium.

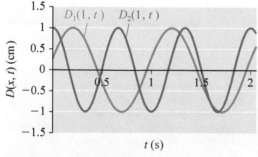

Figure 14-61　Problem 41

42. ✶✶　The displacement function at $x = 2.00$ m for a travelling wave is given by the following equation:

$$D(2.00, t) - (0.75 \text{ m}) \sin(10.0 - 3.50t)$$

where x is in metres, and t is in seconds.
(a) If the phase constant is zero, what are the wavelength, frequency, and speed of the wave?
(b) Write the displacement function for an arbitrary x and t, that is, an equation for $D(x, t)$.

43. ✶✶　The displacement function at $t = 2.00$ s for a travelling wave is given by the following equation:

$$D(x, 2.00) - (0.05 \text{ m}) \sin(10.0x - 15.0)$$

where x is in metres and t is in seconds.
(a) If the phase constant is zero, what are the wavelength, frequency, and speed of the wave?
(b) Write a displacement equation, that is, an equation for $D(x, t)$, for the wave.

44. ✶✶　The displacement function for a travelling wave is given by the following equation:

$$D(x, t) = (0.75 \text{ m}) \sin(10.0x - 3.5t + 0.25)$$

where x is in metres, and t is in seconds.
(a) What are the wavelength, frequency, phase constant, and speed of the wave?
(b) What is the displacement of a point located at $x = 0.10$ m at $t = 2.0$ s?
(c) What is the displacement of a point located at $x = 0.15$ m at $t = 2.0$ s?
(d) At $t = 0.5$ s, what is the phase difference between points located at $x = 1.0$ m and $x = 1.5$ m along the wave?
(e) What is the phase difference between the displacements of a point that is located at $x = 1.0$ m at $t = 1.0$ s and $t = 1.5$ s?
(f) Plot $D(x, 1.0)$ and $D(x, 1.5)$ for the wave. Determine the wave speed from these graphs. Does this speed agree with the speed calculated in part (a)?

45. ✶✶　A wave on a string is described by the displacement equation

$$D(x, t) = (0.02 \text{ m}) \sin\left(\frac{2\pi}{3.0} x + \frac{2\pi}{6.0} t - \frac{\pi}{3}\right)$$

where x is in metres, and t is in seconds.
(a) What are the wavelength, frequency, and speed of the wave?
(b) In which direction is the wave travelling?
(c) Draw a displacement versus time graph for a point located at $x = 0.5$ m for $t = 0$ s to $t = 2.0$ s.
(d) What is the velocity of a segment of the string that is located at $x = 0.5$ m at $t = 2.0$ s?

46. ✶✶✶　A transverse wave on a string is described by the displacement function

$$D(x, t) = (0.20 \text{ m}) \sin(\pi x + 2\pi t)$$

where x is in metres, and t is in seconds.
(a) What is the wave speed, and in which direction is the wave travelling?
(b) What is the transverse velocity of the segment at $x = 0.5$ m at $t = 3.0$ s?
(c) What is the transverse acceleration of the segment at $x = 0.5$ m at $t = 3.0$ s?

Section 14-6 Position Plots and Time Plots

47. ✳ A position plot of a 5.0 Hz wave, moving toward the right, is shown in Figure 14-62.
(a) What is the wavelength of the wave?
(b) What is the speed of the wave?
(c) What is the phase constant of the wave?
(d) Write a displacement equation for the wave.

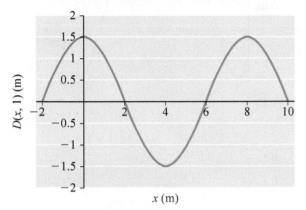

Figure 14-62 Problem 47

48. ✳ The time plot for a wave of wavelength 0.50 m, moving toward the right, is shown at $x = 1.2$ m in Figure 14-63.
(a) What is the angular frequency of the wave?
(b) What is the speed of the wave?
(c) What is the phase constant of the wave?
(d) Write a displacement equation for the wave.

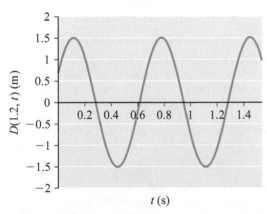

Figure 14-63 Problem 52

49. ✳ A time plot of a wave of at $x = 2.5$ m, moving toward the right with a speed of 4.0 m/s, is shown in Figure 14-64.
(a) What is the frequency of the wave?
(b) What is wavelength of the wave?
(c) What is the phase constant of the wave?
(d) Write a displacement equation for the wave.

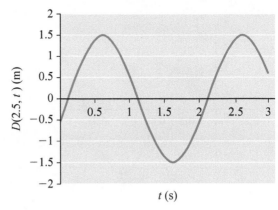

Figure 14-64 Problem 49

50. ✳✳ The position plots for two 50 Hz waves at $t = 1.0$ s are shown in Figure 14-65.
(a) What is the phase difference between the waves if the waves are moving toward the right?
(b) What is the phase difference between the waves if the waves are moving toward the left?

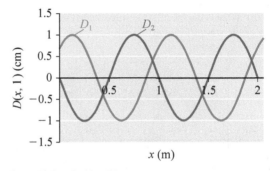

Figure 14-65 Problem 50

51. ✳✳ Figure 14-66 shows a position plot at $t = 1.0$ s and a time plot at $x = 1.2$ m for a travelling wave.
(a) Write a general displacement equation for the travelling wave.
(b) Write an equation for a travelling wave that, when combined with the wave in part (a), produces a standing wave.

(a)

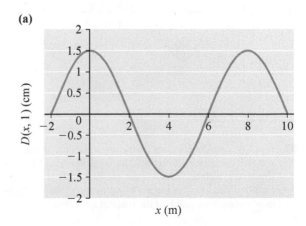

(b)

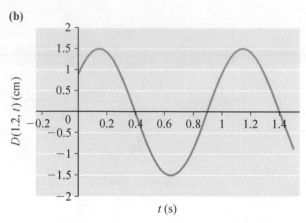

Figure 14-66   Problem 51

52. ✷✷ A one-dimensional wave is travelling along the *x*-axis in the positive direction. The graph in Figure 14-67(a) shows the displacement versus position for the wave at *t* = 1.0 s. The graph in Figure 14-67(b) shows the displacement versus time graph of the wave at *x* = 1.0 m.
   (a) Use the graphs to determine the displacement equation, that is, $D(x, t)$, for the wave.
   (b) What is the speed of the wave, and in which direction is the wave travelling?

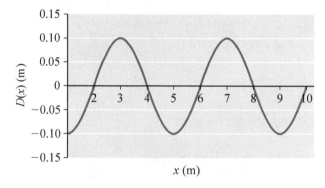

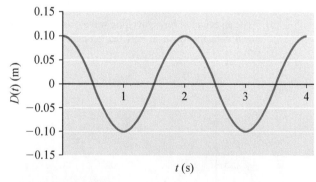

Figure 14-67   Problem 52

53. ✷✷ A one-dimensional wave is travelling along the *x*-axis in the positive direction. The graph in Figure 14-68(a) shows the displacement versus position for the wave at *t* = 1.0 s. The graph in Figure 14-68(b) shows the displacement versus time graph of the wave at *x* = 1.0 m.

   (a) Use the graphs to determine the displacement equation, that is, $D(x, t)$, for the wave.
   (b) What is the speed of the wave?

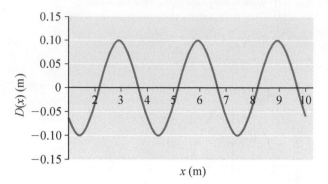

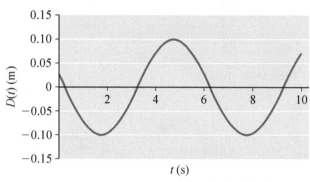

Figure 14-68   Problem 53

Section 14-7   Phase and Phase Difference

54. ✷ The wave function of a wave is given by the equation

$$D(x, t) = (0.2 \text{ m}) \sin(2.0x - 4.0t + \pi)$$

where *x* is in metres, and *t* is in seconds.
   (a) What is the phase constant of the wave?
   (b) What is the phase of the wave at *t* = 1.0 s and *x* = 0.5 m ?
   (c) At a given instant, what is the phase difference between two points that are 0.5 m apart?
   (d) At what speed does a crest of the wave move?

55. ✷✷ The wave function of a wave is given by the equation

$$D(x, t) = (0.1 \text{ m}) \sin(4.0x - 0.5t + 0.2)$$

where *x* is in metres, and *t* is in seconds.
   (a) What is the phase constant of the wave?
   (b) What is the phase of a point located at *x* = 0.5 m at *t* = 2.0 s?
   (c) Does the phase of the point *x* = 0.5 m change with time? Explain your reasoning.
   (d) At *t* = 1.0 s, what is the phase difference between two points that are 0.1 m apart?
   (e) Is the phase difference between these points the same at *t* = 2.0 s?
   (f) What is the phase difference between two points that are one wavelength apart?

56. ✱✱ The wave function of a wave at $x = 2.0$ m is given by the equation

$$D(2.0 \text{ m}, t) = (0.3 \text{ m}) \sin(12.0 + 5.0t)$$

where x is in metres, and t is in seconds.
(a) What is the phase of the point at $t = 1.0$ s?
(b) By how much does the phase at this location change between $t = 1.0$ s and $t = 2.0$ s?
(c) Is the phase constant of the given wave 0 rad? Explain your reasoning.

Section 14-8 Energy and Power in a Travelling Wave

57. ✱ A string of linear mass density 100.0 g/m is under 200.0 N of tension. A sinusoidal wave of frequency 10.0 Hz and amplitude 1.0 cm is propagating along the string. What is the average power carried by the wave?

58. ✱ A sinusoidal wave on a string is described by the equation

$$D(x, t) - (0.10 \text{ m}) \sin(4\pi x - 200\pi t)$$

The linear mass density of the string is 10.0 g/m. Determine the direction of motion and the average power transmitted by the wave.

Sections 14-9 and 14-10 Superposition of Waves and Interference of Waves Travelling in the Same Direction

59. ✱ Figure 14-69 shows position versus displacement plots, at a common time, of two harmonic waves that are travelling in the same medium in the same direction.
(a) Plot the resultant wave.
(b) What are the amplitude and the wavelength of the resultant wave?
(c) Is the speed of the resultant wave the same as the speed of the component waves? Explain your reasoning.
(d) Is the resultant wave a standing wave or a travelling wave?

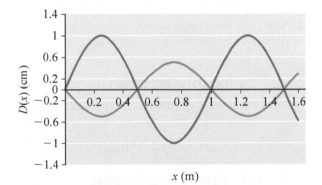

Figure 14-69 Problem 59

60. ✱ Figure 14-70 shows displacement versus position plots for two harmonic waves. Draw the resultant wave obtained by adding these two waves.

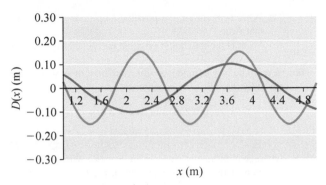

Figure 14-70 Problem 60

61. Figure 14-71 shows plots of two waves at time $t = 0$. Plot the resultant wave on the same graph, and determine the phase constant.

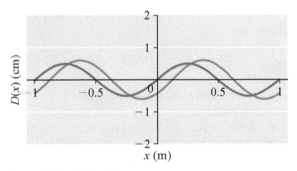

Figure 14-71 Problem 61

62. ✱ Two waves with the same amplitude and frequency arrive at a point P. The waves are $\pi/2$ rad out of phase at this point. What is the resulting amplitude of the waves at P?

63. ✱✱ Two harmonic waves with the same amplitude, wavelength, and frequency are travelling in the same direction. The amplitude of the resultant wave is half the amplitude of the component waves. Find the difference between the phase constants of the two component waves, and the difference between the phase constant of the resultant wave and that of either component wave.

64. ✱ The phase constants of two harmonic waves of same frequency and wavelength, travelling in the same direction, differ by π radians. The amplitude of the waves are A_1 and A_2. Determine the amplitude of the resultant wave.

$$D_1(x, t) = A_1 \sin(kx - \omega t)$$

$$D_2(x, t) = A_2 \sin(kx - \omega t + \pi)$$

Using $\sin(\theta + \pi) = -\sin\theta$, we can write

$$D_2(x, t) = -A_2 \sin(kx - \omega t)$$

The wave function for the resultant wave is,

$$D(x, t) = D_1(x, t) + D_2(x, t) = A_1 \sin(kx - \omega t) - A_2 \sin(kx - \omega t)$$

$$= (A_1 - A_2) \sin(kx - \omega t)$$

Therefore the amplitude of the resultant wave is $A_1 - A_2$

65. ✱ Two waves of amplitudes A_1 and A_2 are simultaneously present in the same medium.

(a) What is the minimum possible amplitude of the resultant wave?

(b) What is the maximum possible amplitude of the resultant wave?

66. ✸✸ The wave functions for two harmonic waves are given by the following equations:

$$D_1(x, t) = (0.2 \text{ m}) \sin(2.0x - 3.0t)$$

$$D_2(x, t) = (0.2 \text{ m}) \sin(2.0x - 3.0t + \pi/4)$$

where x is in metres, and t is in seconds. Write the wave function of the resultant wave when the two waves interfere. What is the amplitude of the resultant wave?

67. ✸✸ The wave functions for two harmonic waves that interfere with each other are given by the following equations:

$$D_1(x, t) = A \sin(kx - \omega t + \phi_1)$$

$$D_2(x, t) = A \sin(kx - \omega t + \phi_2)$$

Show that the wave function $D(x, t)$ of the resultant wave is given by

$$D(x, t) = 2A \cos\left(\frac{\phi_1 - \phi_2}{2}\right) \sin\left(kx - \omega t + \frac{\phi_1 + \phi_2}{2}\right)$$

Section 14-11 Reflection and Transmission of Mechanical Waves

68. ✸ A pulse travelling along a string is reflected from a boundary. The displacement function of the pulse is given by

$$D(x, t) = \frac{1}{(x - 3t)^2 + 4}$$

(a) Write the displacement function of the reflected pulse for a hard reflection.

(b) Write the displacement function of the reflected pulse for a soft reflection.

69. ✸ A harmonic wave travelling along a string is reflected from a boundary. The displacement function of the wave is given by

$$D(x, t) = 0.2 \sin(3x + 4t)$$

(a) Write the displacement function of the reflected wave for a hard reflection.

(b) Write the displacement function of the reflected wave for a soft reflection.

Section 14-12 Standing Waves

70. ✸ A travelling wave is described by the following equation:

$$D(x, t) = (0.5\text{m}) \sin(x - 4t)$$

where x is in metres, and t is in seconds. Which of the following wave(s), when added to the above wave, will result in a standing wave?

(a) $D(x, t) = (-0.5 \text{ m}) \sin(x - 4t)$

(b) $D(x, t) = (0.5 \text{ m}) \sin(x - 4t)$

(c) $D(x, t) = (-0.5 \text{ m}) \sin(x) \cos(4t)$

(d) $D(x, t) = (0.5 \text{ m}) \sin(x + 4t)$

71. ✸ A string oscillates in a standing wave pattern given by the equation (x is in meters and t in seconds)

$$D(x, t) = (1.50 \text{ cm}) \sin((0.6)x) \cos((20\pi)t)$$

(a) Determine the wavelength and frequency of the standing wave.

(b) What is the distance between two consecutive nodes, and between a node and an adjacent antinode?

(c) Write equations for two travelling waves that produce the above standing wave when superimposed.

(d) What is the displacement of a particle located at $x = 0.20$ m when $t = 3.0$ s?

72. ✸ The equation for a sinusoidal wave is given by

$$D(x, t) = (0.50 \text{ cm}) \sin\left(2.0x - 20.0t + \frac{\pi}{4}\right)$$

where x is in metres, and t is in seconds.

(a) Write the equation of a wave that produces a standing wave when added to the given wave.

(b) Find the frequency of the resulting standing wave.

(c) What is the amplitude of a point located at an antinode of the standing wave?

(d) Find the distance between two consecutive nodes of the standing wave.

(e) Find the distance between a node and an adjacent antinode in the standing wave.

73. ✸✸ Two sinusoidal waves, simultaneously travelling along a string, are described by the following equations:

$$D_1(x, t) = (1.0 \text{ mm}) \sin(\pi x - 0.5\pi t)$$

$$D_2(x, t) = (1.0 \text{ mm}) \sin(\pi x + 0.5\pi t)$$

where x is in metres, and t is in seconds.

(a) Determine the equation for the resultant wave.

(b) Starting from $x = 0$, what are the values of x corresponding to the first three nodes and the first three antinodes?

(c) What is the distance between two consecutive nodes?

74. ✸✸ Two sinusoidal waves with frequencies of 100 Hz, wavelengths of 0.500 m, amplitudes of 2.0 cm, and phase constants of 0 are moving in opposite directions along a string interfere to produce a standing wave.

(a) Starting from $x = 0$, what are the values of x corresponding to the first three nodes and the first three antinodes of the standing wave?

(b) What is the amplitude of the standing wave at $x = 2.0$ m?

75. ✸✸ The wave functions for two interfering harmonic waves are given by the following equations:

$$D_1(x, t) = A \sin(kx - \omega t + \phi_1)$$

$$D_2(x, t) = A \sin(kx + \omega t + \phi_2)$$

Show that the wave function $D(x, t)$ of the resultant wave is given by

$$D(x, t) = 2A \sin\left(kx + \frac{\phi_1 + \phi_2}{2}\right) \cos\left(\omega t - \frac{\phi_1 - \phi_2}{2}\right)$$

Section 14-13 Standing Waves on Strings

76. ✸ A stretched string of length L that is fixed at both ends is vibrating in its third harmonic. How far from the end of the string can the blade of a screwdriver be placed against the string and not disturb the amplitude of the vibrations?

77. ✸ A string fixed at both ends is 0.500 m long and has a tension that causes the frequency of the first harmonic (fundamental frequency) to be 260.0 Hz. The tension is increased by 4%. What is the new fundamental frequency of the string?

78. ✷✷ A 1.0 m-long string is fixed at both ends.
   (a) What is the longest wavelength standing wave that can exist on this string?
   (b) The wave speed on the string is 300.0 m/s. What is the frequency of the longest standing wave?
   (c) How should the length of the string be changed to increase the lowest frequency of the string by 5% while keeping the tension the same?

79. ✷✷ A 2.0 m-long string is clamped at both ends and is vibrating so that there are two nodes along the length of the string, in addition to the nodes at the fixed ends.
   (a) What is the longest wavelength standing wave possible on this string?
   (b) When the wave speed is 200.0 m/s, what frequency corresponds to the longest wavelength?
   (c) Can a standing wave of frequency 1.5 times the frequency calculated in part (b) be generated on this string without changing its length or tension?
   (d) Determine the frequencies of the second and fourth harmonics, and sketch the corresponding vibrational pattern of the string.

80. ✷✷ A string is clamped at both ends. The string vibrates at 440 Hz and 660 Hz with no other vibrational frequencies in between. What is the lowest frequency at which this string can vibrate in a standing wave pattern?

81. ✷✷ A string of length 100.0 cm is fixed at both ends and is kept under a tension of 25.00 N. The linear mass density of the string is 0.650 g/m.
   (a) What is the lowest resonant frequency of the string?
   (b) What are the frequencies of the second and third harmonics?
   (c) What is the wave speed for this string?
   (d) What is the lowest resonant frequency when the tension is increased to 35.00 N?

82. ✷✷ One end of a string with a linear mass density of 7.60×10^{-4} kg/m is connected to an oscillator with a frequency of 50.0 Hz. The other end is connected to a hanging variable mass, as shown in Figure 14-72. The string passes over a pulley, and the string between the oscillator and the pulley can vibrate freely.
   (a) The length of the vibrating section of the string is 0.500 m. What mass must be attached so that the string vibrates with (i) one antinode between the pulley and the oscillator and (ii) three antinodes between the pulley and the oscillator?
   (b) What is the oscillation frequency of the string in each case in part (a)?

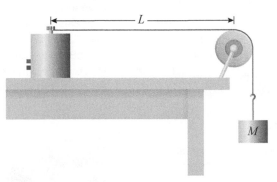

Figure 14-72 Problem 82

Learning Objectives

When you have completed this chapter you should be able to:

1 Describe the physical characteristics of a sound wave.

2 State and use Huygens' principle.

3 Apply Huygens' principle to describe the reflection and refraction of sound waves.

4 Discuss the conditions necessary for standing waves.

5 Sketch a spatial interference pattern.

6 Calculate the sound intensity level from the intensity of a sound, and vice versa.

7 Perform calculations using the Doppler effect equations.

Chapter 15
Interference and Sound

Rock bands use stacks of powerful amplifiers to amplify their music for concerts in large venues (Figure 15-1). Such sound systems typically have a total power of approximately 50 kW. How much of this power reaches your ears when you attend a rock concert? Can such power levels be harmful? In 2009, Ottawa bylaw officers measured a sound intensity level of 136 dB during an outdoor performance by KISS. The officers asked the band to turn the volume down. What is a dB? What are safe sound intensity levels? We will explore the answers to these questions in this chapter.

Figure 15-1 Alberta rock band Nickelback on stage

For additional Making Connections, Examples, and Checkpoints, as well as activities and experiments to help increase your understanding of the chapter's concepts, please go to the text's online resources at www.physics1e.nelson.com.

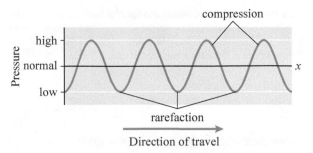

Figure 15-3 Pressure variations in the sound wave illustrated in Figure 15-2.

15-1 Sound Waves

In Chapter 14, we restricted our discussion of wave motion to one dimension. In this chapter, we will examine the properties of sound waves. Sound waves can propagate in three dimensions, and we will devote some attention to those features. We will also consider their propagation in one and two dimensions. Sound waves are a particular type of mechanical wave, so the way they behave is characteristic of many other types of waves. For example, the principles of wave interference described in Section 14-10 apply to all types of waves, including sound waves.

Sound Waves Are Longitudinal Waves

Recall from Chapter 14 that sound is a longitudinal wave: As a sound wave passes through a medium, the molecules of the medium oscillate, alternately creating regions of higher and lower pressure. The molecules are displaced in directions that are either parallel or antiparallel to the direction of propagation of the wave. Figure 15-2 illustrates the direction of propagation, as well as the alternate compressions (darker regions) and rarefactions (lighter regions) in a longitudinal wave.

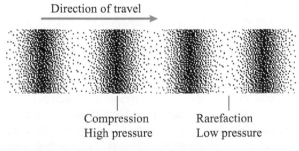

Figure 15-2 Particle positions at an instant of time in a longitudinal wave

Accompanying these particle displacements are pressure variations. In regions in which the medium is compressed, the pressure is elevated above the normal or ambient pressure. Regions in which the medium is "stretched" (rarefaction) correspond to a lowering of the pressure. These pressure variations are illustrated in Figure 15-3. Note that pressure variations are scalar quantities and are not to be confused with the particle displacements, which are either parallel or antiparallel to the direction of propagation.

The Speed of Sound

The speed of a sound wave depends on the properties of the medium through which it is propagating. As you learned in Chapter 14, the speed of waves on strings depends on the linear mass density of the string, μ, and the tension in the string, T_s:

$$v = \sqrt{\frac{T_s}{\mu}} \qquad (14\text{-}5)$$

Note that Equation (14-5) uses a measure of the stiffness of the medium (T_s) and a measure of how much mass is oscillating (μ). The appropriate measure of the string's mass is the linear mass density. For air, and any other medium through which sound can pass, we have to use the density, ρ. What quantity can we use to describe how "stiff" air is?

We know that a spring constant, a measure of the stiffness of a spring, is defined in terms of how much the length of the spring changes when we exert a force on it. For a three-dimensional material, we instead measure by what fraction the volume changes when we change the pressure exerted on the material. The ratio of the change in pressure, Δp, divided by the fractional change in the volume, $\frac{\Delta V}{V}$, is called the **bulk modulus**, B:

$$B = -\frac{\Delta p}{\frac{\Delta V}{V}} = -V\frac{\Delta p}{\Delta V} \qquad (15\text{-}1)$$

The negative sign is included in the definition because the sign of the fractional change in volume is always the opposite of the sign of the pressure change. Paralleling Equation (14-5), the speed of sound is given by

$$v = \sqrt{\frac{B}{\rho}} \qquad (15\text{-}2)$$

This result is quite general and is not restricted to air, or for that matter, even gases. Table 15-1 gives the speed of sound, the bulk modulus, and the density for a variety of media. Note the general trend of increasing sound speed as we move from gases to solids, even though solids tend to be denser than gases. The differences in bulk modulus more than offset the effects of the differences in density.

Table 15-1 Bulk Modulus, Density, and Speed of Sound for Various Media

Medium	Speed of sound (m/s)	Bulk modulus (Pa)	Density (kg/m³)
Air (20°C) (gas)	343	1.01×10^5	1.2
Hydrogen (gas)	1 284	1.51×10^5	0.089
Water (20°C) (liquid)	1 482	2.2×10^9	1 000
Mercury (liquid)	1 450	2.85×10^{10}	13 534
Gold (solid)	3 054	1.8×10^{11}	19 300
Diamond (solid)	11 200	4.42×10^{11}	3 520

As shown in Chapter 14, the relationship between the speed of a sound wave, v, the frequency, f, and the wavelength, λ, is

$$v = f\lambda = \frac{\omega}{k} \qquad (14\text{-}11)$$

EXAMPLE 15-1

Equation for Longitudinal Waves

Write a general expression for the wave function of a longitudinal wave that travels in the z-direction.

SOLUTION

We will still use the variable s to denote the wave function. However, the wave function will depend on the position, z, and the time, t:

$$s(z, t) = s_m \cos(kz - \omega t + \phi)$$

MAKING CONNECTIONS

Wavelength-Dependent Wave Speeds

From Equation (15-2), we can see that the speed of sound in air does not depend on wavelength. However, some types of waves have speeds that are wavelength-dependent. For example, the speed of light in glass varies with the wavelength of the light, which can be demonstrated by the separation of colours of light by a prism. When the wave speed does have wavelength dependence, we say that there is "dispersion" in the medium.

If the speed of sound in air were wavelength-dependent, different musical notes would arrive at slightly different times, and the farther you were from the sound source, the more pronounced this effect would be.

Relationships Between Displacement, Pressure, and Intensity

In addition to having a wave function that describes the displacements of the particles from their equilibrium positions, a sound wave also has pressure variations associated with these particle displacements. The gas that a wave travels through has an ambient pressure. The pressure variations of the wave correspond to increases and decreases in pressure from the ambient pressure. The pressure variations have the same periodicity, in both time and space, as the displacement variations. For a sound wave with a single frequency (a pure tone), the pressure varies sinusoidally. However, the phase relationship between the pressure variations and the displacement is subtle.

We will make our argument in one dimension for clarity and simplicity. One might at first think that the pressure variation is greatest where the displacement is greatest, that is, the two variations are in phase. They are not. If we consider a region of high pressure, it is a region that must have had particles pushed into it from the right and the left. On the left side of a high pressure region, the displacement is positive, while on the right side, it is negative. It must be the case then that the displacement goes through zero right at the point of maximum pressure. A similar argument can be made for the low pressure regions and the point of low pressure is also then a point of zero displacement.

We begin by considering an element of a medium located at the point x, with width Δx and cross-sectional area A. The displacement of this element is in the range $0 < s < s_m$. The volume of medium in the element will

Mathematical Description of the Displacement Amplitude

We can write a mathematical description of a travelling sinusoidal sound wave in one dimension. We will assume that the wave is travelling in the positive x-direction; therefore, the displacement in sound waves is longitudinal in the x-direction. Since we are already using x to denote position, we use s to denote the wave function:

KEY EQUATION
$$s(x, t) = s_m \cos(kx - \omega t + \phi) \qquad (15\text{-}3)$$

In Equation (15-3), s_m is the maximum displacement from equilibrium, $k = \frac{2\pi}{\lambda}$ is the wave number, $\omega = 2\pi f = \frac{2\pi}{T}$ is the angular frequency, and ϕ is the phase constant.

change with x because the displacement of the right side of the element will generally be different from the displacement of the left side. The change in displacement is given by

$$\Delta s = \frac{\partial s}{\partial x} \Delta x \qquad (15\text{-}4)$$

We have used the partial derivative because, in general, s depends on both x and t. To find the change in volume, we multiply Δs by the cross-sectional area A:

$$\Delta V = A\Delta s = A\frac{\partial s}{\partial x} \Delta x \qquad (15\text{-}5)$$

However, the volume of the element is $V = A\Delta x$, to first order, so Equation (15-5) can be rearranged to give

$$\frac{\Delta V}{A\Delta x} = \frac{\Delta V}{V} = \frac{\partial s}{\partial x} \qquad (15\text{-}6)$$

Since we are interested in how the pressure changes, we use the bulk modulus to make the connection between the volume change and pressure change. Solving Equation (15-1) for $\Delta V/V$ and substituting into Equation (15-6), we obtain

$$\frac{\Delta V}{V} = -\frac{\Delta p}{B} = \frac{\partial s}{\partial x} \qquad (15\text{-}7)$$

and

$$\Delta p = -B\frac{\partial s}{\partial x} \qquad (15\text{-}8)$$

We can use Equation (15-3) to calculate the partial derivative of s with respect to x:

$$\frac{\partial s}{\partial x} = \frac{\partial}{\partial x} s_m \cos(kx - \omega t + \phi) = -ks_m \sin(kx - \omega t + \phi)$$

$$(15\text{-}9)$$

Finally, we substitute this result into Equation (15-8) to obtain

KEY EQUATION $\qquad \Delta p = Bks_m \sin(kx - \omega t + \phi) \qquad (15\text{-}10)$

We can identify the amplitude of the pressure variations as the coefficient of the sine factor in Equation (15-10):

$$\Delta p_m = Bks_m \qquad (15\text{-}11)$$

Comparing Equations (15-3) and (15-10), we see that although the wave function has a cosine function and the pressure is a sine function, the arguments are the same in both cases. Thus, they have the same wavelength, period, and wave speed but are $\pi/2$ rad out of phase with each other. These relationships are plotted in Figure 15-4 for a 1000 Hz sound wave with an amplitude that is at the threshold of human hearing—if this sound wave were any quieter, we would not be able to hear it. It is quite incredible that our ear is able to detect displacements on the order of 10^{-11} m, less than the diameter of a hydrogen atom. If our ears were slightly more sensitive, we would hear a constant background noise due to the thermal motion of the air molecules. Our ears have evolved to be as sensitive as possible without subjecting us to constant background noise.

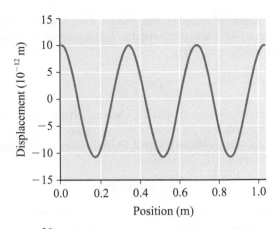

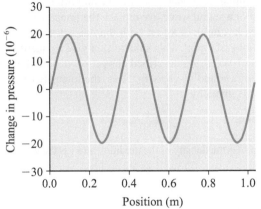

Figure 15-4 Displacement and pressure variations for a 1000 Hz sound wave in air

EXAMPLE 15-2

Wave Amplitude

Calculate the displacement amplitude of a 1000 Hz sound wave whose pressure amplitude is 20.0 μPa.

SOLUTION

We rearrange Equation (15-11):

$$s_m = \frac{\Delta p}{Bk} = \frac{\Delta p}{B\dfrac{2\pi f}{v}} = \frac{0.000\,020\,0\ \text{Pa}}{1.01 \times 10^5\,\text{Pa}\ \dfrac{2\pi(1000\ \text{s}^{-1})}{343\ \text{m}\cdot\text{s}^{-1}}}$$

$$= 1.08 \times 10^{-11}\,\text{m}$$

✔ CHECKPOINT

C-15-1 Sound Wave Amplitudes

The displacement amplitude and pressure amplitude for a sound wave are

(a) in phase;

(b) out of phase by $\pi/2$ rad;

(c) out of phase by π rad;

(d) none of the above.

C-15-1 (b)

We now examine the energy that a sound wave delivers. We define the **intensity**, I, of a wave as the power delivered per unit area:

$$I = \frac{P}{A} \qquad (15\text{-}12)$$

where P is the rate at which the wave delivers energy, and A is the area that the wave is impinging upon.

As was shown in Chapter 14, the power of a mechanical wave in one dimension is

$$P_{\text{avg}} = \frac{1}{2}\mu v \omega^2 A^2 \qquad (14\text{-}32)$$

where μ is the linear mass density, v is the wave speed, ω is the angular frequency, and A is the amplitude of the wave.

For a sound wave, we replace the linear mass density, which has units of kg/m, with the mass density ρ, which has units of kg/m^3. This substitution gives a quantity with units of W/m^2. As described above, we denote the amplitude of a sound wave as s_{m}. Thus, the intensity of a sound wave is given by

$$I = \frac{1}{2}\rho v \omega^2 s_{\text{m}}^2 \qquad (15\text{-}13)$$

where ρ is the density of the medium, and v is the wave speed.

EXAMPLE 15-3

Intensity and Power

Determine the intensity of the 1000 Hz sound wave in Example 15-2. Assuming that a human eardrum has a radius of approximately 0.75 cm, determine how much power is delivered to the ear by such a wave.

SOLUTION

We use Equation (15-13) to determine the intensity:

$$I = \frac{1}{2}(1.2 \text{ kg/m}^3)(343 \text{ m/s})(2\pi)^2(1000 \text{ s}^{-1})^2(1.08 \times 10^{-11}\text{ m})^2$$

$$= 9.48 \times 10^{-13} \text{ W/m}^2$$

To find the power, we solve Equation (15-12) for P:

$$P = IA = (1.5 \times 10^{-13} \text{ W/m}^2)(\pi)(0.0075 \text{ m})^2$$

$$= 1.67 \times 10^{-16} \text{ W}$$

LO 2

15-2 Wave Propagation and Huygens' Principle

Spherical Waves

We begin by considering a single point source of sound. This situation is analogous to dropping a stone in a pool of water. In the water, the wave just travels along the surface

of the water (Figure 15-5(b)). However, the sound wave can propagate in three dimensions and it does so uniformly in all directions (Figure 15-5(a)). Each crest of a sound wave forms a **spherical wave** front centred on the point source.

(a)

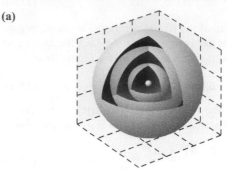

(b)

Figure 15-5 (a) A spherical wave propagates in three dimensions from a point source. (b) The waves created by dropping a stone into a pond propagate uniformly on the water's surface.

The mathematical description of such a wave is much more complex that the description given in Equation (15-3) for a one-dimensional wave. Rather than being a function of just x and t, a spherical wave is a function of x, y, z, and t because the wave propagates in three dimensions. In addition, the amplitude of a spherical wave decreases as it moves outward from the source. We will consider these amplitude changes in more detail in Section 15-6. Christiaan Huygens, a seventeenth-century Dutch scientist, recognized that he could describe the propagation of a wave by considering any point on a wave front to be a source of spherical waves. The resulting wave is then determined by adding all the waves from the point sources. His description is called **Huygens' principle**. Here is a procedure for applying Huygens' principle:

1. Select a set of equally spaced points on a wave front.

2. Draw a set of circles, all of the same radius, each centred at a different point.

3. Construct the resulting wave front by drawing tangents to the circles.

Figure 15-6 illustrates this procedure for a spherical wave.

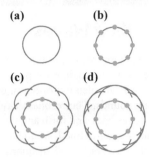

(a)   **(b)**

(c)   **(d)**

Figure 15-6   Applying Huygens' principle. (a) Draw a spherical wave. (b) Draw a set of points along the wave front. (c) Draw a spherical wave from each point. (d) Draw a curve tangent to the individual circular wave fronts.

Plane Waves

We will now consider **plane waves**, which have wave fronts that lie in planes parallel to each other (Figure 15-7). You can create a plane wave in a pond by dropping a long, straight stick horizontally into the pond. You can create a plane sound wave by vibrating a large, flat plate in a direction perpendicular to the surface of the plate. We can use Huygens' principle to show that a plane wave, once created, continues to propagate as a plane wave. In Figure 15-8, notice that the crests and troughs of the wave are parallel to each other.

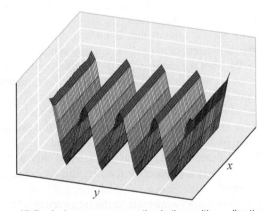

Figure 15-7   A plane wave propagating in the positive *x*-direction. The wave front extends in the *y*-direction, and the vertical axis shows the wave displacement amplitude.

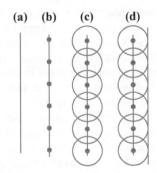

(a)   **(b)**   **(c)**   **(d)**

Figure 15-8   Huygens' principle applied to a plane wave that is travelling from left to right

15-3   Reflection and Refraction

We can use Huygens' principle to help understand what happens to a sound wave when it encounters a boundary between two media where the wave speed changes. At such a boundary, some of the wave can be reflected from the boundary and some transmitted across the boundary into the second medium. The portion of the wave that is reflected travels back into the first medium. As we saw in Chapter 14, the reflected wave may have its phase changed upon reflection. The angle of reflection is the same as the angle of incidence.

The portion of the wave that crosses the boundary into the second medium is called a refracted wave. In general, the direction of propagation of the refracted wave is different from the direction of the incident wave. When the wave speed is slower in the second medium, the wave tends to bend toward the slower medium. When the wave speed is higher in the second medium, the wave tends to bend away from the second medium.

We can use Huygens' principle to understand these effects. Consider a plane wave encountering a planar boundary. To apply Huygens' principle, we place a line of equally spaced dots at the boundary and draw a set of concentric half circles around the dots above and below the boundary, as shown in Figure 15-9. In the figure, the speed of sound in medium 2 is less than the speed of sound in medium 1. We draw only half circles because the speed of the wave is different above and below the boundary. In addition, the circles in the lower half of the figure are drawn closer together because the wave speed is slower in the lower medium.

When we draw our circles, we need to deal with the complication that the wave encounters the boundary at some angle. The incident wave encounters the left-hand points on the boundary before the right-hand points on the boundary. Therefore, the circles we draw around the leftmost points are larger because the wave has had more time to travel from those points. As we move to the right, the circles get successively smaller. Notice that the reflected wave front is perpendicular

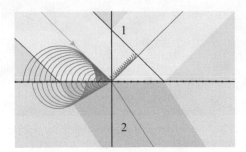

Figure 15-9   Reflection and refraction at a boundary

to the incident wave front. However, the angle of the transmitted wave, measured with respect to the surface normal, is smaller than the incident angle. The wave is bent toward the medium with the lower speed, medium 2. However, if the angle of incidence is 0°, measured with respect to a boundary normal, then both the transmitted wave and the reflected wave leave normal to the boundary.

15-4 Standing Waves

The formation of standing waves in brass, woodwind and stringed instruments is responsible for producing musical notes. Standing waves are also important in other applications not involving sounds. The resonant cavity of a laser reflects the light waves back and forth and creates standing light waves with a well-defined frequency and wavelength. Piezoelectric crystals used in quartz watches are another example where standing waves are used to create an electrical oscillator with a very precise frequency.

We start by considering a long, narrow column of air, such as might be found in a flute. Sound waves can propagate along the length of the column. When they reach the ends of the column, they are reflected, at least partially, and we have the condition for standing waves that we saw in Chapter 14: waves with the same frequency travelling in opposite directions in the same medium.

Similar to strings, air columns can have open (free) and closed (fixed) ends. The closed end of an air column is a displacement node because the air molecules cannot oscillate across such a fixed boundary. If we have an air column that is closed at both ends, the displacement function must have nodes at both ends. The possible standing waves in such a column are then exactly what we would see for a string with fixed ends.

With an air column that is open at both ends, the ends are displacement antinodes, and the standing waves are exactly what we see in a string with open ends. These results are summarized in Table 15-2. Because the pressure amplitude is $\pi/2$ rad out of phase with the displacement amplitude, the displacement amplitude for a column with both ends open has the same shape as the pressure amplitude for a column with both ends closed.

For a column that is open at one end and closed at the other, one end is a displacement node, and the other is a displacement antinode; see Table 15-3.

Standing waves contribute to the rich variety of tones that musical instruments produce. In Table 15-2, we listed the different types of standing waves that are possible for an air column with both ends open or closed. The rich sound of a musical instrument results from the fact that a string or an air column can simultaneously sustain combinations of the possible allowed standing waves. In fact, it is very difficult to avoid having multiple standing waves. The lowest frequency standing wave is called the fundamental and the higher frequencies are called the harmonics.

Figure 15-11 shows the amplitude as recorded on a computer by blowing into one end of a pipe that is open at the other end. The variations in amplitude are periodic but are clearly not simple sine waves. We can create complex periodic waveforms by adding together sine waves with frequencies that are all integer multiples (harmonics) of a fundamental frequency. For example, the wave in

MAKING CONNECTIONS

Refraction of Water Waves

Have you ever wondered why the orientation of waves reaching a beach is always parallel to the beach? For water waves, the speed of the wave depends on the depth of the water when the depth is comparable to the wavelength. As the depth decreases, so does the wave speed. As we saw in Figure 15-9, waves tend to bend toward the region of slower wave speed. Thus, the water waves tend to bend until they are parallel with the beach, as the image in Figure 15-10 illustrates. Despite the complexity of the coastline, the waves are still coming in parallel to the beach.

© Chris Cheadle/Alamy

Figure 15-10 Ocean waves coming in parallel to the beach at Pacific Rim National Park in British Columbia

Table 15-2   Standing waves in an air column with either both ends closed or both ends open

Displacement amplitude		Mode	Wavelength	Frequency
Both ends closed	Both ends open			
		$n = 1$	$\lambda_1 = 2L$	$f_1 = \dfrac{v}{2L}$
		$n = 2$	$\lambda_2 = \dfrac{2L}{2}$	$Lf_2 = \dfrac{2v}{2L}$
		$n = 3$	$\lambda_3 = \dfrac{2L}{3}$	$f_3 = \dfrac{3v}{2L}$
		$n = 4$	$\lambda_4 = \dfrac{2L}{4}$	$f_4 = \dfrac{4v}{2L}$
General case		n	$\lambda_n = \dfrac{2L}{n}$	$f_n = \dfrac{nv}{2L}$

Table 15-3   Displacement amplitudes for a tube that is open at one end and closed at the other end

Displacement amplitude	Mode	Wavelength	Frequency
One end open, one end closed			
	$n = 1$	$\lambda_1 = 4L$	$f_1 = \dfrac{v}{4L}$
	$n = 3$	$\lambda_2 = \dfrac{4L}{3}$	$f_2 = \dfrac{3v}{4L}$
	$n = 5$	$\lambda_3 = \dfrac{4L}{5}$	$f_3 = \dfrac{5v}{4L}$
	$n = 7$	$\lambda_4 = \dfrac{4L}{7}$	$f_4 = \dfrac{7v}{4L}$
General case	only odd n	$\lambda_n = \dfrac{4L}{n}$	$f_n = \dfrac{nv}{4L}$

Figure 15-12 is the sum of sine waves with five different frequencies:

$$s(t) = 0.56\cos(t) + 0.75\cos(3t) + 0.33\cos(5t)$$
$$- 0.16\cos(7t) - 0.05\cos(9t)$$

Notice that the waveform in Figure 15-12 is quite similar to the wave in Figure 15-11. You can also see that all the waves have a frequency that is an integer multiple of the fundamental, $f_0 = \dfrac{1}{2\pi}\,\text{s}^{-1}$. For this particular wave,

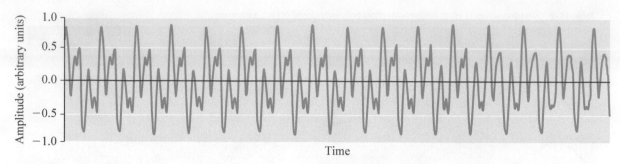

Figure 15-11 Sound wave amplitude for a pipe that was blown into one end, recorded over an interval of approximately 0.1 s

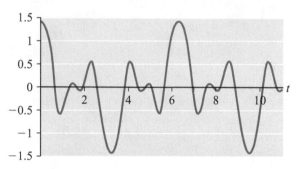

Figure 15-12 A standing wave composed of multiple sine waves with different frequencies and amplitudes

the frequencies are all odd multiples of the fundamental, and some of the terms have a negative sign, which means they are 180° out of phase from a positive term with the same frequency.

Musical instruments produce a sound that is composed of many different frequencies. For instruments that rely on vibrating air columns (e.g., trumpets, flutes, and clarinets), vibrating strings (e.g., violins, pianos, and banjos), and vibrating bars (e.g., xylophones, marimbas, and gamelans), the component frequencies in the sound are integer multiples of the fundamental frequency. However, the sound from instruments based on a vibrating membrane (such as drums) contains multiple frequencies that are not all integer multiples of the fundamental. This is why a drum does not have as distinct a note as, say, a trumpet.

Fourier's Theorem

Figure 15-12 demonstrates that we can generate complicated waveforms by combining sinusoidal waveforms of different wavelengths. In fact, a theorem developed by the French mathematician Jean Baptiste Joseph Fourier (1768–1830) states that *any* periodic waveform can be written as a sum of sine and cosine waves with appropriately chosen frequencies and amplitudes.

According to Fourier's theorem, if a function has a period T, such that $D(t) = D(t + T)$, then

$$D(t) = a_0 + \sum_{n=1}^{\infty}\left[a_n\cos\left(\frac{2\pi nt}{T}\right) + b_n\sin\left(\frac{2\pi nt}{T}\right)\right] \quad (15\text{-}14)$$

Note that the arguments of the sine and cosine functions are all integer multiples of a fundamental frequency that is related to the period by

$$f_0 = \frac{1}{T} \quad (15\text{-}15)$$

In stating the theorem, Fourier also described how to determine the constants a_n and b_n when given $D(t)$. The technique, which involves integral calculus, is beyond the scope of this book.

ONLINE ACTIVITY

Making Waves

The e-resource that accompanies every new copy of this textbook contains an Online Activity using the PhET simulation "Making Waves." Work through the simulation and accompanying questions to gain an understanding of making waves.

LO 5

15-5 Interference

Interference of three-dimensional waves is more complex than interference of one-dimensional waves. In one dimension, waves with the same frequency and wavelength have a fixed phase difference that depends only on the difference between the phase constants of the two waves. The phase difference does not vary with time or position. However, when waves propagate in different directions, their relative phase varies with position.

Figure 15-13 shows two plane waves with the same wavelength and frequency propagating in slightly different directions. We can see that there are points where the two waves are perfectly in phase. At those locations, the amplitudes add, resulting in constructive interference. At the points where the two waves are perfectly out of phase, we observe destructive interference.

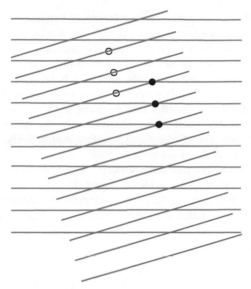

Figure 15-13 Two plane waves with the same wavelength travelling in slightly different directions. The lines represent the crests of the waves, the black circles indicate where the waves are perfectly in phase, and the open circles indicate where the two waves are exactly out of phase

Interference in Space

ONLINE ACTIVITY

Wave Interference

The e-resource that accompanies every new copy of this textbook contains an Online Activity using the PhET simulation "Wave Interference." Work through the simulation and accompanying questions to gain an understanding of interference patterns produced by sound waves.

We will start our exploration of interference in more than one dimension by considering the interference pattern that is produced by two point sources. The two sources are in phase and produce waves with the same frequency and wavelength. Each source produces spherical wave fronts. We can look at the resulting interference pattern in a plane that contains the two sources, as shown in Figure 15-14. This image represents the interference pattern at a particular instant in time.

We can determine mathematically the points where the two waves interfere constructively. At such points, the two waves must be in phase regardless of the elapsed time. Let us consider an instant when both sources are at the peak positive amplitude. If we then locate a point where the waves are interfering constructively, both waves must be at peak positive amplitude at that point. There must be an integer number of wavelengths between each source and the point we are considering. Figure 15-15 shows one such point.

If we consider the instant at which the two sources are at a large negative amplitude, there will still be an integer number of wavelengths between each source and

Figure 15-14 Interference from two identical point sources. The red regions represent large positive amplitude, and the green regions represent large negative amplitude.

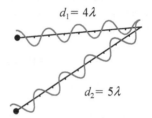

Figure 15-15 The constructive interference of two waves from two point sources

our point; thus, there will still be constructive interference. However, this interference now produces a large negative amplitude. Any point that is an integer multiple of wavelengths from both sources will undergo continuous constructive interference. Defining d_1 as the path length from source 1, and d_2 as the path length from source 2, the condition for constructive interference is

$$d_1 = m\lambda, \quad m = 1, 2, 3, \ldots$$
$$d_2 = n\lambda, \quad n = 1, 2, 3, \ldots \quad (15\text{-}16)$$

where m and n are integers.

These conditions can be made more general by considering the difference between the distances from the two sources to the point of constructive interference:

$$\Delta d = d_2 - d_1 = (n - m)\lambda = p\lambda,$$
$$p = 0, \pm 1, \pm 2, \pm 3, \ldots \quad (15\text{-}17)$$

Thus, the condition for constructive interference is that the path difference between the two sources must be an integer multiple of the wavelength.

The approach we have used to arrive at Equation (15-17) demanded that both paths individually be integer multiples of the wavelength. However, that does not necessarily have to be true for Equation (15-17) to be satisfied. Let us see if the path difference condition is sufficient to produce constructive interference.

As a spherical wave travels away from its source, it oscillates in space and time. However, the amplitude is constant over any spherical surface centred on the source. Thus, the spatial variations can be described as a function of r, the distance from the source. We can write the wave function of a spherical wave as

$$s(r, t) = s_m(r)\cos(kr - \omega t + \phi) \qquad (15\text{-}18)$$

This expression looks very similar to the expression for a one-dimensional wave in Equation 15-3. In the argument of the cosine, we have simply replaced x with r. However, the amplitude for a spherical wave depends on r because, unlike a wave in one dimension, the wave front spreads out over a larger area as it propagates outward; consequently, the amplitude decreases as r increases. We will consider this in more detail in Section 15-6.

For two waves to be in phase, the arguments of the cosine function for each wave must differ by an integer multiple of 2π. Since we have assumed that both waves are in phase, we can neglect the phase constant and write

$$(kd_2 - \omega t) - (kd_1 - \omega t) = k(d_2 - d_1) = n2\pi \qquad (15\text{-}19)$$

$$\therefore \ d_2 - d_1 = n\frac{2\pi}{k} = n\lambda, \quad n = 0, \ \pm1, \ \pm2, \ \pm3$$

From this result, we can see that constructive interference occurs whenever the path difference is an integer multiple of the wavelength. For the special case of $d_2 = d_1 = d$, we can easily add the two waves together to find the resultant wave:

$$\begin{aligned} S(d, t) &= s_m(d)\cos(kd - \omega t + \phi) \\ &+ s_m(d)\cos(kd - \omega t + \phi) \\ &= 2s_m(d)\cos(kd - \omega t + \phi) \qquad (15\text{-}20) \end{aligned}$$

We again consider a situation in which both sources are in phase but will turn our attention to determining the conditions for destructive interference. At an instant when both sources are at a large positive amplitude, we select a point at which the wave from one source is at a large positive amplitude, and the wave from the other source is at a large negative amplitude. One such point is shown in Figure 15-16. At this point, the wave from source 1 has travelled 3 wavelengths, and the wave from source 2 has travelled 3.5 wavelengths. Consequently, the waves are completely out of phase and interfere destructively. The amplitude at this point is quite small.

Destructive interference occurs when one path is an integer number of wavelengths and the other is a half-integer multiple; so, the general condition for destructive interference is that the path difference is a half-integer multiple of the wavelength (i.e., an odd number of half wavelengths):

$$\Delta d = d_2 - d_1 = \frac{(2n + 1)}{2}\lambda = \left(n + \frac{1}{2}\right)\lambda,$$

$$n = 0, \ \pm1, \ \pm2, \ \pm3 \qquad (15\text{-}21)$$

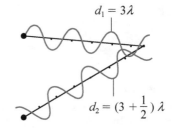

Figure 15-16 The destructive interference of two waves from two point sources

EXAMPLE 15-4

Interference Pattern

Two speakers are placed 2.0 m apart and are emitting sound waves with the same amplitude and phase; the wavelength is 1.0 m. Sketch the interference pattern, and identify points of constructive and destructive interference.

SOLUTION

We make a sketch with concentric circles around the two sources, with each circle having a radius that is one wavelength larger than the previous circle (Figure 15-17).

We can clearly see that there are points where the wave crests overlap—points of constructive interference. However, recognizing that the waves are travelling waves, we should also consider what happens for all times, not just the particular instant shown here.

For all points on the horizontal line that passes midway between the two sources, the path difference is zero.

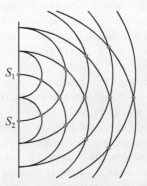

Figure 15-17 Wave fronts for the speakers in Example 15-4

Therefore, that entire line will undergo constructive interference. The other points of constructive interference also appear to lie on lines, and we ask whether these lines are also lines of constructive interference.

Consider the situation in Figure 15-18. To have constructive interference, the path difference must be an integer multiple of the wavelength:

$$d_2 - d_1 = \sqrt{x^2 + (y+1)^2} - \sqrt{x^2 - (y-1)^2} = n\lambda$$

The general solution to this equation is complex. However, for $n\lambda = 1.0$ m, we find that

$$y = \pm \frac{\sqrt{12x^2 + 9}}{6}$$

Figure 15-19 shows the two curves defined by this equation.

We can see that the curves pass through the points of intersection of the circles representing wave fronts 1.0 m apart. Because every point on the two curves satisfies the

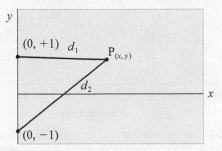

Figure 15-18 The path difference between the two sources and a point (x, y)

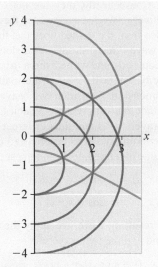

Figure 15-19 The two lines show the solutions corresponding to a path difference of one wavelength.

condition that the path length difference is one wavelength, constructive interference occurs all along the curves. We can infer that there is a curved line of destructive interference somewhere between each of the plotted curves and the x-axis, where the path difference is zero.

✓ **CHECKPOINT**

C-15-2 Interference Patterns

Two sources that are in phase produce an interference pattern like the pattern in Figure 15-14. If the two sources are out of phase by π, we will observe

(a) no interference pattern;

(b) the same interference pattern;

(c) an interference pattern similar to Figure 15-14 but with the destructive interference located where the constructive interference was observed, and vice versa;

(d) none of the above.

C-15-2 (c) One source will emit a crest, and T/2 s later, the other source will emit a crest. The crest from the first source will have travelled half a wavelength in that time. All the places where the in-phase sources had an integer wavelength path difference will now have a half-integer difference in the path and will be locations of destructive rather than constructive interference. A similar argument applies to the locations that experienced destructive interference with the sources in phase, and they will be locations of constructive interference.

Beats

In the patterns described above, regions of constructive and destructive interference are separated by intervals of space. Constructive and destructive interference can also be separated by intervals of time.

When we listen to two waves that have slightly different frequencies, we hear the amplitude increasing and decreasing as a function of time. This type of variation in amplitude is called a **beat**. The rate at which the

amplitude varies is proportional to the frequency difference. However, when the frequency difference is large, we hear the sounds as two distinct tones rather than one tone that varies in intensity.

We can see why this occurs by considering two sources that differ slightly in frequency and imagine we are equidistant from the two. The amplitudes of the two sources and the resultant total amplitude are shown in Figure 15-20.

The sound waves start off interfering constructively. Halfway across the graph, the red wave has completed two full cycles, and the blue wave has only completed one and

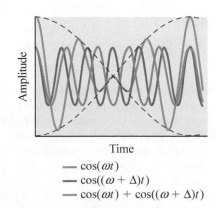

Figure 15-20 Two waves that differ by Δ Hz (red and blue) and the resultant wave (green). The dashed lines outline the amplitude envelope of the resultant wave.

— $\cos(\omega t)$
— $\cos((\omega + \Delta)t)$
— $\cos(\omega t) + \cos((\omega + \Delta)t)$

one-half cycles. Consequently, they are out of phase and interfere destructively. Two red cycles later, they are back in phase. We can see from the dashed lines that the amplitude of the resultant wave (green) strongly and slowly varies in amplitude (is modulated), and we hear this as a beating effect.

For simplicity, we will develop the mathematics of beats in one dimension. Consider two waves that have the same amplitude but different frequencies, ω_1 and ω_2, with different wavelengths and thus different wave numbers. We will set both phases to be zero and consider the resultant amplitude at a fixed position in space, $x = x_o$:

$$s_1(x_o, t) = s_m \cos(k_1 x_o - \omega_1 t)$$
$$s_2(x_o, t) = s_m \cos(k_2 x_o - \omega_2 t)$$

(15-22)

To find the resulting wave, we apply the principle of superposition and add the two waves:

$$\begin{aligned} s_{\text{Total}}(x_o, t) &= s_1(x_o, t) + s_2(x_o, t) \\ &= s_m \cos(k_1 x_o - \omega_1 t) + s_m \cos(k_2 x_o - \omega_2 t) \\ &= s_m[\cos(k_1 x_o - \omega_1 t) + \cos(k_2 x_o - \omega_2 t)] \\ &= 2 s_m \cos \frac{1}{2}(k_1 x_o - \omega_1 t + k_2 x_o - \omega_2 t) \\ &\quad \times \cos \frac{1}{2}(k_1 x_o - \omega_1 t - k_2 x_o + \omega_2 t) \\ &= 2 s_m \cos\left[\left(\frac{k_1 + k_2}{2}\right)x_o - \left(\frac{\omega_1 + \omega_2}{2}\right)t\right] \\ &\quad \times \cos\left[\left(\frac{k_1 - k_2}{2}\right)x_o - \left(\frac{\omega_1 - \omega_2}{2}\right)t\right] \end{aligned}$$

(15-23)

To simplify this expression, we define the following quantities:

$$\varpi = \frac{\omega_1 + \omega_2}{2} \quad \text{Mean angular frequency}$$
$$\Delta\omega = \frac{\omega_1 - \omega_2}{2} \quad \text{Angular frequency difference}$$

(15-24)

With these definitions, and setting $x_0 = 0$, we can rewrite Equation (15-23) as

$$\begin{aligned} s_{\text{Total}}(0, t) &= 2 s_m \cos(-\varpi t)\cos(-\Delta\omega t) \\ &= 2 s_m \cos(\varpi t)\cos(\Delta\omega t) \end{aligned}$$

(15-25)

We have a product of two cosines—one oscillates at the mean angular frequency ϖ, and the other oscillates at the angular frequency difference $\Delta\omega$. When the two waves are close to each other in frequency, we hear a tone that has the mean frequency and whose amplitude varies by the difference in angular frequency. Figure 15-21 shows the result of adding waves with frequencies of 244 Hz and 256 Hz. The red line shows the resultant amplitude, and the blue lines show the envelope of the amplitude modulation. From this figure, we can see that the amplitude modulation has a period of approximately

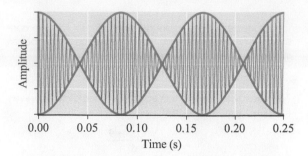

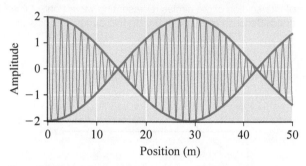

Figure 15-21 (a) The amplitude versus time graph for the sum of a 244 Hz wave and a 256 Hz wave. (b) The amplitude versus position graph for the sum of a 244 Hz wave and a 256 Hz wave. Notice that the spatial modulation of the amplitude has a wavelength of approximately 30 m, which corresponds to a frequency of about 12 Hz.

0.08 s, which corresponds to a frequency of approximately 12 Hz. We hear this modulation as a "beating" with a frequency of 12 Hz. However, when the two waves are not close in frequency, we perceive them as two separate tones (Figure 15-22).

Interestingly, if we listen to two separate tones through headphones, with one frequency in one ear and a different but close frequency in the other, we still hear beats even though the waves do not interfere outside of our head. It is believed that this effect, called binaural beating, results from the way our brain processes signals from our ears rather than from any interference of the sound waves travelling through our skull.

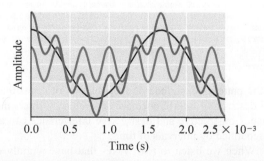

Figure 15-22 Two sound waves with very different frequencies. The black wave has a frequency of 600 Hz, and the blue wave has a frequency of 3000 Hz. In the resultant wave, shown in red, we can clearly see both waves.

WAVES AND OSCILLATIONS

CHECKPOINT

C-15-3 Beat Frequency

A 100 Hz tone and a 105 Hz tone are played simultaneously. What is the beat frequency?

(a) 2.5 Hz

(b) 5 Hz

(c) 10 Hz

(d) 205 Hz

C-15-3 (b) Even though the difference frequency is defined as $\Delta\omega = \dfrac{\omega_1 - \omega_2}{2}$, the effect of this is to modulate the amplitude. The amplitude will be large for both the positive and negative crests; therefore, the amplitude modulations have the same frequency as the difference between the two frequencies.

MAKING CONNECTIONS

Using Beat Frequencies

Beats can indicate that an instrument is out of tune. Musicians can tune an instrument by adjusting the frequency of its notes until the beating stops (Figure 15-23). The beating effect is part of the reason why instruments that are badly out of tune sound so disagreeable. However, a Balinese gamelan orchestra uses pairs of instruments that are intentionally slightly out of tune from each other to add a "shimmering" effect to the music (Figure 15-24).

Figure 15-23 Orchestral musician tuning a violin

Figure 15-24 A Balinese gamelan orchestra

EXAMPLE 15-5

Beat Frequency

The advantage of the beat tuning technique described in the Making Connections feature above is readily apparent when you consider that the smallest change in frequency that the human ear can detect for tones that are played one after the other is about 3 Hz for tones around 440 Hz. What beat frequency would you hear if you played a 440 Hz and a 443 Hz tone simultaneously?

SOLUTION

The amplitude of the resultant wave would be modulated at 3 Hz.

Making sense of the result:

A 3 Hz modulation is very easy to detect. In fact, at 440 Hz it is relatively easy to hear beats with frequency differences as low as 1 Hz.

LO 6

15-6 Measuring Sound Levels

As mentioned previously, the quietest sound that a human ear can detect corresponds to an intensity on the order of 10^{-12} W/m². At the other extreme, the threshold of pain at 1000 Hz corresponds to an intensity of about 10 W/m². This intensity is what you would experience if you were standing about 8 m from a jetliner during takeoff. The human ear has a dynamic range of approximately 10^{13}. For such large ranges, it is often convenient to work with a logarithmic scale. In fact, human perception of sound is nonlinear, much like a logarithmic scale.

Decibels

The **decibel** (dB) is a commonly used unit for logarithmic comparisons of power or intensity. Since logarithmic scales are *relative* scales, we must choose a reference level. For sound, we use the quietest intensity level we can hear at 1000 Hz as our reference intensity. We define the zero of our logarithmic sound intensity scale to correspond to this reference level:

The reference sound intensity level of 0 dB corresponds to an intensity $I_0 = 10^{-12}$ W/m².

(15-27)

To express any other **sound intensity level** (β) in decibels, we use

KEY EQUATION $\beta(I) = 0 \text{ dB} + 10 \log_{10}\left(\dfrac{I}{I_0}\right)$ (15-28)

The decibel is a convenient unit for β because it is about the smallest change in sound level that a human can typically detect. Table 15-4 lists typical β for some common sources. Since decibels indicate the logarithm of the ratio of two levels, "small" differences in decibel measurements indicate much larger differences in the intensity I.

Table 15-4 Typical sound intensity levels for some common sources

Sound source	Sound intensity level (dB)
Chainsaw, 1 m away	110
Lawn mower, 1 m away	107
Dance club speaker, 1 m away	100
Portable stereo, half-volume	94
Transport truck, 10 m away	90
Side of busy road	80
Vacuum cleaner, 1 m away	70
Normal speech, 1 m away	60
Average home	50
Quiet library	40
Quiet bedroom at night	30
Quiet sound studio	20

EXAMPLE 15-6

Sound Intensity Levels

The difference between the β for a lawn mower and a chainsaw is 3 dB. By what factor is the sound intensity of a chainsaw greater than the sound intensity of a lawn mower?

SOLUTION

We use Equation (15-28) to write

$$\beta(\text{chainsaw}) - \beta(\text{lawn mower}) = 3 \text{ dB}$$

$$\therefore \log_{10}\left(\frac{I_{\text{chainsaw}}}{I_{\text{lawn mower}}}\right) = 0.3$$

$$I_{\text{chainsaw}} = I_{\text{lawn mower}} 10^{0.3} \approx 2I_{\text{lawn mower}}$$

So, an increase of 3 dB is actually a doubling of the sound intensity.

Making sense of the result:

This highlights the fact that the decibel scale is a logarithmic scale – the sound intensity level only increases by 3 dB even though the intensity is doubled.

EXAMPLE 15-7

Concert Sound Level

You are at a concert where the sound intensity measures 3.00 W/m². What is the sound intensity level in decibels? Should you be concerned about your hearing?

SOLUTION

We use Equation (15-28) and recall that $I_0 = 10^{-12}$ W/m².

$$\beta = 0 \text{ dB} + 10 \log\left(\frac{I}{I_0}\right)$$

$$= 0 \text{ dB} + 10 \log\left(\frac{3 \text{ Wm}^{-2}}{10^{-12} \text{ Wm}^{-2}}\right) = 125 \text{ dB}$$

This level is close to the threshold of pain and is substantially above the level permitted in a workplace. There is cause for concern about hearing loss, especially since hearing damage is cumulative and permanent.

Making sense of the results:

This sound intensity level is not uncommon at rock concerts.

In Example 15-7, the sound intensity was a modest 3 W/m² at what most of us would consider a loud concert. What happens to all the power from the amplifiers? The loudness of a sound from a point source diminishes as you move away from the source. However, as you move around a point source at a constant distance from the source, the loudness does not vary. In practice, it is very difficult to produce an actual point source for sound waves, but the mathematics of such a source is relatively straightforward.

If we think about conservation of energy, it makes sense that the intensity of a source diminishes as we move away from the source. If we enclose the source with a spherical surface of radius r, all the power radiated by the source must pass through the surface of the sphere. If the power is radiated **isotropically** (uniformly in all directions) by the source, then the intensity over the surface of the sphere will be uniform. Consequently, the total power radiated, P, is equal to the intensity, I, multiplied by the surface area of the sphere:

$$P = I4\pi r^2 \qquad (15\text{-}29)$$

We can rearrange this equation to find

KEY EQUATION
$$I = \frac{P}{4\pi r^2} \qquad (15\text{-}30)$$

The intensity of a sound source that radiates equally in all directions falls off as the inverse square of the distance from the source.

MAKING CONNECTIONS

Hearing Safety

Exposure to loud noise can cause permanent damage to your hearing. In Canada, provincial Occupational Health and Safety legislation restricts the maximum β in a work environment for an 8 h shift to between 85 and 90 dB. If the β is above this limit, either hearing protection must be worn or the exposure time must be reduced to avoid damage to the workers' hearing.

Musicians are particularly concerned about hearing damage. Interestingly, it appears that the incidence of hearing loss is higher among classical musicians than

among rock musicians. High-quality earplugs that reduce the sound level while still allowing a musician to hear the music adequately have been developed to help prevent such hearing loss (Figure 15-25). Ideally, the earplugs should reduce all frequencies equally.

Figure 15-25 High-quality earplugs provide uniform attenuation at all frequencies.

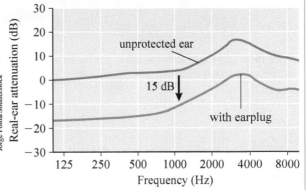

EXAMPLE 15-8

Speaker Power

Assume that you are 50 m from the stage at a concert where the sound intensity is 0.30 W/m². How much power are the speakers radiating, assuming that they direct their sound output toward the audience such that the sound intensity in that direction is 8 times greater than if the speakers were isotropic point sources?

SOLUTION

The sound intensity is 3 W/m². To find the power for an isotropic source, we rearrange Equation (15-29):

$$P = I4\pi r^2 = (0.3 \text{ W/m}^2)(4\pi 50^2) = 9.42 \times 10^3 \text{ W}$$

Since the speakers direct most of their sound toward the audience, their total sound output is about 8 times less than

an isotropic source that produces the same sound intensity in all directions:

$$P = \frac{9.42 \times 10^3 \text{ W}}{8} = 1.2 \text{ kW}$$

Making sense of the results:

We can see that the point source must be quite powerful to produce this level of sound at 50 m. However, as we noted at the beginning of the chapter, amplifier systems at rock concerts typically are capable of delivering approximately 50 kW of electrical power to the speakers. Speakers themselves typically only have an efficiency of about 1%. Thus, if we connect a 100 W amplifier to a speaker, it only radiates about 1 W in sound energy. The rest of the energy is consumed in heating the wires in the speakers and the output stages of the amplifiers.

MAKING CONNECTIONS

Refraction of Sound Waves

If you have ever been out in a canoe on a large lake on a calm day, you may have noticed that it is possible to hear sounds clearly over long distances. The air near the surface of the lake is cooler than the air well above the lake

(Figure 15-26). The speed of sound in air increases as the temperature increases. The sound waves tend to be refracted down to the surface of the lake. Consequently, the power from the sound source is not radiated isotropically.

(continued)

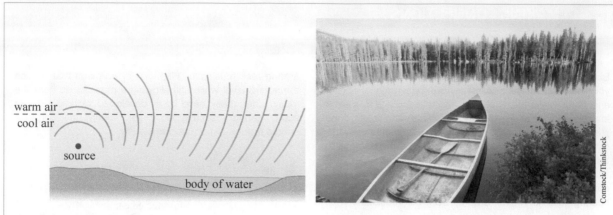

Figure 15-26 On a calm warm day, cool air over the surface of a lake causes sound waves to refract down toward the lake surface, greatly increasing the range of the sound.

Response of the Ear

When we listen to sound, a complex process takes place that converts the sound pressure in the air into our perception of the sound. The outer ear captures the sound waves and directs them to the eardrum, located in the middle ear (Figure 15-27). The sound pressure makes the eardrum vibrate, and three delicate bones connected to the eardrum transmit these vibrations to the inner ear. The inner ear contains the cochlea (shown as a blue coiled object), which is filled with fluid. The vibrations in this fluid stimulate delicate "hairs" on the basilar membrane. Nerve cells in the basilar membrane convert the vibrations of the hairs into electrical impulses that travel to the brain over the auditory nerve.

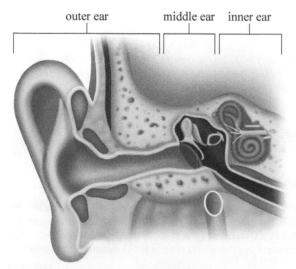

Figure 15-27 A cross section of the human ear

MAKING CONNECTIONS

Ears Can Protect Themselves

The bones in the middle ear are surrounded by muscles that can contract and limit the motion of the bones when the ear is subjected to very loud sounds. Loud percussive noises (Figure 15-28), such as those made when we hammer nails, come too quickly to allow the muscles to respond. Consequently, such sounds can be particularly damaging despite their short duration.

Figure 15-28 Hearing protection can prevent damage from loud percussive noise.

LO 7

15-7 The Doppler Effect

When we stand on the side of a highway and a car goes by, the sound that it generates has a higher frequency when the car is approaching and shifts to a lower frequency as the car passes by. Similarly, if we generate sound as we stand on the side of the road, the people in the car would hear a higher frequency as they approach and a lower frequency as they move away. In general, if there is relative motion between the source of the sound and the receiver of the sound, the frequency at the receiver is different from the frequency that is transmitted. If the two are moving toward each other, the received frequency is higher, and if they are moving apart, the received frequency is lower. This phenomenon is called the **Doppler effect**, named for the Austrian physicist Christian Doppler (1803–53), who first described it.

It is possible to write a single equation for the Doppler effect that covers the three possible scenarios:

- stationary source, moving receiver
- stationary receiver, moving source
- moving source and receiver

We denote the frequency of the source as f_s and the frequency at the receiver as f_r. If the source and receiver are moving along the line that connects them, the general relationship between the two frequencies is

KEY EQUATION
$$f_r = \frac{v \pm v_r}{v \mp v_s} f_s \qquad (15\text{-}31)$$

where v is the speed of sound in air, and v_r and v_s are the speed of the receiver and the source, respectively, relative to the air.

Note that Equation (15-31) has $\pm$ in the numerator but $\mp$ in the denominator. If the receiver is moving toward the source, we observe an increased frequency and use the top sign in the numerator. If the source is moving toward the receiver, we also observe an increase in the frequency and use the top sign in the denominator. In both cases, if the motion is toward, we use the top sign.

We now consider the common special cases, where either $v_r = 0$ or $v_s = 0$.

Moving Source, Stationary Receiver

Figure 15-29(a) shows a stationary source emitting spherical waves, which travel outward at the same speed in all directions. A receiver at any point around this source will detect the same frequency because the wave fronts are all equally spaced and all travel at the same speed. However, if the source is moving to the right, the wave fronts to the right of the source are closer together, and the wave fronts to the left

(a)   (b)

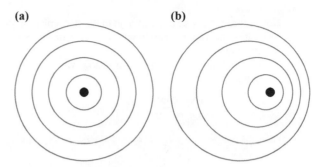

Figure 15-29 (a) A stationary source emitting waves (b) The source moving to the right

of the source are farther apart, as shown in Figure 15-29(b). Since the waves are still all travelling at the same speed, a receiver to the right of the source will detect more waves per second than a receiver to the left of the source will. More waves per second correspond to a higher frequency, and fewer waves per second correspond to a lower frequency.

Another way to understand the Doppler effect is this: Recall that the speed of sound in air at a given temperature is constant. Clearly, the wavelength to the right of the source is shorter, and since $f = v/\lambda$, a shorter wavelength requires a higher frequency.

We can calculate the change in the wavelength. Consider the shortened wavelengths observed to the right of the source. The source emits a wave crest every T seconds, where T is the period of the source. In each period, the wave crest travels a distance $D = vT$, where v is the speed of the wave. In the same time, the source travels a distance $d = v_sT$, where v_s is the speed of the source. Thus, the distance between the source and the last wave crest it emitted is

$$\lambda_r = \frac{v}{f_r} = D - d = (v - v_s)T = \frac{v - v_s}{f_s} \qquad (15\text{-}32)$$

Solving for f_r we find that for a source moving toward a *stationary* receiver:

$$f_r = \frac{v}{v - v_s} f_s \qquad (15\text{-}33)$$

Had we considered the situation with the receiver to the left of the source (i.e., the source moving *away* from a stationary receiver), we would have a plus sign in the denominator of Equation (15-33).

EXAMPLE 15-9

The Doppler Effect

A high-speed train travelling at 250 km/h blows its whistle as it approaches a level crossing. The whistle has a frequency of 660 Hz. What frequency would be perceived by someone standing at the crossing as the train

(a) approaches the crossing;
(b) moves away from the crossing?

SOLUTION

(a) The speed of the source (the train) is 250 km/h = 69.4 m/s, and the speed of the receiver is zero. The speed of sound in air is approximately 343 m/s. Substituting these values into Equation (15-31) gives

$$f_r = \frac{343 \text{ m/s} + 0 \text{ m/s}}{343 \text{ m/s} - 69.4 \text{ m/s}} 660 \text{ Hz} = 828 \text{ Hz}$$

Notice that we chose a minus sign in the denominator: since the train was moving toward the receiver, we use the top sign.

(b) When the train is leaving the crossing, we use the bottom sign:

$$f_r = \frac{343 \text{ m/s} - 0 \text{ m/s}}{343 \text{ m/s} + 69.4 \text{ m/s}} \; 660 \text{ Hz} = 549 \text{ Hz}$$

Making sense of the result:

As the train approaches we hear a higher pitch and as it moves away we hear a lower pitch. This is similar to what we observe when we stand near the side of a highway.

MAKING CONNECTIONS

Hubble and the Expanding Universe

The Doppler effect also occurs with electromagnetic radiation, including visible light. American astronomer Edwin Hubble (1889–1953) made a plot of the Doppler shift of distant nebulas versus their distance from Earth. Hubble noted that almost all the objects had a downward shift in frequency and that the magnitude of the shift is approximately proportional to the distance from Earth to the object. Using these data, Hubble concluded that the universe is expanding. His original data are shown in Figure 15-30. The slope of the plot of recession velocity versus distance is called Hubble's constant. The best current estimate for Hubble's constant is 72.6 ± 3.1 (km/s)/Mpc. A parsec (pc) is a commonly used astronomical unit equal to approximately 3.09×10^{16} m, or 3.26 light years.

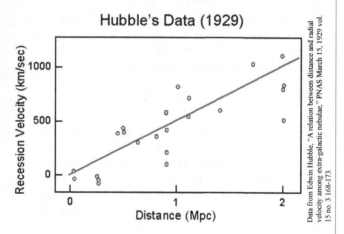

Data from Edwin Hubble, "A relation between distance and radial velocity among extra-galactic nebulae," PNAS March 15, 1929 vol. 15 no. 3 168-173.

Figure 15-30 Hubble's plot of nebula speeds versus distance from Earth

Moving Receiver, Stationary Source

When the source is stationary and the receiver is moving away from it, the receiver detects a reduction in the wave speed, and a receiver moving toward the source detects an increase in the wave speed. First, consider a receiver approaching the stationary source. When the receiver is positioned at a wave crest, the next wave crest is a distance λ away and moving toward the receiver at a speed relative to the receiver of $v + v_r$. The time this second wave crest takes to reach the receiver is

$$T_r = \frac{\lambda}{v + v_r} = \frac{v}{v + v_r}\left(\frac{1}{f_s}\right) = \frac{1}{f_r} \qquad (15\text{-}34)$$

Solving for f_r we find

$$f_r = \frac{v + v_r}{v} f_s \qquad (15\text{-}35)$$

This expression is the special case of Equation (15-31) with $v_s = 0$ and the receiver moving toward the source. For a receiver moving away from a stationary source, we use a minus sign in the numerator of Equation (15-35).

With Equations (15-33) and (15-35), we can readily derive the general result we stated in Equation (15-31). If both the source and the receiver are moving, we replace the stationary source frequency in Equation (15-35) with the stationary receiver frequency from Equation (15-33) to find

KEY EQUATION $\quad f_r = \dfrac{v \pm v_r}{v}\, \dfrac{v}{v \mp v_s}\, f_s = \dfrac{v \pm v_r}{v \mp v_s} f_s \qquad (15\text{-}31)$

The sign conventions use the upper sign when the motion is toward and the lower sign when the motion is away.

Shock Waves

When the speed of the source reaches the wave speed, many wave crests overlap, producing an intense wave front (Figure 15-31(a)). Such wave fronts are called a shock wave.

You may have heard such a shock wave from a supersonic plane. Shock waves can also occur in water, as shown in Figure 15-31(b).

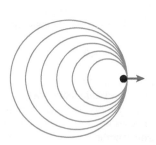

iStockphoto/Thinkstock

Figure 15-31 (a) Overlapping wave crests produce a shock wave. (b) The wake behind a boat is a type of shock wave.

In a sound wave, the particles of the medium are displaced longitudinally. Associated with the displacement of the medium is a pressure wave that is 90° out of phase with the displacement amplitude. Sound waves can travel in three spatial dimensions, and we can describe their propagation qualitatively using Huygens' principle. When a sound wave encounters a boundary, it can reflect and refract. When we confine a sound wave, for example, in a pipe, we get standing waves at specific frequencies. Two sound sources produce a three-dimensional spatial interference pattern. They can also produce a temporal interference pattern if they have different frequencies. Sound intensity levels are expressed in decibels, which are units of a logarithmic scale. The received frequency of a sound wave depends on the relative motion of the source and the receiver.

A sound wave is a longitudinal wave with a wavelength λ and a frequency f, with a displacement amplitude that can be written as

$$s(x, t) = s_m \cos(kx - \omega t + \phi) \qquad (15\text{-}3)$$

where s_m is the maximum displacement, $k = \dfrac{2\pi}{\lambda}$, $\omega = 2\pi f$, and ϕ is the phase constant.

Associated with the displacement amplitude is a pressure wave, which is given by

$$\Delta p = Bks_m \sin(kx - \omega t + \phi) \qquad (15\text{-}10)$$

where Δp is the deviation from the ambient pressure.

The speed of a sound wave is related to the frequency and the wavelength:

$$v = f\lambda$$

The speed of a sound wave in a gas is related to the properties of the gas through the mass density, ρ, and the bulk modulus, B, $v = \sqrt{\dfrac{B}{\rho}}$.

We can use Huygens' principle to qualitatively describe how a wave propagates by considering each point on a wave to be a source of spherical waves. Drawing equally spaced concentric circles around each point illustrates how the wave propagates.

When encountering a boundary, a wave can reflect and refract. The reflected portion of the wave leaves the boundary with the angles of incidence and reflection the same. The refracted portion of the wave passes the boundary and bends at an angle that depends on the wave speeds on each side of the boundary.

For speed of sound v and a pipe of length L, which is open at both ends or closed at both ends, the frequency of the standing waves is given by

$$f_n = \frac{nv}{2L}, \quad n = 1, 2, 3, \ldots$$

For a pipe that is open at one end and closed at the other,

$$f_n = \frac{nv}{4L}, \quad n = 1, 3, 5, \ldots$$

For two sources that are in phase, we get constructive interference when the path difference between the two sources is an integer number of wavelengths. We get destructive interference when the path difference is a half-integer number of wavelengths.

When we have two sources that differ in frequency, the tone we hear has a frequency that is the average of the two frequencies and is amplitude-modulated (beats) at a frequency equal to the difference of the two frequencies.

The intensity of a point sound source a distance r from the source is

$$I = \frac{P}{4\pi r^2} \qquad (15\text{-}30)$$

where P is the total power radiated by the source. The sound intensity level (β) is defined as

$$\beta(I) = 0 \text{ dB} + 10 \log_{10}\left(\frac{I}{I_0}\right) \qquad (15\text{-}28)$$

where the unit is the decibel (dB), and $I_0 = 10^{-12}$ W/m² is the intensity that corresponds to 0 dB.

When a source and a receiver are in motion relative to each other, the frequency detected by the receiver is not the same as the frequency emitted by the source. For sound waves,

$$f_r = \frac{v \pm v_r}{v \mp v_s} f_s \qquad (15\text{-}31)$$

where f_s is the frequency of the source, f_r is the frequency at the receiver, v is the speed of sound in air, and v_r and v_s are the speed of the receiver and the source, respectively, relative to the air. When the motion is toward, we use the top sign, and when the motion is away, we use the bottom sign.

Applications: musical instruments, musical beats, tuning musical instruments, hearing safety, shock waves

Key Terms: beat, bulk modulus, decibel, Doppler effect, Huygens' principle, intensity, isotropic, plane wave, sound intensity level, spherical wave

QUESTIONS

1. A sound wave is a longitudinal wave. True (T) or false (F)?
2. The displacement and pressure amplitudes are
   (a) in phase;
   (b) out of phase by 90°;
   (c) out of phase by 180°.
3. When we double the frequency of a sound wave, by what factor does the wavelength change?
4. When we double the power produced by a source, by what factor does the displacement amplitude of the resultant wave change?
5. The angle of incidence and the angle of reflection are equal when a sound wave reflects from a boundary. True (T) or false (F)?
6. When a sound wave crosses a boundary from a medium of high speed to one of lower speed, the wave
   (a) bends toward the surface normal;
   (b) bends away from the surface normal;
   (c) simply slows without changing direction.
7. We create a standing wave when we confine a wave between boundaries. True (T) or false (F)?
8. The frequencies of standing waves in an air column are all integer multiples of each other. True (T) or false (F)?
9. It is not possible to have multiple standing waves simultaneously in an air column. True (T) or false (F)?
10. We only get an interference pattern when two sources are in phase. True (T) or false (F)?
11. When we double the intensity of a sound source, by what value does the sound intensity level increase?
12. The intensity of a sound source decreases by what factor when we triple the distance from the source?
13. Sound waves can interfere
    (a) spatially;
    (b) temporally;
    (c) both spatially and temporally;
    (d) neither spatially nor temporally.
14. The speed of sound in a gas does not depend on the type of gas. True (T) or false (F)?

PROBLEMS BY SECTION

For problems, star ratings will be used, (✳, ✳✳, or ✳✳✳), with more stars meaning more challenging problems.

Section 15-1 Sound Waves

15. ✳ A wave is observed to have a frequency of 1000 Hz in air. What is the wavelength?
16. ✳✳ When the bulk modulus increases by 10%, by what factor does the speed of sound change?
17. ✳ A depth sounder on a boat sends out a pulse and listens for the echo from the bottom. The water is 30 m deep. How long will it take for the echo to come back?
18. ✳✳ A sound wave is described by $s(x, t) = 2 \times 10^{-9} \cos(37x - 12\,566t)$. Find the wave's
    (a) wavelength;
    (b) frequency;
    (c) speed.
19. ✳ Write an expression for the displacement amplitude of a 60 Hz sound wave.
20. ✳✳ Medical ultrasound machines typically use frequencies in the range of 2 to 18 MHz.

(a) What are the corresponding wavelengths in air?
(b) Given that the human body contains primarily water, estimate the wavelength range in the body.

21. ✳ A 1000 Hz sound wave with a pressure amplitude of 20 N/m² in air is at the threshold of pain for human hearing. It has pressure amplitude of $\Delta p = 20$ N/m². What is the corresponding displacement amplitude?
22. ✳ What is the intensity of a 2000 Hz sound wave in air that has a pressure amplitude of 1.5×10^{-3} N/m²?
23. ✳ When the displacement amplitude of a sound wave doubles, by what factor does the pressure amplitude increase?

Section 15-2 Wave Propagation and Huygens' Principle

24. ✳✳ For the wave front in Figure 15-32, use Huygens' principle to determine the shape of the resulting wave front
    (a) close to the original wave front;
    (b) far from the original wave front.

Figure 15-32 Problem 24

Section 15-3 Reflection and Refraction

25. ✳✳ In Figure 15-33, a plane wave is approaching a rectangular corner. Draw the reflected wave using Huygens' principle.

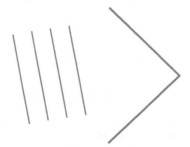

Figure 15-33 Problem 25

Section 15-4 Standing Waves

26. ✳ An air column has a length of 0.75 m and is open at both ends. What is the lowest frequency of a standing wave that can be found in such a column?
27. ✳✳ Standing waves can form in a box. A speaker enclosure has the dimensions $10 \times 15 \times 20$ cm. What are the lowest-frequency standing waves that can form in the three directions?

Section 15-5 Interference

28. ✳✳ Two point sources are 25 cm from each other, and both emit a 2000 Hz tone. Sketch the resulting interference pattern.
29. ✳ Two violins produce frequencies of 294 Hz and 296 Hz. What beat frequency is heard?
30. ✳✳ Two point sources are in phase and emit a 500 Hz tone. What is the phase difference at a point that is 5.0 m from one source and 6.0 m from the other?
31. ✳ In North America, the telephone dial tone consists of a 350 Hz tone and a 440 Hz tone. What is the resulting beat frequency?

445

Section 15-6 Measuring Sound Levels

32. ✱ A food processor produces a sound intensity level of 87 dB at a distance of 1.0 m. How many watts per square metre does this intensity level correspond to?

33. ✱ A single table saw produces a sound intensity level of 95 dB at a distance of 3.0 m. What would the sound intensity level be for two table saws, both 3.0 m away?

34. ✱ What is the displacement amplitude for an 800 Hz sound wave that has an intensity of 20 mW/m²?

Section 15-7 The Doppler Effect

35. ✱ A police siren emits a frequency of 700.0 Hz. When the police car approaches you at a speed of 80.0 km/h, what frequency do you perceive?

36. ✱ The Doppler effect can be used to measure flow rates in blood. A sound wave is transmitted into a vein or an artery, where it is reflected off the moving red blood cells and then detected by a receiver. A transmitter uses a frequency of 10.0 MHz, and the blood flow rate is 0.50 m/s. What is the expected frequency shift?

37. ✱ Some bats use the Doppler shift to detect the direction of motion of prey. The bats use a frequency of about 61 kHz and can detect shifts as small as 35 Hz. To what velocity does this correspond?

COMPREHENSIVE PROBLEMS

38. ✱ At large concerts, it is sometimes disconcerting to observe the musicians moving apparently out of sync with the music. This results from the time it takes the sound to travel from the stage to you. When the musicians are playing at 100 beats/min, at what distance from the stage will they appear to be one full beat behind?

39. ✱ While camping, you observe a lightning bolt, and 4.0 s later you hear the associated thunder. How far away is the strike?

40. ✱ Two sound sources differ in sound intensity level by 20 dB. Find the ratio of their
(a) displacement amplitudes;
(b) pressure amplitudes;
(c) intensities.

41. ✱✱✱ Two sound waves of frequency 200 Hz travel in the x-direction. One has a displacement amplitude of 10 nm, and the other has a displacement amplitude of 20 nm. The phase difference between them is $\pi/4$ rad. What is the resultant amplitude?

42. ✱✱ If we allow the phase difference between the two waves in problem 41 to vary, what are the smallest and largest possible displacement amplitudes?

43. ✱ The sound intensity level a distance of 23 m from an isotropic point source is 65 dB. What is the intensity, and what is the total power radiated by the source?

44. ✱✱✱ B flat played on a tuba has a fundamental frequency of approximately 117 Hz.
(a) Write equations for this fundamental and for second and third harmonics with half the amplitude of the fundamental.
(b) On the same graph, plot the waveform of the fundamental, the two harmonics, and the resultant of these three tones.

(c) Compare the period and amplitude of the resultant to those of the fundamental.

45. ✱✱ Police radar uses the Doppler shift to determine whether you are speeding. The radar unit transmits a wave, which is reflected off your car and Doppler shifted as a result. Rather than detecting the frequency shift directly, the radar receiver mixes the transmitted and received beam and measures the resulting beat frequency. For a radar unit that operates at 10 GHz and for a driving speed of 130 km/h, what beat frequency would result?

46. ✱ An interesting demonstration of standing waves can be done by twirling one end of a flexible, corrugated hose that is open at both ends. The twirling motion causes air to be drawn through the hose, and air rushing over the corrugations creates turbulence in the hose. The turbulence excites waves. It is usually possible to get the three lowest tones from the hose. For a 1.50 m hose, what frequencies will be heard?

47. ✱✱ A bugle is an instrument that has an air column with a fixed length. Different notes can be achieved by exciting different harmonics in the column, not an easy task. The bugle actually acts like a tube with both ends open. For a certain length L, a bugle will have a sequence of four consecutive harmonics with frequencies very nearly equal to the frequencies associated with the notes G_4, C_5, E_5, and G_5. Determine the length of the bugle.

48. ✱✱ A plane wave is normally incident on a boundary where the speed of sound changes. The wave travels from a medium in which the speed of sound is 1200 m/s to a medium in which the speed of sound is 400 m/s. The incident wave has a frequency of 440 Hz.
(a) What is the period of a 440 Hz wave?
(b) How far will a wave front travel in the second medium during the time from part (a)?
(c) What is the wavelength in the second medium?
(d) What is the frequency in the second medium? (Hint: Calculate from the speed and the wavelength from part (c).)

49. ✱✱ The Bay of Fundy has some of the highest tides in the world, with a tidal range as large as 15 m. The bay behaves like a resonator that is open at one end and closed at the other. The Bay of Fundy is approximately 270 km long, and the tidal period is approximately 12.5 h. Assuming that the tide corresponds to the lowest-frequency standing wave, what is the wave speed in the Bay of Fundy?

50. ✱ The air cavity inside stringed instruments is designed to have resonant frequencies that are close to the resonant frequencies that are produced by the instrument. For a 12 cm-deep guitar body cavity, what are the frequencies of the standing waves that could form between the top and back of the guitar?

51. ✱✱ The reverberation time in a performance space is a measure of how long it takes a sound to fade away once it has stopped being produced. The time is defined as the amount of time it takes the sound to drop by 60 dB. For an initial sound intensity of 0.2 W/m², what is the corresponding reduction in intensity after one reverberation time?

52. ✱✱✱ A sound wave loses 10% of its intensity each time it reflects from a particular surface. How many reflections will be required to reduce the sound intensity level by 3 dB?

53. ✶✶ To determine the depth of a well, you drop a stone into the well and hear the splash 2.8 s later. How deep is the well?

54. ✶✶ A jet produces a sound intensity level of approximately 140 dB at a distance of 30 m. Assuming that your ear has an effective radius of about 2.0 cm, what is the total power received by your ear?

55. ✶✶✶ One way of measuring the frequency of a sound is to make measurements of the interference pattern that is created when the sound is produced by two sources. You find that for two sources, the sound is loudest when you are equidistant from the sources. You then determine that there is a minimum in intensity when you are 2.8 m farther from one source than the other. What is the frequency of the source?

56. ✶✶ Standing waves in your music room can cause certain spots in the room to have lower volume levels. For a room with the dimensions $4.00 \times 6.00 \times 2.40$ m, what are the possible standing wave frequencies?

57. ✶✶ Standing waves can be set up in solid rods, and this is the basis for a popular demonstration. The rod, clamped in the middle, is stroked (sometimes a cloth is used) and begins to emit a tone. For a 120 cm aluminum rod in which the speed of sound is 6420 m/s, find
(a) the fundamental wavelength on the rod;
(b) the fundamental frequency.

58. ✶✶ A typical audio amplifier does not have the same gain at all frequencies. An amplifier is rated 100.0 W at 1.0 kHz. The manufacturer specifies that the output falls by 3.0 dB when the frequency output is 20 kHz. What power can the amplifier produce at 20 kHz?

59. ✶✶✶ Show that we still get beats if we add a phase constant to each of the two waves in Equation (15-22).

60. ✶✶ You have been asked to design a set of pipes for an organ. Assuming that the pipes have one open end and one closed end, calculate the lengths you will need to go from 440 Hz (A_4) to 880 Hz (A_5).

DATA-RICH PROBLEM

61. ✶✶✶ You have been hired by an environmental consulting firm to do a noise analysis for a quarry operation. The operation uses two trucks, a drill, and a crusher. The manufacturers of the equipment provided the specifications in Table 15-5 for the sound intensity level of the various pieces of equipment. Local bylaws specify that the maximum sound intensity level at the perimeter of the quarry property not exceed 85 dB. How close to the perimeter can the quarry operate all these devices simultaneously?

Table 15-5 Data for Problem 61

Equipment	Sound intensity level at 10 m
Truck	85 dB
Drill	110 dB
Crusher	110 dB

OPEN-ENDED PROBLEM

62. ✶✶✶ Many of us have heard the effect that can be produced by inhaling helium and speaking. The speaker's voice is shifted to higher frequencies. Discuss the physics behind this effect.

 See the text online resources at www.physics1e.nelson.com for Open Problems and Data-Rich Problems related to this chapter.

THERMODYNAMICS

Learning Objectives

When you have completed this chapter you should be able to:

1 List the three main states of matter and some of their associated physical characteristics.

2 Discuss the state variables for some systems.

3 Describe the relationship between the average speed of a particle in a gas and the pressure that the gas exerts on the walls of its container.

4 Define thermometric property and do calculations related to the linear expansion of solids.

5 Be familiar with some common temperature scales.

6 State the zeroth law of thermodynamics, and describe the concept of thermal equilibrium.

7 Use the ideal gas law to determine various properties of a gas.

8 Describe how a universal temperature scale can be based on an ideal gas thermometer.

9 Discuss the relationship between the temperature of a gas and the average mechanical energy of the atoms in that gas.

10 Qualitatively interpret a probability distribution.

11 Read a phase diagram.

For additional Making Connections, Examples, and Checkpoints, as well as activities and experiments to help increase your understanding of the chapter's concepts, please go to the text's online resources at www.physics1e.nelson.com.

Chapter 16
Temperature and the Zeroth Law of Thermodynamics

The morning landscape in Figure 16-1 contains all the familiar phases of matter: solids in the trees and bushes, liquid in the tiny droplets of water in the fog, and invisible gases in the air.

On such a morning, you would find the grass and bushes covered with heavy dew. In addition to liquid water, fog also contains a lot of water vapour. So, water is present in two phases: liquid and gas. While this is a relatively common occurrence for water, we might wonder whether other substances exhibit this behaviour, and whether water can be present in all three states simultaneously.

Courtesy of Rebecca Dimock

Figure 16-1 Early morning fog in the Annapolis Valley, Nova Scotia

16-1 Solids, Liquids, and Gases

On Earth, matter is usually found in **solid, liquid,** and **gas** phases, as described in Chapter 12. Under suitable conditions, most materials can exist in all three phases. For example, water exists as a solid (ice), a liquid, and a gas (steam). However, there is a fourth phase, called plasma. Plasma is a highly ionized gas and is the predominant phase in the universe. For example, the Sun, which contains over 99% of the mass in the solar system, is in a plasma phase.

Solids tend to be very rigid and do not compress or stretch readily. The atoms in solids tend to remain at fixed positions, but the atoms in a liquid and in a gas can drift around. However, the atoms in solids do vibrate around their equilibrium positions. In crystalline solids, the atoms are arranged in regular periodic arrays, and each atom is bonded to a set of adjacent atoms (Figure 16-2). Crystalline solids are usually composed of a collection of small crystals. Even microscopic individual crystals still contain a huge number of atoms and extend over a very large number of interatomic distances. Such materials are described as having long-range order. In some materials, such as diamond, the individual crystals are often large enough to be seen without magnification (Figure 16-3).

Figure 16-3 A large single crystal of diamond. The faces of the diamond correspond to planes in the crystal structure

liquid

Figure 16-4 A liquid phase does not have long-range order, and the atoms are somewhat more loosely packed than atoms in the solid phase.

solid

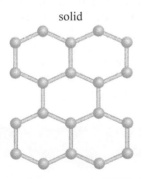

Figure 16-2 A periodic lattice from a solid. The atoms are closely spaced and form a periodic array.

Liquid phases are usually somewhat less dense and more compressible than the solid phase of the same material. Liquids have no long-range order (Figure 16-4). Over short distances, the positions—and sometimes the relative orientations—of adjacent particles in liquids are correlated. As a liquid cools, the extent of this ordering increases, becoming long-range order at the freezing point. Liquids consist of either atoms or molecules.

Gases are much less dense and much more compressible than the liquid and solid phases of the same material. Gases have no long- or short-range order, so the particles are completely free to move around (Figure 16-5). Like liquids, gases are comprised of either atoms or molecules.

gas

Figure 16-5 The atoms in a gas are relatively far apart, and their positions are not related to each other.

16-2 State Variables

The goal of **thermodynamics** is to describe energy transfers in the form of work and thermal energy (heat) from a macroscopic point of view using variables such as temperature and **pressure**. A key concept in thermodynamics is the state of a system. If we know all the forces that act on a particle, and we know the position and velocity of the particle at some instant, then we can, at least in principle, describe the motion of the particle for all time. This description would then specify the state of the particle.

Consider a flask of gas sitting on a lab bench. This system consists of a vast number of individual particles, so single-particle descriptions are not suitable. We can describe the state of a pure gas in terms of the number of particles in the system, the **volume** of space occupied, the pressure that the gas exerts on the walls of the container, and the temperature of the gas. We will discuss these **state variables** for gases in Section 16-6. Other types of systems use different variables. Table 16-1 gives a few examples. Notice that all the variables in Table 16-1 refer to macroscopic properties of the systems. None of the variables relate to properties of specific individual particles.

 CHECKPOINT

C-16-1 Are the Details in the State Variables?

Do state variables depend on all the details of the particles in a system?

C-16-1 Yes. For example, we will see that the pressure in a gas depends on the average speed squared of the particles in the gas. To determine the average, you need to know the speeds of all the particles.

 CHECKPOINT

C-16-2 Can the Details be Extracted from the State Variables?

Can information about specific particles in a system be determined from the state variables?

C-16-2 No. The state variables usually allow us to determine some average values. However, it is not possible to extract information about single particles from the average values.

 PEER TO PEER

State Variables

You can think of state variables as the values you must know in order to recreate a particular thermodynamic system. For any ideal gas system, you only need to know three state variables to know everything about the system. This means that you can calculate any state variable from any three other state variables. For example, if you know the volume, temperature, and number of particles in a system, you can use the ideal gas law to determine the pressure. More complex systems are more difficult to describe and therefore require more state variables to do so.

The fundamentals of thermodynamics were developed during the Industrial Revolution in the late eighteenth and early nineteenth centuries. In fact, one of the first applications of thermodynamics was to improve the efficiency of steam engines (Figure 16-6). The original focus of thermodynamics was the behaviour of gases (steam in particular) and the conversion of thermal energy into mechanical energy.

The four thermodynamic state variables for a gas are the number of gas particles, the volume, the pressure, and the temperature. The number of particles in a gas is a straightforward quantity. Similarly, the volume occupied by a gas is relatively straightforward to understand and is usually simply determined by the size of the container holding the gas. However, the remaining two variables—pressure and temperature—are somewhat more subtle. We will first consider pressure.

LO 3

16-3 Pressure

The pressure that a gas exerts on the walls of its container results from collisions of individual particles with the wall of the container (Figure 16-7). When a particle collides with the walls of its container, it exerts a force on the wall, and the wall exerts a force on the particle.

Table 16-1 State variables for various types of systems

System	State variables			
Gas	Temperature	Volume	Pressure	Number of particles
Magnet	Temperature	Magnetization	Applied magnetic field	Number of magnetic dipoles*
Dielectric	Temperature	Polarization	Applied electric field	Number of electric dipoles*

*Chapters 19 and 23 describe electric and magnetic dipoles, respectively.

Since momentum is conserved, the change in momentum of the container wall must be exactly equal but opposite to the change in momentum of the particle:

$$\Delta\vec{p}_{\text{wall}} = -\Delta\vec{p}_{\text{particle}} = 2mv_x\hat{i} \qquad (16\text{-}2)$$

Equation (16-2) gives us the momentum imparted to the container wall due to a single collision with a single particle. To determine the pressure exerted by the gas, we also need to know the rate at which collisions are occurring.

The time between collisions between a given particle and the vertical wall is the time it takes the particle to cross the length of the container twice:

$$\Delta t = \frac{2L}{v_x} \qquad (16\text{-}3)$$

Dividing Equation (16-2) by Equation (16-3), we obtain the rate of change of momentum of the wall:

$$\frac{\Delta p_{\text{wall}}}{\Delta t} = \frac{2mv_x}{\dfrac{2L}{v_x}} = \frac{mv_x^2}{L} \qquad (16\text{-}4)$$

As we saw in Chapter 7, we can write Newton's second law as $F = \dfrac{\Delta p}{\Delta t}$, so

$$F_x = \frac{\Delta p}{\Delta t} = \frac{mv_x^2}{L} \qquad (16\text{-}5)$$

To determine the total force on the wall, we sum this expression over all the particles to get

$$
\begin{aligned}
F_{x,\text{Total}} &= \sum_i F_{x,i} \\
&= \sum_i \frac{mv_{x,i}^2}{L} \\
&= \left(\frac{mv_{x,1}^2}{L} + \frac{mv_{x,2}^2}{L} + \frac{mv_{x,3}^2}{L} + \cdots + \frac{mv_{x,N}^2}{L}\right) \\
&= \frac{m}{L}\left(v_{x,1}^2 + v_{x,2}^2 + v_{x,3}^2 + \cdots + v_{x,N}^2\right) \qquad (16\text{-}6)
\end{aligned}
$$

where N is the total number of particles in the box.

We note that the average values of the square of the x-components of the velocities is given by

$$\left(v_x^2\right)_{\text{avg}} = \frac{1}{N}\left(v_{x,1}^2 + v_{x,2}^2 + v_{x,3}^2 + \cdots + v_{x,N}^2\right) \qquad (16\text{-}7)$$

Rearranging this equation gives

$$N\left(v_x^2\right)_{\text{avg}} = \left(v_{x,1}^2 + v_{x,2}^2 + v_{x,3}^2 + \cdots + v_{x,N}^2\right) \qquad (16\text{-}8)$$

Thus, the total force exerted on the wall of the container is

$$F_{x,\text{Total}} = \frac{m}{L}N(v_x^2)_{\text{avg}} \qquad (16\text{-}9)$$

To determine the pressure that the gas exerts on the wall, we divide the total force by the area of the wall to obtain

Figure 16-6 Improving the efficiency of steam engines provided much of the impetus for the early work in thermodynamics.

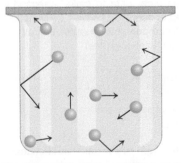

Figure 16-7 Gas atoms collide with the walls of a container.

Consider a gas particle of mass m that is moving with velocity $\vec{v}_{\text{initial}} = v_x\hat{i} + v_y\hat{j} + v_z\hat{k}$ in a cubic container with sides of length L and volume $V = L^3$. We assume that the collisions of the gas particles with the walls of the container are elastic. Then, the particle has the same speed before and after the collision, but one component of the particle's velocity has been reversed. For example, if a particle collides with a vertical wall (perpendicular to the x-axis), the x-component of the velocity is reversed, and $\vec{v}_{\text{final}} = -v_x\hat{i} + v_y\hat{j} + v_z\hat{k}$, the change in momentum of the particle is then

$$\Delta\vec{p}_{\text{particle}} = \vec{p}_{\text{f}} - \vec{p}_{\text{i}} = m\vec{v}_{\text{final}} - m\vec{v}_{\text{initial}} = -2mv_x\hat{i} \qquad (16\text{-}1)$$

THERMODYNAMICS

$$P = \frac{\frac{m}{L}N(v_x^2)_{avg}}{L^2} = \frac{mN(v_x^2)_{avg}}{L^3} \qquad (16\text{-}10)$$

We can generalize this expression to three dimensions. For any set of velocities,

$$(v^2)_{avg} = (v_x^2)_{avg} + (v_y^2)_{avg} + (v_z^2)_{avg} \qquad (16\text{-}11)$$

Because the container is cubical and there is nothing special about either the x-, y-, or z-directions (if we assume there are no external forces acting on the gas), it must be true that

$$(v_x^2)_{avg} = (v_y^2)_{avg} = (v_z^2)_{avg} = \frac{(v^2)_{avg}}{3} \qquad (16\text{-}12)$$

Finally, recalling that $V = L^3$, we can write

 KEY EQUATION $$P = \frac{1}{3}\frac{mN(v^2)_{avg}}{V} \qquad (16\text{-}13)$$

✓ CHECKPOINT

C-16-3 Pressure-Speed Connection in a Gas

If we observe that the pressure in a sample of gas changes, can we conclude that the average speed squared has changed correspondingly?

C-16-3 No. Either the number of particles or the volume that the gas occupies could have changed instead.

You may have questioned the assumption that particles colliding with the container emerge from the collision with their speeds unchanged. Implicit in this assumption is that the velocity of the container did not change as a result of the collision. This situation might appear to violate conservation of momentum. However, gas molecules collide with all the walls of the container. For every particle that collides with the wall on the right, there is a particle with the same speed that collides with the wall on the left. If this were not the case, the total momentum of the gas would be changing, and the centre of mass of the gas would move. In other words, the net momentum transfer resulting from the random motion of the huge number of gas particles averages to zero.

You might also wonder about the assumption implicit in Equation (16-3), which states that the time for the particle to go from one side of the container to the other is equal to the length of the container divided by the speed of the particle. This relationship is only strictly true for a single particle that does not change its velocity during this time. In fact, a particle in a gas collides with other particles, and its velocity does change. However, our assumption is still valid overall because the gas is in a state of equilibrium, with constant volume, pressure, and temperature. To maintain the state of equilibrium, it must be true that for every collision that causes a particle with speed v_x to change to v_x', there must be a corresponding collision that causes a particle with speed v_x' to change to v_x. Otherwise,

 ## EXAMPLE 16-1

How Fast Are the Molecules Going?

At standard temperature (0°C) and pressure (101.325 kPa), or STP, a mole of gas occupies a volume of 22.4 L. The molar mass of an atom of nitrogen is approximately 14 g/mol. Nitrogen is diatomic, that is, its atoms bond in pairs to form N_2 molecules. Estimate the average speed of a nitrogen molecule at STP.

SOLUTION

Since each nitrogen molecule contains two atoms, a mole of nitrogen gas has a mass of approximately 28 g. Solving Equation (16-13) for the speed, we get

$$(v^2)_{avg} = \frac{3PV}{mN}$$

$$\therefore \sqrt{(v^2)_{avg}} = v_{rms} = \sqrt{\frac{3PV}{mN}}$$

where v_{rms} is the root-mean-square velocity, that is, the square root of the average (or mean) of the squares of the individual velocities.

Substituting in the known values, and converting litres to cubic metres (1 L = 0.001 m³):

$$v_{rms} = \sqrt{\frac{3 \times 101.325 \times 10^3\,\text{kg}\cdot\text{m}^{-1}\text{s}^{-2} \times 0.0224\,\text{m}^3}{0.028\,\text{kg}}}$$

$$= \sqrt{2.43 \times 10^5\,\text{m}^2\,\text{s}^{-2}} = 490\,\text{m/s}$$

Making sense of the result:

We can use other data to make a second estimate. It takes 200 kJ of energy to convert 1.0 kg of liquid nitrogen to gaseous nitrogen. If we assume that all of this energy goes into kinetic energy, then

$$v = \sqrt{\frac{2E}{m}} = \sqrt{\frac{2 \times 2 \times 10^5\,\text{J/kg}}{1\,\text{kg}}} = 630\,\text{m/s}$$

Thus our result seems reasonable.

the total momentum of the gas would change, and the gas would not be in a state of equilibrium. *On average*, there is a constant number of particles with a given speed v_x, and that number of particles will, *on average*, cross the container in the time L/v_x.

LO 4

16-4 Temperature and Thermal Expansion

Temperature is something that we all discuss on a daily basis: What is the temperature today? At what temperature should the oven be set to cook a particular dish? In fact, thermometry (the measurement of temperature) was so

THERMODYNAMICS

taken for granted that the law of thermodynamics that deals with it was only developed after the formulation of the other laws. Since these had already been assigned to the first and second laws, thermometry was assigned the zeroth law.

To measure temperature, we use a thermometric property of some object. One common thermometer is the alcohol thermometer, which relies on the thermal expansion of alcohol: as we heat alcohol, it expands. There are many other materials that have thermometric properties that are utilized in thermometers and some of them are listed in Table 16-2.

Table 16-2 Various common temperature-measuring devices and their associated thermometric properties

Temperature-measuring device	Thermometric property
Galinstan*	Thermal expansion
Bimetallic strip	Differential thermal expansion
Thermocouple	Thermoelectric effect

*Galinstan is a liquid alloy used as a substitute for mercury, which is highly toxic.

 **CHECKPOINT**

C-16-4 Wood or Metal: Which Is "Warmer"?

Consider a metal object and a wooden object sitting on a desk. Typically, the metal object feels cool to the touch, but the wooden object does not. Yet, the objects have been sitting side by side in the same room and should be at the same temperature. Why do they not feel equally warm?

C-16-4 When we touch the wood or the metal, thermal energy is transferred from our hand to the object. Wood has a very poor ability to conduct thermal energy, so the spot we are touching quickly warms up to about the same temperature as our hand. Metal, however, is a very efficient conductor, and the energy from our finger rapidly spreads into the entire object. The spot we are touching does not warm up much and therefore feels cooler than the wood.

Thermal Expansion of Solids

Most solids expand when heated. For a long, thin piece of solid material, the expansion occurs principally along the long axis of the material.

If we do not change the temperature of the material too much, we can make the assumption that the amount of expansion is linear, that is, the expansion is directly proportional to the temperature change. Since the molecular bonds in the material expand equally, the total amount of expansion is also proportional to the length of the material. For a given temperature change, the increase in the length of a piece of metal that is 100 cm long is twice the increase in a 50 cm length of the same material.

 MAKING CONNECTIONS

Using Physics in the Kitchen

When we have a jar that is difficult to open, it can help to run hot water over the lid of the jar. The metal in the lid expands, making it easier to unscrew.

For linear thermal expansion, the coefficient of thermal expansion, α, is defined as

KEY EQUATION
$$\alpha = \frac{\Delta L}{L} \frac{1}{\Delta T} \qquad (16\text{-}14)$$

where ΔL is the observed increase in length, L is the original length of the object, and ΔT is the temperature change.

To calculate the change in length of a sample, we rearrange Equation (16-14):

$$\Delta L = L\alpha\Delta T \qquad (16\text{-}15)$$

Because the expansion depends on the details of the bonds between the atoms in the solid, different materials expand to a greater or lesser extent. Table 16-3 lists some coefficients of linear thermal expansion for various materials.

Table 16-3 Coefficients of linear thermal expansion for various materials

Material at 20°C	Coefficient of thermal expansion (10⁻⁶/°C)
Aluminum	23
Brass	19
Carbon steel	10.8
Concrete	12
Copper	17
Diamond	1
Glass	8.5
Gold	14
Iron	11.1
Lead	29
Pine (perpendicular to the grain)	34
Stainless steel	17.3
Water	~69

When designing devices that are made with different materials, it is important to keep in mind these differing coefficients to ensure that the devices function properly across the range of temperatures in which they are expected to operate.

THERMODYNAMICS

EXAMPLE 16-2

Thermal Expansion of a Rod

A long, thin rod of aluminum is measured to have a length of 20.0°C. What will be the change in the length of the rod at 30.0°C?

SOLUTION

According to Table 16-3, the coefficient of linear thermal expansion for aluminum is $23.0 \times 10^{-6}/°C$. We use Equation (16-15) to find the change in length:

$$\Delta L = L\alpha\Delta T$$

$$= (0.250 \text{ m}) (23.0 \times 10^{-6}/°C) (30.0°C - 20.0°C)$$

$$\Delta L = 5.75 \times 10^{-5} \text{ m}$$

Thus, at 30.0°C, the change in length of the rod will be about 0.058 mm.

Note that if we had cooled the rod by 10°C, the change in length would have been negative.

MAKING CONNECTIONS

Sea Level Rise: Melting Ice or Thermal Expansion?

As the temperature of the atmosphere increases, so too does the temperature of the ocean, although it takes some time for the lower levels of the ocean to warm up (Figure 16-8).

As the oceans warm, they also expand. Some researchers predict that this thermal expansion will make the largest contribution to sea level rise caused by global warming.

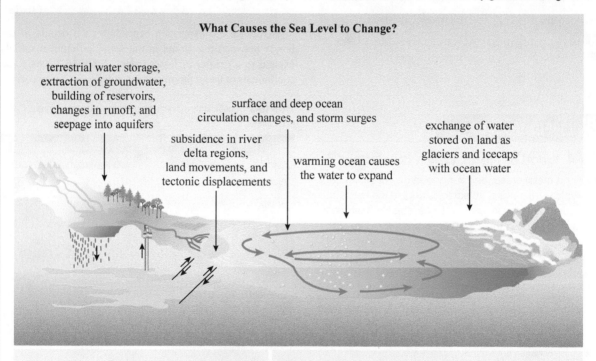

What Causes the Sea Level to Change?

terrestrial water storage, extraction of groundwater, building of reservoirs, changes in runoff, and seepage into aquifers

subsidence in river delta regions, land movements, and tectonic displacements

surface and deep ocean circulation changes, and storm surges

warming ocean causes the water to expand

exchange of water stored on land as glaciers and icecaps with ocean water

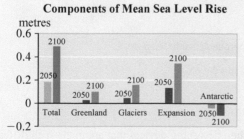

Components of Mean Sea Level Rise

Various assumptions need to be made to model sea level rises. The model depicted to the left assumes rapid economic growth, a global population that peaks in the 21st century and then declines, and the rapid introduction of new and more efficient technologies. The model also assumes a more global economy with a substantial reduction in regional differences in per capita incomes. Another key assumption is what energy sources we rely on, and this model assumes a fossil fuel intensive energy supply.

UNEP/GRID-Arendal

Figure 16-8 Thermal expansion and global warming

THERMODYNAMICS

16-5 Thermometers and Temperature Scales

Temperature-sensitive properties of materials are the basis for various types of thermometers, which convert measurements of quantities such as volume and resistance into temperature readings. In order to quantify the measurements, we must assign a temperature scale to the readings of our thermometer. The most common scale in the world today is the Celsius scale, which assigns a temperature of 0°C to the freezing point of water and 100°C to the boiling point. In practice, we also need to specify the pressure for these calibration points because many of them also depend on the pressure. For example, we need to increase the cooking times for foods boiled at high altitudes because, at higher altitudes, the air pressure is lower and the boiling temperature of water is also lower.

Celsius and Fahrenheit There are several different temperature scales in use today. In addition to the Celsius scale, another common scale is the Fahrenheit scale, which was used in Canada until the 1970s, when the metric system was introduced. Fahrenheit is still used in the United States and Belize. However, even in some countries that have "gone metric," we still see the Fahrenheit scale on some household appliances, most commonly ovens. In the Fahrenheit system, the freezing point of water is 32°F, and the boiling point is 212°F. We can use these reference points to convert temperatures from Celsius to Fahrenheit. In the Celsius system there are 100 increments between freezing and boiling, and in Fahrenheit there are 180 increments. Therefore, the ratio of a Fahrenheit degree to a Celsius degree is 180°F/100°C = 9°F/5°C. We also note that 0°C corresponds to 32°F. Thus,

$$°F = \frac{9°F}{5°C} \times °C + 32°F \quad °C = \frac{5°C}{9°F} \times (°F - 32°F) \quad (16\text{-}16)$$

EXAMPLE 16-3

Temperature Conversion

John, a hungry physics student, wants to cook his favourite recipe, Tandoori chicken. The recipe instructs him to preheat the oven to 175°C. However, the dial on his oven is labelled in Fahrenheit. What setting should John use?

SOLUTION

John needs to find the Fahrenheit equivalent for a Celsius temperature. Using Equation (16-16), we get

$$°F = \frac{9°F}{5°C} \times 175°C + 32°F$$
$$= 315°F + 32°F$$
$$°F = 347°F$$

Kelvin The Kelvin scale was developed by William Thomson (British physicist and engineer, also known as Lord Kelvin, 1824–1907) after he discovered that there is a lower limit for temperature, called absolute zero (about −273.15°C). No object can have a temperature below this limit. The Kelvin scale uses the same degree increments as the Celsius scale, but absolute zero is denoted as 0 K. (Note that SI does not use a degree symbol with kelvin units.) As you will see below, the Kelvin scale is particularly useful for thermodynamic calculations and predicting the behaviour of gases. Table 16-4 lists some common temperatures in Celsius, Kelvin, and Fahrenheit.

Table 16-4 Some common temperatures listed in Celsius, Kelvin, and Fahrenheit

	Celsius	Kelvin	Fahrenheit
Absolute zero	−273.15°C	0 K	−459.67°F
Freezing point of water	0°C	273.15 K	32°F
Typical normal body temperature	37°C	310.15 K	98.6°F
Room temperature	20°C	293.15 K	68°F
Surface temperature of the Sun	~6 000°C	~6 273 K	~10 832°F

16-6 The Zeroth Law of Thermodynamics

Suppose we place a thermometer in thermal contact with an object that is at some steady temperature. We will assume that our thermometer is so small that it does not affect the temperature of the object. The thermometer reading will change when we first establish contact with the object, but the reading will eventually reach a steady state. At this point, we say that the thermometer is in thermal equilibrium with the object. Two objects are said to be in **thermal equilibrium** when placing them in thermal contact with each other causes no change in the temperature of either object—they are at the same temperature.

If we now place the thermometer in thermal contact with a second object and find that the thermometer stabilizes at the same reading, we can say that the thermometer is in thermal equilibrium with the second object.

If we then place the first and second objects in thermal contact with each other, the reading on our thermometer will not change, regardless of which object it contacts. This observation demonstrates the **zeroth law** of thermodynamics: If object A is in thermal equilibrium with object B, and object A is also in thermal equilibrium with object C, then object C is also in thermal equilibrium with object B.

In other words, if we get the same temperature reading on a thermometer from two different objects, then those objects are at the same temperature, and if we put them together, their temperatures will not change.

C-16-5 Holding Hands

When we hold hands with another person, we notice that the other person's hand feels warm, even if the temperature of the other person is exactly the same as ours. Are we in thermal equilibrium with the other person? If so, why does the person's hand feel warm to us?

C-16-5 Our hand is in thermal equilibrium with the other person's hand if that person is at the same temperature as us. Our skin is usually in contact with air that is cooler than our body temperature. As a result, thermal energy is constantly flowing out of our bodies into our surroundings, and our skin is at some temperature that is lower than our body's core temperature. Further, we have evolved to feel comfortable (i.e., neither hot nor cold) at "room temperature," so when we hold hands, the temperature of the surface of our skin is elevated above its normal temperature, and that feels warm to us.

LO 7

16-7 Ideal Gases

One difficulty with the system of thermometry we have described so far is getting two thermometers to agree with each other. While identical thermometers are consistent with each other, two different thermometers typically agree precisely only at common calibration points. Much experimentation has been done to develop thermometers that are accurate over their full range. The discrepancies between various types of thermometers are largely due to differences in the materials used to construct them. The particles in solids and liquids are relatively close together and interact strongly with each other. These interactions differ in the various materials and affect how the materials respond to changes in temperature. Consequently, the materials can differ both in the linearity of their response and in the temperature range in which the nonlinear response becomes significant.

Gases, however, have greater interatomic distances and, consequently, much weaker interactions between particles. The behaviour of gases is much more weakly dependent on the type of gas being studied than is the case for solids and liquids. Table 16-5 compares the densities and interatomic distances for ice, liquid water, and water vapour.

Table 16-5 Densities and interatomic distances for phases of water

Material	Density (g/cm³)	Typical interatomic distance (nm)
Ice at −10°C	0.998	≈0.3–0.4
Water at 40°C	0.992	≈0.3–0.4
Water vapour at 100°C and 1 atm	0.000 59	≈3–4

Much experimentation has been done with gases to determine the relationships between their pressure, temperature, and volume. The motivation for some of these studies was not to develop better thermometers, but to understand and improve the hot-air balloon!

 ONLINE ACTIVITY

The e-resource that accompanies every new copy of this textbook contains an Online Activity using the PhET simulation "Gas Properties." Work through the simulation and accompanying questions to gain an understanding of gas properties.

It was found that all gases exhibit the same behaviour at low densities, regardless of the chemical composition of the gas. This behaviour is described by the **ideal gas law**:

KEY EQUATION
$$PV = NkT \tag{16-17}$$

where P is the pressure, V is the volume, N is the number of particles in the gas, T is the temperature in kelvins, and k is Boltzmann's constant, 1.38×10^{-23} J·K^{-1} (named for the Austrian physicist Ludwig Boltzmann, 1844–1906).

 MAKING CONNECTIONS

Moles versus Number of Particles

Chemists often write the ideal gas law in terms of the number of moles of gas that are present. The amount (number of moles in the gas) is the number of particles in the gas expressed as a multiple of Avogadro's number, 6.02×10^{23}. Avogadro's number, N_A, is defined as the number of atoms in exactly 12 g of carbon-12. Denoting the amount as n, we can then write

$$PV = nN_A kT = nRT$$

where

$$R = N_A k = 8.314\ 472 \text{ J·K}^{-1} \cdot \text{mol}^{-1}$$

EXAMPLE 16-4

Using the Gas Law

The pressure in a 5.00 L flask of gas is 200 kPa, and the temperature is 300 K. Assuming that the gas is ideal, how many gas particles are in the flask?

SOLUTION

We can use the ideal gas law and solve for N:

$$N = \frac{PV}{kT}$$

We can now substitute in the known quantities. However, we must be careful to convert the units properly:

$1\,\mathrm{Pa} = 1\,\mathrm{N/m^2} = 1\,\mathrm{kg \cdot m^{-1} \cdot s^{-2}}$;
$1\,\mathrm{J} = 1\,\mathrm{N \cdot m} = 1\,\mathrm{kg \cdot m^2 \cdot s^{-2}}$; $1\,\mathrm{L} = 1000\,\mathrm{cm^3} = 10^{-3}\,\mathrm{m^3}$

Using these conversions we find that

$$N = \frac{PV}{kT} = \frac{(200\,000\ \mathrm{kg \cdot m^{-1} s^{-2}})\,(5 \times 10^{-3}\ \mathrm{m^3})}{(1.38 \times 10^{-23}\ \mathrm{kg \cdot m^2 s^{-2} K^{-1}})\,(300\ \mathrm{K})}$$

$$= 2.42 \times 10^{23}$$

Making sense of the result:

At first glance, the result seems like a huge number. However, one mole of gas at standard temperature (0°C) and pressure (101.325 kPa) occupies 22.4 L. We have arrived at a result of 2.42×10^{23} particles, which is about one-third of a mole. Our sample is at approximately twice standard pressure but somewhat less than one-quarter of the volume, so we would expect to have approximately 0.5 mol of gas. However, our temperature is higher than STP, so we should have slightly less than 0.5 mol. Thus, our result seems quite reasonable.

Note that real gases behave like ideal gases only at very low densities. Nonetheless, it is somewhat remarkable that all gases do behave the same at low densities, and we will devote some time to understanding why that is so.

Recall that gases are much less dense than liquids and solids and that the typical distance between particles in a gas is large. Consequently, gas particles in a container zoom about with relatively few interactions with the other gas particles. As we reduce the density of the gas, the fraction of the time that a particle spends interacting with other particles becomes smaller and smaller. As a result, the details of those interactions become less and less significant in determining the behaviour of the gas.

Figure 16-9 shows a generic interatomic potential. At large separations, the interaction energy goes to zero. At small separations, the interaction energy is very large and positive: the particles strongly repel each other. The minimum in the potential corresponds to the equilibrium bond length. The particles in a low-density gas spend most

of their time at large interparticle distances. When they do collide, the large repulsive force dominates the interaction. Consequently, a reasonable approximation to this potential, called the hard sphere approximation, is shown in Figure 16-10.

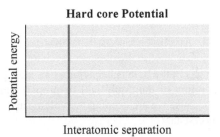

Hard core Potential

Figure 16-10 The hard sphere approximation for the interatomic potential for a dilute gas. The particles experience an infinitely strong repulsive force when they collide but do not interact otherwise.

Even though the approximate potential in Figure 16-10 does not have a minimum, this is not significant for a dilute gas, because the minimum is responsible for the attractive forces that ultimately produce the liquid and solid phases, which are not observed to occur in a dilute gas. Another way of thinking about it is that in a dilute gas, the particles do not spend a lot of time close to each other; thus, the short-range details of the potential are not important.

It is possible to calculate the equation of state for a system of particles that interact through the hard sphere potential, although the details are beyond the scope of this book. The result is the ideal gas law, $PV = NkT$.

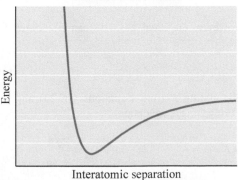

Interatomic Potential

Figure 16-9 A typical interatomic potential

16-8 The Constant-Volume Gas Thermometer

The discovery that all gases have the same behaviour in the low-density limit provided the basis for a universal thermometer against which all other thermometers could be calibrated: the **constant-volume gas thermometer**. This thermometer measures the temperature of a gas-filled bulb. In Figure 16-11, this bulb is shown submerged in a liquid of unknown temperature. The bulb on the right is filled with mercury, and its height can be adjusted so that the mercury in the left-hand tube is brought to the zero mark, keeping the volume of gas constant. The difference in height between the left- and right-hand tubes can then be used to determine the pressure in the gas-filled bulb.

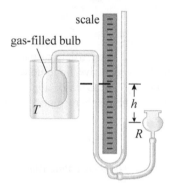

Figure 16-11   A constant-volume gas thermometer

If the gas in the bulb is behaving as an ideal gas, then the temperature of the liquid will be proportional to the pressure because we are keeping the volume and quantity of the gas constant:

$$T = CP \qquad (16\text{-}18)$$

where C is a constant that must be determined.

The pressure in the bulb is given by

$$P = P_{atm} - \rho_{Hg} g h \qquad (16\text{-}19)$$

where P_{atm} is the atmospheric pressure, ρ_{Hg} is the density of mercury, g is the acceleration due to gravity, and h is the height difference between the two mercury columns.

To determine C, we place the bulb at some known temperature, say the triple point of water (see Section 16-13), and we measure the pressure. At this point we find that

$$T_3 = CP_3 \qquad (16\text{-}20)$$

Eliminating C from Equations (16-18) and (16-20), we find

$$T = T_3 \frac{P}{P_3} = (273.16 \text{ K}) \frac{P}{P_3} \qquad (16\text{-}21)$$

Carrying out this procedure with different gases, we find that we get consistent results only at very low densities. By making a series of measurements with successively smaller densities, we can find a limit that gives a precise value for the temperature:

$$T = (273.16 \text{ K}) \lim_{\substack{gas \to 0 \\ density}} \frac{P}{P_3} \qquad (16\text{-}22)$$

It was discovered that for all gases at the low-density limit, the pressure goes to zero at the same temperature: 0 K. Figure 16-12 shows data confirming this discovery.

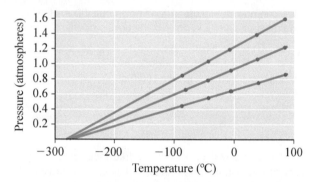

Figure 16-12   The three lines represent running the experiment with successively smaller quantities of gas in the system. The lines all go to zero at approximately −273°C.

16-9 Temperature and Mechanical Energy

We now return to the question of the relationship between the temperature—a macroscopic thermodynamic variable—and the molecular-level details of the system under consideration.

Recall that for a gas we showed that

$$P = \frac{1}{3} \frac{mN(v^2)_{avg}}{V} \qquad (16\text{-}23)$$

However, the ideal gas law can be solved for the pressure:

$$P = \frac{NkT}{V} \qquad (16\text{-}24)$$

Eliminating P from Equations (16-23) and (16-24) and solving for kT, we obtain

$$kT = \frac{VmN(v^2)_{avg}}{3NV} = \frac{1}{3}m(v^2)_{avg} = \frac{2}{3}\left(\frac{1}{2}m(v^2)_{avg}\right) \qquad (16\text{-}25)$$

Note that the product kT has units of energy. The last term in parentheses in Equation (16-25) is the average kinetic energy of a particle in the gas. Rearranging Equation (16-25) yields

KEY EQUATION
$$\frac{3}{2} kT = \frac{1}{2} m(v^2)_{avg} \qquad (16\text{-}26)$$

Equation (16-26) provides us with an elegant interpretation of the temperature of a gas: the **temperature** of a gas is a measure of the average kinetic energy of the gas particles.

Equipartition of Energy

We now consider the factor of 3 that appears on the left-hand side of Equation (16-26). It originated in Equation (16-13), which we derived, arguing that

$$\left(v_x^2\right)_{avg} = \left(v_y^2\right)_{avg} = \left(v_y^2\right)_{avg} = \frac{(v^2)_{avg}}{3} \qquad (16\text{-}12)$$

The factor of 3 results from the fact that there are three directions in which the gas particles can move. If the gas were confined to the xy-plane, we would have concluded that

$$\left(v_x^2\right)_{avg} = \left(v_y^2\right)_{avg} = \frac{(v^2)_{avg}}{2} \qquad (16\text{-}27)$$

In both cases, the energy in the system is equally divided among the "degrees of freedom" that the system has. The number of ways in which a system can absorb energy equals the number of degrees of freedom for the system. For an ideal gas in a container, there are three directions in which the gas particles can move, and the kinetic energy associated with those three directions is equal. The energy is "equally partitioned" between the three directions.

Energy considerations are more complex for a gas in which the particles are molecules instead of single atoms. For example, the atmosphere is composed of approximately 78% nitrogen gas, which consists of nitrogen molecules, N_2. A gas molecule has three degrees of freedom associated with its translational kinetic energy. However, the molecule can also absorb kinetic energy in other ways, which give it more degrees of freedom than a single atom.

A nitrogen molecule, or any diatomic molecule, can rotate and vibrate. In Chapter 8, you learned that the kinetic energy associated with a rotating object is

$$K_{rot} = \frac{1}{2} I \omega^2 \qquad (16\text{-}28)$$

where I is the moment of inertia about the axis of rotation, and ω is the angular velocity.

For a diatomic molecule, the rotational inertia about the axis that joins the two atoms is negligible. However, there is a significant moment of inertia about the two axes that are perpendicular to the axis that joins the two atoms. Since neither of these two rotational axes is preferred, the energy is evenly distributed between them. Thus, a diatomic gas molecule has two additional degrees of freedom associated with the rotational kinetic energy.

There is also energy associated with vibrations in a diatomic gas. The chemical bond that joins the two atoms is not infinitely rigid, and it has a potential as shown in Figure 16-13. This potential curve is called the Lennard–Jones potential. The equilibrium separation corresponds to the minimum in the potential. However, the minimum in the potential is a stable equilibrium point. For reasonably small excursions from the equilibrium point, the potential can be modelled as a parabolic well.

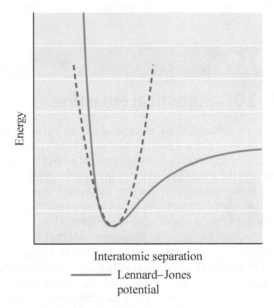

Figure 16-13   The interatomic potential in a diatomic molecule. The solid curve represents the Lennard–Jones potential, and the dashed curve shows a parabolic approximation.

As we know from Chapter 14, a parabolic potential is characteristic of a simple harmonic oscillator. Associated with the oscillator are two energies: the kinetic energy and the potential energy of vibration. Thus, a diatomic gas molecule has two additional degrees of freedom associated with its vibrational motion.

The effects of these additional degrees of freedom are observed in experimental data, but only at sufficiently high temperatures. A full understanding of molecular energy requires quantum theory, which is introduced in Chapter 31.

Lower Dimensions

Are there always three degrees of translational kinetic energy in a gas? Can atoms and molecules be confined to an xy-plane? In fact, there are various techniques for making such systems. One involves creating layered structures of semiconducting materials. At the interface between the two appropriately chosen materials, a large potential well can be created that confines the electrons in a very narrow region at the interface, as illustrated in Figure 16-14.

The electrons are free to move parallel to the interface. In the perpendicular direction, the confinement is such that the laws of classical physics no longer apply, and we must use quantum mechanics. (Strictly speaking, we need to use quantum mechanics in the other direction as well, but the quantization effects are very small.) The result is that the energy for motion in the z-direction becomes quantized, with only certain discrete values allowed. If we cool such a device to very low temperatures, the electrons are all forced into the lowest energy level for motion perpendicular to the interface

while remaining relatively free to move in the xy-plane parallel to the interface. Studies of such a system in strong magnetic fields led to the discovery of the integer quantum Hall effect. In 1985, the Nobel Prize in Physics was awarded to German physicist Klaus von Klitzing for this discovery.

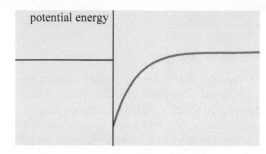

Figure 16-14 Deep potential "well" near the sample surface

LO 10

16-10 Statistical Measures

When we calculated the average speed for particles in a gas, we were using a statistic to describe the state of the gas. How accurately does a statistic such as average speed represent the actual behaviour of the gas? In other words, what is the probability that the theoretical average we have calculated matches the actual behaviour of a sample of gas particles in a laboratory experiment?

Suppose we want to determine whether a coin is "fair," that is, balanced such that there is an equal likelihood of getting either heads or tails when we toss the coin. To quantify the results of the coin toss, we count heads as 1 and tails as 0. If the coin is fair, the average value of a large number of coin tosses should be 0.5. The results of such an experiment are shown in Figure 16-15. We find that the running average gives a reliable description of the coin only when the number of coin tosses is quite large.

Similarly, there must be a large number of particles in a system in order for a statistical thermodynamic description to be applicable. As the number of particles in the sample increases, so does the probability that the

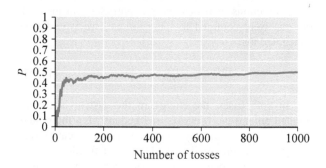

Figure 16-15 Running average value in a coin toss experiment

statistics accurately describes the sample. As the number of particles approaches infinity, the system approaches the **thermodynamic limit**, where the statistical measures are completely accurate. For practical purposes, when the number of particles in a system is of the order of Avogadro's number, the system behaves as though it were at the thermodynamic limit.

However, statistical measures do not give any details about the behaviour of an individual particle. For example, the average speed gives no indication of the proportion of particles that are moving faster or slower than the average at any given moment.

EXAMPLE 16-5

Road Trip Statistics

Suppose that a family drives from Nova Scotia to Manitoba and back for their summer holiday. The total time for the trip is three weeks, with lots of stops along the way to visit and

sightsee. The distance from Halifax to Winnipeg is approximately 3700 km. When they are in the car, they drive at a speed of 120 km/h. What is the average speed? Plot the speed probability distribution.

SOLUTION

To determine the average speed, we divide the total distance travelled, 7400 km, by the total time, 504 h:

$$\text{speed}_{avg} = \frac{\text{total distance travelled}}{\text{total time for trip}} = \frac{7400 \text{ km}}{504 \text{ h}} = 14.7 \text{ km/h}$$

On this basis alone, you might conclude that this family has very cautious drivers, not to mention being fairly annoying to fellow motorists.

To construct the speed probability distribution, we need to calculate the probability that the car is travelling at a particular speed. We already know that the total time for the trip is 504 h. The time spent driving is

$$\frac{7400 \text{ km}}{120 \text{ km/h}} = 61.7 \text{ h}$$

Therefore, the probability of finding the speed to be 120 km/h at any time on the trip would be

$$\text{Prob}(120 \text{ km/h}) = \frac{\text{time at 120 km/h}}{\text{total time for trip}} = \frac{61.7 \text{ h}}{504 \text{ h}} = 0.122$$

The remainder of the trip, 504 h − 61.7 h = 442.3 h, was spent at a speed of 0 km/h. Thus, the probability of the speed being 0 km/h is

$$\text{Prob}(0 \text{ km/h}) = \frac{442.3}{504} = 0.878$$

The probability of finding any other speed is zero, if we neglect the short times it takes to get up to speed and slow to a stop. Figure 16-16 shows a plot of these probabilities.

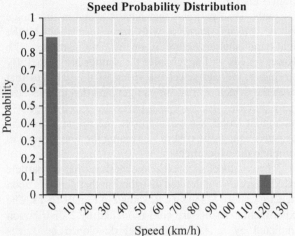

Figure 16-16 Speed probability distribution

Making sense of the result:

Clearly, the speed probability distribution gives us more information about the trip than simply knowing the average speed. For example, we can see that when the family does drive, they travel pretty quickly, but they do not drive that often. Nevertheless, a lot of details are still unknown—how many stops were made and where, for example.

What is the probability that an individual particle in a gas will have a speed within a certain range? An apparatus for measuring the distribution of particle speeds is shown in Figure 16-17(a). The source oven is filled with a gas. The collimating slits ensure that only those gas molecules with horizontal velocity enter the apparatus. The shutter on the left is opened for a brief time, Δt. At some time τ later, the right-hand shutter is also open for a time Δt. The detector produces a signal proportional to the number of particles that impinge upon it. To determine the speeds of the particles that can make it through the apparatus, we construct a space-time diagram as shown in Figure 16-17(b).

As we can see from Figure 16-17(b), the slowest particles detected travel the distance D in a time $\tau + \Delta t$, and the fastest particles can make it in a time $\tau - \Delta t$. Thus, particles that make it through to the detector while the right-hand shutter is open have a speed, s, that falls in the range of

$$\frac{D}{\tau + \Delta t} \leq s \leq \frac{D}{\tau - \Delta t} \qquad (16\text{-}29)$$

(b)

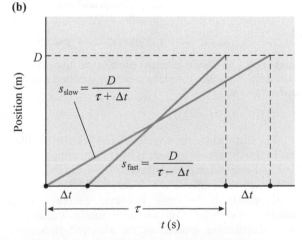

(a)

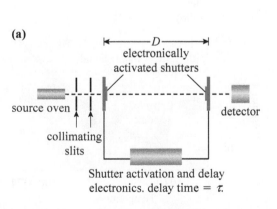

Figure 16-17 (a) An apparatus for measuring molecular speed distributions (b) A space-time diagram for a molecular speed apparatus

THERMODYNAMICS

where D is the distance between the shutters of the apparatus, τ is the delay between the opening of the left- and right-hand shutters, and Δt is the length of time that the shutters are open.

Only those particles that have a speed in this range will be detected. Particles with faster speeds will hit the right-hand shutter before it opens, and particles with slower speeds will arrive after the shutter has closed. By varying τ, we can select the speed of the particles that reach the detector; thus, we can get data that tells us how many particles there are in a particular speed range. By making Δt small, we can improve the resolution of the measurement, at the cost of increasing the time needed to complete the experiment.

The results of such experiments show that the particle speeds have a well defined distribution that depends upon the temperature of the gas, as shown in Figure 16-18. The experimental results are well modelled by a distribution predicted by a statistical model developed by James Clerk Maxwell (Scottish physicist, 1831–79) and Ludwig Boltzmann (Figure 16-19). The equation they derived for this distribution is

$$f(v) = 4\pi \left(\frac{m}{2\pi kT} \right)^{3/2} v^2 e^{-[mv^2/2kT]} \qquad (16\text{-}30)$$

Figure 16-19 Ludwig Boltzmann's grave in Austria. At the top of the marker you can see Boltzmann's equation for entropy, one of the fundamental concepts of thermodynamics.

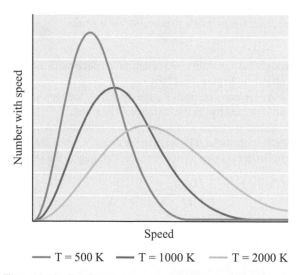

Figure 16-18 The Maxwell-Boltzmann distribution for three different temperatures. It has been shown to be in excellent agreement with experimental data.

The fact that we get a well-defined velocity distribution indicates that the system is near the thermodynamic limit and is in thermal equilibrium. There are many possible ways to distribute the speeds such that the total kinetic energy is the same. However, as the particles collide with each other, they quickly equilibrate, and the speed distribution approaches the Maxwell–Boltzmann distribution.

LO 11

16-11   Phase Diagrams

Now that we have a set of variables to describe a thermodynamic system, we can begin to discuss the behaviour of a particular system as a function of those variables. We can use a **phase diagram**, which is a diagram that shows the phases of a substance at various temperatures and pressures. A phase diagram for water is shown in Figure 16-20.

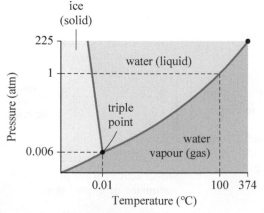

Figure 16-20   A phase diagram for water

We can see that there are distinct regions for each phase. Along the boundaries of each region, different phases can coexist at the same temperature and pressure. In Figure 16-20, there is a broad range of pressures and temperatures where water can coexist in liquid and gas phases along the line extending to the right of the triple point. The conditions for the mist in the image at the beginning of this chapter lie on this line. The triple point, where the three lines intersect in a phase diagram, indicates the pressure and temperature at which the solid, liquid, and gas can all coexist. For water, the triple point occurs at a pressure of 0.006 atm and a temperature of 0.01°C.

Also note that the line separating the solid and liquid phases for water slants upward to the left from the triple point. This slope is a consequence of the fact that water expands when it freezes. If we increase the pressure on solid ice while keeping the temperature constant, we move upward in the phase diagram. Moving upward in the solid region, we cross the boundary into the liquid region, indicating that the ice will melt.

MAKING CONNECTIONS

Water Is Unique

Water is an important exception to the general tendency of liquids and solids to contract as they cool. Water begins to expand as we cool it below about 4°C. The fact that ice floats is a consequence of this expansion.

Lakes cool at the surface due to contact with cold air. If ice were denser than water, bodies of water would fill with ice from the bottom up in freezing weather. Skating on a lake or pond would be possible only after all the water had frozen (Figure 16-21).

Figure 16-21 Outdoor skating

FUNDAMENTAL CONCEPTS AND RELATIONSHIPS

In this chapter, we used state variables, such as volume, pressure, and temperature, to describe a macroscopic assembly of particles. We also established a firm basis for the science of thermometry with the zeroth law of thermodynamics.

We used average values to relate state variables to the properties of the individual particles, recognizing that statistical measures are fully valid only for a system with a very large number of particles.

We can relate the pressure that a gas exerts on the walls of its container to the square of the average speed of the gas particles:

$$P = \frac{1}{3} \frac{mN(v^2)_{\text{avg}}}{V} \qquad (16\text{-}13)$$

At the low-density limit, all gases exhibit the same behaviour, which is described by the ideal gas law:

$$PV = NkT \qquad (16\text{-}17)$$

We can use the universal low-density behaviour of gases to create an ideal gas thermometer:

$$T = T_3 \frac{P}{P_3} = (273.16 \text{ K}) \frac{P}{P_3} \qquad (16\text{-}21)$$

The temperature of a gas is a measure of the average kinetic energy of the molecules in the gas:

$$\frac{3}{2} kT = \frac{1}{2} m(v^2)_{\text{avg}} \qquad (16\text{-}26)$$

The zeroth law of thermodynamics is the foundation for the measurement of temperature: If object A is in thermal equilibrium with object B, and object A is also in thermal equilibrium with object C, then object C is also in thermal equilibrium with object B.

A phase diagram captures the complex behaviour of a material as its temperature and pressure are varied.

APPLICATIONS

Applications: thermometry, gas handling, material response to temperature and pressure changes

Key Terms: constant-volume gas thermometer, gas, ideal gas law, liquid, phase, phase diagram, probability distribution, solid, state variable, temperature, thermal equilibrium, thermodynamic limit, thermodynamics, volume, zeroth law of thermodynamics

QUESTIONS

1. Two different thermometers are calibrated by placing them in the same ice-water bath and then in the same pot of boiling water. The respective readings of each thermometer are assigned to be 0°C and 100°C. The scale on each thermometer is divided into 100 equal increments. If we then use these thermometers to measure the temperature of an object that is at some intermediate temperature, will they agree?

2. A piece of sodium metal and a bath of water are at the same temperature. If we place them in thermal contact, their temperatures do not change. However, if we place them in physical contact, we observe a rather spectacular reaction. Discuss.

3. The maximum in a probability distribution indicates the value that is most likely to be observed. Is the most likely value the same as the average value? Use sketches to illustrate your answer.

4. Most materials expand as they are heated. Discuss why this occurs using the shape of the Lennard–Jones potential. Explain why this expansion would be greatly reduced if the potential were parabolic as shown in Figure 16-13.

5. Using the phase diagram of water in Figure 16-20, discuss whether there are limits on the temperatures at which the three phases can be observed. How would your answers change if liquid water did not expand when cooled?

6. A mercury thermometer uses the fact that mercury expands when heated. The coefficient of thermal expansion for mercury is 6×10^{-5}/°C. This means that if we heat a column of mercury that is 1 m long and cause its temperature to change by 1°C, the column will increase in length by 6×10^{-5} m. However, we are all familiar with mercury thermometers, where the length of the mercury column seems to change by several centimetres with small changes in temperature. How do mercury thermometers work?

7. Two gases are at the same temperature. Do they necessarily exert the same pressure on the walls of their respective containers?

8. Measuring very high temperatures is a challenge. If we try to use a gas, we must measure its pressure and, hence, use a container. However, imagine trying to insert a container of gas into a lava flow. Considering your own experience with the light that is emitted by hot objects, suggest a method of determining the relative temperatures of such objects.

9. Figure 16-22 shows the pressure in a gas sample as a function of temperature, with the volume held constant. Does this gas behave as an ideal gas?

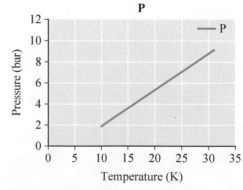

Figure 16-22 Question 9

10. When we derived an expression for the pressure that a gas exerts on the walls of its container, we assumed that the collisions that the gas particles made with the walls were elastic, and the particles emerged from the collisions with the same speed. If that assumption were invalid, what would happen to the gas in the container?

11. Many of us have experienced the phenomenon of a "sea breeze." We wake up on a summer morning near the water and there is no wind. As the Sun rises and the air begins to warm up, we notice that a breeze develops that blows from the water onto the land. Keeping in mind that winds blow from areas of high pressure to areas of low pressure, explain why sea breezes occur.

12. The ideal gas model assumes that there are no attractive interactions between the molecules of a gas, and this is a good approximation at low densities. At densities where the approximation is not valid, would you expect the pressure of a real gas to be higher or lower than the pressure of an ideal gas? Assume the same quantity, volume, and temperature for the real and ideal gases.

13. Many of us have had the experience of pumping up a bicycle tire with a manual pump. You may have observed while doing so that the hose from the pump to the tire becomes warm to the touch, indicating that the pumping process causes the temperature of the air to increase. Clearly, compressing the gas in the pump causes the volume to decrease and the pressure to increase. We might assume, for example, that if we halve the volume we double the pressure. However, if the temperature also increases, is this assumption true? Explain your answer.

PROBLEMS BY SECTION

For problems, star ratings will be used, (✱, ✱✱, or ✱✱✱), with more stars meaning more challenging problems.

Section 16-1 Solids, Liquids, and Gases

14. ✱✱ The density of solid copper is approximately 9.00 g/cm³, and the atomic mass of copper is 63.5 amu.
    (a) How many atoms of copper are in 1.00 cm³ of copper?
    (b) What is the average volume occupied by one copper atom?
    (c) What is the average distance between adjacent copper atoms?

15. ✱ One mole of gas at standard temperature and pressure occupies 22.4 L. What is the average volume occupied by each gas particle? What is the average spacing between particles?

Section 16-2 State Variables

16. ✱ If we know the temperature of a system in Celsius, is the temperature in Kelvin an additional state variable?

17. ✱ The reading of a thermometer in some system fluctuates. Is the system in thermodynamic equilibrium?

Section 16-3 Pressure

18. ✱ The SI unit of pressure is the pascal. Other common units of pressure are the atmosphere, the bar, the torr, and pounds per square inch (psi). Look up the conversions between these units, and express the standard pressure of 101.325 kPa in units of

(a) atmospheres;
(b) torr;
(c) pounds per square inch.

19. ✱✱ Racing bicycles have very narrow tires, which reduce the rolling friction. Assume that such a tire has a diameter of 70.0 cm with a cross-sectional area of 4.00 cm², and that it is inflated to 110 psi.
    (a) Convert the pressure to pascals.
    (b) Determine the volume of the tire.
    (c) Assuming $mv^2 = 1.24 \times 10^{-20}$ J, how many moles of gas are required to inflate the tire?
    (d) What would the mass of this gas be if it were nitrogen? Helium?

Section 16-4 Temperature and Thermal Expansion of Solids

20. ✱ Bridges are usually constructed with expansion joints so that they do not buckle when heated (Figure 16-23). Assume that the expansion of a bridge is dominated by structural steel with a linear thermal expansion coefficient of approximately 1×10^{-5}/°C. Suppose we wish to design expansion gaps in a bridge that will be exposed to temperatures ranging from −40°C to +40°C.
    (a) At what temperature will the gap be the largest?
    (b) How long can a section be if it can expand no more than 3.0 cm?

Figure 16-23 Problem 20

21. ✶ A 2.00 m copper rod has its temperature increased from 20.0°C to 40.0°C. By how much does the length of the rod increase?

22. ✶✶ A thin aluminum ring has a circumference of 20.0 cm. The ring is heated from 10.0°C to 50.0°C. By how much does the diameter of the ring change?

Section 16-5   Thermometers and Temperature Scales

23. ✶ Convert −40°F to Celsius.

24. ✶ On February 3, 1947, Weather Service of Canada workers at Snag, Yukon, filed a notch into the glass casing of an alcohol thermometer because the indicator within fell below the lowest number, −80°F. When they later sent the thermometer to Toronto, officials there determined that the temperature had dropped to −81.4°F, the lowest official temperature ever recorded in North America. Convert this temperature to
   (a) Celsius;
   (b) kelvins.

25. ✶ The highest temperature ever officially recorded is 58°C at Azizia, Libya, in the Sahara desert, recorded in 1922. Convert this temperature to
   (a) Fahrenheit;
   (b) kelvins.

Sections 16-7 and 16-8   Ideal Gases and The Constant-Volume Gas Thermometer

26. ✶✶ A hydrogen gas thermometer is found to have a volume of 100.0 cm^3 when placed in an ice-water bath at 0°C. When the same thermometer is immersed in boiling liquid bromine, the volume of hydrogen at the same pressure is found to be 121.6 cm^3. Assume that the hydrogen gas behaves as an ideal gas.
   (a) Write the ideal gas law for both circumstances.
   (b) Recognizing that the pressure is the same for both cases, solve each equation in part (a) for pressure.
   (c) Set the two equations in part (b) equal to each other, and find the temperature of the bromine.

27. ✶✶ A high-altitude balloon contains helium, whose molar mass is 4.00 g/mol. At the balloon's maximum altitude, its volume is 792 m^3 and the outside temperature and pressure are −53.0°C and 5.00 kPa, respectively. Assume that the helium in the balloon is in equilibrium with the outside air temperature and pressure.
   (a) Use the ideal gas law to find the amount of helium in the balloon at maximum altitude.
   (b) Assuming no loss of helium, what would be the volume of the balloon when it was launched from the ground, where the air temperature and pressure are 20.0°C and 101 kPa, respectively?
   (c) If the volume of the balloon when it was launched was 65.0 m^3, how many moles of helium were lost during the balloon's ascent?

28. ✶ An 18.0 L container holds 16.0 g of oxygen gas (O_2) at 45.0°C.
   (a) Look up the molar mass of oxygen gas, and use that to determine the amount of gas.
   (b) Use the ideal gas law to find the pressure in the container.

29. ✶ All gases behave as ideal gases at sufficiently low densities. Consider one mole of gas at standard temperature and pressure, and assume that it behaves as an ideal gas.

(a) Calculate the volume occupied by the gas.
(b) If the gas is nitrogen, what is the average speed of a nitrogen molecule?
(c) If the gas is oxygen, what is the average speed of an oxygen molecule?
(d) What is the ratio of the average speeds from parts (b) and (c)?

Section 16-9   Temperature and Mechanical Energy

30. ✶✶ Massive amounts of energy are transferred when the atmosphere close to Earth's surface shifts between daytime and nighttime temperatures.
   (a) Calculate the number of atoms in 1.0 km^3 of nitrogen gas at standard temperature and pressure, assuming that nitrogen behaves as an ideal gas.
   (b) Calculate the average kinetic energy of a nitrogen molecule at 25°C.
   (c) Calculate the average kinetic energy of a nitrogen molecule at 15°C.
   (d) How much energy transfers out of 1.0 km^3 of nitrogen gas when it cools by 10°C?

31. ✶ What would the average speed of a nitrogen molecule be at the temperatures in problems 24 and 25?

Section 16-10   Statistical Measures

32. ✶ Consider the function $f(t) = at^2$. What is the average value of f between 0 and T seconds? Hint – you will need to do some integration.

33. ✶✶ Use Equation 16-30 to compute the average speed of a molecule in an ideal gas.

Section 16-11   Phase Diagrams

34. Solid carbon dioxide is sometimes called *dry ice* because it goes directly from the solid to the gas phase when warmed at a pressure of one atmosphere. What can you infer about the triple point of carbon dioxide?

35. On a newly discovered planet, the atmospheric pressure is 500 kPa and the temperature is 200 K. What phase(s) will water, oxygen, nitrogen and carbon dioxide be found in? You will need to look up the phase diagrams for these substances.

COMPREHENSIVE PROBLEMS

36. ✶ We have an ideal gas, and we double the quantity of gas. What other quantities would we need to change, and how, for the temperature of the gas to remain constant?

37. ✶ Your body contains approximately 60% water by mass. Estimate the amount of water in your body.

38. ✶✶ Very cold temperatures cannot be measured with a simple mercury thermometer since mercury freezes at −38.72°C. However, solid-state diodes (described in Chapter 32) can be used as sensors for low temperatures. The relationship between the current, I, in a diode and the voltage, V, across the diode is $I = I_0(e^{q_e V/kT} - 1)$, where I_0 is the saturation current, q_e is the fundamental unit of charge (1.602 × 10^{-19} C), k is Boltzmann's constant, and T is the temperature in kelvins.
   (a) Rearrange this equation to find the relationship between the voltage across the diode and the temperature, assuming a constant current through the diode.

(b) Assuming $I_0 = 1.0 \times 10^{-12}$ A, what voltage will appear across the diode at a temperature of $-190°C$ if a current of 5.0 mA flows through the diode?

39. ✷ The lowest temperatures ever reached in a laboratory are in the order of nanokelvins, 10^{-9} K. Calculate the average speed of a cesium atom in a gas that has been cooled to 1×10^{-9} K.

40. ✷✷ You work for a major theatre chain and have been assigned the job of determining the optimum conditions for running the popcorn machines to get the largest, most uniformly sized popcorn. The popcorn size distributions from two runs of your experiment are shown in Figure 16-24.
(a) Which series has the largest average kernel size?
(b) Which series has the most uniform kernel size?
(c) Which series has the largest number of kernels?
(d) Which series is the more reliable measurement?

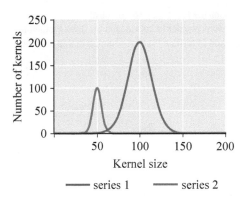

Figure 16-24   Problem 40

41. ✷ A 30 L container at a pressure of 150 kPa contains three moles of an ideal gas. What is the temperature?

42. ✷✷ A sample of ideal gas is compressed until its volume is one-third of its original volume.
(a) If the compression is carried out at constant temperature, what is the resulting pressure?
(b) If the compression is carried out at constant pressure, what is the resulting temperature?

43. ✷✷✷ We showed that the pressure in a gas is given by $P = \frac{1}{3}\frac{mNv_{avg}^2}{V}$. Estimate the number of collisions per second per unit area on the walls of the container. Use data for nitrogen at standard temperature and pressure.

44. ✷✷ You are designing storage tanks that will be filled with helium. The tanks will be filled at 20°C to a pressure of 100.0 atm. What maximum and minimum pressures would you expect to observe in a full tank if the ambient temperature will range from $-20°C$ to 30°C?

45. ✷ Gravity pulls the matter in a star toward the centre. When the star eventually collapses, the volume is reduced, and the temperature and pressure increase. The increase in pressure is eventually sufficient to offset the gravity, and the star is said to be in hydrostatic balance. The increase in temperature is enough to allow nuclear fusion to occur, which further increases the temperature, leading to a new equilibrium. The net result is that in the core of a star we observe very high densities, pressure, and temperatures. At the core of the Sun, the density is approximately 150 000 kg·m^{-3}, and the temperature is as high as 13.6×10^6 K. In this question, we naively treat the Sun as an ideal gas. Calculate
(a) the pressure at the core of the Sun;
(b) the average speed of a hydrogen atom at the core of the Sun.

46. ✷✷✷ An alternative method of measuring the velocity distribution of particles in a gas is to use a rotating cylinder with a helical groove cut in the side, as shown in Figure 16-25. If the cylinder is rotating at an angular speed of ω, what speed of particle will it be selecting for? Express your result in terms of the length and radius of the cylinder, the helix angle, and the angular speed. Discuss how the width of the groove impacts measurements made with this device.

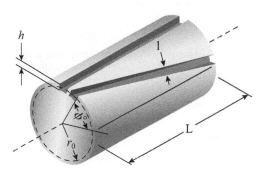

Figure 16-25   Problem 46

47. ✷✷ Table 16-6 contains some observations of the phase that a particular substance is in at various pressures and temperatures. Use the data to construct a rough phase diagram.

Table 16-6   Data for Problem 47

Temperature (°C)	Pressure (atm)	Observed phase
−78	0.5	Gas
−78	1.1	Solid
−78	10	Solid
−57	4	Gas
−57	6	Solid
−57	19	Solid
−50	6	Gas
−50	7	Liquid
−50	15	Solid
−30	7	Gas
−30	8	Liquid
−30	100	Solid

48. ✷✷ Roll a six-sided die, and record the numbers that you roll in a spreadsheet. Calculate the running average value of all of your rolls as you go, and plot your data.

Table 16-7 Data for Problem 49

T (K)	77.0	97.0	117	137	157	177	197	217	237	257	277	297	317
R (Ω)	132.2	134.7	139.5	143.8	146.5	152.0	153.6	157.8	163.7	167.8	168.9	172.2	177.9

49. ✴✴ To construct a thermometer, a physicist decides to use a resistor whose resistance varies as a function of temperature. The physicist and her team construct a circuit that passes a constant current of 1 mA through the resistor; they then measure the voltage across the resistor. Table 16-7 lists the calibration data. Fit the data with a suitable polynomial, and predict the resistance value at a temperature of 400 K.

OPEN PROBLEM

50. ✴✴✴ In problem 43, we found that there is an enormous number of collisions per second with the walls of a container holding a gas. Suppose the pressure sensor in our system has a response time of 1 ms. To what value would we have to reduce the concentration of particles in our system to see fluctuations in the reading of the pressure sensor? State any assumptions you make.

See the text online resources at www.physics1e.nelson.com for Open Problems and Data-Rich Problems related to this chapter.

Chapter 17

Heat, Work, and the First Law of Thermodynamics

The first step in brewing beer is mixing malted barley and water together to form a "mash" and holding them at a temperature that allows enzymes in the barley to convert starches into fermentable sugars (Figure 17-1). How can the brewer determine what initial temperature of the water will produce the desired temperature for the mixture?

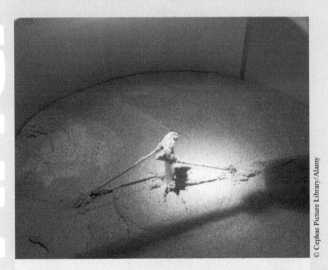

© Cephas Picture Library/Alamy

Figure 17-1   Malted barley and water in the mash tun at a brewery

Learning Objectives

When you have completed this chapter you should be able to:

1   Discuss what is meant by the term "heat."

2   Define specific heat, and perform calculations using heat capacities.

3   Define heats of fusion and vaporization, and use them in calculations.

4   Calculate the work done by a gas during an expansion.

5   State the first law of thermodynamics.

6   Define what is meant by an adiabatic, isothermal, isobaric, and an isochoric process.

7   Describe various heat transfer processes, and perform calculations for radiative and conductive processes.

THERMODYNAMICS

For additional Making Connections, Examples, and Checkpoints, as well as activities and experiments to help increase your understanding of the chapter's concepts, please go to the text's online resources at www.physics1e.nelson.com.

17-1 What Is Heat?

The nature of heat puzzled scientists for centuries. Early theories suggested that there was a "caloric fluid" that flowed from a hot object to a cold object. The rather notorious Anglo-American physicist, inventor, and military officer Count Rumford (Benjamin Thompson, 1753–1814) was the first to dispel this notion and make a connection between motion and heat. He performed an experiment in which he bored out the barrel of a brass cannon with a dulled cutting tool in a water bath. He then monitored the temperature of the water to determine the heat produced. He showed that this mechanical process could produce significant heat continuously for a virtually unlimited time. He also noted that the thermal properties of the shavings produced as a result of the boring appeared to be unchanged. On the basis of these results, Rumford claimed that there could not be a caloric fluid because surely it would be depleted and would limit the amount of heat that could be extracted from an object. In addition, if the heat were some type of fluid, you would expect there to be some observable change in the physical properties of a material if the fluid were removed. He suggested that there is a connection between motion and heat.

In Chapter 16, we noted that the temperature of a simple gas is a measure of the average kinetic energy of the particles that make up the gas. Almost every day, we see that adding heat to an object increases its temperature. Based on these observations, it would seem that heating an object somehow transfers energy into the object and, at least in the case of an ideal gas, the energy is stored in the form of kinetic energy. For more complex systems, the energy may also be stored as potential energy. **Heat** is the energy that is transferred from one object to another when a temperature difference exists between the two objects. Because heat is a form of energy, called thermal energy, it can be measured in joules.

  CHECKPOINT

C-17-1 Thermal Energy and Temperature

Does an object at a higher temperature always contain more thermal energy than an object at a lower temperature?

C-17-1 No. As we will see in quantitative detail in Section 17-2, in general, different materials exhibit different temperature changes when we add the same amount of thermal energy to them.

We can gain some insight into how this energy transfer occurs by considering what happens when we place a hot object and a cold object in contact with each other. The particles in the hot object have greater kinetic energy than the particles in the cold object. When they collide, energy is transferred from the more energetic particles to the less energetic particles. Consequently, the average kinetic energy decreases in the hot object and increases in the cold object, and the temperatures of the objects change correspondingly.

17-2 Temperature Changes Due to Thermal Energy Transfer

We now consider quantitatively how the temperature of an object changes when it absorbs thermal energy. Since temperature is a measure of the average kinetic energy of the particles (at least for an ideal gas), we would expect that the temperature change due to a thermal energy flow into a material would simply be proportional to the amount of thermal energy put into the system:

$$\Delta T \propto Q \qquad (17\text{-}1)$$

where ΔT is the temperature change, and Q is the thermal energy flow into the system.

Notice that when $Q > 0$, $\Delta T > 0$, and when $Q < 0$, $\Delta T < 0$.

Two other factors affect the change in temperature: the mass of the material present and the type of material. The fact that the temperature change depends on the amount of material makes sense. For example, if we double the number of particles, the quantity of kinetic energy added per particle is halved, and we would expect the change in temperature to be halved as well.

 ONLINE ACTIVITY

The e-resource that accompanies every new copy of this textbook contains an Online Activity using the PhET simulation "States of Matter." Work through the simulation and accompanying questions to gain an understanding of how molecular motion changes according to the state of matter.

As demonstrated in the "States of Matter" online activity, the increase in temperature also depends on the number of different ways that particles in the material can move and, hence, have kinetic or potential energy. The atoms of a simple monatomic gas, such as helium, can only have kinetic energy from linear motion. A diatomic gas, such as oxygen, can also have rotational kinetic energy, vibrational kinetic energy, and vibrational potential energy. These different modes are illustrated in Figure 17-2.

As we discussed in Chapter 16, these different ways of absorbing energy are called degrees of freedom, and the absorbed energy is shared equally between all the degrees

THERMODYNAMICS

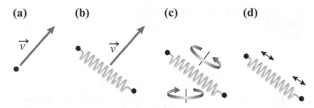

(a)   **(b)**   **(c)**   **(d)**

Figure 17-2 (a) A monatomic particle with translational kinetic energy (b), (c), and (d) A diatomic particle with translational, rotational, and vibrational energy

of freedom. As a result, the absorption of a given amount of thermal energy causes a much smaller increase in translational kinetic energy for a diatomic gas than for an equal mass of monatomic gas.

We can quantify these differences by defining the **heat capacity**, C, of an object as the ratio of the thermal energy added to the object to the resulting temperature change:

$$C = \frac{Q}{\Delta T} \qquad (17\text{-}2)$$

An object with a large heat capacity undergoes a small temperature change for a given thermal energy input compared to an object with a small heat capacity. The units of heat capacity are J/K.

The amount of thermal energy that a sample can absorb depends on both the type and quantity of material in the sample. We define the **specific heat capacity**, c, as $c = C/m$, which is the quantity of thermal energy required to raise the temperature of a unit mass by a unit degree. This allows us to relate thermal energy flows and temperature changes:

KEY EQUATION $Q = mc\Delta T \qquad (17\text{-}3)$

where Q is the amount of thermal energy, m is the mass of the sample, c is the specific heat capacity, and ΔT is the temperature change, $T_{\text{finial}} - T_{\text{initial}}$.

Table 17-1 lists the specific heat capacities of some common substances.

Table 17-1 Specific heat capacities of some common materials at constant pressure (101 kPa)

Material	Specific heat capacity, c (J/(kg·K))
Copper (20°C)	390
Lead (20°C)	129
Glass (20°C)	840
Air (20°C)	1000
Animal tissue (20°C)	3500
Water, ice (−5°C)	2100
Water, liquid (15°C)	4186
Water, steam (110°C)	2010

EXAMPLE 17-1

Temperature Change Due to an Input of Thermal Energy

Find the resultant temperature change when 100 g of water absorbs 1000 J of thermal energy.

SOLUTION

We simply rearrange Equation (17-3) to solve for the temperature change:

$$\Delta T = \frac{Q}{mc} = \frac{1000 \text{ J}}{(0.1 \text{ kg})(4186 \text{ J/(kg·K)})} = 2.39 \text{ K}$$

Making sense of the results:

1000 J is the energy consumed by a 100 W lightbulb in 10 seconds so we would not expect a large increase in T.

MAKING CONNECTIONS

Water Helps Regulate Temperature

You will note that the specific heat capacity of water is the highest value listed in Table 17-1. This property of water helps us regulate our body temperature and plays a significant role in moderating the climate of coastal areas. For example, in Halifax, Nova Scotia, average daily temperatures are −4.4°C in January and 18.6°C in July. In contrast, the corresponding averages for Winnipeg, Manitoba, are −17.8°C and 19.5°C, respectively. The climate in Halifax is moderated by the Atlantic Ocean.

LO 3

17-3 The Flow of Thermal Energy Between Objects

We now consider what happens when a hot object is in thermal contact with a cold object. For simplicity, we will assume that the two objects are completely thermally insulated from their surroundings. Thus, there is no thermal energy flow into or out of the surroundings, any thermal energy that leaves the hot object will go into the cold object, and

$$Q_{\text{hot}} = -Q_{\text{cold}} \qquad (17\text{-}4)$$

We can replace the thermal energy flows for each object with the mass, specific heat, and temperature change from Equation (17-3) to get

$$m_{\text{hot}} c_{\text{hot}} \Delta T_{\text{hot}} = -m_{\text{cold}} c_{\text{cold}} \Delta T_{\text{cold}} \qquad (17\text{-}5)$$

C-17-2 Heat and Temperature

When you bite into a hot, fruit-filled pastry, you can burn your mouth on the filling but not likely on the crust. Does this mean that the filling is at a higher temperature?

C-17-2 No, the entire pastry is at approximately the same temperature. However, the heat capacity of the filling is much higher than the heat capacity of the crust, so there is more energy in the filling. When the pastry contacts your mouth, much more energy is transferred to your mouth from the filling than from the crust; thus, the final temperature of your mouth is increased to the point of pain.

 EXAMPLE 17-2

Mashing

A particular recipe for beer calls for 3.00 kg of crushed barley to be mixed with 14.0 L of water. The desired temperature of the mixture is 66.0°C. The crushed barley is at room temperature, and malted barley has a specific heat of 1600 J/(kg·K). What should the initial temperature of the water be?

SOLUTION

Table 17-1 lists a specific heat of 4186 J/(kg·K) for water. Since 1 L of water has a mass of 1 kg, we have 14 kg of water. Applying Equation (17-5), we get

$$m_{\text{water}} c_{\text{water}} (T_{\text{f,water}} - T_{\text{i,water}}) = -m_{\text{malt}} c_{\text{malt}} (T_{\text{f,malt}} - T_{\text{i,malt}})$$

Since $T_{\text{f,water}} = T_{\text{f,malt}} = 66°C$,

$$T_{\text{i,water}} = \frac{-m_{\text{malt}} c_{\text{malt}} (66°C - T_{\text{i,malt}}) - m_{\text{water}} c_{\text{water}} (66°C)}{-m_{\text{water}} c_{\text{water}}}$$

$$= \frac{-3 \text{ kg} \cdot 1600 \text{ J/(kg·°C)} \cdot (66°C - 20°C) - 14 \text{ kg} \cdot 4186 \text{ J/(kg·°C)} \cdot 66°C}{-14 \text{ kg} \cdot 4186 \text{ J/(kg·°C)}}$$

$$= 69.8°C$$

Note that we have kept our temperatures in Celcius and have expressed the specific heats in Celsius. We can do this because the unit in the specific heat is a change in temperature and a change of 1 K is the same as a change of 1°C.

Making sense of the results:

The initial water temperature is only a few degrees higher than the temperature of the final mixture because there is much more water than malt and the specific heat of water is greater than the specific heat of malted barley.

 PEER TO PEER

When we do these calculations we are really just keeping track of energy. The energy that is removed from the hot object goes into the cold object.

 EXAMPLE 17-3

Using Specific Heat

A 20 g block of copper at a temperature of 100°C is placed in good thermal contact with a 40 g block of copper that is at 10°C. What is the final temperature?

SOLUTION

We once again apply Equation (17-5):

$$m_{\text{hot}} c_{\text{hot}} \Delta T_{\text{hot}} = -m_{\text{cold}} c_{\text{cold}} \Delta T_{\text{cold}}$$

In this case, both blocks are made of the same material, so the specific heat cancels. We also know that $T_{\text{f,hot}} = T_{\text{f,cold}} = T_{\text{f}}$. Therefore,

$$T_{\text{f}} = \frac{m_{\text{hot}} T_{\text{hot}} + m_{\text{cold}} T_{\text{cold}}}{m_{\text{hot}} + m_{\text{cold}}}$$

$$= \frac{0.02 \text{ kg} \times 100°C + 0.04 \text{ kg} \times 10°C}{0.02 \text{ kg} + 0.04 \text{ kg}} = 40°C$$

Making sense of the results:

The final temperature is closer to the temperature of the larger block than to the temperature of the smaller block because the larger block has a larger heat capacity.

LO 4

17-4 Phase Changes and Latent Heat

When you boil water in a pot on a stove, the burner transfers significant energy to the pot, yet the temperature of the liquid water sits at 100°C. The energy being absorbed by the water is being used to break the bonds that exist between the water molecules in the liquid phase so they can move to the gas phase. In some sense, the thermal energy that is being added to the system is "hidden" in that it does not manifest itself as a resulting temperature change. We call these "hidden" heats **latent heats**.

A latent heat called the **heat of vaporization** is associated with the transition from the liquid to the gaseous state. Similarly, a latent heat called the **heat of fusion** is associated with the change from the solid to the liquid state. When a material goes from the solid to the liquid state, the thermal energy responsible for the phase change must be supplied from the surroundings. Conversely, when the material goes from the liquid to the solid state, the thermal energy is liberated to the surroundings. Table 17-2 lists heats of fusion and vaporization for a few common materials. Note that all the values are in J/kg. To determine the total amount of thermal energy required to make a transition, we need to multiply these quantities by the mass of the sample. Thus, the thermal energy required for the phase change is

KEY EQUATION
$$Q_{\text{phase}} = mL \quad (17-6)$$

where m is the mass of the sample, and L is the latent heat.

Table 17-2   Melting points, heats of fusion, boiling points, and heats of vaporization for some common materials

Substance	Melting point (°C)	Heat of fusion (J/kg)	Boiling point (°C)	Heat of vaporization (J/kg)
Water	0	3.33×10^5	100	2.26×10^6
Nitrogen	−210	2.6×10^4	−195.8	2×10^5
Lead	327	2.5×10^4	1750	8.7×10^5
Tungsten	3410	1.84×10^5	5900	48×10^5

 CHECKPOINT

C-17-3 Does the Temperature Always Change When We Add Thermal Energy?

True or false? When we heat a substance, we always observe an increase in temperature.

C-17-3 No. We only observe a temperature change when the substance is not undergoing a change of phase.

 MAKING CONNECTIONS

Reusable Hand Warmers

Reusable hand warmers rely on heat of fusion (Figure 17-3). They contain a supersaturated solution of a salt, typically sodium acetate. Snapping the small metal circle in the package triggers the crystallization of the solution from the liquid to the solid state. You can see that the pack is crystallizing in the image from right to left. As it crystallizes, the heat of fusion is liberated, and the pack feels quite warm to the touch. Once the pack has liberated all the heat, the pack contents can be returned to the liquid state by immersing in boiling water.

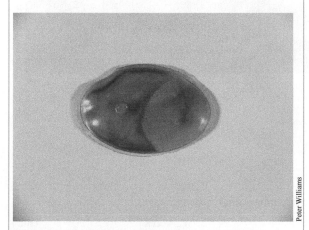

Figure 17-3   A reusable heat pack

When doing thermal energy flow (calorimetric) calculations, we consider whether a substance goes through a phase change. If so, we modify Equation (17-5) as follows:

$$Q = mc_1(T_{\text{phasechange}} - T_i) + mL + mc_2(T_f - T_{\text{phasechange}})$$
(17-7)

where c_1 and c_2 are the specific heats of the two phases, and L is the latent heat per unit mass.

 ### EXAMPLE 17-4

Temperature Changes Due to Thermal Energy When There Is Also a Phase Change

How much thermal energy is required to raise the temperature of 250 g of water from 80.0°C to 120°C?

SOLUTION

For clarity, we will calculate each term in Equation (17-7) separately. We will first heat the water to 100°C. We assume that the specific heat of water in this temperature range is a constant 4186 $\text{J} \cdot \text{kg}^{-1} \text{°C}^{-1}$.

$$Q_{80-100} = (0.25 \text{ kg}) (4186 \text{ J/(kg} \cdot \text{K)}) (100\text{°C} - 80\text{°C})$$
$$= 20\,930 \text{ J}$$

Now, we will determine how much thermal energy is required to convert the water to steam. We refer to Table 17-2 to find the heat of vaporization for steam to write:

$$Q_{\text{water-steam}} = (0.25 \text{ kg}) (2.26 \times 10^6 \text{ J/kg}) = 5.65 \times 10^5 \text{ J}$$

Finally, we will determine how much thermal energy is required to raise the temperature of the steam from 100°C to 120°C. We will assume that the specific heat of the steam is a constant 2010 J/(kg · K):

$$Q_{100-120} = (0.25 \text{ kg}) (2010 \text{ J/(kg} \cdot \text{k)}) (120\text{°C} - 100\text{°C})$$
$$= 10\,050 \text{ J}$$

Thus, the total thermal energy required to heat 250 g of water from 80°C to 120°C is

$$Q_{\text{Total}} = Q_{80-100} + Q_{\text{water-steam}} + Q_{100-120}$$

$$= 20\,903 \text{ J} + 565\,000 \text{ J} + 10\,050 \text{ J} = 596\,000 \text{ J}$$

Making sense of the results:

Since the heat of vaporization is quite large compared to the thermal energy required to warm the water and subsequently the steam, most of the energy required goes in to the phase change.

EXAMPLE 17-5

What Is the Final Phase?

Calculate the final temperature and phase when 100 g of ice at −5.00°C is mixed with 200 g of water at 20.0°C, as well as how much ice will be melted. Assume that no thermal energy is transferred between the mixture and its surroundings.

SOLUTION

There are three possible outcomes for a mixture of ice and water. First, all the ice could melt, resulting in water at some temperature above 0°C. Second, all the liquid could freeze, producing ice at some temperature below 0°C. Finally, the resultant mixture could be part liquid and part solid at a temperature of 0°C. First, we compare the thermal energy transfers required to bring the water and the ice to 0°C:

$$Q_{cool\ water} = (0.2\ \text{kg})(4186\ \text{J/(kg·K)})(0°C − 20°C) = −16\ 744\ \text{J}$$

$$Q_{warm\ ice} = (0.1\ \text{kg})(2100\ \text{J/(kg·K)})(0°C − (−5°C)) = 1050\ \text{J}$$

The minus sign indicates that cooling the water transfers thermal energy *from* the water to the ice. The net transfer

of energy to the ice is 16 744 J − 1050 J = 15 694 J. We now calculate how much ice will be melted by this energy transfer using the heat of fusion from Table 17-2:

$$m = \frac{Q}{L} = \frac{15\ 694\ \text{J}}{3.33 \times 10^5\ \text{J/kg}} = 4.71 \times 10^{-2}\ \text{kg} = 47.1\ \text{g}$$

Thus, 47.1 g of ice at 0°C will be converted to liquid at 0°C. So, the final mixture will be 247 g of liquid water and 52.9 g of solid water, all at a temperature of 0°C.

Making sense of the results:

In this case the final outcome was a solid liquid mixture at 0°C. This is because the energy that was transferred form the water to the ice was only enough to warm the ice to the melting point and then melt some of it. If the ice melted had exceeded 100 g, we would have concluded that the final mixture would have been all water. Conversely, if the energy required to warm the ice had been greater than the energy required to cool the water, some or perhaps all of the water would have frozen.

MAKING CONNECTIONS

Home Ice-Cream Makers

A home ice cream maker (like the one in Figure 17-4) utilizes a mixture of ice and salt water to achieve temperatures below 0°C. Crushed ice and a saturated salt solution are placed in the outer part of the ice cream maker. The saturated salt solution has a freezing temperature of about −21°C. Thermal energy flows from the salt solution into the ice. The solution cools and the ice warms up. However, because the ice dominates the mixture, the final temperature of the mixture is closer to the initial temperature of the ice than that of the salt solution.

Figure 17-4 A home ice-cream maker uses a mixture of ice and salt water to freeze the ice cream.

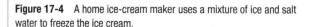

LO 5

17-5 Work

In the previous sections, we examined thermal energy going into and out of a system. We now consider the effects of doing work on a system and having a system do work on its surroundings. We saw in Chapter 6 that doing work on an object changes its kinetic energy. Recall that when a force $\vec{F}$ is exerted on a particle that moves from $\vec{r}_1$ to $\vec{r}_2$, the work done is

$$W_F = \int_{\vec{r}_1}^{\vec{r}_2} \vec{F} \cdot d\vec{r} \qquad (6\text{-}13a)$$

We also recall the work–kinetic energy theorem, which for a single particle states that

$$W_T = \Delta K \qquad (6\text{-}5)$$

where ΔK is the change in kinetic energy of the particle.

How can we do work on an assembly of particles? Let us consider a gas in a cylinder as shown in Figure 17-5. A force is applied to the piston, causing it to move to the left and compress the gas. As the piston moves, the pressure in the gas increases. To determine the work, we consider an infinitesimal motion of the piston, denoted as dx in the figure. Accompanying this motion is an infinitesimal change in the volume of the gas: $dV = A dx$, where A is the cross-sectional area of the cylinder. The work that the applied force does on the piston is given by

$$dW = \vec{F} \cdot d\vec{r} = F dx \qquad (17\text{-}8)$$

because the force and the displacement are parallel.

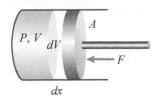

Figure 17-5 A piston compressing a gas

If we assume that the piston either starts and ends at rest, or travels at a constant speed during the process, the kinetic energy of the piston does not change and therefore the net work done on the piston must be zero. Therefore, the gas must do some negative work on the piston to balance off the positive work done by the external force. The pressure of the gas exerts a force to the right on the piston and has a magnitude of $F_{\text{pressure}} = PA$. The work done on the piston by the gas is then

$$dW_{\text{gas on piston}} = -PA dx \qquad (17\text{-}9)$$

The minus sign is included because the force and the displacement are in opposite directions. Since the net work done on the piston is zero,

$$F dx - PA dx = 0 \qquad (17\text{-}10)$$

Thus, the force applied on the piston by the gas has the magnitude PA and is equal but opposite to the applied external force. Newton's third law requires that the piston exert an equal but opposite force on the gas. Consequently, the piston does work on the gas:

$$dW_{\text{piston on gas}} = PA dx = P dV \qquad (17\text{-}11)$$

Consider what happens when the gas particles collide with the walls of the cylinder. If we assume that the collisions are elastic, then a particle emerges from the collision with the same speed and, therefore, the same kinetic energy that it had going into the collision. The only change that occurs is that the component of its velocity vector perpendicular to the face of the piston changes sign. However, in a collision with the moving piston, a gas particle emerges with that component of its velocity increased in addition to having its sign changed. As we saw in Chapter 7, for a perfectly elastic collision in one dimension,

$$u' = \frac{2mv + (M - m)u}{M + m} \qquad (7\text{-}21)$$

and

$$v' = \frac{-2Mu + (M - m)v}{M + m} \qquad (7\text{-}22)$$

where M and m are the masses of the two objects, and u and v are the velocities, with the primes denoting final velocities. In our case, the mass of the piston is much larger than the mass of the gas particle, $M \gg m$. In that limit, Equations (7-21) and (7-22) reduce to

$$V_{\text{f}} = V_{\text{i}} \quad \text{and} \quad v_{\text{f}} = -v_{\text{i}} + 2V_{\text{i}} \qquad (17\text{-}12)$$

Thus, the increase in that component is twice the speed of the piston.

The general expression for calculating the work done in a finite volume change is

KEY EQUATION
$$W = \int_{V_1}^{V_2} P dV \qquad (17\text{-}13)$$

The magnitude of this quantity corresponds to the area under the curve on a **P-V diagram**, as shown in Figure 17-6. Note that the amount work done in going from one point to another depends on the details of the process used.

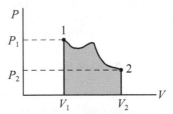

Figure 17-6 A P-V diagram. The shaded area represents the work done in the process of going from point 1 to point 2.

✓ CHECKPOINT

C-17-4 Expansions and Work

Is the following statement true or false: If we carry a gas through an expansion and then compress it so that it returns to its original volume, the net work done by the gas is always zero because there is no net volume change.

C-17-4 The work done depends on the way we carry out the process. The net work is zero only if we carry out the expansion and compression along the same path on the P-V diagram. Otherwise, the area under the two paths will be different and some net work will have been done.

LO 6

17-6 The First Law of Thermodynamics

We saw in the previous two sections that we can add energy to a thermodynamic system in two ways: heating the system and doing work on the system. Applying conservation of energy to a thermodynamic system suggests that the change in energy of the system is simply the sum of the thermal energy put into the system and the work done by the system:

KEY EQUATION
$$\Delta U = Q - W \qquad (17\text{-}14)$$

where U is the internal energy of the system, Q is the thermal energy flow *into* the system, and W is the work done by the system.

Equation (17-14) is the **first law of thermodynamics**, which states that the change in the internal energy of a system is equal to the thermal energy that flows into the system less the work done by the system.

What do we mean by "a system" in this context? In general, a system is simply the set of objects we wish to consider, and we are usually free to define our system in any way that is convenient for us. For example, if we wished to consider what happens when we place a quantity of ice cubes in a glass of water, we could define our system as the water and the ice. This choice is a **closed system** because the total amount of material in the system does not change—no mass enters or leaves the system. Alternatively, we could have chosen just the glass of water as our system. This choice defines an **open system**, which is a system in which mass is added or removed.

Another important distinction is deciding whether or not the system is isolated. In an isolated system, we assume that no thermal energy flows into or out of the system. In general, it is very difficult to make an isolated system. A reasonable approximation might be an insulated cup, which can be treated as an isolated system over short periods of time.

We should also pay some attention to the sign conventions that we have adopted in writing the First Law this way. When thermal energy flows into the system, which raises the internal energy, we consider that to be a positive flow and Q is positive. Conversely, when thermal energy flows out of the system, the internal energy decreases and Q is negative.

✓ CHECKPOINT

C-17-5 Sign of the Thermal Energy Flow

Ice cubes are placed in a glass of water. When we consider the ice and the liquid to be separate systems, which of the following statements is (are) correct?

(a) The thermal energy flow from the ice is negative.
(b) The thermal energy flow from the liquid is negative.
(c) The thermal energy flow from the ice is equal in magnitude but opposite in sign to the thermal energy flow from the liquid.

C-17-5 (b) and (c) Thermal energy flows from the warm liquid to the cooler ice. Thus, for the liquid system, the thermal energy flow is negative (out of the liquid); for the ice system, the thermal energy flow is into the liquid and is positive. The magnitude of the two flows is the same, assuming that the two systems are isolated from all other systems except each other. All the thermal energy that flows from the liquid goes into the ice.

The negative sign in front of the work term in Equation (17-14) is necessary because we are considering the work done *by* the system. When a system does work, it loses energy.

The variables in Equation (17-14) can have quite different properties. For example, in the case of an ideal gas, the internal energy is a function of the temperature of the

EXAMPLE 17-6

Internal Energy and Thermal Energy

Calculate the change in internal energy of an isolated closed system when

(a) 3000 J of thermal energy is added to it, and the system does 4000 J of work;
(b) 3000 J of thermal energy is added to it, and 4000 J of work is done on the system.

SOLUTION

The calculation is a straightforward application of Equation (17-14), but we need to pay careful attention to the sign conventions.

(a) When thermal energy is added to the system, it increases the internal energy of the system and is therefore positive. When the system does work, energy is transferred out of the system. Thus, the net change in the internal energy is

$$\Delta U = Q - W = 3000 \text{ J} - 4000 \text{ J} = -1000 \text{ J}$$

(b) When the system has 4000 J of work done on it, the work done by the system is -4000 J. Then the net change in internal energy is

$$\Delta U = Q - W = 3000 \text{ J} - (-4000 \text{ J}) = 7000 \text{ J}$$

Making sense of the results:

In the first case the system loses more energy by doing work than it received in the form of thermal energy so the net change in internal energy is negative. In the second, work is done ON the system which raises its internal energy.

gas and has a well-defined value at a given temperature. However, as we saw in Section 17-5, the work done in a process that carries a system from one state to another depends on how the process is carried out. Consequently, we cannot associate a quantity of work with the state of a system. In addition, the thermal energy that flows into or out of a system must depend on the way the process is carried out, such that variations in the work done are offset by equal but opposite variations in the thermal energy flow. Otherwise, the first law would not hold for all processes. We will discuss different types of processes in the next section.

✓ CHECKPOINT

C-17-6 Thermal Energy and Work

Is the following statement true or false: The thermal energy that flows and the work that is done in a process depend on the details of the process, but the difference between the thermal energy and the work, $Q - W$, does not, and only depends on the start and end points of the process.

C-17-6 True. By the first law of thermodynamics, $Q - W = \Delta U$, and the internal energy is a state variable. Therefore, the change in the internal energy in any process that has the same start and end points must be the same, regardless of the details of the process.

17-7   Different Types of Processes

When an isolated system undergoes a process, there is no thermal energy flow into or out of the system. We call such a process **adiabatic**. A process carried out at constant temperature is called an **isothermal** process. Similarly, an **isobaric** process is carried out at constant pressure, and an **isochoric** process is carried out at constant volume.

Isothermal Processes

For an ideal gas,

$$PV = nRT \qquad (17\text{-}15)$$

where P is the pressure, V is the volume, n is the number of moles of the gas, R is the gas constant, and T is the temperature in kelvins. When we carry out an isothermal process on a closed system, the product PV must be a constant. Since the internal energy for an ideal gas depends only on the temperature, in an isothermal process there is no change in the internal energy of the system. Thus,

$$\Delta U = W - Q = 0$$
$$\therefore W = Q \qquad (17\text{-}16)$$

Consider the work done in an isothermal expansion of an ideal gas in a cylinder with a movable piston. We immerse the cylinder in a bath of water large enough that any thermal energy flow between the cylinder and the bath has a negligible effect on the temperature of the bath. We also carry out the process slowly enough that the temperature of the gas in the piston is always the same as the temperature of the surrounding water bath. In fact, this process is only **quasi-static** because the piston would have to move infinitesimally slowly to keep the temperature completely constant. We can apply Equation (17-13) to calculate the work done by the gas:

$$W = \int_{V_1}^{V_2} P\,dV = \int_{V_1}^{V_2} \frac{nRT}{V}\,dV = nRT\ \ln\left(\frac{V_2}{V_1}\right) \quad (17\text{-}17)$$

Here, we have used the ideal gas law to replace P in the integral with $\dfrac{nRT}{V}$. The work in such a process is illustrated in Figure 17-7. The solid black line is an **isotherm** (line of constant temperature), and the work done is the shaded area under the curve.

Isobaric Processes and the Constant Pressure Heat Capacity

When we hold the pressure constant while compressing an ideal gas,

$$P = \frac{nRT}{V} = \text{constant} = P_c \qquad (17\text{-}18)$$

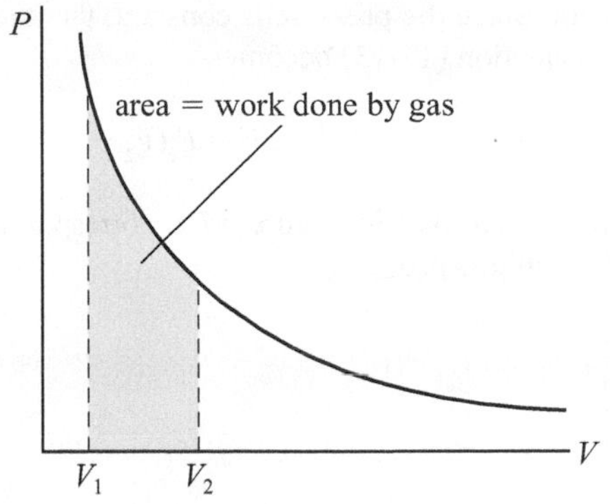

Figure 17-7   A P-V diagram for an isothermal process in an ideal gas

EXAMPLE 17-7

Work and Thermal Energy Flow

One mole of an ideal gas is compressed in an isothermal process at a temperature of 293 K. The initial volume is 22.0 L, and the final volume is 10.0 L. Calculate the work done and the thermal energy flow into the system.

SOLUTION

We use Equation (17-17) to find the work:

$$W = nRT \ln\left(\frac{V_2}{V_1}\right)$$

$$= 1\text{ mol} \times 8.314\text{ J/(K·mol)} \times 293\text{ K} \times \ln\left(\frac{10\text{ L}}{22\text{ L}}\right)$$

$$= -1920\text{ J}$$

We use Equation (17-16) to find the thermal energy flow during the compression:

$$W = Q = -1920\text{ J}$$

The negative sign indicates that thermal energy flowed out of the system.

Making sense of the results:

The gas exerts an outward pressure on the piston, which moves inward to reduce the volume; thus, the work done by the gas is negative.

$$\therefore T = \frac{P_c V}{nR}$$

Consequently, as we change the volume, we must also change the temperature by allowing thermal energy to flow into or out of the system. In Chapter 16, we showed that the average kinetic energy of the particles in an ideal monatomic gas is related to the temperature of the gas. For a monatomic ideal gas, all of the energy is the translational kinetic energy of the gas particles. It therefore follows that the internal energy of an ideal gas is given by

$$U = \frac{3}{2}nRT \qquad (17\text{-}19)$$

Therefore, this isobaric process involves thermal energy flow, work, and a change in the internal energy of

477

the system. Since the pressure is constant, the integral for work in Equation (17-13) becomes

$$W = \int_{V_1}^{V_2} P \, dV = P_c \int_{V_1}^{V_2} dV = P_c(V_2 - V_1) \quad (17\text{-}20)$$

The shaded area in Figure 17-8 corresponds to the work done in this process.

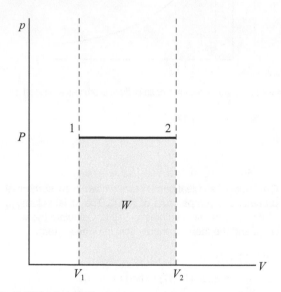

Figure 17-8 An isobaric process

The temperature change can be found using Equation (17-15):

$$\Delta T = T_2 - T_1 = \frac{P_c V_2}{nR} - \frac{P_c V_1}{nR} = \frac{P_c}{nR}(V_2 - V_1) \quad (17\text{-}21)$$

Using this expression for the temperature change with Equation (17-19) gives an equation for the change in internal energy:

$$\Delta U = \frac{3}{2}nR\Delta T = \frac{3}{2}P_c(V_2 - V_1) \quad (17\text{-}22)$$

Rearranging Equation (17-14) to solve for the thermal energy flow, we get

$$Q = \Delta U + W = \frac{3}{2}P_c(V_2 - V_1) + P_c(V_2 - V_1)$$

$$= \frac{5}{2}P_c(V_2 - V_1) \quad (17\text{-}23)$$

When the ideal gas is compressed, all the terms on the right of Equation (17-23) are negative, and thermal energy must flow out of the system. However, if the ideal gas expands, all the terms on the right of Equation (17-23) are positive, and thermal energy must flow into the system. From Equation (17-21), we can see that in the case of an expansion ($V_2 > V_1$), the work done by the gas is positive and, hence, comes at a cost to the internal energy of the gas; for a compression the converse is true.

We can use Equation (17-23) to determine the heat capacity for an ideal gas in an isobaric process. The heat

EXAMPLE 17-8

Work, Thermal Energy, and Internal Energy

One mole of an ideal gas is at a temperature of 293 K and a volume of 22.0 L. It is compressed in an isobaric process to a volume of 10.0 L. Find the work done, the thermal energy flow, and the change in internal energy.

SOLUTION

We can use Equation (17-20) to calculate the work, but first we need to determine the pressure. To do that, we use the ideal gas law and solve for P:

$$P = \frac{nRT}{V} = \frac{1 \text{ mol} \times 8.314 \text{ J/(mol·K)} \times 293 \text{ K}}{0.0220 \text{ m}^3} = 110727 \text{ Pa}$$

We can then determine the work:

$$W = P(V_2 - V_1) = (110\,727 \text{ Pa})\,(0.0100 \text{ m}^3 - 0.0220 \text{ m}^3)$$

$$= -1330 \text{ J}$$

The change in internal energy is

$$\Delta U = \frac{3}{2}P(V_2 - V_1) = \frac{3}{2}W = -1990 \text{ J}$$

and the thermal energy flow is

$$Q = \frac{5}{2}P(V_2 - V_1) = \frac{5}{2}W = -3320 \text{ J}$$

Making sense of the results:

The work done by the gas is negative, as it always is in a compression.

capacity is the ratio of the thermal energy that flows into a system to the resulting temperature change, so

KEY EQUATION
$$C_p = \frac{Q}{\Delta T} = \frac{\frac{5}{2}P(V_2 - V_1)}{T_2 - T_1}$$

$$= \frac{5}{2}\frac{nR(T_2 - T_1)}{T_2 - T_1} = \frac{5}{2}nR \quad (17\text{-}24)$$

The subscript P denotes this is a constant pressure heat capacity.

Isochoric Processes

In an isochoric process, we keep the volume constant and vary the pressure. We can carry out such a process by placing the gas in thermal contact with a thermal energy reservoir and allow thermal energy to flow into or out of the gas. Because the volume is constant, no work is done in such a process, and the first law becomes

$$\Delta U = Q \quad (17\text{-}25)$$

Figure 17-9 shows a P-V graph for an isochoric process. Note that since the volume does not change there is no area under the curve and no work is done.

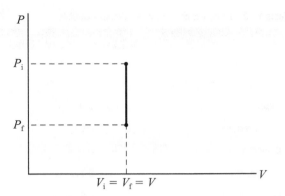

Figure 17-9 In an isochoric process, the pressure and temperature are varied and the volume remains constant.

There is a change in the temperature of the gas during this process, and the change in internal energy is

$$\Delta U = \frac{3}{2} nR(T_2 - T_1) = Q \qquad (17\text{-}26)$$

The heat capacity for an isochoric process can be found from Equation (17-26):

$$C_V = \frac{Q}{\Delta T} = \frac{3}{2} nR \qquad (17\text{-}27)$$

The heat capacity for an isochoric process is less than the heat capacity for an isobaric process. For a given input of thermal energy, we get a larger temperature change in an isochoric process than we get in an isobaric process. This is because in an isochoric process, no work is done by the gas, whereas in an isobaric process, the gas does work and some of the thermal energy is converted into work done by the gas.

Adiabatic Processes

In an adiabatic process, there is no thermal energy flow into or out of the system. Such a process can be accomplished by carefully insulating the system from the surroundings. Carrying out a process very quickly also reduces thermal energy flow to and from the system. In an adiabatic process, the first law of thermodynamics becomes

$$\Delta U = -W \qquad (17\text{-}28)$$

In an adiabatic expansion, work is done and the internal energy of the system changes. For an ideal gas, this change means there is also a change in the temperature of the system and, hence, changes in the pressure and volume of the gas. Since the gas does work as it expands, the change in the internal energy is negative. If the process starts at some temperature T_1 and ends at a lower temperature T_2, the P-V diagram for the process would look like Figure 17-10.

The online supplement for this textbook shows that for adiabatic processes in ideal gases,

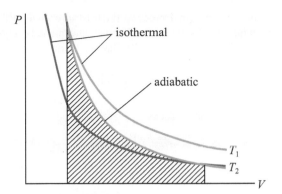

Figure 17-10 An adiabatic expansion

$$PV^\gamma = \text{constant} \qquad (17\text{-}29)$$

where $\gamma = \dfrac{f + 2}{f}$, and f is the number of degrees of freedom, as discussed in Chapter 16.

A monatomic gas only has three translation degrees of freedom and therefore $f = 3$. Therefore, the work done in an adiabatic expansion of a monatomic ideal gas is

$$W = \int_{V_1}^{V_2} P dV = \int_{V_1}^{V_2} \frac{C}{V^\gamma} dV = \frac{1}{1-\gamma} \frac{C}{V^{\gamma-1}} \bigg|_{V_1}^{V_2} \qquad (17\text{-}30)$$

The constant C can be determined from the starting point using the Ideal Gas Law.

State Variables

We now investigate the idea that the internal energy is a state variable, but neither the work nor the thermal energy is. We will compare two different ways of getting from an initial state to a final state for an ideal gas: an isothermal process that takes an ideal gas from (P_1, V_1, T) to (P_2, V_2, T) versus a two-step process with an isochoric stage from (P_1, V_1, T) to (P_2, V_1, T_i), followed by an isobaric stage from (P_2, V_1, T_i) to (P_2, V_2, T). These two different processes are illustrated in Figure 17-11.

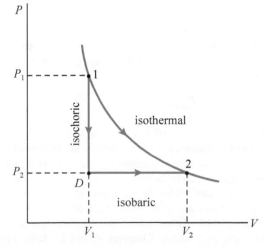

Figure 17-11 An isothermal process and an isochoric/isobaric process

For the isothermal process, there is no change in the temperature and, therefore, no change in the internal energy. We know from Equation (17-17) that the work done is

$$W_{\text{isothermal}} = nRT \ln\left(\frac{V_2}{V_1}\right) = Q \qquad (17\text{-}31)$$

For the isobaric process,

$$W_{\text{isobaric}} = P_2(V_2 - V_1) \qquad (17\text{-}32)$$

$$Q_{\text{isobaric}} = \frac{5}{2} P_2(V_2 - V_1) \qquad (17\text{-}33)$$

and

$$\Delta U_{\text{isobaric}} = Q_{\text{isobaric}} - W_{\text{isobaric}} = \frac{3}{2} P_2(V_2 - V_1) \qquad (17\text{-}22)$$

For the isochoric process, there is no work done but there is a change in temperature and, hence, a change in the internal energy:

$$\Delta U_{\text{isochoric}} = \frac{3}{2} P_2(V_1 - V_2) = Q_{\text{isochoric}} \qquad (17\text{-}34)$$

The net change in internal energy for the two-step isobaric–isochoric process is

$$\Delta U_{\text{isochoric+isobaric}} = \frac{3}{2} P_2(V_2 - V_1) + \frac{3}{2} P_2(V_1 - V_2)$$

$$= 0 = \Delta U_{\text{isothermal}} \qquad (17\text{-}35)$$

However, we can see that the net work in the isothermal process and the two-step process is not the same, and the net thermal energy flows also differ.

LO 8

17-8 Thermal Energy Flow

We will conclude this chapter with a discussion of three different processes by which thermal energy can move from one body to another.

Radiation

All objects whose temperature is not zero emit electromagnetic radiation, which carries energy. An object radiates energy at a rate given by

KEY EQUATION
$$P_{\text{radiate}} = \sigma \varepsilon A T_{\text{object}}^4 \qquad (17\text{-}36)$$

where $\sigma = 5.67 \times 10^{-8}$ W/(m^2·K^4) is the **Stefan–Boltzmann constant**, ε is the emissivity coefficient of the object, A is the surface area of the object, and T is the temperature of the objects in kelvins. The emissivity coefficient is a characteristic of the surface of the object and has a value between 0 and 1. A surface with an emissivity of 1 is called a **blackbody radiator**. The details of blackbody radiation are discussed in Chapters 30 and 31. Table 17-3 lists emissivity coefficients for some common materials.

Table 17-3 Emissivity coefficients for various materials

Surface material	Emissivity coefficient
Aluminum, heavily oxidized	0.2–0.31
Aluminum, highly polished	0.039–0.057
Asphalt	0.93
Brick, red, rough	0.93
Concrete	0.85
Cotton, cloth	0.77
Glass, smooth	0.92–0.94
Gold, polished	0.025
Ice, rough	0.985
Oil paints, all colours	0.92–0.96
Porcelain, glazed	0.92
Paper	0.93
Plastics	0.91
Water	0.95–0.963

If we were to put some typical values into Equation (17-36), such as an emissivity on the order of 1 and temperature, T, on the order of 300 K, we would see that many everyday objects at room temperature radiate energy at a rate of approximately 400 W/m^2. You might think that these objects would cool down rapidly. However, each object is also receiving radiation from its surroundings at a rate that depends on the ambient temperature:

$$P_{\text{absorbed}} = \sigma \varepsilon A T_{\text{ambient}}^4 \qquad (17\text{-}37)$$

The net power radiated by an object is

$$P_{\text{net}} = P_{\text{radiate}} - P_{\text{absorbed}} = \sigma \varepsilon A \left(T_{\text{object}}^4 - T_{\text{ambient}}^4 \right) \qquad (17\text{-}38)$$

We have used the same constant and parameters in Equations (17-37) and (17-38). If these values were not the same, there would be a net flow of energy between an object and its surroundings when they are at the same temperature. Since such flows do not occur, we know that objects are equally efficient at radiating as they are at absorbing.

Conduction

When two objects are in contact with each other, the molecules at the contact surface can collide and transfer energy between the objects. Within a single object that has one end at a higher temperature than the other, the particles that make up the object itself can transfer energy between each other, and this leads to a flow of thermal energy from the hot end to the cold end.

MAKING CONNECTIONS

Where Does the Thermal Energy Go?

Staying warm in the winter in northern climates can be a challenge. In 2008, a total of 920.8 PJ (1 PJ = 10^{15} J) was used for domestic space heating in Canada. This number represents 55% of total domestic energy consumption, excluding energy used in our domestic transportation. The false-colour thermal image in Figure 17-12 was produced by measuring infrared radiation from a house. The red areas have the greatest rate of thermal energy loss. You can also see that there is significant thermal energy loss through the door and windows, under the eaves, and through the roof. Insulating drapes on the windows can reduce thermal energy loss. More dramatic improvements in the windows can be achieved by using double-paned windows with a low thermal conductivity gas between them. It is also possible to use thin-film coatings on the glass to lower the emissivity.

Figure 17-12 A false-colour thermal image of a house showing hot spots in red and cool spots in blue.

EXAMPLE 17-9

Hot Coffee

A ceramic cup of hot coffee is placed on the counter. The liquid is at a temperature of $80°C$. Estimate the rate at which the cup and coffee are radiating energy.

SOLUTION

We will use Equation (17-38), but we will have to make some assumptions about the surface area of the cup, the emissivity of the cup, and the ambient temperature. We can model the cup as a cylinder of radius 5.0 cm and height 10 cm. Then we get a surface area of $A = \pi r^2 + \pi r^2 + 2\pi rh = 0.047 \, \text{m}^2$. From Table 17-3, we see that that the emissivity of glazed porcelain is 0.92 and the emissivity of water is approximately 0.96. So, we can use 0.92 as a reasonable approximation for the cup and coffee system. Assuming that the ambient temperature is $21°C$, and converting to kelvins, we get

$$P = (5.67 \times 10^{-8} \, \text{Wm}^{-2}\text{K}^4) \, (0.92) \, (0.047 \, \text{m}^2)$$

$$\times \, [(353.16 \, \text{K})^4 - (294.16 \, \text{K})^4] = 20 \, \text{W}$$

Making sense of the results:

Putting your hand near a 20 W lightbulb is similar to grabbing a hot cup of coffee. Putting a lid on the cup and insulating it can significantly reduce the surface temperature of the container and reduce the cooling rate.

It is found experimentally that when we place a layer of material between a hot reservoir and a cold reservoir, the rate at which thermal energy is transported from the hot side of the material to the cold side is given by

KEY EQUATION $\quad P_{\text{conduction}} = \dfrac{\kappa A (T_{\text{hot}} - T_{\text{cold}})}{x} \qquad$ (17-39)

where κ is the thermal conductivity, in units of W/(m·K), A is the area, and x is the thickness of the material.

Table 17-4 lists the thermal conductivities of various materials. In general, gases tend to have low thermal conductivities, insulating solids have somewhat higher conductivities, and metals have the highest thermal conductivities (metals are also good conductors of electricity). In metals, both the atoms and some of the electrons contribute to the thermal conductivity. For materials that do not conduct electricity, only the atoms in the material can transport the energy.

Table 17-4 Thermal conductivities of various materials

Material/Substance	Thermal conductivity at 25°C W/(m·K)
Argon (gas)	0.016
Brass	109
Brickwork, common	0.6–1.0
Carbon dioxide (gas)	0.0146
Concrete, stone	1.7
Copper	401
Glass	1.05
Gold	310
Hardwoods (oak, maple, etc.)	0.16
Insulation materials	0.035–0.16
Mica	0.71
Nitrogen (gas)	0.024
Paper	0.05
Snow (temperature < 0°C)	0.05–0.25
Water	0.58

THERMODYNAMICS

EXAMPLE 17-10

Efficient Windows

A 2.0 m² glass window is at a temperature of 20°C on the inside surface and −20°C on the outside. The glass is 5.0 mm thick. At what rate does energy flow out through the window glass? Compare this rate to thermal energy flow through a double-paned window that has a 2.0 cm gap between the panes filled with argon gas.

SOLUTION

For the single-paned window, we use Equation (17-39) and the thermal conductivity of glass from Table 17-4:

$$P_{conduction} = \frac{1.05 \ W/(m \cdot K) \, 2.0 \ m^2 (+20°C - (-20°C))}{0.005 \ m}$$

$$= 16 \ 800 \ W$$

For the double-paned window, we will assume that both surfaces of the inner pane are at 20°C and both surfaces of the outer pane are at −20°C. Then, the rate of thermal energy conduction through the argon between the panes is

$$P_{conduction} = \frac{0.016 \ W/(m \cdot K) \, 2 \ m^2 (+20°C - (-20°C))}{0.02 \ m} = 64 \ W$$

Making sense of the results:

Clearly, investing in double-paned windows can result in a significant reduction in energy consumption. We should also remember that windows allow energy to flow into a house in the form of sunlight and are important elements in a passive solar design.

EXAMPLE 17-11

Insulate the Roof

A cathedral (sloped) ceiling in a house has a total area of 100 m². The ceiling is covered with drywall (RSI 0.45), and the rafters are filled with insulation (RSI 40). The roof is sheathed with ½-in. plywood (RSI 0.63) and asphalt shingles (RSI 0.44). Calculate the rate at which thermal energy is lost, assuming that the interior temperature is 20°C and the exterior temperature is 2°C. Assume that thermal energy losses from conduction through the wood rafters and from ventilation of the roof cavity are negligible.

SOLUTION

$$P_{conduction} = \frac{A}{R} (T_{inside} - T_{outside})$$

$$= \frac{100 \ m^2}{[0.45 + 40 + 0.63 + 0.44] \ m^2/(W \cdot K)} (20°C - 2°C)$$

$$= 43 \ W$$

Making sense of the results:

While this may seem like a modest amount of power, we must remember that this is lost continuously and over time can be quite a loss of energy. Also note that we did not convert the temperatures to kelvins since they appear as a difference and temperature differences are the same in both scales.

Another measure of thermal energy flow through a material that is commonly used when discussing insulation is the R value, a measure of the thermal resistance of a material. The R value combines the thermal conductivity with the thickness of the material, l, and is defined as

$$R = \frac{l}{\kappa} \qquad (17\text{-}40)$$

In the SI system, the units for R are m²/(W·K), often called RSI values. R values simplify calculations for building insulation. If you use a multilayered insulation system, you simply add the R values for the individual layers to find the total R value. To calculate the thermal energy loss of a particular building element, you divide the area of the element by the total R value and multiply by the temperature difference across the element.

Convection

The third thermal energy transport mechanism is convection, which occurs in gases and liquids that are in a gravitational field. When we heat a portion of a gas or liquid, it expands and the density decreases. If there is a cooler, higher-density region above it, there is a buoyant force on

the warmer portion, so the warmer portion rises and the cooler region falls.

You may have experienced getting into a swimming pool or a lake on a hot summer day and noticed that the top layers of the water are nice and warm, but the water farther down is much cooler. Sunlight striking the surface heats the water from the top. The less dense warm water stays at the top, and the cooler water stays below it. There is no circulation; thus, no thermal energy is transported by the water other than by conduction from the top to the bottom of the lake. Figure 17-13 shows temperature

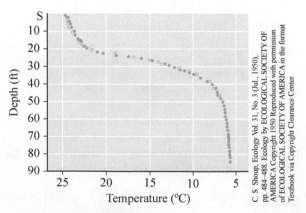

Figure 17-13 Thermal stratification in a lake

C. S. Shoup, Ecology Vol. 31, No. 3 (Jul, 1950), pp. 484–488. Ecology by ECOLOGICAL SOCIETY OF AMERICA Copyright 1950 Reproduced with permission of ECOLOGICAL SOCIETY OF AMERICA in the format Textbook via Copyright Clearance Center.

measurements in a lake on a sunny summer day. Near the surface, the temperature approaches 25°C, and at the deepest point it is only about 6°C. The lake is **thermally stratified**.

However, when a fluid is heated from the bottom, the heated material rises and the cooler material at the top falls. When the cooler material gets to the bottom, it is heated and rises. As long as the rising material loses thermal energy by some process, there will be a continuous **convection cycle**.

Hot water heating systems use both radiation and convection to distribute the thermal energy throughout a room. The radiator through which the hot water is circulated is usually placed on or near the floor close to a wall. The radiator is shaped to have a large surface area to radiate thermal energy, which is absorbed by the air surrounding the radiator. The heated air rises, and the cooler air falls. Because the radiator is placed against one wall, the hot air tends to spread out across the ceiling away from that wall, pushing the cooler air toward the opposite wall where it falls, as shown in Figure 17-14. This cooler air is then drawn across

the floor to the radiator, where it is heated. It then rises to the ceiling, continuing the convection cycle.

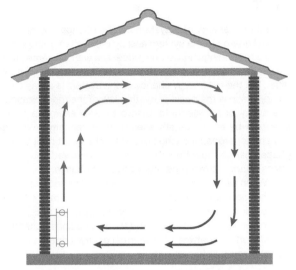

Figure 17-14   Thermal convection in a home heating system

 MAKING CONNECTIONS

Interesting Convection Patterns

A Rayleigh–Bénard cell consists of two parallel plates that are sealed together around the edges with the space between the plates filled with liquid. The bottom plate is heated, and the top plate is cooled. When the temperature difference between the top and bottom plates is quite small, there is simply thermal conduction through the liquid. As the temperature difference increases, however, convection occurs, creating beautiful patterns like the patterns in Figure 17-15. Different patterns emerge as the difference in temperature between plates changes. When the temperature difference is great enough, the convection becomes completely turbulent. This sequence of transitions from a well-ordered system to chaotic turbulence is characteristic of nonlinear dynamical systems.

Convection is also very important in global climate. The Gulf Stream is a large convection current that transports warm water from the Gulf of Mexico to the North Atlantic. The Gulf Stream is responsible for keeping Europe warm. Palm trees can grow in southern England despite it being at about the same latitude as Winnipeg, Manitoba.

Figure 17-15   A time exposure of convection in hexagonal Rayleigh–Bénard cells. The warm fluid is up-welling from the centre of each cell and sinking around the edges. The flow has been made visible by adding small aluminum flakes to the liquid.

FUNDAMENTAL CONCEPTS AND RELATIONSHIPS

The internal energy of a system can be changed by adding thermal energy to the system and by the system doing work on its surroundings. In various processes with the same start and end points, quantities of thermal energy and work may differ, but the net energy change of the system is always the same, which is the basis of the first law of thermodynamics. The ability of a material to absorb thermal energy is characterized by the material's specific heat capacity. When a system undergoes a change of phase, the latent heat of the phase change must be either added or removed. During a phase change, the temperature does not change.

When we heat an object, we transfer energy to the object. The thermal energy required to take a system from one state to another depends on how the process is carried out.

When we heat a system and there is no phase change, the relationship between the thermal energy added and the resulting temperature change is

$$Q = mc\Delta T \qquad (17\text{-}3)$$

where Q is the thermal energy, m is the mass, c is the specific heat, and ΔT is the temperature change.

When we heat a system and it undergoes a phase change, the temperature of the system does not change. Instead, the thermal energy is used to change the phase, and the thermal energy required is given by

$$Q_{\text{phase}} = mL \qquad (17\text{-}6)$$

where m is the mass of the system, and L is the latent heat of the transition.

When a system undergoes a volume change, the system does work that is given by

$$W = \int_{V_1}^{V_2} P\,dV \qquad (17\text{-}13)$$

The work done depends on the path taken between the start and end points.

The first law of thermodynamics states that the difference between the thermal energy that is added to a system and the work done by the system is independent of the path taken to carry out the process. This difference is called the internal energy, U:

$$\Delta U = Q - W \qquad (17\text{-}14)$$

There are countless ways that a system can be taken from one point on a P-V diagram to another. Some common processes are the following:

- Isobaric: constant pressure
- Isochoric: constant volume
- Isothermal: constant temperature
- Adiabatic: no thermal energy flow

Thermal energy transfer mechanisms include radiation, conduction, and convection:

- Radiation: $P_{\text{net}} = \sigma \varepsilon A(T_{\text{object}}^4 - T_{\text{ambient}}^4)$ $\qquad (17\text{-}38)$

where σ is the Stefan–Boltzmann constant, ε is the emissivity, and A is the surface area.

- Conduction: $P = \dfrac{\kappa A \Delta T}{x}$ $\qquad (17\text{-}39)$

where κ is the thermal conductivity, A is the surface area, x is the thickness of the material, and ΔT is the temperature difference.

- Convection: a gas or fluid that is heated tends to expand and rise.

APPLICATIONS

Applications: heating, cooling, moving thermal energy, energy-efficient windows and buildings, temperature control of mixtures

Key Terms: P-V diagram, adiabatic, blackbody radiation, closed system, convection cycle, first law of thermodynamics, heat capacity, heat of fusion, heat of vaporization, heat, isobaric, isochoric, isotherm, isothermal, latent heats, open system, quasistatic, specific heat capacity, Stefan–Boltzmann constant, thermally stratified

QUESTIONS

1. You are asked to heat a sample of gas using the least amount of thermal energy possible. Should you carry out a constant volume process or a constant pressure process?
2. Two boards of insulation are made of the same material. One board is twice as thick as the other.
   (a) Which board has the greater thermal conductivity?
   (b) Which board has the greater R value?
3. Figure 17-16 shows a temperature versus time graph for a sample that is heated from the solid phase to the gas phase. What phases are present in each section?

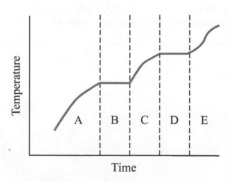

Figure 17-16 Question 3

4. Two objects with the same mass are heated to the same temperature. They are placed into identical insulated water baths that are initially at room temperature. The final temperatures of the water baths with the objects in them are different. Would the bath with the higher final temperature contain the object with the greater or the lesser specific heat?

5. In the winter, even though the air temperature in our homes is 20°C, we usually wear warm clothes. Conversely, on a summer evening, when the inside temperature of the house drops to 20°C, we tend to be quite comfortable in shorts and a light shirt. Why? (Hint: Consider differences in radiant energy absorption by your skin.)

6. Figure 17-17 shows a temperature versus time graph for a sample of ice and a sample of water that are mixed together. What is the final phase?

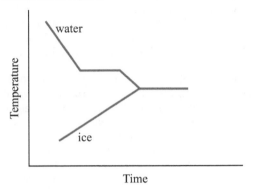

Figure 17-17   Question 6

7. For the *P-V* process for an ideal gas shown in Figure 17-18, the system starts at A and goes to B. Is the work done positive or negative?

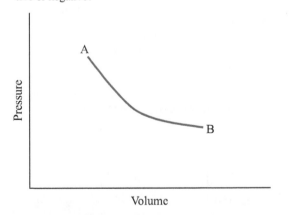

Figure 17-18   Question 7

8. When we double the temperature of a hot object, by what factor does the rate at which it radiates energy increase?

9. When we double the temperature difference across an insulator, by what factor does the thermal energy flow increase?

10. Many animals sweat or perspire in order to keep cool in hot weather. How does perspiring help remove thermal energy from your body?

11. In the cooking technique called *sous-vide*, raw food is sealed in an airtight plastic bag and then immersed in a hot water bath; the water temperature is maintained at the desired final temperature for the food. The types of changes in the food as it cooks are determined primarily by the temperature of the water bath, so the flavour and texture of food cooked *sous-vide* differ from those of the same food cooked by other methods that use higher temperatures. Many types of food can be held in a *sous-vide* water bath for extended periods without overcooking. Explain why this is so.

12. Pressure cookers are used to shorten cooking times for foods that are boiled. How do pressure cookers work?

13. A hot object and a cold object are placed into thermal contact. Thermal energy flows from the hot object to the cool object, and eventually they reach the same temperature. Under what circumstances is the final temperature halfway between the two initial temperatures?

PROBLEMS BY SECTION

For problems, star ratings will be used, (✶, ✶✶, or ✶✶✶), with more stars meaning more challenging problems.

Section 17-2   Temperature Changes Due to Thermal Energy Transfer

14. ✶   A 50.0 g block of copper is at a temperature of 20°C. When 2.00 kJ of thermal energy is added to the block, what temperature will it reach?

15. ✶✶   You wish to make tea, and you put 1.0 kg of water at 5°C into your 500.0 W kettle. How long will it take for the water to boil? Assume no thermal energy loss.

16. ✶✶   Alberta experiences chinooks in the winter, which are winds that can cause the air temperature to rise dramatically. In one instance, the temperature rose from −30°C to 5°C. We assume that the atmosphere is heated by this amount to an altitude of 5.0 km. How much thermal energy per square kilometre is required?

Section 17-3   Thermal Energy Flow Between Objects

17. ✶   A 100.0 g block of aluminum at a temperature of 100°C is placed in a well-insulated water bath at a temperature of 10°C. If the bath holds 500.0 g of water, what is the temperature of the system when it reaches equilibrium?

18. ✶   Twenty grams of milk at a temperature of 2°C is added to 300.0 g of coffee at a temperature of 170°C. The coffee cup has a heat capacity of 109 J/K. What is the equilibrium temperature of the mixture?

19. ✶✶   You determine in an experiment that a 100.0 g sample of an unknown metal experiences a 5.0°C temperature increase when 65 J of thermal energy is added to the sample. What is the specific heat of the metal?

Section 17-4   Phase Changes and Latent Heat

20. ✶✶   A 500.0 W kettle has 1.00 kg of water in it at a temperature of 100°C. How long will it take for the kettle to boil dry?

21. ✶✶   Twenty grams of ice at a temperature of −4°C is placed into 100.0 g of water at a temperature of 30°C. Find the final temperature of the mixture.

Section 17-5   Work

22. ✶✶   One mole of an ideal gas is compressed isobarically from a volume of 40.0 L to a volume of 30.0 L at a pressure of 110.0 kPa. How much work is done?

23. ✶   Three moles of an ideal gas is expanded isothermally at a temperature of 100°C. During the expansion, the volume increases by 30%. How much work is done?

24. ✶   A gas undergoes an isochoric process in which the pressure increases from 50 kPa to 100 kPa. How much work is done?

Section 17-6   The First Law of Thermodynamics

25. ✳ An ideal gas undergoes an isothermal process in which the gas does 100 J of work. How much thermal energy flows into the gas?

26. ✳✳ A gas undergoes an isothermal expansion from a certain pressure, temperature, and volume. The volume of the gas doubles. The same gas undergoes an adiabatic expansion from the same starting point, and again the gas doubles in volume. In which process is more work done? In which process is there a thermal energy flow?

Section 17-8   Thermal Energy Flow

27. ✳ Comet Ikeya-Seki was observed to reach a temperature of approximately 650°C at its closest approach to the Sun. At what rate does the comet radiate energy per square metre? Assume an emissivity of 0.50.

28. ✳ A particular insulating material is 10 cm thick and has an RSI value of 40. What is the thermal conductivity of the material?

29. ✳✳ A 0.50 cm-thick piece of copper has an area of 4.0 cm². One side of the copper is at a temperature of 100°C, and the other side is at 400°C. At what rate does thermal energy flow through the copper?

COMPREHENSIVE PROBLEMS

30. ✳✳ The Sun delivers radiant energy to Earth at a rate that averages approximately 1.3 kW per square metre of the cross-sectional area of Earth. If we assume that all of this energy is absorbed by Earth (i.e., that Earth has an emissivity of 1), at what rate must Earth radiate energy if Earth's temperature is to remain constant? What would that constant temperature be? (Note: The actual temperature of Earth is somewhat higher due to the greenhouse effect of the atmosphere.)

31. ✳ A researcher heats a 100.0 g sample using a 400.0 W resistive heater. The temperature of the sample increases at a rate of 3.0°C/min. Find the heat capacity and the specific heat of the sample.

32. ✳✳ (a) The Sun has a surface temperature of approximately 6000 K and a diameter of 14×10^5 km. What is the total power that is radiated by the Sun?
(b) The Sun derives its energy from fusion reactions in which $E = mc^2$, where E is the energy released, m is the mass converted to energy, and c is the speed of light. At what rate does the Sun lose mass as radiated energy?

33. ✳✳ Athletes can lose significant amounts of water from their bodies as a result of perspiring during a competition. During an intense 2.0 h tennis match, a player lost 4.0 kg. Estimate the rate at which her body was using energy to evaporate all of that perspiration. List any assumptions you make for the estimate.

34. ✳✳ Three moles of an ideal gas are at a temperature of 273 K and a pressure of 100.0 kPa. The gas is compressed isothermally to half of its original volume. It is then expanded adiabatically to its original state. How much work is done in the compression? In the expansion? What is the highest temperature reached by the gas? The lowest?

35. ✳✳ One and a half moles of an ideal gas is at standard temperature and pressure. The gas is isothermally compressed to one-third of its original volume. How much work is done?

36. ✳✳ The compression in problem 35 is carried out adiabatically. How much work is done in that case?

37. ✳✳ The Etruscan shrew is a small rodent that has a very high metabolic rate: its heart beats over 1000 times/min. The animal is approximately 4 cm long and has a body temperature of approximately 36°C. Estimate the rate at which it loses thermal energy due to radiation.

38. ✳✳✳ The ice on the surface of a lake in winter helps insulate the lake. For these questions, assume that the air temperature is −10°C and the ice is 10 cm thick.
(a) How much thermal energy per square metre is lost from the water to the atmosphere?
(b) Using the rate in part (a), determine how long it will take another centimetre of ice to form.

39. ✳✳ An ideal gas is in a cylinder that has a light frictionless piston. The pressure outside the cylinder is 101 kPa. You slowly add 1000.0 kJ of thermal energy to the gas and observe the volume to increase from 10.0 m³ to 15.0 m³. Determine the work done by the gas and the change in the internal energy of the gas.

40. ✳✳ An ideal gas is at an initial pressure of 3000.0 kPa and a volume of 7.0 L. The gas is expanded at constant pressure to a volume of 11.0 L. It is then cooled at constant volume until the temperature returns to its original value. Calculate the work done, the change in internal energy, and the thermal energy flow into or out of the gas.

41. ✳✳ Consider an ideal gas that initially has a pressure of 1909 kPa and a volume of 11.0 L. The gas is compressed at constant pressure to a volume of 7.0 L. Thermal energy is then added at constant volume until the pressure reaches 3000.0 kPa. Calculate the work done, the change in internal energy, and the thermal energy flow into or out of the gas.

42. ✳✳ An ideal monatomic gas undergoes the increases in pressure and volume shown in the graph in Figure 17-19. The initial temperature of the gas is 297 K.
(a) Calculate the work done.
(b) Calculate the temperature of the gas at the end of the process.
(c) Calculate the change in internal energy in this process.
(d) Calculate the quantity of thermal energy added to or removed from the gas during this process.

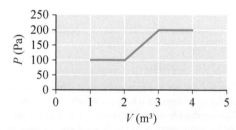

Figure 17-19   Problem 42

43. An ideal gas is carried through a thermodynamic cycle consisting of two isobaric and two isothermal processes, as shown in Figure 17-20. Calculate the total work done in the entire cycle when $V_1 = 0.50$ m³, $V_2 = 2.0$ m³, $P_1 = 1.00 \times 10^5$ Pa, and $P_2 = 1.50 \times 10^5$ Pa.

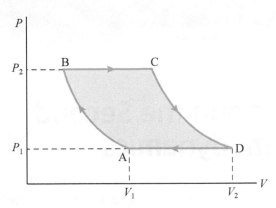

Figure 17-20 Problem 43

Table 17-5 Data for Problem 47

Time (s)	Temperature (°C)
0	4.79
40	6.53
80	8.04
120	9.86
160	11.05
200	12.50
240	13.56
280	14.94
320	15.80
360	17.04
400	17.83
440	18.90
480	19.98
520	20.90
560	21.92
600	22.51
640	23.63
680	24.28
720	24.94
760	26.00
800	26.31
840	27.03
880	27.71
920	28.62
960	28.98
1000	29.88
1030	30.06

44. ✶✶ One mole of an ideal gas is at a temperature of 50°C and a pressure of 100.0 kPa. The gas is expanded isothermally to three times its initial volume. Find the final temperatures, the thermal energy input to the gas, the work done by the gas, and the change in the internal energy of the gas. Draw the process on a *P-V* diagram.

45. ✶✶ One mole of an ideal gas is at a temperature of 50°C and a pressure of 100.0 kPa. The gas is expanded isobarically to three times its initial volume. Find the final temperature, the thermal energy input to the gas, the work done by the gas, and the change in the internal energy of the gas. Draw the process on a *P-V* diagram.

46. ✶✶ One mole of an ideal gas is at a temperature of 50°C and a pressure of 100.0 kPa. The gas is expanded adiabatically to three times its initial volume. Find the final temperatures, the thermal energy input to the gas, the work done by the gas, and the change in the internal energy of the gas. Draw the process on a *P-V* diagram.

DATA-RICH PROBLEM

47. Table 17-5 lists temperatures recorded as a sample of water was heated from about 5°C to 30°C. The ambient temperature in the room was 20.0°C. The thermal energy was supplied by a 10.0 W heater. Determine the mass of the water being heated.

OPEN PROBLEM

48. ✶✶✶ We can reduce the cost of heating our homes by increasing the amount of insulation we put into them. Of course, there is a cost to installing the insulation. Suppose that insulation costs $X per square metre per centimetre and that thermal energy costs $Y per joule. Determine the thickness of insulation one should use when building a house that minimizes the total cost of insulating and heating the house. You will have to make some assumptions about how long the house will be lived in, the thermal conductivity of the insulation, and the average inside–outside temperature difference.

 See the text online resources at **www.physics1e.nelson.com for Open Problems and Data-Rich Problems related to this chapter.**

Learning Objectives

When you have completed this chapter you should be able to:

1 Describe the operation of a refrigerator and a heat engine.

2 Sketch a Carnot cycle on a *PV* diagram.

3 Describe entropy in terms of macroscopic variables.

4 State the second law of thermodynamics in several ways.

5 Discuss the circumstances under which thermodynamics is applicable.

6 Discuss some of the consequences of the second law of thermodynamics.

7 Describe entropy from a microscopic perspective.

Chapter 18
Heat Engines and the Second Law of Thermodynamics

What do the refrigerator and car engine in Figure 18-1 have in common? In the refrigerator, we supply electrical energy to power a motor, which does work on a gas. The gas is circulated in such a way that thermal energy is transferred from the interior of the refrigerator to the coils on the back. Work is done on the system, and thermal energy is moved from a lower to a higher temperature.

In the car engine, fuel is burned at a high temperature inside cylinders with pistons. The hot gases do work on the pistons, pushing them outward. The gases cool somewhat during this expansion, and are then ejected through the exhaust pipes. The moving pistons rotate the crankshaft, which is connected to a gearing system that drives the wheels. In this engine, thermal energy is moved from a higher temperature to a lower temperature, and the system does work on the car.

One of the goals of this chapter is to understand how heat pumps and heat engines work. As you will see, the study of these devices was a key step in the formulation of the second law of thermodynamics. The second law of thermodynamics introduces the concept of entropy. We will first describe entropy in terms of macroscopic quantities: thermal energy and temperature. Later in the chapter, we will look at a microscopic description. We begin our approach to the second law by examining in more detail how heat engines and refrigerators (and heat pumps) work.

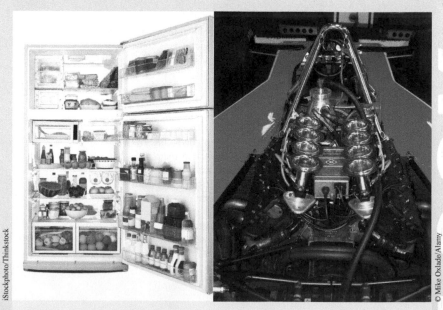

Figure 18-1   A refrigerator and an internal combustion engine

For additional Making Connections, Examples, and Checkpoints, as well as activities and experiments to help increase your understanding of the chapter's concepts, please go to the text's online resources at www.physics1e.nelson.com.

18-1 Heat Engines and Heat Pumps

Heat Engines

A **heat engine** is a device that performs work while allowing thermal energy to flow from a high-temperature reservoir to a low-temperature reservoir, as shown in Figure 18-2. A **reservoir** is an object whose heat capacity is so large that it can be considered to be at a constant temperature. The reservoir from which we are removing thermal energy is called the source, and the reservoir that we are putting thermal energy into is called the **sink**.

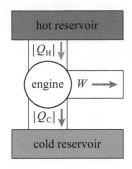

Figure 18-2   Schematic illustration of a heat engine

A high-temperature source provides thermal energy to the engine. For a heat engine, the temperature of the source of thermal energy is always greater than the temperature of the heat sink. The engine does some work and also delivers a quantity of thermal energy to the low-temperature sink. Since energy must be conserved, for a heat engine,

KEY EQUATION $$|Q_{\mathrm{H}}| = |Q_{\mathrm{C}}| + |W| \qquad (18\text{-}1)$$

In Equation (18-1), we have used absolute value signs to avoid any ambiguity that could arise from the fact that the thermal energy flows can be either positive or negative, depending on our reference point. For example, Q_{H} is a negative thermal energy flow from the perspective of the high-temperature source but positive from the perspective of the engine.

Steam engines are a common type of heat engine. While obsolete for railway locomotives, a steam engine is the mechanism that converts thermal energy into mechanical energy in thermal power plants (both fossil fuel-fired and nuclear). In a steam engine, water is heated to produce steam at a high temperature. In a fossil fuel-fired plant, hot combustion gases from burning fuel heat the water flowing through tubes in the boiler. In a nuclear

C-18-1 Sign Conventions for the Flow of Thermal Energy

The flow of thermal energy is
(a) positive from the perspective of the engine and negative from the perspective of the low-temperature sink;
(b) positive from the perspective of both the engine and the low-temperature sink;
(c) negative from the perspective of the engine and positive from the perspective of the low-temperature sink;
(d) negative from the perspective of both the engine and the low-temperature sink.

C-18-1 (c) Thermal energy flows out of the engine into the low-temperature sink.

plant, the reactor core has coolant pumped through it to capture the thermal energy from the fission reaction, and that thermal energy is used to generate the steam. Such plants use a steam turbine to drive an electrical generator, as shown in Figure 18-3.

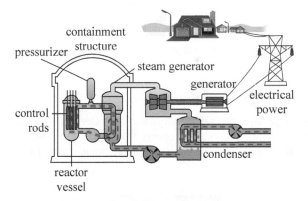

Figure 18-3   A steam turbine in a nuclear power plant

Often, pressurized water is circulated through the reactor core to both control the temperature of the core and capture the thermal energy that is generated by the nuclear reactions in the core. Since the water is under high pressure, it can be heated to a high temperature (well above 100°C) without boiling. The pressurized water circulates through pipes in the steam generator, where thermal energy from these pipes vaporizes water in the turbine loop to form high-pressure steam. The reactor coolant, which has been exposed to intense radiation in the core, circulates in a closed loop of piping and does not mix with the water in the turbine loop. The steam is fed to the turbine, causing it to turn. The turbine drives the generator, producing electrical power. The steam expands and cools as it passes through the turbine into the condenser. However, not all the thermal energy in the high-temperature steam is converted into work in the turbine. To increase the efficiency of the engine, we cool the steam in the condenser so that the thermal energy that is ejected from the engine is at as low a temperature as possible. We will examine heat engine efficiencies in Section 18-2.

A heat engine always generates some waste thermal energy. In a car, that thermal energy is dissipated through the radiator, although some of this thermal energy can be used to warm the passenger compartment on cold days. The fact that waste thermal energy is unavoidable leads to the **Kelvin–Planck statement** of the **second law of thermodynamics**: No heat engine can convert thermal energy completely into work.

The device shown in Figure 18-4 is not possible because the sole effect of the device is to convert thermal energy completely into work.

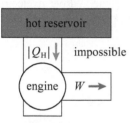

Figure 18-4 This heat engine process is impossible, a fact summarized by the Kelvin–Planck statement of the second law of thermodynamics.

Heat Pumps and Refrigerators

Heat pumps and refrigerators are both devices that use energy to transfer thermal energy from a low-temperature source to a high-temperature sink, as shown in Figure 18-5.

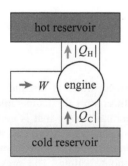

Figure 18-5 A heat pump and a refrigerator

When we apply conservation of energy to a heat pump, we find that the thermal energy expelled to the high-temperature reservoir must equal the thermal energy

extracted from the low-temperature reservoir plus the work input:

KEY EQUATION
$$|Q_H| = |W| + |Q_C| \qquad (18\text{-}2)$$

We are all familiar with household refrigerators, which transfer thermal energy from a cool interior compartment to the warmer surroundings. With heat pumps, the focus is on the thermal energy that is delivered to the higher-temperature area. For example, a heat pump for home heating extracts thermal energy from the cool ground outside and delivers it to the warmer interior of the house. A heat pump operates on the same physical principles as a refrigerator, and both transfer thermal energy from a cooler body to a warmer body. The distinction between the two is based on their purpose.

Figure 18-6 shows the basic operation of a refrigerator. The pressure in the internal heat exchange coil (on the left side of the diagram) is lower than in the external heat exchange coil (right side). When gas passes through the expansion valve, the gas expands and cools to a temperature that is lower than the temperature of the interior of the refrigerator. Thermal energy in the air in the refrigerator is then absorbed by the cool gas. The gas is then compressed as it is pumped into the external heat exchange coil. This compression raises the temperature of the gas above the ambient temperature in the room, so thermal energy transfers from the coil on the back of the refrigerator to the room. Thus, the refrigerator transfers thermal energy from its cool interior to the warmer room. To accomplish this transfer, energy must be supplied to the compressor, which does work on the gas.

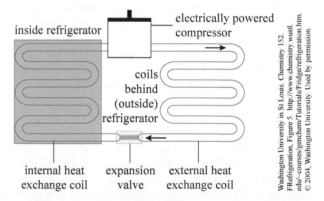

Washington University in St Louis. Chemistry 152. FRefrigeration, Figure 5. http://www.chemistry.wustl.edu/~courses/genchem/Tutorials/Fridge/refrigeration.htm. © 2004, Washington University. Used by permission.

Figure 18-6 Schematic of a refrigerator

We find that thermal energy does not flow from a cooler body to a warmer body unless we do work on the system. This observation leads to the **Clausius statement** of the second law of thermodynamics: No device can transfer thermal energy from a low-temperature reservoir to a high-temperature reservoir without work input. The heat pump shown in Figure 18-7 violates the Clausius statement because it has no work input.

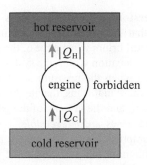

Figure 18-7 This heat pump process is impossible, a fact summarized by the Clausius statement of the second law of thermodynamics.

LO 2

18-2 Efficiency and the Carnot Cycle

We have seen that heat engines cannot completely convert thermal energy into work and that heat pumps (e.g., refrigerators) cannot simply move thermal energy from a low temperature to a high temperature without work being done. Of course, when designing a heat engine, we would like to extract as much work as possible from a given quantity of thermal energy. Similarly, we would like to minimize the energy that a heat pump requires to transfer a given quantity of thermal energy. Now, we consider the efficiencies of these devices.

Heat Engine Efficiency

For a heat engine, we are primarily interested in maximizing the amount of work done for a given thermal energy input. In so doing, we are also minimizing the thermal energy output, which is typically lost energy. We define the **efficiency**, η, of a heat engine as the ratio of the work done, $|W|$, to the thermal energy input, $|Q_H|$:

KEY EQUATION
$$\eta \equiv \frac{|W|}{|Q_H|} = \frac{|Q_H| - |Q_C|}{|Q_H|} \qquad (18\text{-}3)$$

We used Equation (18-1) to replace W in the numerator with $|W| = |Q_H| - |Q_C|$.

✓ CHECKPOINT

C-18-2 Limits of Efficiency

A manufacturer of a heat engine claims that the engine has an efficiency of 110%. Is this claim reasonable?

EXAMPLE 18-1

Work, Thermal Energy, and Efficiency

An automobile engine has an efficiency of 20.0%. When the engine radiates thermal energy at a rate of 100 kJ/s, how much mechanical work is being extracted?

SOLUTION

Looking at Equation (18-3), we can see that to determine W, we need to know Q_H. We are given Q_C, so we start by solving for Q_H:

$$|Q_C| = |Q_H| - \eta|Q_H|$$

$$\therefore |Q_H| = \frac{|Q_C|}{1 - \eta} = \frac{100 \text{ kJ/s}}{1 - 0.2} = 125 \text{ kJ/s}$$

We can now determine the work to be

$$W = |Q_H| - |Q_C| = 125 \text{ kJ/s} - 100 \text{ kJ/s} = 25 \text{ kJ/s}$$

Making sense of the results:

The work done is certainly less than the total thermal energy input, so we know that the engine has not violated the Kelvin–Planck statement of the second law of thermodynamics.

Efficiency of Heat Pumps and Refrigerators

For heat pumps and refrigerators, the definition of efficiency centres on the amount of work done per unit of thermal energy transferred. For a refrigerator, we are primarily interested in the thermal energy transferred from the cool interior. Therefore, we define the **coefficient of performance of a refrigerator**, CP_R, as

KEY EQUATION
$$CP_R = \frac{|Q_C|}{|W|} \qquad (18\text{-}4)$$

In contrast, our primary focus with a heat pump is the thermal energy delivered to the higher temperature. We define the **coefficient of performance of a heat pump**, CP_H, as

KEY EQUATION
$$CP_H = \frac{|Q_H|}{|W|} \qquad (18\text{-}5)$$

✓ CHECKPOINT

C-18-3 Too Good to Be True?

A manufacturer of home-heating systems claims that its heat pump can deliver up to 3 J of thermal energy for every joule of electrical energy it consumes. Is this even possible?

C-18-3 Yes. An examination of Figure 18-5 shows that the thermal energy delivered to the high-temperature reservoir is, in general, greater than the energy supplied in the form of work. As we will see later in this section, the actual ratio depends on the difference between the inside and outside temperatures.

C-18-2 No. Getting more work out than energy you put in would violate conservation of energy. In addition, we can see from 18-3 that the only way to get an efficiency greater than 1 is to have $Q_C < 0$ which is not possible.

THERMODYNAMICS

Getting More Than You Pay For

Heat pumps can provide more energy than they consume, but most other forms of home heating cannot (Figure 18-8). Electrical heating systems convert the electrical energy directly into thermal energy, producing a thermal energy output equal to the electrical energy input. Home-heating systems that burn a fuel, such as natural gas, oil, or wood, always lose some thermal energy with exhaust gases that go up the chimney. These losses can be significant—around 20% for a typical midefficiency furnace. However, the latest high-efficiency gas furnaces can reduce the waste thermal energy to as little as 4%.

Figure 18-8 In this heat pump installation, the coils that absorb the thermal energy from outside of the house are installed underground. Heat pump efficiencies decrease substantially when the source reservoir temperature is low compared to the interior temperature. The underground installation helps guard against low air temperatures.

The Carnot Cycle

French scientist Nicolas Léonard Sadi Carnot (1796–1832) devoted considerable attention to maximizing the efficiencies of heat engines. His studies introduced some of the fundamental concepts of thermodynamics.

As is often done in physics, Carnot chose to study a simple model system under ideal conditions. He considered a heat engine that used an ideal gas in a reversible process. A **reversible process** is a process that is carried out so slowly that the system is always in equilibrium. A system is in equilibrium when none of the state variables vary with time or location within the system. An ideal gas is in equilibrium when the pressure and temperature are the same everywhere in the gas, the number of particles in the gas is constant, and the volume of the gas is constant.

All real processes are *irreversible*. For example, if we suddenly compress a gas, we create a pressure wave that travels throughout the gas. The wave is a region of high pressure, and the compression reduces the volume of the gas, so clearly the gas is no longer in equilibrium. As the wave propagates, the molecules in the gas collide with each other. The wave spreads out, and eventually the pressure becomes uniform. At this point, the gas has returned to equilibrium.

Imagine breaking up the sudden compression into smaller, sudden steps. At each step, the departure from equilibrium is smaller than for the entire compression, and the time taken to return to equilibrium is shorter. If we break up the process into a series of infinitesimal steps, the departure from equilibrium becomes infinitesimally small, and the time away from equilibrium becomes infinitesimally short. In the limit, the process maintains equilibrium throughout. However, it will take an infinite amount of time to carry out the process. Nonetheless, this infinite process can be a useful theoretical construct.

Figure 18-9 illustrates the **Carnot cycle** for a heat engine. As indicated by the arrows, we proceed clockwise around the cycle. We could also proceed counterclockwise to describe a cycle for a heat pump or refrigerator.

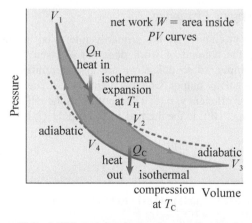

Figure 18-9 A *PV* diagram for a heat engine operating on a Carnot Cycle

The Carnot cycle consists of four reversible processes. Starting in the upper-left-hand corner of the cycle, we first carry out an isothermal expansion at T_H. During this process, thermal energy $|Q_H|$ is absorbed by the gas. We then carry out an adiabatic expansion where no thermal energy flows into or out of the gas, but the temperature is lowered to T_C. Once we reach that temperature, we carry out an isothermal compression, and thermal energy $|Q_C|$ is removed from the system. The final step is an adiabatic compression that raises the temperature back to T_H. During the two expansions, the gas does work on its surroundings. During the two compressions, the surroundings do work on the gas. The net work done on the surroundings is represented by the pink shaded area in Figure 18-9.

We can determine how much work is done in this cycle by calculating all the thermal energy flows. In Chapter 17, we showed that for isothermal processes

$$W = \int_{V_1}^{V_2} P dV = \int_{V_1}^{V_2} \frac{nRT}{V} dV = nRT \ln\left(\frac{V_2}{V_1}\right) \quad (18\text{-}6)$$

Thus, for the Carnot cycle we have

$$W_{1-2} = nRT_H \ln\left(\frac{V_2}{V_1}\right) \quad (18\text{-}7)$$

Since this is an isothermal process, there is no change in the internal energy of the gas. The first law of thermodynamics then reduces to $W = Q$ (Equation (17-16)) and we can write:

$$|Q_H| = nRT_H \ln\left(\frac{V_2}{V_1}\right) \quad (18\text{-}8)$$

All the same arguments apply to the isothermal compression. Therefore,

$$|Q_C| = nRT_C \ln\left(\frac{V_3}{V_4}\right) \quad (18\text{-}9)$$

Recall that $|Q_C|$ is the magnitude of the thermal energy absorbed by the low-temperature reservoir. To make that quantity positive, we used $\frac{V_3}{V_4}$ as the argument of the logarithm.

For the adiabatic portions of the cycle, we can eliminate the volumes and express the thermal energy flows simply in terms of the temperatures. In Chapter 17, we showed that for an ideal gas undergoing an adiabatic process,

$$PV^\gamma = \text{Constant} \quad (18\text{-}10)$$

where $\gamma = \dfrac{f+2}{f}$, and f is the number of degrees of freedom.

Thus we can write

$$P_2 V_2^\gamma = P_3 V_3^\gamma$$

$$P_4 V_4^\gamma = P_1 V_1^\gamma \quad (18\text{-}11)$$

Since we are using an ideal gas, we can apply the ideal gas law to give

$$\frac{P_2 V_2}{T_H} = \frac{P_3 V_3}{T_C} \quad (18\text{-}12)$$

$$\frac{P_4 V_4}{T_C} = \frac{P_1 V_1}{T_H}$$

In Equation (18-12), all four ratios are equal, but we have written the pairings to mirror those in Equation (18-11). Dividing the corresponding pairs in Equation (18-11) by those in Equation (18-12), we obtain

$$T_H V_2^{\gamma-1} = T_C V_3^{\gamma-1}$$

$$T_H V_1^{\gamma-1} = T_C V_4^{\gamma-1} \quad (18\text{-}13)$$

Dividing the upper row by the lower row in Equation (18-13) we find

$$\left(\frac{V_2}{V_1}\right)^{\gamma-1} = \left(\frac{V_3}{V_4}\right)^{\gamma-1} \quad (18\text{-}14)$$

Therefore,

$$\ln\left(\frac{V_2}{V_1}\right) = \ln\left(\frac{V_3}{V_4}\right) \quad (18\text{-}15)$$

Using this result in Equations (18-8) and (18-9), and combining those two equations, we find (finally!)

$$\frac{|Q_H|}{|Q_C|} = \frac{T_H}{T_C} \quad (18\text{-}16)$$

Rearranging gives

$$\frac{|Q_H|}{T_H} = \frac{|Q_C|}{T_C} \quad (18\text{-}17)$$

We can use this result to determine the efficiency of a heat engine operating on a Carnot cycle:

KEY EQUATION

$$\eta = \frac{|Q_H| - |Q_C|}{|Q_H|} = 1 - \frac{|Q_C|}{|Q_H|} = 1 - \frac{T_C}{T_H} \quad (18\text{-}18)$$

It is important to remember that these temperatures are expressed using the Kelvin scale. Note that the efficiency of such an engine is always less than 1 for an engine operated at finite temperatures. Since the Kelvin–Planck statement of the second law indicates that a heat engine cannot have an efficiency of 100%, T_C must be greater than 0 K.

Carnot was able to prove that this efficiency applies for any reversible engine operating between two given temperatures. He was also able to show that any irreversible engine has a lesser efficiency. This result is called **Carnot's theorem** and can stated as follows: The efficiency of any reversible engine operating between two fixed temperatures is the same, and any irreversible engine operating between the same two temperatures is not as efficient.

Thus, Carnot's theorem establishes the benchmark against which all real (irreversible) engines are measured.

As mentioned earlier, we can run the Carnot cycle in reverse. The coefficients of performance for a heat pump and for a refrigerator operating on a Carnot cycle are as follows:

Heat pump: $CP_H = \dfrac{|Q_H|}{|W|} = \dfrac{|Q_H|}{|Q_H| - |Q_C|}$

KEY EQUATION

$$= \frac{1}{1 - \dfrac{|Q_C|}{|Q_H|}} = \frac{1}{1 - \dfrac{T_C}{T_H}} = \frac{T_H}{T_H - T_C} \quad (18\text{-}19)$$

Refrigerator: $CP_R = \dfrac{|Q_C|}{|W|} = \dfrac{|Q_C|}{|Q_H| - |Q_C|} = \dfrac{1}{\dfrac{|Q_H|}{|Q_C|} - 1}$

 $= \dfrac{1}{\dfrac{|T_H|}{|T_C|} - 1} = \dfrac{T_C}{T_H - T_C}$ (18-20)

Note that in both cases the temperature difference $T_H - T_C$ appears in the denominator. As a consequence, the coefficients of performance of these devices decrease as the temperature difference increases. Thus, we can reduce the energy consumption of a refrigerator by placing it in a cool place. Similarly, heat pumps work more effectively when the low-temperature reservoir is not too cold.

☑ **CHECKPOINT**

C-18-4 Practical Limits to Efficiency

Can a heat engine operating between 400 K and 300 K have an efficiency of 24%?

[rotated, upside-down answer text:]

C-18-4 The maximum possible efficiency for a reversible engine operating between the given temperatures is

$$u = 1 - \dfrac{T_C}{T_H} = 1 - \dfrac{300\ K}{400\ K} = 0.25 = 25\%$$

Thus, an efficiency of 24% is theoretically possible. However, this high an efficiency is very unlikely because the best practical heat engines only approach approximately 75% of the maximum possible efficiency.

LO 3

18-3 Entropy

Now, we consider the direction of the thermal energy flow during a Carnot cycle in a heat engine. Q_H is the thermal energy absorbed by the engine at the high temperature, and our convention is that a heat flow into the engine is a positive quantity. Therefore, $|Q_H| = Q_H$. However, Q_C is the thermal energy transferred from the engine to the low-temperature reservoir and is therefore a negative quantity. Thus $|Q_C| = -Q_C$. Therefore, we can write Equation (18-17) as

$$\dfrac{Q_H}{T_H} = -\dfrac{Q_C}{T_C} \qquad (18\text{-}21a)$$

or

$$\dfrac{Q_H}{T_H} + \dfrac{Q_C}{T_C} = 0 \qquad (18\text{-}21b)$$

☑ **CHECKPOINT**

C-18-5 Sign Conventions

How would Equation (18-21a) change if we considered a heat pump instead of a heat engine?

[rotated, upside-down answer text:]

equation identical to Equation (18-21a).

Thus, Equation (18-17) would become $-\dfrac{Q_H}{T_H} = \dfrac{Q_C}{T_C}$, and multiplying both sides by −1 gives an

C-18-5 There would be no change. We would replace $|Q_H|$ with $-Q_H$, and $|Q_C|$ with Q_C.

So far, we have demonstrated only that Equation (18-21a) applies to a Carnot cycle. We now consider whether this relationship applies to any reversible process for an ideal gas. Any such process takes us from one point on a PV diagram to another, and can be accomplished with a combination of a reversible adiabatic process and a reversible isothermal process. The adiabatic process gets us from the starting temperature to the isotherm that the end point lies on. Then the isothermal process takes us to the end point, as shown in Figure 18-10. We could also start with an isothermal process and then carry out an adiabatic process.

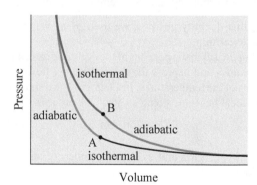

Figure 18-10 We can get from point A to point B by either an adiabatic process (green) followed by an isothermal process (red) or an isothermal process (black) followed by an adiabatic process (blue).

We could break up any arbitrary reversible process that takes the system from A to B into a large number of reversible adiabatic and isothermal steps. The greater the number of steps, the better the approximation. Similarly, we could approximate any arbitrary reversible process that returns the system to point A from point B with a large number of reversible adiabatic and isothermal steps. Thus, any reversible process over a closed path (a path that returns the system to its starting point) can be approximated as a series of small adiabatic and isothermal steps. In the limit of an infinite number of steps, we would have a perfect approximation. Thus, any closed path on a *PV* diagram can be approximated with arbitrary accuracy with a series of Carnot cycles. Such a process is illustrated in Figure 18-11.

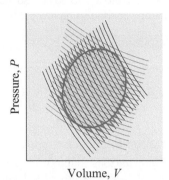

Figure 18-11 Approximating an arbitrary process with a series of Carnot cycles

We now turn our attention to the thermal energy flows that occur in such a step-wise process. We can generalize Equation (18-21b) to

$$\sum_{\text{reversible}} \frac{Q_i}{T_i} = 0 \qquad (18\text{-}22)$$

where Q_i denotes the thermal energy flow for each successive step, and the temperature at which it occurs is T_i.

In the limit of an infinite number of steps, the summation becomes an integral:

$$\oint_{\text{reversible}} \frac{dQ}{T} = 0 \qquad (18\text{-}23)$$

Since the integral sums to zero over a closed path, regardless of the shape of the path, there must be a state variable associated with this integral. Recall that a state variable has a well-defined value that is unique to the state of the system. For example, when we specify the temperature, pressure, and volume for an ideal gas with a fixed number of particles, we have specified the state of the gas. We saw in Chapter 17 that for an ideal gas the internal energy is a state variable, but the work and the thermal energy are not.

We call this new state variable **entropy** and denote it with the symbol *S*. For a reversible process going from *a* to *b*,

$$S_b - S_a = \Delta S = \int_a^b \frac{dQ}{T} \qquad (18\text{-}24)$$

From Equation (18-24) we can identify

`KEY EQUATION` $$dS = \frac{dQ}{T} \qquad (18\text{-}25)$$

Entropy has units of J/K.

18-4 Entropy and the Second Law of Thermodynamics

We have several statements of the second law of thermodynamics so far, and it is not immediately apparent how they are related. As you will see, these statements are all linked by the concept of entropy.

Clausius Statement Violated: The Clausius statement says that thermal energy cannot flow from a cold temperature to a hot temperature without work being done. If such a process were to occur, the net effect would be a flow of thermal energy out of the low-temperature reservoir while the same amount of thermal energy flows into the high-temperature reservoir. The net change in entropy of the system comprised of the two reservoirs is:

$$\Delta S = -\frac{Q}{T_C} + \frac{Q}{T_H} < 0 \qquad (18\text{-}26)$$

Thus, if the Clausius statement of the second law is violated, the entropy of the system will decrease.

Kelvin–Planck Statement Violated: The Kelvin–Planck statement of the second law says that no process can completely convert thermal energy into work. If such a process were possible, the net effect would be a removal of thermal energy from a high-temperature reservoir. Thus, the change in entropy would be:

$$\Delta S = \frac{-Q}{T_H} < 0 \qquad (18\text{-}27)$$

Thus, if the Kelvin Planck statement of the second law is violated, the entropy will decrease.

Carnot Theorem Violated: Suppose we have a heat engine that is more efficient than a Carnot engine. For example, the efficiency would be greater if the thermal energy flow at the lower temperature were somewhat reduced with a corresponding increase in the amount of work done by the engine. We denote the reduced low-temperature thermal energy flow as Q'_C. We are assuming that $Q'_C < Q_C$, where Q_C is the heat flow for the engine that does not violate Carnot's Theorem. It must also be true that $\frac{Q'_C}{T_C} < \frac{Q_C}{T_C} = \frac{Q_H}{T_H}$. The change in entropy will then be:

$$\Delta S = \frac{Q'_C}{T_C} - \frac{Q_H}{T_H} < \frac{Q_C}{T_C} - \frac{Q_H}{T_H} = 0 \qquad (18\text{-}28)$$

Thus, there would also be a decrease in entropy if a heat engine was more efficient than a Carnot engine.

THERMODYNAMICS

Summary: We can see that in each of these processes that violate one of the statements of the second law, the entropy of the system *decreases*. However, we have restricted ourselves to considering reversible processes so far, and all real processes are irreversible. For a real heat engine, we find that we are able to extract less work than with a reversible process. Since energy is conserved, the thermal energy delivered to the low-temperature reservoir must be greater than for a reversible process. The change in entropy for an irreversible process is

Efficiency less than Carnot efficiency:

$$\Delta S = \frac{-Q_H}{T_H} + \frac{Q''_C}{T_C} > 0 \qquad (18\text{-}29)$$

where $Q''_C > Q_C$ is the increased amount of thermal energy.

Thus, the entropy either remains constant (reversible processes) or increases (all real processes). The most general statement of the **second law of thermodynamics** is as follows: **The total entropy never decreases. The result of all real processes is that the total entropy increases.**

EXAMPLE 18-2

Entropy Change in a Reversible Process

Find the change in total entropy when a sample of liquid water at 0°C is placed in thermal contact with a large reservoir that has a temperature infinitesimally lower than 0°C.

SOLUTION

Thermal energy will slowly flow out of the liquid into the reservoir, and the liquid will freeze. The change in entropy of the liquid is

$$\Delta S_{\text{liquid}} = \frac{-mL}{273.15 \text{ K}}$$

where m is the mass of the liquid, and L is the latent heat associated with the liquid–solid transition.

We can see that the entropy of the liquid decreases. However, we must also consider the change in the entropy of the reservoir. The thermal energy that comes out of the liquid transfers to the reservoir. Since the reservoir is large and its temperature is only infinitesimally lower than the temperature of the liquid water, the amount of thermal energy transferred to the reservoir is so small that the temperature of the reservoir does not change. The change in the entropy of the reservoir is

$$\Delta S_{\text{reservoir}} = \frac{mL}{273.15 \text{ K}}$$

Thus, the total change in the entropy of the system consisting of the liquid sample and the reservoir is zero.

Making sense of the result:

The process is reversible, so we would expect the change in the entropy to be zero.

PEER TO PEER

Entropy is a state variable, so the change in entropy between two points is independent of the process used to go between those two points. So, you can calculate the entropy change using any process you like.

EXAMPLE 18-3

Entropy Change in an Irreversible Process

Find the change in the total entropy when a small sample of liquid water at 10.00°C is placed in thermal contact with a large thermal reservoir at exactly 0°C.

SOLUTION

Thermal energy flows from the warmer sample into the cooler reservoir until the sample cools to 0.000°C. The thermal energy that flows out of the water sample is

$$Q_{\text{water}} = -mc\Delta T$$

where m is the mass of the liquid, c is the specific heat, and $\Delta T = 283.15 \text{ K} - 273.15 \text{ K} = 10.00 \text{ K}$.

The same amount of thermal energy flows into the reservoir. Since the reservoir is large, its temperature does not change, and the change in entropy of the reservoir is

$$\Delta S_{\text{reservoir}} = \frac{mc\Delta T}{273.15 \text{ K}} = 0.0366 \, mc > 0$$

However, the temperature of the liquid is not constant during the thermal energy transfer. Recall from Chapter 17 that the thermal energy flow in a process where there is a finite temperature change is

$$Q = mc\Delta T \qquad (17\text{-}3)$$

For an infinitesimal temperature change, Equation (17-3) becomes

$$dQ = mc \, dT$$

We can use this result to determine the change in entropy for the liquid:

$$\Delta S_{\text{liquid}} = \int \frac{dQ}{T} = \int_{283.15 \text{ K}}^{273.15 \text{ K}} \frac{mc \, dT}{T}$$

$$= mc \ln\left(\frac{273.15 \text{ K}}{283.15 \text{ K}}\right) = -0.0360 \, mc$$

The decrease in the entropy of the liquid is less than the increase in the entropy of the reservoir. Thus, the total entropy has increased by $0.0007 \, mc$.

Making sense of the result:

The change in the total entropy is positive, as we would expect for an irreversible process.

18-5 The Domain of the Second Law of Thermodynamics

The second law of thermodynamics establishes a direction for all processes. Only those processes that lead to a net increase in entropy are observed to occur.

In Example 18-3, we examined a situation in which an object was placed in contact with a cooler object. Thermal energy flowed from the warmer object to the cooler object, and the total entropy increased. Imagine two bodies at the same temperature—placed in contact—and having thermal energy flow from one to the other so that a temperature difference is created. Such a process would not violate conservation of energy because the exact amount of thermal energy that left one body would move to the other body. However, this flow of thermal energy would result in a net decrease in entropy and, hence, violate the second law. Such processes are never observed.

Let us consider some other natural processes for which the reverse process never occurs. Place a full glass of water on the counter in a sealed room. Over the period of a few days, the water evaporates from the glass, and the room becomes somewhat more humid. However, placing an empty glass on the counter never results in the room becoming less humid and the glass filling with water. Similarly, when we open a bottle of perfume in a closed room, the scent spreads throughout the room and remains uniformly distributed after all the perfume has evaporated from the bottle. We do not observe the perfume refilling the bottle, and the scent does not even concentrate in any one part of the room. In both of these examples, the water in the glass and the perfume in the bottle are at the same temperature as the room. It is not obvious how thermal energy has flowed from the room into the perfume or the water to cause it to evaporate. However, we can see that the degree of organization or "order" of the system has changed. Having all the water in the glass is a state that is more ordered than having the water molecules dispersed throughout the room. Conversely, there is more disorder when the water or perfume molecules are dispersed throughout the room than there is when they are all in a container. Such examples lead us to another statement of the second law: All natural processes tend toward states of greater disorder.

So, we can interpret the entropy of a system as a measure of the disorder in the system. Increasing disorder equates to increasing entropy.

We can think of many examples of things that we see happen on a regular basis that are never observed to occur in reverse. If a glass falls off a counter and shatters on the floor, we would be startled to see the shards come back together and the reformed glass then leap back up onto the counter. In fact, we usually find it amusing to watch motion pictures run backward because of the absurdity of seeing such processes occur in reverse.

However, we can think of certain events that are readily reversible. For example, consider a perfectly elastic collision between two particles, as shown in Figure 18-12. In the collision, both momentum and energy are conserved. We can see that the collision in Figure 18-12(b) is simply the reverse of the collision in Figure 18-12(a). If we were to observe either of these collisions, we would not in any way be able to tell if the "clock" was running backward or forward.

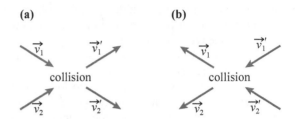

Figure 18-12 Two perfectly elastic collisions between two point particles

What about entropy? Doesn't the second law of thermodynamics apply to all processes? One important difference between the processes illustrated in Figure 18-12 and the processes that we discussed previously is the number of particles involved.

Consider a vessel full of some ideal gas and equipped with pressure and temperature sensors. As we saw in Chapter 16, the pressure is created by collisions between the gas molecules and the walls of the container. Similarly, the temperature is measured by allowing the thermometer to experience collisions with the gas molecules. These collisions transfer energy back and forth between the gas molecules and the material the thermometer is made of. Eventually, the net transfer of energy back and forth is zero, and the gas and the thermometer are in thermal equilibrium. At this point, the thermometric property of the thermometer stops changing and we can read the temperature. For nitrogen gas at standard temperature and pressure, there are approximately 10^{25} molecular collisions with the walls of the container per square metre per second. As a result, the readings on the pressure and temperature sensors appear constant.

Now consider what happens when we remove some of the gas molecules. We would expect the reading on the pressure sensor to decrease because there would be fewer gas molecules colliding with the sensor. If we are careful about how we remove molecules, it is possible to keep the temperature constant. We would need to ensure that the average kinetic energy of the molecules we remove is equal to the average kinetic energy of the molecules in the gas. However, as we reduce the number of molecules further and further, we would find that the frequency of collisions with the pressure sensor would become so low that instead of a steady reading, we would see spikes in the readings as individual molecules hit the sensor.

Our thermometer readings would also start to become peculiar as the number of gas molecules got very low. As we saw in Chapter 16, not all the molecules have the same speed in a gas; hence, there is a distribution of kinetic energies, as well. As the frequency of collisions becomes very small, we would expect fluctuations in the temperature reading, assuming we have an "ideal" thermometer that responds instantaneously to each collision. When a very energetic molecule hits the sensor, it deposits a lot of energy and makes the sensor reading spike up. Conversely, when a low-speed molecule collides with the sensor, we would detect a transfer of energy from the sensor to the molecule, and the reading would spike down.

Once the number of molecules gets so low that the readings of our pressure sensor and thermometer are no longer constant, these variables are no longer well-defined, and we begin to question whether a thermodynamic description of the system is appropriate.

How would we approach thinking about the entropy of a system that contained a single molecule? It would be very difficult to discuss, for example, the degree of disorder of a single gas molecule. Suppose we had two molecules. As they moved around the container we would not be terribly surprised if at some point they passed very close to each other. However, if we have several moles of gas in the container, we would be completely surprised if all of those molecules congregated in the same place at the same time.

All of these state variables—temperature, pressure, and entropy—are based on averages, and these averages are meaningful only when taken over a large number of particles. The number of particles needed to make a thermodynamic approach valid is not precisely known. Strictly speaking, thermodynamics only applies perfectly in the limit of an infinite number of particles—the **thermodynamic limit**. As we reduce the number of particles, we begin to see fluctuations in the readings of thermometers and pressure sensors. Once the fluctuations become comparable to the readings themselves, the variables are no longer well-defined. Then we are no longer in the thermodynamic limit, and the concept of entropy no longer applies.

✅ CHECKPOINT

C-18-6 When Does Thermodynamics Apply?

Which of the following systems can we describe using thermodynamics?

(a) a gas sample containing several moles of gas
(b) a hydrogen atom
(c) a transistor
(d) a galaxy

C-18-6 All but (b) should be amenable to a thermodynamic analysis, because these systems contain a very large number of particles.

18-6 Consequences of the Second Law of Thermodynamics

Absolute Zero

In Section 18-2, we found that the efficiency of a heat engine operating on a Carnot cycle is

$$\eta = \frac{|Q_H| - |Q_C|}{|Q_H|} = 1 - \frac{|Q_C|}{|Q_H|} = 1 - \frac{T_C}{T_H} \quad (18\text{-}18)$$

The Kelvin–Planck statement of the second law asserts that no heat engine can convert all the thermal energy into work. Consequently, $|Q_C| > 0$ and $\eta < 1$. It follows that, for any finite T_H, $T_C > 0$. Thus, there is a lower limit on temperature below which one cannot go. That limit temperature is called **absolute zero**. Absolute zero is 0 K on the Kelvin scale and −273.15°C on the Celsius scale. The lowest temperatures that have been produced in laboratories are on the order of 10^{-10} K.

This lower limit rests on the assumption that T_H is finite. Since temperature is a measure of the average kinetic energy of the particles in a system, we could have an infinite temperature only if we had an infinite amount of energy.

🧩 MAKING CONNECTIONS

Additional Limits from Quantum Mechanics

As we will see in Chapter 31, one of the foundations of quantum physics is the uncertainty principle. One of the uncertainty relations states that the precision with which we can simultaneously measure the position, x, and the momentum, p, of a particle are limited such that

$$\Delta x \Delta p \geq \frac{h}{4\pi} \quad (18\text{-}30)$$

where $h \cong 6.62 \times 10^{-34}$ J·s is Planck's constant.

Because of this limit, confining a particle to a region of space results in some uncertainty about the momentum of the particle and, hence, about its kinetic energy. Therefore, we cannot reduce the kinetic energy of such a particle to zero.

Heat Death

The second law of thermodynamics provides all real processes with a direction, which some call the "arrow of time." We now consider in which direction the arrow points and the ultimate fate of the universe.

All processes either move thermal energy from a high temperature to a lower temperature and do some work, or require some work input to move thermal energy from a lower temperature to a higher temperature. The second law establishes that to perform work, we must have temperature differences because there is no heat engine that can solely convert thermal energy into work. Consider a simple model universe that consists of a high-temperature reservoir and a low-temperature reservoir. If we run a heat engine between these two reservoirs, the engine will perform work *and* transfer thermal energy from the high-temperature reservoir to the low-temperature reservoir. The temperature of the high-temperature reservoir

decreases as we remove thermal energy from it, and the temperature of the low-temperature reservoir increases as thermal energy is deposited into it. At some point, the two reservoirs reach the same temperature, making it impossible for the heat engine to produce any more work. This universe will have reached a state where the energy is uniformly distributed throughout the universe, and no further energy flow is possible. Such a universe will be "dead" in the sense that all matter will be uniformly distributed and no life possible.

MAKING CONNECTIONS

You Are a Thermodynamic Fluctuation

The fact that the universe is not in heat death tells us something about the structure of the universe at its beginning. Because there is structure in the universe today, there must have been structure in the universe when it began. This structure is responsible for our existence. Evidence of this structure can be seen in two ways. The way that matter is distributed in the universe on a very large scale—how galaxies are distributed, for example—reflects the structure in the early universe. Other signs of structure are the fluctuations in the cosmic microwave background (CMB) radiation throughout the universe.

As we saw in Chapter 15, the Doppler shift of the light emitted from distant galaxies is evidence that the universe is expanding. Observations suggest that all matter in the universe was primarily hydrogen, and most of the other elements were formed by fusion processes in stars that coalesced from the hydrogen gas. If we extrapolate the expansion backward to before the formation of stars, the universe becomes denser, and it becomes hotter if we model the hydrogen as a simple ideal gas (a reasonable approximation). At some earlier time, the temperature would have been hot enough that all the hydrogen atoms

would have been ionized into electrons and protons. Light scatters from charged particles, and any radiation in the universe at that point would have been interacting significantly with the matter in the universe.

When we have light that is in thermal equilibrium with matter at a fixed temperature, the light has a characteristic pattern of frequencies called a blackbody spectrum. Analysis of such a spectrum reveals the temperature of the object. The CMB radiation is microwave radiation that appears almost uniformly throughout the sky. A detailed analysis of this radiation shows that it is consistent with an object that was in thermodynamic equilibrium with the radiation, and the characteristic temperature of the radiation is now 2.725 K. The "photon gas" that makes up the CMB radiation has also expanded and has been shifted to longer wavelengths and lower frequencies as a result. Consequently, it appears to be at this low temperature from a temperature of approximately 3000 K at which the electrons and protons combined. Although the radiation is remakably uniform, it is not perfectly so, and these fluctuations are believed to result from small deviations from equilibrium that existed very early in the universe (Figure 18-13).

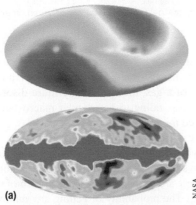

Figure 18-13 (a) Fluctuations in the cosmic microwave background. False-colour images of the temperature inferred from CMB radiation measurements made by the COBE satellite. In the upper image, red corresponds to 2.729 K, and blue corresponds to 2.721 K. The dominant structure is due to the Doppler shift that results from Earth's motion. When that is removed, we are left with the bottom image, where the difference between red and blue is only 0.002 K. (b) Large-scale structure in the universe. A near-infrared image of the sky, which reveals the distribution of galaxies beyond our own. Galaxies that are coloured blue are closest, and galaxies that are coloured red are farthest away. The band across the centre is the Milky Way galaxy.

Such a universe would not necessarily have zero energy. The energy will simply be uniformly distributed. Fortunately, this process will take a very long time to occur. The universe is estimated to be approximately 14 billion years old, and there is still plenty of thermal energy flowing around. In fact, there is even some debate over whether this heat death will be the ultimate fate of our universe, because it depends on some things that we do not know, such as whether the universe will continue to expand or whether the expansion will stop and the universe will collapse back in on itself.

LO 7

18-7 A Microscopic Look at Entropy

We begin our comparison between the macroscopic definition and the microscopic details of a system by examining the concepts of macrostate and microstate.

A "macrostate" of an ideal gas is specified by knowing the pressure, volume, temperature, and number of particles in the system. We have no knowledge of the positions or speeds of any of the individual particles in the system. We have knowledge of the average speeds of the particles because, as we saw in Chapter 16, we can relate the temperature and the pressure of an ideal gas to those quantities. We also know that the average velocity of all the particles is zero if the gas has no net momentum.

A "microstate" of an ideal gas is specified by stating how many particles have a particular speed and the volume of the gas. Knowing this, we can then calculate the total number of particles, the average speed, and, hence, the temperature and pressure.

Let us consider how many different microstates correspond to a given macrostate. The two quantities that must be conserved are the number of particles and the average speed of the particles. One way to create the macrostate is to give each particle the same speed equal to the average speed. There is only one way to do this, however. We say that such a microstate is highly ordered.

Suppose we increase the speed of one particle and decrease the speed of another particle in such a way that we do not change the average speed. There are many ways to do this simple modification because we can change the speed by an arbitrary amount. If we extend this to changing the speeds of three particles, we get an even larger number of ways to do it. These microstates would be less ordered, or more disordered, because the number of particles with changed speeds increases.

The order of a state is a reasonably intuitive concept. For example, we say that a solid material is more ordered than its liquid state. The solid consists of an array of atoms and/or molecules that are in well-defined positions relative to one another. The liquid state does not have well-defined positions for the atoms and/or molecules, and

even the shape of the liquid is less defined than the solid, as illustrated in Figure 18-14.

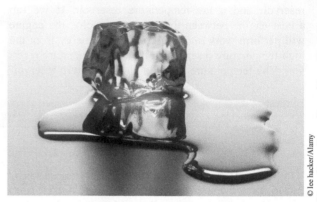

Figure 18-14 A melting ice cube. The solid is in a more ordered state than the liquid.

Similarly, the gas is less ordered than the liquid, and we expect the water from the melted ice cube to eventually evaporate into the air. The natural tendency is toward states of greater disorder.

✓ CHECKPOINT

C-18-7 Thermal Energy and Order

Which is more ordered, a cold block of metal and a hot block of metal, or the two blocks pressed together at some intermediate equilibrium temperature?

C-18-7 The separate hot and cold blocks represent a more ordered state because we have two distinct sets of particles with different average kinetic energies. Once the temperatures are the same, the two sets of particles are indistinguishable in terms of their average kinetic energy.

It appears as though systems naturally evolve toward states of higher disorder, and from our consideration of the microstates of a gas, there are many more disordered microstates corresponding to a particular macrostate than there are ordered microstates.

A fundamental concept in statistics is that the probability of an event occurring is proportional to the number of different ways that the event can occur. For example, if we roll a die, there are six possible outcomes, each of which is equally likely. The probability of getting any number on the roll is then 1/6. Suppose we roll two dice, one red and one blue. There are now 36 possible outcomes for the roll because, for each number on the first die, there are 6 possible outcomes for the second die. If we categorize the rolls based on the sum of the two dice, we get the results in Table 18-1. The most probable outcome is that the dice sum to 7 because that result can be achieved in the largest number of ways.

Table 18-1 Outcomes for rolling two dice

Sum of two dice (macrostate)	Possible outcomes (microstates)						Number of outcomes (microstates)
2	1, 1						1
3	1, 2	2, 1					2
4	1, 3	2, 2	3, 1				3
5	1, 4	2, 3	3, 2	4, 1			4
6	1, 5	2, 4	3, 3	4, 2	5, 1		5
7	1, 6	2, 5	3, 4	4, 3	5, 2	6, 1	6
8	2, 6	3, 5	4, 4	5, 3	6, 2		5
9	3, 6	4, 5	5, 4	6, 3			4
10	4, 6	5, 5	6, 4				3
11	5, 6	5, 6					2
12	6, 6						1

We make the analogy between the dice and the thermodynamics systems as follows: Knowing the sum of the two dice is equivalent to specifying the macrostate of the system. Specifying the result of each die is specifying the microstate.

The important point to note is that, in general, there is more than one way to achieve the same outcome. The same is true of a thermodynamic system. In general, there are many microstates that correspond to the same macrostate.

EXAMPLE 18-4

States for Coin Tosses

Enumerate the possible macrostates (number of heads and tails) and their corresponding microstates (head and tail for each coin) when 4 coins are tossed.

SOLUTION

The possible macrostates are 4 tails, 3 tails and 1 head, 2 tails and 2 heads, 1 tail and 3 heads, and 4 heads.

There is only one microstate each corresponding to all heads (H, H, H, H) and all tails (T, T, T, T).

For the case of 3 tails and 1 head, we can have (T, T, T, H), (T, T, H, T), (T, H, T, T), and (H, T, T, T). The case of

1 tail and 3 heads can be arrived at by switching all heads for tails, and vice versa. We then obtain (H, H, H, T), (H, H, T, H), (H, T, H, H), and (T, H, H, H).

The case of 2 tails and 2 heads can be arrived at with (H, H, T, T) and (T, T, H, H), (H, T, T, H) and (T, H, H, T), and (H, T, H, T) and (T, H, T, H). Note that each pair is the complement of each other. We summarize our results in Table 18-2.

Making sense of the result:

We find that the most number of microstates corresponds to the most "disordered" state: 2 heads and 2 tails.

Table 18-2 Results for Example 18-4

Macrostate	Microstates						Number of microstates
4 tails	T, T, T, T						1
3 tails, 1 head	H, T, T, T	T, H, T, T	T, T, H, T	T, T, T, H			4
2 tails, 2 heads	T, T, H, H	H, H, T, T	T, H, H, T	H, T, T, H	T, H, T, H	H, T, H, T	6
1 tail, 3 heads	T, H, H, H	H, T, H, H	H, H, T, H	H, H, H, T			4
4 heads	H, H, H, H						1

The results of a coin toss can be generalized to a large number of coins. For example, if we toss 100 coins, there are on the order of 10^{30} possible microstates. On the one hand, the probability of getting all heads or tails is thus about 10^{-30}, a very small number. On the other hand, approximately 8% of the states are states in which there are the same number of heads and tails. Seventy-three percent of the states lie between 45% and 55% being all heads. As we increase the number of coins, more and more of the probability shifts toward the same number of heads and tails, and in the limit of an infinite number of coins, the probability of getting 50% tails and 50% heads is 1.

The same thing happens in a thermodynamic system. All microstates are occupied with equal probability, but the macrostate that is observed is the one that has the largest number of microstates associated with it. In the thermodynamic limit of an infinite number of particles, this corresponds to a single macrostate. If we prepare a system in another macrostate that has fewer microstates, it will evolve toward the macrostate with the largest number of microstates, and this is also the most disordered state.

A study of the efficiency of heat engines and heat pumps led to the development of an idealized, reversible process called the Carnot cycle. The expression obtained for the efficiency of a Carnot cycle suggests the existence of a state variable called entropy. For reversible processes, the entropy remains constant; for all other processes, the entropy increases. This led to the most general statement of the second law of thermodynamics: entropy always increases or stays the same. The idea of entropy also leads to an understanding of the "arrow of time." Processes that are always observed to occur in one direction but never in reverse are processes in which the entropy increases.

A heat engine is a device that moves thermal energy from a high temperature to a low temperature and performs some work:

$$|Q_H| = |Q_c| + |W| \qquad (18\text{-}1)$$

The Kelvin–Planck statement of the second law of thermodynamics: No heat engine can convert thermal energy completely into work.

A heat pump is a device that uses work input to move thermal energy from a low temperature to a high temperature:

$$|Q_H| = |W| + |Q_c| \qquad (18\text{-}2)$$

The Clausius statement of the second law of thermodynamics: No heat pump can move thermal energy from a low temperature to a higher temperature without work input.

A reversible process is an idealized process that occurs so slowly that the system remains in equilibrium throughout the process.

The most efficient heat engine possible is one that uses reversible processes. The processes used are adiabatic and isothermal processes, and such an engine is said to operate on a Carnot cycle. For all Carnot cycles operating between a low and a high temperature, we find that

$$\frac{Q_H}{T_H} + \frac{Q_c}{T_c} = 0 \qquad (18\text{-}21b)$$

Entropy is a state variable defined by

$$dS = \frac{dQ}{T} \qquad (18\text{-}25)$$

The second law of thermodynamics states that the total entropy never decreases. The result of all real processes is that the total entropy increases.

Entropy is a measure of the disorder of a system. The second law of thermodynamics can be stated as all natural processes tend toward states of greater disorder.

Thermodynamics only applies to systems where the number of particles is large enough that statistical measures of their behaviour are valid.

APPLICATIONS

Applications: engine design, home heating, refrigeration, power generation, improving energy efficiency of devices,

Key Terms: absolute zero, Carnot cycle, Carnot's theorem, Clausius statement, coefficient of performance of a heat pump, coefficient of performance of a refrigerator, efficiency, entropy, heat engine, heat pump, Kelvin–Planck statement, reservoir, reversible process, second law of thermodynamics, sink, thermodynamic limit

QUESTIONS

1. Two physics students are trying to stay cool on a hot day. They read in their textbook that a refrigerator and an air conditioner are both heat pumps. They come up with the idea of opening the door of their refrigerator to try to cool their kitchen. Will this idea work? Explain.
2. Discuss how the order of a system changes as it is cooled from the gas phase to the solid phase.
3. An inventor claims to use the waste thermal energy from a heat engine as thermal energy input to the engine and achieve an efficiency greater than the Carnot efficiency. What is wrong with this claim?
4. When a liquid evaporates from a glass into a room, it is not completely clear how the latent heat of evaporation is supplied to the water for it to evaporate, because, if the water is in equilibrium with the room, the water is at the same temperature as the room and thermal energy only flows spontaneously from high to low temperatures. Can this truly be an equilibrium process?
5. An inventor has designed a heat pump to recover waste thermal energy from a heat engine and use it to resupply the engine. The inventor claims that adding this heat pump will improve the efficiency of any heat engine. Is this claim valid? Explain.
6. Which is the most effective way to increase the efficiency of a Carnot heat engine: increase the temperature of the high-temperature reservoir or decrease the temperature of the low-temperature reservoir by the same amount?
7. Two keen physics students tell their parents they do not want to clean their rooms because they feel they will be violating the second law of thermodynamics. Are they correct? Explain.
8. Describe a process that respects conservation of energy but violates the second law.

9. In Section 18-5, we considered an elastic collision between two particles and claimed that we are unable to distinguish between the collisions depicted in Figure 18-12 on thermodynamic grounds. Suggest another example.

PROBLEMS BY SECTION

For problems, star ratings will be used, (∗, ∗∗, or ∗∗∗), with more stars meaning more challenging problems.

Section 18-1   Heat Engines and Heat Pumps

10. ∗ A heat engine exhausts 1100 J of thermal energy in the process of doing 400 J of work. What is the thermal energy input to the engine?

11. ∗ A car has a fuel consumption of 5.8 L/100 km while travelling at 100 km/h. The power output of the engine is 40 kW. When we burn gasoline, the thermal energy released is approximately 33 MJ/L
(a) How many litres of gasoline does the engine burn per second?
(b) How much thermal energy is released per second as a result?

12. ∗ A heat engine absorbs 20 kJ of energy from a high-temperature reservoir and ejects 15 kJ of energy to a low-temperature reservoir. How much work does the engine perform?

13. ∗∗ A camper has a propane-powered refrigerator. The refrigerator consumes approximately 0.50 kg of propane per day. When propane is burned completely, it releases approximately 50 MJ/kg. The interior of the refrigerator is maintained at an average temperature of 2°C, and the exterior temperature is 22°C. Assume that the refrigerator has a surface area of 2.0 m² and that the insulation has an R factor of 3.2 K·m²/W.
(a) At what rate does thermal energy flow into the refrigerator? (Hint: See Example 17-9.)
(b) At what rate must the refrigerator remove thermal energy from its interior?
(c) How much energy is released per second from the propane combustion?
(d) How much thermal energy is delivered to the room?

14. ∗ A heat pump operates between the outside and the inside of a house. The pump absorbs 10 kJ of energy from the cool exterior and delivers 30 kJ to the interior. How much work is required to accomplish this?

Section 18-2   Efficiency and the Carnot Cycle

15. ∗ A heat engine operates between a high temperature of 1000.0°C and a low temperature of 500.0°C. Assume it operates on a Carnot cycle.
(a) What is the efficiency of the engine?
(b) When the heat engine absorbs 100.0 kJ from the high temperature reservoir, how much work does it do?

16. ∗ A heat pump operates on a Carnot cycle between the cool exterior of a house and the warm interior. When the exterior temperature is 2°C and the interior is 20°C, what is the coefficient of performance for the heat pump?

17. ∗∗ The heat pump in problem 16 delivers 63 MJ of energy to the interior of the house. How much energy input is required in the form of work?

Section 18-3   Entropy

18. ∗∗∗ Suppose we have two identical samples, one at a temperature of T_2 K and the other at a temperature of $-T_1$ K. The two samples are placed in thermal contact.
(a) Determine the final temperature of the two samples.
(b) What is the change in entropy of the sample that was at temperature T_2?
(c) What is the change in entropy of the sample that was at temperature $-T_1$?
(d) What is the net change in entropy?
(e) Are negative temperatures possible?

Section 18-4   Entropy and the Second Law of Thermodynamics

19. ∗∗ An ideal gas is placed in good thermal contact with a reservoir, which is held at a constant temperature of 200.0 K. The temperature of the gas is initially 210.0 K. The volume of the gas is held constant.
(a) How much thermal energy, per mole of gas, flows from the gas to the reservoir to cool the gas to 200.0 K?
(b) What is the change in the entropy of the reservoir?
(c) What is the change in the entropy of the gas? (Hint: You will need to do an integral.)
(d) What is the overall change in the entropy of the reservoir–gas system?
(e) Is this a reversible process?

20. ∗∗ A 20.0 g piece of copper at an initial temperature of 50.0°C is placed in an insulated container that holds 300.0 g of water at an initial temperature of 20.0°C.
(a) What is the final temperature of the system?
(b) How much thermal energy flows out of the copper?
(c) How much thermal energy flows into the water?
(d) What is the change in the entropy of the copper?
(e) What is the change in the entropy of the water?
(f) What is the total change in the entropy of the copper–water system?

21. ∗∗∗ A heavy-duty restaurant pot is placed on a hot burner on a stove. The temperature of the burner is 500.0°C, and the pot is filled with tap water at a temperature of 10.0°C. The bottom of the pot is steel with a thickness of 1.0 cm and an area of 0.030 m².
(a) At what rate does thermal energy flow through the bottom of the pot?
(b) What is the change in entropy per second of the burner?
(c) What is the change in entropy per second of the water?
(d) What is the net rate of change of entropy?

Section 18-5   The Domain of the Second Law of Thermodynamics

22. ∗∗ Suppose you toss six coins, which can land either heads or tails. If we treat each coin as being unique, there are 64 unique results for the toss. However, in terms of the number of heads and tails possible, there are only 7 possible outcomes for a toss, ranging from all heads and no tails to all tails and no heads.
(a) How many ways can you toss all heads?
(b) How many ways can you toss all tails?
(c) How many ways can you toss only one head?
(d) Only two heads?
(e) Three heads?
(f) Four heads?

(g) Five heads

(h) Plot a histogram showing the number of ways each result can be achieved.

23. ✳ There is a seemingly never-ending reduction in the size of electronic devices. However, at some point, transistors will become small enough that they will no longer be in the thermodynamic limit. How many atoms would be in a cube of silicon that is 20 nm on edge?

24. ✳✳ In a coin toss experiment, there are only two possible outcomes, and the probability of getting a certain outcome is given by the binomial probability distribution. When we toss N coins, the probability of getting n heads is given by

$$P(n, N) \frac{N!}{n!(N-n)!} 0.5^N$$

(a) Use software (such as a spreadsheet) to calculate $P(n, N)$ for tossing 1000 coins for $0 \le n \le 1000$.

(b) What is the probability of getting 500 heads?

(c) What is the probability of getting between 450 and 550 heads?

Section 18-6 Consequences of the Second Law of Thermodynamics

25. ✳✳ A 200.0 g block of copper is cooled to very low temperatures. Calculate the reduction in the entropy of the block in going from

(a) 100.0 K to 99 K;

(b) 10.0 K to 9.0 K;

(c) 2.0 K to 1.0 K;

(d) 0.020 K to 0.010 K.

For simplicity, assume that the specific heat of copper does not change as the block is cooled. (In fact, the specific heat is proportional to T^3.)

26. ✳ Helium atoms interact very weakly with each other—helium only liquefies at 4.2 K at standard pressure.

(a) What is the kinetic energy of a helium atom at 100 K?

(b) At 10 K?

27. ✳✳ A suggested strategy to avoid heat death is to use the work from a heat engine to drive a heat pump to restore the original temperature difference. Is this possible?

COMPREHENSIVE PROBLEMS

28. ✳✳ Human beings are highly ordered arrangements of molecules. Do we violate the second law of thermodynamics?

29. ✳ You have been asked to review a patent application for a new design of a heat engine. The inventor has supplied the following data on the application:

• The engine takes in 4.50 kW of thermal energy at a temperature of 400 K.

• The engine expels 2.00 kW of thermal energy at a temperature of 200 K.

• The engine performs work at the rate of 2.5 kW.

Do you approve or reject the application? Explain.

30. ✳✳✳ A reasonable approximation to the cycle in an internal combustion engine is the Otto cycle, which is depicted in Figure 18-15. At point 1, the piston is at the bottom of its stroke, and the cylinder is filled with exhaust gases. The exhaust valve was opened at point 4. The exhaust valve remains open, and the piston moves up in the cylinder, pushing the exhaust gases out. This is depicted by the blue horizontal line in the figure. Once the piston reaches the top of the stroke, the exhaust valve closes and the intake valve opens. The piston moves down and draws the fuel–air mixture into the cylinder. This is represented by the green horizontal line, and we have returned to point 1. The intake valve closes and the piston moves up, compressing and raising the temperature of the fuel–air mixture along the yellow line, which we approximate as an adiabatic process ending at point 2. The spark plug then ignites the fuel–air mixture, leading to a sharp increase in temperature and pressure as indicated by the red segment going from 2 to 3, which we approximate as an isochoric process. This increase in pressure forces the piston downward along the red line from 3 to 4, and the pressure and temperature both decrease in this adiabatic process. At point 4, the exhaust valve opens, and the pressure and temperature drop along the blue line from 4 to 1 in another isochoric process. Compare the work done in this process with the work that would be done by a Carnot cycle operating between temperatures T_1 and T_3.

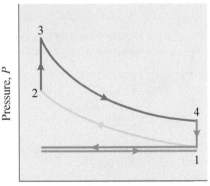

Figure 18-15 Problem 30. The Otto cycle

31. ✳✳✳ We want to investigate how the reading on a sensor in a system becomes more stable as we add more particles to the system. We will consider a very simple one-dimensional system of length L. All the particles in the system travel with the same speed, v m/s. A detector that has the width $\Delta L \ll L$ is placed in the middle of the system; the detector measures some property of the particle. When a particle is in the detector, the detector reads a value Y_0, and when there is no particle in the detector, it reads 0. If we have a single particle in the system, it will travel from one end to the other in time $T = \dfrac{L}{v}$. During that time, it will pass through the detector once and spend time $\Delta T = \dfrac{L}{v}$ in the detector. As a result, the reading on our sensor will look something like that shown in Figure 18-16. Clearly, the sensor reading is not very stable.

Assuming that the particle enters the sensor at time t_0, the average value of the sensor reading is

$$\bar{Y} = \frac{1}{T} \int_0^T Y(t)dt = \frac{1}{T} \int_{t_0}^{t_0 + \frac{\Delta L}{v}} Y_0 dt = \frac{1}{T} \frac{Y_0 \Delta L}{v}$$

$$= \frac{v}{L} \frac{Y_0 \Delta L}{v} = \frac{Y_0 \Delta L}{L}$$

We will characterize the "wobble" in the sensor with a statistical measure called the standard deviation, σ_Y. It is defined as the positive square root of the following expression

$$\sigma_Y^2 = \frac{1}{T}\int_0^T (Y(t) - \bar{Y})^2\, dt$$

Notice that this measure adds up the square of the deviations from the average value and is zero when the sensor reading is constant. For our sensor, with a single particle, we calculate this as follows. We will simplify the integrations by assuming $t_0 = 0$:

$$\sigma_Y^2 = \frac{1}{T}\int_0^T \left(Y(t) - \frac{Y_0 \Delta L}{L}\right)^2 dt$$

$$= \frac{1}{T}\left[\int_0^{\frac{\Delta L}{v}} \left(Y_0 - \frac{Y_0 \Delta L}{L}\right)^2 dt + \int_{\frac{\Delta L}{v}}^0 \left(\frac{Y_0 \Delta L}{L}\right)^2 dt\right]$$

$$= \frac{Y_0^2 \Delta L}{L}\left(1 - \frac{\Delta L}{L}\right)$$

We skipped a few steps at the end. If we examine the ratio of σ_Y to $\bar{Y}$, remembering that $\Delta L \ll L$, we find

$$\frac{\sigma_Y}{\bar{Y}} = \frac{\sqrt{\dfrac{Y_0^2 \Delta L}{L}\left(1 - \dfrac{\Delta L}{L}\right)}}{\dfrac{Y_0 \Delta L}{L}} \cong \sqrt{\frac{L}{\Delta L}} \gg 1$$

Generalize this result to N particles, and show that the ratio gets small in the limit of large N.

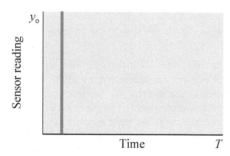

Figure 18-16   Problem 31

32. ✹✹   A block of wood is sliding across a rough surface. The block has a mass of 10 kg and is moving with an initial speed of 10 m/s. It slides across the surface and comes to rest. The kinetic energy of the block has been converted into thermal energy. Assume that the block and the surface are at a constant temperature of 293 K.
   (a) What is the kinetic energy of the block?
   (b) How much thermal energy has been deposited in the block and the surface?
   (c) What is the change in entropy?
   (d) Is this a reversible process?

33. ✹✹   A kettle is put onto the stove and brought to a boil. One litre of liquid water is converted to steam. What is the change in the entropy of the litre of water if the initial temperature of the water was 22°C?

34. ✹✹   A typical shower uses approximately 6 L of water per minute. When the temperature of the water is 40°C and the water ends up in the sewer at 5°C, how much does the entropy of the water change during a 5 min shower?

35. ✹✹   Wood burns at a temperature of approximately 500°C. One cord of seasoned (dry) firewood releases approximately 20×10^9 J of energy when it is burned. All the energy is delivered to the room at approximately 20°C. What is the change in the entropy of the wood? The air in the room? The total change in entropy of the wood–air system?

36. ✹✹   An air conditioner removes thermal energy from the interior of a house at 20°C and releases it to the exterior of the house at 35°C. Assume that the system operates on a Carnot cycle. How much energy is removed from the interior per 1000 J of energy supplied to run the device?

37. ✹✹   The temperature of 10.0 mol of an ideal monatomic gas is increased reversibly from 300.0 to 400.0 K. During the process, the volume is held constant. What is the change in the entropy of the gas?

38. ✹✹   Two constant-temperature reservoirs are kept in thermal contact with a 1.00 kg aluminum bar. One reservoir is at 100.0°C, and the other is at 50.0°C. After 1000.0 J of energy has flowed from the hotter reservoir to the cooler reservoir, determine the change in entropy of
   (a) the hotter reservoir;
   (b) the cooler reservoir;
   (c) the bar.

39. ✹✹   Heat pumps for home heating commonly have a coefficient of performance, CP_H, of about 3.5 when the temperature inside the home is 20°C and the temperature outside is 10°C. Estimate the coefficient of performance for such a heat pump when the outside temperature is −10°C, assuming that the change in the coefficient is determined primarily by the difference between the inside and outside temperatures.

40. ✹   Coal releases approximately 30 MJ/kg of thermal energy when it is burned. A power plant produces approximately 1500 MW h while burning 7.0×10^5 kg of coal. How efficient is the plant?

41. ✹✹✹   Ice on the surface of a lake forms an insulating barrier that slows further freezing. Assume that the temperature profile shown in Figure 18-17 is valid for this question.
   (a) For a given thickness of ice, at what rate will thermal energy go from the water to the air?
   (b) At what rate will the entropy of the air change?
   (c) At what rate will the entropy of the water change?
   (d) At what rate will the thickness of the ice grow?
   (e) Put your answers to (a) and (d) together to show that the thickness of the ice is proportional to the square root of time.

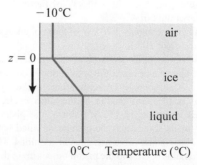

Figure 18-17   Problem 41

42. ✱✱ When dealing four cards from a full deck of cards, determine the probability of dealing
    (a) four cards of the same face value, (e.g., four 3's or four queens);
    (b) four completely different cards (i.e., four cards all with different face values and different suits).
    (Hint: Consider how many different ways (microstates) each outcome (macrostate) can be generated.)

43. ✱✱ An object of mass m falls from a height h and lands on the ground, coming to rest at temperature T. What is the change in the entropy of the system? Discuss the change in entropy in terms of order and disorder.

44. ✱✱ You clean your room and put all of your clothes and books away in an orderly arrangement. The disorder of your room has decreased. Have you violated the second law of thermodynamics? Note: This is *not* a good excuse for refusing to clean your room.

45. ✱✱ As we mentioned in the Making Connections feature called "You Are a Thermodynamic Fluctuation," there was some structure in the early universe. The fluctuations have been amplified by gravity over time to result in the large-scale structure we observe in the universe today: galaxies, etc. Thus, it seems that the universe has evolved from a very nearly uniform state (high disorder) to one of significant structure (low disorder). Does this violate the second law of thermodynamics?

46. ✱✱ The coefficient of performance of a heat pump decreases as the temperature difference increases, so heat pumps become less efficient on very cold days. A clever scientist came up with the idea of using two heat pumps in a home heating system. One transfers heat from the surroundings of the house to a heat reservoir, and runs only when the exterior temperature is high enough for the process to be reasonably efficient. The second transfers heat from the reservoir to the interior of the house, and runs when the house needed to be heated. Discuss the practicality of such an arrangement. What would be a good material to use for the heat reservoir? How large would the reservoir need to be? Clearly state any assumptions.

47. ✱✱✱ Assume that a Carnot heat engine operates between reservoirs T_2 and T_1. The work derived from the heat engine is used to drive an irreversible heat pump that operates between the same two reservoirs. Consider the following three cases:
    (a) The efficiency of the irreversible heat pump is less than the efficiency of the Carnot engine.
    (b) The efficiency of the irreversible heat pump is the same as the efficiency of the Carnot engine.
    (c) The efficiency of the irreversible heat pump is greater than the efficiency of the Carnot engine.
    Prove Carnot's theorem by showing that only one of these cases is possible.

48. ✱✱ Five moles of an ideal monatomic gas is at STP. Calculate the work done and the change in the entropy of the gas when the gas doubles in volume during an
    (a) adiabatic expansion;
    (b) isothermal expansion.

OPEN PROBLEM

49. ✱✱ People have long looked to natural temperature gradients to exploit for driving heat engines. Examples include differences in sea water temperature between the surface and the depths, and differences in the temperature between Earth's surface and core. Estimate the efficiencies of these two examples, and identify any adverse environmental effects that a scheme using these differences may pose.

 See the text online resources at www.physics1e.nelson.com for Open Problems and Data-Rich Problems related to this chapter.

Chapter 1—Introduction to Physics

1. c

3. $\mu = 4$ m, $\sigma = 0.9$ m

5. a

7. c

9. The graph does not state the difference between the measurement points represented by dots and those represented by circles. Since both distance and velocity measurements have associated uncertainties, the corresponding x and y error bars should be included. There is no indication on the graph as to what the two solid and dashed lines represent.

11. The duration of 9 192 631 770 periods of the radiation corresponding to the transition between the two hyperfine levels of the ground state of a caesium 133 atom at rest at 0 K.

13. While the ampere is equal to a current of one coulomb passing a point in one second, it is defined formally as follows: "The ampere is that constant current which, if maintained in two straight parallel conductors of infinite length, of negligible circular cross section, and placed 1 meter apart in vacuum, would produce between these conductors a force equal to 2×10^{-7} newton per meter of length." (e.g. http://www.physics.nist.gov)

15. d

17. d

19. 4

21. (a) 4   (b) 2   (c) 4   (d) 2   (e) 3   (f) 4

23. (a) 2.452×10^{-3} (b) 5.95×10^{-1} (c) 1.2×10^{4}
(d) 4.5×10^{-5}

25. 25.8

27. (a)

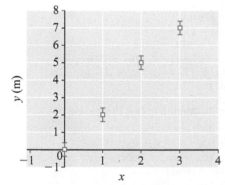

(b)

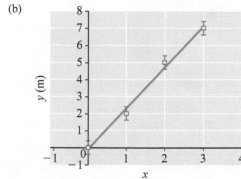

(c) 7.2 m (or ± 3.6 m)

29. mgh

31. V = IR

33. kg/s

35. About 7.5×10^{11} assuming the space between the cells is negligible

37. Roughly 50–150 thousand (varies with assumptions made)

39. 0.0012 pg

41. Years 1990–1999: mean win percentage 0.513 with SDOM of 0.020
Years 2000–2009: mean win percentage 0.497 with SDOM of 0.012
Although the teams in years 1990–999 had a somewhat better win-loss record, the difference is within the range of the SDOMs, so these data do not prove any significant difference between the two sets of teams.

43. (a) 3.9×10^{-8} kg m^{-2} y^{-1} (b) 350

45. (a) $\sqrt{\dfrac{Gh}{c^5}}$ (b) 5.39×10^{-44} s

Chapter 2—Scalars and Vectors

1. (a) $r = 30$ N, $\theta = 315°$
(b) $(21.21, -21.21)$ N
(c) $21.21\hat{i} - 21.21\hat{j}$

3. $(2.53, -7.59)$, $(-2.53, 7.59)$ There are two possible answers due to quadratic equation obtained from the Pythagorean theorem.

5. 5.74 m

7. (a) $(4, -2, 11)$
(b) $(-4, 2, -11)$
(c) $(-24, 11, 22)$
(d) $(-20, 5, 10)$

9. c. Vector magnitudes are invariant under coordinate transformation.

11. Student 2: Consider a vector in each plane directed such that they both form a 45° angle with the line of intersection of the two orthogonal planes. These two vectors are not perpendicular to each other. Another counterexample is vectors in each of the intersecting planes that are parallel to the line of intersection of the orthogonal planes. Such vectors are either parallel or antiparallel to each other, so they are clearly not orthogonal.

13. (a) For $\vec{A}$: $(2, 0, 0)$, $(0, 2, 0)$ and $(0, 0, -2)$ For $\vec{B}$: $(-1, 0, 0)$, $(0, 3, 0)$ and $(0, 0, -2)$
(b) 8
(c) $2\hat{i} + 6\hat{j} + 8\hat{k}$

15. Disagree; if more than one component equals 1, the magnitude will be greater than 1.

17. $\vec{F}_1 = 1.00\hat{i} - 3.00\hat{j} + 1.00\hat{k}$
$\vec{F}_2 = -2.00\hat{i} - 4.00\hat{j} + 3.00\hat{k}$
$\vec{F}_3 = 1.00\hat{i} + 2.00\hat{j} + 0\hat{k}$
$|\vec{F}_1| = \sqrt{11.00}$ N, $|\vec{F}_2| = \sqrt{29.00}$ N, $|\vec{F}_3| = \sqrt{5.00}$ N
$\vec{F}_R = -5.00\hat{j} + 4.00\hat{k}$, $|\vec{F}_R| = \sqrt{41.00}$ N

19. Disagree. The magnitude of the resultant is not equal to the sum of the radius vectors of individual vectors. Similarly, the resultant angle is not equal to the sum of the individual vector angles.

21. Mathematically, head-to-tail rule is equivalent to addition of vectors in Cartesian form. The associative property of vector addition, that is, $\vec{V}_1 + \vec{V}_2 + \vec{V}_3 = (\vec{V}_1 + \vec{V}_2) + \vec{V}_3 = \vec{V}_1 + (\vec{V}_2 + \vec{V}_3)$, proves the head-to-tail rule for more than two vectors.

23. The scalar product of two vectors is defined as the product of the magnitudes of the two vectors and the cosine of the angle between them. Because the definition involves the product of three real numbers, and the product of real numbers is commutative, the scalar product of two vectors is commutative.

   The vector product of two vectors is not commutative because the direction of the cross product is defined using the right-hand rule, and the direction is opposite if the two vectors are multiplied in opposite order.

25. $4.33\hat{i} + 3.83\hat{j} + 2.50\hat{k}$ N

27. $2\hat{i} + 2\hat{j} + 4\hat{k}$. It is impractical to add the vectors by hand, as it is difficult to accurately draw the vectors head-to-tail in three dimensions.

29. 21.9 m

31. (a) $\vec{A}$: $19.3\hat{i} - 5.18\hat{j}$, $\vec{B}$: $12.3\hat{i} + 8.60\hat{j}$, $\vec{C}$: $-14.3\hat{i} + 20.5\hat{j}$
   (b) $17.3\hat{i} + 23.9\hat{j}$, $(29.5, 54.1°)$
   (c) $(201.1, 51.3°)$

33. (a) $(3\hat{i} - 5\hat{j} + 6\hat{k})/\sqrt{70}$
   (b) Any unit vector $\hat{u}$ that satisfies the condition $\hat{u} \cdot (3\hat{i} - 5\hat{j} + 6\hat{k}) = 0$ is perpendicular to the given vector. This question has infinite number of answers, such as $(3\hat{i} + 3\hat{j} + \hat{k})/\sqrt{19}$
   (c) There are infinite number of unit vectors parallel to the plane. Any unit vector $\hat{u} = a\hat{i} + b\hat{j} + c\hat{k}$ satisfying the relation $3a + 2b - 4c = 0$ is a vector parallel to the given plane.
   (d) $\dfrac{1}{\sqrt{29}}(3\hat{i} + 2\hat{j} - 4\hat{k})$

35. (a) $F_x = -\sqrt{2}$, $F_y = 3$, $F_z = 5$
   (b) $3\hat{i} - 5\hat{j} + F_x\hat{i} - 2F_y\hat{j} + F_z\hat{k} - 3\hat{k} = 0$.

37. The two roots of the quadractic equation for the magnitude give unit vectors that are antiparallel: $\dfrac{2\sqrt{13}}{13}\hat{i} + \dfrac{3\sqrt{13}}{13}\hat{j}$ and $-\dfrac{2\sqrt{13}}{13}\hat{i} - \dfrac{3\sqrt{13}}{13}\hat{j}$.

39. 46.98

41. $-\sqrt{5}$

43. $6\hat{i} - 22\hat{j} - 14\hat{k}$

45. $|\vec{A} \times \vec{B}| = (10)(5)\sin(20°) = 17$ pointing in $+y$ or $-y$ direction depending on the order of vector product.

47. (a) $F_x = 2\sqrt{2}$; $F_y = 14$ (b) $F_x = -\dfrac{20}{\sqrt{3}}$; $F_z = -\dfrac{80}{\sqrt{3}}$

49. (a) $\begin{pmatrix} \cos(60°) & -\sin(60°) & 4 \\ \sin(60°) & \cos(60°) & 5 \\ 0 & 0 & 1 \end{pmatrix}\begin{pmatrix} x' \\ y' \\ 1 \end{pmatrix} = \begin{pmatrix} x \\ y \\ 1 \end{pmatrix}$
   (b) $x = x'\cos(60°) - y'\sin(60°) + 4$;
   $y = x'\sin(60°) + y'\cos(60°) + 5$

51. 1. This fact can be proved by definition of the vector product—see Equation 2-26 and Figure 2-18.
   2. This fact can also be proved using the algebraic definition of cross product and the fact that the scalar product of two orthogonal vectors is zero.

53. $\Rightarrow A^2B^2 - (A_xB_x + A_yB_y)^2$
   $= A^2B^2(1 - \cos^2\theta) = A^2B^2 \sin^2\theta$

55. 0

57. Since the magnitudes of vectors $\vec{r}$ and $\vec{R}$ are both equal to the radius of the circle, $\vec{A} \cdot \vec{B} = 0$. Thus, the vectors $\vec{A}$ and $\vec{B}$ are perpendicular to each other, and so the triangle containing these vectors as its sides must be a right triangle.

59. $\Rightarrow C^2 = A^2 + B^2 - 2\vec{A} \cdot \vec{B}$
   $\Rightarrow C^2 = A^2 + B^2 - 2AB\cos\theta$

61. (a) $v = \sqrt{v_x^2 + (v_{0y} - gt)^2}$
   (b) Assuming that v_{0x} is positive, a graph of the magnitude of the velocity will have a V-shape similar to the example shown below:
   For $v_{0x} = 10$ m/s and $v_{0y} = 30$ m/s

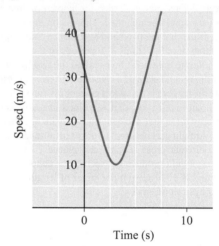

   (c) $t = \dfrac{v_{0y}}{g}$
   (d) The motion of a projectile, assuming that air resistance is negligible.

63. (a) 178 m/s²
   (b) $\gamma = 141.8°$
   (c) $\vec{a} = 110\hat{j} - 140\hat{k}$
   (d) $110\hat{j}$, $-140\hat{k}$, $110\hat{j} - 140\hat{k}$

65. $\cos(\beta - \alpha) = \cos\alpha\cos\beta + \sin\alpha\sin\beta$

67. 109.5°

Chapter 3—Motion in One Dimension

1. (a) Both objects have same average velocity.
   (b) B at $t = 30$ s; A at $t = 90$ s
   (c) Yes, when the slopes of their position-time graphs are equal, around $t = 55$ s.

3. (a) No
   (b) Yes, at $t = 0$.
   (c) A. Maximum accelerations: A, about 100 m/s²; B, 25 m/s²

5. (a) B
   (b) B
   (c) Same magnitude for both (opposite signs)
   (d) A
   (e) Somewhere between $t = 50$ and 100 s the objects have the same displacement, and they have the same acceleration around $t = 130$ s

7. (a) Yes, at the maxima and minima of the curve
   (b) Highest at $t = 200$ s, lowest at $t = 30$ s and $t = 140$ s

9. b

11. c

13. b

15. b

17. d

19. Yes, for example, an object moving at a constant acceleration has equal average and instantaneous accelerations.

21. Yes, for example, the average velocity is equal to the instantaneous velocity for an object moving with constant velocity.

23. a

25. b, if we consider only the present moment; d, if the acceleration continues after the direction of motion reverses, since the speed will then start to increase.

27. Only momentarily. If you are moving one direction but have acceleration in the opposite direction, you will have zero speed for just the instant when you come to a stop before starting to move in the direction of the acceleration.

29. d

31. d

33. a

35. c

37. (a) 2400 m
(b) −200 m

39. $d = 230$ cm, $\vec{d} = 80$ cm [down]

41. 4.1 m/s, 1 m [down], 0.1 m/s [down]

43. $v_{av} = 1.3$ m/s, $\vec{v}_{av} = 1.3$ m/s [down]

45. 6.2 h

47. 29.8 km/s

49. 1.04 km

51. (a) 5 km [forward]
(b) 15 km
(c) (i) 40 km/h [forward], (ii) 120 km/h

53. (a) 105 m, 70 m, 40 m
(b) −10 m/s²

55. (a) 8.3 min
(b) 8.6 years
(c) about 100 000 years

57. $v = 800$ m/s, $\vec{v} = -800$ m/s

59. 12.6 m/s

61. (a) 14.2 m/s
(b) 22.3 m/s

63. (a) 10.70 m/s² (b) 1713 km

65. We have (for simplicity we will leave out arrows to represent vectors)

$$v_{12} = v_{012} + a_{12}t \qquad (1)$$

$$\Delta x_{12} = v_{012}t + \frac{1}{2}a_{12}t^2 \qquad (2)$$

$$(1) \Rightarrow t = \frac{v_{12} - v_{012}}{a_{12}}$$

Substituting the expression for t into (2) gives

$$\Delta x_{12} = v_{012}\left(\frac{v_{12} - v_{012}}{a_{12}}\right) + \frac{1}{2}a_{12}\left(\frac{v_{12} - v_{012}}{a_{12}}\right)^2$$

Simplifying this we get

$$v_{12}^2 - v_{012}^2 = 2a_{12}\Delta x_{12}$$

67. 9.8 m/s²

69. 1.95 s

71. $\vec{a}_{av} = 9.8$ m/s² [down], $v_{av} = 16$ m/s, $\vec{v}_{av} = 0$

73. (a) 11.29 m/s²
(b) 381.3 m
(c) 20.0 s

75. (a) 1.4 s (b) 7.4 m/s (c) 8.4 m/s, 14.5 m

77. 38 m/s, 84 ms

79. 180 m/s

81. (a) 2.818 km/s (b) 3816 km (c) 480 km

83. 117 s

85. $v_{av} = 17.2$ m/s, $\vec{a}_{av} = 9.8$ m/s² [down], $d = 120$ m, $\vec{d} = $ zero

87. (a) 12 m/s (b) 12 m/s

89. (a) −45 km/h (b) 45 km/h

91. 33.1 m/s

93. (a) 1.22 s (b) 1.83 m, no (c) 1.22 s

95. 77.6 m

97. (a) $\dfrac{n\pi}{2}$; $n = 1, 3, 5, \ldots\ldots$

(b) (i)

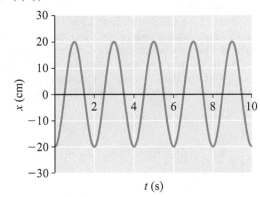

(ii)

(iii)

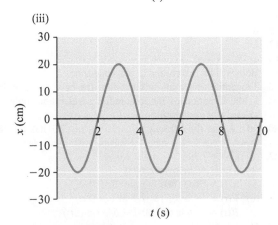

(c) $v = -x_0\omega\sin(\omega t + \phi)$
(d) $a = -x_0\omega^2\cos(\omega t + \phi)$
(e) $a = -\omega^2 x$

99. (a) 5.1 min (b) 5.71 km toward the dock, 14.29 km
101. 51.1 m, 31.6 m/s
103. 66.3 m/s
105. 0.9 s
107. 37 m/s
109. (a) Less than 0.5 km/s per kilometre of altitude at 120 km; about 2.9 km/s per kilometre of altitude at 92 km
(b) Since the meteor fragment is slowing, the force of atmospheric friction must be somewhat greater than the force of Earth's gravity at an altitude of 120 km, and much greater at an altitude of 90 km.
(c) The density of the atmosphere increases substantially as your descend in altitude from 120 to 85 km.
111. (a) About 30 m/s², 20 m/s², and 25 m/s²
(b) About 13 m/s²
(c) about 15 m/s², 12.5 m/s², and 12.3 m/s²
113. Approximate shape shown below. Since this acceleration graph is based on values estimated from a small velocity graph, the resulting graph does not exactly match Figure 3-41 and does not show the details of the sudden changes in acceleration between the stages of the launch.

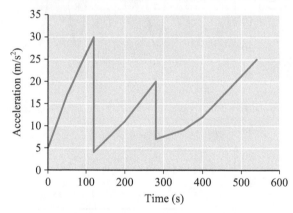

Chapter 4—Motion in Two and Three Dimensions

1. b
3. c
5. b
7. b
9. d
11. d
13. d
15. d
17. a
19. a
21. Both will hit the ground at the same time.
23. Tangential acceleration: tangent to the bowl and pointing in the direction of motion, radial acceleration: pointing towards the center of the bowl
25. (a) 1.4 m/s (b) 1.9 m/s (c) 1.8 m/s (d) 2.0 m/s (e) No, these average speeds apply for only part of the time elapsed.
27. (a) $v_x = -7.99r$, $v_y = -0.23r$
(b) $8.0r$ [1.6° below $-x$ axis]
$\vec{x}(t) = r\cos(8t)\hat{i} + r\sin(8t)\hat{j}$
(c) $\vec{v}(t) = -8r\sin(8t)\hat{i} + 8r\cos(8t)\hat{j}$
$\vec{a}(t) = -64r\cos(8t)\hat{i} - 64r\sin(8t)\hat{j}$
29. Hyperbola with x- and y-axes as asymptotes.
31. (a) $166.9\,\text{m}\,\hat{i} + 121\,\text{m}\,\hat{j} + 34.6\,\text{m}\,\hat{k}$, assuming that the ramp is in the xz plane, and that the car is moving in the y-direction after turning.

(b) $9.43\,\text{m/s}\,\hat{i} + 6.84\,\text{m/s}\,\hat{j} + 1.95\,\text{m}\,\hat{k}$
(c) $0.40\,\text{m/s}^2\hat{j}$
33. 5.9 m/s
35. 17 m/s
37. (a) 2 h (b) 1285 km/h
39. 417 m
41. $\vec{a} = 3.63\,\text{m/s}^2$ [43.2° with respect to radial direction]
43. About 21 krad/s
45. $\vec{a} = 16\,\text{m/s}^2$ 13.1° with respect to radial direction
47. 1.1 s, 0.08 turns
47. 0.18 s
49. The acrobat will land in the car; 7.9 m
51. 0.83 m/s
53. 5.0 m
55. 11 m/s
57. For the given angle, there is no speed that will get the stone to the required height and horizontal displacement at the same time.
59. 29°
61. 16.8 m/s, 22.2 m/s
63. (a) 796 m
(b) 579 m/s, 1430 m
(c) Right before hitting the ground
65. $x = 52.0t$, $y = 43.6t$, $z = 122t - 4.9t^2$
67. (a) 26.6 m (b) 125.2 m (c) 36.1 m/s
69. (a) 954 km/h
(b) 13 h forward when both cities are on standard time
71. About 1200 km/h or 1600 km/h depending on whether you have to adjust your watch forward or back.
73. (a) $2\hat{i} + 3\hat{j}$
(b) 0.85 m/s
(c) $(0.2\hat{i} + 0.2\hat{j} + 1.2\hat{k})\,\text{m/s}$
(d) $(-0.01\hat{i} + -0.01\hat{j} + 0.8\hat{k})\,\text{m/s}^2$
(e) $(0.21\hat{i} + 0.21\hat{j} + 2.42\hat{k})\,\text{m}$
(f) The particle oscillates with a displacement magnitude of 0.5 m on a plane 45° between the x- and y-axes while accelerating uniformly the z direction.

75. $\vec{x}_{rel} = -\dfrac{17r}{6}\hat{i} + \sqrt{\dfrac{5}{6}}r\hat{j}$,
$\vec{v}_{rel} = -\sqrt{30}\pi r\hat{i} + \pi r\hat{j}$, $\vec{a}_{rel} = -6\pi^2 r\hat{i} - 36\pi^2 r\sqrt{\dfrac{5}{6}}\hat{j}$

77. $h = 12.6$ km, $R = 8.13$ km, $v = 301$ m/s, $a = 20.0\,\text{m/s}^2$
79. (a) 3.0 m
(b) 15.2 m/s
(c) The acorn lands at an angle of 72° with respect to the horizontal plane. The projection of the acorn's velocity makes an angle of 7.7° with the road.
81. A graphing calculator was used to find that there is no solution. Any angle between 36.9° and 60.2° will result in the cannon ball clearing Sir Burnalot.
83. (a) $(7.97\hat{i} + 0.67\hat{j})\,\text{m/s}$
(b) $\vec{v}_{avg} = (-4.66\hat{i} - 0.13\hat{j})\,\text{m/s}$,
$\vec{a}_{avg} = (7.5 \times 10^{-3}\hat{i} + 1.83\hat{j})\,\text{m/s}^2$
85. 31.2° north of east
87. 85.2° up from the ground
89. (a) 0.36 km/s
(b) 0.42 km/s, 0.14 km/s, 2.0 km/s, 2.5 km/s
(c) The small increase in the slope of the graph for the next 250 s indicates some radial acceleration; after that the curve is almost straight, so there is little or no radial acceleration after 1750 s.
(d) Since the graph shows altitude versus time, the slope does not reflect any tangential component of the velocity.

91. 23.4 m/s; yes, the negative root of the quadratic is spurious since it corresponds to a negative speed.

Chapter 5—Forces and Motion

1. a
3. b
5. b
7. b
9. 16 kg mass accelerates down while 4 kg mass accelerates up
11. d
13. b
15. d
17. c
19. c
21. c
23. b
25. (a) 841 N (b) 13.0 kN (c) 600 N (d) 600 N (e) 19.0 kN
27. a
29. c
31. a
33. a
35. b
37. (a) $6N$: $\vec{F} = 5.14\hat{i} + 3.09\hat{j}$

$8N$: $\vec{F} = -4.93\hat{i} + 6.30\hat{j}$

$5N$: $\vec{F} = -1.87\hat{i} - 4.64\hat{j}$

(b) 5 N $\angle 109°$
39. (a) $(-\hat{i} + 3\hat{j} - 3.5\hat{k})$ N

(b) $(6.2\hat{i} + 1.1\hat{j})$ N

(c) $(-5\hat{i} + 16\hat{j} + 6\hat{k})$ N

(d) $(22.6\hat{i} - 0.2\hat{j} + 25.9\hat{k})$ N
41. 20.1 m/s
43. 0.54 m
45. (a) 21.5 kN [radially inward]

(b) 1.05 kN [radially outward]

(c) 19.3 kN [radially inward]
47. (a) along the slope, $v_f = v_i + gt\sin\theta$

(b) along the slope, $v_f = v_i + gt(\sin\theta - \mu\cos\theta)$

(c) toward the centre of the horizontal circular path of the pendulum bob $v = 2\pi r/t$, where r is the radius of the path and t is the time taken for one circuit $t = 2\pi\sqrt{\dfrac{L\cos\theta}{g}}$, where L is the length of the pendulum string and θ is the angle of suspension measured from the z-axis

(d)

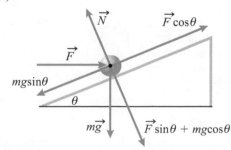

(e) parallel to the plane,

$$v_f = v_i + \frac{Ft}{m}\cos\theta - \mu t\left(g - \frac{F}{m}\sin\theta\right)$$

(f) (i) $\vec{a} = \dfrac{T}{m} + g$ [down]   (ii) $\vec{a} = \dfrac{T}{m} - g$ [up]

(g) radial acceleration directed towards the centre, $v = \sqrt{rg\tan\theta}$

(h) $v = \sqrt{rg\dfrac{(\sin\theta + \mu_s\cos\theta)}{\cos\theta - \mu_s\sin\theta}}$

49. (a) 820 N [up]

(b) 23.2 kN [up]

(c) 1.6 kN [down]

(d) 22.6 kN [down]
51. 35 N [down]
53. 2°
55. 0.71
57. (a) $T = \dfrac{M_1 M_2 g}{M_1 + M_2}$, $M_2 > M_1$ (b) no change   (c) 16.8 N
59. (a) $g/2$ (b) 0
61. 8.2 cm
63. 272 N/m
65. (a) 3.6 m/s   (b) 0
67. (a) 62.3 km/h   (b) 300 N [radially inward]
69. 0.51
71. 4.5×10^2 N
73. 4.7 m/s
75. 2.7°
77. 12.4 m/s
79. (a) m_1: 2.45 m/s, m_2: -2.45 m/s

(b) 127 N
81. $\sqrt{\dfrac{3(9.81\,\text{m/s}^2)^2}{4} + \dfrac{(2.3\,\text{m/s})^4}{R^2}}$ when the marble is at a height of $R/2$
83. $(M + m)g\tan\theta$
85. $\dfrac{(M + m)(g + a\sin(\theta))}{\cos(\theta) - \mu\sin(\theta)}(\mu\cos(\theta) + \sin(\theta)) + Ma\cos(\theta)$
87. $3g(M + 4m)$
89. (a) 59.9 N

(b) $\vec{F}_A = 12.8$ N [right]

$\vec{F}_B = 37.6$ N [right]

$\vec{F}_C = 9.41$ N$\hat{i}$ + 3 N$\hat{j}$

The arm that supports the pulley exerts a backward force on Block B.
91. (a) $Mg\cos\theta$

(b) $\dfrac{Mg}{\cos\theta}$

(c) $\dfrac{mg}{\cos(\theta) - \mu_s\sin(\theta)}$

(d) Downward parallel to the surface, towards the center, towards the center, resolved components are used to determine the net force
93. 16.6 kN
95. $9.81\,\text{m/s}^2 \times \dfrac{5.76\dfrac{M}{m} + 4.53}{\left(2\dfrac{M}{m} + 0.72\right)^2}$

97. $F = (M + m) g \dfrac{\sqrt{R^2 - h^2}}{h}$

99. ($a_1 = 7.28\,\text{m/s}^2$, $a_2 = 3.64\,\text{m/s}^2$, $v_f = 16.6\,\text{m/s}$)

101. (a) $1.5 \times 10^2\,\text{N}$
   (b) $6.8 \times 10^2\,\text{N}$
   (c) $1.5 \times 10^2\,\text{N}$

103. 0.40

Chapter 6—Energy

1. Yes. The work-kinetic energy theorem was derived using equation 3-20. Question 106 in chapter 3 showed that equation 3-20 is valid for nonconstant forces.

3. T since work is being done on the spring to stretch it.

5. Negative, since the spring applies the force in a direction opposite to the direction of motion.

7. Yes, if the displacement is in a direction opposite to the applied force

9. It depends on the *net* work done on the car by yourself and other forces. For example, if you push a moving car on level ground, it will speed up. If you push a moving car up a hill, it may slow down. Although you did positive work, gravity did even more negative work.

11. zero, zero

13. zero

15. The weight began with zero speed, and ended with zero speed, so:
$W = K_f - K_i = 0 - 0 = 0$

17. d

19. b; The projectile still has some kinetic energy at the apex due to its horizontal velocity.

21. b

23. Since the child is moving at constant speed, we must have
$\mu mg \cos \theta = mg \sin \theta$
$\Rightarrow \mu = \tan \theta$
Since the child is moving at constant speed, the work done by gravity must be equal and opposite to the work done by friction.

25. a and c

27. e

29. c and d.

31. c

33. e

35. (a) $W_g = \vec{F} \cdot \Delta \vec{y} = -mgh$
   $= -(0.400\,\text{kg})(9.81\,\text{m/s}^2)(0.45\,\text{m}) = -1.77\,\text{J}$
   (b) $W_{hand} = \vec{F} \cdot \Delta \vec{y}$
   $= mgh = (0.400\,\text{kg})(9.81\,\text{m/s}^2)(0.45\,\text{m}) = 1.77\,\text{J}$
   (c) $W_{total} = W_g + W_{hand} = 0$

37. $-18\,\text{J}$

39. (a) As the ball drops 50 cm, the total work done by you and gravity is enough to bring the kinetic energy to zero.
   $W_{Total} = \dfrac{1}{2} mv_f^2 - \dfrac{1}{2} mv_i^2 = 0 - \dfrac{1}{2} (8)(3)^2 = -36\,\text{J}$
   $W_g = \vec{F}_g \cdot \Delta \vec{y} = -(8.0\,\text{kg} \times 9.81\,\text{m/s}^2) (-0.50\,\text{m})$
   $= 39.2\,\text{J}$
   $W_{you} = W_{total} - W_g = -36\,\text{J} - 39.2\,\text{J} = -75.2\,\text{J}$
   (b) As calculated in (a), the work done by gravity is 39.2 J.

41. (a) Using conservation of mechanical energy:
   $v_1 = \sqrt{2gh} = \sqrt{(2)(9.81)(7 - 1.5)} = 10.4\,\text{m/s}$
   $W = U_f + K_f = mgh + 0 = (20.0\,\text{kg})(9.81\,\text{m/s}^2)(700\,\text{m})$
   $= 1.37\,\text{kJ}$

   (b) $W_g = \vec{F}_g \cdot \Delta \vec{y}$
   $= -mgh = -(20.0\,\text{kg})(9.81\,\text{m/s}^2)(1.5\,\text{m}) = -294\,\text{J}$

43. (a) $v = \sqrt{4.91\,\text{m}^2/\text{s}^2 - 9.38 \times 10^{-3}\,\text{m}^2/\text{kg} \times k}$
   (b) $h = (M - m) \dfrac{g}{k}$
   (c) $v = 2.0\,\text{m/s}$

45. (a) Force balance gives $F = f$, where $f = \mu mg$ is the force of friction
   $\Rightarrow \mu mg = F$
   $\Rightarrow \mu = \dfrac{F}{mg}$
   (b) $W = Fd$
   (c) If the same force is applied over the same distance, the work is the same as found in part (b). The work creates gravitational potential energy instead of heat.
   (d) Again, if the force and distance are the same, the work is the same. In this case, the work creates a mix of gravitational potential energy and kinetic energy.
   (e) The work done is the same. The work creates a mix of gravitational potential energy, kinetic energy, and heat.

47. $W = -176\,\text{J}$, $F = 180\,\text{N}$

49. $\mu_k = 0.14$

51. (a) $v = 24\,\text{m/s}$
   (b) $W_f = -80.5\,\text{J}$, $W_g = -43.2\,\text{J}$, $v = 18\,\text{m/s}$

53. mgh

55. 1.09 m

57. (a) 1293 kJ
   (b) 1.9 kJ

59. 13.7 cm

61.

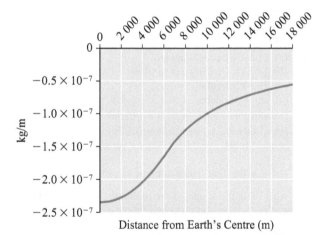

Distance from Earth's Centre (m)

63. (a) $\sqrt{\dfrac{2KQq}{m}\left(\dfrac{1}{R} - \dfrac{1}{R + \Delta r} \right)}$
   (b) 107 m/s

65. 800 kN

67. 4.1 kN

69. (a) 0.54°
   (b) 1.17 m/s

71. $\dfrac{5}{2} r$

73. $h = \dfrac{2}{3} r$, $N = 3mg$

75. 45 cm

77. (a) $T = mg \cos(23°)\left(\dfrac{2L}{r} + 1\right)$

    (b) $mgL \cos(23°)$

    (c) zero

79. (a) $-\dfrac{mg}{k}$

    (b) $v = g\sqrt{\dfrac{m}{k}}$

81. $a_{max} = 2.26\,\text{m/s}^2$, No energy lost due to friction

83. (a) 0.94 m/s

    (b) 0.74 m/s²

85. (a) $\dfrac{A}{2}\left(\dfrac{1}{r_1^2} - \dfrac{1}{r_2^2}\right)$

    (b) $\dfrac{A}{2}\left(\dfrac{1}{r_1^2} - \dfrac{1}{r_2^2}\right)$

    (c) $-\dfrac{A}{2}\left(\dfrac{1}{r_1^2} - \dfrac{1}{r_2^2}\right)$

    (d) attractive

    (e) $8.14\ \text{m}^{-1}\sqrt{A}$

87. $-\dfrac{14r}{(r^2 + 6)^2}\hat{r},\ -\dfrac{14r}{m(r^2 + 6)^2}\hat{r}$

89. 2.3 m

91. Force points in the direction of steepest descent

93. (a) $\theta = 43.3°$

    (b) $\theta = 33.5°$

95. (a) $1.87 \times 10^8\,\text{W}$

    (b) 6.46 m/s²

    (c) $3.22 \times 10^{12}\,\text{J}$

Chapter 7—Linear Momentum, Collisions, and Systems of Particles

1. d

3. c

5. d

7. 1000 N [away from the wall]

9. b

11. The centre of mass moves down at an acceleration of 9.81 m/s²

13. Yes, if air resistance is negligible

15. 44.1 N

17. d

19. c

21. a

23. a

25. b

27. c

29. b

31. T

33. a

35. c

37. a

39. 1.2 kg m/s

41. (a) 0.31 m/s [forward]

    (b) 3.2 kN [forward]

43. (a) 3.5 N [forward]   (b) 151 N [forward]

45. (a) 16 ms

(b)

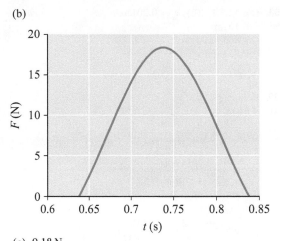

(c) 9.18 N

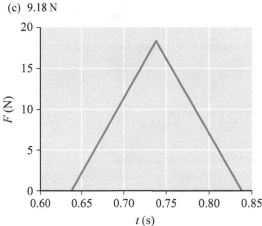

    (d) 18.4 N

47. $\dfrac{1}{15}R$

49. $(0.5\,L,\ 0.3\,L)$

51. 1.1 m

53. 0.99 m away from the child

55. 4.1 m/s

57. 8.7 km/s

59. $1 - \dfrac{\cos^2\theta}{\cos^2\theta'}$

61. $0.66\,v$

63. 9.0 m/s

65. 5:1

67. $1{:}0.66,\ \dfrac{1}{2}v$

69. 660.5 m/s, 792.5 N

71. (a) 7.9 cm/s²   (b) 1.15 m/s   (c) 9.6 cm/s²

73. 7×10^7 km, 5×10^7 km

75. 39.6 cm

77. 0.016 m/s, 9.8 m/s² [down]

79. (a) 97 cm   (b) zero

81. 5.9 m/s

83. 26 cm

85. 35.4 m/s

87. 10.66 m, twice the distance between her and the centre of mass

89. $\dfrac{3}{2}v$

91. $A\sqrt{k\left(1 + \dfrac{M}{m}\right)}$

93. (a) 52.7 J   (b) $v_c = 0.201$ m/s
95. (a) 15 μm
    (b) 803 nm/s
    (c) 803 nm/s
97. $\dfrac{11}{18}L$
99. 0.96 m/s
101. 4.61 m
103. (a) $L + v_1\sqrt{\dfrac{m_1 m_2}{k(m_1 + m_2)}}$ (b) 51.7 cm
105. (a) $(v + u)\sqrt{\dfrac{mM}{k(m + M)}}$
    (b) $\dfrac{mv - Mu}{m + M}$
    (c) $v' = \dfrac{m - M}{m + M}v - \dfrac{2M}{m + M}u,$
       $u' = -\dfrac{M - m}{m + M}u + \dfrac{2m}{m + M}v$
    (d) At max compression, mass M has the same speed as mass m.
107. $v' + \sqrt{\dfrac{k}{2m}}x$
109. (a) 10.0 kg   (b) zero
111. (a) 0.48 m/s   (b) 0.77 m

Chapter 8—Rotational Dynamics

1. Spherical Shell, its moment of inertia is larger
3. All three quantities for the point on the rim are higher since the radius is larger.
5. The smaller disk has the greater angular speed and radial acceleration on the circumference. The two disks have the same speed on their circumferences.
7. Increase
9. Shorter rod
11. The equation is still valid if other units (such as degrees) are used for all of the angular quantities.
13. (a) No.   (b) The dot has radial acceleration but no tangential acceleration.
15. The total kinetic energy of the figure skater never changes. Therefore she does zero work bringing her arms out, and zero work bringing her arms in. Because energy is conserved, her angular speed in the end is the same as her initial angular speed.
17. A thin cylindrical shell can be considered as a number of rings stacked on top of one another. Now, since moment of inertia depends on how mass is distributed about the axis of rotation, the number of rings in the stack will not change the moment of inertia.
19. zero
21. Tilting the axis of rotation changes the direction of the angular velocity vector, and hence changes the angular momentum vector.
23. Tension is the same when there is no acceleration.
25. Yes, at the moment when the angular acceleration reverses the direction of rotation of the wheel
27. Yes, as long as they are not pushing the car straight toward the tree
29. No, torque is defined about a pivot. Force is a vector quantity, so its magnitude is independent of the frame of reference.
31. 393 m/s

33. 46.4 cm
35. 1.45×10^{-4} rad/s
37. (a) 69 rad/s
    (b) 0.43 km/s²
    (c) 2.8 m/s²
    (d) 2.8 m/s²
39. 8.4 rad/s
41. (a) 0.41 rad   (b) 1.0 rad/s
43. (a) 46.3 m/s²   (b) 250 m/s²
45. $(-10.8\hat{i} + 5.4\hat{j} + 5.8\hat{k})$ Nm
47. $(28\hat{i} + 105\hat{j} + 67\hat{k})$ Nm
49. 806 N·m
51. $(-28\hat{i} + 13\hat{j} - 19\hat{k})$ Nm, (36 Nm, 121.6°, 150.1°)
53. 0.48 kg m²
55. 0.23 kg m²
57. $\dfrac{2MR^2}{5}$
59. $\alpha = \dfrac{m_1 g \sin\theta}{r(m_1 + m_2)}$
61. 5.07×10^5 kg m²/s, 5.07×10^5 kg m²/s
63. 25°
65. (a) 0.27 rad/s   (b) -2.6×10^{-4} J
67. $-28\hat{i} + 39\hat{j} - 20\hat{k}$
69. (a) 12.1 m/s   (b) 38.8 kJ, 120 km/h
71. 0.32 rad/s
73. 1.94 s, 104.2 rad
75. 32.2 kJ
77. (a) $I = mL^2$
    (b) Yes.
    (c) All of the mass is the exact same distance from the axis of rotation.
    (d) $2\,mL^2$
79. $\dfrac{mL^2}{12}$; $\dfrac{M}{12}(L^2 + T^2)$ if T not negligible
81. (a) 1.2 m/s
    (b) 1.2 m/s
    (c) 1.2 m/s
    (d) 2.3 rad/s
    (e) 0.33
83. (a) (i) 6.9 m/s,   (ii) 9.8 m/s   (b) 8.8 m/s
85. 2.00 m/s
87. 3.6 N
89. $\left[\dfrac{2d(m_2 g - m_1 g \sin\theta - \mu m_1 g \cos\theta)}{m_1 + m_2 + I/r^2}\right]^{1/2}$
91. (a) 11 rad/s   (b) 1.7×10^2 J
93. (a) 6.6 rev/min   (b) 6.6 rev/min
95. The smallest angular speed ever reached is 33 rad/s, so the flywheel will never stop.
97. (a) 49 rad/s   (b) 6.3 m/s²
99. (a) Positive
    (b) increase
    (c) 1.3 rad/s
    (d) decrease
    (e) −47 kJ

Chapter 9—Rolling Motion

1. Rolling Ring
3. (a) solid cylinder   (b) The sphere will have a higher angular speed
5. To prevent locking up of wheels that may cause skidding

7. Static
9. Sphere
11. $2mr^2$
13. Both disks will reach the bottom of the hill at the same time.
15. Both shells will reach the bottom of the hill at the same time.
17. Not possible.
19. $m_r = \dfrac{40}{7} m_s$
21. Friction and air resistance
23. Opposite to the direction of motion.
25. (a) 30 m/s, -608 m/s$^2\hat{i}$ (b) 21.2 m/s, 608 m/s^2; 0, 608 m/s^2
27. (a) 18 m/s (b) no
29. (a) -5.68 m/s^2 [in the x direction] (b) 117 m/s^2, $-1.7°$ wrt the vertical.
31. The apparent acceleration is due to change in frame of reference and not a real effect.
33. $\cos^{-1}\left(\dfrac{r}{R}\right)$
35. $\dfrac{g}{2}$
37. F_1 and F_3 right; F_2 and F_4 left

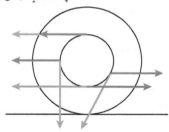

39. 2.5 N, will remain the same
41. (a) 3.8 N (b) 3.8 N
43. $\dfrac{4F}{4M - 3m}$
45. (a) No, since $f > \mu_s N$ (b) 60.8 rad/s^2 (c) 1.37 J
47. 24.8°
49. 6.45 N
51. (a) $K_{rot} = 13$ J, $K_{trans} = 27$ J (b) $K_{rot} = 20$ J, $K_{trans} = 20$ J
53. (a) 5 m
(b) Yes, the ball spins (not rolls) as it moves up the incline.
(c) 78.0 rad/s
55. (a) $\dfrac{5}{7} h$, energy is conserved
(b) zero
(c) $\dfrac{2}{7} mgh$
(d) A ring of the same mass has a higher moment of inertia, so more of the total energy will go towards rotational motion as it moves down. Less translational kinetic energy means that the ring will move up to a smaller height on the right hand side as compared to the solid sphere.
57. (a) 30.4 kN (b) 41.4 kN (c) 1.1×10^2 rad/s
59. (a) $\dfrac{2M_c g \sin \theta_2 - M_s g \sin \theta_1}{\left(6M_c + \dfrac{4I_2}{r_2^2} + 4m + \dfrac{4I_1}{r_1^2} + \dfrac{7}{5}M_s\right)}$
(b) $\dfrac{2M_c gv_s \sin \theta_2 - M_s gv_s \sin \theta}{\left(\dfrac{7}{5}M_s + 4\dfrac{I_1}{r_1^2} + 4m + 4\dfrac{I_2}{r_2^2} + 6M_c\right)}$

61. (a) In the direction of the force
(b) $\alpha = \dfrac{3F(r + R)}{4mr^2}$, where $r = \dfrac{R}{3}$
63. -6.02 m/s
65. (a) $\dfrac{mg}{5}(11 - 6\cos(72°))$
(b) $mg(3 - 2\cos(72°))$
(c) With Rolling 12.0 N, Without Rolling 15.7 N
67. (a) Once a ball starts rolling without slipping, we can view rolling as rotation about the pivot. There is no torque about the pivot, so the sphere does not accelerate once it is rolling on a horizontal surface.
(b) 1.87 s
(c) 24.5 J
69. $\dfrac{mg}{m + \dfrac{3}{8}m_c + \dfrac{I}{r^2}}$
71. Only if the object is already rolling without slipping. Otherwise the object will slip because there is no frictional torque to change the angular speed.
73. If the marble a perfectly rigid sphere and the horizontal surface is totally frictionless, the marble will not lose any energy to the surface. However, air resistance will cause the marble to gradually come to a stop.
75. $\dfrac{1}{2}I_{cm}\omega^2 + \dfrac{1}{2}mv_{cm}^2$
77. (a) $0.35Mg\left(1 + \dfrac{4}{3}\dfrac{M}{m}\right)$
(b) $0.35\dfrac{4}{3}\dfrac{M}{m}g$
79. $F\left(\dfrac{\beta - 1}{\beta + 1}\right)$
81. (a) $\sqrt{g\dfrac{R - r}{R - h}}\sqrt{2Rh - h^2}$
83. (a) 0.67 rad/s^2 (b) 43.0 N

Chapter 10—Equilibrium and Elasticity
1. No, even when the object is at its highest point an unbalanced force (gravity) acts on it.
3. c
5. c
7. d
9. No, an unbalanced centripetal force is acting on the object.
11. Yes. For example, the centre of gravity of a V-shaped object lies outside the object.
13. no
15. $\dfrac{6}{5} r$ from the centre of the star
17. No
19. Steel
21. Compression
23. c
25. a and c
27. b
29. a
31. 0.42
33. 2.7 kN
35. $F_1 = F_2 = 6.87 \times 10^4$ N
37. 0.44
39. $T_2 = 3240$ N, $T_1 = 2770$ N

41. 0.16 kN

43. $\dfrac{L}{2} - \dfrac{120 - e^{-L/120}\,(L + 120)}{1 - e^{-L/120}}$

45. $(49.3\hat{\imath} + 98.6\hat{\jmath})\,\mathrm{m}$

47. 1.4 kN

49. $m_2 = \frac{1}{2}m_1$

51. $8.1 \times 10^2\,\mathrm{N}$

53. $A_x = -35480\,\mathrm{N}$, $A_y = -20315\,\mathrm{N}$; $B_x = 35480\,\mathrm{N}$, $B_x = 47783\,\mathrm{N}$; $C_x = 35480\,\mathrm{N}$, $C_y = 40916\,\mathrm{N}$

55. $F_x = 5002\,\mathrm{N}$, and $F_y = 2511\,\mathrm{N}$; Considering that the only two forces that act on each beam in the horizontal direction are the force of friction and the horizontal contact force: $f = F_x = 5002\,\mathrm{N}$

57. 1574 kg

59. 25440 kg

61. 2.9 MPa

63. (a) $7.2 \times 10^4\,\mathrm{kg}$ (b) $1.3 \times 10^5\,\mathrm{kg}$ (c) 4.0 cm

65. 0.016

67. 31550

69. (a) $2.31 \times 10^5\,\mathrm{J}$
    (b) 1180 kg

71. Balancing torque hinges $N = 8985\,\mathrm{N}$; For reaction forces, we balance horizontal and vertical forces, $F = 7782\hat{\imath} + 6298\hat{\jmath}\,\mathrm{N}$

73. 8.2°

75. 0.53 kg

77. $T = 15\,\mathrm{kN}$, $R_x = 8805\,\mathrm{N}$, $R_y = -2961\,\mathrm{N}$

79. $r(1 + \mu)$

81. 0.175

83. 1.44

85. (a) 0.31 kN
    (b) Since $\tau_F < \tau_G$, the refrigerator will not tip first.
    (c) 112 cm

87. 28 N

89. (a) $Y = 1.3\,\mathrm{GPa}$
    (b)

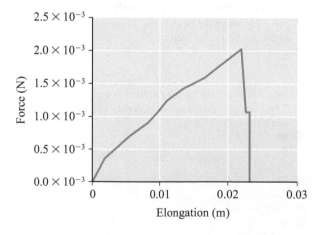

    (c) $1.26 \times 10^{-8}\,\mathrm{J}$
    (d) around (1.5%, 200 MPa)
    (e) 0.94 m/s, yes

91. 1.22 N/m

93. 1.44 MPa, 7.2×10^{-6}

95. 2.5 mm

97. (a) $5.99 \times 10^4\,\mathrm{m}$ (b) 5.8 mm

99. Longitudinal; compression $1.4 \times 10^5\,\mathrm{N}$, tension $9.5 \times 10^4\,\mathrm{N}$
    Transverse; compression $2.1 \times 10^6\,\mathrm{N}$., shear $4.6 \times 10^4\,\mathrm{N}$ to $5.0 \times 10^4\,\mathrm{N}$

101. (a) The ultimate tensile strength increases as the steel passes through subsequent cycles. From the graph we see that the UTS of the steel was about 850 MPa. The UTS increased to about 950 MPa after first pass and to about 1150 MPa after four passes.
     (b) The pressing process increases the ultimate tensile strength of the steel, and hence increases the energy required to break the steel.

Chapter 11—Gravitation

1. b

3. Weight is a force and should be measured in units such as newtons.

5. b

7. c

9. a

11. increase

13. b

15. c

17. c

19. $1.4 \times 10^{-11}\,\mathrm{N}$

21. $2.21 \times 10^{-6}\,\mathrm{N/kg}$

23. $3.7\,\mathrm{m/s^2}$

25. $9.801\,\mathrm{m/s^2}$

27. (a) $8.0\,\mathrm{m/s^2}$
    (b) $20.5\,\mathrm{m/s^2}$
    (c) $7.68 \times 10^{24}\,\mathrm{kg}$

29. $2.04 \times 10^6\,\mathrm{J}$

31. $F_x = 0, F_y = -2ky, F_z = 0$

33. 2.4 km/s

35. (a) 7.5 yr
    (b) 2400 km/s

37. 1680 m/s

39. (a) 449 km from centre of the Sun
    (b) 0.000645

41. (a) 23.4 s
    (b) $6.47 \times 10^{23}\,\mathrm{kg}$

43. (a) $1.18\,\mathrm{kg \cdot m^2}$
    (b) $2.65 \times 10^{-4}\,\mathrm{N \cdot m}$
    (c) $1.45 \times 10^{-7}\,\mathrm{N}$
    (d) $1.31 \times 10^{-7}\,\mathrm{N \cdot m}$
    (e) $9.89 \times 10^{-4}\,\mathrm{rad}$

45. (a) $5.0 \times 10^{23}\,\mathrm{kg}$ (b) 3.9 km/s

47. 42.1 km/s

49. $1400\,\mathrm{kg/m^3}$

51. See solutions manual.

53. (a) 73000 km/s
    (b) 4.7 ms

55. The simulation shows enhanced numbers of asteroids that have orbital radii and periods similar to Jupiter, but travel approximately 60° ahead or behind Jupiter. These locations are two of the Lagrange points (L4 and L5).

57. (a) The planet with the 10 day period is closer.
    (b) The one with the 10 day period is slightly more massive.

59. 2.3 km/s

61. $\dfrac{3kMm}{r^4}$

63. (a) $5.1 \times 10^{41}\,\mathrm{J}$
    (b) $4.2 \times 10^7\,\mathrm{yr}$

65. (a) 22 m/s (b) 396.78097 to 396.78103 nm

67. Estimates will vary. The answers given assume a comet about 1/20 the size of Halley's comet.

Chapter 12—Fluids

1. c

3. d

5. ABC

7. (a) When the iron cube is in the boat, the water displaced is due to the weight of the cube. However, when it is dropped in the water, it displaces water equal to its volume. Therefore the water level will fall.

9. c

11. b

13. Yes, the pressure at the bottom depends on the height of the water, not on the surface area at the bottom.

15. a

17. 3.1×10^{-8} cm

19. 69 kN/m^2

21. (a) 4.10×10^{20} kg/m^3
(b) 4.1×10^{15} N, absolutely not

23. about 1100 kg/m^3; no

25. 57 kN

27. 107 kPa

29. 349 N

31. a

33. 27 m

35. 880 kg/m^3

37. 0.20 N

39. 2.78 g/cm^3

41. (a) 0.69
(b) 27
(c) Yes, some of the weight of the survivors would be off-set by the buoyant force corresponding to the volume of water displaced by their legs.

43. 6.2 m/s^2. Acceleration will decrease as drag from the water increases with speed. Although drag is the dominant factor, the density of the water increases slightly with depth due to the greater pressure. This increase in density increases the buoyance force, and decreases the acceleration correspondingly.

45. 650 m^3

47. 1.3 m

49. (a) 3.3×10^{-5} m^3/s
(b) 0.11 m/s
(c) 2.0 m/s
(d) 1.7×10^{-5} m^2

51. (a) 8.0 m/s
(b) 172 kPa; since the height does not change, by Bernoulli's principle.

53. (a) 13 kPa (b) -1.5%

55. (a) 47 kPa (b) 5.5 m

57. Bernoulli's equation indicates that the pressure increases when the speed of water is reduced.

59. $v_1 = A_2 \sqrt{\dfrac{2(P_t - P_o) + 2\rho g (y_2 - y_1)}{\rho (A_2^2 - A_1^2)}}$, where P_o is the pressure in the opening at the side of the tank

61. See solutions manual.

63. 6.3 kPa

65. See solutions manual.

67. 1.9 N • m

69. 1.9 N • m

Chapter 13—Oscillations

1. $4A$, zero

3. Since force is proportional to distance from equilibrium point, acceleration increases with amplitude, keeping the period constant.

5. Increased

7. The system has to be treated as a physical pendulum, and the period will be $2\pi \sqrt{\dfrac{I}{MgR}}$.

9. a

11. Decreases

13. c

15. a, b, and c

17. d

19. (a) 0.3 m
(b) 0.47 m/s
(c) 0.75 m/s^2
(d) $x(t) = 0.3 \cos(1.58t)$

21. (a) $\omega = 7.7$ rad/s, $f = 1.2$ Hz
(b) 15.0 cm
(c) 0
(d) $x(t) = -(0.15\,\text{m}) \cos(7.7t)$

23. (a) $x(t) = 0.40 \cos\left(\pi t + \dfrac{\pi}{2}\right)$
(b) 1.3 m/s, 4.0 m/s^2
(c) 0.079 J

25. No, since $x(0) = A \cos\phi = -x(T/2)$ we have only one independent equation, so we cannot solve for the two unknowns.

27. (a) CBDAE
(b) EADBC
(c) BDACE
(d) EADBC

29. 7.27×10^{-5} rad/s, 24 h

31. DABEFC

33. (a) 32 rad/s, 0.20 s
(b) 6.3 m/s, 0.20 km/s^2
(c) 2.0 J

35. (a) 200 N/m
(b) 1.6 s^{-1}
(c) $x(t) = 0.050 \cos(10t + 1.283)$
(d) 0.49 m

37. (a) 0.31 s
(b) No change since period does not depend on spring compression.

39. (a) 4.7×10^4 N/m (b) 1.8 Hz

41. (a), (b), (c): See solutions manual.
(d) No. At the instant when the oscillating mass passes through its initial position, both springs are in their equilibrium positions. When the mass is at any other position, it exerts a force on spring 1, which in turn exerts a somewhat lesser force on spring 2. These forces either stretch both springs or compress both of them.

43. 6.2 cm

45. (a) 9.9 N/m
(b) 0.012 J
(c) 0.64
(d) $y(t) = 0.050 \cos(\pi t)$

47. (a) 63 J (b) 65 m/s

49. $\dfrac{1}{14.1} \sqrt{\dfrac{k}{m}}$

51. ADBCEF

53. (a) 200.6 s   (b) 200.9 s, yes
55. (a) See solutions manual.   (b) increase by $L/360$
57. See solutions manual.
59. (a) 0.6 m, 3 s

(b) $\dfrac{\pi}{2}$

(c) $\dfrac{7\pi}{6}, \dfrac{11\pi}{6}$

(d) $x(t) = 0.6 \cos\left(\dfrac{2\pi}{3}t + \dfrac{\pi}{2}\right)$

(e) 0.63 m/s
(f) -2.3 m/s^2

61. (a) 0.20 m, 3.0 s
(b) 1.0 m
(c) $\pi/3$ rad

(d) $x(t) = (0.20 \text{ m}) \cos\left(\dfrac{2\pi}{3}t + \dfrac{\pi}{3}\right) + 1.0$ m

(e) $x(t) = (0.20 \text{ m}) \cos\left(\dfrac{2\pi}{3}t + \dfrac{5\pi}{6}\right) + 1.0$ m

$\qquad = (0.20 \text{ m}) \sin\left(\dfrac{2\pi}{3}t + \dfrac{4\pi}{3}\right) + 1.0$ m

(f) 0.88 m/s^2

63. (a) 312   (b) 132   (c) 321   (d) 1, 3, 2
65. (a) 6.4×10^{-3} kg/s   (b) 15.8   (c) No
67. ACB

69. $\phi = \tan^{-1}\left[-\dfrac{x(t = T/4)}{x(t = 0)}\right]$, $A = \dfrac{x(t = 0)}{\cos\phi}$

71. 42.2 min. The period of the motion is $T = 2\pi\sqrt{\dfrac{R_{Earth}^3}{GM_{Earth}}}$, which is independent of the mass of the object.

73. See solutions manual.

75. (a) $\dfrac{2m}{m + M}v$ (b) $\dfrac{2mv}{m + M}\sqrt{\dfrac{M}{k}}$

Chapter 14—Waves

1. (a) F. For a transverse wave, the particles of the medium vibrate perpendicular to the direction of wave. For a longitudinal wave, the particles move back and forth in the direction of the wave.
(b) F. Wave speed of a mechanical wave in a medium depends on the elastic and inertial properties of the medium.
(c) T. The speed of waves in a medium often varies somewhat with wavelength.
(d) F. Mechanical waves depend on the movement of particles of the medium through which they pass.
(e) F. Since the speed of the wave depends on the properties of the medium, the speed will be constant in a uniform medium.
(f) F. A traveling wave always carries energy.
(g) F. The maximum amplitude cannot be greater than the sum of the amplitudes of individual waves.
(h) F. To produce a standing wave, the waves must be traveling in opposite directions.
3. Light wave
5. No
7. The wave propagates outward from the point where the stone hits the water surface. As the wavefront expands the energy per unit length decreases, and therefore the amplitude decreases.

9. c for a mechanical wave
11. The two waves must have same frequency and amplitude to produce a standing wave. If the two waves have same amplitude but different frequencies, interference will occur but standing waves will not be produced. (b) If the two waves have same frequency but different amplitudes, a pattern similar to a standing wave will be produced but without any nodes.
13. No, a difference in linear mass density could be offset by a difference in tension.
15. (a) π rad/m   (b) 20π rad/s   (c) 20 m/s
17. (a) 143 Hz   (b) 6.98 ms   (c) 1.05 m
19. 17 mm to 17 m
21. 10 mm to 6 m in ocean water; 2.3 mm to 1.4 m in air
23. $D(1.5, 2.0) = 6.9$ mm, $D_{min} = 0$, $D_{max} = 25$ cm
25. (a) 3.0 m/s, traveling left
(b) -23.6 mm
(c) -0.50 m
(d) 1.2 cm/s
(e)

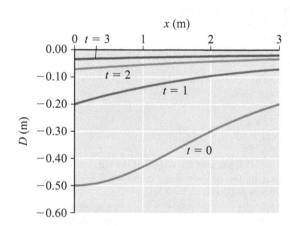

27. (a) The numbers on the vertical scale of the graph should be increased by a factor of 10, so that the range is from -2.5 to $+2.5$.

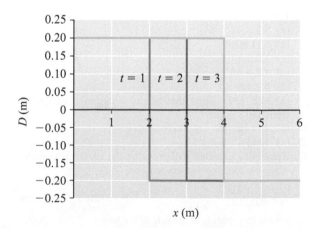

(b) 1.0 m/s, right
(c) 2 m

(d) $D(x, t) = \begin{cases} +2 \text{ m} & \text{if } |x + t| \leq 1 \\ -2 \text{ m} & \text{if } |x + t| > 1 \end{cases}$

29. (a)

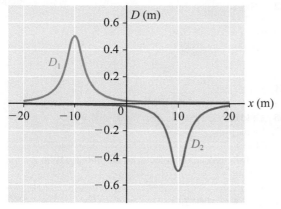

(b) D_1: −10.0 m, D_2: 10.0 m
(c) $t - 2$ s
(d) Yes, at $x = 0$

31. 25 N

33. 0.22 km/s

35. 13.2 m from the far end of string B

37. (a) See solutions manual.

(b) $T = \sqrt{\dfrac{2\pi\lambda}{g}} = \dfrac{2\pi v}{g}$

(c) 19 m/s

39. $\lambda_b < \lambda_a = \lambda_c < \lambda_d$, $f_a = f_b = f_d < f_c$; all waves moving toward increasing x.

41. $D_1(x, t) \approx \sin\left(\dfrac{4\pi}{3}x - 2\pi t - 1.4\right)$

$D_2(x, t) = \sin\left(2\pi x - 3\pi t + \dfrac{\pi}{2}\right)$

43. (a) 0.626 m, 1.19 Hz, 0.750 m/s
(b) $D(x, t) = 0.05 \sin(10.0x - 7.50t)$

45. (a) 3.0 m, $\dfrac{1}{6}$ Hz, 0.50 m/s

(b) Toward decreasing x.

(c)

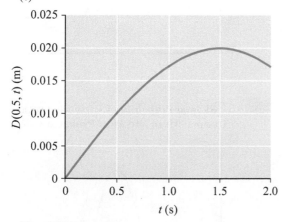

(d) −1.0 cm/s

47. (a) 8.0 m
(b) 40 m/s

(c) $\dfrac{21\pi}{2}$, which is equivalent to $\dfrac{\pi}{2}$

(d) $D(x, t) = 1.5 \sin\left(\dfrac{\pi}{4}x - 10\pi t + \dfrac{21\pi}{2}\right)$

49. (a) 0.50 Hz
(b) 8.0 m
(c) 1.5 rad

(d) $D(x, t) = (1.5 \text{ m}) \sin\left(\dfrac{\pi}{4}x - \pi t + 1.5\right)$

51. (a) $D(x, t) = 1.5 \sin\left(\dfrac{\pi}{4}x - 2\pi t + \dfrac{\pi}{2}\right)$

(b) $D(x, t) = (1.5 \text{ m}) \sin\left(\dfrac{\pi}{4}x + 2\pi t + \dfrac{\pi}{2}\right)$

53. (a) $D(x, t) = (0.10 \text{ m}) \sin\left(\dfrac{\pi}{2}x - \dfrac{\pi}{3}t + 5.0\right)$

(b) 0.50 m/s

55. (a) 0.2 rad
(b) 1.2 rad
(c) Yes, the phase changes linearly with time.
(d) 0.4 rad
(e) Yes
(f) 2π rad, which is equivalent to 0 rad

57. 0.88 W

59. (a)

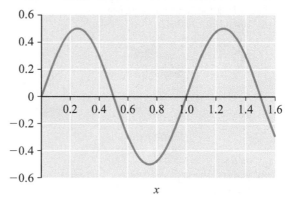

(b) $A = 0.50$, $\lambda = 1.0$
(c) The two component waves have the same frequency and wavelength, so these waves have the same speed. Consequently, the resultant wave will also have the same speed.
(d) Travelling

61. −0.42 rad

63. 2.636 rad, 1.318 rad

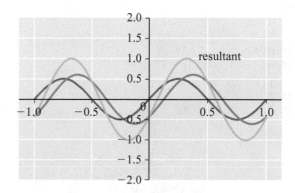

65. (a) $|A_1 - A_2|$ (b) $A_1 + A_2$

67. (a) $D(x,t) = \dfrac{-1}{(x + 3t)^2 + 4}$

(b) $D(x,t) = \dfrac{1}{(x + 3t)^2 + 4}$

69. (a) $D(x, t) = 0.2 \sin(3x - 4t + \pi)$

(b) $D(x, t) = 0.2 \sin(3x - 4t)$

71. (a) $\dfrac{10\pi}{3}$ m, 10 Hz

(b) $\dfrac{5\pi}{3}, \dfrac{5\pi}{6}$

(c) $D_1(x, t) = 0.0075 \sin(0.6x - 20\pi t)$,
$D_2(x, t) = 0.0075 \sin(0.6x + 20\pi t)$

(d) 1.8 mm

73. (a) $D(x, t) = (2.0 \text{ mm}) \sin(\pi x) \cos(0.5\pi t)$

(b) Nodes: ±1.0 m, ±2.0 m, ±3.0 m,
Antinodes: ± 0.5 m, ± 1.5 m, ±2.5 m

(c) 1.0 m

75. $D(x,t) = 2A \cos\left(\omega t - \dfrac{\phi_1 - \phi_2}{2}\right) \sin\left(kx + \dfrac{\phi_1 + \phi_2}{2}\right)$

77. 265 Hz

79. (a) 4.0 m

(b) 50.0 Hz

(c) No, all frequencies of standing waves are integer multiples of the fundamental frequency.

(d) 100.0 Hz, 200.0 Hz

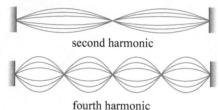

second harmonic

fourth harmonic

81. (a) 98.0 Hz

(b) 196 Hz, 294 Hz

(c) 196 m/s

(d) 116 Hz

Chapter 15—Interference and Sound

1. T

3. 0.5

5. T

7. T

9. F

11. 3.01 dB

13. c

15. 34.32 cm

17. 0.040 s

19. $s(x, t) = s_m \cos(1.1x - 377t)$

21. 1.08×10^{-5} m

23. 2

25.

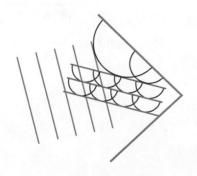

27. 1.7 kHz, 1.1 kHz, 0.86 kHz

29. 2 Hz

31. 90 Hz

33. 98 dB

35. 748 Hz

37. 9.8 cm/s

39. 1.4 km

41. 28 nm

43. 3.2×10^{-6} W/m², 21 mW

45. 2.4 kHz

47. 1.3 m

49. 24 m/s

51. Intensity reduces by a factor of 10^6 to 0.2 μW/m².

53. 35.6 m

55. 61 Hz

57. (a) 2.4 m   (b) 2.68 kHz

59. See solutions manual.

61. 252 m

Chapter 16—Temperature and the Zeroth Law of Thermodynamics

1. Only if both thermometers are linear between the calibration temperatures

3. The most probable and mean values are the same only for symmetric distributions with a central peak. If a distribution is skewed, the mean differs from the most probable value.

(a) **Symmetric Probability Distribution with Equal Mean and Most Probable Values**

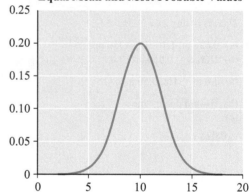

(b) **Skewed Probability Distribution with Unequal Mean and Most Probable Values**

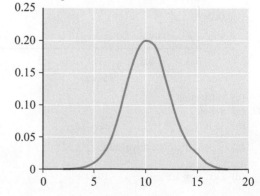

5. The triple point of water exists at exactly 0.01°C or 273.16 K. If liquid water did not expand when cooled, the melting point of ordinary ice would not decrease as a function of pressure and the triple point would be the limit below which water would not exist in liquid state.

7. No, the gases would have to have equal low-density concentrations (N/V).

9. No

11. As the Sun rises, the temperature of the land increases more rapidly than that of the water, and the air just above the land becomes warmer than the air above the water. As a result the air pressure on land is lower than the pressure above water. This pressure difference forces the cooler air above the water to move toward the land.

13. For an ideal gas the relationship between pressure, volume and temperature is given by $PV = NkT$. Therefore, if the temperature increases while the volume is reduced by half, the pressure will more than double.

15. $3.71 \times 10^{-26} \ m^3$, 3.34 nm

17. No, when a system is in thermodynamic equilibrium, all macroscopic properties, including temperature, remain constant.

19. (a) 758 kPa
    (b) $8.80 \times 10^{-4} \ m^3$
    (c) 0.267 mol
    (d) 7.51 g nitrogen, 1.07 g helium

21. 0.68 mm

23. $-40°$ C

25. (a) 136°F   (b) 331 K

27. (a) 8.65 kg   (b) 52.2 m³   (c) 530 moles

29. (a) 0.0224 m³   (b) 493.1 m/s   (c) 461.4 m/s   (d) 1:1.069

31. 433 m/s, 0.54 km/s

33. $\sqrt{\dfrac{8kT}{\pi m}}$

35. Water, oxygen and carbon dioxide in solid phase; nitrogen in liquid phase

37. 0.60 m, for example, 48 kg of water for a body mass of 80 kg

39. 4×10^{-4} m/s

41. 180 K

43. $3.3 \times 10^{27} \ m^{-2} \ s^{-1}$

45. Answers given are for hydrogen in diatomic molecules. For separate atoms of hydrogen, the pressure is about 1.7×10^{17} Pa and the RMS speed is 5.8×10^5 m/s.

47.

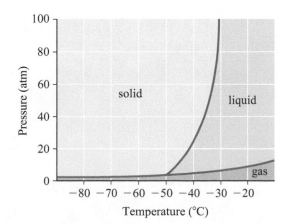

49. $R = 193 \ \Omega$ at $T = 400$ K

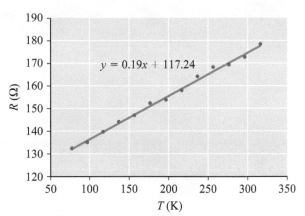

Chapter 17—Heat, Work, and the First Law of Thermodynamics

1. Constant volume

3. A solid; B solid and liquid; C liquid; D liquid and gas; E gas

5. All surfaces, including our skin, lose heat through radiant heat transfer with rates that differ in winter and summer months. The reason is the unequal temperature differentials that exist between objects and their surroundings in winter and summer months. Hence in winter months we lose heat at a faster rate than in summer months.

7. Positive

9. 2

11. The food has reached an equilibrium state.

13. The heat capacities of the objects are equal (i.e., the product cm is the same).

15. 8.0×10^2 s

17. 13.8°C

19. 130 J kg⁻¹ K⁻¹

21. 11°C

23. 2.4 kJ

25. 100 J

27. 21 kW/m²

29. 9.6 kW

31. 8.0 kJ/K, 80.0 kJ kg⁻¹ K⁻¹

33. 1.3 kJ/s assuming that all of the perspiration was evaporated by heat from the athlete's body

35. -3.74 kJ

37. About 0.15 W (assuming 0.5 emissivity and cylindrical shape with 1 cm radius)

39. 505 kJ, 495 kJ

41. Change in internal energy is zero.

43. 61 kJ

45. 696°C, 13.4 kJ, 5.37 kJ, 8.06 kJ

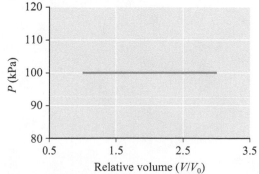

47. about 0.10 kg

Base SI Units

Unit name	SI symbol	Measures
Ampere	A	Electric current
Candela	cd	Luminous intensity
Kelvin	K	Temperature
Kilogram	kg	Mass
Metre	m	Length
Mole	mol	Amount of a substance
Second	s	Time

The Most Common Derived SI Units

Unit name	SI symbol	Measures	In base units
Coulomb	C	Electric charge	$A \cdot s$
Farad	F	Electric capacitance	C/V
Henry	H	Electromagnetic inductance	$V \cdot s/A$
Hertz	Hz	Frequency	$/s$
Joule	J	Energy and work	$N \cdot m$
Newton	N	Force	$kg \cdot m/s^2$
Ohm	Ω	Electric resistance	V/A
Pascal	Pa	Pressure	N/m^2
Radian	rad	Angle	dimensionless
Tesla	T	Magnetic field strength	$V \cdot s/m^2$
Volt	V	Electric potential difference	J/C
Watt	W	Power	J/s

Common Prefixes

Prefix	Abbreviation	Power
Zepto-	z	10^{-21}
Atto-	a	10^{-18}
Femto-	f	10^{-15}
Pico-	p	10^{-12}
Nano-	n	10^{-9}
Micro-	μ	10^{-6}
Milli-	m	10^{-3}
Centi-	c	10^{-2}
Deci-	d	10^{-1}
Kilo-	k	10^{3}
Mega-	M	10^{6}
Giga-	G	10^{9}
Tera-	T	10^{12}
Peta-	P	10^{15}
Exa-	E	10^{18}
Zetta-	Z	10^{21}

Appendix C
GEOMETRY AND TRIGONOMETRY

Arc Length and Angle

$$s = \text{arc length} = r\theta$$

Angle θ is in radians.

$$2\pi \text{ rad} = 360°$$

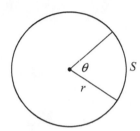

Trigonometric Functions
Right-angled Triangle

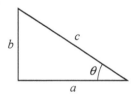

Pythagoras' Theorem: $a^2 + b^2 = c^2$

$$\sin\theta = \frac{\text{opposite}}{\text{hypotenuse}} = \frac{b}{c}; \quad \csc\theta = \frac{\text{hypotenuse}}{\text{opposite}} = \frac{c}{b}$$

$$\cos\theta = \frac{\text{adjacent}}{\text{hypotenuse}} = \frac{a}{c}; \quad \sec\theta = \frac{\text{hypotenuse}}{\text{adjacent}} = \frac{c}{a}$$

$$\tan\theta = \frac{\text{opposite}}{\text{adjacent}} = \frac{b}{a}; \quad \cot\theta = \frac{\text{adjacent}}{\text{opposite}} = \frac{a}{b}$$

General Triangle

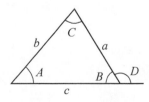

Angles: A, B, C Opposite sides: a, b, c

$$A + B + C = 180° \quad A + C = D$$

Law of cosines $c^2 = a^2 + b^2 - 2ab\cos C$

Law of sines $\dfrac{\sin A}{a} = \dfrac{\sin B}{b} = \dfrac{\sin C}{c}$

Trigonometric Identities

$$\tan\theta = \frac{\sin\theta}{\cos\theta}; \quad \sin^2\theta + \cos^2\theta = 1$$

$$\sin(A \pm B) = \sin A \cos B \pm \cos A \sin B$$

$$\cos(A \pm B) = \cos A \cos B \mp \sin A \sin B$$

$$\tan(A \pm B) = \frac{\tan A \pm \tan B}{1 \mp \tan A \tan B}$$

$$\sin(\theta \pm \pi/2) = \pm\cos\theta \quad \cos(\theta \pm \pi/2) = \mp\sin\theta$$

$$\sin(\theta \pm \pi) = -\sin\theta \quad \cos(\theta \pm \pi) = -\cos\theta$$

$$\sin(2\theta) = 2\sin\theta\cos\theta \quad \cos(2\theta) = \cos^2\theta - \sin^2\theta$$

$$\sin A \pm \sin B = 2\sin\left(\frac{A \pm B}{2}\right)\cos\left(\frac{A \mp B}{2}\right)$$

$$\cos A + \cos B = 2\cos\left(\frac{A + B}{2}\right)\cos\left(\frac{A - B}{2}\right)$$

$$\cos A - \cos B = -2\sin\left(\frac{A + B}{2}\right)\sin\left(\frac{A - B}{2}\right)$$

Expansions and Approximations
Trigonometric Expansions

$$\sin x = x - \frac{x^3}{3!} + \frac{x^5}{5!} - \frac{x^7}{7!} + \dots \quad (x \text{ is in radians})$$

$$\cos x = 1 - \frac{x^2}{2!} + \frac{x^4}{4!} - \frac{x^6}{6!} + \dots \quad (x \text{ is in radians})$$

$$\tan x = x + \frac{x^3}{3} + \frac{2x^5}{15} + \frac{17x^7}{315} + \dots \quad (x \text{ is in radians})$$

Small-angle Approximation

If $x \ll 1$ rad, then $\sin x \approx \tan x \approx x$. The small-angle approximation is generally good for $x < 0.2$ rad.

Appendix D
KEY CALCULUS IDEAS

Derivatives

Given a function $f(x)$, the derivative is defined to be

$$f'(x) = \frac{df}{dx} = \lim_{h \to 0} \frac{f(x+h) - f(x)}{h}$$

The derivative can be interpreted as the instantaneous rate of change of the function at x. We can also think of the derivative as the slope of the tangent to the curve as shown in the figure.

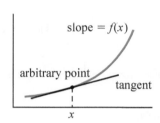

From a physics perspective, if the function $f(t)$ describes the position of an object, the derivative is the velocity of the object.

Basic Properties

For two functions $f(x)$ and $g(x)$ whose derivatives exist, the following properties apply:

1. $\dfrac{d}{dx}(cf(x)) = c\dfrac{df}{dx}; \quad c = \text{constant}$

2. $\dfrac{d}{dx}(f \pm g) = \dfrac{df}{dx} \pm \dfrac{dg}{dx}$

3. $\dfrac{d}{dx}(fg) = f\dfrac{dg}{dx} + g\dfrac{df}{dx} \quad \text{(Product Rule)}$

4. $\dfrac{d}{dx}\left(\dfrac{f}{g}\right) = \dfrac{\dfrac{df}{dx}g - f\dfrac{dg}{dx}}{g^2} \quad \text{(Quotient Rule)}$

5. $\dfrac{d}{dx}f(g(x)) = \dfrac{df}{dg}\dfrac{dg}{dx} \quad \text{(Chain Rule)}$

Common Derivatives

1. $\dfrac{d}{dx}x = 1$

2. $\dfrac{d}{dx}x^n = nx^{n-1}$

3. $\dfrac{d}{dx}\sin x = \cos x$

4. $\dfrac{d}{dx}\cos x = -\sin x$

5. $\dfrac{d}{dx}\tan x = -\sec^2 x$

6. $\dfrac{d}{dx}\sec x = \sec x \tan x$

7. $\dfrac{d}{dx}\csc x = -\csc x \cot x$

8. $\dfrac{d}{dx}\cot x = -\csc^2 x$

9. $\dfrac{d}{dx}\sin^{-1}x = \dfrac{1}{\sqrt{1-x^2}}$

10. $\dfrac{d}{dx}\cos^{-1}x = -\dfrac{1}{\sqrt{1-x^2}}$

11. $\dfrac{d}{dx}\tan^{-1}x = \dfrac{1}{\sqrt{1+x^2}}$

12. $\dfrac{d}{dx}a^x = a^x \ln(a)$

13. $\dfrac{d}{dx}e^x = e^x$

14. $\dfrac{d}{dx}\ln x = \dfrac{1}{x}$

Integrals

Definition: Suppose that $f(x)$ is continuous on the interval $[a, b]$. We divide $[a, b]$ into N subintervals of width $\Delta x = \dfrac{b-a}{N}$, with centre $x_i = \left(i + \dfrac{1}{2}\right)\Delta x$. We then define the definite integral as follows:

$$\int_a^b f(x)\,dx = \lim_{N \to \infty} \sum_{i=0}^{N} f(x_i)\Delta x$$

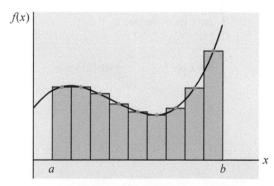

We define the antiderivative of $f(x)$ as $g(x)$ such that $\dfrac{dg}{dx} = f$.

The indefinite integral is defined as $\displaystyle\int f(x)\,dx = g(x) + C$, where

$$\frac{dg}{dx} = f \quad \text{and} \quad C = \text{constant}$$

We interpret the integral as the area bounded by the function $f(x)$ and the x-axis, where the areas above the axis are positive, and the areas below the axis are negative.

Some Properties of Integrals

$$\int (f(x) \pm g(x))\, dx = \int f(x)\, dx \pm \int g(x)\, dx$$

$$\int c f(x)\, dx = c \int f(x)\, dx; \quad c = \text{constant}$$

$$\int_a^b f(x)\, dx = -\int_b^a f(x)\, dx$$

$$\int u \frac{dv}{dx}\, dx = uv - \int v \frac{du}{dx}\, dx$$

Some Indefinite Integrals

$$\int b\, dx = bx + c; \quad b, c \text{ are constants}$$

$$\int x^n\, dx = \frac{1}{n+1} x^{n+1} + c$$

$$\int \frac{dx}{x} = \ln(x + c)$$

$$\int e^x\, dx = e^x + c$$

$$\int \sin x\, dx = -\cos x + c$$

$$\int \cos x\, dx = \sin x + c$$

$$\int \tan x\, dx = \ln|\sec x| + c$$

$$\int \sin^2 x\, dx = \frac{1}{2}x - \frac{1}{4}\sin(2x) + c$$

$$\int \frac{dx}{\sqrt{x^2 + a^2}} = \ln\left(x + \sqrt{x^2 + a^2}\right) + c$$

$$\int \frac{x\, dx}{(x^2 + a^2)^{3/2}} = -\frac{1}{(x^2 + a^2)^{1/2}} + c$$

$$\int \frac{dx}{(x^2 + a^2)^{3/2}} = \frac{x}{a^2 (x^2 + a^2)^{1/2}} + c$$

$$\int \frac{x\, dx}{x + a} = x - a \ln(x + a) + c$$

$$\int e^{-ax}\, dx = -\frac{1}{a} e^{-ax} + c$$

$$\int x e^{-ax}\, dx = -\frac{1}{a^2}(ax + 1)e^{-ax} + c$$

$$\int x^2 e^{-ax}\, dx = -\frac{1}{a^3}(a^2 x^2 + 2ax + 2)e^{-ax} + c$$

Partial Derivatives

We sometimes need to take the derivative of a function that depends on more than one variable. For example, the displacement amplitude of a wave depends on both position and time, and in one dimension is thus a function of both x and t. When we need to find the derivatives of such functions, we can use partial derivatives.

Consider a function $f(x, y, z)$. We denote the partial derivatives by $\frac{\partial f}{\partial x}, \frac{\partial f}{\partial y}$ and $\frac{\partial f}{\partial z}$ (pronounced die f by die x, etc.). We calculate a partial derivative by differentiating with respect to the variable of interest while treating all other variables as constants. For example, for the function $f(x, y, z) = ax^2 + bxy + cz$, the partial derivatives are given by

$$\frac{\partial f}{\partial x} = 2ax + by, \quad \frac{\partial f}{\partial y} = bx, \quad \text{and} \quad \frac{\partial f}{\partial z} = c.$$

Appendix E

USEFUL MATHEMATICAL FORMULAS AND MATHEMATICAL SYMBOLS USED IN THE TEXT AND THEIR MEANING

Lengths, Areas, and Volumes

Rectangle

$$\text{Area} = ab$$

$$\text{Perimeter} = 2(a + b)$$

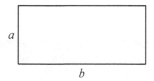

Rectangular Cuboid

$$\text{Surface Area} = 2(ab + bc + ac)$$

$$\text{Volume} = abc$$

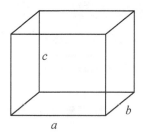

Triangle

$$\text{Area} = \frac{1}{2}bh$$

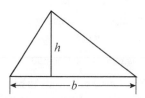

Circle

$$\text{Circumference} = 2\pi r$$

$$\text{Area} = \pi r^2$$

Sphere

$$\text{Surface area} = 4\pi r^2$$

$$\text{Volume} = \frac{4}{3}\pi r^3$$

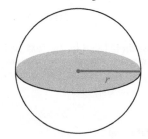

Cylinder

$$\text{Surface area} = 2\pi rh + 2\pi r^2$$

$$\text{Volume} = \pi r^2 h$$

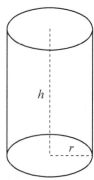

Quadratic Equation

An equation of the form $ax^2 + bx + c = 0$ has as its solution

$$x = \frac{-b \pm \sqrt{b^2 - 4ac}}{2a}$$

Linear Equation

A linear equation is of the form $y = mx + b$, where b is the intercept of the line with the y-axis, and m is the slope of the line;

$$m = \frac{y_2 - y_1}{x_2 - x_1} = \frac{\Delta y}{\Delta x}$$

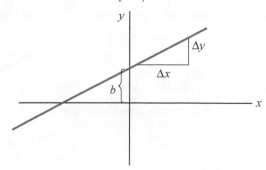

Logarithms

These identities apply to all bases:

$$\log a + \log b = \log (ab)$$

$$\log a - \log b = \log\left(\frac{a}{b}\right)$$

$$\log a^b = b \log a$$

Expansions

$$(1 + x)^n = 1 + \frac{nx}{1!} + \frac{n(n-1)x}{2!} + \cdots \quad (x^2 \ll 1)$$

$$e^x = 1 + x + \frac{1}{2!}x^2 + \frac{1}{3!}x^3 + \cdots$$

$$\ln(1 + x) = x - \frac{1}{2}x^2 + \frac{1}{3}x^3 - \frac{1}{4}x^4 \quad (|x| < 1)$$

$$\sin\theta = \theta - \frac{1}{3!}\theta^3 + \frac{1}{5!}\theta^5 - \cdots$$

$$\cos\theta = 1 - \frac{1}{2!}\theta^2 + \frac{1}{4!}\theta^4 - \cdots$$

Mathematical Symbols Used in the Text and Their Meaning

Symbol	Meaning	Sample expression	How it is read
$+$	Addition sign	$7 + 14 = 21$	7 plus 14 equals 21 The sum of 7 and 14 equals 21.
$\cdot$ $\times$	Multiplication signs	$7 \cdot 3 = 21$ $7 \times 3 = 21$	7 times 3 equals 21 The product of 7 and 3 equals 21
$\cdot$	Dot product sign	$\vec{A} \cdot \vec{B} = 32$ $(1, 2, 3) \cdot (4, 5, 6) = 32$	Vector A dot vector B equals 32 The dot (scalar) product of vector A and vector B equals 32.
$\times$	Cross product sign	$\vec{A} \times \vec{B} = \vec{C}$ $(1, 2, 0) \times (3, 4, 0) = (0, 0, -2)$	Vector A cross vector B equals vector C The cross (vector) product of vector A and vector B equals vector C
$-$	Subtraction sign Minus sign	$7 - 3 = 4$ $-(3 + 5) = -8$	7 minus 3 equals 4 The difference of 7 and 3 equals 4 Negative of three plus five equals minus 8 The opposite of three plus five equals minus 8
$\div$ $/$ $-$	Division signs	$21 \div 3 = 7$ $^3/_4 = 3/4 = 0.75$ $\frac{3}{4} = 0.75$	21 divided by 3 equals 7 The quotient of 21 and 7 equals 3 3 over 4 equals three quarters equals point 75
%	Percent symbol	$58\% \equiv 0.58$ $30\%(\$200) = \$60$	58 percent equals point 58 30 percent of \$200 equals \$60
:	Ratio	$1 : 5 = 100 : 500$	1 is to 5 as 100 is to 500
$\pm$ $\mp$	Plus/minus sign Minus/plus sign	$50\ \Omega \pm 10\%$ $(a \pm b)^2 = a^2 \pm 2ab + b^2$ $\cos(\alpha \pm \beta) = \cos^2\alpha \mp \sin^2\beta$	Fifty Ohms plus minus 10 percent a plus minus b squared equals... cos alpha plus minus beta equals cos squared alpha minus plus sin squared beta
$=$	Two values are the same	$-(-7) = 7$ $x + 3y = 57$	The opposite of minus seven equals seven
$\equiv$	A definition	$\sqrt{x} \equiv x^{1/2}$	Square root of x is defined as x to the power of one half
$\neq$	Two values are different	$-1 + 5 \neq -6$	Minus 1 plus 5 is not equal to minus 6
$\approx$	Two values are close to each other (of the same order of magnitude)	$x + y \approx z;\ 2^{10} \approx 1000$	x plus y is approximately equal to z 2 to the power of 10 is approximately equal to 1000

~	Two objects are geometrically similar	$\triangle ABC \sim \triangle ABD$	Triangle ABC is similar to triangle ABD.
$\propto$	Two variables change in direct proportion	$y \propto x$	y is proportional to x
$>(<)$	greater than (less than)	$2^{10} > 10^2$ and $2^3 < 3^2$	Two to the power of 10 is greater than ten to the power of two. Two cubed is less than three squared.
$\gg (\ll)$	much greater than (much less than)	$2^{20} \gg 20^2$ and $10^{-5} \ll 10^5$	Two to the power of twenty is much greater than twenty squared. Ten to the power of minus five is much less than ten to the power of five.
$\geq (\leq)$	greater than or equal to (less than or equal to)	$x \geq 10^2$ and $y \leq 2^3$	x is greater than or equal to ten squared. y is less than or equal to two cubed
∞	Infinity	$x \rightarrow \infty$	x approaches infinity
$\ldots$	Continuation	$1 + \dfrac{1}{2} + \dfrac{1}{4} + \dfrac{1}{8} + \cdots + \dfrac{1}{1024}$ $1 + \dfrac{1}{2} + \dfrac{1}{4} + \dfrac{1}{8} + \cdots$	... and so on up to ... ... and so on indefinitely
$\Rightarrow$	Logical implication	$A \Rightarrow B$	A implies B If A then B
$\Leftrightarrow$	Logical equivalence	$A \Leftrightarrow B$	A is logically equivalent to B
$\therefore$	Logical following	$A = B$ and $B = C$ $\therefore A = C$	A equals B and B equals C, therefore (it follows that), A equals C
() [] {}	Parentheses Square brackets Curly brackets	$[(a + b)^2 + 3c]^2$ $\vec{A} = (3, 4, 5)$ $\{a_i\}_{i=1}^{100} = \{a_1, a_2, a_3, \ldots, a_{100}\}$	... quantity, list, set of coordinates, open interval ( ) or closed interval [ ], set { }
$|x|$	Absolute value (a magnitude of a vector). It is always a non-negative quantity.	$|\vec{F}| = 5\,\text{N}$ $|-37| = 37$	The magnitude of force F equals 5 Newtons The absolute value of negative 37 equals 37
$\sqrt{\ }$	Square root symbol	$\sqrt{25} = 5$	The square root of twenty five equals five
$\sqrt[n]{\ }$	Root symbol	$\sqrt[4]{81} = 3$	The fourth root of eighty one equals three
!	Factorial	$3! = 1 \times 2 \times 3 = 6$	Three factorial equals six
°	Degree symbol	$\angle ABC = 60°$ $\cos(90°) = 0$	Angle ABC has a value of sixty degrees
$\perp$	Perpendicularity symbol	$\vec{A} \perp \vec{B}$	Vector A is perpendicular to vector B
$\parallel$	Parallel symbol	$\vec{A} \parallel \vec{B}$	Vector A is parallel to vector B
$\angle$ or $\sphericalangle$	Angle symbols	$\angle ABC, \sphericalangle ABC$	Angle ABC
$\rightarrow$	An object is a vector	$\vec{F}, \vec{v}$	Vector F, vector v
$\sum\limits_{i=1}^{N} x_i$	Finite summation	$\sum\limits_{i=1}^{N} x_i = 345$	The summation of N quantities x_i equals 345. The sum of all quantities x_i from $i = 1$ to $i = N$ equals 345.
$\sum\limits_{i=1}^{\infty} x_i$	Infinite summation	$\sum\limits_{n=1}^{\infty} \dfrac{1}{2^n} = 1$	The summation of an infinite number of quantities, ... the sum of all quantities x_i from $i = 1$ to $i = \infty$ (infinity)

(continued)

Mathematical Symbols Used in the Text and Their Meaning *(continued)*

Symbol	Meaning	Sample expression	How it is read
Δ	The change in ... The vertices of a triangle	Δx ΔABC	Delta x (the change in x) Triangle ABC
$\rightarrow$	Approaching	$\Delta x \rightarrow 0$	Delta x approaches zero
$\dfrac{d\ldots}{dt}$	Time derivative	$\dfrac{dx}{dt} ; \dfrac{d\vec{v}}{dt}$	The derivative of x with respect to time The derivative of vector v with respect to time
$\dfrac{\partial\ldots}{\partial t}$	Partial time derivative	$\dfrac{\partial f(x,t)}{\partial t}$	The partial derivative of function f with respect to time
$\int$	Indefinite integral	$\int x^2 dx = \dfrac{x^3}{3} + \text{const}$	The integral of x squared dx equals ...
$\iint$	Indefinite double integral	$\iint f(x,y)dxdy$	The double integral of of $f(x, y)\,dxdy$
$\iiint$	Indefinite triple integral	$\iiint f(x,y,z)dxdydz$	The triple integral of of $f(x, y, z)\,dxdydz$
$\int_a^b$	Definite integral	$\int_0^2 x^3 dx = 2\dfrac{2}{3}$	The integral of x squared dx equals 2 and 2/3.
$\oint$	Line (path) integral	$\oint \vec{B} \cdot d\vec{l}$	The line integral of $B\,dl$
$\oint$	Line integral around a closed path of	$\oint_L \vec{B} \cdot d\vec{l}$	The line integral of $B\,dl$ around a close path L
$\oiint$	A surface integral over a closed surface of	$\oiint_s \vec{E} \cdot d\vec{A}$	The surface integral of $E\,dA$ over a closed surface S

Periodic Table of the Elements

Legend (sample):
Uranium — 92 ---- Atomic number — U ---- Symbol

Key:
- MAIN GROUP METALS
- TRANSITION METALS
- METALLOIDS
- NON-METALS

Period 1
Group 1: Hydrogen 1 H
Group 18: Helium 2 He

Period 2
- Lithium 3 Li
- Beryllium 4 Be
- Boron 5 B
- Carbon 6 C
- Nitrogen 7 N
- Oxygen 8 O
- Fluorine 9 F
- Neon 10 Ne

Period 3
- Sodium 11 Na
- Magnesium 12 Mg
- Aluminum 13 Al
- Silicon 14 Si
- Phosphorus 15 P
- Sulfur 16 S
- Chlorine 17 Cl
- Argon 18 Ar

Period 4
- Potassium 19 K
- Calcium 20 Ca
- Scandium 21 Sc
- Titanium 22 Ti
- Vanadium 23 V
- Chromium 24 Cr
- Manganese 25 Mn
- Iron 26 Fe
- Cobalt 27 Co
- Nickel 28 Ni
- Copper 29 Cu
- Zinc 30 Zn
- Gallium 31 Ga
- Germanium 32 Ge
- Arsenic 33 As
- Selenium 34 Se
- Bromine 35 Br
- Krypton 36 Kr

Period 5
- Rubidium 37 Rb
- Strontium 38 Sr
- Yttrium 39 Y
- Zirconium 40 Zr
- Niobium 41 Nb
- Molybdenum 42 Mo
- Technetium 43 Tc
- Ruthenium 44 Ru
- Rhodium 45 Rh
- Palladium 46 Pd
- Silver 47 Ag
- Cadmium 48 Cd
- Indium 49 In
- Tin 50 Sn
- Antimony 51 Sb
- Tellurium 52 Te
- Iodine 53 I
- Xenon 54 Xe

Period 6
- Cesium 55 Cs
- Barium 56 Ba
- Lanthanum 57 La
- Hafnium 72 Hf
- Tantalum 73 Ta
- Tungsten 74 W
- Rhenium 75 Re
- Osmium 76 Os
- Iridium 77 Ir
- Platinum 78 Pt
- Gold 79 Au
- Mercury 80 Hg
- Thallium 81 Tl
- Lead 82 Pb
- Bismuth 83 Bi
- Polonium 84 Po
- Astatine 85 At
- Radon 86 Rn

Period 7
- Francium 87 Fr
- Radium 88 Ra
- Actinium 89 Ac
- Rutherfordium 104 Rf
- Dubnium 105 Db
- Seaborgium 106 Sg
- Bohrium 107 Bh
- Hassium 108 Hs
- Meitnerium 109 Mt
- Darmstadtium 110 Ds
- Roentgenium 111 Rg
- Copernicium 112 Cn
- Ununtrium 113 Uut
- Ununquadium 114 Uuq
- Ununpentium 115 Uup
- Ununhexium 116 Uuh

Lanthanides
- Cerium 58 Ce
- Praseodymium 59 Pr
- Neodymium 60 Nd
- Promethium 61 Pm
- Samarium 62 Sm
- Europium 63 Eu
- Gadolinium 64 Gd
- Terbium 65 Tb
- Dysprosium 66 Dy
- Holmium 67 Ho
- Erbium 68 Er
- Thulium 69 Tm
- Ytterbium 70 Yb
- Lutetium 71 Lu

Actinides
- Thorium 90 Th
- Protactinium 91 Pa
- Uranium 92 U
- Neptunium 93 Np
- Plutonium 94 Pu
- Americium 95 Am
- Curium 96 Cm
- Berkelium 97 Bk
- Californium 98 Cf
- Einsteinium 99 Es
- Fermium 100 Fm
- Mendelevium 101 Md
- Nobelium 102 No
- Lawrencium 103 Lr

At the date of publication, elements 113–118 had not been named and have been given temporary names.

Atomic Masses of the Elements[†] (IUPAC 2007), Based on relative atomic mass of $^{12}C = 12$ exactly

Name	Symbol	Atomic Number	Atomic Mass	Name	Symbol	Atomic Number	Atomic Mass
Actinium*	Ac	89	(227)	Molybdenum	Mo	42	95.96(2)
Aluminum	Al	13	26.9815386(8)	Neodymium	Nd	60	144.242(3)
Americium*	Am	95	(243)	Neon	Ne	10	20.1797(6)
Antimony	Sb	51	121.760(1)	Neptunium*	Np	93	(237)
Argon	Ar	18	39.948(1)	Nickel	Ni	28	58.6934(4)
Arsenic	As	33	74.92160(2)	Niobium	Nb	41	92.90638(2)
Astatine*	At	85	(210)	Nitrogen	N	7	14.0067(2)
Barium	Ba	56	137.327(7)	Nobelium*	No	102	(259)
Berkelium*	Bk	97	(247)	Osmium	Os	76	190.23(3)
Beryllium	Be	4	9.012182(3)	Oxygen	O	8	15.9994(3)
Bismuth	Bi	83	208.98040(1)	Palladium	Pd	46	106.42(1)
Bohrium	Bh	107	(272)	Phosphorus	P	15	30.973762(2)
Boron	B	5	10.811(7)	Platinum	Pt	78	195.084(9)
Bromine	Br	35	79.904(1)	Plutonium*	Pu	94	(244)
Cadmium	Cd	48	112.411(8)	Polonium*	Po	84	(209)
Cesium	Cs	55	132.9054519(2)	Potassium	K	19	39.0983(1)
Calcium	Ca	20	40.078(4)	Praseodymium	Pr	59	140.90765(2)
Californium*	Cf	98	(251)	Promethium*	Pm	61	(145)
Carbon	C	6	12.0107(8)	Protactinium*	Pa	91	231.03588(2)
Cerium	Ce	58	140.116(1)	Radium*	Ra	88	(226)
Chlorine	Cl	17	35.453(2)	Radon*	Rn	86	(222)
Chromium	Cr	24	51.9961(6)	Rhenium	Re	75	186.207(1)
Cobalt	Co	27	58.933195(5)	Rhodium	Rh	45	102.90550(2)
Copernicium	Cn	112	(285)	Roentgenium	Rg	111	(280)
Copper	Cu	29	63.546(3)	Rubidium	Rb	37	85.4678(3)
Curium*	Cm	96	(247)	Ruthenium	Ru	44	101.07(2)
Darmstadtium	Ds	110	(281)	Rutherfordium	Rf	104	(267)
Dubnium	Db	105	(268)	Samarium	Sm	62	150.36(2)
Dysprosium	Dy	66	162.500(1)	Scandium	Sc	21	44.955912(6)
Einsteinium*	Es	99	(252)	Seaborgium	Sg	106	(271)
Erbium	Er	68	167.259(3)	Selenium	Se	34	78.96(3)
Europium	Eu	63	151.964(1)	Silicon	Si	14	28.0855(3)
Fermium*	Fm	100	(257)	Silver	Ag	47	107.8682(2)
Fluorine	F	9	18.9984032(5)	Sodium	Na	11	22.98976928(2)
Francium*	Fr	87	(223)	Strontium	Sr	38	87.62(1)
Gadolinium	Gd	64	157.25(3)	Sulfur	S	16	32.065(5)
Gallium	Ga	31	69.723(1)	Tantalum	Ta	73	180.94788(2)
Germanium	Ge	32	72.64(1)	Technetium*	Tc	43	(98)
Gold	Au	79	196.966569(4)	Tellurium	Te	52	127.60(3)
Hafnium	Hf	72	178.49(2)	Terbium	Tb	65	158.92535(2)
Hassium	Hs	108	(270)	Thallium	Tl	81	204.3833(2)
Helium	He	2	4.002602(2)	Thorium*	Th	90	232.03806(2)
Holmium	Ho	67	164.93032(2)	Thulium	Tm	69	168.93421(2)
Hydrogen	H	1	1.00794(7)	Tin	Sn	50	118.710(7)
Indium	In	49	114.818(3)	Titanium	Ti	22	47.867(1)
Iodine	I	53	126.90447(3)	Tungsten	W	74	183.84(1)
Iridium	Ir	77	192.217(3)	Ununhexium	Uuh	116	(293)
Iron	Fe	26	55.845(2)	Ununoctium	Uuo	118	(294)
Krypton	Kr	36	83.798(2)	Ununpentium	Uup	115	(288)
Lanthanum	La	57	138.90547(7)	Ununquadium	Uuq	114	(289)
Lawrencium*	Lr	103	(262)	Ununtrium	Uut	113	(284)
Lead	Pb	82	207.2(1)	Uranium*	U	92	238.02891(3)
Lithium	Li	3	6.941(2)	Vanadium	V	23	50.9415(1)
Lutetium	Lu	71	174.9668(1)	Xenon	Xe	54	131.293(6)
Magnesium	Mg	12	24.3050(6)	Ytterbium	Yb	70	173.054(5)
Manganese	Mn	25	54.938045(5)	Yttrium	Y	39	88.90585(2)
Meitnerium	Mt	109	(276)	Zinc	Zn	30	65.38(2)
Mendelevium*	Md	101	(258)	Zirconium	Zr	40	91.224(2)
Mercury	Hg	80	200.59(2)				

[†]The atomic masses of many elements can vary depending on the origin and treatment of the sample. This is particularly true for Li; commercially available lithium-containing materials have Li atomic masses in the range of 6.939 and 6.996. The uncertainties in atomic mass values are given in parentheses following the last significant figure to which they are attributed.

*Elements with no stable nuclide; the value given in parentheses is the atomic mass number of the isotope of longest known half-life. However, three such elements (Th, Pa, and U) have a characteristic terrestrial isotopic composition, and the atomic mass is tabulated for these. http://www.chem.qmul.ac.uk/iupac/AtWt/

Source: © 2009 IUPAC, Pure and Applied Chemistry 81, 2131–2156. Note: The atomic mass referred to in this table is the abundance averaged atomic mass of all the isotopes of the particular element. This quantity is sometimes referred to as the atomic weight of an element.

Index

International Space Station (ISS), 177–78, 187, 288, *288f*
  gravity at, 288–89
International System of Units (SI). *See* SI
ion propulsion rockets, 533
ionizing radiation, 909–10
  dosages, 910–11, *911f*
  radiation weighting factors, 910, *910t*
iPad, *564f*
irradiance, 711
irrotational flow, 323
isobaric processes, 477, 478, *479f*, 484
isobars, 887, 900
isochoric processes, 477, 478–79, *479f*, 484
isomers, 905
isothermal processes, 477, *479f*, 484, *492f*
isotones, 887
isotopes, 887, *888t*, 913
  of carbon, *887t*, 901, 902
  of helium, *888t*
  medical, 614, 938
  of potassium, 903
  radioisotopes, 614, 938
  of uranium, 77, 612, *612f*, 908
isotropic radiation, 438, 440–41

Japan Aerospace Exploration Agency, 714
Joint Attosecond Science Laboratory (JASLab), 770–771
joule, 10, 132, 470
  converting kilowatt hours to, 579
  converting volts to, 532
  energy orders of magnitude, 134
  petajoule, 156
Joule, James Prescott, 578
Joule–Lenz law, 578, 595
junction points
  in parallel circuit, 578, 583, 587, 595
Jupiter, 299–300
  asteroids, 283
  moons of, 752

Kao, Charles Kuen, 3
Kapoor, Anish, 726
kelvin, 9, *455ft*
  SI unit, 455
Kelvin, Lord (William Thomson), 455
Kelvin scale, 455, 498
Kelvin–Planck statement, 490–91, 498, 503
  violated, 495
Kepler, Johannes, 294
Kepler's laws of planetary motion, 97, 294–96, 301
  1st, 294–95, *295f*
  2nd, 294–95, *295f*
  3rd, 294, 295–96
  periods in, 294, 295–96, 300, 301
Kepler space telescope, 299
Kerckhove, Diane Nalini de, 2
kilowatt hour, 134, 156
kinematics
  definition, 42
  equations, general, 51–52, 61
  equations, constant acceleration, 53–58, 131
  equations, for rotation, 201
kinematics equations, *220t*
  for constant acceleration, 53–56, 58–59, 144
  for rotation with constant acceleration, 201

kinetic energy, 131–32, *134t*, 301, *358f*, 895
  hydroelectric energy, 131, 157, 653
  minimum, 843
  relativistic, 786, 794–96, 803
  of rolling object, 238–42
  rotational, and work, 211–15
  SI unit, 322
  and special relativity (relativistic), 786
  and temperature, 452–53, 470
  in a wave, 400–02
  work-kinetic energy theorem, 131–32, 148, 158, 220
Kirchhoff, Gustav, 582
Kirchhoff's laws, 581–82
  applications of, 587–89
  loop rule, 582–83, 587–88, 595, 683–84, 687–88
  parallel circuits, 583
  series circuits, 582–83
  sign conventions for, *588t*
Kleist, Ewald Georg von, 552
Klitzing, Klaus von, 460
Kronig, Ralph, 630

laminar flow, 324
Large Hadron Collider, 809, *809f*, *918f*
  discovery of Higgs particle, 918
Large Magellanic Cloud, 71
laser cooling of atoms, 185, 217
Laser Interferometer Gravitational-Wave Observatory (LIGO), 766, 799
laser printers, 514
lasers, 727, 764, *771f*
  helium-neon, 707–08
  standing waves in, 414, 430
latent heat, 472–73, 484
lattice parameter, 865–86
Laue, Max Theodor Felix von, 833
Lawrence, E.O., 613
laws of motion (Newton)
  first, 94, 97, 172
  second, 94–97, 114, 119, 131, 205, 234, 285, 320
    applied to fluids, 320, 331
  third, 97, 100, 105, 284, 331
LC circuit, 685–86, *685f*, 694
Lenz, Heinrich, 578, 648, 659
Lenz's rule, 648–50, 668
  Joule–Lenz law, 578, 595
leptons, *919t*, 920, *924f*, 930,942
  lepton number, conservation of, 929
light and light waves
  interference
    constructive, 765–69, *769f*, *773f*, 774, 833
    double slit, 766–68, *766f*, *767f*, 780
    single slit, 773–76, 780
    thin film, 771–73, 780
light bulbs, 580–81
  incandescent light bulb, 576, 580, 727
  relative brightness of, 584, 587
  tungsten, 580
light-emitting diodes (LEDs), 580
  internal reflection in, 742
lightning, 508, *508f*
  plasma and, 308
LIMPET (Land Installed Marine Powered Energy Transformer), 403, *403f*

line integrals, 624
linear accelerator, 938–40
linear momentum, 168–89
  conservation of, 172–75, 177–78, 183–84, 186
  rate of change, 170–71
linear quantities, 220
liquid crystal display (LCD) technology, 816
liquids, *308f*, 449, *449f*
  density of, *309t*
  See fluids
Livingston, M.S., 613
longitudinal waves, 382–83, *383f*, 415, 425–26, 444
  equation for, 426
Lorentz, Hendrik, 605
Lorentz force, 605, 634
lumen, *10t*
lux, *10t*

macroscopic model of electric conductivity and resistivity, 574–75, 595
maglev trains, 644, *644f*, 667
magnetic boots, 219
magnetic dipole moments, 619, 630
magnetic domains, 632–33, *632f*
magnetic field lines, *208f*, 604, *604f*, 634, *716f*
magnetic fields, 604–05
  Ampère's law for, 624–25, 634, 669–702, *699t*, 703
  Earth's, *604f*, 605
  energy stored in, 666–67, 668
  Gauss's law for, 699, 705
  charged particle motion in, 608–16
  strengths of, 605–06
  of solenoids, *627f*, *628f*
magnetic flux, *10t*, 646–53, 668, 699, 703, 705, *705f*
  rate of change, 646–53
magnetic force, 604–05, *605f*, 634
  Ampère's law, 624–25, 634, 703
    applications of, 625–29
    for electromagnetic waves, 706–07
    for magnetic fields, 669–702, *699t*
    for solenoids, 661–62
  Biot–Savart law, 620–23, 634
  on charged particle, 604–07
    motion of, 608–16, *609f*
  crossed, 606–07
  on current-carrying wire, 616–17, 622–23
    straight, 625–27
  direction of, 606–07, 608–16
    handedness rules, 606–07, 618, 622
  torque on current loop, 617–19, *618f*, 620
  between two parallel conductors, 629–30
magnetic force microscope (MFM), 880
magnetic levitation (maglev), 644, 667
magnetic permeability/susceptibility, *631t*, 632
magnetic properties of materials, 630–33, *631t*, *633t*
  Bohr magneton, 630
  diamagnetism, 632
  electron spin, 630
  ferromagnetism, 632–33, *633t*
  hysteresis, 633
  paramagnetism, 630–31
  relative permeability, 631–32, *61t*
magnetic quantum number, 852–53

timing circuits, 563
Titan, Saturn satellite, 296
Titanic, 170, *170f*, 267
tokamaks, 606
Tolman–Stewart experiment, 572
torque, 202–04, 220, 259, 261, 262
   and angular motion, 211–12, 216
   calculating, 203
   direction of vector, 202
   in equilibrium, 255, 272
   and moment of inertia, 205–06, 234
   right-hand rule, 32, 202, 216
   without contact, 203–04
total angular momentum number, 850
total orbital energy, 297–98, 301
   and escape speed, 293
   negative, 294, 298
   positive, 298
touch-sensitive
   lamps, 594
   screens, 564
towing tension, 106–07
transducers, 316
   capacitors in, 563–64
transistors, 564
translational and rotational motion, 232–34
translucent materials, 727
transmission electron microscope (TEM), 879
transparent materials, 727, 771, 773
transverse speed, 387–88
transverse velocity, 396–97
transverse waves, 382, *382f*, *383f*, 385–88,
   390, 415
   along a string, 408–09
Tri-University Meson Facility (TRIUMF),
   525, *525f*, 533, 613–14, *613f*, 936–37,
   937f
triangle construction rule, 23, 35
triboelectric
   charging, 509, *509t*, *510f*
   series, 509, *509t*, 520
trough, 390, 399, 407
   parallel, 429
true errors
   definition, 4, 16
tsunami, 137, 886
tuned-mass damper
   in earth quake-resistant structures, 368
tungsten light bulbs, 580
Tupolev Tu114, 202
turboprop plane, 202
turbulent flow, 323
twin paradox, 790–91

Uhlenbeck, George, 630, 857
ultimate tensile strength (UTS), 268, 269
ultraviolet (UV) light, 709, *709f*
   UV index, 709
umbra, 727
unbound orbits, 293
   bound orbit, 293
   hyperbolic orbit, 297
uncertainty principles, 498, 835–38
underdamped oscillators, 66–68
uniform circular motion
   and simple harmonic motion, 351, *351t*,
   371, *371f*
unit conversion, 11–12, 16

universal gravitational force, 301
   Newton's law of, 99, 284–85, 301
universe, structure in, 499–500
uranium, 612, *631t*, 892, 906–07
   in nuclear reactors, 77, 612

valence band, 872
valence electrons, 572, 810, 866
van der Waals forces, 702
vaporization, 472, *473t*
variable mass, 10
   rocket propulsion and, 186–88, 189
vectors
   addition rules, 23–27
   algebraic method of addition of, 24, 35
   cross product of, 32–35, 202
   definition, 21, 35
   geometric method of addition of, 23, 34–35
   products of, 32–35, 72
   resolution, 22
   resultant force, 25–26
   subtraction of, 26–27
   in two and three dimensions, 29, 34, 72–74
   unit, 28
velocity
   average, 44, 49, 72–73
   components of, 73–74
   instantaneous, 45, 48–49
   of recession and distance, 3–4
   and speed, 44–45, 47
velocity selector, 610–11, *610f*
Venturi meter, 329
Vesta, 533
virtual photon, 921–22, *922f*
viscosity, 323, 332–35, *333t*
   effect on volume flow rate, 333–35
   measuring, 332, *332f*
   Poiseuille's law for viscous flow, 333
visible light, 709, *769f*
vision, human, 750–53, *751f*
   spectacles/eyeglasses for, 750, 752, *752f*,
   753f
Volta, Alessandro, 571
volume flow rate, 324–25, 336

walking, 361
Walton, Ernest, 896–97
water, 463
   refraction of, 430
   helps regulate temperature change, 471
   waves of, 381, *381f*
wave crest, 390, 407, 441–43
   of sound wave, 428, 434–35
   of plane wave, 429, *429f*, *433f*
   shock wave, 443, *443f*
wave function, 386–88, 391, 415, 427
   of interference, 406–07
   of longitudinal waves, 426
   of a spherical wave, 434
   of standing waves, 409–11, *411t*
   of superposition, 404
   of travelling harmonic wave, 394–95,
   398–99
   in opposite directions, 409
wave number, 391–93, 415, 436
   in wave function, 391, 426
wave optics. *See* physical optics
wave packet/wave group, 835

waveform, 381, *382f*, 385–86
   equations, 387
wavelength, *383f*, 391, *391f*
   range of various, 43
   and wave speed, 393
waves, 381
   amplitude
      constructive interference of, 407
   crest, 390, 407, 441–43
      of sound wave, 428, 434–35
      of plane wave, 429, *429f*, *433f*
   cycle, *382f*
   energy and power in travelling, 400–03
   equilibrium position, *381f*
   Fourier's theorem, 432
   frequency, 392–93
   interference of waves, 404
      constructive, 407, 433, 434, 444
      destructive, 407, 432, 434, 444
      of three dimensional waves, 433–35
      travelling in the same direction, 406–07
   longitudinal, 382–83, *383f*, 415,
      425–26, 444
   mechanical, 381, 407
   nodes and antinodes, 410–11, 413
   ocean, 402–03
   period of, 382, 392–95, 399, 402–03, 411,
      426
   phase constant, 395
   phase difference, 399–400, *399t*
   position plots of, 397–98
   reflection, 408–09
   resultant, 404
   shock wave, 443, *443f*
   standing, 409–11, 430, *431t*
      in air columns, 430–32, *431f*, *432f*, 444
      amplitude of, 410, *410f*
      electromagnetic, 414
      in lasers, 414, 430
      nodes and antinodes, 410–11, 413
      on strings, 412–14
   superposition of, 404–05, 406, 409, 436
   time plots of, 381, 384–85, 398–99
   transmitted, 401, 409
   transverse, 382, *382f*, *383f*, 385–88, 390,
      415
      along a string, 408–09
   transverse speed, 387–88
   transverse velocity, 396–97
   trough, 390, 399, 407
      parallel, 429
   water, 381, *381f*
      refraction of, 430
   waveform, 381, *382f*, 385–86
      equations, 387
   wavelength, *383f*, 391, *391f*
      range of various, 43
      and wave speed, 393
   wave number, 391–93, 415
   wave speed, 382, 387–89
   See also pulse; string, waves on; sound
      waves; wave function
weak force, 889, 900, 920, *922f*, 924
   as fundamental force, 920, 935
   and decay, 920
weber (Wb), *10t*, 646
Weber, Wilhelm, 540
Wehrlie/Standard Solar Spectrum, *824f*

Pedagogical Colour Chart

MECHANICS AND THERMODYNAMICS

Displacement and position vectors

Displacement and position component vectors

Linear ($\vec{v}$) and angular ($\vec{\omega}$) velocity vectors

Velocity component vectors

Force vectors ($\vec{F}$)

Force component vectors

Acceleration vectors ($\vec{a}$)

Acceleration component vectors

Energy transfer arrows

W_{eng}

$Q_?$

$Q_?$

Process arrow

ELECTRICITY AND MAGNETISM

Electric fields

Electric field vectors

Electric field component vectors

Magnetic fields

Magnetic field vectors

Magnetic field component vectors

POSITIVE CHARGES

Protons

NEGATIVE CHARGES

Electrons

Neutrons

LIGHT AND OPTICS

Light ray

Focal light ray

Central light ray

Converging lens

Diverging lens

LIGHT AND OPTICS continued

Linear ($\vec{p}$) and angular ($\vec{L}$) momentum vectors

Linear and angular momentum component vectors

Torque vectors ($\vec{\tau}$)

Torque component vectors

Schematic linear or rotational motion directions

Dimensional rotational arrow

Enlargement arrow

Springs

Pulleys

Resistors

Batteries and other DC power supplies

Switches

Capacitors

Inductors (coils)

Voltmeters

Ammeters

AC Sources

Lightbulbs

Ground symbol

Current

Mirror

Curved mirror

Objects

Images

Some Physical Constants

Gravitational constant	G	6.673×10^{-11} N·m/kg²
Speed of light	c	2.998×10^{8} m/s
Fundamental charge	e	1.602×10^{-19} C
Avogadro's number	N_A	6.022×10^{23} particles/mol
Gas constant	R	8.315 J/mol·K 1.987 cal/mol·K
Boltzmann's constant	$k = R/N_A$	1.381×10^{-23} J/K 8.617×10^{-5} eV/K
Unified mass unit	$u = (1/N_A)$	1.661×10^{-24} g
Coulomb's constant	$k = \dfrac{1}{4\pi\varepsilon_0}$	8.988×10^{9} N·m²/C²
Permittivity of free space	ε_0	8.854×10^{-12} C²/N·m²
Permeability of free space	μ_0	$4\pi \times 10^{-7}$ N/A² 1.257×10^{-6} N/A²
Planck's constant	h $\hbar = h/2\pi$	6.626×10^{-34} J·s 4.136×10^{-15} eV·s 1.055×10^{-34} J·s 6.582×10^{-16} eV·s
Mass of electron	m_e	9.109×10^{-31} kg 511.0 keV/c²
Mass of proton	m_p	1.673×10^{-27} kg 938.3 MeV/c²
Mass of neutron	m_n	1.675×10^{-27} kg 939.6 MeV/c²
Bohr magneton	$m_B = e\hbar/2m_e$	9.274×10^{-24} J/T 5.788×10^{-5} eV/T
Nuclear magneton	$m_n = e\hbar/2m_p$	5.051×10^{-27} J/T 3.152×10^{-8} eV/T
Magnetic flux quantum	$\phi_0 = h/2e$	2.068×10^{-15} T·m²
Quantized Hall resistance	$R_K = h/e^2$	2.581×10^{4} Ω
Rydberg constant	R_H	1.097×10^{7} m⁻¹
Josephson frequency–voltage quotient	$2e/h$	4.836×10^{14} Hz/V
Compton wavelength	$\lambda_c = h/m_e C$	2.426×10^{-12} m

Physical Data Often Used

Mass of Earth, M_E	5.98×10^{24} kg
Radius of Earth, R_E, mean	6.37×10^6 m
Mass of Moon, M_M	7.35×10^{22} kg
Radius of Moon, R_M	1.74×10^6 m
Mass of Sun, M_S	1.99×10^{30} kg
Radius of Sun, R_S	6.96×10^8 m
Earth–Sun distance, $D_{E\text{-}S}$, mean	1.496×10^{11} m
Earth–Moon distance, $D_{E\text{-}M}$, mean	3.844×10^8 m
One light year (the distance light travels in one year)	9.47×10^{15} m
Acceleration of gravity on Earth, g (standard value)	9.81 m/s^2
At sea level, at equator	9.78 m/s^2
At sea level, at poles	9.83 m/s^2
Acceleration of gravity on Moon, g_M	1.62 m/s^2
Orbital speed:	
Earth	2.98×10^4 m/s
Moon	1.02×10^3 m/s
Period:	
Earth (sidereal year)	365.26 days
Moon (sidereal month)	27.3 days
Escape velocity:	
Earth, $v_{E\text{-}esc}$	1.12×10^4 m/s
Moon, $v_{M\text{-}esc}$	2.4×10^3 m/s
Solar constant	1.35 kW/m^2
Time it takes light to travel the distance:	
from Earth to Sun	498 s = 8.3 min
from Earth to Moon	1.28 s
around Earth	0.13 s
equal to one light year	1 year = 3.16×10^7 s
Milky Way Galaxy data:	
Diameter	100 000 to 120 000 light years ($\approx 10^{21}$ m)
Thickness (average)	1 000 light years ($\approx 10^{19}$ m)
Number of stars	100 to 400 billion
Estimated mass	$5.8 \times 10^{11} \, M_S$
Useful ratios	$\dfrac{M_S}{M_E} = 3.33 \times 10^5; \ \dfrac{M_E}{M_M} = 81.3$
	$\dfrac{R_S}{R_E} = 109; \ \dfrac{R_E}{R_M} = 3.66$
	$\dfrac{D_{E\text{-}S}}{D_{E\text{-}M}} = 389; \ \dfrac{D_{E\text{-}M}}{R_E} = 60.4; \ \dfrac{D_{E\text{-}S}}{R_E} = 2.35 \times 10^4;$
	$\dfrac{g_E}{g_M} = 6.1; \ \dfrac{v_{E\text{-}esc}}{v_{M\text{-}esc}} = 4.67$

Solar System Data

	Mass (kg)	Planetary radius (m)	Semi-major axis (m)	Eccentricity
Mercury	3.30×10^{23}	2.44×10^{6}	5.76×10^{10}	0.206
Venus	4.87×10^{24}	6.05×10^{6}	1.08×10^{11}	0.00676
Earth	5.97×10^{24}	6.37×10^{6}	1.50×10^{11}	0.0167
Mars	6.42×10^{23}	3.38×10^{6}	2.28×10^{11}	0.0933
Jupiter	1.90×10^{27}	7.07×10^{7}	7.79×10^{11}	0.0488
Saturn	5.68×10^{26}	6.03×10^{7}	1.43×10^{12}	0.0557
Uranus	8.68×10^{25}	2.53×10^{7}	2.88×10^{12}	0.0444
Neptune	1.03×10^{26}	2.46×10^{7}	4.55×10^{12}	0.0112
Ceres (d)	9.43×10^{20}	4.80×10^{5}	4.13×10^{11}	0.0791
Pluto (d)	1.31×10^{22}	1.15×10^{6}	5.87×10^{12}	0.249
Eris (d)	1.67×10^{22}	1.2×10^{6}	1.46×10^{13}	0.436

(d) = dwarf planet

Some Prefixes for Powers of 10

Zepto	(z)	10^{-21}
Atto	(a)	10^{-18}
Femto	(f)	10^{-15}
Pico	(p)	10^{-12}
Nano	(n)	10^{-9}
Micro	(μ)	10^{-6}
Milli	(m)	10^{-3}
Centi	(c)	10^{-2}
Deci	(d)	10^{-1}
Kilo	(k)	10^{3}
Mega	(M)	10^{6}
Giga	(G)	10^{9}
Tera	(T)	10^{12}
Peta	(P)	10^{15}
Exa	(E)	10^{18}
Zetta	(Z)	10^{21}

Standard Abbreviations and Symbols for Units

Unit name	Symbol	Measures	Equal to
Ampere (*)	A	Electric current	C/s
Atmosphere	atm	Pressure	1.01×10^5 Pa
Becquerel	Bq	Radioactive decay	decays/s
Candela (*)	cd	Luminous intensity	
Coulomb	C	Electric charge	A · s
Day	d	Time	8.64×10^4 s
Decibel	dB	Logarithmic ratio of intensities	
Degree	°	Angle	π/180 rad
Degree Celsius	°C	Temperature	K − 273
Farad	F	Electric capacitance	C/V
Henry	H	Electromagnetic inductance	V · s/A
Hertz	Hz	Frequency	cycle/s
Hour	h	Time	3600 s
Joule	J	Energy or work	N · m
Kelvin (*)	K	Temperature	
Kilogram (*)	kg	Mass	
Kilowatt hour	kW·h	Energy	kW · h
Litre	L	Volume	10^{-3} m^3
Lumen	lm	Luminous flux	cd · sr
Lux	lx	Luminous flux per unit area	l m/m^2
Light year	ly	Distance light travels in a year	9.46×10^{15} m
Metre (*)	m	Length	
Mole (*)	mol	Avogadro's number of units	
Newton	N	Force	kg · m/s^2
Ohm	Ω	Electric resistance	V/A
Pascal	Pa	Pressure	N/m^2
Radian	rad	Angle	m/m
Second (*)	s	Time	
Steradian	sr	Solid angle	
Tesla	T	Magnetic field strength	Wb/m^2
Volt	V	Electric potential difference	J/C
Watt	W	Power	J/s
Weber	Wb	Magnetic flux	V·s or T·m^2
Year	yr	Time	3.156×10^7 s

Base SI units are marked with *. The equivalents given in the last column are not the only possibilities. See Chapter 1 for equivalents expressed in terms of the seven SI base units.

Conversions

$1 \text{ rev} = 2\pi \text{ rad} = 360°$

$1 \text{ rad} = 57.3° = \pi/180°$

$1 \text{ d} = 8.64 \times 10^4 \text{ s}$

$1 \text{ yr} = 3.16 \times 10^7 \text{ s}$

$1 \text{ angstrom (Å)} = 1.00 \times 10^{-10} \text{ m}$

$1 \text{ light year (ly)} = 9.46 \times 10^{15} \text{ m}$

$1 \text{ astronomical unit (au)} = 1.50 \times 10^{11} \text{ m}$

$1 \text{ mile} = 1.61 \times 10^3 \text{ m}$

$1 \text{ litre (L)} = 1.00 \times 10^{-3} \text{ m}^3$

$1 \text{ unified mass unit (u)} = 1.66 \times 10^{-27} \text{ kg}$

$1 \text{ tonne (t)} = 1.00 \times 10^3 \text{ kg}$

$1 \text{ mile/h (mph)} = 0.447 \text{ m/s}$

$1 \text{ litre (L)} = 1.00 \times 10^{-3} \text{ m}^3$

$1 \text{ dyne} = 1.00 \times 10^{-5} \text{ N}$

$1 \text{ lb} = 4.45 \text{ N}$

$1 \text{ erg} = 1.00 \times 10^{-7} \text{ J}$

$1 \text{ kilowatt hour (kW·h)} = 3.60 \times 10^6 \text{ J}$

$1 \text{ cal} = 4.18 \text{ J}$

$1 \text{ Cal} = 4.18 \times 10^3 \text{ J}$

$1 \text{ electron volt (eV)} = 1.60 \times 10^{-19} \text{ J}$

$1 \text{ British thermal unit (Btu)} = 1.05 \times 10^3 \text{ J}$

$1 \text{ horsepower (hp)} = 746 \text{ W}$

$1 \text{ Btu/h} = 0.293 \text{ W}$

$1 \text{ atmosphere (atm)} = 1.01 \times 10^5 \text{ Pa} = 760 \text{ mm Hg}$

$1 \text{ torr} = 133 \text{ Pa}$

$1 \text{ gauss (G)} = 1.00 \times 10^{-4} \text{ T}$

$1 \text{ curie (Ci)} = 3.70 \times 10^{10} \text{ Bq}$

$1 \text{ rem} = 0.0100 \text{ Sv}$

$K = °C + 273.15 \text{ (for temperatures)}$

Greek Letters Commonly Used in Physics

alpha	α
beta	β
delta	δ
epsilon	ε
eta	η
gamma	γ
lambda	λ
mu	μ
nu	ν
omega	ω
phi	ϕ
pi	π
psi	ψ
rho	ρ
sigma	σ
tau	τ
theta	θ
Delta	Δ
Gamma	Γ
Omega	Ω
Phi	Φ
Sigma	Σ